Complex Digital Control Systems

Complex Digital Control Systems

GUTHIKONDA V. RAO, Ph. D.
Consulting Systems Analyst
Beltsville, Maryland

VNR VAN NOSTRAND REINHOLD COMPANY
NEW YORK CINCINNATI ATLANTA DALLAS SAN FRANCISCO
LONDON TORONTO MELBOURNE

Van Nostrand Reinhold Company Regional Offices:
New York Cincinnati Atlanta Dallas San Francisco

Van Nostrand Reinhold Company International Offices:
London Toronto Melbourne

Library of Congress Catalog Card Number: 78-26041
ISBN: 0-442-20110-9

Manufactured in the United States of America

Published by Van Nostrand Reinhold Company
135 West 50th Street, New York, N.Y. 10020

Published simultaneously in Canada by Van Nostrand Reinhold Ltd.

15 14 13 12 11 10 9 8 7 6 5 4 3 2 1

Library of Congress Cataloging in Publication Data

Rao, Guthikonda V
Complex digital control systems.

Bibliography: p. 496
Includes index.
1. Digital control systems. I. Title.
TJ216.R36 629.8′3 78-26041
ISBN 0-442-20110-9

To
my sisters, Kanakrathnam and Dr. Subaima,
and grandpa Nandam Virswami

Preface

This tutorial control engineering book on the complex sampled-data/digital control systems, presently used in modern industrial applications, is primarily concerned with the study and exposition of the correlation between theory and technology at the juncture of the latest computerized industrial revolution. These applications belong to the category of broadcast high-quality quadruplex color videotape recorders, missile attitude control, automatic pilot, high-precision radar, TT&C (tracking, telemetry, and control of synchronous and interplanetary satellites), simulation and signal processing, manufacturing plant-automation, high-precision numerical machine control, etc. Automation, computation and control, and TDM (time-division multiplex) data storage and real-time data communications via synchronous satellites, fiber-optic optical communications, and microwave links are, all-in-all, the order of the day. There is a proliferation in developments in theory, components, and systems, and the book modestly aims at briefly acquainting the reader with the latest trends in the implementation of the modern control systems. A sophisticated multiloop interacting sampled-data control system is comprehensively analyzed at length to illustrate the extent of complexity encountered in some of the major applications in the field.

A brief survey of the theoretical aspects of analysis and synthesis in automatic control is presented in Chapter 1 under four different classifications for both continuous and digital control systems. (Most systems are actually hybrid in these respects, and *digital filters* are exclusively used for compensation in digital control systems.) These classifications are:

1. The frequency response techniques.
2. The statistical design technique.
3. Modern control theory with state-space techniques.
4. Miscellaneous techniques such as the extension of frequency response methods to nonlinear systems as quasi-linear approximations, graphical phase-plane tra-

jectories for display of transient response in low-order linear/nonlinear systems, Lyapounov's Second or Direct Method, and general ac carrier servo systems.

Chapter 2 is a comprehensive introduction to the complex digital control system referred to in Chapter 3. Actual worked-out examples in Chapter 3 illustrate the method of application of the various theoretical concepts in each classification to simple mathematical models, as far as the transfer functions or pulse-transfer functions go. Realistic, hard-to-derive, highly complex transfer functions are generally beyond practical consideration in the case of the more complex, multivariable physical control systems using the latest components. The simplified mathematical models used for analysis may be therefore more pertinent to the various subsystems or grossly approximated subloops in a modern control-system complex. The manner in which these examples are explained, as an illustration of the theoretical concept involved in each technique, is conducive to a rapid and meaningful understanding of the subject to the undergraduate or graduate student and the practical electrical engineer, because a physical image of a substantial control element with a certain function is immediately available as a reference in the complex sampled-data control system chosen for exposition in Chapters 2 and 4. It is generally anticipated in this context that the student and the practicing control engineer will have had some elementary knowledge in Laplace and Z-transforms, and differential and difference equations. At the same time, some insight into the heuristic methods normally resorted to in the laboratory for the successful implementation of these complex sampled-data control systems is provided by some of the theoretical examples chosen in this method of approach.

Most textbooks in the control engineering field usually devote extensive space to the peripheral mathematics related to the theoretical concepts, and give less importance to the applicability of the concept to some physical system, which is invariably nonlinear and thus unduly complex to solve for the *necessary and sufficient conditions of stability*. Some textbooks specialize and concentrate on a rigorous treatment of a few methods for a strictly limited range of applications. This control engineering book, on the other hand, attempts to introduce in principle the status of practically every classification of control technique and its applicability at the present time. It will therefore serve as an introduction to the comparative study of the various techniques, past and present. It will provide an overall perspective to establish guidelines by the examples chosen for a particular technique. Without this all-around theoretical orientation, the engineer working on a heuristic basis may perhaps waste time and effort on unrealizable objectives in his or her particular application. The latest achievements—in both theory and practice—in modern control systems should mutually contribute to one another by regular intercommunication and coordination as far as possible, instead of drifting further apart.

Chapter 4 introduces to the student and exponents of theory a typical example of a successful, modern, high-precision digital control system-complex, and its mostly unsolvable stability criterions as far as the theory is concerned. The technological sophistication of the control system and the closely related system electronics expose students of control theory to the physical reality of the control system, and give them a substantial impression of the elaborate electronics circuit design and signal processing involved in each and every feedback loop in the three interdependent sampled-data control systems of a high-quality quadruplex color videotape recorder. The *measure-and-optimize* technique introduced by the author for the stage-by-stage and loop-by-loop development of the interacting nonlinear digital control systems in the laboratory on a heuristic basis is considered appropriate for the following reasons. The development and design of the associated digital controllers are not along the lines of conventional *trial-and-error* technique usually employed in feedback control systems for the analysis and synthesis of suitable compensating networks to meet the requirements of parameters such as stability, bandwidth, noise immunity and gain. The measure-and-optimize technique is more in line with the developments and trends in the state-space techniques of modern control theory, where the measured variables through the various stages of the system for desired coefficients of the system equation, and the optimization of the system parameters for the desired performance-index share a prominent role. As an incidental development, a comprehensive, up-to-date background is presented on the technology of compatible color television since magnetic tape recording and color television are inseparable in modern telecasting.

The technological exposition of the system chosen is profusely illustrated with detailed block diagrams from both electronics and control engineering aspects. The theoretical representation and interpretation of the complex interacting digital control systems of the quadruplex color videotape recorder will provide a direct channel of communication between theory and practice to avoid possible controversy. In high-precision nonlinear control systems, the loop-by-loop compression of the performance-index (the limit-cycle or residual phase-jitter of a nonlinear control system) on an automatic adaptive basis, by interloop transfer from the lower to the higher rates of sampling, may be perhaps a revealing example for introduction in a control engineering textbook. With the staggered higher rates of sampling used in the recorder, the system organizes itself toward self-optimization in terms of nanosecond control-timing patterns in arriving at the final result, viz., the reproduction of true color from videotape within an undetectable $\pm 2.5^\circ$ phase-jitter at the color-subcarrier frequency of 3.579545 MHz in approximately 2 sec when the start push-button is pressed. The performance-index is of course entirely beyond the scope of a continuous feedback control system; this is how the modern digital control systems have revolutionized the accomplishments in the control field as far as the accuracy is concerned.

Along with the latest revolution of the silicon monolithic large-scale integration in microelectronics, complex digital control systems are just about to take a new turn altogether. In view of this timely development, Chapter 5 is devoted to the application of microprocessors and solid-state RAM and ROM memories to digital control systems. By employing auxiliary BIFET and BIMOS analog-to-digital and digital-to-analog LSI converter chips (built-in or external), the low-cost microcomputers are all set to play a major role, in conjunction with LSI digital filters, in simplifying the processing of present and future digital control systems as an alternative approach (and in displacing some of the former techniques).

Three appendixes, on the subjects of the local-oscillator frequency synthesis, the high-speed phase-lock loop, and some basic concepts of feedback control systems, present a useful supplement to the subject under consideration. The last two appendixes are incidental to the subject of the book.

The author extends his sincere appreciation to professor Cornelius N. Weygandt of the University of Pennsylvania and A. C. Luther, Jr., Chief Engineer, Broadcast & Communications Division, Radio Corporation of America, Camden, New Jersey for their continuous interest and encouragement during the early stage of this project. Thanks to Maxine Moore of AMECOM (Litton Systems, Inc.) and Margaret Pitchalonis of the Moore School of Electrical Engineering for typing the manuscript in record time. Grateful acknowledgments are due to the many authors who originally published the information (which is summarized in Chapters 1 and 5) in the voluminous literature on the subjects surveyed. The bibliography gives a partial list of these authors and their works.

The author considers himself fortunate to have been a member of the RCA team responsible for the development and design of the *first* broadcast solid-state quadruplex color videotape recorder in the world market.

Guthikonda V. Rao, Ph.D.
Beltsville, Maryland
(Washington, D.C.)

Introduction

The author admires the influence of students, control engineers, and control theorists (of a diversity of backgrounds in physical sciences and humanities) who exult in open, enlightened, tolerant, flexible, and harmonious communication with one another. In the theoretical and technological context of the book, the author understands that validated theory and *de facto* successful technological practice—via short-term heuristic and/or long-term computer simulation techniques—contribute to one another to assure optimal results when sheer complexity is encountered in systems.

The 10-billion neuron masterpiece of evolution, the human brain, and its neural communications supernetwork and the physiological glandular hormones and enzymes, form the most complex bioelectrochemical digital feedback control system in nature. As a parallel state-of-the-art achievement, 1 to 10 million functional building blocks of microelectronics will soon be incorporated on a minute chip as a result of creative research and development in modern science and technology. High-technology very-large-scale integration (VLSI) currently fabricates 100,000 functions on a silicon monolithic microcomputer chip!

The free individual's self-realization via persistent experimentation, that the brain-complex can be self-controlled and optimized in performance (or behavior) by positive constructive education and training via audio and visual aids and hence by strength of character, is the vital step toward achieving long-term stability in thought and/or action, as a precondition to inner psychological tranquility. Self-discipline with a broad perspective (as opposed to excesses of gain or unrestrained greed) is a simulation equivalent to the negative-feedback control systems of a stable and effective broad-band system organization. In particular, the unpredictable instigator of problems is the addictive life-style of indulging in drug- and alcohol-induced "highs" that intermittently interfere with the normal gain and sensitivity of the metabolism, which is both involuntarily and voluntarily (through learning and/or iterative programming) controlled by complex and fragile bioelectrochemical processes in the system-architecture of the brain. It is synonymous with disturbing the most important parameter "gain" of an otherwise optimally stable combined analog and digital feedback control system.

Constructive guidelines, based on an intelligent and tenacious effort along the *automatic* stabilized patterns in the brain and its network, merge to insure survivability, health, and happiness of the optimistic, secure individual in a sociological environment for ultimately maintaining peace on an expanding global scale. Peace, as a close parallel to stability in a nonlinear control system, is a continually converging process from a collective conglomerate made up of self-disciplined individuals in the body politic. "Blessed are the peacemakers" who righteously strive for the emergence and establishment of unity of purpose in life amid natural diversity of backgrounds. "They shall be called the children of God" (or, in classic Sanskrit, the *brahmanas*). As far as the memory and development of the cerebral cortex are concerned, there is no virtue that supersedes the acquisition of a broad spectrum of constructive knowledge that enhances a well-adjusted day-to-day participation of the individual in a harmonious community.

Contents

Complex Digital Control Systems

1
Survey of Analog and Digital Control Theory

1.1 z-TRANSFORMS AND THEIR APPLICATION TO SAMPLED-DATA/ DIGITAL CONTROL SYSTEMS

The *sampling*, as used in pulse systems, is a *time-quantization* process, inherently capable of improving stability by improving immunity to interference. The effect of sampling as carried out by a pulse modulator is described by a *constant-coefficient linear difference equation*, which is commonly solved by a recursion formula in *numerical approximation theory*. Relay systems, in which level-quantization is used, are more difficult to account for, since they are described by nonlinear difference equations.

Figures 1-1 through 1-3 illustrate a sampled-data feedback control system, the sampling process in time-domain, and the quantization error, respectively.

When the input to a quantizer is a sampled-data signal, the *mean-square of the quantization error* is given by

$$\overline{Y}^2 = \frac{1}{e/m} \int_{-e/2m}^{e/2m} (mt)^2 \, dt = \frac{e^2}{12}$$

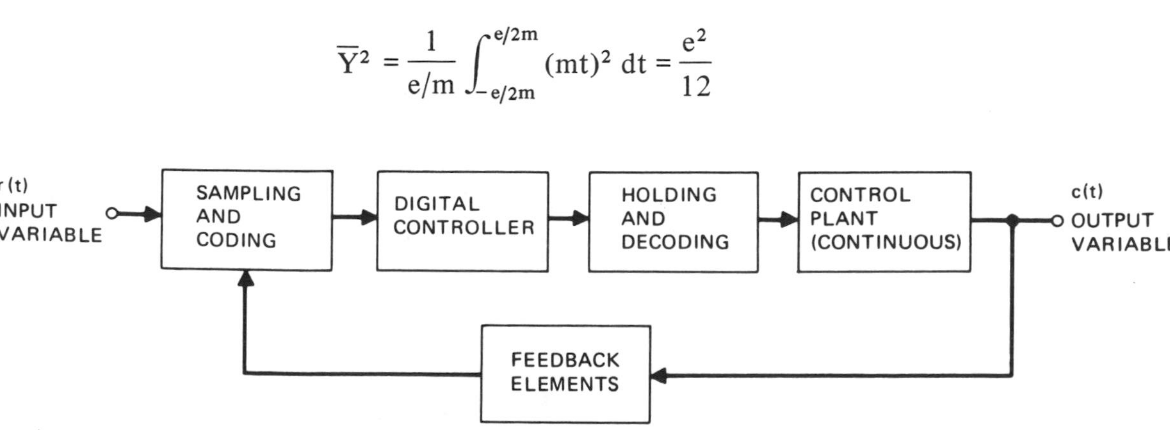

Fig. 1-1. Sampled-data feedback control system.

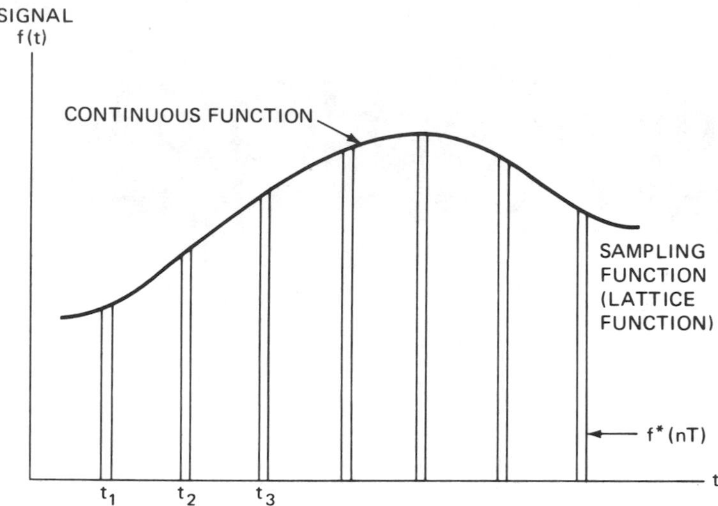

Fig. 1-2. Sampling process in time-domain.

where the line-segment of the error signal e is Y = mt and m = the slope. The equivalent statistical sampler for the quantizer can be indicated as a device consisting of a summer (Σ) and a sampler (∗). In terms of the *probability distribution* of the input, to make it an area-sampler, the mean-square quantization error is also given by

$$\overline{Y}^2 = \frac{d^2}{du^2} F_n(x)\Big|_{x=0}$$

where

$$F_n(x) = e \frac{(\sin ex/2)}{(ex/2)} = \frac{e^2}{12}$$

As a pulse-code modulator (PCM), level- and time-quantization are used for accuracy and immunity to interference in a digital computer, respectively. It reduces to a relay system for the smaller values of the signal being quantized and to a pulse system for the larger values. As a pulse system, the digital computer is described by the difference equation, which represents the algorithmic operation-sequence of the computer software program. In the case of a complex digital control system such as a quadruplex color videotape recorder (Q-CVTR), the digital controllers amount to the various pulse transfer-functions operating on the various variables. Then, the linear difference equations, expressed in the alternative form of z-transforms, are in order for the examination of the various sampled-data control systems.

The application of the *z-transformation in digital or pulse sampled-data control systems* is closely analogous to the application of *Laplace transformation in continuous feedback control systems.* For systems having lumped constants such as inductors, capacitors, and resistors—that is, those described by constant-coefficient linear difference equations—the z-transformation gives expressions that are *rational polynomial ratios in the complex-variable z*, as defined by the relation $z = e^{st}$, where

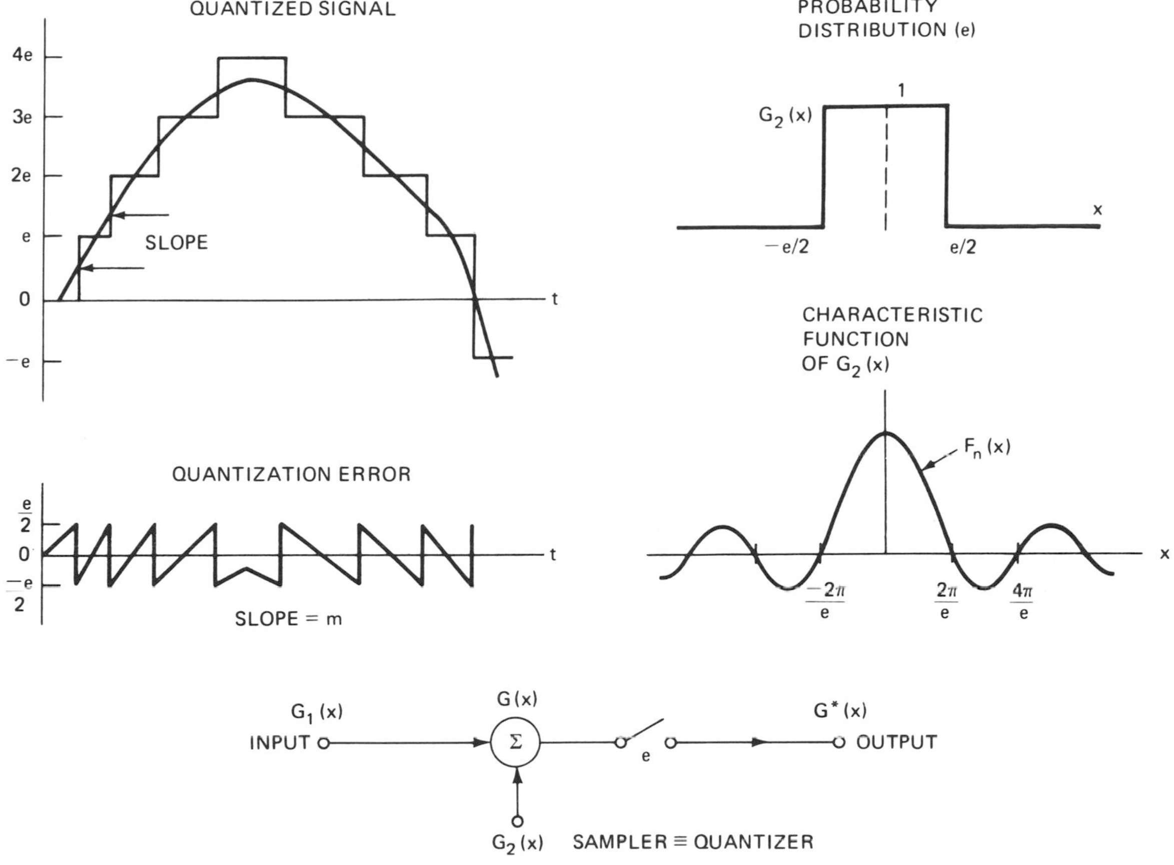

Fig. 1-3. Quantization error.

s is the complex-variable used in Laplace transforms

$$s = \sigma + j\omega$$

The complex z-plane is used to study the behavior of the sampled-data transfer-function in z-transforms, as a parallel to the complex s-plane that is used for the study of the transfer-functions in the continuous system by $\mathcal{L}$-transforms. All the points comprising the $j\omega$ imaginary-axis in the s-plane lie on the unit-circle in the z-plane.

$$z = e^{j\omega T} \text{ for } s = j\omega$$

Therefore, z is a complex number whose magnitude is unity and whose phase angle is ωT. The z-plane is a *Rieman surface* with an infinite series of superimposed planes, as it repeats the periodicity of e^{sT}. All points in the right half of s-plane map outside the unit-circle in the z-plane, and those to the left map inside the unit-circle. The properties of transfer-functions, mapping, inversion, etc., hold good in the theory of z-transforms also. Networks definable by $\mathcal{L}$-transforms can be restated in z-transform notation by implying (1) a sampling-switch in series with the network and (2) a sampling-period T. The spectrum F(z) or F*(s) or F*($j\omega$) of the amplitude-modulated pulse-train f(nT) is obtained by summing the spectra of the individual pulses. If $\delta_T(t)$ is a unit impulse-train,

$$f(nT) = \int_{-\infty}^{\infty} f(t)\delta(t - nT)\,dt$$

and

$$\delta_T(t) \triangleq \sum_{n=-\infty}^{\infty} \delta(t - nT) \tag{1}$$

$$f^*(t) \triangleq f(t) \cdot \delta_T(t) = f(nT) \sum_{n=-\infty}^{\infty} \delta(t - nT) \tag{2}$$

where

$\delta(t - nT) \triangleq 0$ for $t \neq nT$.

$$F^*(j\omega) = F(z) = \mathcal{L}[f^*(t)] \triangleq \sum_{n=0}^{\infty} f(nT)e^{-nsT} \tag{3}$$

if $\text{Re}(s) > 0$ (Re(s) = Real part of s)

Example:

$$\mathcal{L}(1/s + a) = e^{-aT}$$

$$F(z) \triangleq \sum_{n=0}^{\infty} f(nT) \cdot z^{-n} = \frac{1}{(1 - e^{-aT} z^{-1})} \qquad \text{since } f(nT) = e^{-naT}$$

Symbol $\triangleq$ stands for *by definition* and *equal to.*

The z-transform expansion is, as shown, conventionally given in the closed-form. The sum of the sampling-function f(nT) plays the same role in sampled-data (digital or pulse) systems, as the integral plays in the continuous (analog or linear) systems. The sampling-function is sometimes called the *lattice-function.* For conventional transfer-functions, the z-transforms in the closed-form can be directly read from tables as for $\mathcal{L}$-transforms. The z-transform is rational in z whenever F(s) is rational in s, so as to allow the *expression of a z-transform in a closed-form for a pulsed linear network.* In short, the $\mathcal{L}$-transform of a network F(s) is the $\mathcal{L}$-transform of its impulse-response, f(t), while the z-transform (or the discrete $\mathcal{L}$-transform) of a network F(z) is the $\mathcal{L}$-transform of the *sampled impulse-response.*

1.1.1 Impulse Sampling. The effect of the sampling process is to introduce a succession of spurious spectra proportional to the signal spectrum. The repeated spu-

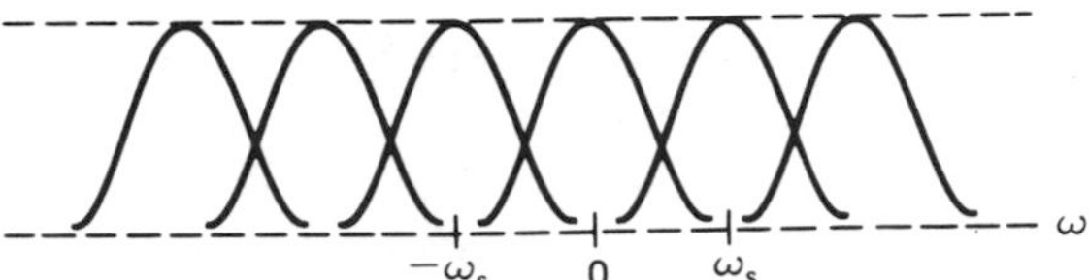

Fig. 1-4. Spectra of impulse.

rious spectra are all equal in magnitude and are periodically repeated at a frequency $\omega_0 = 2\pi/T$.

$$F^*(j\omega) = \frac{1}{T} \sum_{k=-\infty}^{\infty} F[j(\omega - k\omega_0)] \tag{4}$$

In practice, the impractical impulse (or unit-pulse) sampling is actually approximated by a finite-width pulse-sampling process. The repeated spectra in this case diminish in relative amplitude. See Figs. 1-4 and 1-5 for the spectra of a unit-pulse and pulse of finite-width, respectively.

$$F^*(j\omega) = \sum_{k=-\infty}^{\infty} xF[j\omega - \omega_0 k] \tag{5}$$

The response of a linear system to a finite-width pulse is the sum of the impulse-response of the system transfer function G(s) and its various derivatives. The sampled function c(t) is given by:

$$c(t) = f(nT)\gamma \left[g(t) - \frac{\gamma}{2!} g'(t) + \frac{\gamma^2}{3!} g''(t) - \cdots \right] \tag{6}$$

where nT is the sampling instant of duration γ.

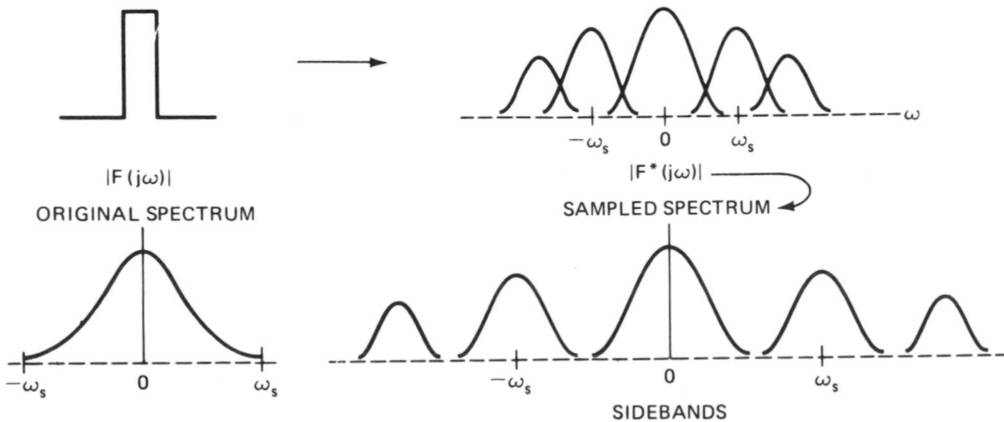

Fig. 1-5. Spectra of a pulse of finite-width.

As a typical example, for the simple transfer-function $(1/s + a)$, if the duration of the sampling pulse is 10% of the time-constant $(1/a)$ of this system, the response according to (6) differs from impulse-response by 5%.

The sampled-data control systems have a unique property of producing a small periodic output component called *ripple*, due to the intermittency in the signal caused by sampling. Means must be therefore used to reduce or control the magnitude of this component. The impulse-response of the most popular and simple *zero-hold network used for clamping the sampled input function* takes the following form:

$$\mathcal{L}[g_h(t)] = \mathcal{L}[u(t) - u(t - T)]$$

$$G_h(s) = \frac{1}{s} - \frac{1}{s} e^{-sT} = \frac{1 - e^{-sT}}{s} \tag{7}$$

$$G_h(j\omega) = \frac{1}{j\omega}(1 - e^{-j\omega T}) = T\frac{\sin(\omega T/2)}{(\omega T/2)}, \ \underline{/-\omega T/2} \tag{8}$$

The preceding expression for $G_h(s)$, suitable for reproducing *step-input* function, makes a regular appearance in sampled-data control systems, although at times the more complex *first-order hold-network suitable for reproducing the ramp-function* is used. The latter is not frequent since it introduces higher noise-level and a phase-shift of $-280°$ as compared to $-180°$ of the zero-order hold. In practice, other special networks termed *data extrapolators* are also included to control the ripple, noise, bandwidth, and transient-response of sampled-data feedback control systems. The hold-network incidentally provides a smooth analog output, since most sampled-data control systems are basically hybrids of continuous and pulse/digital portions. The data reconstruction is rendered by the network through a process of extrapolation by means of the preceding set of samples, as shown in Fig. 1-6.

Evaluation of the Pulse-Transfer Function: F(z) is conveniently evaluated by using the *convolution integral.*

$$F(z) = \mathcal{L}[f(t)\delta_T(t)]$$

$$F(s) = \mathcal{L}[f(t)] = \int_0^\infty e^{-sT} f(t)\, dt$$

$$F(z) = \sum_{n=-\infty}^{\infty} \int_{-\infty}^{\infty} e^{-snT} f(t)\delta(t - nT)\, dt$$

$$= \sum_{n=-\infty}^{\infty} f(nT)z^{-n}$$

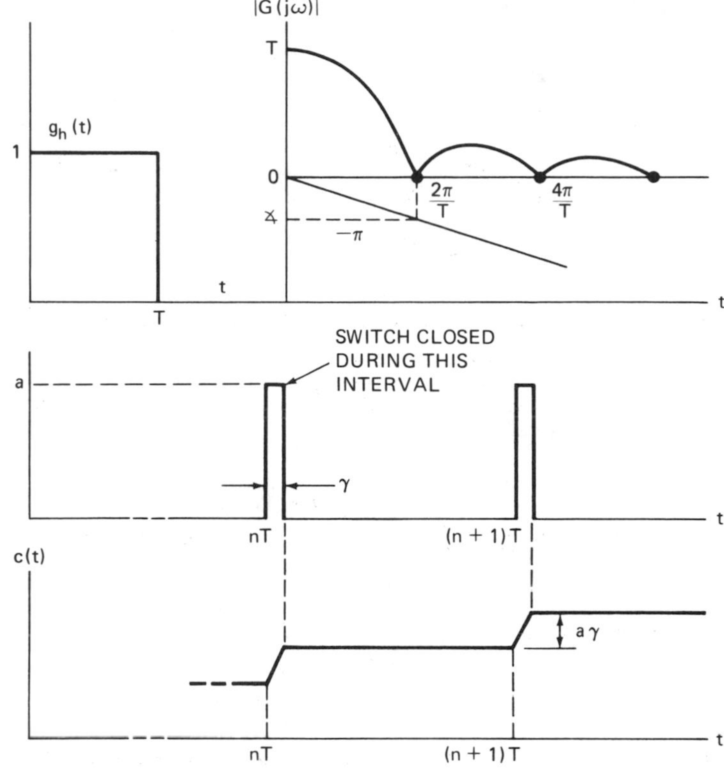

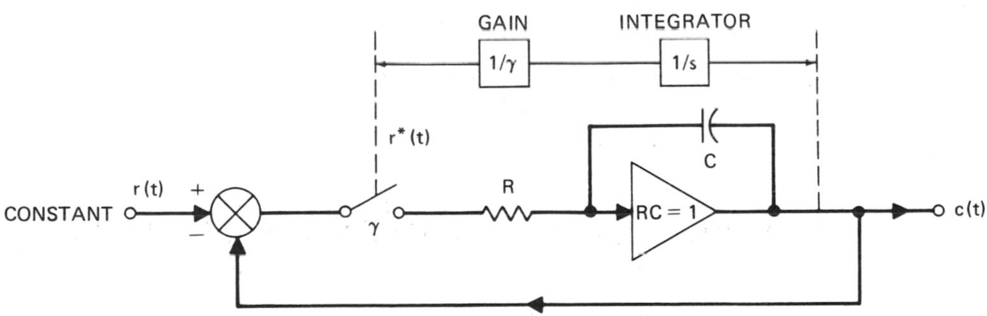

Fig. 1-6. Zero-order hold, switch, and integrator.

$$
\begin{cases}
\delta c(t) = \dfrac{a}{\gamma}\, t\Big|_{nT}^{nT+\gamma} = a \text{ V/sec} \\
G_h(s) = \dfrac{1}{s}\,(1 - e^{-sT})
\end{cases}
$$

From convolution theorem:

$$
\mathcal{L}[f_1(t)f_2(t)] = \frac{1}{2\pi j}\int_{c-j\infty}^{c+j\infty} F_1(s-\lambda)F_2(\lambda)\,d\lambda
$$

$$
\mathcal{L}[f_1(t)] = \mathcal{L}[\delta_T(t)] = 1 + e^{-sT} + e^{-2sT} + \cdots = 1/(1 - e^{-sT})
$$

$$\mathcal{L}[f_2(t)] = \mathcal{L}[f(t)] = F(s)$$

$$\therefore F(z) = \mathcal{L}[f(t)\delta_T(t)]$$

$$= \frac{1}{2\pi j}\int_{c_2-j\infty}^{c_2+j\infty} \frac{F(\lambda)\, d\lambda}{1 - e^{-sT(s-\lambda)}}$$

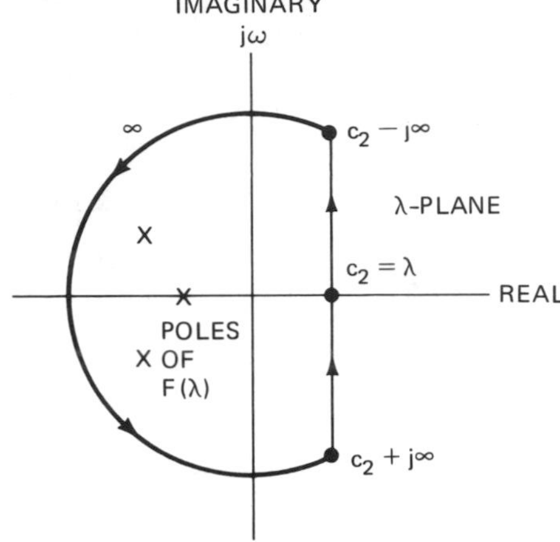

The preceding equation evaluated by *contour integration* along a line $\lambda = c_2$ and a semicircle of ∞-radius in the left-hand plane encloses the poles of $F(\lambda)$ only when (1) $c_2 < s$, and (2) the poles of $F(\lambda)$ are located to the left of c_2. Then F(z) is given by the sum of residues as follows:

Example: $F(s) = 1/(s + a)$: to obtain the output of a pulse transfer-function of a network F(s):

$$F(z) = \sum_{\text{poles of } F(\lambda)} \text{residue of } F(\lambda)/1 - 4^{-T(s-\lambda)} \quad \text{or } F(\lambda)/(1 - z^{-1}e^{\lambda T})$$

$$F(z) = \frac{1}{2\pi j}\int_{c-j\infty}^{c+j\infty} \frac{1}{(\omega + a)}\frac{1}{[1 - e^{-(s-\omega)T}]}\, d\omega$$

In the left-hand plane, the pole at $(\omega = -a)$ only contributes to the integral.

The residue at $\omega = -a$: $F(z) = 1/(1 - e^{-aT}e^{-sT}) = 1/(1 - e^{-aT}z^{-1})$

Modified z-Transform. For plants or transfer-functions exhibiting *propagation or transportation-lag*, this version of the z-transform is applicable. The behavior of the sampled function between the sampling instants can be evaluated by using a pulse-sequence derived from the time functions. The latter must be delayed by nonintegral multiples of the sampling-frequency.

$$F(s) \triangleq \sum_{n=0}^{\infty} f(nT)z^{-n}$$

Another form: If λ is an integer, $F(z, \Delta)$ = Delayed z-transform $= Z[e^{-\lambda sT}F(s)e^{msT}]$

Advanced or modified z-transform: $F(z, m) \triangleq z^{-1} \sum_{n=0}^{\infty} [f(n + m)T] z^{-n}$

Advanced z-transform: $F(z, m) = Z[e^{msT}F(s)]$, and $F(z, \Delta) = z^{-\lambda}F(z, m)$

F(z, m) can be evaluated using the contour integral as before:

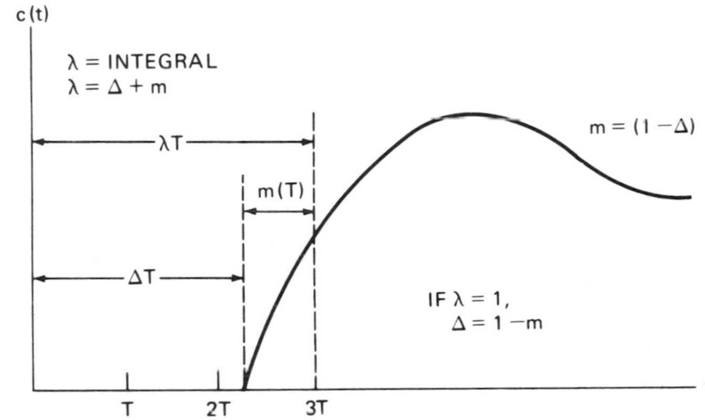

$$F(z, m) = z^{-1} \sum_{\text{poles of } F(\lambda)} \text{residue} \left[\frac{F(\lambda)e^{m\lambda T}}{1 - z^{-1}e^{\lambda T}} \right] \text{ at } \lambda = c_2$$

The following three examples illustrate the principle of deriving z- and modified z-transforms in sampled-data control systems:

Example 1.

$$G(s) = \frac{1}{(s + a)} : G(z, m) = z^{-1} \sum_{n=0}^{\infty} g(nT + mT)z^{-n}$$

$$g(nT + mT) = e^{-a(n+m)T}, \qquad \text{since } g(t) \left.\begin{matrix} = e^{-at}; t \geqslant 0 \\ = 0 \quad ; t < 0 \end{matrix}\right\}$$

$$G(z, m) = z^{-1} \sum_{n=0}^{\infty} e^{-a(n+m)T} z^{-n}$$

$$= z^{-1} [e^{-amT} + e^{-a(1+m)T} z^{-1} + e^{-a(2+m)T} z^{-2} + \cdots]$$

$$G(z, m) = e^{-amT}/(z - e^{-aT});$$

put $m = (1 - \Delta)$ if delay z-transform $F(z, \Delta)$ is required.

Using inversion integral: $G(z, m) = \dfrac{z^{-1}}{2\pi j} \displaystyle\int_{c-j\infty}^{c+j\infty} \frac{e^{mT}}{(\omega + a)} \frac{1}{1 - e^{-(s-\omega)T}} \, d\omega$

In left-hand plane, residue at ($\omega = -a$) only is required.

$$G(z, m) = z^{-1}e^{-amT}/(1 - e^{-aT}z^{-1})$$

If r(t) = unit-step, u(t),

$$R(z) \triangleq 1/(1 - z^{-1})$$

$$G(z, m) = \frac{z^{-1}e^{-amT}}{(1 - z^{-1})(1 - e^{-aT}z^{-1})}$$

Now, response

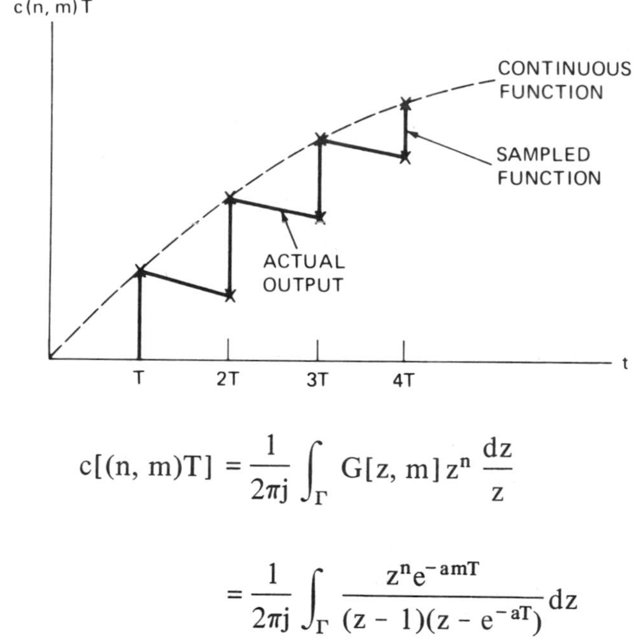

$$c[(n, m)T] = \frac{1}{2\pi j}\int_{\Gamma} G[z, m]\, z^n \frac{dz}{z}$$

$$= \frac{1}{2\pi j}\int_{\Gamma} \frac{z^n e^{-amT}}{(z - 1)(z - e^{-aT})} dz$$

With residues at $z = 1$ and $z = e^{-aT}$,

$$c[(n, m)T] = \frac{e^{-amT}(1 - e^{-anT})}{(1 - e^{-aT})}$$

Example 2. The output and the modified-output pulse-transfer functions of a typical sampled-data feedback control system are given by:

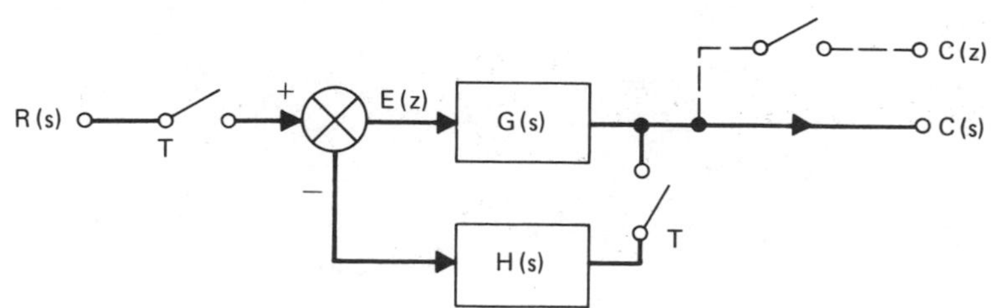

$$C(s) = E^*(s)G(s) \text{ and } E^*(s) = R^*(s) - H(s)C^*(s)$$

$$C(s) = R^*(s)G(s) - GH(s)C^*(s) \tag{1}$$

$$C^*(s) = R^*(s)G^*(s) - GH^*(s)C^*(s) \tag{2}$$

If two pulse-transfer-functions are separated by a sampler, then—and then only—the resultant pulse-transfer-function is the product of the two pulse-transfer-functions; without the sampler, the resultant pulse-transfer-function is the z-transform of the product of the two transfer-functions.

$$C^*(s) \triangleq C(z) = R(z)G(z)/[1 + GH(z)]$$

From (1): $C(s, m) = R^*(s)G(s, m) - GH(s, m)C^*(s)$

$$C^*(s, m) = R^*(s)G^*(s, m) - GH^*(s, m)[R^*(s)G^*(S)/(1 + GH^*(s)]$$

$$C(z, m) = R(z)\left[G(z, m) - \frac{G(z)GH(z, m)}{1 + GH(z)}\right]$$

Example 3. The direct inversion of the z-transformation of a pulse-sequence for a power-series output is given by:

$$C(z) = 1/(1 - 1.2z^{-1} + 0.2z^{-2}) = 1 + 1.2z^{-1} + 1.24z^{-2} + \cdots$$

Thus, long-division gives: $c^*(t) = 1.0\delta(t) + 1.2\delta(t - T) + 1.24\delta(t - 2T) + \cdots$

This impulse-sequence can be organized into a numerical-routine for a desk-calculator or a program of a digital controller, to obtain the transient-response of the pulse-transfer-function C(z). It may be noted that the inverse-term z^{-n} is the delayed-impulse $\delta(t - nT)$.

The *initial* and *final value theorems* applicable in $\mathcal{L}$-transform techniques take the corresponding form in z-transform analysis as shown:

$$\text{If } F(z) = f(0)z^0 + f(T)z^{-1} + \cdots f(nT)z^{-n} + \cdots$$

Initial-value theorem: $\lim_{z\to\infty} F(z) = f(0)$; corresponding to: $\lim_{t\to 0} f(t) = \lim_{s\to\infty} sF(s)$

Final-value theorem: $\lim_{z\to 1} (1 - z^{-1})F(z) = F(\infty)$; corresponding to:

$$\lim_{t\to\infty} f(t) = \lim_{s\to 0} sF(s)$$

where F(z) has no poles on or outside the unit-circle.

1.2 SOLUTIONS OF LINEAR DIFFERENCE EQUATIONS WITH CONSTANT-COEFFICIENTS

Since time-quantization is a predominant feature of the pulse/digital systems, the numerical linear difference equation provides a natural means for the formulation of the digital processes.

If we start with a function f(x), then g(x), given by the following form, is called the *backward difference.* The latter is used in real-time problems.

$$\nabla f(x) = g(x) = [f(x) - f(x-1)] \text{ for } x = n, n+1, \ldots, n+k; \text{ and } f(x) = 0 \text{ for } n < 0$$

In z-transforms: $z[g(x)] = (1 - z^{-1})G(z)$; from the definition of *shift-theorem*, viz., if $z[f(x)] = F(z)$, then $z[f(x-1)] = F(z)/z$; g(x) is called the *forward difference* if $g(x) = \Delta f(x) = f(x+1) - f(x)$. From the shift-theorem in this case, $z[g(x)] = (z - 1)F(z) - zf(0)$. The second forward difference $\Delta^2 f(x)$ is given by the first forward difference of g(x):

$$\Delta^2 f(x) = \Delta f(x+1) - \Delta f(x) = g(x+1) - g(x)$$

and, in general,

$$\Delta^k f(x) = \Delta^{k-1} f(x+1) - \Delta^{k-1} f(x)$$

Thus, a linear difference equation is formed by the linear relationship of the sampling-function and its various higher-order differences, the simplest example being $\Delta f(x) = 0 =$ operator Δ, and $(1 - \Delta) = E^{-1}$. By giving the next values 3, 4, . . . to k, the expression can be written in the general linear difference equation form:

$$\Delta^k f(x) = \sum_{n=0}^{k} (-1)^n \binom{k}{n} f[x + k - n]$$

where $\binom{k}{n} = k!/n!(k-n)!$ are the binomial coefficients $1 \leqslant n \leqslant k$.

Example: For the sampling-function $f(x) = e^{ax}$:

First linear difference equation: $\Delta f(x) = e^{x(n+1)} - e^{x(n)} = e^{ax}(e^x - 1)$

*k*th difference equation: $\Delta^k f(x) = (e^x - 1)^k e^{ax}$

Difference operator form: $\Delta^k = (1 - E^{-1})^n = [1 - \binom{n}{1}E^{-1} + \binom{n}{2}E^{-2} + \cdots]$

where E^{-k} is z^{-k} in z-transforms.

The process is similar to obtaining the derivative of an exponential function. The linear difference equation with constant-coefficients can also be written in the form:

$$f(x) = a_k \Delta^k y(x) + a_{k-1} \Delta^{k-1} y(x) + \cdots a_0 y(x)$$

or

$$f(x) = a_k y(x+k) + a_{k-1} y(x+k-1) + \cdots a_0 y(x)$$

where f(x) is the given known function, and y(x), the unknown, is the solution for the difference equation of the k-th order. The difference equation is thus a recurrence relationship, and, if y(0), y(1), etc., are given, we can obtain $y(x+k)$ for $x = 0, 1, 2, \ldots$ successively. It is this property of the sampled-data/pulse/digital system that distinguishes a difference equation from the differential equation of a continuous control system.

If $f(x) = 0$, the difference equation is homogeneous. The difference equations arise also in the analysis of physical systems with some timewise discrete-variables, such as equivalent-filters to transmission lines and voltage dividers. The constant-coefficient linear difference equations can be solved by using the z-transforms and their properties—(1) linearity, (2) original translation, and (3) image of difference—in the same way as the $\mathcal{L}$-transforms are used for the solution of differential equations. The algebraic z-transform equations thus formed contain all the boundary conditions and hence can be directly solved.

As an alternative, the difference equations can be solved by using the latest *state-variable* techniques along the following lines.

A second-order difference equation, $f(x) = a_1 f(x-1) + a_2 f(x-2)$, can be put in the form of the first-order difference equations. If $g(x) = f(x-1)$: $f(x) = a_1 f(x-1) + a_2 g(x-1)$

$$g(x) = f(x-1)$$

In matrix form,

$$\begin{bmatrix} f(x) \\ g(x) \end{bmatrix} = \begin{bmatrix} a_1 & a_2 \\ 1 & 0 \end{bmatrix} \begin{bmatrix} f(x-1) \\ g(x-1) \end{bmatrix}$$

f(x) and g(x) are the state-variables of the system of equations given in general by

$$\overline{Y}(x) = \overline{A}\,\overline{Y}(x-1)$$

where $\overline{Y}(x)$ and $\overline{Y}(x-1)$ are *state-vectors*, and $\overline{A}$ is *transition-matrix*. (The state-vector is represented by a column matrix.) It can be shown by iteration that the solution of this equation, in terms of the initial states of f(0), g(0), . . . , when $\overline{A}$ is a *square-matrix* of constants, is given by

$$\overline{Y}(k) = \overline{A}^k\,\overline{Y}(0),\ k = 0, 1, 2, \cdots \tag{1}$$

The *necessary and sufficient condition* that the above be a *stable solution* is that the *eigenvalues* or the roots λ_i of the characteristic equation $[(\overline{A} - \lambda\overline{I}) = 0]$ be less than one. $\overline{I}$ is the *unit identity-matrix* and $\overline{A}^k$ can be expressed as:

$$\overline{A}^k = p\overline{I} + q\overline{A} + r\overline{A}^2$$

in terms of the roots λ_i of the *characteristic equation*. By solving the above system of equations for p, q, and r, the solution is obtained from Eq. 1.

1.3 SIGNIFICANCE OF DIGITAL CONTROLLER (OR DIGITAL COMPUTER OR DIGITAL FILTER) IN STABILIZING PULSE-TRANSFER-FUNCTION OF SAMPLED-DATA SYSTEM

A digital controller should receive a sequence of digits (pulses) at equal time-intervals and process them in real-time into a command signal. The digital controller may be either an active or passive compensating device; it is normally preceded and followed by synchronous samplers as shown in Fig. 1-7 in the block-diagram of a sampled-data feedback control system.

The overall pulse-transfer-function K(z), describing the performance of a typical error-sampled digital-controller, can be a purely numerical process in which there is necessarily no associated physical system. The pulse processors in the sampled-data control systems of a complex digital control system, such as the quadruplex color videotape recorder, can be included under the classification of digital-controller. This method of including a digital-controller in general, in an otherwise continuous system makes it possible to obtain an overall system-stabilizing characteristic in the sampled-data system. These discrete controllers are usually classified under (1) general-purpose digital computers, (2) computers with digital storage and analog arithmetic operations, and (3) *digital filters.*

The desired overall pulse transfer-function K(z) in the system shown is given by:

$$D(z), \text{ in terms of } K(z) = \frac{1}{G(z)}\,\frac{K(z)}{1 - K(z)}$$

$$K(z) = \frac{D(z)G(z)}{1 + D(z)G(z)} \tag{1}$$

where

$$D(z) = \frac{E_2(z)}{E_1(z)} = \frac{a_0 + a_1 z^{-1} + \cdots a_n z^{-n}}{1 + b_1 z^{-1} + \cdots b_n z^{-n}}$$

The missing constant b_0 in the denominator is restricted to unity for *physical realizability*, since the expansion of D(z) into a power-series must not contain terms of positive powers in z. Terms a_0 and a_1 in the numerator frequently disappear. The

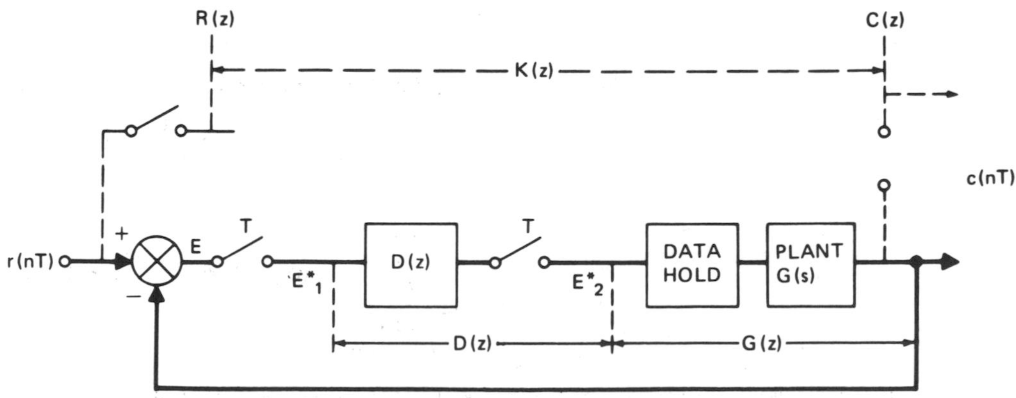

Fig. 1-7. Single-loop error-sampled feedback control system.

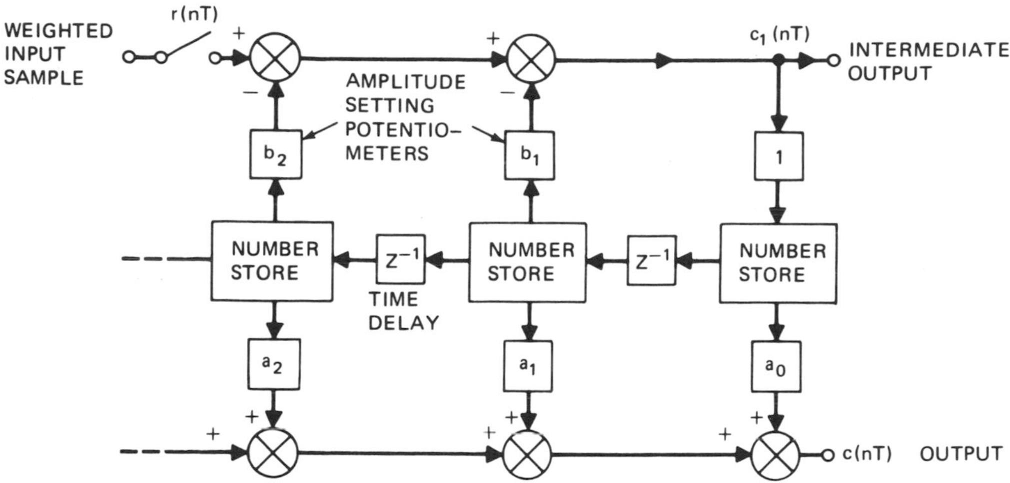

Fig. 1-8. A practical form of a digital controller.

linear program of a digital-controller is thus normally expressed as the ratio of two polynomials in the variable z. If $D(z) = D_1(z) \cdot D_2(z)$ in two factors or sections, the pulse-transfer-function D(z) can be implemented, for example, as shown in Fig. 1-8.

A linear computation that can be performed in a digital computer will be of the linear difference equation form:

$$c(nT) + b_1c[(n-1)T] + \cdots b_{n-1}c[(n-N+1)T]$$

$$= a_0r(nT) + a_1r[(n-1)T] + \cdots a_{n-1}r[(n-N+1)T]$$

where a_i and b_i are either constants or functions of the independent variable (t = nT). The input function r(t), being sampled at t = nT, is then stored and delayed (shifted) in time by integral values of T. In the direct method of programming, the operation for implementing the above difference equation will be directly arithmetic (additions, subtractions, multiplications) and data-handling (data transfer and storage, and constant storage).

The pulse-transfer-function K(z) must satisfy the following requirements if the synthesis of a closed-loop system by using digital compensation specifies a *minimal-response function* for obtaining an overall stable control system.

1. K(z) must contain as its zeros all the zeros of G(z) that lie on or outside the unit-circle in the z-plane.
2. [1 - k(z)] must contain as its zeros all the poles of G(z) that lie on or outside the unit-circle in the z-plane.
3. K(z) must be of the form:

$$K(z) = \frac{M_m z^{-m} + \cdots M_p z^{-p}}{N_0 + N_1 z^{-1} + \cdots N_q z^{-q}}$$

as obtained from Equation 1-1 with zero cancelling terms. In this equation, N_0 is not zero but unity for physical realizability, and m is the lowest order of

G(z) in z^{-1}. G(z) is given by:

$$G(z) = (P_m z^{-m} + \cdots P_n z^{-n})/(q_0 + q_1 z^{-1} + \cdots q_i z^{-i})$$

Examples of pulse-transfer-functions K(z) for various types of inputs R(z):

For step,

$$\frac{R(z) \longrightarrow K(z)}{1/(1 - z^{-1}) \longrightarrow z^{-1}}$$

For ramp,

$$Tz^{-1}/(1 - z^{-1})^2 \longrightarrow (2z^{-1} - z^{-2})$$

For acceleration,

$$T^2 z^{-1}(1 + z^{-1})/(1 - z^{-1})^3 \longrightarrow (3z^{-1} - 3z^{-2} + z^{-3})$$

4. K(z) is specified so that the *steady-state error* from the application of an input of the above form $[R(z) = A(z)/(1 - z^{-1})^m]$ is zero. A(z) is a polynomial with no terms of the form $(1 - z^{-1})$.

$$E_1(z) = R(z)[1 - K(z)]$$

Applying the final-value theorem,

$$e_1(\infty) = \lim_{z \to 1} \{(1 - z^{-1})R(z)[1 - K(z)]\}$$

If $[1 - k(z)] = (1 - z^{-1})^m F(z)$ . . . where F(z) is an unspecified ratio of the polynomials in z^{-1}, the steady-state error for the inputs of the above form will be zero. For m = 1, the system will respond with zero steady-state error for the step input. For m = 2, zero steady-state order is obtained for the ramp input.

If F(z) is equal to 1, a *minimal-prototype response function* is produced and the order of K(z) in z^{-1} is a minimum. This type of minimal-function is effective only if K(z) has only a numerator polynomial in z^{-1}, as shown in the preceding examples, and does not contain for stability any zeros on or outside the unit-circle in the z-plane. *Finite settling-time* sampled-data systems also belong to this category, but they must be *ripple-free*. Special techniques are available for this purpose.

1.4 SURVEY OF ANALOG AND DIGITAL FEEDBACK CONTROL TECHNIQUES FOR ANALYSIS AND SYNTHESIS

1.4.1 Comparison of Two Basic Original Analytical Techniques for Sampled-Data Control Systems. Most of the available steady-state frequency-response methods of the original control-theory approach, such as (1) Nyquist criterion, (2) logarithmic Bode and Nichols plots, and (3) Evan's root-locus method, used for continuous feedback control systems have been extended to cover sampled-data control systems on a modified basis, since the introduction of the z-transform calculus to this branch by Hurewicz and Barker in 1950. The frequency-response methods, involving *deterministic* or known functions of time as input signals, emphasize the treatment on the *transient-response* basis, without any consideration of the effects of noise, and disturbances on the system. On the other hand, random or *stochastic* input signals—such as (1) noise and low-frequency disturbances in digital control systems of the complexity of the quadruplex color videotape recorder (Q-CVTR) and (2) wind and waves, respectively, in a gyro used in an aircraft autopilot control system and a ship —can be mathematically described as a random process governed by the concepts of probability distribution-functions and probability density-functions of various orders. This approach of control theory to the analytical synthesis of compensation in the form of a *digital filter*, without a specific need for the complex definition of a transfer-function of the overall control system, is *statistical* in character, since it is based entirely on the rejection or the immunity to the random noise and disturbances.

The statistical method is substantiated by the design techniques of Wiener's *least-mean-squared-error (LMSE) criterion* in minimizing the system response to noise, by using the auto- and cross-correlation functions formulated from the mathematical concepts of random processes. With this technique, the desired system transfer-function is expressed in terms of the signal and the noise power-spectra. The statistical method is presently applicable to the continuous and sampled-data control systems on a separate basis, although in practice most sampled-data control systems are basically hybrid in this respect with a continuous plant. Modern *state-variable* techniques, however, make an attempt on a generalized basis. The Wiener's criterion ignores the aspect of the transient-response too, by assuming a nominal optimum transient-response from experience. Therefore, the sampled-data control systems designed on this basis have a qualitative tendency to become underdamped systems with an oscillatory transient-response. Pronounced overshoots used to cause scaling problems in the more conventional fixed-point digital computers of the last generation, and underdamped filters make an overall digital-analog control system occasionally sensitive to parameter changes for reliable stability on a long-term basis. The LMSE approach has another demerit because it emphasizes large errors as shown in Fig. 1-9a. Weighting in Fig. 1-9b is more realistic.

In view of the shortcomings inherent in both the preceding basic approaches in the sampled-data control theory, the latest *computer-aided methods* attempt to develop an analytically manageable *composite criterion* for the design of a digital-

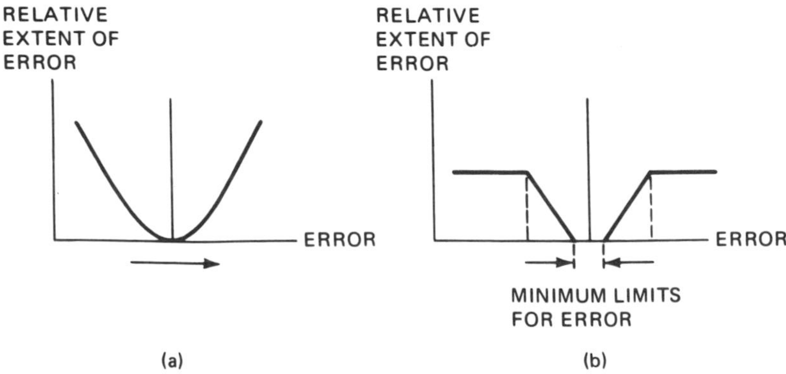

Fig. 1-9. *a*, Least-mean-squared error-weighting. *b*, Nonlinear error-weighting.

filter that gives a maximum rejection of noise, consistent with some control over the overshoot of a step-input signal. With this approach, the input is considered as a signal, belonging, respectively, to either (1) a *stationary random process* or (2) a *deterministic-function of time*, the signal in this case being represented by a polynomial of a finite degree. Noise of a stationary random character is treated as a common factor in both the cases, and the synthesis aims at separating the noise from the signal to perform the *linear* or *nonlinear operation* on the signal.

However, direct syntheses of transient performance have been developed and presented in charts for a few realistic but restricted cases. The application of the Direct (Second) Method of Lyapounov, involving state-variables and vector equations, to the design of both linear and nonlinear continuous and sampled-data control systems, is an outstanding example of the application of classical mechanics to the control systems by using the *calculus of variations* as a primary tool. Modern control theory has produced still another special technique, the *Pontryagins Maximum Principle* (also of classical mechanics), for the optimization of the analog and digital systems. Some control systems are classified in modern control theory as (1) *optimal* and (2) *adaptive control systems*, and a distinction between these two types of systems will follow the various techniques mentioned. Before we attempt to apply some of the preceding techniques to comparable sampled-data control systems in the color videotape recorder, a brief exposition of the basic concepts governing these techniques is given in this section to acquaint the digital-circuits engineer concerned with the development and design of the various nonlinear sampled-data control systems in the Q-CVTR. The final control system-complex evolved in the laboratory is essentially reliable for all practical purposes. Actually, it is the final successful result of a heuristic development-and-design approach, using the *measure-and-optimize* techniques on a step-by-step and loop-by-loop basis.

It may be noted that the problems in linear and nonlinear automatic control systems can be treated, in general, under (1) *analysis* where it is necessary *to predict the behavior* of a specified system for system-stability and transient-response; (2) *synthesis for the design* of a system with the necessary compensation in the form of a *digital-filter* to meet the desired specification; and (3) design of a *self-adaptive learning system.* Modern techniques with sampled-data control systems require the use of a general-purpose digital computer and/or *built-in microprocessors* for applying the algorithms arrived at after painstaking *iterative debugging effort* for the solution of the large amount of numerical-analysis and computation involved; programmable read-only memories that could be electronically alterable (EAROMs) will be convenient for this function of the software.

The conventional block-diagram approach requires the transfer-characteristics of the system components, and the determination of the overall transfer-function, so that the designer could then choose the controller of compensation that would meet both the steady-state static and the dynamic (transient) performance specifications. The *modern state-space approach* of analysis and synthesis, on the other hand, characterizes the system by a number of first-order differential equations representing the *state-variables* of the system (such as the generalized coordinates in the classical mechanics), whereas the *initial-state conditions* are represented by the appropriate state-transition equations.

1.4.2 Frequency-Response Techniques for Linear Control Systems. Sampled-data control systems may be very roughly approximated to continuous systems at high sampling rates to enable in principle the application of the conventional frequency-response techniques, for example, by replacing the sampler and the *first-order* hold network by unity; but the production of the harmonics or sidebands produced by the sampling operation make the design far from satisfactory. The continuous network approximation to the more common *zero-order* hold (or clamp) and the sampler is actually more difficult. Therefore, special techniques are formulated, from the point of view of transient-response, to allow the application of the frequency-response methods to sampled-data control systems.

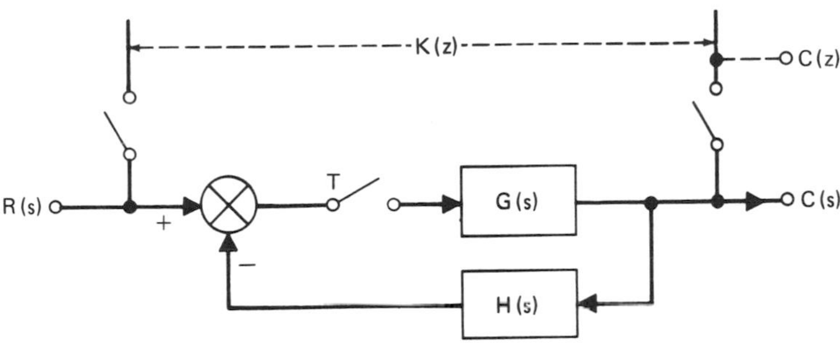

The overall closed-loop pulse-transfer-function of a sampled-data feedback control system consisting of one or more samplers, as shown in the error sampled feedback control system above, is directly obtained as:

$$K(z) = \frac{C(z)}{R(z)} = \frac{G(z)}{1 + GH(z)} = \frac{N_1(z)D_2(z)}{[D_1(z)D_2(z) + N_1(z)N_2(z)]}$$

where

$$G(z) = \frac{N_1(z)}{D_1(z)} = \frac{\text{zeros}}{\text{poles}}$$

$$H(z) = \frac{N_2(z)}{D_2(z)}$$

The zeros of the characteristic equation make the poles of K(z). (The denominator is equated to zero.)

The stability of a feedback control system is uniquely determined by those values of z that satisfy the characteristic equation $1 + GH(z) = 0$, appearing in the closed-loop pulse-transfer-function. All methods of stability-analysis confine themselves to the investigation of this equation one way or the other by techniques that (1) obtain the *explicit value of the roots* of the equation, as in the case of Evan's root-locus method, and (2) obtain conclusive information about the *bounded-regions wherein all the roots lie*, as in the case of the frequency-response methods such as the Nyquist criterion, the Bode-Nichols plots (in an indirect manner), and the Routh-Hurwitz criterion.

In the case of the sampled-data pulse-transfer-function, Mobic's *bilinear-transformation* is applied to the characteristic equation in z, so as to transform (1) the area *outside the unit-circle in the z-plane back to the right-half of a corresponding w-plane* (back, since it simulates the complex s-plane) and (2) the area *inside the unit-circle to the left-half of the w-plane*, if the w-plane is by definition related to the z-plane by the transformation,

$$z = \frac{(1 + w)}{(1 - w)} \quad \text{or} \quad \frac{(1 + z)}{(1 - z)} = w$$

1.4.3 Nyquist Criterion. For a continuous linear control system, the primary purpose of the Nyquist test is to show the existence of any zeros of the characteristic equation $1 + GH(s) = 0$, in the right half-plane of the s-plane, since a single such zero makes the system unstable. If all the roots are negative and all complex roots have negative real-parts, it is *absolute stability*. The manipulation, *based on Cauchy's principle of argument, is a conformal-mapping procedure* of the imaginary-axis (in the s-plane) to a polar-plot defined by the loop transfer-function GH(s) in the GH-plane. The roots within the right half-plane are interpreted by the *counterclockwise encirclement of point* (*-1, 0*) in the GH(s) closed-contour plot containing poles and zeros.

CCW revolutions in positive sense: N = [number of GH(s) poles in the right-half of s-plane] - [number of zeros of $1 + GH(s) = 0$]. The criterion is based on the fact that *the frequency-response of the open-loop transfer-function GH(s) indicates the stability characteristics of the closed-loop system.*

In the case of the sampled-data control systems, the contour of the pulse-transfer-function GH(z) in the polar-plane is plotted for magnitude, using the expression

$$z = e^{j\omega t} = (\cos\theta + j\sin\theta)$$

for various values of θ. The z-transformation transfers the right half-plane to the region outside the unit-circle. If GH(z) and hence $[1 + GH(z) = 0]$ do not contain poles outside the unit-circle, then the open-loop pulse-transfer-function GH(z) is stable. If, however, a zero does lie outside the unit-circle, the closed-loop system will become unstable. A pole at the origin of the s-plane for a continuous transfer-function corresponds to a pole at (1, 0) on the unit-circle in the case of the pulse-transfer function. It is *detoured* (as in the continuous case), so that the pole is included in the unit-circle as shown. While the proximity to the critical-point (-1, 0) is avoided in the continuous systems for satisfactory transient-response, by allowing adequate phase and gain margins, the practice is not important in sampled-data control sys-

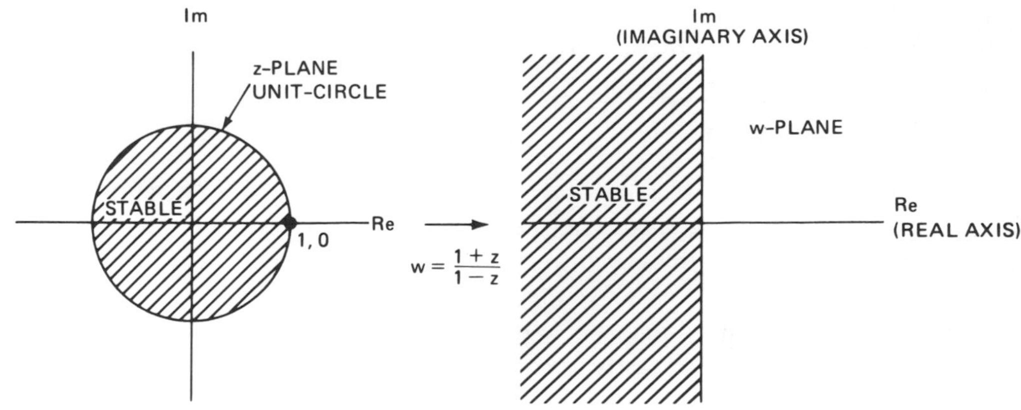

Fig. 1-10. Use of w-transform for analyzing the pulse-transfer function.

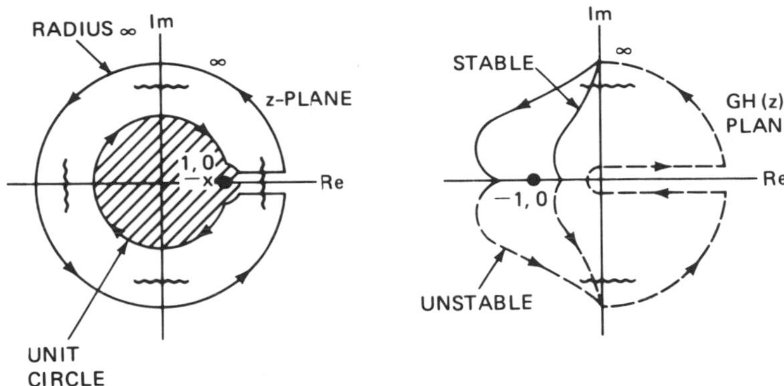

Fig. 1-11. Nyquist stability test with the polar-plot in GH(z) plane.

tems, because the transient-response is not tied to these margins in quite the same way, although the degree of damping does change to some tolerable extent. The use of the Nyquist pulse-transfer-locus is not common in sampled-data control systems since the time-domain performance characteristic of the system is more easily obtained by taking the inverse-transform of the closed-loop pulse-transfer-function.

1.4.4 Bode-Nichols Plots. For well-behaved, minimum-phase GH(s), the negative-feedback closed-loop system is stable if, at the frequency where the log-magnitude of $G(j\omega) \cdot H(j\omega)$ is equal to zero, its phase-angle is less than $-180°$. For minimum-phase transfer-functions or networks with no poles or zeros in the right half-plane, magnitude and phase are uniquely defined; the specification of one will specify the other. At the gain crossover, a positive phase-margin indicates an unstable system. A gain-constant (K), a differentiator and an integrator $(s^{\pm 1})$, a simple lead or lag network $(s + \omega_0)^{\pm 1}$, and a quadratic lead or lag network, $(s^2 + 2\zeta\omega_0 s + \omega_0^2)^{\pm 1}$ having a damping-constant ζ, make the basic building-blocks for the magnitude-versus-frequency Bode plots $(\omega_0 = 2\pi f_o)$. *The closed-loop response can be obtained from a gain-phase plot of the open-loop GH(s) transfer-function* by superimposing and inspecting the points of intersection of this gain-phase characteristic on the Nichols curves in the complex-plane. The curves in the Nichols chart correspond in rectangular coordinates to the loci of *constant-magnitude M-circles* and *constant-phase α-circles* in a closed-loop feedback control system.

The Bode-plot technique is easily extended to the sampled-data control systems by (1) applying the bilinear-transformation $z = (1 + w)/(1 - w)$ to the open-loop GH(z), and then (2) obtaining the log-magnitude $20 \log_{10}|GH(jv)|$, vs v plot for the GH(jv) transfer-function, to arrive at the gain and phase-margin figures. The complex variable $w = u + jv$.

The variable jv is considered a fictitious frequency in this particular technique. The plots can be then extended to the *Nichols chart for the closed-loop response*, making sure that due modification is made to GH(jv)-response on the Nichols chart by 1/H(jv) for the overall closed-loop response (since the M and the α-loci curves in the Nichols chart are by definition based on the assumption of unity feedback).

1.4.5 Routh-Hurwitz Criterion. The Routh-Hurwitz method is an algebraic method for finding whether a polynomial representing the characteristic equation of the overall closed-loop feedback control system has roots with positive real-parts in the right half-plane, because a single such root will make the system unstable. Thus, the

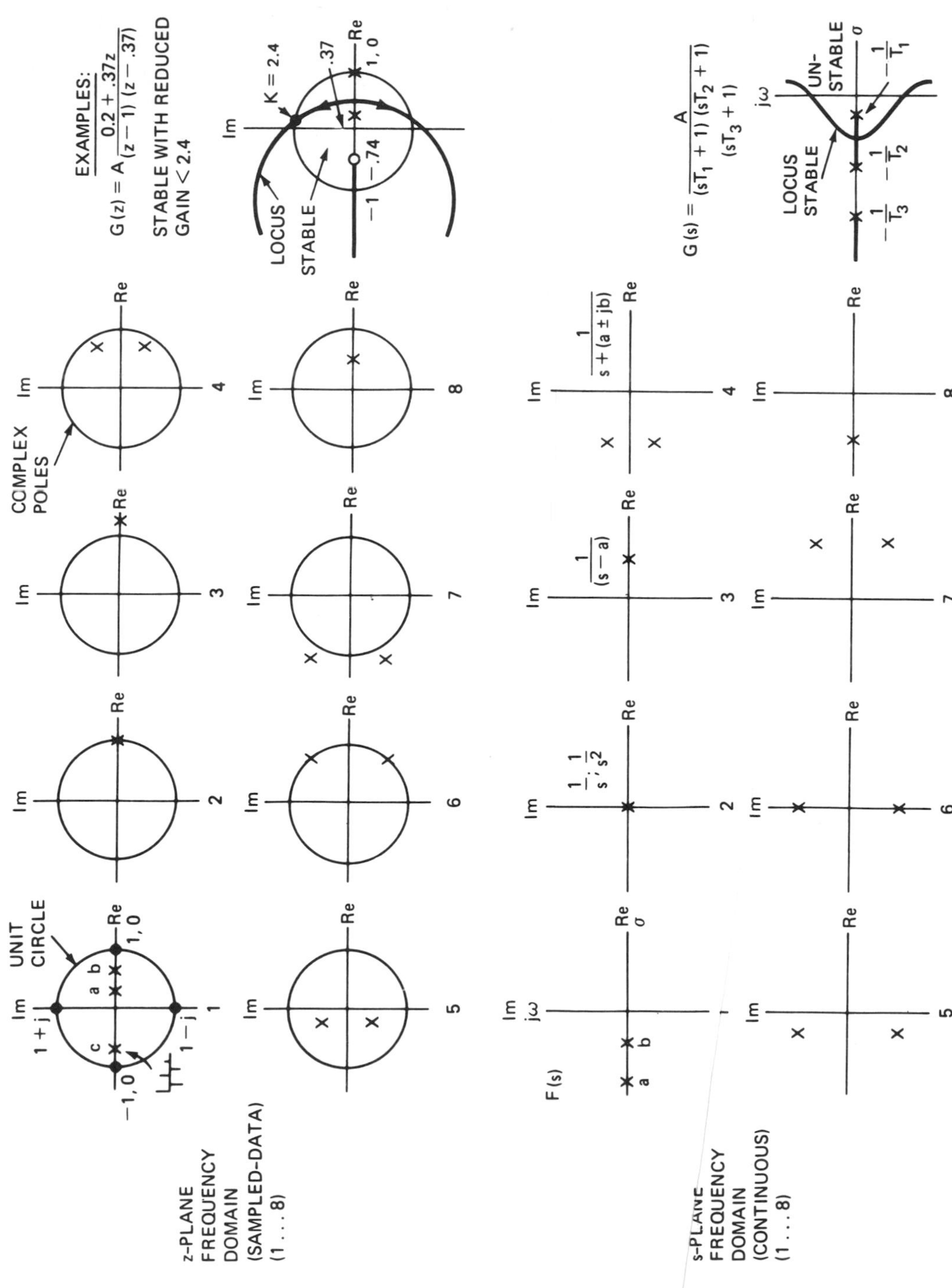

z-PLANE FREQUENCY DOMAIN (SAMPLED-DATA) (1 . . . 8)
UNIT CIRCLE
1 + j
1 − j
−1, 0
1, 0
Im
Re
COMPLEX POLES
EXAMPLES:
G (z) = A (0.2 + .37z)/((z − 1) (z − .37))
STABLE WITH REDUCED GAIN < 2.4
K = 2.4
LOCUS
STABLE
−1
−.74
.37
s-PLANE FREQUENCY DOMAIN (CONTINUOUS) (1 . . . 8)
F (s)
jω
σ
1/s , 1/s²
1/(s − a)
1/(s + (a ± jb))
G (s) = A/((sT₁ + 1) (sT₂ + 1) (sT₃ + 1))
UN-STABLE

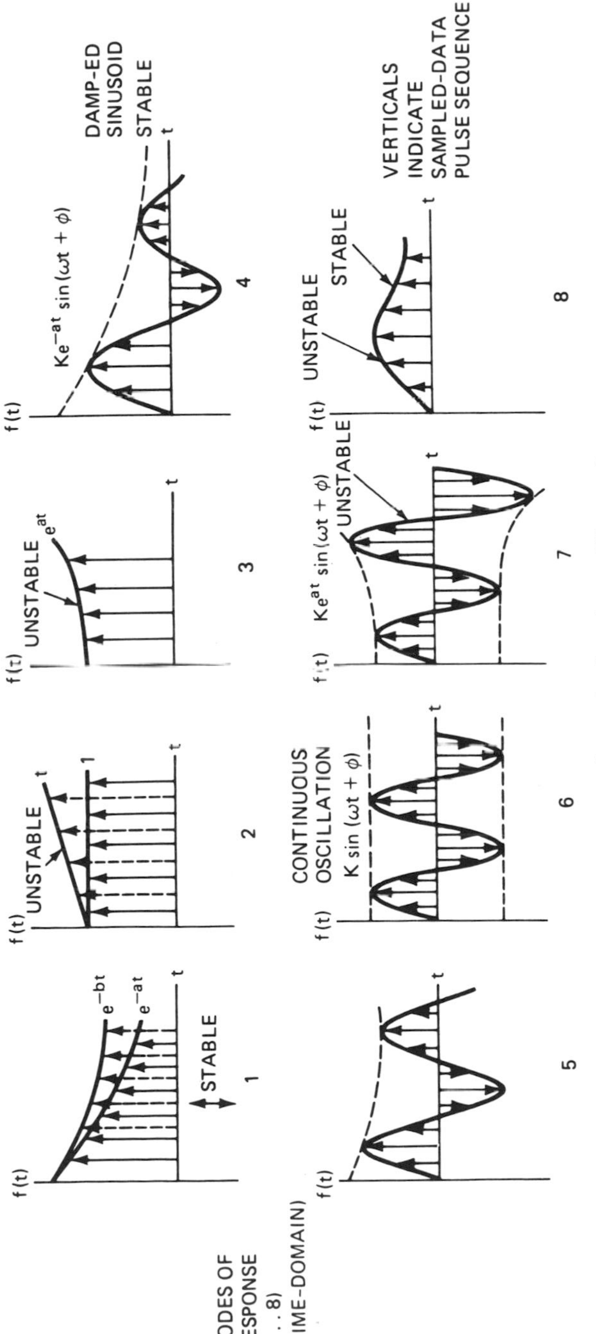

Fig. 1-12. Plotting of root loci: pole-position and response.

method indicates merely the *bounded-regions where the roots lie* in the complex s-plane. A triangular array is formed in terms of the constant-coefficients in the polynomial ($s^n + a_1 s^{n-1} + \cdots a_{n-1}s + a_n = 0$). The number of changes in the sign of the terms in the first column of the triangular array determine the number of roots with the positive real-parts, when the system is unstable; the system is stable if no change of sign occurs.

$b_1 = (a_1a_2 - a_3)/a_1$	s^3	1	a_2	a_4	a_6:	(Array:
$b_3 = (a_1a_4 - a_5)/a_1$	s^2	a_1	a_3	a_5		coefficient
$c_1 = (b_1a_3 - a_1b_3)/b_1$	s^1	b_1	b_3			1 of s^3
etc.	s^0	c_1				may be a_0.)

No change of sign for stability

In the case of the sampled-data control systems, the characteristic equation of the closed-loop pulse-transfer-function in z is changed over to the w-transform by bilinear-transformation, as in the case of the Bode plot, and the triangular array is formed out of the polynomial in w to indicate the bounded-region wherein the roots lie. The constants of the original system may appear differently, as complex-numbers in the w-transform polynomial, to directly correlate the original coefficients. The criterion, therefore, serves merely as a check on the stability conditions obtained by other mapping procedures. The method is laborious for the higher-order systems.

The Routh-Hurwitz stability criterion may be more conveniently stated in the determinant form. In the characteristic equation in s (for continuous) or w-transform (for sampled-data), the roots of the equation will have negative real-parts if the constants are all positive. A necessary and sufficient condition for stability is that each of the following determinants formed out of the polynomial must be positive.

$$\Delta_1 = \begin{vmatrix} a_1 \end{vmatrix} > 0; \quad \Delta_2 = \begin{vmatrix} a_1 & 1 \\ a_3 & a_2 \end{vmatrix} > 0; \quad \Delta_3 = \begin{vmatrix} a_1 & 1 & 0 \\ a_3 & a_2 & a_1 \\ a_5 & a_4 & a_3 \end{vmatrix} > 0, \text{ etc.}$$

Schur-Cohn stability criterion and Jury's stability criterion are two other special techniques along this line for sampled-data control systems. Both of these methods are based on the algebraic manipulation of the determinants formed out of the coefficients of the characteristic equation in z directly. (The application of the Schur-Cohn stability criterion is explained in Section 3.3.)

1.4.6 Root-Locus Method. Evan's root-locus method is a *graphical technique for finding the roots of the characteristic equation of a closed-loop control system, as functions of the open-loop system-gain.* The *loci of the roots are plotted* from the open-loop pole-zero configurations in the GH(s) complex plane. The values of the roots, along with their exact locations in the complex plane at a certain gain, clearly indicate the system-stability and the modes of the transient-response. If the system behavior is unsatisfactory, the locations of the poles in the plot, as a function of the gain, indicate possible methods of compensation. The plotting of the root-loci is a time-consuming procedure; mechanical drafting devices such as a spirule aid the vector-angle summation and the locus-calibration.

The basic rules for plotting the root-locus are:

1. The loci on the real axis are plotted by applying the rule that only those parts of the real-axis lying on the left of the odd number of poles and zeros, in that order, may be root-loci.

2. The asymptotes of the complex root-loci are graphically plotted by applying the formulas: (a) the angle that the asymptotes make with the real-axis is $h\pi/(m - n)$, where $h = \pm 1, \pm 3, \ldots$ for $n \neq m$; n = number of poles; and m = number of zeros; and (b) the crossover of the asymptotes on the negative real-axis is at a distance from the origin equal to

$$\frac{\Sigma n_i - \Sigma m_i}{n - m}$$

n_i and m_i being the sum of the real-parts of the poles and zeros.

3. The points of intersection of these loci with the imaginary axis are obtained by applying the Routh-Hurwitz criterion to the closed-loop characteristic equation.

In the case of the sampled-data control systems, the transient component of the output-response $c(nT)$ is determined by the poles and zeros of the closed-loop pulse-transfer-function $K(z)$. In general,

$$K(z) = \frac{AG(z)}{1 + \overline{AGH}(z)} = \frac{a_0 + a_1 z^{-1} + a_2 z^{-2} + \cdots a_n z^{-n}}{1 + b_1 z^{-1} + b_2 z^{-2} + \cdots b_n z^{-n}}$$

If the zeros of the characteristic equation are z_i, then $K(z)$ can be expanded into partial fractions:

$$K(z) = a_0 + \frac{A_1}{1 - z_1 z^{-1}} + \frac{A_2}{1 - z_2 z^{-1}} + \cdots$$

For impulse input, the time-domain sequence is then given by

$$c(nT) = a_0 + A_1(z_1)^n + A_2(z_2)^n + \cdots$$

It is obvious from the preceding expression that if the poles of the pulse-transfer-function $\overline{AGH}(z)$ have magnitudes greater than unity (for $n > 1$), the impulse-response increases without bounds to make the system unstable.

Since it is difficult to obtain the poles of the overall closed-loop pulse-transfer-function $K(z)$ for the higher-order systems, the root-loci in the complex plane are plotted from the roots of the open-loop $\overline{AGH}(z)$ pulse-transfer-function, as in the case of the continuous systems.

The root-locus, described by the relationship [angle $\overline{AGH}(z) = \pi \pm 2\pi n$], satisfies the phase-relationships implicit in finding all those roots of z in the complex-plane that satisfy the condition $\overline{AGH}(z) = 1/(\pi \pm 2\pi n)$. The phase-relationships are satisfied by using the gain A, which represents the gain-constants of the system, as an adjustable parameter. The only difference in the case of the analyses of the sampled-data control systems is that the behavior of the poles is examined with respect to the unit-circle in the complex z-plane, as against the imaginary-axis in the case of the complex s-plane. If it is the synthesis of the closed-loop system that is required, the effect of the desired compensation with respect to the location of the poles in the graphical plot is examined. For example, the closer that the poles approach the magnitude unity, the larger the transient decay-time when the system is subjected to an abrupt command.

For *physical realizability and causality*, a network function F(s) must be a positive real-function as defined by (1) the function must have no poles in right half-plane, (2) the function may have only simple poles on $j\omega$-axis with real and positive residues, and (3) Re $F(j\omega) \geqslant 0$ for all ω. In the case of a transfer-function G(s), which is a rational function with real coefficients, the order of the numerator may exceed the order of the denominator by 1, but the numerator order may be any order less than that of the denominator. Poles and zeros may be either real, or located on the $j\omega$-axis or complex-conjugate. G(s) is measured unlike a driving-point impedance at two different pairs of terminals (2-port). The zeros can lie anywhere in the complex plane, but the denominator polynomial must be Hurwitz. Once F(s) satisfies these requirements, its z-transform equivalent for the sampled-data system should meet similar requirements with respect to the unit-circle.

1.4.7 Linvill's Method of Sampler Approximation. In Linvill's method, a polar-plot of the loop transfer-function $G(j\omega) \cdot H(j\omega)$ is plotted using the Nyquist method, and then a few harmonic terms of the sampled transfer-function, given by the following expression, are vectorially added for approximation.

$$\frac{1}{T} \sum_{-\infty}^{\infty} G(j\omega + jn\omega_s)H(j\omega + jn\omega_s) \cdots$$

$$\omega_s \text{ corresponding to the sampling-frequency} \quad (1)$$

A compensating network $N(j\omega)$ is then chosen, using one of the preceding analog techniques; the compensated loop transfer-function, with the new product-term of $(j\omega + jn\omega_s)$ in (1), is replotted with the same number of harmonic terms as before. The method is merely an extension of the continuous analog system techniques and includes due approximation for the sampler effects. The z-transform analysis is then in order to check the performance of the system.

1.5 STATISTICAL DESIGN TECHNIQUES BY MEANS OF THE LEAST-MEAN-SQUARED ERROR APPROACH

Wiener's criterion for the successful optimization of the parameters of a particular configuration of compensation is basically a synthesis procedure for a *digital filter.* The solution specifying the form of compensation required, with or without some *power constraint*, when random noise and disturbances limit the system performance objectives, is an *explicit mathematical model* of the desired system transfer-function. It is expressed in relation to the *signal and noise spectra.* The LMSE-value aimed at in sampled-data control systems is a statistical property of the sampled-variable for *minimum noise-transfer.* It can be derived from a knowledge of the auto- and cross-correlation functions encountered in *random/stochastic processes.*

1.5.1 Auto- and Cross-Correlation Functions and Wiener's Criterion. If x is a variable having values $x_1, x_2, \ldots, x_n$ with respective probabilities of $Pr(x_1), Pr(x_2), \ldots,$

$Pr(x_n)$, and $\Sigma Pr(x_i) = 1$, then x is called a *random* or *stochastic-variable.* A random signal-process is represented by the mathematical concepts of *probability distribution*, and *probability density functions* of various orders. For example, the first-order probability density function f(x,t) associated with a random signal X(t), stationary in time, is thus designated,

$$Pr[a \leqslant X(t) \leqslant b] = \int_a^b f_1(x,t)\, dx$$

The dc content or roughly the moment of the first-order random-variable signal is given by:

$$E[X(t_1)] = \int_{-\infty}^{\infty} f_1(x_1,t)\, dx$$

The second-order density-function $f_2(x_1 t_1; x_2 t_2)$ gives a more detailed description of the random signal.

The generalized second-moment of a random variable of second-order probability density function is defined as the *autocorrelation function of a stationary random process*, and is obtained by averaging in time (ensemble-average):

$$\phi_{xx}(t_1,t_2) = \overline{X(t_1)X(t_2)} = \phi_{xx}(\tau) = \int_{-\infty}^{\infty}\int_{-\infty}^{\infty} x_1 x_2 f_2(x_1, x_2, \tau)\, dx_1\, dx_2$$

where

$$\tau = t_2 - t_1$$

On time-basis,

$$\phi_{xx}(\tau) = \lim_{T \to \infty} \frac{1}{2T} \int_{-T}^{T} x(t)x(t+\tau)\, dt = E[X(t)X(t+\tau)]$$

$$= E[X^2(t)] \text{ when } t_2 = t_1$$

$E[X^2(t)]$ is the mean-square value of the random-variable at t_1. The auto-correlation function can be interpreted as a measure of the extent to which the value of f(t) at any given time can be used to predict f(t) at a time-interval τ later.

If, on the other hand, two stationary random signals are under consideration such as signal $f(x,t_1)$ and noise $f(y,t_2)$, the term *cross-correlation* is used to describe the process. (In the case of noise, the average value is zero.)

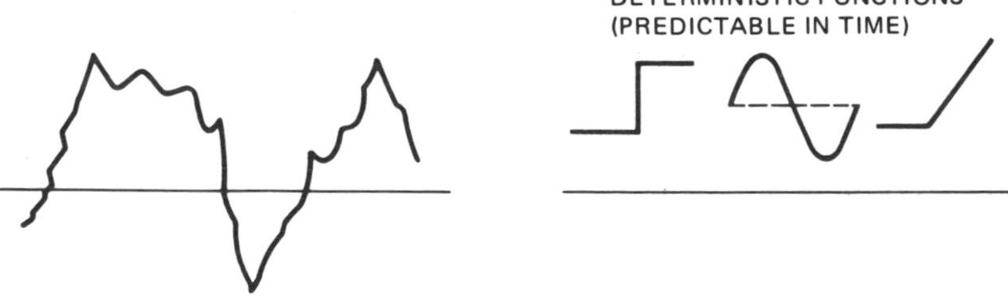

Fig. 1-13. Random or stochastic signal.

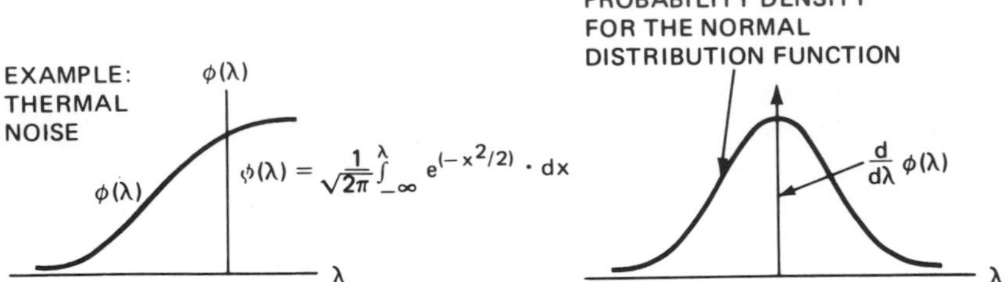

Fig. 1-14. Normal or Gaussian distribution function, $\phi(\lambda)$.

On an ensemble-basis,

$$\phi_{xy}(t_1,t_2) = \phi_{xy}(\tau) = \int_{-\infty}^{\infty}\int_{-\infty}^{\infty} xyf_{11}(x,y,\tau)\,dx\,dy$$

On a time-basis,

$$\phi_{xy}(\tau) = \lim_{T\to\infty} \int_{-T}^{T} x(t)x(t+\tau)\,dt$$

In more familiar terms analogous to the Fourier-transform and its inverse,

The auto-correlation function:

$$\phi_{xx}(\tau) = \frac{1}{2\pi}\int_{-\infty}^{\infty} \Phi_{xx}(s)e^{j\omega t}\,d\omega \tag{1}$$

The auto-correlation spectrum:

$$\Phi_{xx}(s) = \int_{-\infty}^{\infty} \phi_{xx}(\tau)e^{-j\omega\tau}\,d\tau \tag{2}$$

The latter is also called the power spectral-density; it is a measure of the mean-power in signal/unit-bandwidth.

$$\Phi_{xx}(s) = \Phi_{xx}^{+}(-s) + \Phi_{xx}^{+}(s)$$

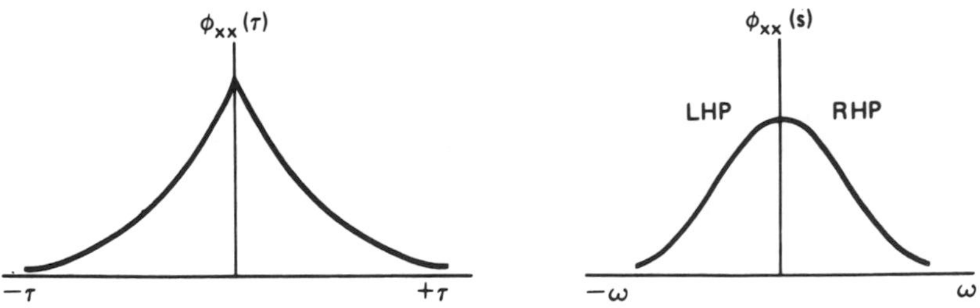

Fig. 1-15. Auto-correlation function and spectral density.

Also

$$\Phi_{xx}(s) = \underset{\text{LHP}}{\Phi^{+}_{xx}(s)}\underset{\text{RHP}}{\Phi^{-}_{xx}(s)}$$

The spectral-density has symmetric characteristics as given in the preceding equation. It is called *spectrum factorization.*

The Weiner-Hopf integral equation that follows is derived for a linear system by the method of the calculus-of-variations by taking into account the auto- and cross-correlation functions, respectively, representing the input and ideal output and actual output. If w(t) is the system weighting function,

$$\int_{-\infty}^{\infty} w(t)f[\phi_{xx}(\tau - t)]\, dt - f[\phi_{xy}(\tau)] = 0$$

The explicit time-domain solution of this equation gives the *system weighting function that minimizes the mean-squared error* when the system configuration is free. Under semifree configurations and constraints such as saturation and minimum bandwidth, the formulation of the integrand and hence the solution are subject to modification.

$$\text{System weighting function } W(s) = \left[\frac{\Gamma(s)}{\Delta^{-}(s)}\right]_{+} \Big/ \Delta^{+}(s)$$

where $\Delta(s)$ and $\Gamma(s)$ are functions of auto- and cross-correlation power-spectrums.

Example:

$$\Phi_{xx}(s) = -\frac{1}{s^2}; \Phi_{yy}(s) = \frac{15}{16 - s^2}$$

Noise and data are not correlated:

$$[\Phi_{xy}(s) = \Phi_{yx}(s) = 0] : s = j\omega$$

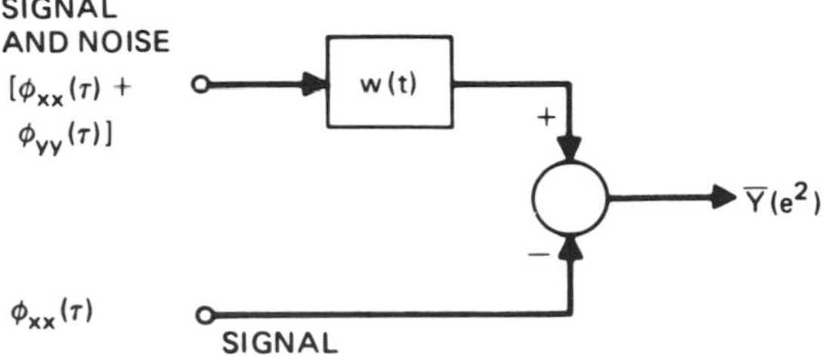

This simple example illustrates the method of design for a physically realizable digital-filter network, W(s), that is suitable for the minimization of the mean-squared error:

If $W_d(s)$ = Desired transfer function in the absence of noise = 1,

$$W(s) = \left[\frac{\Gamma(s)}{\Delta^{-}(s)}\right]_{+} \Big/ \Delta^{+}(s)$$

where

$$\Gamma(s) = W_d(s)[\cancel{\Phi_{xy}(s)} + \Phi_{xx}(s)] = \Phi_{xx}(s)$$

$$\Delta(s) = \Phi_{xx}(s) + \Phi_{yy}(s) + \cancel{\Phi_{xy}(s)} + \cancel{\Phi_{yx}(s)}$$

$$\Delta(s) = W_d(s)[\Phi_{xx}(s) + \Phi_{yy}(s)] = 1\left[-\frac{1}{s^2} + \frac{15}{16 - s^2}\right] = \frac{16(1+s)(1-s)}{[-s(4-s)][s(4+s)]}$$

$$\Delta^-(s) = 16(-s+1)/[-s(-s+4)] \text{ and } \Delta^+(s) = (-s+4)/s(s+4)$$

$$\frac{\Gamma(s)}{\Delta^-(s)} = -\frac{1}{(s)(s)}\,\frac{(-s)(-s+4)}{16(-s+1)} = \frac{-s+4}{16s(-s+1)} = \frac{1}{4s} + \frac{3}{16(-s+1)}$$

using partial fractions.

Desired filter,

$$W(s) = \frac{(s+4)\cancel{s}}{4(s+1)\cancel{s}}$$

The above transfer-function is of the network form $= \dfrac{T_1\left(s + \dfrac{1}{T_1}\right)}{T_2\left(s + \dfrac{1}{T_2}\right)}$

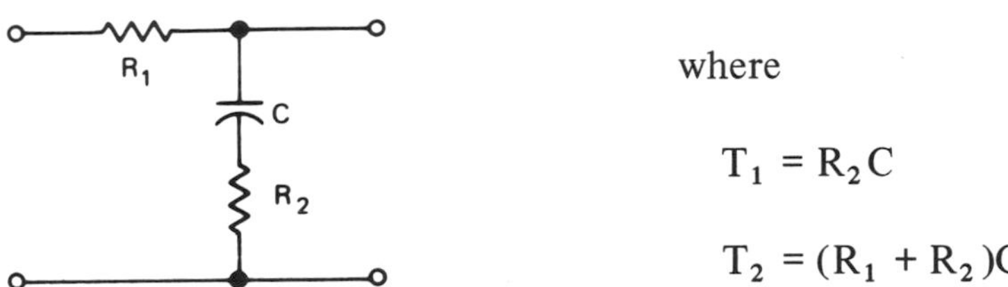

where

$$T_1 = R_2C$$

$$T_2 = (R_1 + R_2)C$$

The mean-squared error referred to in the Weiner-Hopf equation is defined by the equation:

$$\overline{e}^2(t) \triangleq \lim_{T\to\infty} \frac{1}{2T} \int_{-T}^{T} e^2(t)$$

The right-hand side, integral-squared error can be evaluated by making use of the complex convolution theorem:

$$\mathcal{L}[f(t)g(t)] = \frac{1}{2\pi j} \oint F(\lambda')G(s - \lambda')\, d\lambda'$$

$$\text{If } \lambda' = \left(\frac{s}{2} + \lambda\right)$$

$$= \frac{1}{2\pi j} \oint W\left(\frac{s}{2} + \lambda\right) W\left(\frac{s}{2} - \lambda\right) d\lambda$$

$$\mathcal{L}[e^2(t)] = \frac{1}{2\pi j} \oint^{LHP} \frac{b_1\lambda^{2n-2} + b_2\lambda^{2n-4} + \cdots b_n\lambda^0}{[a_0\lambda^n + a_1\lambda^{n-1} + \cdots a_n\lambda^0][f(-\lambda)]} d\lambda$$

Then,

$$\overline{e}^2(t) = \int_0^\infty e^2\, dt = \lim_{s \to 0} \left\{ \frac{(-1)^{n+1}}{2a_0} \begin{vmatrix} b_1 & a_0 & 0 \\ b_2 & a_2 & a_1 \\ b_3 & a_4 & a_3 \end{vmatrix} \Bigg/ \begin{vmatrix} a_1 & a_0 & 0 \\ a_3 & a_2 & a_1 \\ a_5 & a_4 & a_3 \end{vmatrix} \right\}$$

Hurwitz determinants

In practice, it is simpler to compute mean-square error directly from solution tables as explained on p. 35. The above *integral-squared error* is the most common criterion used in *statistical design techniques for linear dynamic systems.* There are, for example, other criteria like time-weighted absolute and integral-squared errors:

$$\int_0^\infty t|e|\, dt, \int_0^\infty te^2\, dt, \text{etc.}$$

In sampled-data control systems such as a coded search-radar sweep for the target echos, an aircraft autopilot for precision attitude-gyro, and the digital control systems of a quadruplex color videotape recorder, the *auto-correlation function* can be alternatively expressed as a *sequence-average:*

$$\phi_{xx}(nT) = \phi^*_{xx}(nT) \triangleq \lim_{N \to \infty} \frac{1}{2N+1} \sum_{m=-N}^{N} x(mT)x[(n+m)T]$$

This is obtained by the impulse-modulation of $\phi_{xx}(\tau)$:

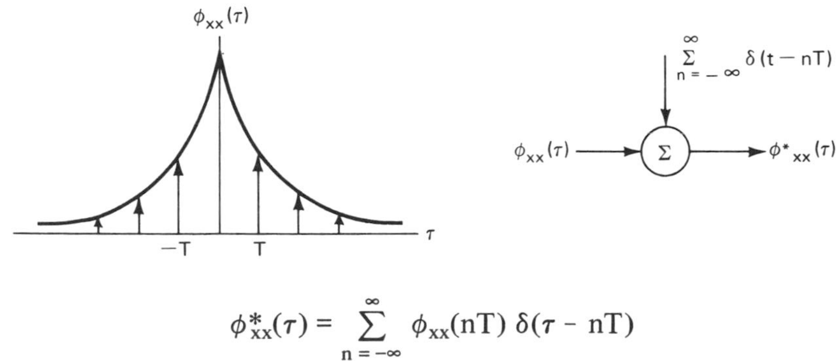

$$\phi^*_{xx}(\tau) = \sum_{n=-\infty}^{\infty} \phi_{xx}(nT)\, \delta(\tau - nT)$$

The auto-correlation function of the sampled signal, in the form of an impulse-modulated function as shown in the accompanying diagram, allows the application of the two-sided Laplace transforms to preserve the totality of the information contained in the correlation function. Hence, the *bilateral impulse-modulated auto-*

correlation function or the power spectral-density is given by:

$$\Phi^*_{xx}(s) = \int_{-\infty}^{\infty} \Phi^*_{xx}(\tau)e^{-s\tau}\, d\tau$$

With the z transforms,

$$(z = e^{sT}): \Phi_{xx}(z) = \sum_{n=-\infty}^{\infty} \phi_{xx}(nT)z^{-n}$$

$$= \sum_{n=-\infty}^{-1} \phi_{xx}(nT)z^{-n} + \sum_{n=0}^{\infty} \phi_{xx}(nT)z^{-n}$$

In the closed-form,

$$\Phi_{xx}(z) = [\Phi_{xx}(z)]_- + [\Phi_{xx}(z)]_+$$

Since $\phi_{xx}(\tau) = \phi_{xx}(-\tau)$, it can be shown,

$$[\Phi_{xx}(z)]_- = [\Phi_{xx}(z^{-1})]_+$$

For the direct use of the z-transform tables,

$$\Phi_{xx}(z) = \underset{(1)}{[\Phi_{xx}(z^{-1})]_+} + \underset{(2)}{[\Phi_{xx}(z)]_+}$$

The poles of the expression (1) will lie outside the unit-circle, while those of the unilateral z-transform (2) will lie inside the unit-circle (for $t < 0$, $\phi_{xx}(\tau) = 0$).

The *discrete equivalent of the Weiner-Hopf equation* takes the following format:

$$2\sum_{k=0}^{\infty} \{w(kT)f[\phi^*_{xx}(j-k)T] - f[\phi^*_{xy}(jT)]\} = 0; j = 0, 1 \ldots$$

As a basis for the system evaluation, the mean-square value is a statistical property of the sampled-variable, and it is derivable from the spectral-density function $\Phi_{xx}(z)$, and hence by the use of the inversion-integral. The mean-square-value of the sampled variable,

$$\bar{r}^2(mT) = \left[\frac{1}{2\pi j}\oint_C \Phi_{xx}(z)z^{n-1}\, dz\right]_{n=0} = \frac{1}{2\pi j}\oint_C \Phi_{xx}(z)\frac{dz}{z}$$

C stands for the contour on the unit-circle. Therefore, for the transmittance of the power spectral-density function through a linear digital-filter:

$$\bar{c}^2(nT) = \frac{1}{2\pi j}\oint_C G(z)G(z^{-1})\Phi_{xx}(z)\frac{dz}{z} \ldots$$

where C encloses the poles inside the unit-circle.

As distinguished from the preceding equation, it can be shown by applying the advanced z-transforms that the mean-square value of a continuous system-response $\overline{c}^2(t)$ is the upper bound of $\overline{c}^2(nT)$.

R(z) $\Phi_{xx}(z)$ → G(z) → C(z) $\Phi_{yy}(z)$

Since it is not simple enough in control system applications to find the roots of the closed-loop characteristic equation and to study the effect of the system parameters on the solution, frequent use is made of tabular mathematical solutions, as expressed in terms of the polynomial coefficients in the integrand of the definite integral:

$$I_n = \frac{1}{2\pi j} \oint_{-j\infty}^{j\infty} \frac{G(w)G(-w)}{d(w)d(-w)} dw$$

where

$$G(w) = \sum_{k=0}^{n-1} G_k w_k ; \; d(w) = \sum_{k=0}^{n} d_k w_k$$

and the zeros of d(w) are restricted to the left-half plane LHP.

Thus, in w-transform,

$$\overline{c}^2(mT) = \frac{1}{2\pi j} \oint_C \Phi_{xx}(z) \frac{dz}{z} = \frac{1}{2\pi j} \oint_C \Phi_{cc}(w) \frac{2dw}{1 - w^2} \cdots \text{with } z = \frac{1+w}{1-w}$$

where

$\Phi_{cc}(w) = \Phi_{xx}(w) \cdot G_hG(w) \cdot G_hG(-w)$, G_h and G stand for hold and pulse-transfer function, respectively.

Then, by making use of the solution tables for the definite integral, the *physically realizable linear digital-filter* that could generate the *desired mean-square response* at the output, is directly computed as a *statistical method of synthesis.* The statistical method with the least-square criterion is further developed for optimization in terms of a more realistic mathematical model. The optimization process with the digital compensation may impose an ultimate performance limitation such as (1) saturation, (2) finite-memory on the digital process, and (3) power capability of the plant in a hybrid control system. Under these circumstances, the relevant parameter will be considered as a restraint on the optimization of the hybrid D(z), G(s) (digital-analog) control system. For example, an *optimum design, with an accompanying sacrifice in performance due to a constraint* such as power-capability, may require the use of a weighting factor λ (called the *Lagrangian multiplier*) in a *performance-index* of the form, $q = \overline{e}^2(nT) + \lambda \overline{i}^2(nT)$.

1.5.2 Composite Criterion for Both Rejection of Noise and Control of Step-Response. In digital processes, it is possible to develop a composite criterion based on the design of a discrete-filter giving maximum rejection of noise and consistent with the control of transient-response to a step input, as a compromise between the two basic control-system design approaches. This rather highly complex and time-consuming procedure is developed with a sampled unit-step at the input of the discrete-filter, as a combination of (1) Graham and Lathrop's discrete normalized-equivalent of the *analog integral-error criterion* and (2) *the statistical minimum-squared error criterion, under the constraint of linearity.* The problem under consideration, and the solutions arrived at for the composite design of the discrete-filter are enumerated in the following with a short introduction to show the complexity of the procedure involved.

The statistical design in this case is performed (1) for dimensional homogeneity and (2) according to the degree of importance attached to the noise-reduction as against the transient-behavior. The two parameters of this criterion should exhibit minima of some magnitude. The details of a control problem in this respect are thus stated:

1. The process:

$$c_n = \sum_{k=0}^{N-1} a_k r_{n-k}$$

The process is of homogeneous dimensions.

2. Transient control with a minimized variance reduction-factor:

$$\frac{\sigma_c^2}{\sigma_r^2} = \sum_{k=0}^{N-1} a_k^2 \,*$$

3. Subject to a constraint of linearity:

$$\sum_{k=0}^{N-1} a_k = 1; \sum_{k=0}^{N-1} k^q a_k = (-\alpha)^q \text{ for all } q \leqslant (N-1)*$$

Now, corresponding to the integral $\int_0^\infty e^2(t)\,dt$ in a continuous system, the sampled-data system uses the approximately equivalent expression:

$$\sum_{n=0}^{N-1} (U_n - \sum_{k=0}^{n} a_k U_{n-k})^2 \,*$$

where

$U_n = 1, n = 0, 1, 2, \cdots$

$U_n = 0, n < 1$

under the constraint of linearity; $a_0 = 1$, $a_j = 0$, $j \neq 0$ are the trivial minimum. The constraints of (1) homogeneity and (2) minima of some relative magnitude for noise-

*From *Digital Processes for Sampled Data Control Systems* (1962) by Alfred J. Monroe. Courtesy of John Wiley & Sons.

reduction and transient-behavior are obtained by dividing the error-criterion with a normalizing function:

$$\frac{\sum_{n=0}^{N-1} (U_n - \Sigma a_k U_{n-k})^2}{\sum_{n=0}^{N-1} U_n^2}$$

The third constraint, the specification of the degree of importance of the noise-reduction and transient-behavior, is realized by introducing a special parameter,

$$\beta(\leqslant 1, \geqslant 0)$$

$$f[a_k, \beta] = (1 - \beta) \sum_{k=0}^{N-1} a_k^2 + \left[\beta \sum_{n=0}^{N-1} \left(U_n - \sum_{k=0}^{n} a_k U_{n-k}\right)^2\right] \Big/ \sum_{k=0}^{N-1} U_n^2$$

For $\beta = 0$, the mean-squared error-criterion alone is effective. By solving (1) the linear difference-equations by the successive-difference technique and (2) the first-order difference-equations for the linear constraint, the complete solution for the a_j as a function of β and N is as follows:

$$a_{i+2} - \left(2 + \frac{\beta}{N(1 - \beta)}\right) a_{i+1} + a_i = 0 \tag{1}$$

the difference equation satisfying the first and third constraints.

The eigenvalues λ_0^i and λ_1^i of the preceding equation, expressed as a set of first-difference equations in matrix form (using the state-variable approach), are obtained to provide the solution to the given problem:

If

$$\lambda_0 = \frac{1}{2}\left(2 + \frac{\beta}{N(1 - \beta)}\right) + \frac{1}{2}\left[\frac{\beta}{N(1 - \beta)}\right]^{1/2} \left[4 + \frac{\beta}{N(1 - \beta)}\right]^{1/2} \tag{2}$$

$$\lambda_1 = \frac{1}{2}\left(2 + \frac{\beta}{N(1 - \beta)}\right) - \frac{1}{2}\left(\frac{\beta}{N(1 - \beta)}\right)^{1/2} \left(4 + \frac{\beta}{N(1 - \beta)}\right)^{1/2} \tag{3}$$

$$\beta_0 \triangleq \frac{\lambda_0^N - 1}{\lambda_0 - 1} \tag{4}$$

$$\beta_1 \triangleq \frac{\lambda_1^N - 1}{\lambda_1 - 1} \begin{cases} \gamma_0 \triangleq [(N - 1)\lambda_0^{N+1} - N\lambda_0^N + \lambda_0]/(\lambda_0 - 1)^2 \\ \gamma_1 \triangleq [(N - 1)\lambda_1^{N+1} - N\lambda_1^N + \lambda_1]/(\lambda_1 - 1)^2 \end{cases} \tag{5}$$

Then

$$a_0 = \frac{\gamma_1 - \gamma_0}{\beta_0 \gamma_1 - \beta_1 \gamma_1} \quad \text{and} \quad a_1 = \frac{\lambda_0 \gamma_1 - \lambda_1 \gamma_0}{\beta_0 \gamma_1 - \beta_1 \gamma_0} \tag{6, 7}$$

1.6 LINEAR AND NONLINEAR SYSTEMS

1.6.1 Continuous (Analog) Signal Processing. In continuous *linear* feedback control systems of various degrees of complexity, the analysis and design of the systems are concerned with the solution of the related linear constant-coefficient differential equations of the respective orders. The characteristic equation of the control system, presented as a polynomial in s of Laplace transformation, is solved to determine its roots, because the location of the roots in the complex s-plane determine the transient-response of the system. The *principle of superposition holds*, and the Fourier integral gives the formal relationship between the time-domain and the frequency-domain to justify the application of the methods of frequency-response, and the design on that basis.

1.6.2 Sampled-Data (Digital) Signal Processing. In a *nonlinear* (NL) system, described by a *nonlinear differential equation*, there is no concept of roots, and hence there is no defining of any transient-response and any formal relationship between the time and the frequency-domains. With approximated frequency-response methods, the relationships are purely empirical as some linear equivalent. Since the potentially best performance can be achieved in most control systems by nonlinear operation (either of an incidental, or of an imposed slow-speed on-off character), the signal-level and the operating-range of all the simultaneous inputs must be specified in order to define the system performance. The synthesis of the nonlinear systems in practice is mostly a trial-and-error process tempered with experience and common sense. The amplifier saturation, the motor sensitivity limits due to coulomb-friction and hysteresis, and the motor velocity and acceleration limits due to magnetic-saturation, granularity in helipot, dead-space, backlash in mechanical linkages (gear-trains), variable gain and variable delay-time, and other dynamic slow- and fast-changing nonlinear phenomena constitute the common nonlinearities in control systems.

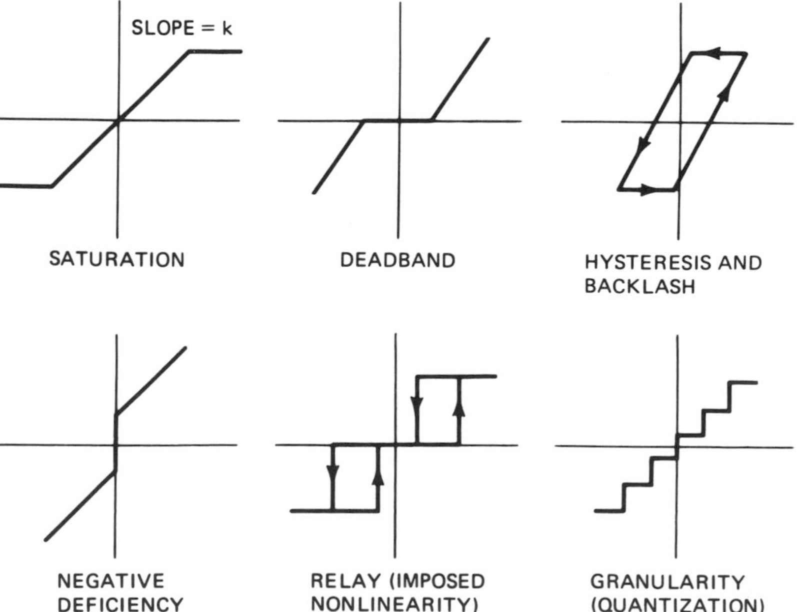

Fig. 1-16. Common nonlinearities in control systems.

These various phenomena produce in control systems such effects as bounded-oscillation or *limit-cycle*, *jump-resonances*, *subharmonic generation*, and *intermodulation effects* on gain.

A system represented by a differential equation with time-varying coefficients can be linear, but the Laplace and frequency-response methods are applicable to the constant-coefficient linear differential equations only. Therefore, corresponding computer methods, using the analog-computer and the state-space techniques, are suitable for such systems.

1.6.3 Control Techniques for Nonlinear Systems. 1. *Linear approximation* for nonlinear systems is not uncommon since many practical so-called linear control systems may contain a small amount of nonlinearity. It is then represented by a characteristic equation of the form:

$$A_n \frac{d^n y}{dt^n} + A_{n-1} \frac{d^{n-1} y}{dt^n} + \cdots A_0 y + \epsilon f\left(y, \frac{dy}{dx}\right) = x(t)$$

For simple nonlinearities, an expansion of the solution to this differential equation can be written in a power series of ϵ of the form:

$$y(t) = \underbrace{[y_{(0)}(t)]}_{\text{Linear part}} + \underbrace{[\epsilon y_{(1)}(t) + \epsilon^2 y_{(2)}(t) + \cdots]}_{\text{Nonlinear part (neglected if } \epsilon \text{ is very small)}}$$

Piecewise linear approximation is also considered using linear segmented-intervals and relatively simple linear differential equations for the segmented-intervals.

2. The *describing-function* (G_D) method makes an attempt to extend the transfer-function approach of a linear system to a quasilinear system, if it is a simple *time-invariable* nonlinear-element. G_D is defined as the ratio of the fundamental component of the output-response of a nonlinear device to the sinusoidal input signal. However, in practice, the performance of a nonlinear system may be entirely different to the step-response as compared to the sinusoidal response. The G_D representing a nonlinear-element can be directly substituted into the system-equation of the nonlinear characteristic in order to quasi-linearize the frequency-response equation. Since G_D is a function of amplitude, the system frequency-response will be a function of both frequency and amplitude to allow the application of the Nyquist and gain-phase plots.

$$\frac{C(j\omega)}{R(j\omega)} = \frac{G_D G(j\omega)}{1 + G_D G(j\omega)}$$

where

$$G_D = \frac{N(j\omega)}{M(j\omega)}$$

The characteristic equation:

$$1 + G_D G(j\omega) = 0$$

$$G(j\omega) = -1/G_D$$

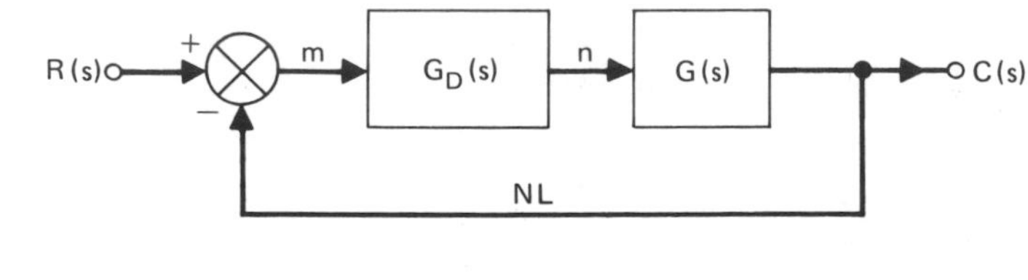

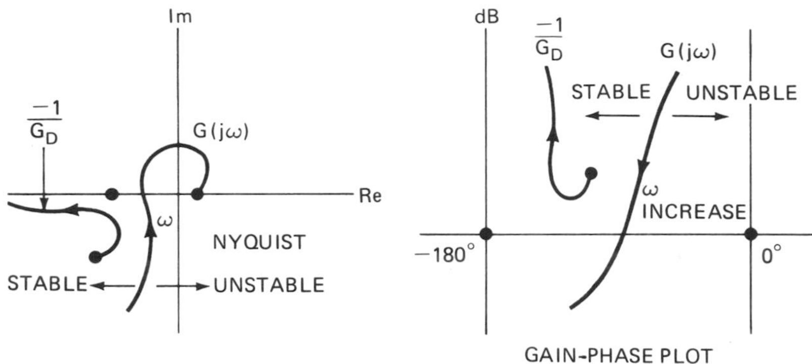

Fig. 1-17. Describing function method.

or

$$20 \log_{10}|G(j\omega)| = 20 \log_{10}\left|\frac{1}{G_D}\right| \quad \text{and} \quad \underline{/G(j\omega)} = -180\text{-Angle } G_D$$

If the locus of the various values of $(-1/G_D)$ is plotted, the system will be stable if the $(-1/G_D)$ locus does not intersect the $G(j\omega)$ locus when the $G(j\omega)$ locus is traversed in the direction of increasing frequency. This is of course the common Nyquist plot.

The describing-function method has two extensions for improved approximation:

a. Klotter's *new describing-function* method for describing the steady-state performance of a nonlinear system assumes the solution as a series of *orthogonal-functions*, and then computes the *coefficients according to the conditions of the Ritz-Galerkin method.*
b. The Taylor-Kauchy transform method assumes a power-series expansion of the highest derivative and integrates it until a series for the dependent-variable is obtained. The second method, however, does not guarantee the convergence of the solution.

3. The *phase-plane technique* is a graphical method for analyzing the transient-response of a linear or nonlinear system to the *step-input under various initial con-*

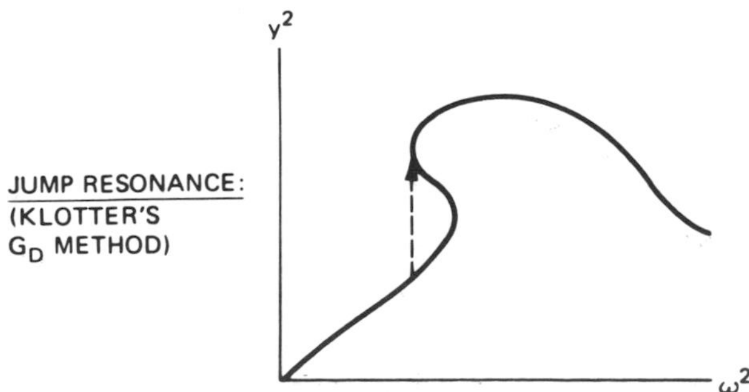

ditions. A series of these velocity (ordinate) versus position (abcissa) plots, called the *phase trajectories*, giving the locus of the solution to the differential equation, show the effect of the parameter variations, so that the plots indicate the trends, and hence lead to the optimal conditions for design. Although the independent-variable time is not explicit in the plot, the accurate transient-response from the *phase-portrait* can be computed in order to obtain a specific solution from the general solution. Unlike the describing-function method, the technique is limited to the basic systems of the second-order with a single degree of freedom, such as

$$\frac{d^2x}{dt^2} + A_1(x, \dot{x})\frac{dx}{dt} + A_2(x, \dot{x})x = 0$$

where

$$\dot{x} = \frac{dx}{dt}$$

If

$$\frac{dx}{dt} = y$$

$$\frac{dx}{N(x, y)} = \frac{dy}{D(x, y)}$$

where

$$N(x, y) = \frac{dy}{dt}$$

$$D(x, y) = \frac{dx}{dt}$$

If the trajectories plotted in the phase-plane for the preceding nonlinear equation converge on the node, the system is stable; if they diverge, the system is unstable. If adjacent trajectories converge to a closed-path, called the *limit-cycle*, the phenomenon describes a characteristic oscillation about the point of convergence; the limit-cycle is symbolic of a nonlinear system.

Lines in the phase-plane corresponding to the slopes of the phase-portrait are called *isoclines*; they are merely the *loci of points from which the segments of*

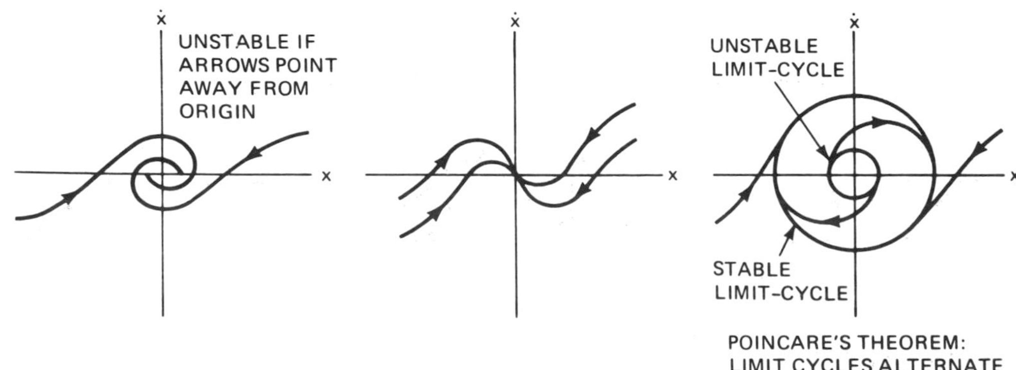

Fig. 1-18. Typical phase trajectories.

constant-slope are initiated: f(x, y) = m. In general, this equation represents a family of curves. If a series of short segments, each with its center on a curve, are drawn to produce a series of tangents and if the process is repeated for different values of m, the *phase-trajectories* can be sketched as solutions from these tangents. This is the *method of isoclines* associated with the phase-plane technique.

For complex control systems of the higher order, the phase-plane method is replaced by the phase-space techniques; these come under the category of the "state-variable" concepts explained in the next section.

1.6.4 Nonlinear Sampled-Data Control Systems. In the case of a sampled-data system containing a single nonlinear element, more than one mode of limit-cycle can be excited with periods that are integral-multiples of the sampling-period, because the nonlinearity interacts with the sampling process in a more complex way, as compared to the continuous nonlinear system. Hence, it is more likely that the *statistical methods* with the state-variable techniques may prove more suitable in the case of the nonlinear sampled-data control systems.

The describing-function method in the sampled-data case becomes highly complex since the process of sampling—and the higher harmonics—make the describing-function take a multidimensional character, with its dependence on any particular limit-cycle in effect at any instant.

As for the phase-plane method, it is modified with an arbitrary set of initial conditions to an *incremental phase-plane technique*; this is usually limited to the nonlinear second-order systems. The solution takes the form of a discrete set of points in the phase-plane, without the concept of a continuous trajectory. The successive solution-points are joined by a straight-line segment called the *solution-path.* For a given set of initial conditions, the solution path that provides the time information directly is graphically computed by the *method of isoclines.* For a nonlinear system, a coordinate-transformation is then developed to (1) simplify the system, (2) obtain a means for synthesizing a useful feedback compensation, and (3) detect the existence of potential intersample ripple that is not seen in e and Δe *incremental phase-plane coordinates.* For example, the homogeneous difference equation for the closed-loop system is derived in terms of the error-variable of the form

$$\Delta^2 e(n) + b\,\Delta e(n) + ce(n) = 0 \tag{1}$$

where Δ^2 and Δ are the second and first differences, respectively.

$$\text{In z-transform domain: } [(z-1)^2 + b(z-1) + c]\,E(z) = 0 \tag{2}$$

Comparing Eq. 2 with the general characteristic equation of a second-order system, system,

$$(z^2 + Bz + C)E(z) = 0 \tag{3}$$

and equating like terms in Eqs. 2 and 3, Eq. 1 takes the form:

$$\Delta w(n) + bw(n) + ce(n) = 0 \tag{4}$$

The locus of a set of points in the e-w-plane represents the solution to this equation, since the coordinates of the incremental phase-plane are then defined by e and Δ.

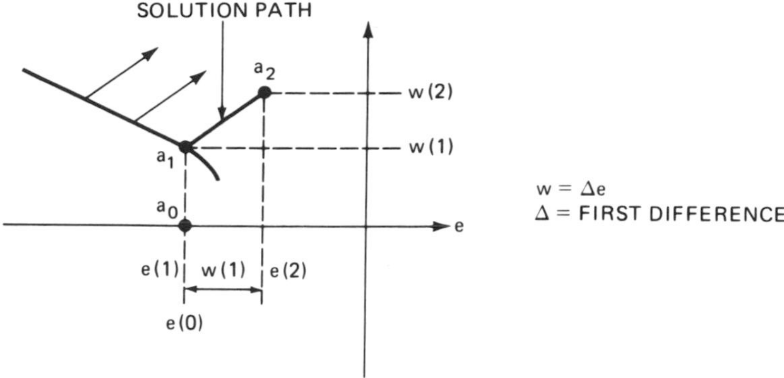

Fig. 1-19. Isocline plotting in incremental phase-plane.

From Eq. 4, we obtain with $\frac{\Delta w}{\Delta e} = k$, the isocline equation,

$$k = -b - c\frac{e}{w} \tag{5}$$

From Fig. 1-19, a point a_2 can be derived from a_1, if isocline at a_1 is known. The e-axis corresponds to an isocline representing a k of infinity. The isocline is defined in this case also as the *locus of the points from which the segments of constant-slope are initiated.* The solution-path must be along a vertical line for infinite k.

Special modern control design techniques for nonlinear sampled-data control systems follow the concepts of the state-variables explained in the following section.

1.7 STATE-VARIABLES IN MODERN TECHNIQUES OF ANALYSIS AND SYNTHESIS FOR LINEAR AND NONLINEAR CONTROL SYSTEMS

The performance of a control system is expressed by *state-transition equations* formed with **state variables**, instead of attempting to find a solution for a complex higher-order differential equation. The state-variables are written in *vector-matrix form as state-vectors* (or column-matrices). They represent the coordinates of points in a *multidimensional* **state-space** or *vector-space*, and provide a complete description of the dynamic-state of the plant in terms of added variables, in contrast to the transfer-function, characterizing only an output-input response relationship of a system, that is, initially at the zero-state (or equilibrium). The state-variables can be identified as **phase variables** when each element in a column-matrix is the *time-derivative* of the element above it. The solution of the state-transition equations in the appropriate form can then give a geometrical interpretation for visualizing the control process. In this mathematical analysis, the state-variables are subjected to *linear-transformation* to become **principal variables**, when the square-matrix of the system transition-equation is reduced to *a diagonal-matrix of eigenvalues* by means of the *Vander-Monde matrix.* For stability, the complete set of the eigenvalues must have negative real-parts. In some system-optimization techniques, a set of adjoint-equations are then used to decide what boundary conditions are desired at the zero initial-time for the system-variables to reach a certain set of values some time later.

The concept of the state-variable techniques has been originally introduced from the classical mechanics by using the calculus-of-variations as the primary tool. The state-space approach readily lends itself to the consideration of the initial conditions of the system as part of the solution. This modern approach is applicable in the case of most nonlinear and time-varying systems of both continuous and sampled-data configurations. Therefore, it can be classified as a more or less *generalized* approach.

In the case of the continuous analog systems, the state-transition equations describe the system in terms of a set of first-order differential equations, while in the case of the sampled-data control systems the state-transition equations describe the system in terms of a set of first-order difference equations.

1.7.1 Introduction to Application of State-Variables

1.7.1.1 Continuous Control Systems. The approach is based on the characterization of the system by state-variables. For example, as a parallel to the phase-plane technique, the variables can be position and velocity. But the characterization is not unique, although *measurable* quantities must be chosen. Another pair of variables can be chosen in a *multidimensional state-space by a simple rotation of axes.* As an example, the linear nonhomogeneous differential equation with constant-coefficients in a three-dimensional case is given by:

$$a_0\dddot{x} + a_1\ddot{x} + a_2\dot{x} + a_3 x = e \qquad (1)$$

This can be rewritten in the state-variable, *vector-matrix form.* The vector-matrix form is depicted by the diacritical mark above the letter.

$$\dot{\bar{x}} = \bar{A}\bar{x} + \bar{b}e \qquad (2)$$

where

$$\dot{\bar{x}} = \begin{bmatrix} x_1 \\ x_2 \\ x_3 \end{bmatrix}; \qquad \bar{A} = \begin{bmatrix} 0 & 1 & 0 \\ 0 & 0 & 1 \\ \dfrac{-a_3}{a_0} & \dfrac{-a_2}{a_0} & \dfrac{-a_1}{a_0} \end{bmatrix}; \qquad \bar{b} = \begin{bmatrix} 0 \\ 0 \\ \dfrac{1}{a_0} \end{bmatrix}$$

Column-Matrix (State-Vector)

Coefficient-Matrix (Jordan-Canonical form) -square matrix

As a linear equation, it is $\mathcal{L}$-transformable (symbol; underline):

$$s\underline{x} - \underline{x}(0^+) = \bar{A}\underline{x} + \bar{b}\underline{e} \cdots$$

where $\underline{x}(0^+)$ is the initial value of the state-variable.

Writing λ for s,

$$\underline{x} = [\lambda\bar{I} - \bar{A}]^{-1}[\underline{x}(0^+) + \bar{b}\underline{e}]$$

where $\bar{I}$ is the unit or identity matrix.

$$\begin{bmatrix} 1 & 0 & 0 \\ 0 & 1 & 0 \\ 0 & 0 & 1 \end{bmatrix}$$

In time-domain, we find the roots λ_1, λ_2, λ_3 as characteristic, or eigenvalues, of the matrix $\bar{A}$ for the characteristic equation

$$\text{Det}\,|\lambda\bar{I} - \bar{A}| = 0 \tag{3}$$

The variables are now subjected to linear-transformation by using the Vander-Monde matrix,

$$\bar{P}^{-1} = \begin{bmatrix} 1 & 1 & 1 \\ \lambda_1 & \lambda_2 & \lambda_3 \\ \lambda_1^2 & \lambda_2^2 & \lambda_3^2 \end{bmatrix}$$

to a new set of variables given by:

$\bar{Y} = \bar{P}\bar{x} \cdots$ *the roots are invariant under this transformation of similiarity.* (4)

where $\bar{P}$ is a constant and a nonsingular matrix such that $\bar{P}\bar{A}\bar{P}^{-1} = \bar{D}$, and D is a diagonal-matrix with the eigenvalues:

$$\bar{D} = \begin{bmatrix} \lambda_1 & 0 & 0 \\ 0 & \lambda_2 & 0 \\ 0 & 0 & \lambda_3 \end{bmatrix}$$

Elements above the diagonal elements can be other than zero, if they are adjacent to two equal eigenvalues.

Inserting (4) in (2),

$$\dot{\bar{Y}} = (\bar{P}\bar{A}\bar{P}^{-1})\bar{Y} + \bar{P}\bar{b}e \tag{5}$$

Then, Eq. 5 can be decomposed into n(=3) dependent first-order equations having elementary exponential solutions for Y:

$$Y_n = Y_k e^{\lambda_k^t} \cdots \tag{6}$$

Then, $\bar{x}$ is determined from Eq. 4:

$$\bar{x} = \bar{P}^{-1}\bar{Y}$$

For stability of the control system, all the eigenvalues must have negative real-parts. That is, a system *enforcing-function* is formed that will bring the plant to zero-state or *equilibrium in finite-time.*

1.7.2.2 Sampled-Data Control Systems. In the case of linear time-invariant discrete sampled-data systems using uniform sampling, Kalman has shown that a plant, which is controllable in finite settling-time before sampling, remains so after sampling (that is, without any cancellations in the transfer-function of the plant) if the plant is free of poles that contain *equal real and imaginary parts.* The poles are subject to the condition that they must be *separated by an integral-multiple of the sampling-frequency*.

During the successive sampling-periods, the dynamic system behaves as a continuous system. Therefore, the sampled-data system under consideration can be represented by a set of first-order linear homogeneous differential-equations with constant-coefficients in the place of the first-order linear difference-equations.

In vector-matrix form, the system dynamics can be characterized by

$$\dot{\bar{x}}(\lambda) = \bar{A}\bar{x}(\lambda) \tag{1}$$

where $\lambda = (t - nT)$, $0 < \lambda \leqslant T$, $\overline{x}$ is the *state-vector* of the whole system, with the state-variables functioning as the input-variables of the system. Since the state-variables change from one set of values to another at each sampling-instant, the *initial conditions* for the state differential equation (Eq. 1) vary from one sampling-instant to the next and can be depicted by the transition-equation:

$$\overline{x}(nT^{+}) = \overline{B}\,\overline{x}(nT) \tag{2}$$

$\overline{A}$ and $\overline{B}$ are square-matrices, and they are found by inspection of the system state-variable diagram. Equations 1 and 2 describe a conventional sampled-data system during the sampling-period; the system is analyzed by solving the first equation with the values of n derived from the second equation. When $T \to \infty$, the sampled-data system becomes a continuous system.

$\mathcal{L}$-transformation Eq. 1:

$$\underline{\overline{X}} = [s\overline{I} - \overline{A}]^{-1}\overline{x}(0^{+})$$

The solution to Eq. 1 is then:

$$\overline{x}(\lambda) = \overline{M}(\lambda) \cdot \overline{x}(0^{+}) \tag{3}$$

If $\overline{M}(\lambda) = \mathcal{L}^{-1}[s\overline{I} - \overline{A}]^{-1}$; and $\overline{x}(\lambda) = e^{\overline{A}\lambda}$

overall
transition
matrix

During the sampling-interval,

$$nT < t \leqslant (n + 1)T$$

The state-vector $\overline{x}(t) = \overline{N}(t - nT)\overline{x}(nT)$ if $\overline{N}(t - nT) \triangleq \overline{M}(t - nT)\overline{B}$ (4)

At the $(n + 1)th$ sampling-instant,

$$t = (n + 1)T$$

and

$$\overline{x}[(n + 1)T] = \overline{N}(T)\overline{x}(nT) \tag{5}$$

Equation 5 gives the recurrence relationship for obtaining the values of the state-vector at the successive sampling-instants. Iteration of Eq. 5 gives:

$$\overline{x}(nT) = \overline{N}^{n}(T)\overline{x}(0) \tag{6}$$

$X(z) \triangleq \sum_{n=0}^{\infty} x(nT)z^{-n}$ according to the definition of z-transform:

Equation 6 in z-transform: $X(z) = [\overline{I} - z^{-1}\overline{N}(T)]^{-1}\overline{x}(0)$ (7)

Now the inverse-transform of Eq. 7 gives the state-vector $\overline{x}(nT)$:

$$\overline{x}(nT) = Z^{-1}\{[\overline{I} - z^{-1}\overline{N}(T)]^{-1}\}\overline{x}(0) \tag{8}$$

Then, the dynamic performance of the system is described by:

$$\overline{x}(t) = \overline{N}(t - nT)Z^{-1}[\overline{I} - z^{-1}\overline{N}(T)]^{-1}\overline{x}(0) \tag{9}$$

Thus, the dynamic performance of a sampled-data control system is analyzed by the use of the Eqs. 8 and 9. The procedure involved the determination of the overall transition-matrix $\overline{M}$ and the $\overline{B}$ square-matrix of the system.

$\overline{N}(T)$ is partitioned according to the dimensions of the state-variables x_1 and x_2:

$$N(T) = \begin{bmatrix} \overline{a}(T) & 0 \\ \overline{c}(T) & \overline{b}(T) \end{bmatrix} \tag{10}$$

where $\overline{a}(T)$ is a square-matrix

Since $\overline{X}(z)$ is a column-vector,

$$\overline{X}_1(z) = [\overline{I} - z^{-1}\overline{a}(T)]^{-1}\overline{x}_1(0) \tag{11}$$

$$\overline{X}_2(z) = z^{-1}[\overline{I} - z^{-1}\overline{b}(T)]^{-1}\overline{c}(T)\overline{X}_1(z) + [\overline{I} - z^{-1}\overline{b}(T)]^{-1}\overline{x}_2(0) \tag{12}$$

At the initial zero-state, the z-transform of the plant state-vector is given by

$$\overline{X}_2(z) = z^{-1}[\overline{I} - z^{-1}\overline{b}(T)]^{-1}\overline{c}(T) \cdot \overline{X}_1(z) \tag{13}$$

where the determinant $[\overline{I} - z^{-1}\overline{b}(T)] = 0$ is the system characteristic equation. (14)

Then the system is stable if the roots of Eq. 14 lie inside the unit-circle in the z-plane. Other stability criteria may be applied to Eq. 14 to find the location of the roots—after the appropriate partitioning of $\overline{N}(T)$ to obtain the matrix $\overline{b}(T)$ of the characteristic equation.

1.7.2 State Variables. In sampled-data control systems, the state-variables can be defined in an alternative way by the transfer-function approach. The sampled-data control system shown can be represented in a nonunique fashion as shown in Fig. 1-20. Measurable quantities are chosen as the state-variables in this approach.

Example:

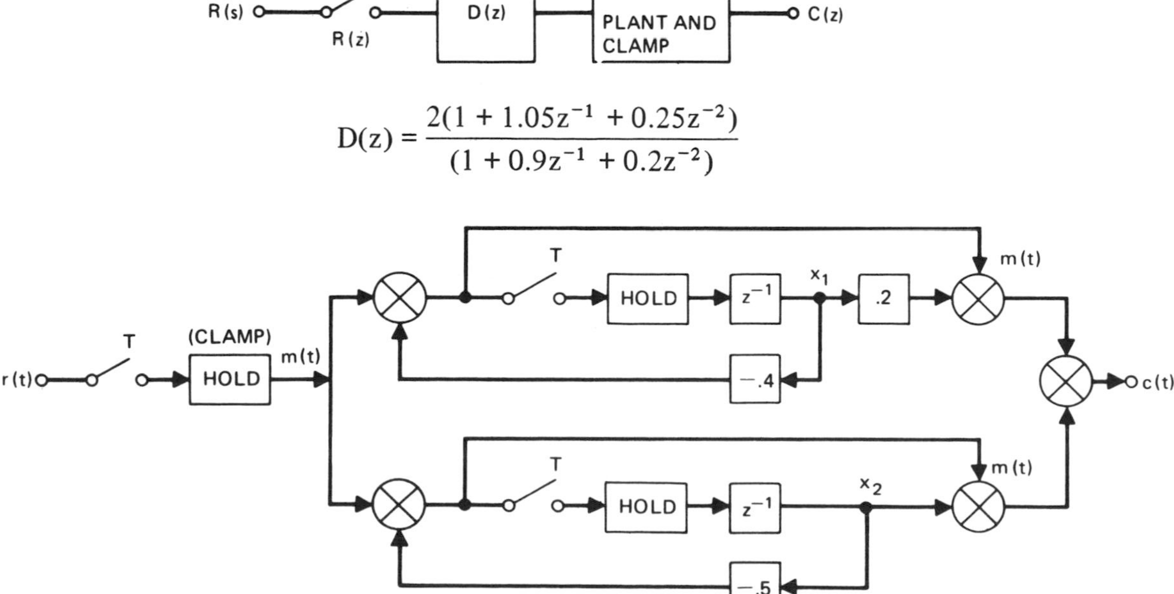

$$D(z) = \frac{2(1 + 1.05z^{-1} + 0.25z^{-2})}{(1 + 0.9z^{-1} + 0.2z^{-2})}$$

Fig. 1-20. Representation of state-variables in block diagram.

Using partial fractions for parallel-programming,

$$D(z) = \frac{1 + .2z^{-1}}{1 + .4z^{-1}} + \frac{1 + z^{-1}}{1 + .5z^{-1}}$$

The output is obtained from the above block-diagram as

$$c(t) = 0.2x_1(t) + x_2(t) + 2m(t)$$

In this specific case, the programming can be actually done (1) directly, (2) in parallel as above, (3) in tandem by factoring and iteration, or (4) for more complex systems (such as variable-rate, multirate, random nonsynchronized, pulse-width modulated, pulse-frequency modulated, time-variant, and nonlinear sampled-data control systems), the state-variable techniques are presently developed on a theoretical basis. These techniques are also simultaneously interspersed with statistical approaches. The modern optimal and adaptive control systems and the Lyapounov "direct method" approach make use of state-variable techniques; the basic principles of these specific cases are explained later on in this chapter.

1.7.3 Multivariable Systems. A *multivariable* system is one where the system to be controlled has *more than a single degree of freedom*, or in other words, more than one initiating enforcing equation and more than one variable in each equation. A voltage, a pressure, or a torque constitute a force. The quadruplex color videotape recorder is a typical example. (Some sources define systems with more than two degrees of freedom as *multiloop* systems, since the access to several outputs of the plant is essential for more degrees of freedom.) A typical *signal flow-graph* is illustrated in Fig. 1-21.

The equations governing the response of the system-variables to the applied forces are determined either by mathematical analysis of the physical hardware in the system, or by experimental analysis. In most cases, these equations will turn out as nonlinear ordinary differential equations. If the system parameters are distributed (as in heat-propagation), they will take the form of partial differential equations. Nonlinear differential equations are linearized to the best first-approximation with the aid of Taylor's power-series, and formulated in Laplace-transform to represent the relationship between the various interacting variables ($x_1, x_2, \ldots$) and the applied forces due to the interaction of the final control elements ($u_1, u_2, \ldots$). Most of the difficulty lies in the generation of these equations from mathematical and experimental data. The following example of a system with two variables uses two linear algebraic equations to show the procedure:

$$H_{11}x_1(s) + H_{12}x_2(s) = A_{11}u_1(s)$$

$$H_{21}x_1(s) + H_{22}x_2(s) = A_{22}u_2(s)$$

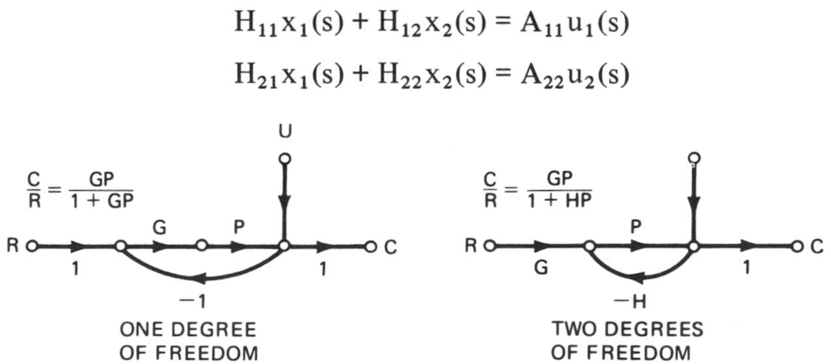

Fig. 1-21. Signal flow-graph.

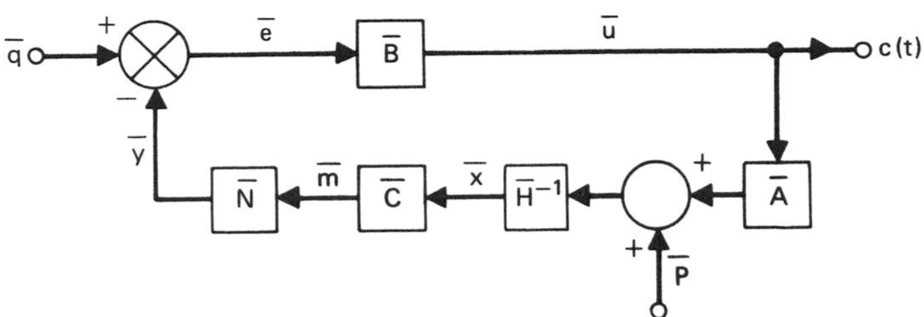

Fig. 1-22. Typical multivariable system.

$\overline{H}^{-1}$: Controlled system and a part of the final control element.
$\overline{A}$: Rest of the final control element.
$\overline{x}$: System variables.
$\overline{B}$: The desired controller matrix. It relates the manipulated variables to the error signals.
$\overline{P}$: The disturbance matrix.
$\overline{e}$: $-\overline{y} + \overline{q}$ = Error matrix: (sum of the control variables and the desired values of the *control variables or set points.*)
$\overline{C}$: Matrix to convert $\overline{x}$ to $\overline{m}$ (matrix of measured variables).
$\overline{N}$: Matrix to describe the *measuring transducers* for converting $\overline{m}$ (such as speed in rpm) to control variables $\overline{y}$ (such as voltage). *Example:* speed, measured as the voltage from a tachometer. The tachometer sensitivity in V/rpm makes an element of $\overline{N}$.

where

$$\overline{H} = \begin{bmatrix} H_{11} & H_{12} \\ H_{21} & H_{22} \end{bmatrix}; \qquad \overline{x}(s) = \begin{bmatrix} x_1(s) \\ x_2(s) \end{bmatrix}$$

$$\overline{A} = \begin{bmatrix} A_{11} & 0 \\ 0 & A_{22} \end{bmatrix}; \qquad \overline{u}(s) = \begin{bmatrix} u_1(s) \\ u_2(s) \end{bmatrix}$$

In the matrix-form of the state-variables,

$$\overline{H}\overline{x}(s) = \overline{A}\overline{u}(s)$$

Matrices developed as above for each component in the system and its control can be described by an overall matrix block-diagram as shown in Fig. 1-22.

The matrix equations formed from Fig. 1-22 are solved for $\overline{m}$ to formulate the multivariable control problem. Then the problem is to design physically realizable elements of $\overline{B}$, so that the elements of $\overline{m}$ are related to the elements of $\overline{q}$ (set points) in the desired manner, in the presence of the disturbances and noise $\overline{P}$.

One design technique enables the choice of $\overline{B}$ by specifying $D[=\overline{C}\overline{H}^{-1}\overline{A}\overline{B}\overline{N}]$ in the above example as a diagonal matrix, and obtaining the elements of $\overline{m}$ in terms of the elements of $\overline{D}$ in the following form:

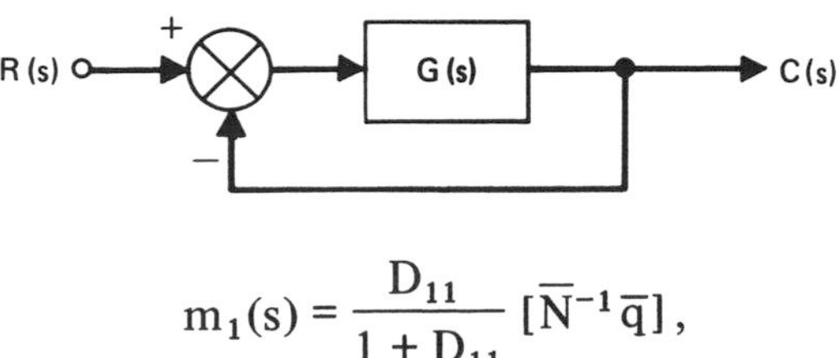

$$m_1(s) = \frac{D_{11}}{1 + D_{11}} [\overline{N}^{-1}\overline{q}],$$

The preceding equation corresponds to the output of a single-loop servo:

$$C(s) = \frac{G(s)}{1 + G(s)} R(s)$$

that is, the frequency-response methods can be directly applied to synthesize the elements of $\overline{D}$, so as to meet the requirements of $\overline{m}$ for the final solution of $\overline{B}$, which, in the above example, takes the form

$$\overline{B} = \overline{A}^{-1}\overline{H}\overline{C}^{-1}\overline{D}\overline{N}^{-1}$$

In the case of the sampled-data multivariable control systems, the whole procedure starts with linear difference equations and z-transforms in the place of the differential equations and the $\mathcal{L}$-transforms. Matrix-manipulation will then produce for the elements of $\overline{m}(z)$ single-loop pulse-transfer-functions of the form

$$C(z) = \frac{G_h G(z) R(z)}{1 + G_h G(z)}$$

where G_h corresponds to a hold-clamp in an equivalent loop.

1.7.4 Optimal Control Systems. The design of control systems, which are "optimal" in the sense of performance or otherwise, is a complex problem, and the expanding theoretical concepts in the field, developed originally in 1950 by MacDonald, have yet to reach the stage of successful control applications. Optimal control for internal combustion engines, minimization or optimization of fuel consumption, and minimization of transient response-time are some examples. A feedback control system with a phase-margin of 60° and gain-margin of 6-dB resulting in a control system of *optimum* transient-response is the nearest practical analogy to a synthesis of the optimal strategy. The system should be theoretically designed, starting with the *desired performance-index* such as transient-response. This theory, at times referred to as *time-optimal control*, is, however, potentially applicable to a great variety of control problems—linear, nonlinear, and time-variable—for continuous and sampled-data control systems, including their stochastic and statistical variations. Some linear and nonlinear systems have been synthesized according to the concept of *variable-gain* in the case of the sampled-data control systems, the gain-parameter having different values at different sampling-instants.

Besides the conventional state-space approach, the theory of optimal control is based on two other principles: (1) dynamic programming, which is concerned with *minimizing the performance-index criterion*, and (2) Pontryagin's **maximum principle**, which is concerned with *maximizing a scalar* H. (H is the Hamiltonian of the energy concept). The latter will also result in minimizing the performance-index criterion.

1.7.4.1 Dynamic Programming by direct approach is a systematic iterative scheme of solving control problems for minimization of the performance-index criterion by using the mathematical theory of *multistate decision* processes.

The minimization of cost of control y^2 over a period T^2 makes a simple example in this case for a total cost equal to

$$\int_0^T (x^2 + y^2)\, dt$$

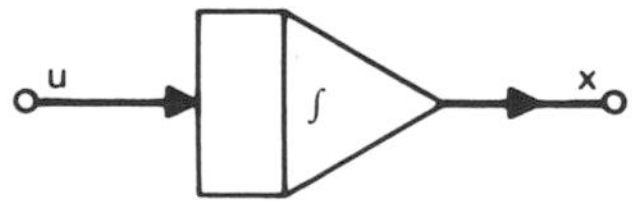

with $\dot{x} = y$.

An optimal control process (or plant) whose dynamics are of the form, $\dot{\overline{x}} = f(\overline{x}, \overline{e}, t)$, must have (1) limitations in input $\overline{e}$ and/or plant $\overline{x}$, (2) a reference signal $\overline{r}(t)$ producing the desired output-response c(t), and (3) a performance-index given by

$$I = \int_0^T F[\overline{c}(t), \overline{e}(t), \overline{r}(t), t]\ dt$$

The preceding description states a typical *time-optimal control* problem. To take a simple example, for a simple unity position-control feedback control system,

$$I = \int_0^T [(|\overline{c}-\overline{r}|)^2 + \lambda(\overline{e})^2]\ dt$$

where $|\overline{c}-\overline{r}|$ is the magnitude of the error-vector, and λ is a *tolerance-factor* for the appropriate error-amplitude. Input $\overline{e}$, over the operating interval t = 0 to t = T, is called the *control law*. The integrand, which is a loss-function, is a measure of the instantaneous change from the ideal performance. The optimal control problem is concerned with finding the control-input $\overline{e}$ that minimizes the performance-index criterion I, subject to any constraints on $\overline{c}$ and $\overline{x}$. If the input $\overline{e}$ is manipulated so that it minimizes I, then $\overline{e}_0(t) = f[\overline{c}(0), \overline{r}(t); t]$ is called the *open-loop policy* (for the initial-state of output); $\overline{e}_0(t) = f[\overline{c}(t), r(t), t]$ is called the **closed-loop policy.**

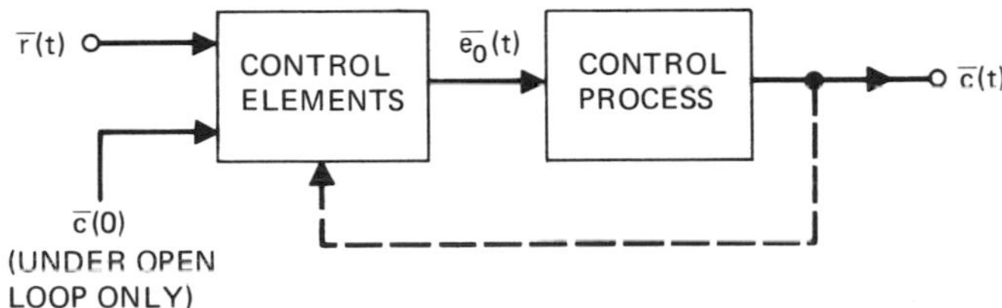

The basic optimal theory, generally applicable to a variety of control problems, takes the following form in the sampled-data control systems by using the difference equations in the vector form. The dynamics of plant,

$$\dot{\overline{x}}_{k+1} = f(\overline{x}_k, \overline{e}_k) = \overline{x}_k + (\Delta t) \cdot f(\overline{x}_k, \overline{e}_k)$$

In the case of "stochastic" optimal control,

$$\dot{\overline{x}}_{k+1} = f(\overline{x}_k, \overline{e}_k, \overline{a}_k)$$

where $\overline{a}_k$ is the "environmental" parameter due to random disturbance or noise. The performance-index

$$I = \sum_{k=0}^{n-1} F(\overline{c}_k, \overline{e}_k, \overline{r}_k)$$

and

$$I = \sum_{k=0}^{n-1} |\overline{c}_k - \overline{r}_k|^2, \text{ etc.}$$

Input for open-loop policy:

$$\overline{e}_k^0 = f(\overline{C}_0, \overline{r}_k)$$

Input for feedback policy:

$$\overline{e}_k^0 = f(\overline{C}_k, \overline{r}_k)$$

There are presently three approaches for defining an optimal control problem to meet the requirements of an objective, an *adequate model*, a *control policy, and stability*: (1) direct approach, (2) dynamic programming, and (3) Pontryagin's maximum-principle. A system designed on the basis of optimal control theory is, however, not necessarily stable. As an example of the state-space direct approach in the sampled-data control systems, for a simple regulator problem with $\bar{r} = 0$, if the controlled process is assumed as a first-order linear plant, $[C_{k+1} = AC_k + Be_k]$ defines the dynamics.

If

$$I = \sum_{k=0}^{2} (C_k)^2 = (C_0)^2 + (C_1)^2 + (C_2)^2$$

and if the controlled-input satisfies the limitation $-1 \leqslant e_k \leqslant 1$, and if the optimal input-sequence e_0^0, e_1^0 is required during the interval $0 < k < 2$

$$\left.\begin{aligned} C_1 &= AC_0 + Be_0 \\ C_2 &= AC_1 + Be_1 \end{aligned}\right\} \quad C_2 = A^2C_0 + ABe_0 + Be_1 \text{ and } I = (C_0)^2 + (AC_0 + Be_0)^2 + (A^2C_0 + ABe_0 + Be_1)^2$$

By taking the partial-derivatives of I with respect to e_0 and e_1 and setting them equal to zero, e_0 and e_1 that minimize I are found by tabulation and comparison of results. Thus, with the *direct approach*, the optimization process involves *two-dimensional minimization.*

1.7.4.2 **But with the dynamic programming approach,** the optimization process involves minimization with respect to one variable at a time in iteration. The concept of dynamic programming as developed by R. Bellman is based on (1) the invariant embedding principle relating to fixed initial states and fixed operation intervals and (2) the optimality principle relating to optimization over an interval and its subinterval.

The difficult problem of solving the partial differential equations, subject to two-point boundary conditions, is eliminated by adapting the dynamic programming approach. Also, the *availability of constraints* in this technique aids the computation of the solution.

The fundamental principle of dynamic programming is the *principle of optimality* or the optimal policy. This is based on the concept of *invariant-embedding*, which allows the substitution of a *multidimensional* optimization process by a sequence of *single-dimensional* optimization processes. In mathematical terms, the *principle of invariance or invariant embedding* in multistage decision problems yields the following result for maximizing the total return.

The final solution S_N of an N-dimensional problem can be expressed by the functional-equation:

$$S_N = f_i(\bar{x}', m_1) + f_{N-1}[g(\bar{x}', m_1)]$$

where $\bar{x}'$, is the state-vector (with dimensional one), characterizing a physical system at any instant, and m_1 is a single-stage decision or policy. The first term in the preceding equation represents the initial solution, and the second term represents the optimum overall solution, or the maximum return from the final $(N - 1)$ dimensional problem. The optimal solution is expressed by the equation:

$$f_N(\bar{x}')]_{N \geqslant 2} = \underset{m_1}{\text{Max.}}\{f(\bar{x}', m_1) + f_{N-1}[g(\bar{x}', m_1)]\}$$

If N = 1, optimum solution is

$$f_1(\overline{x}') = \underset{m_1}{\text{Max.}}\{r(\overline{x}', m_1)\}$$

This is the *principle of invariance*, which thus allows the breakup of an unsolvable N-dimensional control process to N number of one-dimensional control processes. They can be then solved in a systematic iterative manner on a digital computer. The method can be extended to time-varying problems too.

1.7.5.3 Pontryagin's Maximum Principle for Optimal Control. Pontryagin's maximum principle for optimum control is closely related to dynamic programming. It states that a control input $\overline{e}$ that minimizes the performance-index,

$$I = \sum_{i=0}^{n} C_i x_i(T)$$

maximizes on an average the Hamiltonian Scalar-function,

$$H = \sum_{i=0}^{n} P_i f_i(\overline{x}, \overline{e})$$

where P_i is the momentum-vector, and

$$\dot{P}_i \triangleq \sum_{j=0}^{n} \left.\frac{\partial f_j}{\partial x_i}\right|_{\overline{e}=\overline{e}^0}^{P_j} \qquad i = 0, 1, 2, \ldots, n$$

and, if $\overline{e}^0$ is the input-restraint for the maximization of H,

$$\dot{\overline{x}}_i \triangleq f_i(\overline{x}, \overline{e}^0)$$

The maximum principle can be modified by including the minimization of a performance-index

$$I = \int_0^T F_1[\overline{x}, \overline{e}, \overline{r}, t]\, dt$$

Then, it will simultaneously maximize a new Scalar

$$H^1 = \sum_{i=0}^{n} [P_i f_i(\overline{x}, \overline{e}) - F_2(\overline{x}, \overline{e}, \overline{r})]$$

$$\text{if } P_i = \frac{-\partial H^1}{\partial x_i} \qquad i = 1, 2, \ldots, n$$

In the case of the sampled-data control systems, the maximum principle assumes the form:

$$I = \sum_{i=0}^{n} C_i x_{i_{N-1}}$$

and

$$\overline{x}_{k+1} - \overline{x}_k = f(\overline{x}_k, \overline{e}_k) \qquad k = 0, 1, \ldots, n$$

The control input-sequence e_k, which minimizes the performance-index, must maximize the Hamiltonian,

$$H = \sum_{i=0}^{n} P_{i_k} f_j(\overline{x}_k, \overline{e}_k)$$

assuming $\dot{x}_k \triangleq f_j(\overline{x}_k, \overline{e}_k^o)$ and for maximization,

$$P_{i_{k+1}} - P_{i_k} = -\sum_{j=0}^{n} \frac{\partial f_j}{\partial x_i}\bigg|^{P_{j_k}}$$

$(e_k = e_k^0)$ is the restraint. The optimal control theory is yet in its developmental stages. Since it is a powerful technique, the principle of optimality is extended to adaptive control systems.

1.7.5 Adaptive Control Systems. Most feedback control systems are theoretically not capable of precision performance in the presence of (1) the effect of the external disturbances on the operating parameters and (2) the effect of the changes in the control system parameters. This limitation can be avoided by using an adaptive control system, which is, in principle, an extension of the optimal control system. An adaptive control system is generally characterized by the availability of automatic devices that *measure the actual dynamics of the controlled system, continuously compare them with a predetermined optimum figure-of-merit* or operating point, and continuously adjust the parameters of the controlled plant to minimize the error and maintain the precision. If a direct measurement of the figure-of-merit (performance-index) is feasible, the system can be figured as an optimal-control system at the same time. The measurement of the index to be optimized is termed *identification*, and the generation of the *command signal* required for the control of a system parameter to optimize the system is termed *actuation.*

The type of adaptive-control that is most effective for best precision and reliability depends entirely on the parameter one decides to optimize, such as (1) the desired performance-criterion, (2) the minimization of, for example, the integral time absolute-error, (3) the maximization of the probability of a successful beam-sight at a moving-target in the presence of a disturbance, for example, in a radar tracking system.

The high-precision headwheel servo in the quadruplex color videotape recorder can be included in the classification of the adaptive-control systems. The random variations in the 60-Hz vertical synchronizing pulse-rate are continuously measured and compared in phase in an automatic discrete tape-vertical-alignment loop (TVA). The resultant error at an appropriate bandwidth is continuously added to the precision 15.75 kHz automatic horizontal line-lock digital feedback control-loop in order to maintain the performance index of line-lock servo subsystem within an optimized horizontal precision or tolerance of ±0.07 μsec of residual phase-jitter. This figure corresponds to a motor speed-regulation of 0.002%. The monochrome and color correction systems in this recorder can be also classified in this category, because the monochrome automatic tape-horizontal-alignment loop, and the color fine tape-horizontal-alignment loop are likewise taken into consideration. (In the context of the adaptive-control system-applications, it may be mentioned that originally the flight-control fuel-economy problem for obtaining the maximum cruising-range of long-range aircraft, per pound of fuel used in their internal-combustion engines, stimulated the development of the first adaptive-control systems that operated on the parameters of the throttle-setting and the fuel-air mixtures. The manifold pressure-ratio was used for identification, and these systems in fact did raise the cruising-range by as much as 25%.)

The adaptive-control systems are in practice subdivided into four classifications:

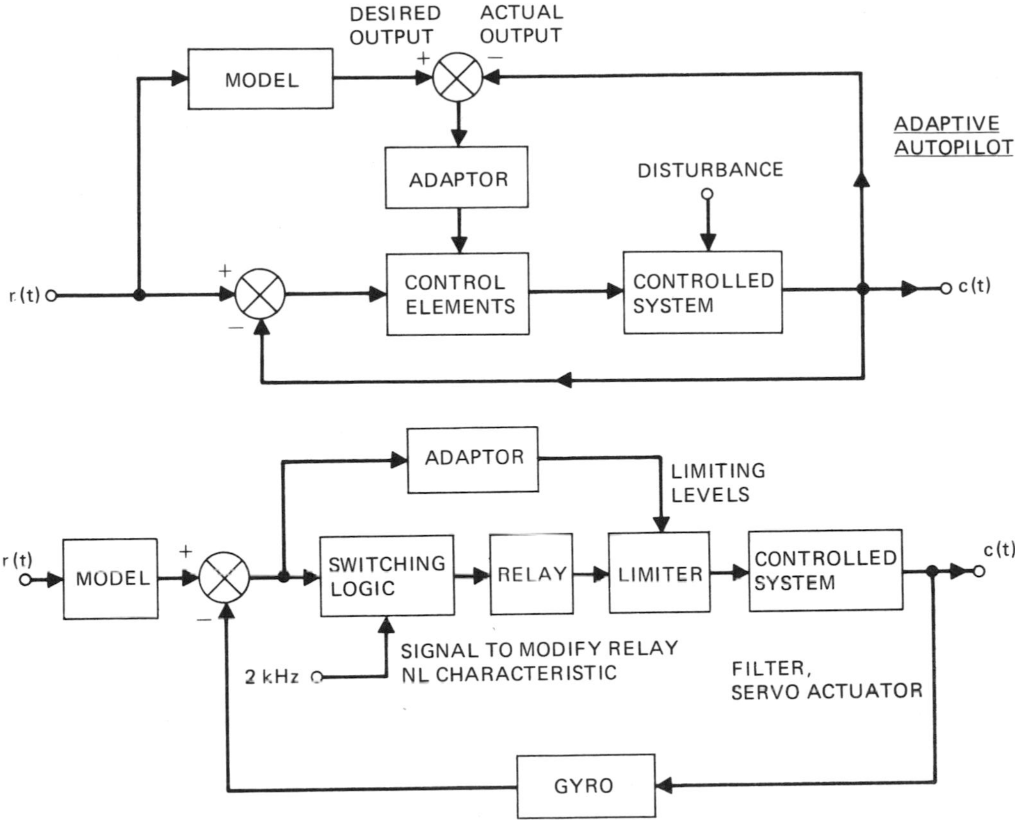

Fig. 1-23. Typical model-reference and nonlinear adaptive control systems.

(1) model-reference, (2) nonlinear, (3) impulse-response, and (4) digital-computer control. The control subsystems of the quadruplex color videotape recorder can be placed under the second and last classifications. Two typical examples of diagrams under the categories of model-reference and nonlinear adaptive control systems are shown in Fig. 1-23.

The adaptor attenuates the relay-output when the error is below a prescribed level only. In an adaptive-control system, if the machine is set back to some nonoptimum parameter, the system will initiate a *search-procedure* in a logical fashion and change its parameters automatically. If the system makes use of its *past experience* under the direction of an exclusive built-in logic to proceed more directly to an optimum-setting, the machine can be classified as a *learning machine*. Digital computers make an important contribution in this respect. For example, at start, the built-in sensing logic in the color videotape recorder enables the machine to reproduce a stabilized color picture from the videotape in 1 to 3 sec. The actual pull-in time mostly depends upon the relevant operating modes applicable at any instant. In the line-lock-only (LLO) mode, with a system-disturbance of the nature of a *drop-out* and a momentary loss of synchronizing pulses, the built-in logic allows the precise choice of the nearest single sample-pulse of two adjacent pulses, for an instantaneous horizontal lock-up of the color picture at half the subcarrier-frequency rate (1.79 MHz) without going through the whole cycle of the primary feedback-loops in the head-wheel sampled-data control system-complex. Such a constraint presents a semblance of "learning" techniques used in adaptive- and optimal-control systems.

1.7.6 Principles of Lyapounov's Direct Method. With Lyapounov's Direct (or Second) Method, a definite solution for the system equation is not required. It is based on the concept of total-energy V contained in a system, and its steady decrease with time at the incidence of a perturbation. For a control system, the method enables the possible choice of a *Lyapounov function* $V(x_1, x_2, \ldots)$, which is *positive-definite*. Then, if the V-function is applicable to the control system, the system will be *asymptotically stable* if $\dot{V}(x_1, x_2, \ldots)$ is *negaitve-definite*. There is no unique method for synthesizing the scalar V-functions; however, a few methods, such as (1) Schultz's *variable-gradient method* for linear and nonlinear systems, (2) Routh's and Weygandt-Puri's *cononical-matrix method* for linear systems, (3) Lurie's *transformation method*, and (4) matrix-formulation method with what is called *Aizermann's conjecture*, are available, although these procedures are strictly restrictive in the type of nonlinearity they can handle. Nevertheless, in the case of nonlinear systems and adaptive-control systems, it is simpler to generate a satisfactory V-function than to find a unique solution to a nonlinear differential equation. The method is extended to the first-order difference equations of the state-space concept in the case of the sampled-data control systems by Hahn, Kalman, and Bertram.

(*Lyapounov's first method* involves the determination of an *explicit solution*, general or particular, of a set of linear difference or differential equations, especially when the system equation is represented by an infinite power-series. He called the systems stable according to his first method, if "the motions which are once near-together, remain near each other for all time." Lyapounov's methods are of course based on the concepts of original classical physics.)

A stationary, autonomous (time-invariant), force-free control system is *asymptotically stable*, if at the incidence of a small disturbance under a particular set of initial conditions, it returns to its original state of equilibrium as the time extends to infinite. If this statement holds good for *any initial conditions*, the system is *globally asymptotically stable* or asymptotically stable at large. Time does not appear explicitly in an *autonomous system.* Linear systems are *globally stable.* Stability in the conventional sense means that the solution of the characteristic differential equation must be *bounded for a bounded disturbance*, without requiring the clause that the solution approach zero as the time $t \longrightarrow \infty$. For example, a control system *with phase-jitter* (or a phase-plane limit-cycle at the origin) is "stable" in the conventional sense, since the *jitter* is bounded, but it is *not asymptotically stable.* In the case of the Lyapounov's *Direct Method for a nonlinear control system*, if a V-function meeting the stated requirements for stability does exist, the system is stable in the Lyapounov sense for a certain *range of values of the system parameters*, but if such a V-function cannot be found, the system may be *either stable or unstable.* Thus, for nonlinear systems, Lyapounov's stability criterion with the possible choice of an appropriate V-function is only *a sufficient condition, but not a necessary condition*, for global asymptotic stability about the origin. Specific applications of Lyapounov's Direct Method date back to 1946 in Russian control literature.

The Direct Method explicitly *includes the linear-case* (the Routh-Hurwitz stability criterion mentioned earlier, and Lurye and Letov made important contributions on this basis to the stability analysis of a few nonlinear systems, working either under direct proportional-control or under *indirect control*, the latter *if the error-signal is used to achieve a rate of displacement of the input variable.* This test procedure is based on a comparison of the stability conditions defined by Routh-Hurwitz *after the nonlinearity is replaced by its upper-bound.* The stability conditions are obtained by Lyapounov's Direct Method for generating the appropriate V-function by using the matrix-formulation techniques. Then it can be shown that for a given linear plant, the overall system will be stable if the nonlinearity is restricted to certain bounds or limits. As an example, for a *third-order linear system* with one single-valued nonlinearity, it can be tested with Lyapounov criterion, according to the so-called *Aizermann's conjecture*, that the open-loop poles of the plant remain in the left-half complex-plane, while the *nonlinearity is restricted* to the first and third quadrants.

1.7.7 Lyapounov's Direct Method for Continuous and Sampled-Data Control Systems. In the case of a linear analog/continuous system, the state-variable technique is applied to reduce the system equation to a set of first-order linear differential equations:

$$a_n \frac{d^n y}{dt^n} + a_{n-1} \frac{d^{n-1} y}{dt^{n-1}} + \cdots a_1 \frac{dy}{dt} + a_0 y = 0$$

$$y = x_1; \qquad \frac{dy}{dt} = x_2 = \dot{x}_1; \qquad \frac{d^2y}{dt^2} = x_3 = \dot{x}_2; \qquad \text{and } \dot{\bar{x}} = \bar{A}\bar{x} + \bar{b}e$$

for a *force-free system.* Now, some appropriate function $V(\bar{X})$ is chosen (or generated) containing all the above state-variables so that $V(\bar{X})$ is *positive-definite* and is equal to zero only when all the state-variables are equal to zero.

It is known that, for the vector equation $V(\bar{X}) = \bar{X}'\bar{Q}\bar{X}$, matrix $\bar{X}'$, being the *transpose* of $\bar{X}$, is a positive-definite function. It is then a *necessary and sufficient condition that $\bar{Q}$ be a positive-definite symmetric-matrix.* If the positive-definite quadratic-form is the Lyapounov function for the above force-free system, then

$$\left.\begin{aligned} V(\bar{X}) &= \bar{X}'\bar{Q}\bar{X}, \quad \bar{Q} > 0 \\ \dot{V}(\bar{X}) &= (\dot{\bar{X}}'\bar{Q}\bar{X}) + (\bar{X}'\bar{Q}\dot{\bar{X}}) \\ \dot{V}(\bar{X}) &= (\bar{X}'\bar{A}'\bar{Q}\bar{X}) + (\bar{X}'\bar{Q}\bar{A}\bar{X}) \end{aligned}\right\} \text{Since } \dot{\bar{X}} = \bar{A}\bar{X} \text{ for the free system, and } \dot{\bar{X}}' = \bar{X}'\bar{A}'$$

$$\dot{V}(\bar{X}) = \bar{X}'(\bar{A}'\bar{Q} + \bar{Q}\bar{A})\bar{X} = \textit{a negative-definite quadratic-form,}$$

$$\text{only if } \bar{A}'\bar{Q} + \bar{Q}\bar{A} = -\bar{M} \cdots \bar{M} > 0$$

That is, only if $\bar{M}$ is a positive-definite symmetric matrix. Thus, the *conditions for stability are obtained in terms of the coefficients of the variables* in the system equation, without any need for solving the differential equation of a higher-order system, whether linear or nonlinear.

In the case of a nonlinear control system, the Lyapounov criterion is equivalent to plotting a chosen V-function, which appears as a closed curve in the *phase-plane*, and making sure that all possible system trajectories pass *from the outside to the inside* of the circle. The closed curve chosen is a circle in this case. The system is unstable if the trajectories do the reverse, although the phase-plane plot indicates a final convergence to the origin. The principles underlying the generation of such scalar

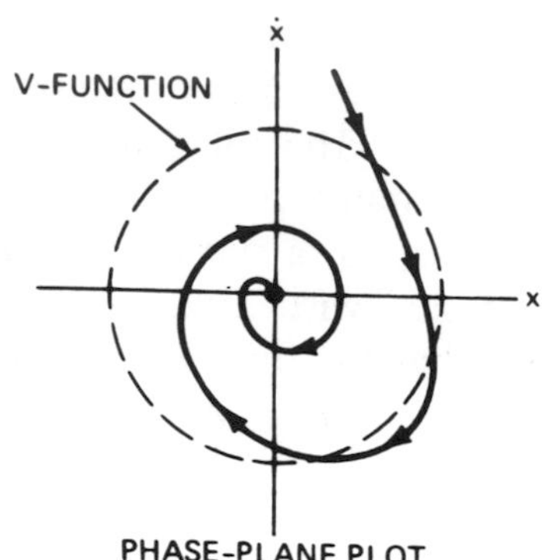

V-functions for a few restricted cases of nonlinearities are mentioned in the previous section.

In the case of a sampled-data control system, the method is entirely analogous. A linear pulse-transfer function is considered, and it is reduced to its canonical form:

$$\frac{C(z)}{R(z)} = \frac{\sum_{k=0}^{N-1} a_k z^{-k}}{1 + \sum_{j=1}^{M-1} b_j z^{-j}}$$

the denominator being the characteristic equation.

$$C(z) = \sum_{k=0}^{N-1} a_k z^{-k} D(z) \tag{1}$$

where

$$D(z) = \frac{R(z)}{1 + \Sigma b_j z^{-j}}$$

or

$$D(z) = R(z) - \sum_{j=1}^{M-1} b_j z^{-j} D(z) \tag{2}$$

Expanding Eqs. 1 and 2,

$$c(n) = a_0 D(n) + a_1 D(n-1) + \cdots + a_k D(n-k) \tag{3}$$

where

$$k = N - 1$$

and

$$D(n) = r(n) - b_1 D(n-1) - \cdots - b_j D(n-j) \tag{4}$$

where

$j = M - 1$

Substituting

$$D(n) = x_1(n); \qquad D(n-1) = x_2(n)$$

From Eq. 4,

$$x_1(n) = D(n) = r(n) - b_1 x_1(n-1) - \cdots - b_j x_j(n-1)$$

and

$$x_2(n) = x_1(n-1); \qquad x_3(n) = x_2(n-1)$$

$$\begin{bmatrix} x_1(n) \\ x_2(n) \\ x_3(n) \\ \cdots \\ \cdots \\ x_j(n) \end{bmatrix} = \begin{bmatrix} -b_1 & -b_2 \cdots -b_j \\ 1 & 0 \ \cdots \ 0 \\ 0 & 1 \ \cdots \ 0 \\ \vdots & \vdots \quad\quad \vdots \\ 0 & 0 \ \cdot\cdot 1 \ \ 0 \end{bmatrix} \begin{bmatrix} x_1(n-1) \\ x_2(n-1) \\ x_3(n-1) \\ \cdots \\ \cdots \\ x_j(n-1) \end{bmatrix} + r_n \tag{5}$$

$$\overline{x}_n = \overline{A}\overline{x}_{n-1} + \overline{d}r_n \cdots$$

where

$\overline{d}r_n = 0$ for a force-free system,

and

$$c(n) = a_0 x_1(n) + a_1 x_2(n) + \cdots + a_k x_{k+1}(n) \cdots k = N - 1$$

or

$$c_n = \overline{a}'\overline{x}(n) \qquad \text{if } \overline{a} = \begin{bmatrix} a_0 \\ a_1 \\ \vdots \\ a_k \end{bmatrix}$$

For nonlinear control systems, there will be no z-transform, and the system equation will be known only in a difference equation form.

For a force-free linear sampled-data control system, the system state-space equations are given by:

$$\overline{X}_n = \overline{A}\overline{X}_{n-1} \tag{6}$$

and

$$c_n = \overline{a}'\overline{X}_n \tag{7}$$

Choosing a positive-definite quadratic form for the Lyapounov function,

$$V(X_n) = \overline{X}'_n\overline{Q}\overline{X}_n \tag{8}$$

where $\overline{Q}$ is any *positive-definite* symmetric matrix.

The system will be stable if the first backward difference of $V(X_n)$ is *negative-definite*, according to the Lyapounov criterion.

$$\nabla V(X_n) = [V(\overline{X}_n) - V(\overline{X}_{n-1})] < 0 \text{ for } \overline{X} \neq 0$$

$$\nabla V(X_n) = \overline{X}'_n\overline{Q}\overline{X}_n - \overline{X}'_{n-1}\overline{Q}\overline{X}_{n-1} \text{ and } \overline{X}_n = \overline{A}\overline{X}_{n-1}$$

$$\nabla V(X_n) = \overline{X}'_{n-1}\overline{A}'\overline{Q}\overline{A}\overline{X}_{n-1} - \overline{X}'_{n-1}\overline{Q}\overline{X}_{n-1} = \overline{X}'_{n-1}(\overline{A}'\overline{Q}\overline{A} - \overline{Q})\overline{X}_{n-1} \tag{9}$$

$\nabla V(X_n)$ will be negative-definite, only if, $-\overline{M} = \overline{A}'\overline{Q}\overline{A} - \overline{Q} \cdots \overline{M} > 0$

The result is *identical* to that obtained for the continuous system. This condition, which is both *necessary and sufficient* in the case of linear systems only, must be therefore satisfied if a system of difference equations, given by Eqs. 6 and 7, is to be stable.

1.8 POPOV'S GENERAL STABILITY CRITERION FOR NONLINEAR CONTINUOUS AND SAMPLED-DATA CONTROL SYSTEMS

Popov's stability criterion for nonlinear continuous control systems has been extended to sampled-data control systems by Tsypkin. For a system with one nonlinear element of any shape, if f(e) satisfies the two conditions,

1. $f(e) = 0$, and

$$f(e)/e \begin{cases} < k \\ > 0 \end{cases} \quad \text{for } e \neq 0$$

2. For a stable linear part G(z) of the system, a sufficient condition for the stability of the force-free nonlinear system is given by

$$\mathrm{Re}[G(e^{j\omega T})] > \frac{-1}{k}$$

where k is the slope of the nonlinearity at the origin, and Re = real-part.

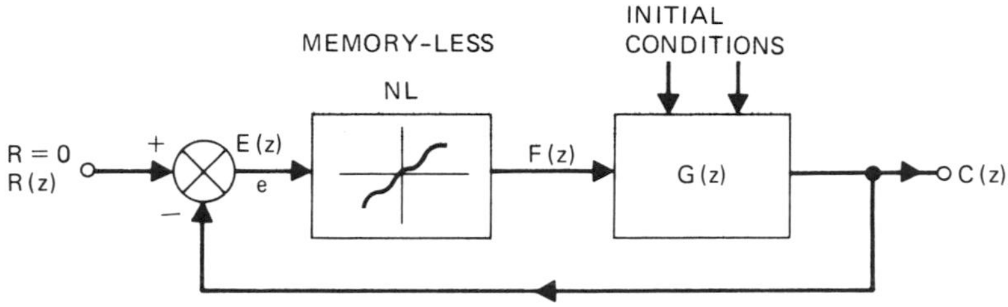

Continuous or sampled-data control system.

According to these conditions, the nonlinearity lies within the first and the third quadrants. The graphical solutions satisfying the *sufficiency condition for stability* and no sufficiency are shown in Fig. 1-24a, assuming that the linear part has no poles on or outside the unit-circle.

For the generalization of the nonlinearity, a *coordinate transformation* is then performed to *replace the unstable plant by a stable plant*, to which the stability analysis applies.

The new nonlinear function $f^1(e)$ and the new linear part are indicated by the two inner loops in Fig. 1-24b. The two paths through k_1 from the output are identically cancelled for equality to the system in the original block-diagram. The function k_1 is chosen so that the poles of $G^1(z)$, the new linear part, are inside the unit-circle.

The secondary-slope chosen with k_1 will have a narrowing effect on the sector occupied by the nonlinearity.

$$G^1(z) = G(z)/1 + k_1 G(z)$$

Then, if some linear gain K is chosen so that $K > f^1(e)/e$, the system will be stable if

$$\operatorname{Re} G^1(e^{j\omega T}) > -\frac{1}{K}$$

From the diagram,

$$f^1(e) = f(e) - k_1 e$$

The transformed nonlinearity must now satisfy condition 1 of the Popov criterion:

$$\left.\frac{f^1(e)}{e}\right\} \begin{matrix} < K \\ > 0 \end{matrix}$$

$$k_1 < [f(e)/e] < (k_1 + K) \qquad \text{or} \left.\frac{f(e)}{e}\right\} \begin{matrix} < k \\ > k_1 \end{matrix}$$

This result gives the condition for generalizing the stability criterion for any shape of nonlinearity. If G(z) contains a pole at $z = 1$, an infinitesimally small k_1 will do to bring the pole inside the unit-circle to make $G(z) = G^1(z)$ and avoid the above *coordinate-transformation.*

Above equations from *Theory of Sampled Data Control Systems* (1965) by David P. Lindorff. Courtesy of John Wiley & Sons.

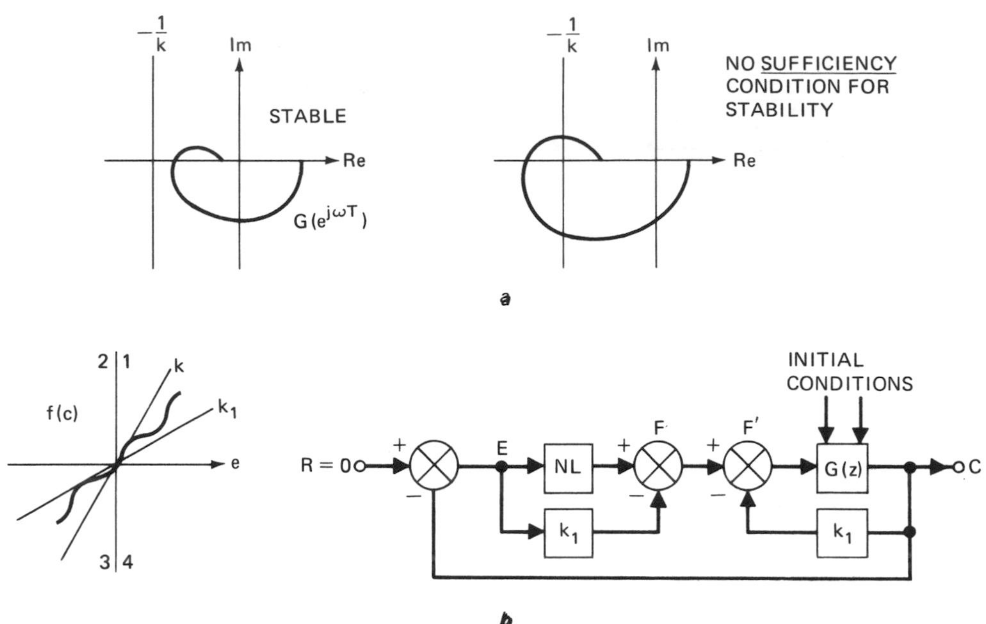

Fig. 1-24. *a*, Popov's criterion. *b*, Unstable plant replaced by stable plant for analysis by transformation. (From *Theory of Sampled Data Control Systems* (1965) by David P. Lindorff. Courtesy of John Wiley & Sons.)

1.9 Multirate Sampled-Data Control Systems. Digital controllers that have *at least two signals at two different sampling rates*, are *multirate* sampled-data systems. The multirate sampling is used:

1. for improving the response of the systems receiving data at lower rates
2. to meet the requirements of the intrinsic nature of the digital systems for higher sensitivity and special purposes, as in the Q-CVTR (color video tape recorder)

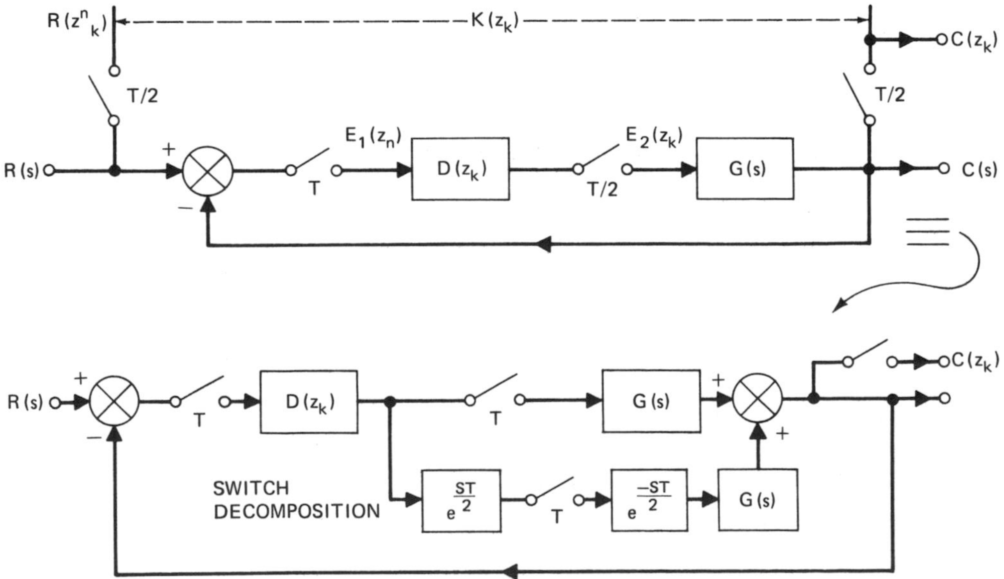

Fig. 1-25. Multirate sampled-data control system. Switch-decomposition technique works irrespective of the sequence of high and low sampling-rates in the system.

3. to read the intersample ripple at the output of a system to detect possible "hidden oscillations" in between the sampling intervals, and
4. to include remote data-transmission links at different rates in the same control loop.

A multirate sampled-data control system with sampling-periods T and T/k can be reduced to an *equivalent multiloop single-rate problem* by realizing the pulse-transfer-function of the system by "switch-decomposition" method. Then, the multirate system can be realized to produce an output-response to a step-input identical to that of a single-rate system with a T/k sampling-period.

For a higher-rate sampler following the lower sampling-rate,

$$C(z_k) \triangleq \sum_{m=0}^{\infty} c\frac{mT}{k} z_k^{-m} \tag{1}$$

and

$$c(t) = \sum_{n=0}^{\infty} r(nT)g(t - nT)$$

$$C(z_k) = \sum_{m=0}^{\infty}\sum_{n=0}^{\infty} r(nT)g\left[\frac{mT}{k} - nT\right] z_k^{-m} \tag{2}$$

where m and n are integers. Hence, (m − nk) = j, an integer. For large z_k, the series converges uniformly, and the order of summation can be interchanged.

$$C(z_k) = \sum_{n=0}^{\infty} r(nT) \sum_{m=0}^{\infty} g\left[\frac{mT}{k} - nT\right] z_k^{-m} \tag{3}$$

$$= R(z_k^n)G(z_k)$$

on further simplification. As an example, if

$$r(t) = e^{-at}, \qquad R(z_k^n) = \frac{1}{1 - e^{-at}z_k^{-n}}$$

$$C(z_k) = \frac{R(z_k^n)D(z_k)G(z_k)}{1 + Z[D(z_k)G(z_k)]} \tag{4}$$

Now, the pulse transfer function,

$$K(z_k) = \frac{C(z_k)}{R(z_k)}$$

and

$$K(z) \triangleq Z[K(z_k)]$$

$$K(z_k) = \frac{D(z_k)G(z_k)}{1 + Z[D(z_k)G(z_k)]} \tag{5}$$

The digital controller $D(z_k)$ is then obtained by inversing Eq. 5 and substitution:

$$D(z_k) = \frac{1}{G(z_k)} \frac{K(z_k)}{1 - K(z_k^n)} \tag{6}$$

Thus, given an overall pulse-transfer-function, the multirate controller for the desired response can be obtained from Eq. 6.

Example of Switch Decomposition:

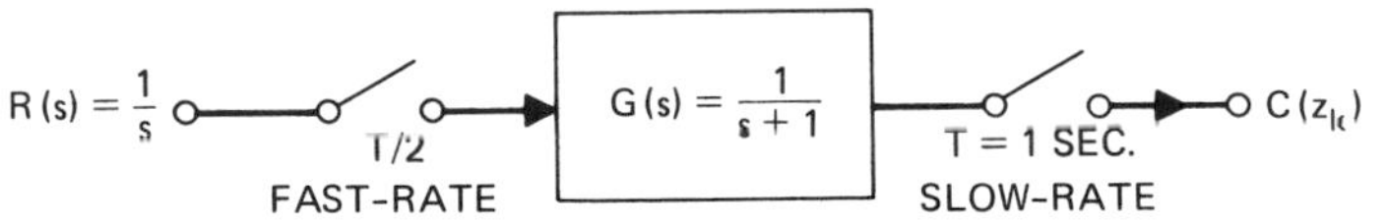

R (s) = 1/s — T/2 FAST-RATE — G (s) = 1/(s + 1) — T = 1 SEC. SLOW-RATE — C (z_k)

$$C(z_k) = Z\left(\frac{1}{s}\right) \cdot Z\left(\frac{1}{s+1}\right) + Z\left(\frac{e^{s/2}}{s}\right) \cdot Z\left(\frac{e^{-s/2}}{s+1}\right)$$

$$= \frac{1}{(1 - z^{-1})(1 - e^{-1}z^{-1})} + \frac{e^{-1/2}z^{-1}}{(1 - z^{-1})(1 - e^{-1}z^{-1})}$$

1.10 AC CARRIER-SERVO SYSTEMS

The suppressed ac carrier-servo system is an example of one type of sampled-data control system. If the input-signal changes little in a single-period of the carrier wave (sampling-period), the equations reduce to those of the equivalent continuous servo system; however, as the carrier-frequency decreases relative to the signal-bandwidth, the approximation becomes less valid, and a more comprehensive sampled-data technique will be found necessary. A carrier system is Type 1, if both the input and output signals are modulated carriers (with the motor operating as an ideal demodulator); it is Type II, if the input alone is the signal-frequency; and Type III, if the output alone is the signal-frequency. A regular demodulator is required in the final case.

In an ac carrier-servo system, the control information takes the form of the modulation on a carrier, while, in a conventional continuous control system, the control signals are proportional to the instantaneous amplitude. The carrier systems are quite common. The Q-CVTR does use an ac carrier-servo as a partial facility. They allow sensitive transducers and comparatively less expensive and easily reproduced amplifiers; the power elements need less maintenance, and the motors incidentally act as perfect demodulators. In general, synchros and choppers act as system modulators.

For a carrier and sinusoidal input signal-functions $f_c(t)$ and $f_s(t)$,

1. *Amplitude-modulation* function

$$f_m(t) = f_s(t) \cdot f_c(t)$$

where

$$f_c(t) = \cos \omega_c t, \text{ and } f_s(t) = K[a + kE_s \cos \omega_s t] \cdots k = \text{degree of modulation.}$$

a = 1 for *unbalanced-modulation* and a = 0 for *balanced-modulation.*

$$f_m(t) = e_m(t) = k[a + kE_s \cos \omega_s t] \cos \omega_c t$$

$$e_m(t) = aK \cos \omega_c t + \frac{kK}{2} E_s[\cos(\omega_c + \omega_s)t + \cos(\omega_c - \omega_s)t]$$ {with *carrier* and *upper* and *lower sidebands.*

2. Representing the input signal as a general periodic-function,

$$f_s(t) = \sum_{n=1}^{M} E_n \cos \omega_n t$$

$$e_m(t) = aK \cos \omega_c t + \frac{kK}{2} \sum_{n=1}^{M} [\cos(\omega_c + \omega_n)t + \cos(\omega_c - \omega_n)t] \cdots \omega_n = \frac{2\pi n}{T}$$

3. If input is nonperiodic,

$$e_m(t) = aK \cos \omega_c t + kK \int_{-\infty}^{\infty} e_s(t) \cos \omega_c t e^{-j\omega t} \, dt$$

If $E_s(j\omega)$ is the Fourier-transform of $e_s(t)$:

$$e_m(t) = aK \cos \omega_c t + \frac{kK}{2} [E_s(j\omega + j\omega_c) + E_s(j\omega - j\omega_c)]$$

4. If the system is nonlinear, or a linear time-variant system, new frequency components are produced. For a time-variant linear case, if $S_h(t)$ is a switching function

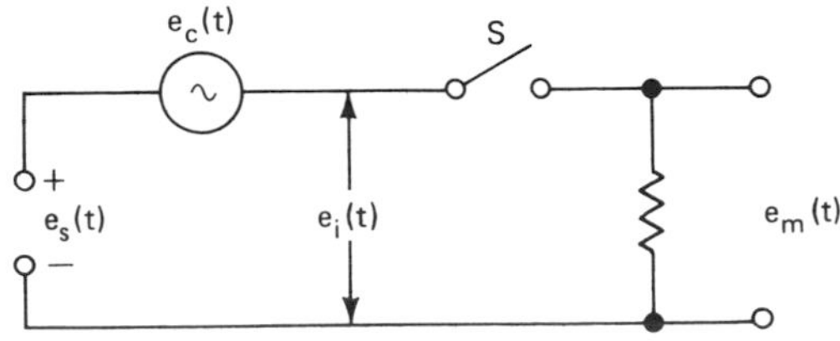

$$e_i(t) = e_s(t) + \cos \omega_c t$$

$$f_m(t) = e_m(t) = e_i(t)S_h(t) = S_h(t)[f_s(t) + \cos \omega_c t]$$

Above equations from *A. C. Carrier Control Systems* (1964) by Keith A. Ivey. Courtesy of John Wiley & Sons.

where

$$S_h t \left\} \begin{array}{l} = 1, \text{ when S is closed} \\ = 0, \text{ when S is open} \end{array} \right.$$

The Fourier Series for

$$S_h(t) = \frac{1}{2} + \sum_{n=1}^{\infty} \frac{2}{n\pi} \sin \frac{n\pi}{2} \cos n\omega_c t$$

If a filter is used for the output spectrum, $(\omega_c - \omega_m) < \omega < (\omega_c + \omega_m)$ with an attenuation-factor α, and if the amplifier gain-constant of $S_h(t)$ is K_0, and if ω_m (the upper limit of the spectrum of $e_s(t)$) $< \dfrac{\omega_c}{2}$

$$f_m^1(t) = e_m^1(t) = \frac{\alpha K_0}{2} [1 + ke_s(t)] \cos \omega_c t$$

where

$$k = \frac{4}{\pi}$$

Most control-system modulating devices are of this character.

5. For a nonlinear case (e.g., square-law), in the region of operation about $e_i = 0$

$$e_m(t) \approx a_1 e_i + a_2 e_i^2$$

$$f_m(t) = e_m(t) \approx a_1 [e_s(t) + \cos \omega_c t] + a_2 [a_s(t) + \cos \omega_c t]^2 + \cdots$$

With a sinusoidal carrier signal,

$$e_m(t) \approx a_1 \left[1 + \frac{2a_2}{a_1} e_s(t)\right] \cos \omega_c t + a_1 e_s(t) + a_2 e_s^2(t) + a_2 \cos^2 \omega_c t$$

With filter,

$$(\omega_c - \omega_m) < \omega < (\omega_c + \omega_m)$$

where

$$\omega_m < \frac{\omega_c}{2}, \text{ and attenuation-factor } \alpha,$$

$$\text{output } e_m^1(t) = \alpha a_1 \left[1 + \frac{2a_2}{a_1} e_s(t)\right] \cos \omega_c t$$

Above equations from *A. C. Carrier Control Systems* (1964) by Keith A. Ivey. Courtesy of John Wiley & Sons.

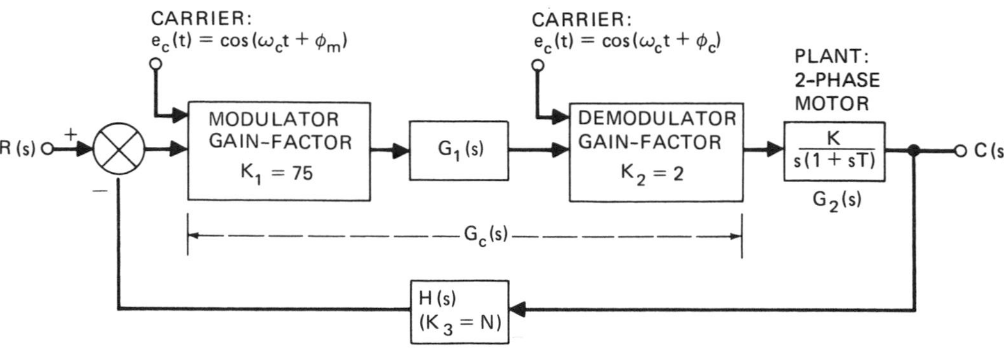

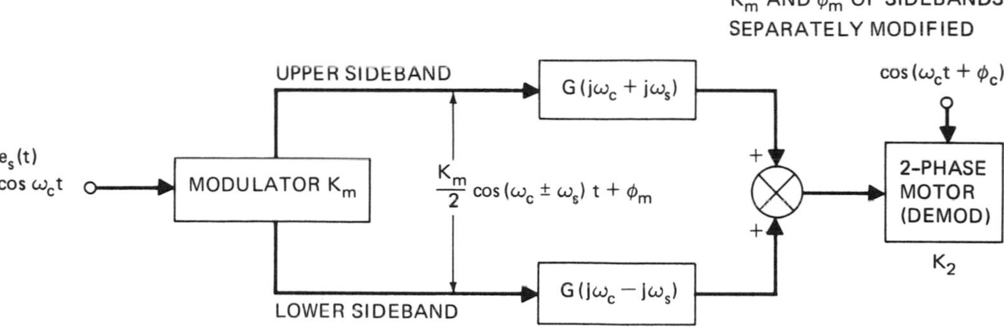

Fig. 1-26. Typical ac carrier-servo. (Courtesy of John Wiley & Sons. From *A.C. Carrier Control Systems* by Keith A. Ivey, 1964)

where

$$\alpha a_1 = K; \qquad \frac{2a_2}{a_1} = k$$

6. System analysis of a suppressed-carrier feedback control system:

$$G(s) = \frac{C(s)}{R(s)} = \frac{G_c(s)G_2(s)}{1 + G_c(s)G_2(s)H(s)}$$

where

$$G_c(s) = \frac{{}^*2K_1K_2}{4} \, \mathrm{Re}_j[\overset{(+)\,\rightarrow}{G}(s + j\omega_c)e^{-j(\phi_c - \phi_m)}]$$

$$= \frac{K_1K_2}{2} \{[G(s + j\omega_c)] \cos(\phi_c - \phi_m) - j[G(s + j\omega_c) \sin(\phi_c - \phi_m)]\}$$

By expansion of the term (+),

$$G_c(s) = [\underbrace{G(s + j\omega_c)e^{-j(\phi_c - \phi_m)}}_{\mathrm{Re}_j} + \underbrace{G(s - j\omega_c)e^{-j(\phi_c - \phi_m)}}_{\text{deleted}^*}] \frac{K_1K_2}{4};$$

Above equations from *A. C. Carrier Control Systems* (1964) by Keith A. Ivey. Courtesy of John Wiley & Sons.

$$G_2(s) = \frac{K}{s(1 + sT)} \text{ and } H(s) = N$$

may serve as examples in the above system, G(s).

The above transfer-function G(s) gives the basic relationship of the carrier control-system analysis. When G(s) is the ratio of two low-order polynomials, the algebraic manipulation of $G_c(s)$ will be fairly simple, because with the low-order complex-roots there are simplified trigonometric techniques. The demodulator is considered ideal when it is assumed to suppress the harmonic-content generated by the demodulation process. The frequency response techniques can then be applied to G(s) to obtain the closed-loop transient-response and the stability conditions. As far as the compensation is concerned, the capacitors and inductors used in the RLC networks of the dc systems after demodulation are replaced as shown below by the series- or shunt-resonant *equivalent-transformations* (the ac networks in the carrier systems prior to the demodulation by the ac servo motor). Parallel and bridged-T networks are most common for ac compensation.

Example:

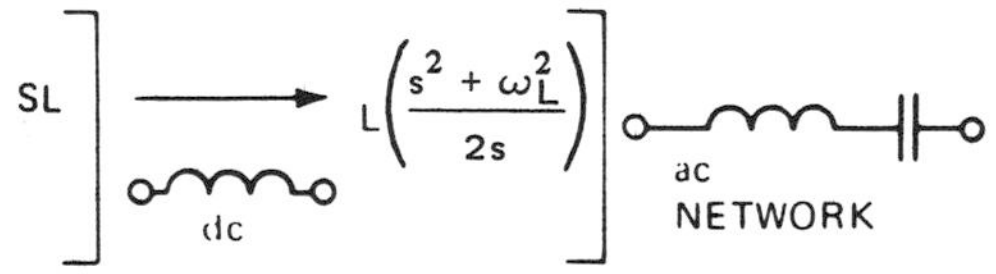

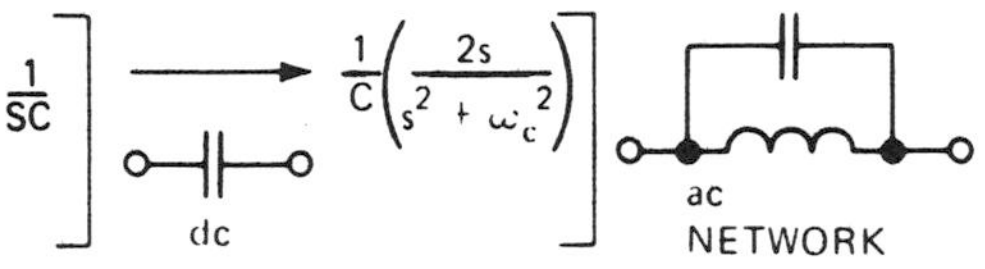

$$\left.\begin{array}{c} SL \\ \text{dc} \end{array}\right] \longrightarrow L\left(\frac{s^2 + \omega_L^2}{2s}\right) \left[\begin{array}{c} \text{ac} \\ \text{NETWORK} \end{array}\right.$$

$$\left.\begin{array}{c} \frac{1}{SC} \\ \text{dc} \end{array}\right] \longrightarrow \frac{1}{C}\left(\frac{2s}{s^2 + \omega_c^2}\right) \left[\begin{array}{c} \text{ac} \\ \text{NETWORK} \end{array}\right.$$

1.11 ANALOG COMPUTER SIMULATION OF CONTROL SYSTEMS

The analog computer performs its operation by setting up analog parameters, such as voltages, currents, and angles of displacement by integration, attenuation, summation, poles, zeros, and nonlinearities, for the variables involved. By operating on these analog parameters in a *continuous* fashion, the results are computed by the processing of electrical signals in order to solve an equation or process a mathematical algorithm. The behavior of the computer is thus made directly comparable (or *simulated*) to the behavior of the physical or mathematical system under consideration. Some analog machines make use of mechanical shaft-positions, signal frequencies, light intensities, and hydraulic elements. The *precision* of the results depends mainly on the actual *fabrication* of the device; and the uniformity with which it is operated in practice depends on calibrated scales and human-errors inherent in the use of measuring apparatus. A device such as a link-trainer, that simulates an air-flight for the ground instruction of a pilot, is a good example.

Since an analog computer is a working model of the control function being performed, the design of the system may require only a straightforward connection of the components (e.g., by means of patch-cords), thereby rendering the programming process easier and faster (unlike that of a digital computer). The continuous nature of the device enables a rapid functioning of the device in a closed-loop feedback control system. A differential analyzer is an example of such a general-purpose analog computer, and it is used for the rapid analysis of equations or functions that represent physical systems. The system to be analyzed is simulated by differential equations or by transfer-functions, and then analyzed by the analog computer for transient response, etc. The system elements of an analog computer are described in the following sequence:

1. **Potentiometers** are used to adjust the coefficients in an equation.

$+E_1$ —○— $+E_2$

Transfer Function: $E_2 = aE_1 \bigg\} \dfrac{E_2(s)}{E_1(s)} = a$

2. **Operational amplifier:** $R_f/R_i = 1$, to change sign as an inverter, and $R_f/R_i \neq 1$, to set gain as an operational amplifier. μ = amplification factor.

$$\left.\begin{aligned} E_1 - I_iR_i &= E_g \\ E_g + I_fR_f &= -E_2 \\ E_2 &= -\mu E_g \end{aligned}\right\} \frac{E_2(s)}{E_1(s)} = \frac{-R_f}{R_i}$$

3. A **summing amplifier** is used for addition, subtraction, and multiplication by a constant, and for simulation of error-detectors in feedback control systems.

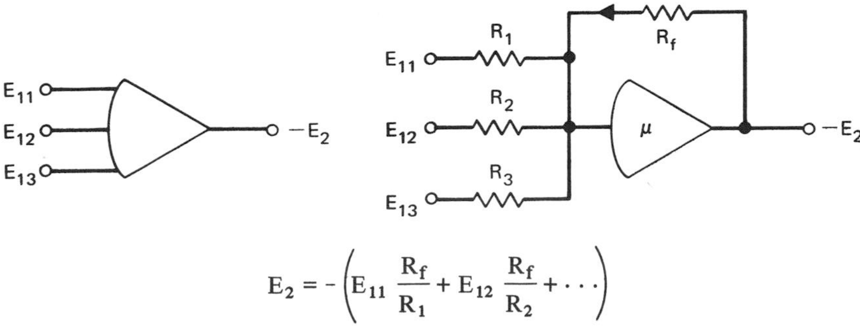

$$E_2 = -\left(E_{11}\frac{R_f}{R_1} + E_{12}\frac{R_f}{R_2} + \cdots\right)$$

4. The **integrator** is the most important device for setting up the analog of a differential equation. Initial conditions are set by charging the feedback capacitor. It simultaneously operates as a summing device.

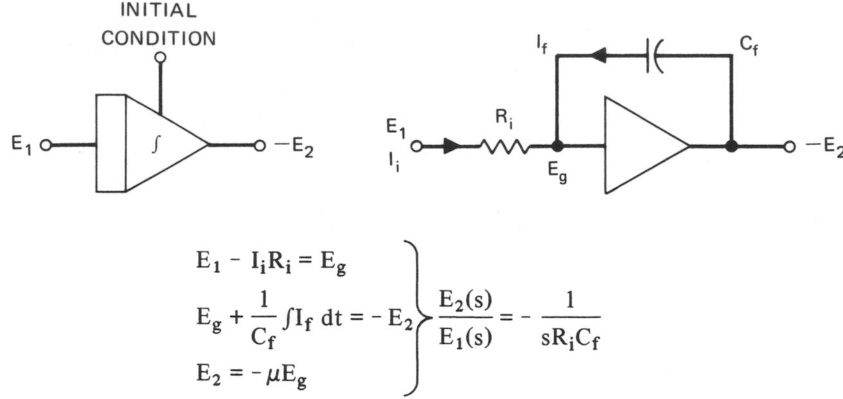

$$\left.\begin{aligned} &E_1 - I_i R_i = E_g \\ &E_g + \frac{1}{C_f}\int I_f\,dt = -E_2 \\ &E_2 = -\mu E_g \end{aligned}\right\} \frac{E_2(s)}{E_1(s)} = -\frac{1}{sR_iC_f}$$

5. A **differentiator** is usually avoided because of noise problems and a tendency to saturate:

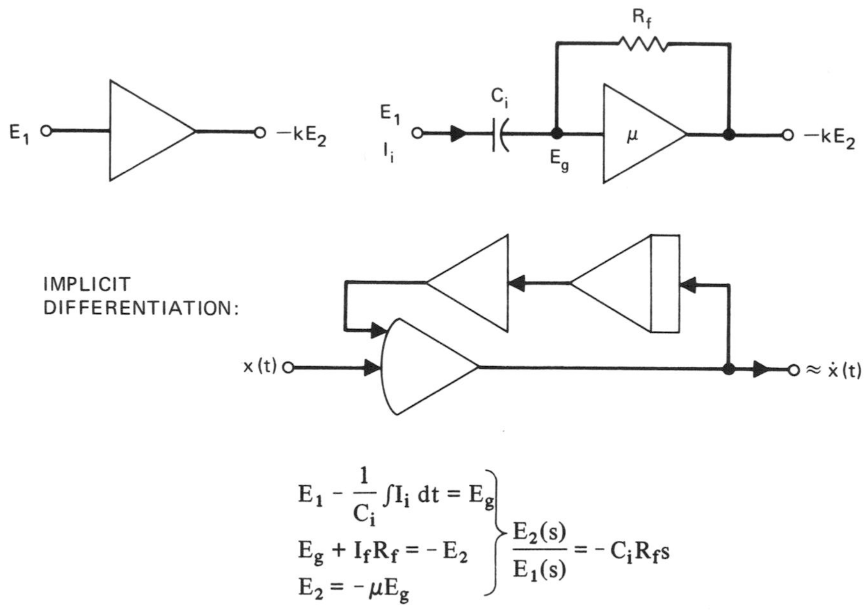

$$\left.\begin{aligned} &E_1 - \frac{1}{C_i}\int I_i\,dt = E_g \\ &E_g + I_f R_f = -E_2 \\ &E_2 = -\mu E_g \end{aligned}\right\} \frac{E_2(s)}{E_1(s)} = -C_i R_f s$$

6. **Simple pole:**

$$\frac{E_2(s)}{E_1(s)} = \frac{-R_1}{R}\,\frac{1}{(R_1C_1s + 1)}$$

7. **Simple zero:**

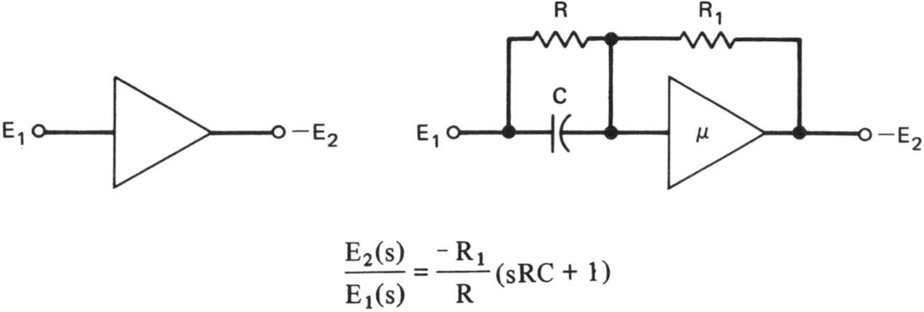

$$\frac{E_2(s)}{E_1(s)} = \frac{-R_1}{R}(sRC + 1)$$

8. **Quadratic complex pole:**

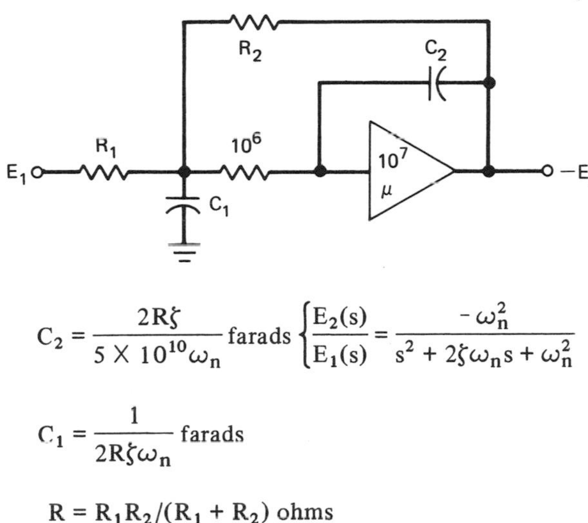

$$C_2 = \frac{2R\zeta}{5 \times 10^{10}\omega_n} \text{ farads} \quad \left\{ \frac{E_2(s)}{E_1(s)} = \frac{-\omega_n^2}{s^2 + 2\zeta\omega_n s + \omega_n^2} \right.$$

$$C_1 = \frac{1}{2R\zeta\omega_n} \text{ farads}$$

$$R = R_1R_2/(R_1 + R_2) \text{ ohms}$$

Examples:

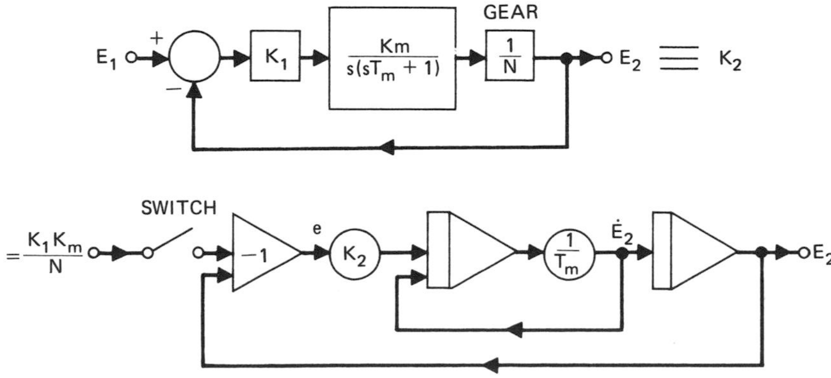

1. A simple *position-control* system:
2. *Bang-bang* servo (*predictor-control*) was the basic control system that was introduced prior to Pontryagin's *maximum principle* for the optimal control of a nonlinear system in the sense of minimal-time or maximum-range. A com-

mand is used to change the dynamic-state of the controlled system from one set of values to another in minimal time with minimal overshoot.

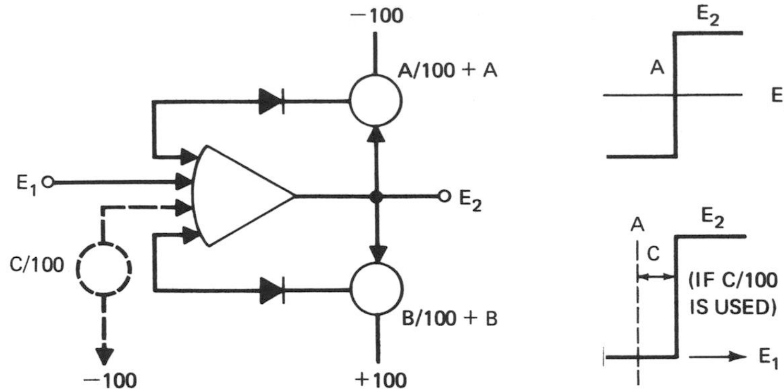

3. An analog device for the *simulation of a polynomial:*

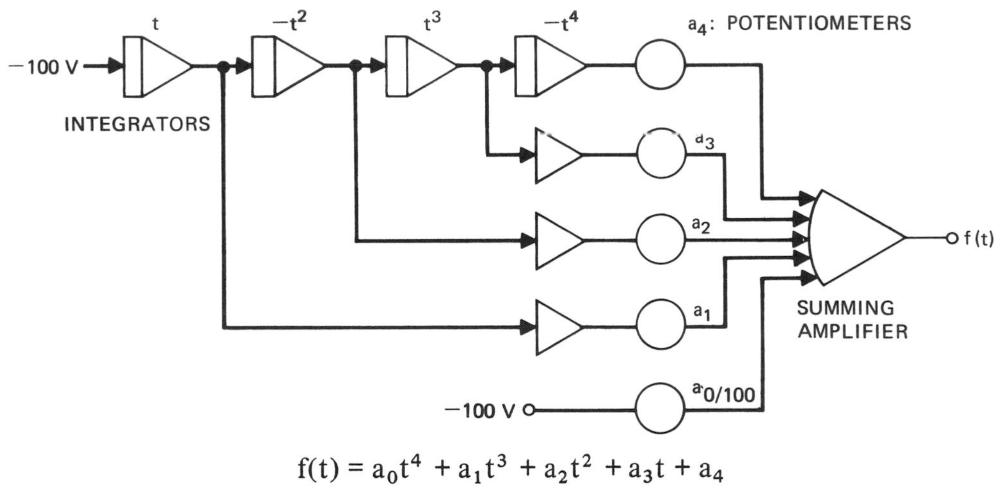

$$f(t) = a_0t^4 + a_1t^3 + a_2t^2 + a_3t + a_4$$

1.12 MASON'S TECHNIQUE FOR FEEDBACK CIRCUIT ANALYSIS

Mason's technique for feedback circuit analysis is a *signal flow-graph* technique; it replaces the transfer-function blocks in the conventional feedback block-diagrams by branches marked by arrows. The junctions of these *branch-transmittances* T_{ij} become *nodes*, and the node signals make the system variables. In short, the signal-flow-graph is a *topological representation* of the equations that describe a system operation. Each dependent node signal is the *algebraic* sum of the incoming branch signals at that node. One *minus* the loop-transmittance is termed the *return-difference.* A node is termed a *sink* if the branches enter only; it is termed a *source* if only outgoing branches emanate at that node. Mason's technique of analysis is an improvement on the basic flow-graph *reduction technique*, which corresponds to the conventional block-diagram reduction in control systems.

Mason's formula for the *analysis of linear feedback circuits* using signal flow-graph representation is expressed in the following form:

$$C(s) = \frac{1}{\Delta_g} \sum_{j=1}^{m} T_j \Delta_j$$

where

$$\Delta_g = 1 - \sum_n T_n + \sum_{n,p} T_n T_p - \sum_{p,n,q,} T_n T_p T_q + \cdots$$

Δ_j = determinant for that part of the graph, not touching the *jth* forward path

T_j = gain or transmittance between the input and the output of the *jth* forward path

m = No. of forward paths; ΣT_n = sum of independent loop-gains

$\Sigma T_n T_p$, etc. = product of the loop-gains of 2 (3, etc.) at a time for nontouching feedback loops only.

Example 1: three-stage amplifier

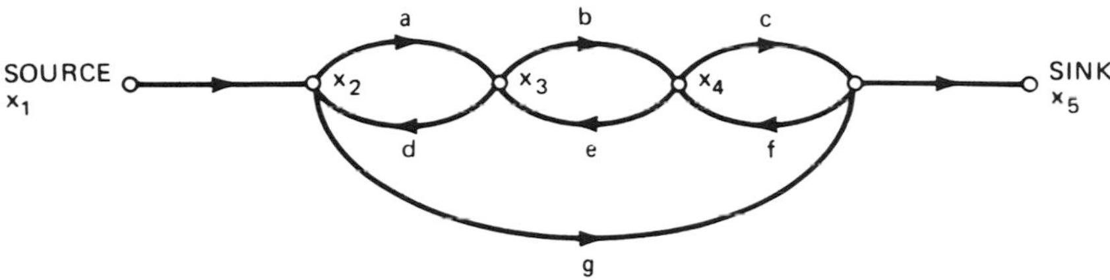

T_j: G_1 = abc; G_2 = g

Signal flow-graph chart.

Δ_j: $\Delta_1 = 1$ (All loops touch the signal-path from the output to the input)

$\Delta_2 = 1 - bc$

(determinant of that part of the graph not touching the second forward path, g)

$$\sum_n T_n = (ad + be + cf + gfed)$$

$$\sum_{n,p} T_n T_p = (ad)(cf)\text{: the nontouching feedback loops, two at a time.}$$

$$G(s) = \frac{x_5}{x_1} = \frac{G_1 \Delta_1 + G_2 \Delta_2}{1 - \Sigma T_n + \Sigma T_n T_p}$$

$$= \frac{abc + g(1 - be)}{(1 - ad - be - cf - gfed + adcf)}$$

Example 2: single-stage transistor feedback amplifier (see Fig. 1-27).

$$h_{11} = \left.\frac{E_1}{I_1}\right]_{E_o=0} = \text{Input impedance} \approx r_b + \frac{r_e}{1 - a}$$

$$h_{12} = \left.\frac{E_1}{E_0}\right]_{I_o=0} = \text{Voltage feedback ratio} \approx \frac{r_e}{r_d}$$

$$h_{21} = \left.\frac{I_0}{I_1}\right]_{E_0=0} = \text{Current amplification} \approx \frac{a}{1 - a}$$

$$h_{22} = \left.\frac{I_0}{E_0}\right]_{I_1=0} = \text{output admittance} \approx \frac{1}{r_d}$$

where

$$r_d = (1 - a)r_c$$

$$\begin{bmatrix} E_1 \\ I_0 \end{bmatrix} = \begin{bmatrix} h_{11} & h_{12} \\ h_{21} & h_{22} \end{bmatrix} \begin{bmatrix} I_1 \\ E_0 \end{bmatrix}$$

The overall *indefinite-admittance matrix* is directly written from the equivalent circuit:

$$\begin{array}{r} 1. \\ 2. \\ 3. \\ 4. \end{array} \begin{bmatrix} \dfrac{1}{h_{11}} + Y_p & -Y_p & -\dfrac{1}{h_{11}} & 0 \\ -Y_p + \dfrac{h_{21}}{h_{11}} & Y_p + h_{22} + g_L & -\left(h_{22} + \dfrac{h_{21}}{h_{11}}\right) & -g_L \\ -\left[\dfrac{h_{22}}{h_{11}} + \dfrac{1}{h_{11}}\right] & -h_{22} & h_{22} + \left[\dfrac{h_{22}}{h_{11}} + \dfrac{1}{h_{11}}\right] + Y_s & -Y_s \\ 0 & -g_L & -Y_s & Y_s + g_L \end{bmatrix}$$

Node 1:

$$I_1 = i + E_1 Y_p - E_0 Y_p \tag{1}$$

$$E_2 = E_1 - i h_{11} \tag{2}$$

$$(E_1 - E_0)Y_p = h_{21} i + (E_0 - E_2)h_{22} + E_0 g_L \tag{3}$$

$$i(h_{21} + 1) = -E_0 h_{22} + (E_1 - i h_{11})(h_{22} + Y_s) \tag{4}$$

Node 2:

$$i = \frac{-h_{22}}{x} E_0 + \frac{y}{d} E_1 \tag{5}$$

where

$$x = (h_{21} + 1) + h_{11}(h_{22} + Y_s)$$
$$y = (h_{22} + Y_s)$$

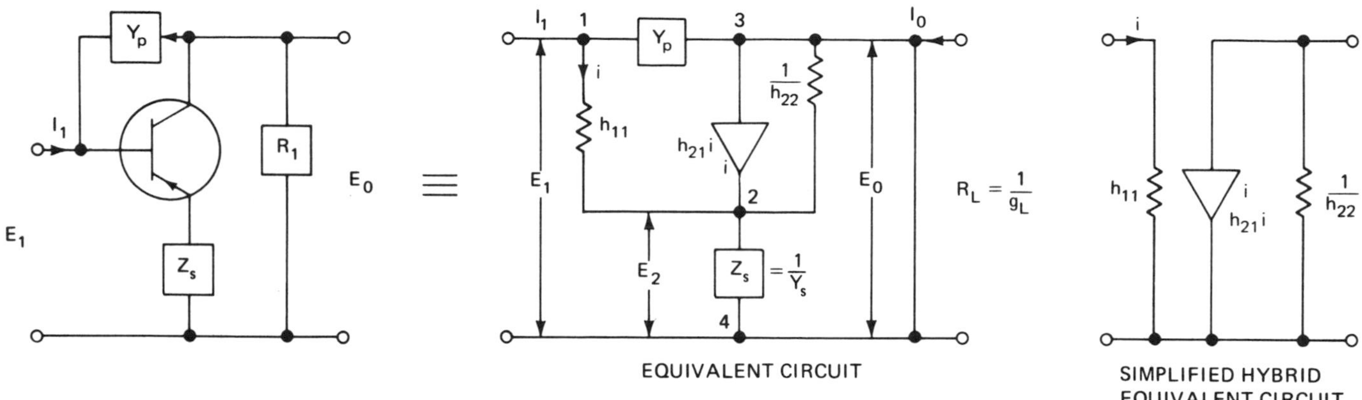

Fig. 1-27. Transistor feedback amplifier.

Substituting Eq. 2 in Eq. 3,

$$E_1Y_p - E_0Y_p - h_{21}i - E_0h_{22} + E_1h_{22} - ih_{11}h_{22} - E_0g_L = 0$$

$$-E_0(Y_p + h_{22} + g_L) + E_1(Y_p + h_{22}) - i(h_{21} + h_{11}h_{22}) = 0 \quad (6)$$

Node 3:

$$E_0 = \frac{b}{a}E_1 - \frac{c}{a}i \quad (7)$$

where

$$a = Y_p + h_{22} + g_L$$
$$b = Y_p + h_{22}$$
$$c = h_{21} + h_{11}h_{22}$$

To determine *the gain* with Eqs. 1 through 6, the selected nodes E_1 and E_0 in the *hybrid equivalent-circuit* are modified to formulate the terminations in the signal flow-graph illustrated in Fig. 1-28.

There are two forward paths:

$$G_1 = -Y_p^2yc/xa; \qquad \Delta_1 = 1$$

$$G_2 = b/a; \qquad \Delta_2 = 1$$

$$\Sigma T_n = (Y_py/x) + (ch_{22}/xa) + (Y_p^2yc/xa) \quad (8)$$

$$\Sigma T_nT_p = 0$$

$$\text{Gain} = \frac{E_0}{E_1} = \frac{G_1\Delta_1 + G_2\Delta_2}{1 - \Sigma T_n} = \frac{\dfrac{-Y_p^2yc}{xa} + \dfrac{b}{a}}{1 - \left[\dfrac{Y_pya + ch_{22} + Y_p^2yc)}{ax}\right]} \quad (9)$$

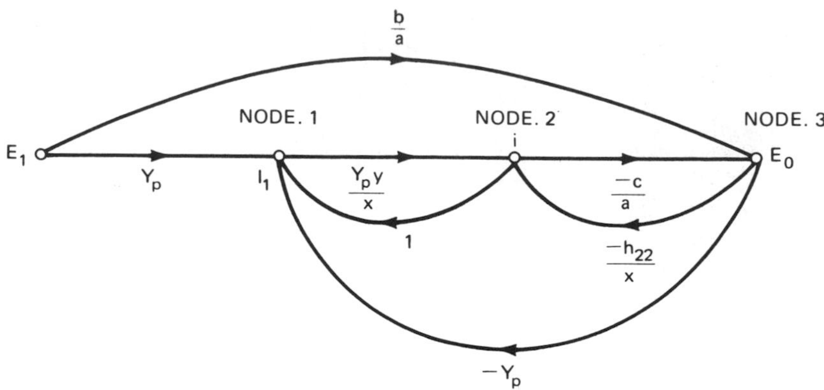

Fig. 1-28. Signal flow-graph of a transistor feedback amplifier.

where

$$a = (Y_p + h_{22} + g_L)$$
$$b = (Y_p + h_{22})$$
$$c = (h_{21} + h_{11}h_{22})$$
$$x = (h_{21} + 1) + h_{11}(h_{22} + Y_s)$$
$$y = (h_{22} + Y_s)$$

As another example, if the *output impedance* ($z_{out} = E_0/I_0$) is required, the above signal flow-graph will be rearranged, and the equations are reformulated in terms of E_1, E_0, I_1, I_0, and i. The feedback signal flowchart for this purpose will then have the output node at E_0 and the input node at I_0. Then the application of Mason's formula is repeated as above to obtain the output impedance.

Table 1-1. z and modified z-transforms for the basic common transfer-functions.

G(s)	g(t)	G(z)	G(z, m)
1	$\delta(t)$ impulse	$1(=z^{\circ})$	0
$\dfrac{1}{s}$	$u(t)$ unit-step	$\dfrac{z}{(z-1)}$	$\dfrac{1}{(z-1)}$
$\dfrac{1}{s^2}$	t ramp	$\dfrac{Tz}{(z-1)^2}$	$\dfrac{mT}{(z-1)} + \dfrac{T}{(z-1)^2}$
$\dfrac{1}{\left(s - \dfrac{1}{T}\ln a\right)}$	$a^{t/T}$	$\dfrac{z}{(z-a)}$	$\dfrac{a^m}{(z-a)}$
$\dfrac{1}{(s+a)}$	e^{-at}	$\dfrac{z}{(z-e^{-aT})}$	$\dfrac{e^{-amT}}{(z-e^{-aT})}$
$\dfrac{1}{(s+a)^2}$	te^{-at}	$\dfrac{Tze^{-aT}}{(z-e^{-aT})^2}$	$\dfrac{Te^{-amT}[e^{-aT} + m(z-e^{-aT})]}{(z-e^{-aT})^2}$
$\dfrac{a}{s(s+a)}$	$1 - e^{-at}$	$\dfrac{(1-e^{-aT})z}{(z-1)(z-e^{-aT})}$	$\dfrac{1}{z-1} - \dfrac{e^{-amT}}{z-e^{-aT}}$
$\dfrac{a}{s^2(s+a)}$	$t - \dfrac{1-e^{-at}}{a}$	$\dfrac{Tz}{(z-1)^2} - \dfrac{(1-e^{-aT})z}{a(z-1)(z-e^{-aT})}$	$\dfrac{T}{(z-1)^2} + \dfrac{mT - \dfrac{1}{a}}{(z-1)} + \dfrac{e^{-amT}}{a(z-e^{-aT})}$
$\dfrac{a}{s^2+a^2}$	$\sin at$	$\dfrac{z \sin aT}{z^2 - 2z\cos aT + 1}$	$\dfrac{z \sin maT + \sin(1-m)aT}{z^2 - 2z\cos aT + 1}$
$\dfrac{s}{s^2+a^2}$	$\cos at$	$\dfrac{z(z-\cos aT)}{z^2 - 2z\cos aT + 1}$	$\dfrac{z \cos maT - \cos(1-m)aT}{z^2 - 2z\cos aT + 1}$
$\dfrac{a}{s^2-a^2}$	$\sinh at$	$\dfrac{z \sinh aT}{z^2 - 2z\cosh aT + 1}$	$\dfrac{z \sinh maT + \sinh(1-m)aT}{z^2 - 2z\cosh aT + 1}$
$\dfrac{b-a}{(s+a)(s+b)}$	$e^{-at}e^{-bt}$	$\dfrac{z}{z-e^{-aT}} - \dfrac{z}{z-e^{-bT}}$	$\dfrac{e^{-amT}}{z-e^{-aT}} - \dfrac{e^{-bmT}}{z-e^{-bT}}$
$\dfrac{a}{(s+b)^2+a^2}$	$e^{-bt}\sin at$	$\dfrac{ze^{-bT}\sin aT}{z^2 - 2ze^{-bT}\cos aT + e^{-2bT}}$	$\dfrac{[z \sin maT + e^{-bT}\sin(1-m)aT]e^{-amT}}{z^2 - 2ze^{-bT}\cos aT + e^{-2bT}}$
e^{-ksT}	$\delta(t-kT)$	z^{-k}	z^{m-1-k}
$e^{-ksT}G(s)$	$g(1-k)T$	$z^{-k}G(z)$	$z^{m-1-k}G(z, m)$
$G(s+a)$	$e^{-at}g(t)$	$G(e^{aT}z)$	$G(e^{aT}z, m)$.

2
Typical Complex Digital Control System: Quadruplex Color Videotape Recorder

2.1 HISTORICAL BACKGROUND OF VIDEOTAPE RECORDING

Magnetic videotape recording and color television were pioneered during the 1930s through the 1950s. While monochrome (black-and-white) and color television were passing through their infancy, so too was magnetic tape recording.

A field-sequential color television system was adopted rather prematurely in 1950 by the Federal Communications Commission (FCC) in the United States. This early electromechanical technique involved a rotating color filter-disk of the three primary colors (green, red, and blue) at both the camera and the receiver; hence, it was a transreceiving system for color television alone, and excluded black-and-white. As a result, the FCC revised its decision in 1953 and adopted the NTSC (National Television Systems Committee) all-electronic compatible color television. It was *compatible* in the sense that color television transmissions meant for color receivers could nevertheless be directly received in black-and-white receivers as black-and-white. This basic compatibility of the NTSC color television system, with the incorporation of an exclusive in-band *subcarrier* for the color information, facilitated the gradual development of the monochrome quadruplex videotape recorder to its refinement of color recording during the late 1950s. The actual breakthrough of the quadruplex videotape recorder as a recording medium for television occurred in the mid-1950s, prior to the development of color VTR.

During the early stages of color introduction, the sampled-data control systems in the Q-CVTR were primarily basic. They involved a large amount of residual *servo phase-jitter*, and color had to be separately handled by a special non-phase double-heterodyne electronic technique in order to counteract the effects of phase-jitter. This original color television recording system had several shortcomings, and the results were rather marginal (and poor in maintenance). The feedback digital control systems of the Q-CVTR gradually matured with the advent of the solid-state transistor technology during the early 1960s. Color is presently an inseparable part of the digital control system in the modern Q-CVTR, and color reproduction from tape, after repeated rerecordings, is remarkably true and faithful up to the fourth generation at least. It is currently difficult to distinguish the live color camera signal from the videotape signal when the recorder is equipped with control accessory refinements like the monochrome automatic timing corrector (MATC), color automatic timing corrector (CATC), and velocity error compensator (VEC). The latter is meant for countering the effects of *color-banding due to the four-magnetic-head system*, sometimes noticed with worn-out magnetic heads or old videotapes. There was a serious effort in France to replace the technique of the simultaneous NTSC-compatible color television by an alternative version, because the rendition of color according to the early heterodyne recording technique was not up to the mark. The successful integration of color into the videotape control system has finally altered the situation in Europe (except for France and eastern Europe) and around the world in favor of either the NTSC color television system or a slightly modified version, phase-alternation by line (PAL).

2.2 PRINCIPLES OF NTSC COLOR TELEVISION SYSTEM AND Q-CVTR

The Q-CVTR is one of the essential and indispensable pieces of equipment in a television broadcast studio—as essential as the intricate color television camera. During the late 1950s in the United States, the unique significance of the Q-CVTR in television broadcasting (or telecasting) was recognized the instant the machine went into production at Ampex and RCA as a vacuum-tube version. RCA produced a compact solid-state version in the early 1960s. The videotape recorder facilitates rehearsals, retakes of scenes, on-the-spot editing, and coast-to-coast programming schedules. Television newscasters from New York and Washington (or anywhere else around the world) are simultaneously seen in color and heard throughout the continental United States. The Q-CVTR, the microwave links, and the Intelsat satellites have made this possible. During President Carter's European trip in 1978, the Q-CVTR and the Comsat and Intelsat synchronous satellites extended these up-to-the-minute real-time international colorcasts from New York and the West Coast to London and to the Eurovision television network as if it were a routine day-to-day telecast. An automated television studio is currently made feasible by including the latest microcomputer systems. This is all routine because the intricate, interacting, nonlinear, multirate sampled-data control systems in the Q-CVTR do their task of color reproduction with a high degree of reliability—that is, within 5 nsec of unnoticeable phase-jitter in the color television receiver. A large number of nonlinear elements naturally show up throughout the whole control system, as seen in Figs. 3-22 through 3-24. Adaptive control, for instance, has made this possible, in a way, because a 20-nsec jitter in place of 5-ns would turn the faces of the newscasters and performers either greenish-yellow or crimson-purple. The specification of a 5-nsec phase-jitter component corresponds to $\pm 2.5°$ at 3.58 MHz, the color subcarrier used in the NTSC-compatible color television system.

2.3 COLOR TELEVISION

The color television camera proper and the associated signal processing system have three high-resolution video channels for green, red, and blue signals, along with an independent or a processed high-resolution monochrome channel. The monochrome signal can be obtained by matrixing the G-R-B color signals. The three signals are matrixed according to the principles of colorimetry and encoded in a colorplexer by means of a color frequency subcarrier at 3.579545 MHz to produce a compatible color television signal, as an alternative to the conventional black-and-white television camera signal. (The corresponding frequency of the European PAL color television system is 4.4296875 MHz.) The colorplexer employs a two-channel combined amplitude and phase modulation technique for, respectively, the saturation and hue of the color components in the whole visible color spectrum. These components, synthesized with respect to two axes I and Q of the color-triangle, are suitably restricted in bandwidths for the most pleasing color presentation according to the principles of colorimetry. The color information appears alongside the monochrome video information, on a spectrum *frequency-interleaving* basis, when the encoded signal is transmitted on a VHF, UHF, or microwave transmitter with a video bandwidth of 4.2 MHz (Fig. 4-6a and b).

This signal is presently transmitted in a digital format at 43 megabit/sec for the purpose of satellite data communications or optical communications. In practice, an individual colorplexer makes a regular accessory of the signal processing equipment internal to the color television camera. However, it is possible to use one colorplexer per several color cameras, if a more complex high-precision color video switching system is available to meet the appropriate differential gain and phase requirements through the full video bandwidth. With the latest linear medium-scale integrated chips, the colorplexer is merely a subassembly of a compact one-piece color television camera.

Every single piece of signal-generating and -processing equipment (including the color television cameras and the Q-CVTRs) is kept in proper synchronism by a discrete multirate pulse-train from a *sync generator.* This basic pulse-train, usually designated as "sync," is a complex train of (digital) pulses at both horizontal and vertical rates. The rates involved are 31.5 kHz, 15.75 kHz, and 60 Hz. The half-line distinction between interlaced alternate odd and even fields at 60-Hz can also produce a 30-Hz component, if a special digital processor is specifically designed for generating this low-frequency pulse-train. The line and field sync intervals are timed to occur between corresponding blanking intervals, as determined by another pulse-train called *blanking.* This, in turn, will serve to define the *retrace* intervals (in the television receiver), during which no picture information is transmitted. The sync mentioned previously is added, along with a *reference color synchronizing burst* of about eight cycles at the beginning of each line-interval, to the color video signal in the colorplexer itself; this provides the timing and phase control for the deflection and color processing circuits, respectively, in color video monitors and in color or black-and-white television receivers. The sync is *phase-locked* in a special digital device to the reference color frequency standard of 3.58 MHz used in the colorplexers as a color subcarrier. This will enable *dot-interlace* of the subcarrier information in black-and-white television receivers for minimum visibility of the subcarrier pattern. The *line-interlace* of two fields per frame will aid in the formation of a high-resolution picture with minimum possible video-bandwidth. Local and remote camera signals can be fully synchronized using a device called *genlock;* however, it is not suitable for the videotape recorders due to the servo phase-jitter components.

2.4 Q-CVTR

The advances in the state-of-the-art of the black-and-white and color television broadcast media are primarily influenced by the Q-CVTR. Four wide-band magnetic heads in quadrature, on a rotating headwheel (HW), make the basis of the quadruplex video recording. Monochrome and color television signals recorded on the magnetic videotape in an FM format can be immediately reproduced on air on a high-quality basis, since unlike film, no intermediate image processing is involved. Where programs are of a routine rehearsal character, economy is maintained by master-erasing and reusing a tape several times. The video master-erase system is automatically energized during the recording process only.

The Q-CVTR employs a lateral transverse-scanning technique (as compared to the longitudinal technique in the case of the audio) by using a rotating head-wheel (or drum) to obtain the very high writing speeds required to record wide-band color and monochrome video signals. The 2-in.-wide magnetic-tape used in the high-quality broadcasting industry is illustrated with the FM information on the magnetic tracks in Figs. 4-2a and b. Picture signals are recorded by the four rotating magnetic heads in transverse video-tracks, as the tape moves at a speed of 15 or $7\frac{1}{2}$ in./sec against the preset penetrating pressure of the rotating headwheel and a vacuum-guide assembly. The three servo systems—headwheel, capstan, and vacuum-guide—interact right at the instantaneous contact of the headwheel. Each track contains approximately 16 horizontal lines of picture information, both color and monochrome. One of the three longitudinal audio-tracks is reserved for the control-track (CT) signal that is required for playback synchronization. The corresponding head is situated at the base of the vacuum-guide. The control-track signal consists of the basic 240-Hz control information for the use of the capstan servo during the playback.

A typical transport panel for the tape is illustrated in Fig. 4-27c. The illustration shows the detachable headwheel panel with the associated vacuum-guide mechanism. A part of the guide drive-mechanism is situated under the headwheel panel. The videotape is cupped to conform to the circumference of the rotating headwheel by means of the vacuum-guide (shoe), which employs the interface of vacuum to maintain a firm hold on the moving tape. The driving force, which pulls the tape from the supply-reel to the take-up reel, is imparted by a rotating capstan in conjunction with a solenoid-operated pinch-roller. The four FM-signal currents from the four video-heads of the rotating headwheel located at one end of the motor-shaft are drawn either through a brush-and-slip-ring assembly or preferably a rotary transformer. A tonewheel at its other end furnishes a 240-Hz feedback tone-wheel signal for the headwheel servo that regulates the speed and the phase of the headwheel motor.

Figure 4-31 presents the complete functional system block-diagram of the quadruplex color videotape recorder. As a solution to the nonlinearity problem of the tape medium, the FM record-playback signal-processing used in the four individual independently controlled FM-channels is illustrated with essential details. The various erase, record, and playback magnetic-heads are shown in their proper location in the path of the tape, with respect to the vacuum-guide, headwheel, and capstan. The major details of the frequency modulation and demodulation techniques are briefly explained in Chapter 4. Suffice it to say here that the FM technique provides a better signal-to-noise ratio and eliminates the need for very-low-frequency equalization for the video by translating the entire signal upward to a higher frequency-band. For monochrome and color recording, the normal FM-deviation characteristics are stated in Chapter 4 under the specification of a typical Q-CVTR. Figure 4-8a shows the FM-versus-video and the side-band energy distribution. As an example, in the case of the monochrome recording, the band actually recovered from the tape extends from 0.5 to 7 MHz, a range broad enough to cover the actual carrier-deviation, and the first-order lower side-band energy. The missing upper side-band energy is restored by the special limiting signal-processing used during the playback. The low deviation-index restricts the significant signal-energy within the first-order side-band spectrum. The latest FM recording techniques, as will be explained in Chapter 4, allow a wide-band deviation in high-band FM to give fine results with better signal-to-noise ratio for both color and monochrome.

During playback, the four FM-channels, containing the overlapping FM information in each video-track, are properly combined in a single channel by electronic switching during the horizontal blanking interval. Potential switching transients in the picture are thus eliminated. The clean FM output signal so obtained is reverted to a monochrome or color composite-video signal by the demodulation process (Fig. 4-8). The composite-color signal is reprocessed in a video signal-processor to furnish the clean video and color information and the regenerated sync and color synchronizing burst. The tape-sync, separated from the composite signal in this accessory, provides the necessary pulse sequences for the control systems. Built-in monitoring devices provide a continuous check on the quality and conventional performance-indices of the color picture on a color video monitor. Operation of the color videotape recorder is fairly simple, and most basic operations can be conveniently controlled from a remote location.

2.5 INTERACTING DIGITAL CONTROL SYSTEMS OF QUADRUPLEX COLOR VIDEOTAPE RECORDER (Q-CVTR)

The feedback control systems used in a television system—whether they are of the hybrid sampled-data (digital) and servomechanism type that automatically controls a power plant like a motor and its associated gear or load-mechansim, or (2) purely of the sampled-data (digital) type that automatically controls electronic system parameters like gain or frequency or phase or delay—are all inherently discrete. It is not by choice as in the application of a regular digital controller in, for instance, industrial automation. The reason is obvious. The implied amplitude or frequency or phase or delay of the control input/output variable concerned can be scrutinized only at predetermined sampling instants, at those sampling-rates that bear a divisible relation to the various rates of the synchronizing pulse-sequences or color frequency-standard in a black-and-white or color television system.

As briefly noted in the previous section, there are three complex interacting sampled-data feedback control systems in the Q-CVTR. All three are multirate and multiloop with several nonlinear elements: (1) headwheel servo, (2) capstan servo, and (3) vacuum-guide servo make an interesting control engineering example of a successful, extremely complex, industrial application for further study. These complicated nonlinear systems were expeditiously developed and designed in the laboratory on a heuristic *measure-and-optimize* basis, stage by stage, and facility by facility, from the commercial viewpoint of urgency and competition. The objective was simply the feasible minimization of the performance-index (minimum extent of phase-jitter) according to the state-of-the-art. The whole development took place during a period of about 6 years to reach the highly reliable, high-precision final product as a parallel to the refinements of color television itself. The early monochrome version using vacuum-tubes took about 3 years of this period (AMPEX and RCA). By 1964, the performance-index had reached the incredible limit $< \pm 2.5$ nsec of horizontal phase-jitter on the screen of a color television receiver. The corresponding 2.5° figure-of-merit at the color subcarrier-frequency is satisfactory from the viewpoint of the threshold of color perception by the human eye.

A simplified block-diagram of the three interacting control systems is illustrated in Fig. 2-1 with a few salient features.

The complexity of these systems can be seen in the control system-block-diagrams of Figs. 3-22 through 3-24, for headwheel, capstan, and vacuum-guide servos, respectively. Some of the system characteristics, including the types of nonlinearities encountered, are enumerated below:

1. Multiloop system of a multivariable character
2. Multirate sampled-data feedback loops
3. Pulse-width modulation as an adjunct to an ac carrier servo
4. Three-phase ac carrier servo using amplitude modulation; two-phase ac carrier servo using phase modulation; two-phase ac carrier servo using amplitude modulation and phase-reversal
5. Mutual interaction of input, output, and other system variables of three individual sampled-data feedback control subsystems
6. Mutual interaction due to mechanical linkages, subassembly placement, and interfering tolerances on a common video-plus-audio tape transport assembly
7. Interacting system electronics in the video and FM sections, and the tape-magnetics
8. Noise and disturbances from one control system to the other
9. Adaptive control of three incremental orders at various sampled-data rates
10. Pulse transfer-functions based on digital and analog computer techniques.

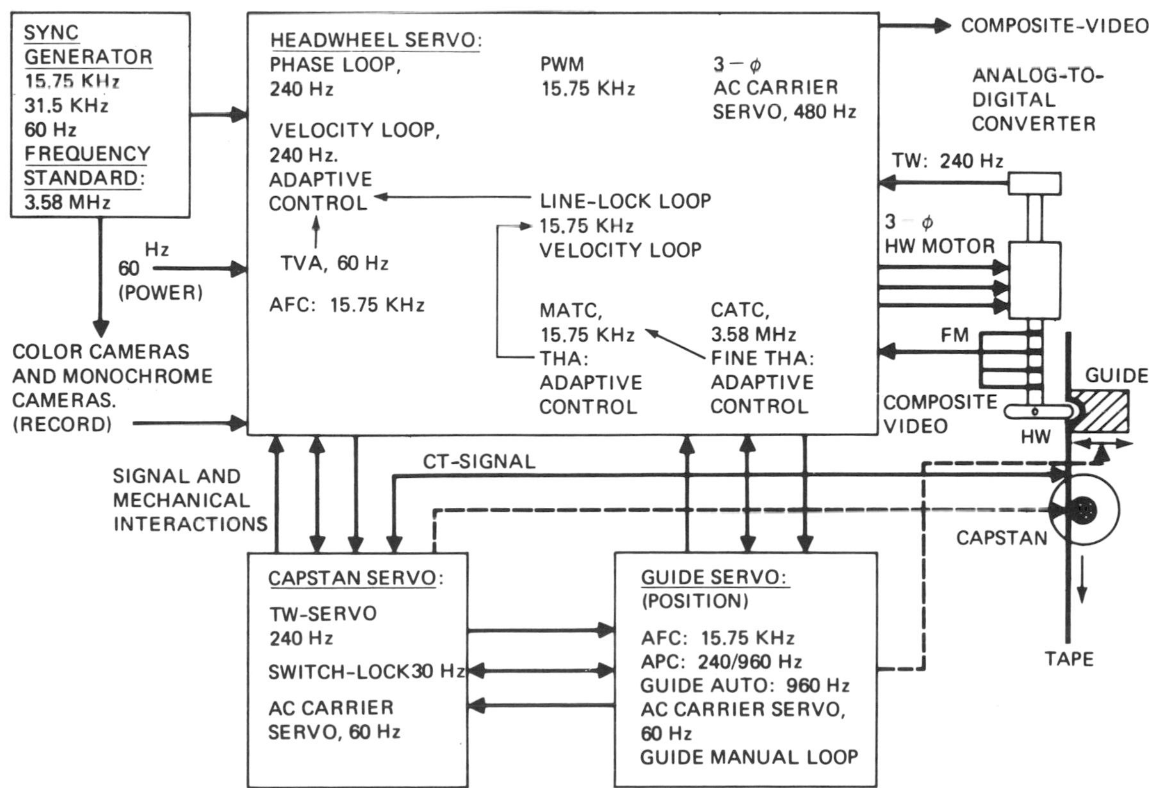

Fig. 2-1. Three interacting digital control systems of the Q-CVTR.

Pulse-formers, pulse-shapers, differentiators, integrators, dc error-holding extrapolation networks, compensation networks for noise-rejection and suitable transient-response, bandpass, high-pass and low-pass filters, flip-flop memories, multiplexers, shift registers, binary counters, decoders, comparators, reference dc clamps, diode matrixing and logic, pulse-gating, matching networks, phase discriminators, differential amplifiers, limiters, and predictor logic, etc.

11. Mechanical linkages like gear-trains, belt-drives, pinch-roller, eccentric and other couplings
12. Tens of factory-preset control settings
13. Air-cooled power transistors on heat-sinks
14. Vibration from air-bearing pump; air circulation for air-guides; head-wheel air-blower
15. Reel stabilizing-arms
16. Complex vacuum-guide mechanism, and scallop-error corrector for quadrature magnetic-heads
17. Automatic stability-checking indicator system for the servo subsystems
18. Following nonlinear elements:

Saturation, practically in every loop
Electrical hysteresis of three-phase headwheel, and two-phase capstan asynchronous synchronous (HAS) motors
Pulse-width modulation in headwheel servo consists of two nonlinearities, saturation of pulse-width, and signal amplitude
The nonlinear pull-in characteristics of the automatic frequency-and phase-control loops
Saturation in power amplifiers
Stiction in the motors due to constant-velocity operation
On-off relay switching with rotary solenoids
Mechanical hysteresis or back-lash in the various gear-trains of the vacuum-guide mechanism in manual and auto modes of operation
Granularity of helipot in the manual mode of the vacuum-guide servo
Minimal dead-space to eliminate possible mechanical jitter of the vacuum-guide servo; the minimal dead-space results in a stable limit-cycle
Variable gain
Variable time-delay in the various loops
The nonlinear characteristic of the varicaps used in the *automatic* delay lines
Other dynamic slow- and fast-changing nonlinear phenomena in the solid-state pulse and FM-video circuitry, subharmonic generation, intermodulation effects, stable multiple limit-cycles seen as constantly-varying phase-jitter due to multi-nonlinear phenomena
Random phase-jumps in the reference power-supply frequency, 60 Hz

2.6 BASIC PRINCIPLES OF HEADWHEEL SERVO

The headwheel sampled-data feedback control system, consisting of several digital controllers and the associated variables, is not independent as a single unconnected servo system regulating the speed of a nonlinear hysteresis asynchronous synchronous (HAS) motor. The other two complex sampled-data control systems in the quadruplex color videotape recorder are closely interacting as a result of the multirate sampling input excitation-variables, output controlled-variables, and in some cases intermediate product-variables, all mostly appearing as pulse-sequences. Instability in any one of the three servo subsystems will invariably result in the instability of the other two during the reproduction of the color or monochrome picture from the videotape. This can be demonstrated by a simple example: if a picture recording is made on a typical quadruplex color videotape recorder at an insufficient vacuum-guide pressure on the headwheel, and the tape is immediately played back, both the headwheel and the capstan servos fail; thus, the whole picture breaks up into a disorderly pattern of noise. The vacuum-guide pressure will be "automatically" insufficient since the original recording is made accordingly. Actually, the magnetic recording-process would be faulty because of the nonoptimum penetration of the headwheel into the videotape during the recording process. Thus, a deficiency on the part of the vacuum-guide servo would destroy the correct incidence of the pulse-sequences for the headwheel and the capstan servos. (This particular maintenance fault may occur as an accidental error in the manual setting of the guide-pressure at the time of recording.)

The headwheel servo, in principle, controls the speed and the phase of the three-phase HAS motor as shown in the simplified block diagram of Fig. 2-2. While recording, the major objective of the headwheel servo is to assure the writing of the transverse pattern of FM video-tracks on the tape, in which each television field at the rate of 60 fields/sec requires exactly 16 tracks (provided that the capstan servo is appropriately phased). The specific pattern would then alone allow the interchange and the intersplicing of the videotapes on different quadruplex color videotape recorders situated at different locations. The feedback variable consists of a 240-Hz pulse-train derived from a miniature tone-generator (or transducer) mounted adjacent to a slot on the rotating tone-wheel. The phase of the pulse-train is compared against that of the vertical 60-Hz synchronizing component of the composite color or monochrome signal being recorded. The dc error-component and the subsequent motor-drive circuits keep the output-variable (240 Hz) phase-locked to the fourth-harmonic of the 60-Hz reference-pulse, as a multirate sampled-data feedback control system.

During playback, the 60-Hz reference pulse-train for the headwheel servo is derived from an external source, such as a studio sync generator or 60-Hz power line.

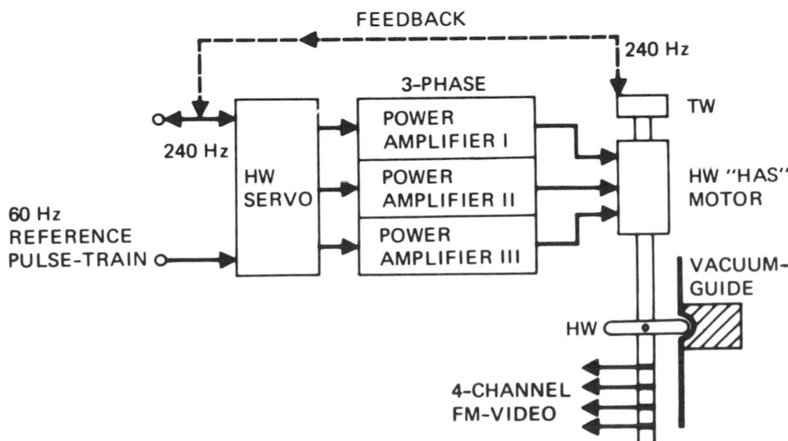

Fig. 2-2. Simplified block diagram, headwheel servo.

The final phase-jitter component due to the nonlinear elements in the control subsystem extends to a limit-cycle up to a maximum limit of 1 μs (0.25 in. across on the 17-in. diagonal video-monitor). The duration of one horizontal-line is 63.6 μs. As far as the regular television receiver is concerned, this phase-jitter component is not objectionable, since the receiver deflection-circuits follow the relative phase-jitter of the synchronizing information and produce a stable picture for the viewer. But for handling videotape signals in television studios as freely as camera signals—that is, for mixing and performing "special effects" between the videotape signals and the television or film-camera signals—the modern color videotape recorder, with a PLL line-lock subcontrol system at a 15.75-kHz rate and a slow-acting tape vertical-alignment (TVA) at 60-Hz rate, can minimize the above limit-cycle due to phase-jitter to a duration of ±0.1 to ±0.07 μsec in the case of monochrome signals. For the correct reproduction of color from the videotape, this limit-cycle, in the wake of so many nonlinearities, must be reduced to ± 2.5 nsec. This fine figure-of-merit (0.0017% with respect to 240 rps) for the present highly sophisticated headwheel servo subsystem is feasible in a reliable way due to the third-order adaptive control presently used. The desired extra high-precision is achieved by the addition of open-loop, monochrome and color automatic timing-corrector subsystems (MATC and CATC) to the high-precision line-lock servo subsystem. A PLL fine-tape-horizontal-alignment (FTHA), a PLL tape-horizontal-alignment (THA), and a PLL line-lock tape-vertical-alignment (TVA) make the "third-order adaptive control" against the unpredictable variations of the basic 60-Hz reference. These three loops are naturally closed-loop feedback type, and they are closely associated with the CATC, MATC, and line-lock subsystems, respectively. The technological aspects of these subsystems in the headwheel servo are briefly explained in Chapter 4 (Figs. 4-18, 4-22, and 4-17, respectively).

2.7 BASIC PRINCIPLES OF CAPSTAN SERVO

To maintain a fixed timing-relationship between the tape-movement and the transverse video-head scanning during the *record* process, the speed of the two-phase asynchronous synchronous capstan motor is synchronized or phase-locked to that of the headwheel motor (Fig. 2-3). In the record mode, the capstan motor is driven by a 60-Hz built-in power-supply derived from an oscillator, which is phase-locked to the 240-Hz tonewheel signal. For synchronization purpose, the recording speed of the capstan is, in turn, imprinted on the videotape as a 240-Hz control-track signal. Each cycle corresponds to exactly four video-tracks (Figs. 4-7a and b). In the

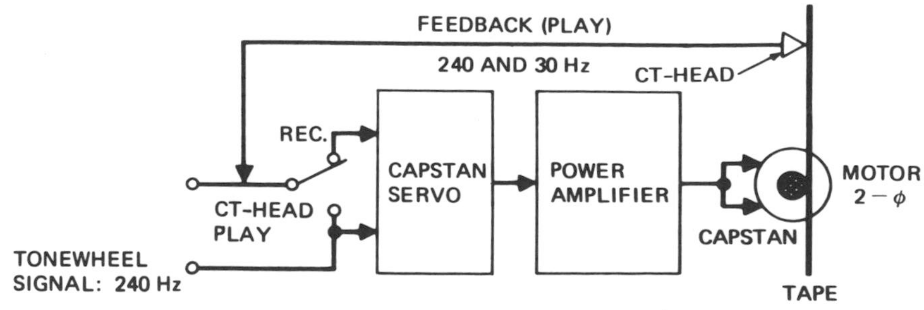

Fig. 2-3. Simplified block diagram, capstan servo.

playback mode, the 240-Hz rate control signal derived from the control-track is phase-locked to the 240-Hz rate tone-wheel signal from the headwheel assembly by means of the capstan servo. The dc error-signal from the phase-lock loop (PLL), if any, assures that the speed of the capstan-motor is controlled by feedback so that exactly four video-tracks are moved past the headwheel during each revolution. For best results in respect to S/N ratio, each video-head must be accurately centered over a track on the moving tape, and this is in practice done by a manual phase-control adjustment during playback. For all practical purposes, this is actually achieved on a preset basis.

It has been stated earlier that the line-lock system of the headwheel servo is required for the "special effects" between the camera signals and the video tape-reproduced signals. This facility necessarily requires that the capstan be *switch-locked* at a 30-Hz rate. The two facilities, the 15.75-kHz line-lock and the 30-Hz switch-lock (at the picture or frame rate), in combination, make what is called the "pixlock" for the complete synchronization of the signals from the quadruplex color videotape recorder to the other camera signals in the television studios. A *line-lock only* (*LLO*) mode is also used by the television networks, without the switch-lock, for minimum effect of disturbances on a picture frame in a special color transmission from videotape. The principles of these techniques are explained in Chapter 4 (Fig. 4-9).

2.8 VACUUM-GUIDE POSITION SERVO

The vacuum-guide servo automatically sets the degree of penetration of the headwheel into the moving-tape during playback, to that originally used for recording each part of the edited program on a long-duration tape-reel. The position-error appears at 960-Hz rate; this is the head-to-head switching rate. The feedback path involves practically the whole FM and video system, since the tape-sync carries the error information due to the *skew* distortion caused by an improper headwheel penetration into the tape in the playback mode (Fig. 4-26).

The major features of the vacuum-guide sampled-data position feedback control system are illustrated in Fig. 2-4. (See Figs. 3-24, 4-26, 4-27, and 4-29 for further details.) It will be seen that the control system is directly interacting with the headwheel servo, by way of the input reference and output feedback variables. Since the headwheel and capstan are synchronized, the capstan does have an indirect influence on the accuracy of the vacuum-guide position in the manual mode. The moving tape transfers a slight mechanical influence in terms of friction (or traction) between the capstan and the vacuum-guide assembly, as noticed by preset-setting under manual record at times.

Four sampling rates are involved:

1. the 15.75 KHz sampled-data automatic frequency control loop
2. the multirate 240/960-Hz sampled-data automatic phase control loop
3. the 960-Hz sampled-data vacuum-guide position feedback-control system with its exclusive digital controller
4. the amplitude-modulated 60-Hz suppressed-carrier ac servo.

The compensation networks used in the associated AFC and PLL circuits are not naturally optimized from the vacuum-guide servo point of view, since the other loops

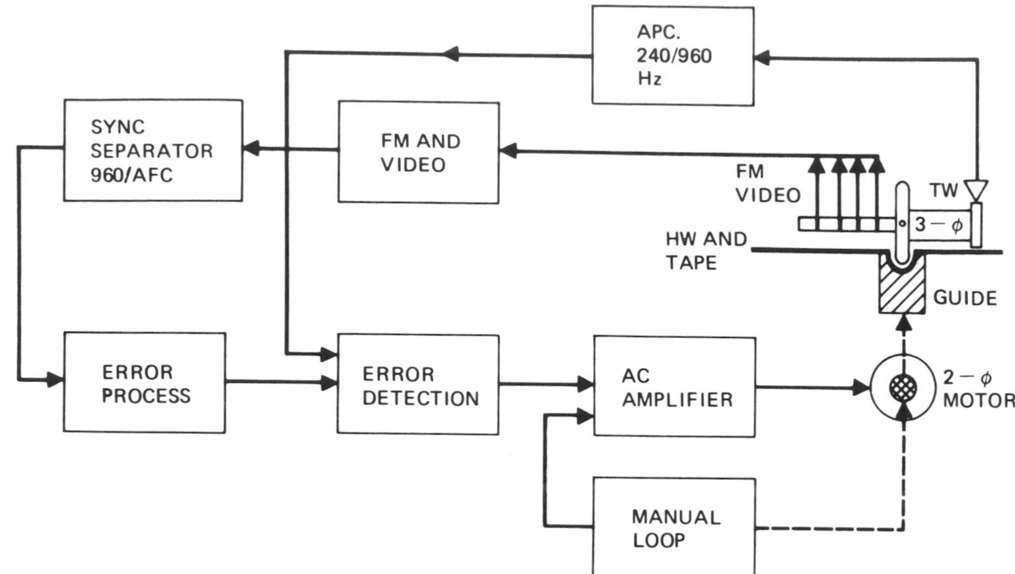

Fig. 2-4. Simplified block diagram, vacuum-guide servo.

are expected to meet the high-speed high-precision requirements of the line-lock loop and the FM switching system, respectively. So, the vacuum-guide servo must by specification rely on its own digital controller, and its exclusive dc and ac compensation networks for noise rejection and transient response. Two elements of backlash appear in the ac carrier servo of the control plant, a two-phase induction motor. In the manual mode, three backlash elements and one granularity of the helipot make the nonlinear elements. Hence, the vacuum-guide servo is more reliable and accurate in the automatic playback mode.

Three preset trimpots are used:

1. a sensitivity control for the dc error-signal from the 960-Hz automatic loop, at the output of the dc compensation
2. a sensitivity control for the manual/auto ac carrier servo loop, to meet frictional variations in vacuum-guide mechanisms
3. a zero-setting for balancing the positive and negative excursions of the error-signal with respect to the sampling process.

The performance-index obtained is an undetectable $\pm 0.02\ \mu s$ on the video monitor screen. Due to the theoretical implication of the several nonlinear elements involved, a small amount of phase-jitter (as mechanical jitter) corresponding to the stable limit-cycle, is inevitable. This is inadmissible in the picture, as it takes the appearance of a minute ripple in vertical lines. Therefore, the sensitivity at the output of the 960-Hz error-detector is optimized, with adequate gain-margin, so that (1) a dead-zone of an unnoticeable $\pm 0.02\ \mu s$ is the effective operational-index, and (2) the system is sensitive enough to respond only to error signals above this tolerance. Then the system will settle down dead-beat within the dead-zone with a minimum of overshoot and the suppression of the vertical-ripple mentioned. The $0.02\ \mu s$ corresponds to 0.02 mil in position error. For the best reproduction of color using the MATC and the CATC accessories (without any trace of color-banding at the head-switching rate), the correct operation of the vacuum-guide automatic position servo within its figure-of-merit is a prerequisite. The CATC accessory will minimize the above error tolerance of $\pm 0.02\ \mu$sec to ± 2.5 nsec.

2.9 COMPLEX DIGITAL CONTROL SYSTEMS: THEORY vs. PRACTICE

Complex digital control systems of the category of missile attitude control, high-precision radar, interplanetary satellites, etc., using digital computers, give rise to immense theoretical and software problems for direct overall application of available control theory. The enormous number of variables, the large number of nonlinear elements involved, and the multidimensional character of some of these systems bring about this problem. The interacting control systems of the Q-CVTR, described as typical examples of control subsystems in a complex digital control system in this work, belong to this complex classification. The author considers that the stage-by-stage compression of the performance-index (limit-cycle or phase-jitter) on an automatic adaptive control basis from a coarse 1-μs tolerance to a fine 5-ns tolerance, on a loop-by-loop basis, is a successful new example in the control field. With the staggered high rates of sampling used, the system organizes itself with self-optimization in terms of nanosecond behavior patterns in arriving at the final result—that is, the reproduction of color from videotape within an undetectable tolerance of $\pm$ 2.5° of phase-jitter at a 3.58-MHz color subcarrier frequency. The process is entirely beyond the capabilities of former continuous feedback control systems for direct or indirect application. That is, nonlinear continuous feedback control systems can be made more and more accurate by using the time-wise self-organized, PLL sampled-data control techniques, such as those used in the quadruplex color videotape recorder on an adaptive control basis.

Preliminary simulation studies with a mathematical model on an analog computer and subsequent computations on a digital computer are conducive to a better understanding. But the correlation between the simulating process and a physical system of the character of the quadruplex color videotape recorder will remain vague and impractical from the point of view of implementation. Thus, a multidimensional system is an unsolvable problem on a theoretical basis, unless the control system is much simpler, possibly with some parallel. Modern control theory and its broad state-space characterization, involving state and phase variables, and optimization to a performance-index, are aimed at a generalized approach to solving simpler control problems, but actual application to inordinately complex physical systems is not a feasible commercial proposition. Therefore, a *channel of communication*—involving well-documented features of interest between applicable theoretical concepts in terms of simple mathematical models, and system and development techniques in successful control engineering practice—is generated in this treatise and considered as a substantial contribution of the author. The side-by-side exposition of theory and practice will enable the reader to realize the immensity of the interface problem involved in producing a commercially successful and reliable complex digital control system.

Using time-consuming, highly expensive software along simplified algorithms on a heuristic basis, and supporting hard-wired design of computer interface logic may be one potential system-design approach in defense and aerospace applications, but it is entirely out of consideration for the competitive, profit-oriented commercial applications, where a meager, purely nominal, research-and-development funding is the basic policy of most industrial budgets. The designated products must be produced according to schedules at minimal cost, and the preliminary field problems are debugged as an essential part of product development during the usual *learning stage.* Thereupon, the final success of the product is assured on a gradual basis. In some applications, the product itself may be rendered obsolete by, technologically, a more successful lower-cost or higher-quality replacement from a competitor.

With the amazing evolution of the latest low-cost large-scale integrated (LSI) microprocessors/microcontrollers and microcomputers, and I^2L (integrated-injection logic) bipolar PLL chips using linear differential amplifier techniques, these complex digital control systems can be further developed on a commercial basis to introduce the application of modern digital-filter techniques. That is, the active analog input/output variables from the micro-PLL chips will be first converted to the digital format by using, on a systems-design basis, high-resolution analog-to-digital converter chips to enable the application of the latest LSI microcomputer systems and the digital-filter techniques, if adequate research-and-development funding is invested in the industry. International technological competition and high demand will gradually lower production costs and minimize the cost of the final products from the viewpoints of space, scheduling, and improved reliability. These trends definitely point in the direction of successful implementation of complex digital control systems in the near future. Chapter 5 is concerned with the digital/data aspects of this approach.

2.10 TECHNIQUES OF ANALYSES AND SYNTHESES IN CONTROL THEORY

There is presently a trend to refer to control theory under four different classifications, for both continuous/analog and sampled-data systems. As briefly surveyed in Chapter 1, these are:

1. The **frequency response methods**, such as the Bode, Evan's root-locus, Nyquist, and Routh-Hurwitz procedures, are used for the design of phase-lock loops (PLLs) in linear control systems with the aid of the *open-loop transfer function* and the *closed-loop characteristic equation.* They are in practice successfully designed by the so-called trial-and-error techniques, as far as the determination of the transient response and the compensation network are concerned.
2. The **statistical design technique**, by means of the mean-squared-error criterion for stochastic inputs and the integral-squared-error criterion for transient inputs, is generally classified as an analytical technique for linear systems. The solution, specifying the form of compensation required, with or without some power-constraint (such as saturation or the minimization of bandwidth) when random noise and disturbances limit the desired system objectives, is an explicit mathematical-model for the desired system transfer-function. It is expressed, in relation to the signal and noise, as *auto- and cross-correlation functions in the time-domain* and the corresponding *power-spectrums in the frequency-domain.* The explicit time-domain solution of the **Weiner-Hopf integral equation** gives for a free-configuration the system weighting-function that minimizes the mean-squared error. The solution is subject to modification by a *Lagrange multiplier* for a semifree-configuration in the presence of minimum- or nonminimum-phase fixed elements. (The statistical techniques are extended to the **modern optimal control** to combine the control-system

aspects of the noise-rejection and the transient response.) Since control systems designed exclusively along the statistical approach tend to be oscillatory, as some quarters point out, a composite criterion giving maximum rejection of noise, consistent with control of transient response, has been suggested by Graham and Lathrop for linear systems.

3. The **extension of frequency response techniques to nonlinear systems as quasi-linear approximations** are covered under the *describing-function techniques.* The graphical *phase-plane method* for analyzing and plotting the transient response of a nonlinear system (of the first and the second orders) in the form of *phase trajectories* and their convergence to a *limit-cycle* are also covered under this category. A special technique like **Lyapounov's direct method** for linear and nonlinear systems may belong to this category, since it is one of the classical methods. The **ac carrier servo systems** can be also included in this class, since the class-2 compensation used in these systems is a matter of approximation to the dc compensation used under frequency-response techniques.
4. **Modern control theory** is exclusively formulated around the characterization of the control system by *state-variables in a multidimensional state-space approach.* This approach to complex control systems avoids the need of solving a complex higher-order differential equation, linear or nonlinear. The performance of the control system is expressed, as a *generalized approach* to linear, nonlinear, continuous, and sampled-data control systems, by *state-transition equations* formed with the state-variables. The response of the system is then expressed in terms of the excitation or *measured variables* through the system and the state-transition equations. The state-variables form a set of first-order differential equations that determine the dynamic behavior of the system. The technique readily lends itself to the consideration of the initial conditions of the system as part of the solution. Since during successive sampling periods, the dynamic system behaves like a continuous system, the sampled-data system under consideration can be represented by a set of first-order linear homogeneous differential equations with constant coefficients (in the place of the first-order linear difference equations). Modern control theory extends to **optimal control systems** under *direct approach* and *dynamic programming* (involving the invariant-embedding principle), and **Pontryagin's maximum principle.** The techniques of analysis and synthesis include the consideration of both the transient response to a step-input and the statistical aspects of noise-rejection. **Self-optimizing adaptive control** can be classified under optimal control, since the major distinction between the adaptive and optimal control systems, is that, in the case of the latter only, a *direct measurement of the performance-index* is feasible, as an additional dividend.

2.11 MEASURE-AND-OPTIMIZE TECHNIQUE USED FOR THE QUADRUPLEX COLOR VIDEOTAPE RECORDER

The *measure-and-optimize* technique on a heuristic basis, as used for the development and design of the hybrid sampled-data and continuous control systems, and the digital control systems in the Q-CVTR, is in line with the developments and trends in modern control theory, where the *measured variables* and the analytical techniques of *optimization to a performance-index* share an outstanding role. The measure-and-optimize technique is a natural solution to the expeditious evolution of a complex digital control system. The design of the digital controllers and the digital compensation is accomplished on an actual real-time operational basis, with the performance-index on display either (1) on a color video monitor, (2) a high-precision waveform CRT display equipment, or (3) some unique, specially designed measuring setup or instrument. Some ingenuity is required on the part of the systems and electronics engineers, who are directly responsible for the development and design of the digital controllers and the associated extrapolation and compensation networks. Their tools are mainly digital circuit design techniques with the latest hardware, solid-state or otherwise, and the knowledge of digital and analog computer engineering techniques. Conclusions are drawn, however, on the basis of their knowledge in control theory.

The laboratory measure-and-optimize technique is highly successful and reliable in experienced and skilled hands. The technique requires (1) the determination of the relevant parameters, digital and continuous, for a particular control-process, (2) judgment, based on the engineer's experience and knowledge of control theory, to choose with a sense of proportion and then optimize the right parameters (out of a host of less important parameters), and (3) choice of the appropriate test signals (for the transient and steady-state tests) and the right stage where the performance-index or a specific waveform needs careful examination. Theoretical techniques of analyses and syntheses are usually oriented around the synthesis of the appropriate compensation for the desired response or performance-index. But this aspect of the design is only a minor part of the development and design of the solid-state active and passive circuits associated with the various pulse transfer functions in the forward and feedback paths of the numerous phase-lock loops. In the case of the quadruplex color videotape recorder, there is not available a precise or even an approximate transfer-function for the nonlinear three-phase hysteresis asynchronous synchronous motor, which is used for the headwheel and the tonewheel. Thus, heuristic measure-and-optimize techniques in the laboratory, with reference to a performance-index, have a substantial lead over rigorous analytical techniques, when complex interacting control systems of the Q-CVTR classification are encountered in engineering practice. There is absolutely nothing undignified about this results-oriented approach from the theoretical standpoint.

As an analogy, essentially the display of the concrete self-explanatory display-waveforms take the place of the equations of the mathematical-models in theory. The waveforms directly indicate what is happening to the stability or transient response of the control system. The optimization of a pertinent parameter is a gradual procedure to satisfy a condition for stability. It is similar to working on the coefficients of either a characteristic equation or mathematical model, when the necessary or sufficient conditions for stability are required in a theoretical aspect. In the measure-and-optimize technique, the process is indeed expeditious, because the requisite performance-index is directly accessible on the waveform display-equipment or a special measuring instrument. With a purely algorithm-based theoretical approach in a complex system, a digital programming or some other time-consuming technique is necessary to see what is going to happen to the final performance objectives, when some changes are contemplated, out of necessity, half-way through the mathematical process. With computer software, final debugging is also a time-consuming and highly expensive procedure.

The electronics development engineer conceives a "gut" feeling of confidence, out of experience, intuition, or some logical reasoning based on theoretical knowledge, how the various waveforms at different stages in a digital controller contribute to the final stability and performance objectives of the system. This corresponds to the variations in algorithmic procedures we initiate as we analyze or synthesize half-way through the mathematical model, when we aim at a particular solution to a complex theoretical problem.

Complex digital control systems and heuristic measure-and-optimize techniques in the laboratory have a common ground—generally, the waveform-display provided by the electron-beam. So, the technique of interpretation is essentially *graphical*, as the development progresses toward the result. (The more expensive, *interactive computer graphics* procedure using *light-pen* is a parallel approach.) The graphical aspect especially presents a direct link between the performance-index, the process, and the doer; hence, the eventual success of the technique is assured, even if repeated efforts are necessary. The experienced and intelligent human brain is the most trustworthy sampled-data control system that there is. The latest research in bionics and the architecture of the brain point toward this finding. Above all, in the case of the complex control systems like those in the quadruplex color videotape recorder, the heuristic measure-and-optimize technique in the laboratory is directly practical. It has always been an achievement based on specialized test procedure, depending in turn upon the state-of-the-art at any time. The prerequisite is that the development and design engineer have a thorough knowledge of (1) video and digital techniques, (2) digital and analog computers, and (3) control theory.

2.12 EXAMPLES OF THE HEURISTIC LABORATORY MEASURE-AND-OPTIMIZE TECHNIQUES

As measure-and-optimize suggests, specialized measuring equipment, test signal generators, and waveform CRT-display equipment play the basic role in arriving at the desired performance index. Some of the test and measuring instruments used in a television laboratory at the time of the development and design of the videotape recorder are listed below:

1. Dual-trace oscilloscopes, Tektronix 535 (DC-15 MHz), 547 (DC-50 MHz), 585A (DC-85 MHz)
2. Transistor characteristic-curve tracer, Tektronix 575
3. Tektronix color-vectorscope for measuring differential-phase and magnitude-versus-phase of the color vector-components in the NTSC color television signals
4. Low-pass, high-pass unit for measuring differential-gain (gain-vs.brightness level) of color signals to an accuracy of ±0.1%, using a dual-trace high-gain preamplifier like Tektronix 1A2.
5. Television color signal analyzer and stair-step burst generator with 10, 50, and 90% average-picture-level (APL) signals
6. Sine-squared pulse-and-bar test signal generator
7. Multiburst test signal generator
8. Television synchronizing pulse distribution
9. Digital pulse repetition frequency read-out equipment
10. Square and rectangular pulse generators

11. Time-marker generators
12. 20-MHz video FM-sweep generators
13. Wandel and Goltermann group or envelope-delay measuring set
14. Intermodulation test-set developed and designed by the author.
15. Noise generators
16. Wow-meter, distortion measuring set-up
17. Low-frequency function generators
18. Special Lab-designed test equipment, special k-factor graticules, and trace-recording cameras for comparing waveforms, RF signal generators, etc.

2.12.1 Capstan Multirate Sampled-Data Control. An engineering research-and-development assignment to analyze and design a product, as a first of its kind and scope, makes it an advance development project—as the conventional statement goes. The project's purpose is to develop a 240/30-Hz multirate capstan sampled-data feedback control system and to compare its performance with an earlier single-rate design at 240/240-Hz sampling-rate. The electronics engineer is expected to compare the performance of both systems from the point of view of stability, transient response, speed of pull-in into phase-lock, and extent of limit-cycle (which usually determines the performance-index in nonlinear control systems). The speed of pull-in and transient response are closely interrelated.

The performance index for the capstan servo is readily obtained by checking the FM-video output of the 4:1 demultiplexer (switcher) on an oscilloscope. During playback of the videotape, the peak-to-peak FM-video output is adjusted to a nominal 1 V, when the phasing adjustment is optimized to center each video-head of the quadruplex headwheel over a single transverse-track on the tape. Then the percentage of the peak-to-peak amplitude-jitter or variation of the CRT-waveform gives a direct idea of the transient-response when some sudden disturbance is simulated around the capstan mechanical assembly. The time of pull-in is easily measured by starting a stopwatch when the play push-button is pressed. The settling-down of the set-point (indicated by a high-intensity dot) on the slope of the trapezoid waveform, as observed on the oscilloscope, indicates the phase lock-up of the capstan sampled-data feedback control system. The stability of the set-point determines the extent of the limit-cycle too. The set-point is the result of sampling with narrow sharp pulse-trains of opposite polarity against the phase of the reference pulse-train, which is shaped as shown in Fig. 2-5 to form a trapezoid waveform. The narrow sampling-pulse is the practical equivalent of a unit-pulse.

The control-track signal shown is converted by means of a signal processor consisting of filters, pulse-shapers, differentiators, delay multivibrators, integrators, pulse amplifier-limiters, etc., to generate a clean trapezoid pulse-train such as that shown. The 240-Hz tonewheel signal from the headwheel motor assembly is similarly processed by another signal processor to generate a square wave pulse-train at 240 Hz. For the requirement of this particular PLL, the 240-Hz pulse-train is applied to three serial binary-counters to obtain a 30-Hz square pulse-train. This waveform is differentiated and clipped to generate two sharp sampling pulse-trains of opposite polarity. An impulse generating transistor stage, consisting of a pulse transformer in its collector circuit, is employed for this function. The transformer used is center-tapped on the secondary side to output sharp pulse-trains of opposite polarity at the secondary terminals. When the feedback phase-lock loop is closed, the phase of the

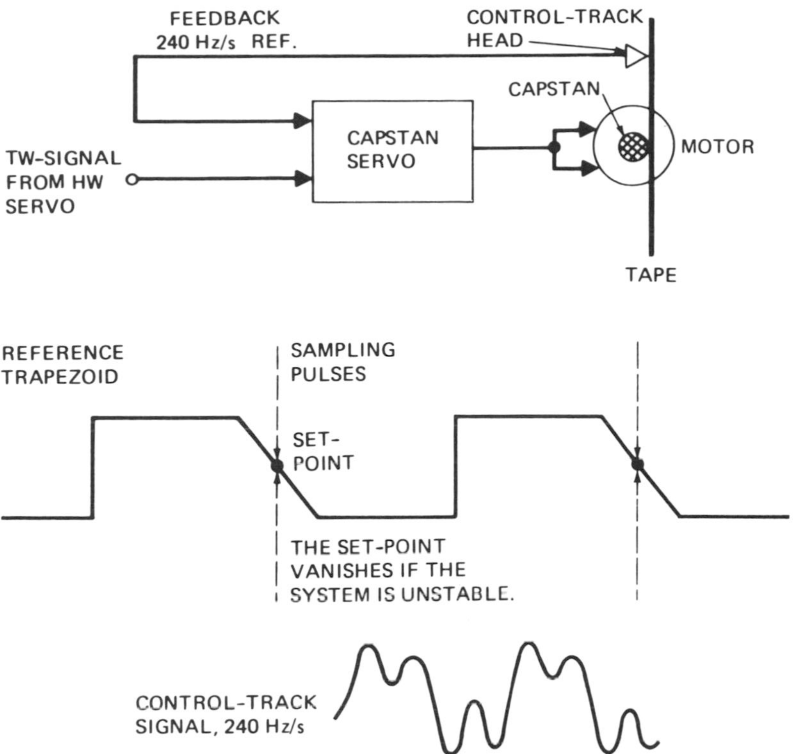

Fig. 2-5. Phase-lock set-point.

240-Hz rate trapezoid and that of the 30-Hz rate sampling-pulses are compared in a phase-discriminator. A silicon fast-recovery diode-bridge is used for this function. The ± dc error-signal on a simple "hold" capacitor at the output is further processed through a "chopper" to generate an ac phase-modulated carrier. A servo amplifier with sufficient power output drives the two-phase hysteresis asynchronous-synchronous capstan servo motor, having a complement of a quadrature phase-shift capacitor between the control and reference windings.

The extrapolation and compensation networks are finally optimized for the best results concerning (1) minimum of phase-jitter in the FM-signal, (2) minimum up-and-down excursions of the phase-lock set-point, and (3) minimum time of pull-in when the machine is repeatedly started. In some cases, an ac compensation network in the carrier part will contribute to better transient response. In practice, the amount of development work involved around the optimization of compensation hardly amounts to 10% of the overall system design effort, which is based on analog and digital circuit design. Delay optimization in the signal-process of the sampling pulse-train by means of a monostable delay multivibrator may contribute toward an improved performance-index, because it would boost the maximum allowable-gain and hence faster pull-in for producing the desired effective transient response. The design engineer must be extremely observant, and logical in the conclusions drawn from the logged data as several variations in circuit parameters are experimented on an iterative basis.

The control system engineer might discover that the multirate 240/30-Hz capstan sampled-data feedback control system would give improved results in respect to stability and transient response as compared to those of the single-rate 240/240-Hz control system. A mathematical example of the type presented in Chapter 3 on a multirate sampled-data control system would provide an excellent theoretical support to this finding. As far as the overall system is concerned, in practice several problems and incidental disturbances regarding stability occur during the development work on the signal processors associated with the sampled-data digital controllers in the reference and feedback paths, as a result of the imminent interactions from other feedback-loops and the overall system dynamics.

2.12.2 Switch-Lock and Sharp Sampling Pulses. *Problem:* With the original 240/240-Hz/s single-rate capstan sampled-data control system, when a videotape is played back on transmission, the phase of the tape-reproduced picture frame will not lock with that of the studio camera signals at the instant of switching. The probability of concurrence is one in eight, since the picture-frame rate is 30 Hz/sec. Therefore, when the television studio switches on a videotape signal in succession to a studio camera signal, a vertical rollover of the picture frame is imminent as a disturbance in the distant television receiver, because of the phase discrepancy.

Solution: If the capstan sampled-data control system is phase-locked to the studio 30-Hz synchronizing pulse-train, the vertical picture rollover in the distant television receiver is avoided when a camera-to-tape program-changeover takes place, because the cameras are already synchronized to the studio sync. Therefore, a 30-Hz reference *edit* pulse-train is derived from the multirate complete-sync by means of a special 30-Hz/s frame pulse generator and recorded along with the 240-Hz control-track signal. During the playback, if the tape 30-Hz edit pulse-train can be successfully separated from the playback waveform shown, its phase can be compared with that of the 30-Hz pulse-train derived from the studio sync (Fig. 2-6). The dc error from this subsidiary PLL can then be incorporated for system effectiveness into the main capstan servo to ascertain that the tape signal is synchronized to the camera signals at the 30-Hz frame-rate. Of course, for the best pull-in and transient-response characteristics, new optimized extrapolation and compensation networks were mandatory at the 30-Hz rate. It was the very first successful minisize solid-state servo developed and designed for the former vacuum-tube RCA videotape recorder. (The author has also had this opportunity.) The recorder had no spare room available in the cabinet-rack for an alternative, large vacuum-tube version. Incidentally, it was just the very beginning of introducing compact, reliable solid-state designs into the television systems at that time (1958).

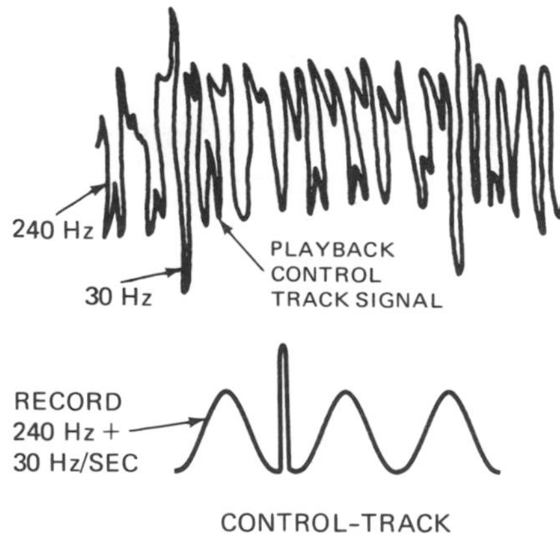

Fig. 2-6. Tonewheel control-track signal.

For sampling, a 30-V peak-to-peak impulse pulse-train was used. It was 100 nsec wide at 30-Hz rate, and it enabled the best possible results regarding reliability of stability, fast pull-in, and minimum of phase-jitter (the limit-cycle due to the nonlinearities in the system), as indicated by the set-point on the trapezoid nonlinear slope. Since 30-Hz/sec is the lowest pulse-rate in the quadruplex color videotape recorder, the picture frame lock-up time of the recorder, at start, is mainly determined by this particular PLL. So the fast pull-in of this servo at the 30-Hz rate, with a satisfactory transient response, is a very important requirement. Since practical sampled-data control systems are capable of handling sequences of input functions or waveforms of any shape, the reference trapezoid-waveform used in this particular subsystem was intentionally shaped for a nonlinear-slope, as shown in Fig. 2-7. It was actually the result of the particular optimum compensation-network that was used for the high-speed pull-in with minimum of overshoot. The pull-in time of this PLL was 1 to 2 sec after the "start"push-button was hit. This very first switch-lock servo was incorporated in eight RCA videotape recorders in a New York television studio, and the chief engineer was very pleased with the results. Two months later, a different version of the switch-lock subsystem was designed and tried on one of these machines in the presence of the author. The second version used a linear-slope trapezoid and low-amplitude sampling pulses of about 10-μsec pulse-width in the place of the 100-nsec, 30-V peak-to-peak sampling pulses of the first version. With the compensation used under the condition of the wide and short sampling pulses, the pull-in time increased to 2 to 3 sec. The performance-index, namely, the phase-jitter, indicated by the up-and-down movement of the set-point, was much larger.

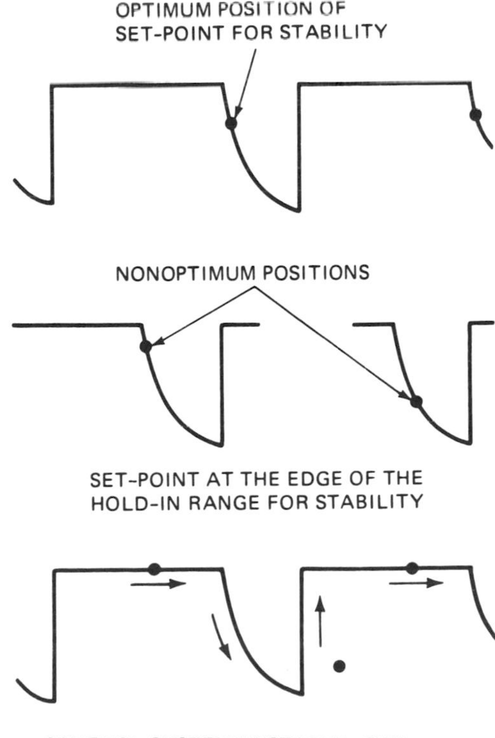

Fig. 2-7. "Switch-lock" PLL waveforms.

Naturally, the chief engineer preferred the original switch-lock servo and rejected the alternative modified version. This heuristic measure-and-optimize example is theoretically supported by the mathematical proof of the impulse versus finite-width pulse-sampling given in Chapter 3. As a result of the demonstrated substantial improvement in performance, the original switch-lock unit was finally chosen for regular production.

The chief engineer casually remarked that some old videotapes arrive at his studio without a 30-Hz edit-pulse on the control-track, since the switch-lock accessory to the recorder at that time was a new innovation. The switch-lock device, developed on the basis of the *presence of the edit-pulse* in the recorded control-track, naturally would not furnish the new switch-lock facility for the old videotapes that had no edit-pulse. Therefore, the original version of the switch-lock accessory was slightly modified by the author to meet the new requirement so that it could remain automatically effective with or without the need of the above 30-Hz edit pulse-train in the control-track of the tape. Using additional switching logic, a special 30-Hz synchronizing pulse-train is derived from the tape-sync via the FM composite-video since the standard edit-pulse is not available on the tape. It was discovered that, at the start, the headwheel and capstan servos work in synchronism; the headwheel servo also partially contributes toward the pull-in of the capstan at the 30-Hz rate. Based on this finding, with the appropriate logic for automatic switching and lock-in, the new switch-lock facility, without the former complementary edit-pulse, was demonstrated by the author for the first time as an advance development project. Switch-lock, it may be noted, is essentially required for the *pixlock* mode of playback in conjunction with *line-lock* to fully synchronize the videotape and live camera signals for enabling *special effects* between the two signals in *real-time*. As a result, the previous practice of adding (during recording) a 30-Hz rate edit-pulse to the 240-Hz rate control-track signal was eventually discarded.

2.12.3 A Guide Servo Problem and a Solution by Measure-and-Optimize Signal Processing. 1. The waveform of the "skew" distortion error-signal, as received from the 15.75 kHz AFC-loop (automatic frequency control), is of the shape shown in Fig. 2-8. The error-signal appears at the 960-Hz FM-video head-switching rate. As one of the requirements, the vacuum-guide servo should not respond to minute headwheel mechanical errors of the nature of *scallops* and *quadratures*, which occur at the same rate. (For details of guide servo, see Figs. 4-26 through 4-31, and 3-24.)

2. The error-signal is processed in a digital controller to the trapezoidal waveform by means of filters, I^2L (integrated injection logic) pulse-shapers and clippers, integrators, etc. (Incidentally, the I^2L technique, using discrete PNP/NPN transistors, was originated and first used by the author for signal processing in several projects.)

3. Two RC filters are optimized in the above digital controller for noise-rejection, and low-level low-frequency disturbances from the other interacting control systems. Noise can be easily injected into the vacuum-guide sampled-data feedback control system by slightly misphasing the capstan servo. The magnetic-heads on the headwheel are thereby diverted away from the video-tracks slightly, and the signal-to-noise ratio can be directly controlled, as desired, via the path of the tape-sync.

4. The dc error from a bidirectional sampler is dc-coupled to a bilateral-transistor chopper (modulator), after the necessary extrapolation and compensation.

The transient response is indicated by the control output signal shown in Fig. 2-9;

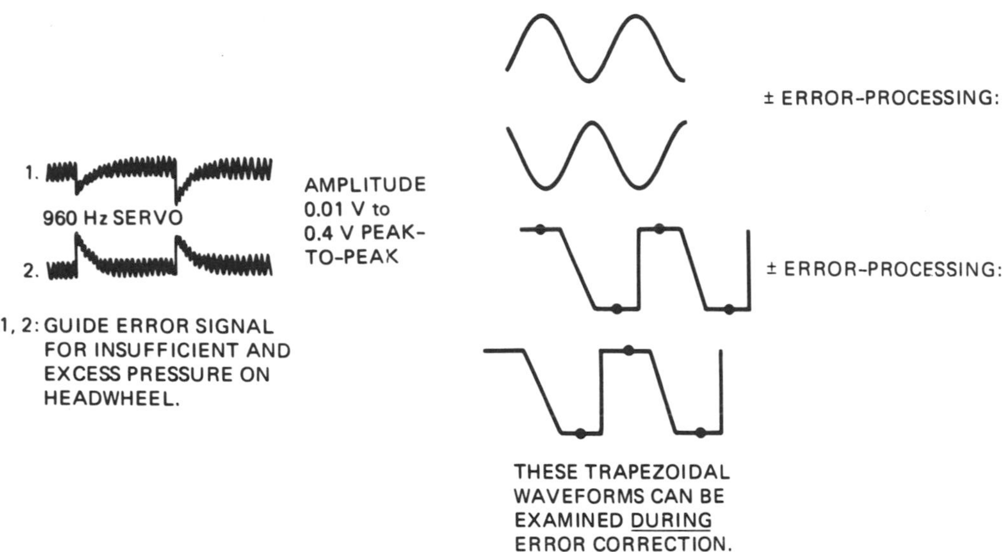

Fig. 2-8. Signal-processing of the vacuum-guide input error-signal.

it is displayed on the oscilloscope at a slow time-base rate by simulating a *step* input-error via the associated manual servo-loop (waveform A). Waveform B represents the start of the actual automatic correction process; the feedback signal is processed through the whole system via the FM-video from the tape. By direct-coupling of the chopper to the dc error-signal, the transient response, as indicated by the output signal, was improved. For the necessary optimized gain-setting and minimum cor-

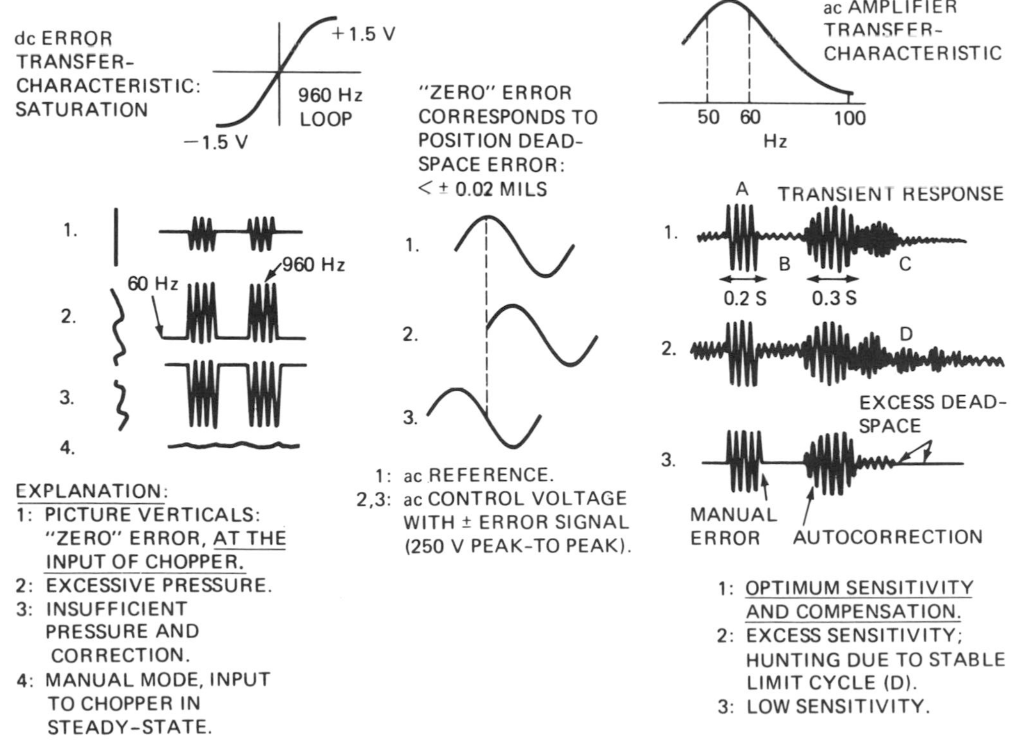

Fig. 2-9. Signal-processing waveforms—vacuum-guide.

rection time, the transient response was worse when the chopper was isolated by a high input-impedance emitter-follower; the emitter-follower was therefore eliminated to minimize the input impedance at this stage and obtain the optimized transient response. All dc and ac compensations are carefully optimized to minimize lobe-C in the transient-response waveform shown in Fig. 2-9.

Problem. A typical problem from the field (as a feedback from a television studio) during the early stages is described in the following example. Some second-generation videotapes in the field, containing a faulty vertical sync-timing (due to a certain temperature-sensitive drift in the television signal-processing system in the microwave links), would disturb the normal stability of the vacuum-guide servo. With faulty videotapes of this character, the input error-waveform from the 15.75 kHz AFC-loop would appear as shown in Fig. 2-10.

The large-amplitude 60-Hz rate recurring pulse-disturbance over the normal 960-Hz error-signal would appear on a part of the program on a reel of videotape, due to the usual program-editing procedure. That is, the disturbance mentioned may not be present at all on 90% of the program, in which case the vacuum-guide would remain absolutely stable.

Solution. The problem is solved by the measure-and-optimize process. After the modification shown in Fig. 2-11, which consists of two new germanium crystal diodes, the whole instability problem, with the specific fault mentioned, was solved once and for all. The crystal-diodes speed up the inverter switching process somewhat similar to the latest Schottky diodes in high-speed I^2L logic.

Explanation. The positive-half of the abnormally large 60-Hz vertical sync disturbance in the AFC-loop, from occasional faulty tape programs, would drive stage A_1 transistor well into saturation region. (The 15.75 kHz AFC-loop, from which the error is obtained, is *not* compensated to meet the requirements of the vacuum-

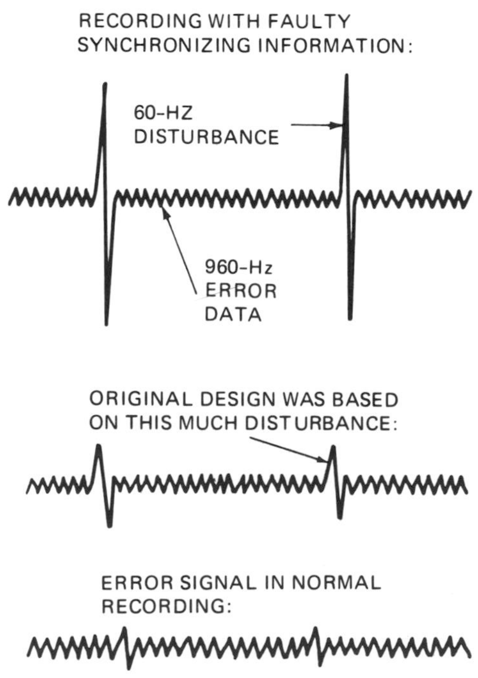

Fig. 2-10. Input error-signal—vacuum-guide.

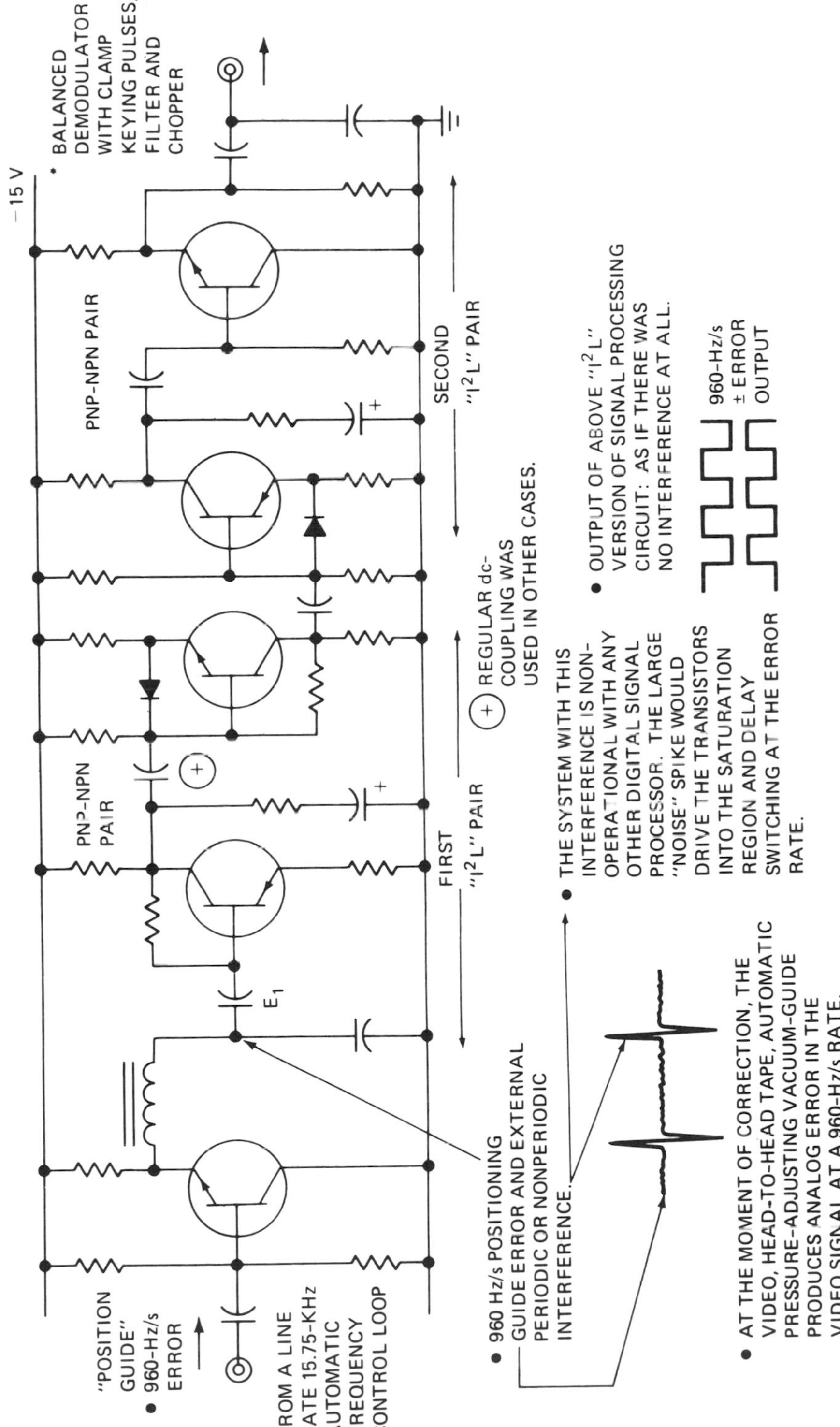

−15 V
"POSITION GUIDE"
• 960-Hz/s ERROR
FROM A LINE RATE 15.75-KHz AUTOMATIC FREQUENCY CONTROL LOOP
PNP-NPN PAIR
E_1
PNP-NPN PAIR
• BALANCED DEMODULATOR WITH CLAMP KEYING PULSES, FILTER AND CHOPPER
FIRST "I²L" PAIR
SECOND "I²L" PAIR
(+) REGULAR dc-COUPLING WAS USED IN OTHER CASES.
• 960 Hz/s POSITIONING GUIDE ERROR AND EXTERNAL PERIODIC OR NONPERIODIC INTERFERENCE.
• THE SYSTEM WITH THIS INTERFERENCE IS NON-OPERATIONAL WITH ANY OTHER DIGITAL SIGNAL PROCESSOR. THE LARGE "NOISE" SPIKE WOULD DRIVE THE TRANSISTORS INTO THE SATURATION REGION AND DELAY SWITCHING AT THE ERROR RATE.
• OUTPUT OF ABOVE "I²L" VERSION OF SIGNAL PROCESSING CIRCUIT: AS IF THERE WAS NO INTERFERENCE AT ALL.
960-Hz/s ± ERROR OUTPUT
• AT THE MOMENT OF CORRECTION, THE VIDEO, HEAD-TO-HEAD TAPE, AUTOMATIC PRESSURE-ADJUSTING VACUUM-GUIDE PRODUCES ANALOG ERROR IN THE VIDEO SIGNAL AT A 960-Hz/s RATE.

Fig. 2-11. High-speed I^2L signal-processing with clamping-diodes.

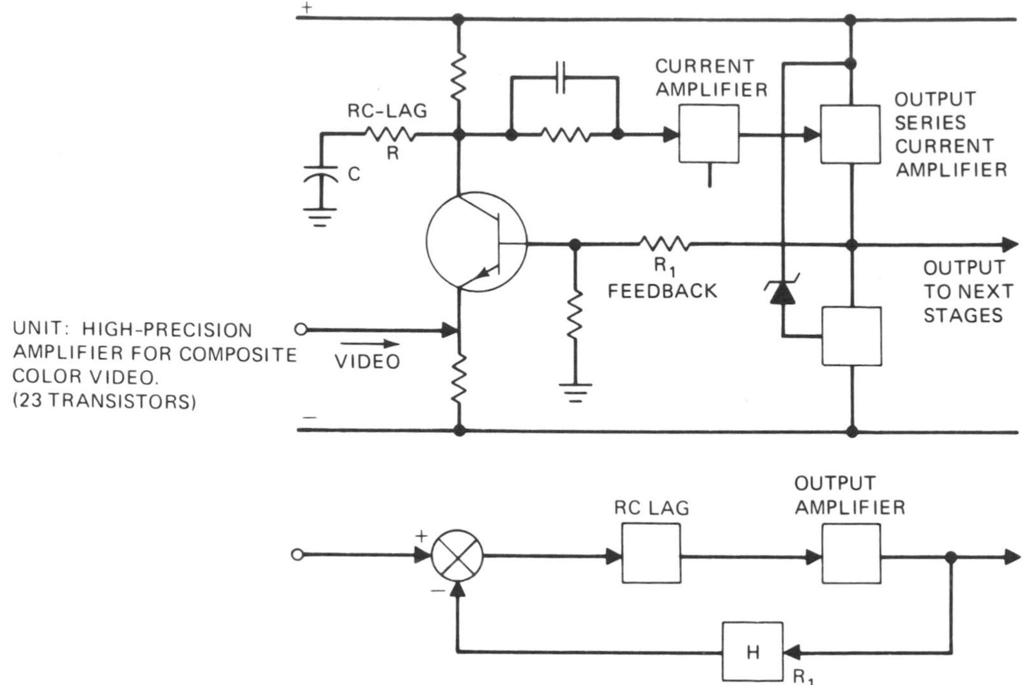

Fig. 2-12. Video amplifier–RC-lag compensation.

guide servo, because the same AFC-loop should serve the fast-acting 15.75 kHz line-lock loop. On the other hand, if the AFC-loop is compensated for the specific requirements of the vacuum-guide servo, the damped residual 60-Hz disturbance would be harmless in the 960-Hz guide servo-loop. At the time of the original development of this control system, the AFC-loop had a compensation that *was* more suitable for the vacuum-guide servo, from the viewpoint of the 60-Hz disturbance in this loop.) In the above circuit, the transistor saturation storage-effects will then restrict the normal 960-Hz switching-rate with the normal error-signal; with the disturbance shown, the rate of switching in this stage goes far below, down to 60 Hz. So the complete vacuum-guide error-sampling process would become erratic, and the system would become unstable. A limiting germanium diode-clamp, as shown in the lower circuit, provides the necessary solution to the saturation delay-effect. The clamp-diode conducts and prevents the collector-to-emitter voltage from falling below the base clamping-potential, when the large 60-Hz disturbance arrives. The diode, besides clamping, provides some degenerative feedback to minimize the effect of the overdrive. The PNP transistor provides the load-current for the switch-on process; and it minimizes the pulse-stretch effect during the switch-off process. Similarly, the large negative-excursion of the 60-Hz disturbance affects the next stage, and an identical diode-clamp at stage B_2 is effective in totally eliminating the effect of the periodic 60-Hz disturbance. The interfering pulse-train is merely clipped off, and the 960-Hz vacuum-guide digital controller is left undisturbed for correct automatic operation when such faulty videotapes are encountered. The dc compensation is optimized all over again to maintain the transient response with minimum of overshoot. The solution is equally effective for nonperiodic impulse noise.

2.13 A Simple Case in the Video Section of the Color Videotape Recorder. It has been stated earlier that the video and FM sections of the recorder are closely inter-

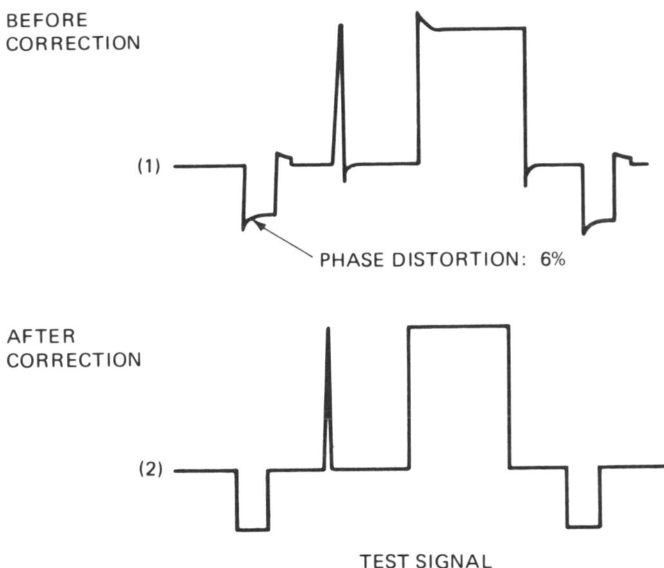

Fig. 2-13. Composite sine-squared pulse-and-bar.

related to the sampled-data control systems in the Q-CVTR because the tape-sync used for the servo systems is processed in those areas. See Fig. 2-12.

Here is a measure-and-optimize technique that eliminated a difficult problem. A color video amplifier was designed earlier with the best performance-index for a certain video gain. The index was measured by means of the sine-squared pulse-and-bar test signal shown in Fig. 2-13. The performance is indicated in terms of a k-factor. This particular color signal accessory must have an excellent characteristic in this respect (k = ±0.1%). Due to certain revised system requirements, the gain had to be doubled expeditiously. Normally, a redesign of the whole amplifier on a stage-by-stage basis was mandatory because of the high-precision performance involved. But, in this particular case, the problem was solved without a customary redesign, by the use of the measure-and-optimize process, in the following way.

The current feedback was first extrapolated by the change of the feedback resistor R1 to double the gain, while it was directly measured on the oscilloscope. From feedback control theory, we know that the transient response would be adversely affected by this abrupt increase in sensitivity. The overshoot, as measured with the sine-squared pulse-and-bar test signal shown in Fig. 2-13, increased to about 6% at the adjusted revised sensitivity. A video frequency-sweep and group-delay measuring setup indicated that the response was affected, with the associated phase-distortion, around the 1-to-2-MHz area. Then the sine-squared pulse-and-bar test signal was applied, and an RC-lag compensation at the collector of the transistor was chosen to correct the output waveform (1) to its original shape and (2) to within ±0.1%. The expeditious measure-and-optimize technique as a commercial design option would normally require in this complex circuit configuration an hour or two to solve this problem. A theoretical background in control theory is essential for the application of this particular idea. As far as the distortion components are concerned, the input common-base video amplifier itself would act as the differential amplifier for error detection. Hence, the lag compensation was simultaneously effective right at the collector of that stage. Three units made for a special project gave identical results regarding the desired performance-index. The abovementioned 6% of sync (and picture) distortion, if not corrected, would eventually appear during playback as a time-modulation of sync (and picture "reflections") and affect the final ± 2.5-nsec performance-index for the color and the control system-complex.

2.14 POSSIBLE EXTENSION OF HEURISTIC SAMPLED-DATA MEASURE-AND-OPTIMIZE TECHNIQUES TO AUTOMATION

As a typical example of possible extension of measure-and-optimize techniques to automation, the capstan servo of the quadruplex color videotape recorder may be considered as a case of simulation for the rate at which the material is moved in an industrial process, while the plant servo conducting a high-precision multiprocessing function of the unit (such as the headwheel servo), is synchronized to the servo moving the material. Sensitive transducers take the place of the tonewheel and the headwheel data. A specially coded disk encoder may be used, for instance, as an analog (shaft-position) to digital converter. As an additional requirement, a vacuum-guide servo can be emulated by devising an x-y coordinated positioning system for the material-disposition with respect to the primary process-control. The overall system-complex can be simulated as a sampled-data process-control via analog-to-digital converters, phase-locked to a central synchronizing multirate digital pulse generator. The latter can serve, in fact, as a synchronizing medium for several diverse processes located at that particular plant, or plants elsewhere (by phase-locking the various loops in a data format via microwave or optical communication links). The digital computer presently used on an individual basis in automation does partly come under this category, but the central reference sync generator and the pulse processing digital controllers designed along the lines of those used in the quadruplex color videotape recorder (using measure-and-optimize techniques) will add a new dimension of adaptability to the place of the conventional digital computer in automation for greater economy, flexibility, speed, and accuracy. The electronic sampled-data phase-lock techniques and the digital pulse logic involved would obviously eliminate a portion of the complex bulky mechanical and hydraulic interlocks and hardware involved in present processing systems.

3 Application of Theoretical Concepts to a Complex Digital Control System (Q-CVTR)

3.1 HEADWHEEL SAMPLED-DATA FEEDBACK CONTROL SYSTEM

As seen from Fig. 3-22 and as explained in Chapter 2, the headwheel servo system is highly complex. It is a multirate, multiloop, multivariable sampled-data feedback control system with several nonlinear elements, some unavoidable due to the characteristics of the system components, and most purely intentional, such as the nonlinear saturation of a reference trapezoid input pulse-waveform for the sampled high-speed lockup of the feedback phase-lock loops. The fine nonlinear *hold* quantization-errors inherent in sampled-data control systems may be considered negligible in view of the comparatively high rates of sampling used in the Q-CVTR. The system, consisting of several digital controllers and the associated system-variables, is not independent as a single servo system; the other two complex sampled-data feedback control systems in the Q-CVTR–the capstan and the vacuum-guide servos–are closely interacting as a result of the mutual exchange of the multirate sampled (1) input excitation-variables, (2) output controlled-variables, (3) system-variables, and, in some cases, (4) intermediate product-variables. Many of these variables are accessible and effective only via the complex FM-color-video system associated with the tape-data. *Instability in any one of the three servo systems will invariably result in the virtual instability of the other two as an imminent outcome.* (There is, however, an exception in the *automatic* vacuum-guide servo if the system instability occurs elsewhere in the signal processing system of the Q-CVTR, since an alternative *manually adjustable feedback loop* is available for vacuum-guide position-setting; the automatic mode is changed over to manual on an automatic basis, as a means of protection to the quadruplex headwheel.) Thus, from the theoretical standpoint, the system on a commercial basis is too complex to attempt, at the present state of the art, any known systematic application procedures for the independent analysis and synthesis of the various digital subsystems for arriving at the *necessary and sufficient conditions for stability*. Therefore, a theoretical interpretation will follow on some of the significant features in this control system, mostly in the form of applicable and instructive examples, dealing with *simple mathematical models.* These application examples serve toward a clearer understanding of the principles governing the commercial implementation of sampled-data control systems of this complex classification, when similar systems are developed and designed along *heuristic measure-and-optimize techniques* in the laboratory.

The solid-state video and control systems of the quadruplex color videotape recorder were developed and designed with solid-state discrete transistors during the early 1960s, and hence the heuristic situation cited was a necessary and valid step on a competitive commercial basis. If such complex control systems are contemplated at the *present state of the art*, silicon monolithic, I^2L (integrated-injection logic), large-scale, integrated (LSI) phase-lock loops (PLL) would be an alternative to simplify the design effort. The more sophisticated choice would be a regular low-cost microcomputer system to process and control the whole system-complex. The analog input variables from the input transducers, after the requisite small-signal amplification and quantization (or high-resolution analog-to-digital conversion), would be processed by *individual microcontrollers and digital filters for the subsystems* under the real-time master-control of a microprocessor serving as a central processing unit. Software would be in the form of electronically programmable read-only-memory (EPROM) firmware; solid-state LSI RAM memory would function as the real-time working memory. The output digital data would be reverted back to the analog format to activate the power plant of the control system. The idea is elaborated in Chapter 5.

3.2 HEADWHEEL HYSTERESIS MOTOR

The nucleus of the control system in the Q-CVTR is the quadruplex FM-video headwheel (HW), run by the speed-regulated, three-phase, self-starting hysteresis asynchronous-synchronous (HAS) motor. The basic theory of the HAS-motor is explained briefly as a start.

The HAS-motor combines the advantages of the synchronous and induction motors, without the demerits of either. Voltage regulation (by regulation of the reactive power) of the HAS-motor is fairly simple, as compared to that of an induction motor. It retains its characteristics in both steady-state and applicable transient states. It can operate synchronously if the law of excitation versus regulation only exists; however, it can operate asynchronously within known limits of the rotor-slip relative to the synchronous-speed. The slip of the motor is the rate at which the rotor lags behind the starter rotating-field and is measured in revolutions per second (or percentage of synchronous-speed).

$$\text{Slip of motor in cycles/sec, } \ell = (\text{rps of field}) - (\text{rps of rotor})$$
$$= (f/p) - 240$$

where p is the number of pole-pairs on the stator winding, and f is the supply-frequency.

With three-phase supply, the rotating stator-field produces on rotor the same effects as if it were surrounded by a system of rotating poles of constant strength. In practice, the efficiency of the HAS-motor at 60 Hz ranges from 20% to 40%, and it is the quietest of all motors. It develops torque as a result of residual-flux in a rotor of high-permeability, low-loss, soft magnetic steel. The flux-density lags the magnetizing-current by an angle, as a characteristic of the nonlinear hysteresis loop, and torque is actually developed due to the hysteresis-loss per cycle (time-lag) in the magnetic steel (Alnico V). The motor is unstable at lag-angles close to $\pi/2$ and above. It exhibits induction-motor torque below synchronism due to the eddy currents in the rotor-ring. The lag-angle varies with the shaft-torque and vanishes at zero-slip. Like the squirrel-cage rotor of the induction motor, the HAS motor develops the highest torque in a given size and the highest torque-inertia ratio. But it has the inherent capacity to give the highest speed-accuracy product when the phase of the operating frequency (60 Hz) is regulated by an appropriate servo-system that is capable of delivering the requisite voltage to give a constant flux-density and hence a constant uniform torque. (An induction motor is no match in this respect.) When phase-locked, the ratio of the applied voltage to frequency is a constant. Then, the power transmitted from the stator to the rotor is constant due to the asynchronous operation.

$$\text{Rotor hysteresis loss, } W_h = kf\,(\text{volume} \times \text{area of loop})$$

where k is a constant of units.

The power developed at the shaft for a constant torque $= W_h(1 - \ell)$, where ℓ = slip. It is only necessary that the stator ampere-turns meet the requirement for necessary degree of magnetization. The hysteresis motor hunts, but the rotor-losses tend to damp and minimize the hunting to a small phase-jitter, which the complex head-wheel servo system minimizes still further. Since the sinusoidal magnetomotive force (mmf) distribution around the air-gap is influenced by the stator-slot openings, the HAS motor selected is one of the closed-slot type (with minimum slot-openings) to reduce the pulsating magnetization and hence the minor hysteresis loops along the periphery of the main loop. The torque is obviously reduced by the area of the minor loops. The efficiency is actually increased by allowing adequate air-gap for a cushioning-effect on the preceding parasitic losses.

The power developed versus the air-gap flux is highly nonlinear in an asynchronous motor (unlike that of a synchronous version with rated voltage and frequency). The asynchronization of a synchronous machine follows the criterion that the magnetic-flux receive an additional rotation at the speed of the rotor-slip, relative to the synchronous speed, to give an excellent starting acceleration on high-inertia load. The magnetic-flux is no longer oriented along the direct rotor-axis; when there is slip, it shifts in relation to the rotor at the slip-speed, as the magnitude of the flux is maintained constant by uniform magnetic reluctance as far as the rotor is concerned. Unlike the synchronous motor, the steady induced rotor-field is of the same multiphase character as that of the applied stator-field. The slip-dependent asynchronous component F_a of the torque F has the same shape as the torque of an induction motor. The second component F_s of the torque due to the presence of the steady rotor-field induced in the rotor-ring does not depend on the slip after the condition of asynchronization is achieved, since the magnitude of the rotor-slip (and not the angle) determines the torque F. The asynchronization allows minimum speed-variation too, since the rotor-speed varies between the limits set by F_a of torque. Hence, at start, as the motor accelerates to asynchronization, the stability is assured by seeing that the component F_s of the torque F is dependent on the rotor-slip; and by regulating F_s within the limits of the slip, the best characteristic of the torque as a function of the slip is obtained in the asynchronous mode of operation (Fig. 3-1).

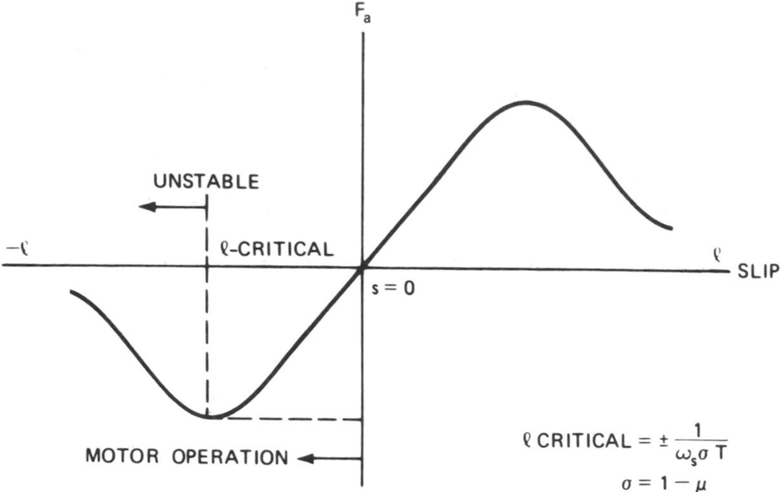

Fig. 3-1. Asychronous torque as a function of slip.

A multiphase system can be simplified as a two-phase system to write the equations of the asynchronous motor. In the place of the permanent-magnet rotor, two simulated excitation windings are assumed for the rotor to write the voltage-current relationships in the rotor (Fig. 3-2).

Under steady-state conditions with constant-slip operation, F is independent of the angle θ. When θ undergoes a constant change, the steady-state conditions are possible with the rotor-slip, $\ell = \frac{\omega - \omega_p}{\omega_p}$:

$$\theta = \omega_p \int_o^t \ell \, dt = \ell_o \omega_p t + \theta_{initial}$$

If V is the voltage of the power system,

$$F = e_d i_q - e_q i_d$$

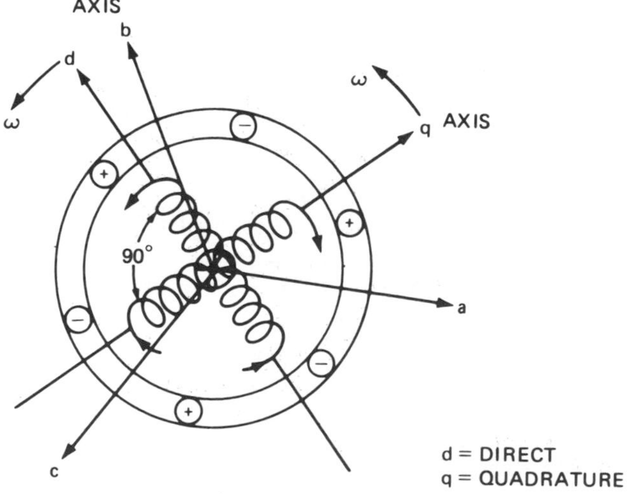

Fig. 3-2. Assumed ⊕directions of the axes, HAS motor.

(F is the torque, and ω_p is the synchronous angular-velocity). Then, the **Park-Gorev set of differential equations** for the operating conditions of the synchronous machine are given by:

Stator:

$$\frac{1}{\omega_p} xsi_d + \frac{1}{\omega_p} se_d + (1 + \ell)\, xi_q + (1 + \ell)e_q = V \sin\theta \tag{1}$$

$$(1 + \ell)xi_d + (1 + \ell)e_d - \frac{1}{\omega_p} xsi_q - \frac{1}{\omega_p} se_q = V \cos\theta \tag{2}$$

Rotor:

$$kTxsi_d + (1 + sT)e_d = E_d, \text{ the emf applied to the rotor} \tag{3}$$

$$kTxsi_q + (1 + sT)e_q = E_q \tag{4}$$

T = time-constant of excitation
s = complex-variable
x = synchronous reactance of the system

The torque or motion is a nonlinear differential equation:

$$Js\ell + e_d i_q - e_q i_d = M_o \tag{5}$$

where M_o is the external torque, and J is the mechanical inertia constant of the rotor in sec. The above equations are further modified for the HAS motor. A general solution can be found only if a steady rotor-field with a full compensation for the delay in the rotor circuit is available. Therefore, no attempt will be made in this work to obtain the solutions and the constants required for writing the transfer-function of the control-plant. Since the HAS motor is, in principle, a *hybrid induction-synchronous motor*, the standard transfer-function of an induction motor will be used while considering this control-plant in the interpretive theoretical examples.

The **Ashland $\frac{1}{20}$-hp three-phase HAS motor** used for the headwheel servo is rated for 100 V at a synchronous frequency of 300 Hz, but it is operated asynchronously by means of the headwheel servo system at 240 rps. The torque at the rated voltage and frequency is 3.5 in-oz minimum at 14,400 rpm. The nonlinear voltage/speed versus torque characteristics at the synchronous frequency are shown in Fig. 3-3. The stall-torque at the rated voltage and frequency is 110 in-oz minimum. The asynchronous speed of 14,400 rpm is attained in less than 1 sec at start. In operation, the HAS motor requires a ceiling voltage of 86 V (delta, phase-to-phase) at start, as it races toward synchronization; it drops to 36 V (phase-to-phase) when it achieves the phase-lockup in the regulated running mode. Fluid (air) bearings are used for a perfect lubrication without metal-to-metal contact. Since rotational velocities are low, a pump is essential for maintaining a perfect air-lubrication.

As a matter of theoretical interest, the equations of the AS-machine are given in the following form, by introducing a set of conditions into the regulation principle:

1. $E_{d1} = \beta si_d\, E_{q1} = \beta si_q$
2. The regulation is adjusted so that $\beta = \mu Tx$ where μ is the magnetic coupling-coefficient.
3. $E_{d2} = (1 + sT)E_d$; $E_{q2} = (1 + sT)E_q$ . . . with two split-components for each rotor voltage.

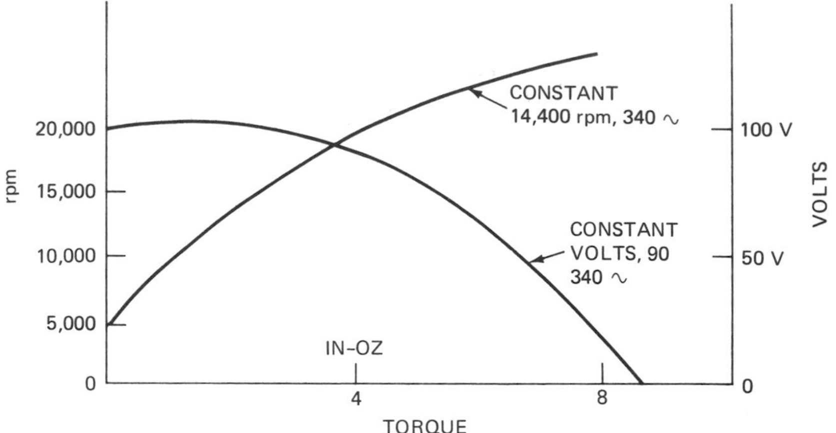

Fig. 3-3. HAS motor characteristics.

Equations of asynchronous-synchronous machine:

$$\frac{1}{\omega_p} xsi_d + \frac{1}{\omega_p} se_d + (1 + \ell)xi_q + (1 + \ell)e_q = V \sin \theta \quad (1)$$

$$(1 + \ell)xi_d + (1 + \ell)e_d - \frac{1}{\omega_p} xsi_q - \frac{1}{\omega_p} se_q = V \cos \theta \quad (2)$$

$$(e_d - \epsilon_d)(1 + sT) = 0 \quad (3)$$

$$(e_q - \epsilon_q)(1 + sT) = 0 \quad (4)$$

$$Js\ell - e_d i_q - e_q i_d = M_o \quad (5)$$

The hysteresis component of the particular HAS motor used can be treated for analytical purposes as a separate nonlinear element of the *electrical backlash*. It is unfortunate that at the present stage of the development of these devices in heavy electrical machinery, the manufacturers concerned do not appear to have any specialized research facilities to determine the constants involved in the transfer-function of each motor before it is supplied with the usual specifications for feedback control purposes.

3.3 GENERAL THEORETICAL CONCEPTS APPLICABLE TO HEADWHEEL SERVO

3.3.1 Sampling Systems in the Q-CVTR. In the feedback phase-error detection used in the various loops of the Q-CVTR, impulse sampling is approximated by commonly sampling a reference trapezoid pulse-transfer-function by means of sharp sampling pulses of opposite polarity. The *hold* dc error signal (with a nonlinear saturation characteristic) is obtained with respect to a suitable reference dc error potential. The sharp sampling pulses are commonly generated by a pulse transformer in a solid-state active circuit. The analysis shown in the following pertinent example clearly illustrates the theoretical necessity of extremely sharp sampling pulses for reproducing the frequency spectrum-components of the error lattice-function f*(t) as faithfully as possible. See Fig. 3-4.

In the place of the complex transfer-function of the trapezoid, f(t) in the time-domain is taken, for example, as 10 cos ωt. The sampling pulse-train p(t) and the pulse lattice-function f*(t) are represented by the following expressions. The sampling-period T is assumed 1 sec, and ω_o is assumed 0.5 rad/sec. Two cases will be examined: first a finite-width sampling, 0.1 sec wide, and then an impulse, according to the theoretical definition.

$$f(t) = 10 \cos \omega t: \tag{1}$$

$$p(t) = 5 \sum_{k=-\infty}^{\infty} [u(t - kt) - u(t - kt - m)] \tag{2}$$

$$f^*(t) = E_o \cos \omega_o t + E_1 \left\{ \cos\left[\left(\frac{2\pi}{T} + \omega_o\right) t + \theta_1\right] + \cos\left[\left(\frac{2\pi}{T} - \omega_o\right) t + \theta_1\right]\right\}$$
$$+ E_2 \left\{ \cos\left[\left(\frac{4\pi}{T} + \omega_o\right) t + \theta_2\right] + \cos\left[\left(\frac{4\pi}{T} - \omega_o\right) t + \theta_2\right]\right\} + \cdots \tag{3}$$

where E_0, E_1, E_2, etc., are the amplitudes of the frequency spectrum components.

Finite Pulse-Width Sampling. The sampling pulse-train p(t), which is periodically repeated, is expressed in Fourier series:

$$p(t) = 5 \left[a_o + \sum_{k=1}^{\infty} \left(a_k \cos \frac{2\pi k}{T} t + b_k \sin \frac{2\pi k}{T} t \right) \right] \tag{4}$$

where

$$a_o = \frac{1}{T} \int_{-T/2}^{T/2} p(t)\, dt$$

$$a_k = \frac{2}{T} \int_{-T/2}^{T/2} p(t) \cos \frac{2\pi k}{T} t \cdot dt$$

$$b_k = \frac{2}{T} \int_{-t/2}^{t/2} p(t) \sin \frac{2\pi k}{T} t \cdot dt$$

$$\frac{2\pi}{T} = \omega_s$$

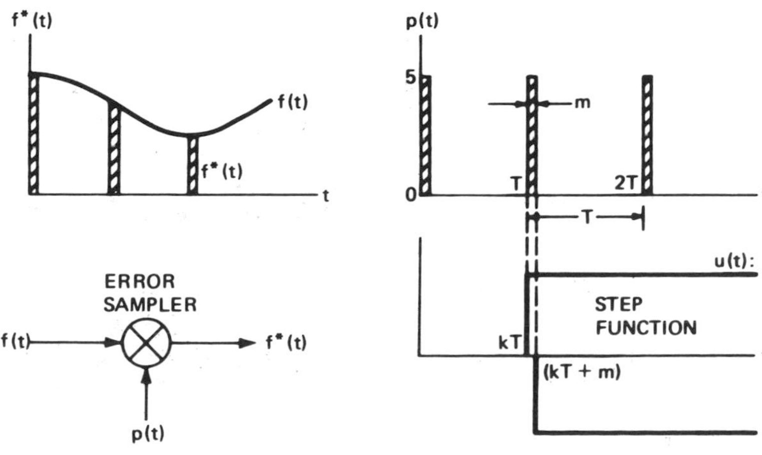

Fig. 3-4. Finite pulse-width sampling.

In the exponential form, the pulse-train:

$$p(t) = 5 \sum_{-\infty}^{\infty} c_k e^{j\omega_s kt} \tag{5}$$

where

$$c_k \triangleq \frac{1}{T} \int_0^T p(t) e^{-j\omega_s kt}\, dt$$

But, it is given that

$$p(t) = 5 \sum_{-\infty}^{\infty} [u(t - kt) - u(t - kt - m)]$$

$$\text{i.e., } p(t) = \begin{cases} 5 \text{ for } kt < t < kt + m \\ 0 \text{ at other instants.} \end{cases}$$

$$c_k = \frac{1}{T} \int_0^m 5e^{-j\omega_s tk}\, dt = \frac{-5}{jk\omega_s t} e^{-jk\omega_s t}\Big|_0^m = \frac{5}{jk\omega_s t}(1 - e^{-jk\omega_s m})$$

$$p(t) = \sum_{k=-\infty}^{\infty} \frac{5}{jk\omega_s t}(1 - e^{-jk\omega_s m}) e^{jk\omega_s tk} \tag{6}$$

Now c_k, when $k = 0$:

$$c_0 = \frac{5}{jk\omega_s t}(1 - e^{-jk\omega_s m}) = \frac{0}{0}$$

Apply L'Hôpital's rule:

$$\left(\frac{dNr.}{dk} \Big/ \frac{dDr.}{dk}\right)\Bigg|_{k=0} = \frac{5}{j\omega_s t}(j\omega_s m \cdot e^{-jk\omega_s m})_{k=0}$$

$$c_0 = \frac{5m}{T}$$

From (6):

$$p(t) = c_0 + \sum_{k=1}^{\infty} \frac{5}{jk\omega_s T} \{[e^{jk\omega_s t} - e^{jk\omega_s(t-m)}] - [e^{-jk\omega_s t} - e^{-jk\omega_s(t-m)}]\} \tag{7}$$

by splitting $\sum_{-\infty}^{\infty} = \sum_{k=0} + \sum_{k=1}^{\infty} + \sum_{k=-1}^{-\infty}$ $\sin x = (e^{jx} - e^{-jx})/2j$

$$p(t) = \left\{\frac{5m}{T} + \sum_{k=1}^{\infty} \frac{j2 \times 5}{jk\omega_s T} [\sin k\omega_s t - \sin k\omega_s(t - m)]\right\}$$

But $f^*(t) = f(t) \cdot p(t)$.

$$f^*(t) = \frac{50m}{T} \cos \omega_o t + \sum_{k=1}^{\infty} \frac{100}{k\omega_s t} \cos \omega_o t\, [\sin k\omega_s t - \sin k\omega_s(t - m)]$$

using the trigonometric identities: $\sin A - \sin B = 2 \sin \frac{1}{2}(A - B) \cos \frac{1}{2}(A + B)$, where

$A = k\omega_s t$
$B = k\omega_s(t - m)$

$$C = \frac{A + B}{2}$$

$$D = \omega_o t$$

$$\cos C \cos D = \tfrac{1}{2} \cos (C + D) + \tfrac{1}{2} \cos (C - D)$$

$$f^*(t) = \frac{50m}{T} \cos \omega_o t + \sum_{k=1}^{\infty} \frac{100}{k\omega_s t} \sin \frac{k\omega_s m}{2} \left\{ \cos \left[(k\omega_s + \omega_o)t - \frac{k\omega_s m}{2} \right] + \cos \left[(k\omega_s - \omega_o)t - \frac{k\omega_s m}{2} \right] \right\} + \cdots$$

The preceding expression and the given expression for $f^*(t)$ are compared to obtain the spectrum component amplitudes E_0, E_1, E_2, etc.

$$\frac{\omega_s m}{2} = \theta_1 \text{ for } k = 1;\ \omega_s m = \theta_2 \text{ for } k = 2;\ \frac{2\pi}{T} = \omega_s;\ E_0 = 50;\ \omega_o = 2.5 \text{ rad/sec}$$

$$\frac{k\omega_s m}{T} = 0.1\pi \text{ for } k = 1, \text{ and } 0.2\pi \text{ for } k = 2.$$

$$E_0 = \frac{50m}{T};\ E_1 = \frac{100}{\omega_s T} \sin \frac{\omega_s m}{2};\ E_2 = \frac{100}{2\omega_s T} \sin \left(\frac{2\omega_s m}{2} \right) \cdots$$

Putting m = 0.1 sec, t = 1 sec:

$$E_0 = 50 \times 0.1 = 5$$

$$E_1 = \frac{100}{2\pi} \sin 18^\circ = 4.92$$

$$E_2 = \frac{50}{2\pi} \sin 36^\circ = 4.66$$

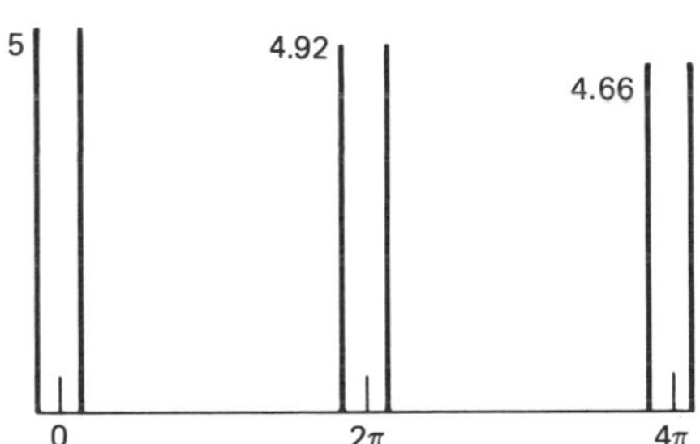

FREQUENCY SPECTRUM COMPONENTS

In this case, the amplitudes (coefficients) are determined by the **reference f(t) and the sampling pulse dimensions.**

Impulse-Sampling. The unit-impulse (Dirac or delta-function) $\delta_T(t)$ is defined thus:

$$p(t) = \delta_T(t) = \sum_{k=-\infty}^{\infty} \delta(t - kt)$$

where $\delta(t - kt)$ represents an impulse of unit area at time kt; theoretically, Em = 1 as $m \longrightarrow 0$.

$$p(t) = \sum_{k=-\infty}^{\infty} c_k e^{j\frac{2\pi k}{T} t}$$

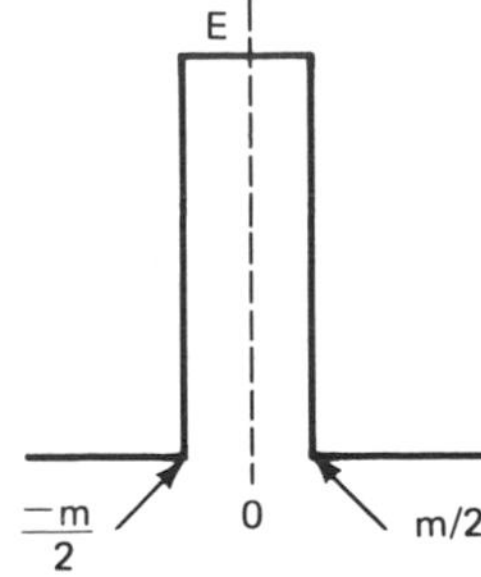

c_k can be obtained by deriving the Fourier series for the rectangular pulse, extending from $-m/2$ to $m/2$, and putting $Em = 1$, as $m \longrightarrow 0$.

$$c_k = \frac{1}{T}\int_{-T/2}^{T/2} p(t)e^{-j\frac{2\pi k}{T}t}\,dt$$

$$\text{and let } p(t)\begin{cases} = E \text{ through } \dfrac{m}{2} \text{ to } \dfrac{-m}{2}. \\ = 0 \text{ at other times.}\end{cases}$$

$$= \frac{1}{T}\int_{-m/2}^{m/2} Ee^{-j\frac{2\pi k}{T}t}\,dt = \frac{E}{T}\frac{T}{-j2\pi k}\,e^{-j\frac{2\pi k}{T}t}\Bigg|_{-m/2}^{m/2}$$

$$= -\frac{E}{j2\pi k}\left(e^{-j\frac{2\pi k}{T}\frac{m}{2}} - e^{j\frac{2\pi k}{T}\frac{m}{2}}\right) = \frac{E}{\pi k}\sin\frac{\pi km}{T}$$

As $m \longrightarrow 0$, $\sin\dfrac{\pi km}{T} \longrightarrow \dfrac{\pi km}{T}$

$$c_k\Big|_{m\to 0} - \frac{E}{\pi k}\frac{\pi km}{T} = \frac{Em}{T} = \frac{1}{T} \text{ since } Em = 1 \text{ as } m \longrightarrow 0.$$

$$p(t) = \delta_T(t) = \sum_{k=-\infty}^{\infty}\frac{1}{T}e^{j\frac{2\pi k}{T}t} = \frac{1}{T} + \frac{2}{T}\left[\frac{(e^{j\frac{2\pi t}{T}} + e^{-j\frac{2\pi t}{T}})}{2} + \frac{(e^{j\frac{4\pi t}{T}} + e^{-j\frac{4\pi t}{T}})}{2} + \cdots\right]$$

$$= \frac{1}{T} + \frac{2}{T}\left[\cos\left(\frac{2\pi t}{T}\right) + \cos\left(\frac{4\pi t}{T}\right) + \cdots\right]$$

$$f^*(t) = f(t)\cdot p(t).$$

$$= \frac{10\cos\omega_o t}{T} + \frac{20}{T}\left[\cos\left(\frac{2\pi t}{T}\right)\cos\omega_o t + \cos\left(\frac{4\pi t}{T}\right)\cos\omega_o t\right] + \cdots$$

Since $2\cos A\cdot\cos B = \cos(A + B) + \cos(A - B)$

$$f^*(t) = \frac{10}{T}\cos\omega_o t + \frac{10}{T}\left[\cos\left(\frac{2\pi}{T} + \omega_o\right)t + \cos\left(\frac{4\pi}{T} + \omega_o\right)t + \cos\left(\frac{2\pi}{T} - \omega_o\right)t + \cos\left(\frac{4\pi}{T} - \omega_o\right)t + \cdots\right]$$

where $T = 1$.

Comparing this expression with given $f^*(t)$ as before,

$$E_0 = E_1 = E_2 = \cdots = 10$$

(determined by the **amplitude of the reference f(t) only**). Thus, it is seen that the theoretical impulse-sampling reproduces the error frequency-spectrum with absolute accuracy. Hence, the requirement of the sharp sampling pulses is met as far as practically possible in the Q-CVTR.

3.3.2 Pulse-Transfer-Functions in Q-CVTR and Their Interactions. If the number of open-loop pulse-transfer-functions (ptf's) involved is n, the equations of these pulse systems can be written in the following form, when in series in-phase and out-

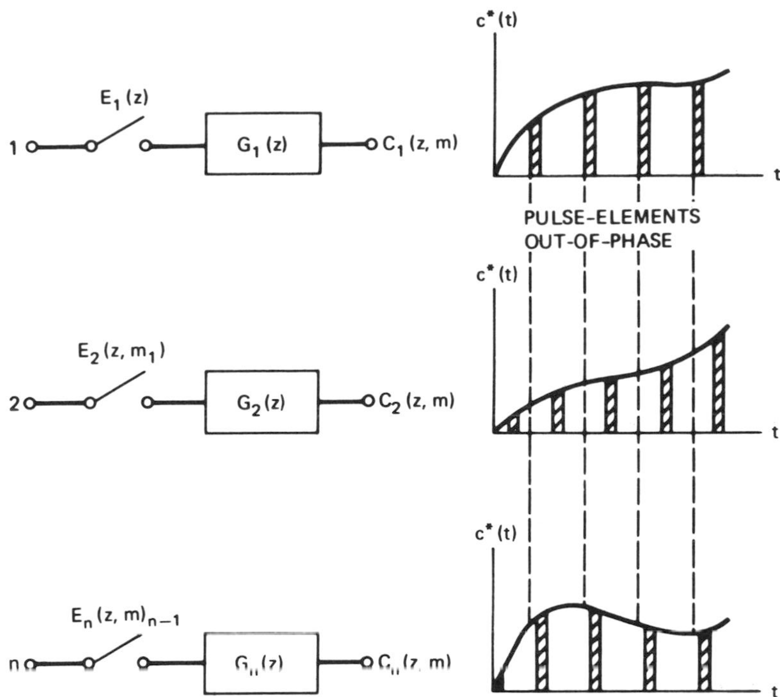

Fig. 3-5. Representation of pulse-transfer-functions.

of-phase, and when in parallel in-phase and out-of-phase. The block-diagrammatic representation of the three sampled-data control systems in the Q-CVTR merely shows for simplicity the overall ptf's (with logic and internal feedback in pulse circuitry) as R(z), H(z), and G(z), leaving aside the power-plant. Each ptf is a miniature digital controller.

1. $C_1(z, m) = G_1(z, m)E_1(z)$
   $C_2(z, m) = G_2(z, m - m_1)E_2(z, m_1)$
   $C_n(z, m) = G_n(z, m - m_{n-1})E_n(z, m_{n-1})$

2. If the above ptf's are **connected in series,** the overall ptf, C(z, m), is given by:

$$C(z, m) = \left[\prod_{r=1}^{n-1} G_r(z, m_r - m_{r-1})\right] G_n(z, m - m_{n-1})$$

If pulse elements are in-phase,

$$C(z, m) = \left[\prod_{r=1}^{n-1} G_r(z,m_r - m_{r-1})\right] G_n(z, m)$$

For **parallel-connected** ptf's,

$$\textbf{Out-of-phase:}\, C(z, m) = \sum_{r=1}^{n} G_r(z, m)$$

$$\textbf{In-phase:}\, C(z, m) = \sum_{r=1}^{n} G_r(z, m - m_{r-1})E(z, m_{r-1})$$

A simple example of determining the overall output ptf for the multiloop sampled-data feedback control systems, when the pulse elements are in-phase, follows.

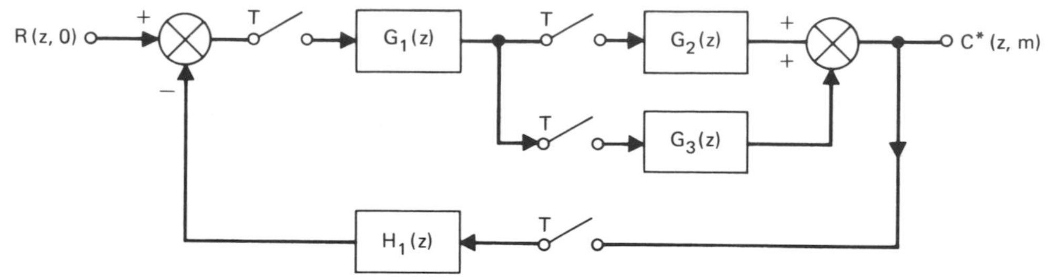

$$C(z, m) = \frac{R_1(z, o)G_1(z, o)[G_2(z, m) + G_3(z, m)}{1 + G_1(z, o)[G_2(z, o) + G_3(z, o)H_1(z, o)]}$$

3. The interconnection of out-of-phase ptf's that commonly occur in the Q-CVTR overall-control system presents a very cumbersome problem as seen from the following simple example of a **two-loop system with respective input and output variables interacting on each other in out-of phase** as represented below by $E_1(z, o)$ and $E_2(z, m_1)$. *Note:* $E_1(z, o) = z^{-1}E_1(z)$.

If the system is disconnected at the input of the pulse elements, the output ptf in general is expressed thus:

$$C_k(z, m) = \sum_{r=1}^{n} G_{kr}(z, m - m_{r-1})E_r(z, m_{r-1}) \tag{1}$$

where k = 1, 2, . . . , n

$$E_k(z, m_{k-1}) = R_k(z, m_{k-1} - C_k(z, m_{k-1})$$

$$= R_k(z, m_{k-1}) - \sum_{r=1}^{n} G_{kr}(z, m_{k-1} - m_{r-1})E_r(z, m_{r-1}) \tag{2}$$

These expressions give a system of equations, which can be manipulated in matrix form, to describe the processes in a closed interacting sampled-data control system with n pulse elements:

$$[1 + G_{11}(z, o)E_1(z, o)] + G_{12}(z, -m_1)E_2(z, m_1) + \cdots = R_1(z) \tag{3}$$

$$G_{21}(z, m_1)E_1(z, o) + [1 + G_{22}(z, o)E_2(z, m_1)] + \cdots = R_2(z) \text{ etc.} \tag{4}$$

Applying this technique to the simple two-loop case:

$$E_1(z, o) = \left[\frac{1 + G_{22}(z, o)}{\Delta(z)} R_1(z, o)\right] - \left[\frac{G_{12}(z, -m_1)R_2(z, m_1)}{\Delta(z)}\right] \tag{5}$$

$$E_2(z, m_1) = \left[\frac{1 + G_{11}(z, o)}{\Delta(z)} R_2(z, m_1)\right] - \left[\frac{G_{21}(z, m_1)R_1(z, o)}{\Delta(z)}\right] \tag{6}$$

where

$$\Delta(z) = [1 + G_{11}(z, o)]\,[1 + G_{22}(z, o)] - [G_{21}(z, m_1)G_{12}(z, -m_1)] \tag{7}$$

Then,

$$C_k(z, m_{k-1}) = R_k(z, m_{k-1}) - E_k(z, m_{k-1}) \tag{8}$$

$$C_1(z, m) = R_1(z, o) - E_1(z, o)$$

$$C_2(z, m) = R_2(z, m_1) - E_2(z, m_1)$$

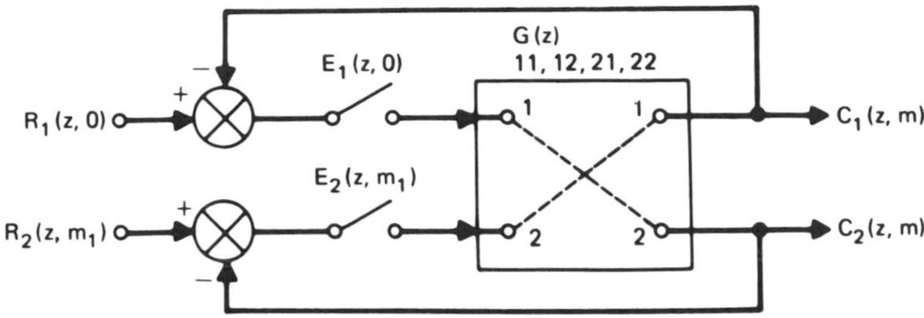

Fig. 3-6. Two-loop interacting sampled-data control system.

With each ptf as a digital controller, the highly complex nature of the output ptf's in the Q-CVTR with three basic interacting feedback control systems can be imagined from the preceding expressions for a very simple interacting case in Fig. 3-6.

4. The **equivalent transfer-function** techniques, shown in either **block-diagram** representation or the alternative **flowchart** representation, find frequent application in reducing a multiloop system to the final mathematical model of a single feedback-loop sampled-data control system. It is seen that elements in-and-out of a loop can be interchanged during this system manipulation. In the Q-CVTR, this is not a readily feasible proposition. The various nonlinear elements in the forward and feedback paths of the various loops, and the multirate character of the loops from 30 Hz through 1.79 MHz make the reduction (with possibly describing-function approximations) of the transfer-function by interchange extremely cumbersome and impractical. In fact, the extraordinary character of the multirate samplings in the Q-CVTR interacting control systems strictly restricts the analysis or synthesis to a loop-by-loop basis, and that with undefined constraints on the input and output variables. Where the system reduction is applicable, these basic principles in block-diagram and flowchart are very important. Flowchart approach is more common in circuit analysis (Fig. 3-7).

3.3.3 Multirate Sampling Systems Used in Control Systems of Q-CVTR. The multirate sampling techniques and the valid sampling rates used for the various loops in the three interacting control systems in the Q-CVTR can be primarily attributed to the inherent nature of the multirate ptf variables involved in the television tape-signal processing system. The situation is entirely different to that deliberately employed with digital controllers used in automatic control for improving system performance. As seen from the control-system block-diagrams in Figs. 3-22 through 3-24, there are seven different rates of sampling, varying from 30 Hz through 1.79 MHz. With the theoretical techniques available such as the switch-decomposition method, the computations involved in analyzing the various multirate sampling systems of the order encountered in the Q-CVTR will be excessively high with respect to the overall control system. Therefore, a simple example is given below to illustrate the common theoretical technique employed for analyzing a multirate sampled-data control system.

The **switch decomposition technique** introduced in Chapter 1 will be applied to the basic control-loop of the **headwheel control system** (Fig. 3-8), where the feedback signal is derived from a 240 Hz (T/4) digital controller, while the forward-path functions at 60 Hz (T) with a 60 Hz reference trapezoid input. For simpler computation and as a matter of illustration, T is assumed as 1 sec, along with the following mathematical models.

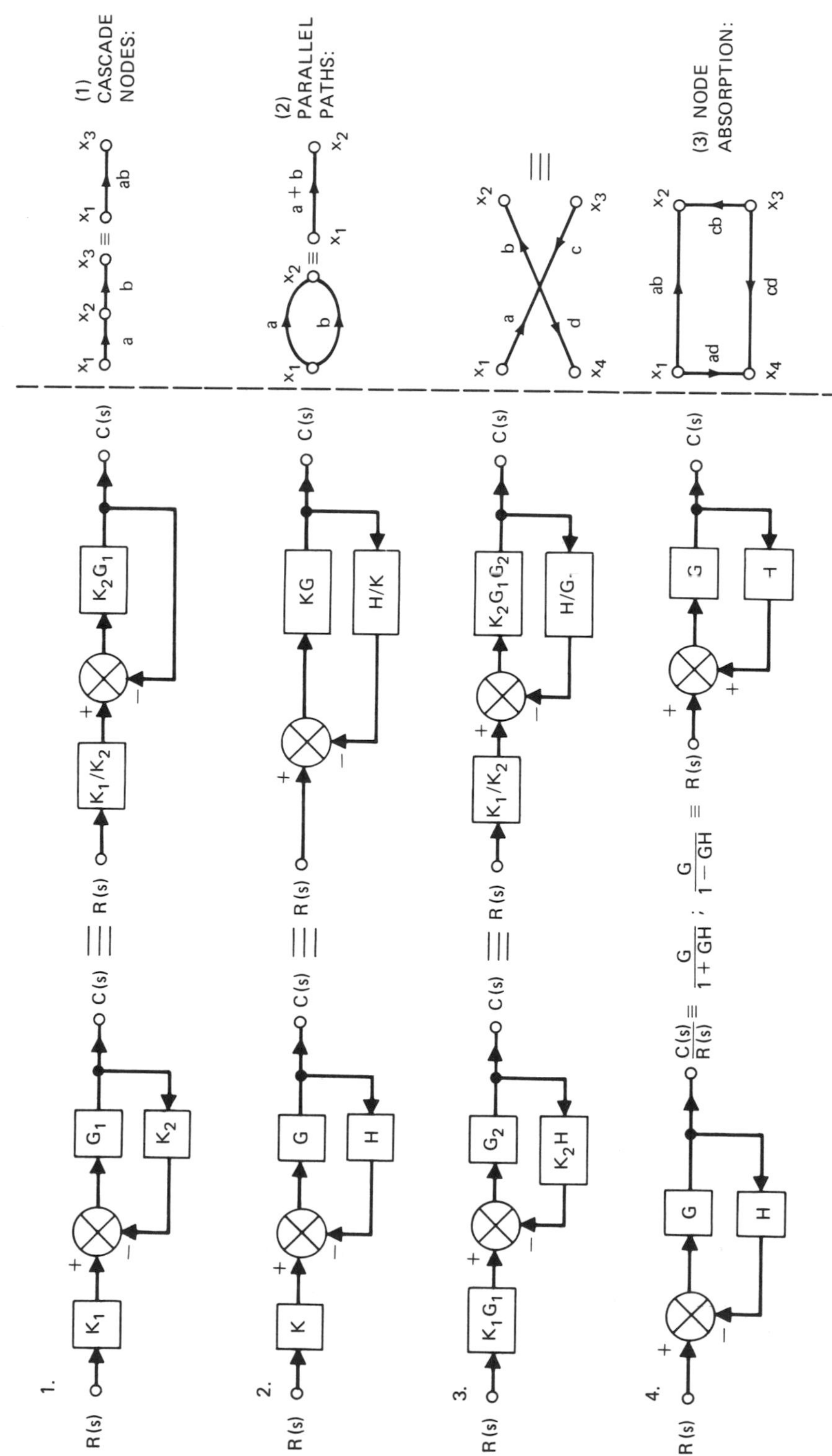
1.
R(s)
K1
G1
K2
C(s)
≡
R(s)
K1/K2
K2G1
C(s)
2.
R(s)
K
G
H
C(s)
≡
R(s)
KG
H/K
C(s)
3.
R(s)
K1G1
G2
K2H
C(s)
≡
R(s)
K1/K2
K2G1G2
H/G
C(s)
4.
R(s)
G
H
C(s)/R(s) ≡ G/(1 + GH) ; G/(1 − GH) ≡
R(s)
G
H
C(s)
x1 a x2 b x3 ≡ x1 ab x3
(1) CASCADE NODES:
x1 a b x2 ≡ x1 a + b x2
(2) PARALLEL PATHS:
x1 a b x2 x3 c d x4
≡
x1 ab x2 cb x3 cd x4 ad
(3) NODE ABSORPTION:

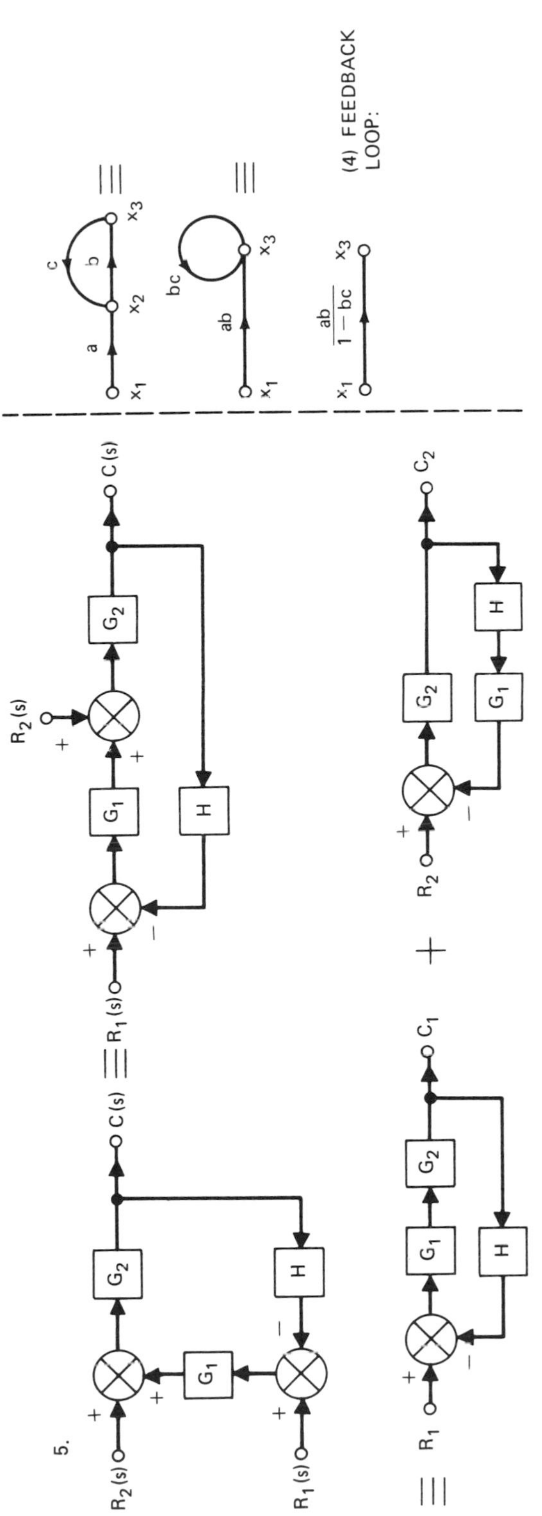

$R_2 = 0$: $\dfrac{C_1}{R_1} = \dfrac{G_1G_2}{1 + G_1G_2H}$; $R_1 = 0$: $\dfrac{C_2}{R_2} = \dfrac{G_2}{1 + G_1G_2H}$ $\qquad C = \dfrac{G_1G_2R_1 + G_2R_2}{1 + G_1G_2H}$

Fig. 3-7. Equivalent transfer-functions: block-diagram and flowchart methods.

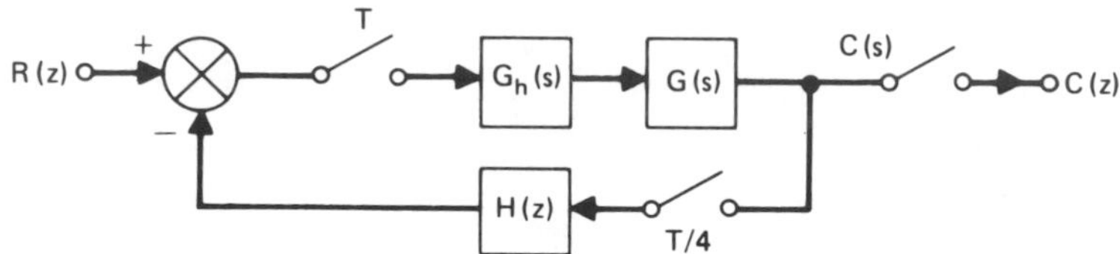

Fig. 3-8. Basic multirate HW sampled-data control.

$$G_h(s) = \text{zero-hold} = (1 - e^{-s})/s = (1/s)(1 - z^{-1})$$

$$G(s) = K/(s + 1) \text{ and } H(s) = 1/s$$

Now, if the z-transform with respect to the sampling-period T/n is $G(z_n)$:

$$C(z)_n = G(z)_n R(z)$$

1. Also, $C(z_n) \triangleq G(z_n)Rz_n^n)$.

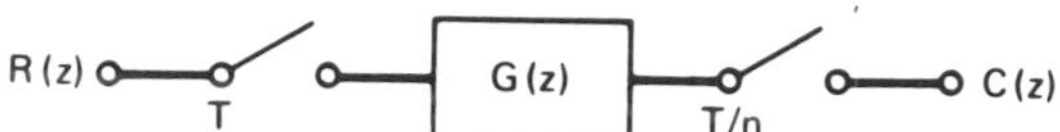

2. The relationship between the z-transform G(z) with respect to T, and $G(z_n)$ with respect to T/n is given by the following expression:

$$G(z)_n = G(z) + \sum_{p=1}^{n-1} z^{1-\frac{p}{n}}\left(z, \frac{p}{n}\right)$$

where p stands for the *pth* delay-sampler in switch-decomposition.

3. It can be shown, for the multirate sampled-data system in Figs. 3-8 and 3-9 with T/n sampling-rate in the feedback-path, that the output ptf is given by the expression:

$$C(z) = \frac{G(z)R(z)}{1 + G(z)H(z) + \sum_{p=1}^{n-1} Z[e^{pst/n}G(s)]\, Z[e^{-pst/n}H(s)]}$$

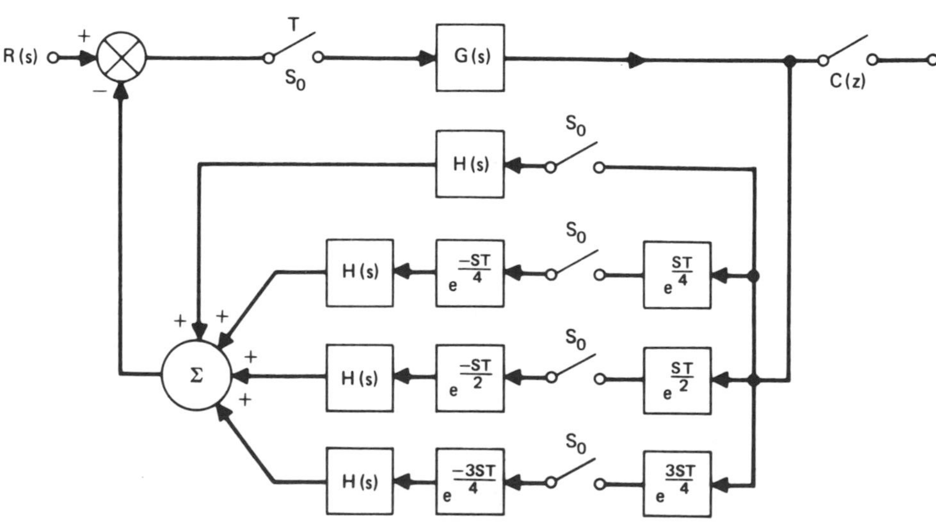

Fig. 3-9. Equivalent switch-decomposition block-diagram for the given multirate system.

and since C(z, m) = G(z, m) · E(z), only the numerator G(z) is replaced by G(z, m) for obtaining the modified z-transform of the output C(z, m). For n = 4,

$$C(z) = \frac{G(z)R(z)}{1 + G(z)H(z) + Z[e^{st/4}G(s)]\,Z[e^{-st/4}H(s)] + Z[e^{2st/4}G(s)]\,Z[e^{-2st/4}H(s)] + Z[e^{3st/4}G(s)]\,Z[e^{-3st/4}H(s)]}$$

To obtain C(z), the z-transforms required in the above equation are determined in the following manner:

$$G(z) = Z\left(\frac{1 - e^{-st}}{3}\,\frac{K}{s+1}\right)$$

$$= \left.\frac{K(1 - e^{-T})}{(z - e^{-T})}\right]_{T=1}$$

$$= \frac{0.632K}{(z - 0.368)}$$

$$G(z, m) = \frac{K(1 - e^{-m})z - (e^{-m} - 0.368)}{z(z - 0.368)}$$

$$H(z) = Z[e^{-st/4}H(s)] = Z\left[\frac{e^{-st/4}}{s}\right]$$

From z-transform tables:

$$\frac{Ka}{s(s+a)} \xrightarrow{G(z)} \frac{(1 - e^{-at})z^K}{(z - 1)(z - e^{-at})}$$

$$\frac{Ka}{s(s+a)} \xrightarrow[G(z,m)]{} \left[\frac{1}{(z - 1)} - \frac{e^{-amt}}{(z - e^{-at})}\right]K$$

where a = 1

$$\frac{1}{s} \xrightarrow[G(z)]{} \frac{z}{z - 1}$$

$$\frac{1}{s} \xrightarrow[G(z,m)]{} \frac{1}{z - 1}$$

Now,

$$\left.\begin{aligned} G(z) &= Z[e^{pst/n}G(s) = ZG(z, m)]_{m=p/n} \\ H(z) &= Z[e^{-pst/n}H(s)] = H(z, m)_{m=1-p/n} \end{aligned}\right\}$$ from the relationship between z and (z, m) transforms. p = (n − 1) = 3.

$$H(z) = Z\left[\frac{e^{-st/4}}{s}\right] = \left.\frac{1}{z - 1}\right]_{m=3/4} = \frac{1}{z - 1}$$

If $G_1(s) = K/s(s + 1)$, $G(s) = G_1(s)(1 - e^{-st})$.

$$Z[e^{st/4}G(s)] = zG_1(z, m)]_{m=1/4} - G_1(z, m)]_{m=1-3/4}$$ {From the above relationships

$$= (z - 1)G_1(z, m)]_{m=1/4}$$

$$= (z - 1)\left[\frac{K}{z - 1} - \frac{Ke^{-mt}}{z - e^{-T}}\right]_{m=1/4}$$

$e^{-.5} = .548$
$e^{-.75} = .472$
$e^{-1} = .368$
$e^{-.25} = .779$

$$= K\left[1 - \frac{(z - 1)e^{T/4}}{z - e^{-T}}\right]_{T=1} = \frac{0.221Kz + .411K}{z - 0.368}$$

Similarly,

$$Z[e^{st/2}G(s)] = [(z - 1)G_1(z, m)]_{m=2/4}$$

$$= K\left[1 - \frac{(z - 1)e^{-T/2}}{z - e^{-T}}\right]_{T=1} = \frac{0.452Kz + 0.182K}{z - 0.368}$$

$$Z[e^{3st/4}G(s)] = (z - 1)G_1(z, m)]_{m=3/4}$$

$$= K\left[1 - \frac{(z-1)e^{-3T/4}}{z - e^{-T}}\right]_{T=1} = \frac{0.528Kz + 0.104K}{z - 0.368}$$

$$Z[e^{-2st/4}H(s)] = H(z, m)]_{m=2/4} = \left.\frac{1}{z-1}\right]_{m=3/4} = \frac{1}{z-1}$$

$$Z[e^{-3st/4}H(s)] = H(z, m)]_{m=3/4} = \frac{1}{z-1}$$

Substituting these ptf's in the output C(z),

$$C(z) = \frac{\dfrac{0.632K}{z-0.368}R(z)}{1 + \left(\dfrac{0.632K}{z-0.368}\right)\left(\dfrac{1}{z-1}\right) + \left(\dfrac{0.221Kz + .411K}{z-0.368}\right)\left(\dfrac{1}{z-1}\right) + \left(\dfrac{0.452Kz + 0.182K}{z-0.368}\right)\left(\dfrac{1}{z-1}\right) + \left(\dfrac{0.528Kz + 0.104K}{z-0.368}\right)\left(\dfrac{1}{z-1}\right)}$$

$$C(z) = \frac{0.632K(z-1)R(z)}{z^2 + z(1.20K - 1.368) + (1.329K + 0.368)}$$

The characteristic equation of this multirate sampled-data system is therefore,

$$A(z) = z^2 + z(1.201K - 1.368) + (1.329K + 0.368) = 0$$

Just as the Routh-Hurwitz stability criterion is commonly employed in analytically testing continuous systems, the **Schur-Cohn stability criterion** is suitable for sampled-data or digital feedback control systems. According to this technique, the test for finding the presence of any root of the characteristic equation, outside the unit-circle in the z-plane, involves the expression of its coefficients and their conjugates in the form of a determinant. For the **second-order system,**

$$\Delta_{k=2} = \begin{vmatrix} a_o & 0 & a_n & a_{n-1} \\ a_1 & a_o & 0 & a_n \\ \bar{a}_n & 0 & \bar{a}_o & \bar{a}_1 \\ \bar{a}_{n-1} & \bar{a}_n & 0 & \bar{a}_o \end{vmatrix}$$

and

$$\Delta_{k=1} = \begin{vmatrix} a_o & a_n \\ \bar{a}_n & \bar{a}_o \end{vmatrix}$$

The system is stable provided that $\Delta_k < 0$ for k odd, $\Delta_k > 0$ for k even.

$$A(z) = a_o + a_1 z^1 + \cdots a_n z^n = 0\big|_{n=2}$$

The above condition is further simplified for a quadratic polynomial with real coefficients, and unity coefficient for z^2. Then, the *necessary* and *sufficient conditions* for the roots of the characteristic equation to lie inside the unit-circle (for stability) are simply given by:

$$(1)\quad |A(0)| < 1$$

$$(2)\ A(1) > 0$$

$$(3)\ A(-1) > 0$$

Therefore, applying this criterion to the characteristic equation A(z),

$$(1)\ 1.329K + 0.368 < 1$$

$$(2)\ 2.53\ K > 0$$

$$(3)\ 0.128K + 3.736 > 0$$

Hence, the maximum allowable gain for stability from condition (1) is

K = 0.48 for a stable multirate control system.

The multirate sampled-data control system is now compared with the same system as a single-rate loop.

$$C(z) = \frac{G(z)R(z)}{1 + G(z)H(z)} = \frac{\dfrac{0.632KR(z)}{(z - 0.368)}}{1 + \left(\dfrac{0.632K}{z - 0.368}\right)\left(\dfrac{z}{z - 1}\right)}$$

$$= \frac{0.632K(z - 1)R(z)}{z^2 - z(0.632K + 1.368 + 0.368)}$$

$$A(z) = z^2 - z(0.632K + 1.368) + 0.368 = 0$$

Applying the Schur-Cohn stability criterion, from conditions $A(\pm 1) > 0$, it is seen that the conditions for stability fail, since K becomes a negative quantity.

Thus, it can be concluded from the above analytical examples, that a multirate system has a maximum allowable gain, higher than that of the corresponding single-rate system. In this particular case, it is seen that an unstable single-rate system has been turned into a stable system by applying the technique of multirate sampling. And this is a theoretical conclusion that favors the multirate sampling techniques used in the headwheel and capstan sampled-data control systems of the Q-CVTR, wherein actually the multirate sampling is done as a necessity to meet the requirements of the pulse-rates in the television signal-processing system.

To obtain the unit step-function response of the above multirate system at the sampling instants, put K = 0.3 and R(z) = z/(z - 1).

$$R(s) = u(t) \xrightarrow{R(z)} z/(z - 1).$$

$$C(z) = \frac{0.632Kz}{z^2 + z(1.201K - 1.368) + (1.329K + .368)} = \frac{0.19z}{z^2 - 1.01z + 0.767}$$

Expanding the above expression into a power-series in z^{-1} (by long division),

$$\begin{array}{lllllll}
z^{-1} & z^{-2} & z^{-3} & z^{-4} & z^{-5} & z^{-6} & z^{-7} \\
0.19 & .192 & & & & &
\end{array}$$

$$\begin{array}{l}
1 - 1.01 + .767\,\overline{)\,0.19} \\
\quad\underline{0.19 - 0.192 + 0.146} \\
\qquad 0.192 - .146 \\
\qquad \underline{0.192 - .192 + .148} \\
\qquad\qquad 0.056 - .148
\end{array}$$

$$\begin{array}{l} \qquad\qquad\quad 0.19z^{-1} + 0.192z^{-2} + 0.056z^{-3} - 0.132z^{-4} - 0.174z^{-5} - 0.07z^{-6} + 0.06z^{-7} + .1z^{-8} - .07z^{-9} \\ 1 - 1.01 + .767\overline{).056 - .148} \\ \qquad\qquad\quad \underline{.056 - .016 + .043} \\ \qquad\qquad\qquad -.132 - .043 \\ \qquad\qquad\qquad \underline{-.132 + .131 - .1} \\ \qquad\qquad\qquad\qquad -.174 + .1 \\ \qquad\qquad\qquad\qquad \underline{-.174 + 0.17 - 0.13} \\ \qquad\qquad\qquad\qquad\qquad -.07 + .13 \\ \qquad\qquad\qquad\qquad\qquad \underline{-.07 + .07 - .05} \\ \qquad\qquad\qquad\qquad\qquad\qquad .06 + .05 \\ \qquad\qquad\qquad\qquad\qquad\qquad \underline{.06 - .06 + .04} \end{array}$$

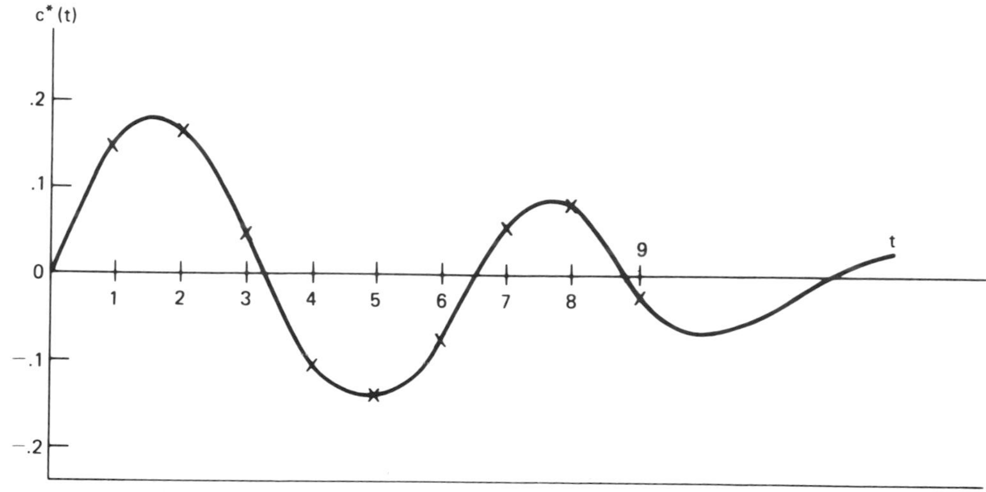

$$C^*(t) = 0.19\delta(y - 1) + 0.192\delta(t - 2) + 0.056\delta(t - 3) - 0.132\delta(t - 4)$$
$$- 0.1748\delta(t - 5) - 0.07\delta(t - 6) + 0.06\delta(t - 7) + 0.1\delta(t - 8) - \cdots$$

The transient-response of this particular multirate example at the gain $K = 0.3$ is not quite satisfactory since the system is altogether unstable (with continuous oscillation) as a single-rate system. As seen from the trend of the response, it will settle down after a decaying oscillation (*ringing*), thus confirming, as a case of extension, that a multirate sampler in principle does improve transient-response if the system is stable to start with.

3.3.4 Delay Elements in Control Systems of Q-CVTR. In the pulse systems of the three sampled-data control systems, the delay elements are a constantly recurring phenomena. A pure-delay element is represented by $e^{-st} = z^{-1}$, but in the active pulse circuitry of the system, the delay is often represented by the delayed z-transform of a fictitious simple-lag element, corresponding to the transfer-function $K/(s + a)$ with the corresponding time-function Ke^{-at}. As a physical example, the transfer-function of the automatic-delay lines in the headwheel sampled-data control system with about 80 reactive elements in sequence will be extremely complex.

The following basic concepts concerning the pure-delay element are hence particularly important. A simple example, illustrating the method of analysis, will clarify its significance by indicating the effect of delay on system stability.

Pure time-delay, e^{-st} can be represented in a Bode-plot as shown:

$$20 \log e^{-st}\big|_{s=j\omega} = 20 \log 1 - j\omega t = j\omega t(-)$$

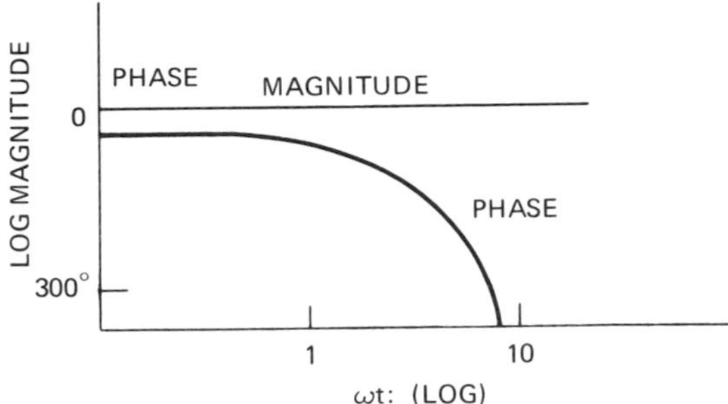

The magnitude is independent of frequency; phase is linearly related to frequency.

The modified z-transform Z_m will be used to analyze a simple error-sampled feedback loop with a delay element $e^{-\lambda s}$ and system transfer-function K/s(s + 1) with T = 1 sec and λ = 0.25 sec.

$$G(z) = Z\left[\frac{Ke^{-\lambda s}}{s(s+1)}\right] = Z_m\left[\frac{K}{s(s+1)}\right]_{m=1-\lambda}$$

From the modified z-transform table for the above Z_m,

$$G(z) = \left[\frac{K}{z-1} - \frac{Ke^{-(1-\lambda)}}{z-e^{-1}}\right]$$

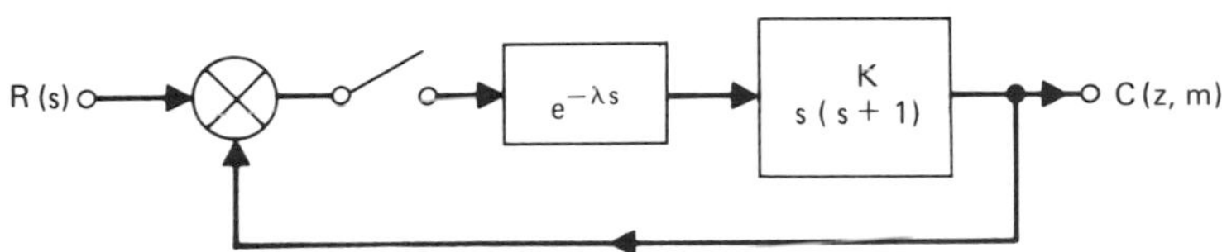

The characteristic equation A(z) = 1 + G(z) = 0.

$$A(z) = z^2 - z(1 - e^{-1} + Ke^{-(1-\lambda)} - K) + (1 - K)e^{-1} + Ke^{-(1-\lambda)} = 0$$

Applying the Schur-Cohn criterion for the quadratic polynomial,

$$|A(o)| = K[e^{-(1-\lambda)} - e^{-1}] + e^{-1} < 1 \quad (1)$$

$$A(1) = K[1 - e^{-1}] > 0 \text{ or } 0.632K > 0 \quad (2)$$

$$A(-1) = 2(1 + e^{-1}) - K[1 + e^{-1} - 2e^{-(1-\lambda)}] > 0. \qquad e^{-.75} = 0.472 \quad (3)$$

with λ = 0.75: From (1): 0.104K + 0.368 < 1 or 0.104K < 0.632

From (3): 2.736 - 0.424K > 0

From (1) K = 6. For this maximum value of K, with a pure-delay of 0.25 sec, all three necessary and sufficient conditions are satisfied for the stability of the system. To examine the effect of the delay elements on the stability of a closed-loop sampled-data control system, the procedure is now repreate without the delay element:

$$G(z) = Z\left[\frac{K}{s(s+1)}\right] = \frac{K(1-e^{-1})z}{(z-1)(z-e^{-1})}$$

the Laplace to z-transform from table in Chapter 1.

$$A(z) = 1 + G(z) = z^2 + Z[K(1 - e^{-1}) - 1 - e^{-1}] + e^{-1} = 0$$

Applying the Schur-Cohn conditions, the system is stable if

$$|A(0)| = e^{-1} < 1 \tag{1}$$

$$A(1) = K(1 - e^{-1}) > 0 \tag{2}$$

$$|A(-1)| = 1 - K(1 - e^{-1}) + (1 + e^{-1}) + e^{-1} > 0 \tag{3}$$

$$K < \frac{2(1 + e^{-1})}{1 - e^{-1}} = 4.32$$

The maximum allowable gain in this case, without the delay element, is 4.32 only, compared to 6 with a delay of 0.25 sec. If the delay, for example, is 1 sec in the first case, then from condition (1) in that case:

$$K(e^{-(1-1)} - e^{-1}) + e^{-1} < 1$$

$$0.632K < 0.632 \text{ or } K = 1$$

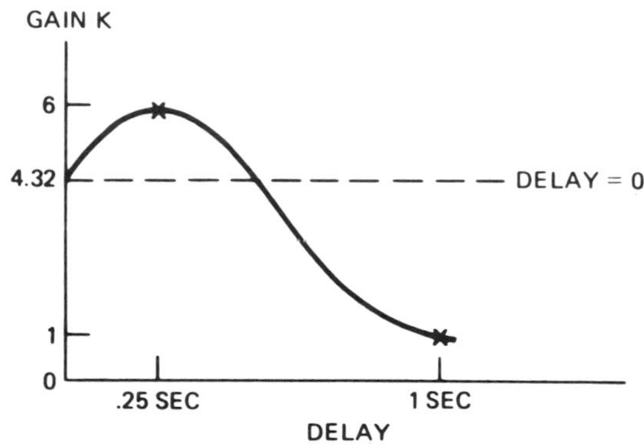

It is interesting to note that the delay element does have an *optimum* value for stability with a gain-figure higher than that possible without any delay element whatsoever. Thus, an appropriate value of delay in the forward-path of this loop is conducive to greater stability in a sampled-data feedback control system and an element in the forward-path of the loop can have a corresponding equivalent in the feedback-path. From practical experience with the sampled-data control systems in the Q-CVTR, this is one of the most noteworthy theoretical conclusions, since, in the heuristic measure-and-optimize techniques employed in the design of the complex Q-CVTR control systems, pulse delay-optimization in the forward or feedback paths, against the parameters of gain and stability, is one of the most common pulse circuit-development techniques.

3.3.5 Pulse-Transfer-Function of Rectangular Pulse-Train. A rectangular pulse-train is one of the most frequent occurrences in the Q-CVTR control system. Its pulse-transfer-function is derived in the following manner.

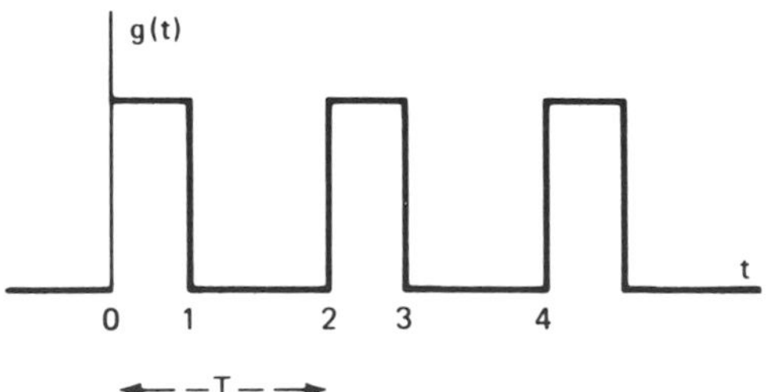

$$g(t) = u(t) + \sum_{k=1}^{\infty} (-1)^k u(t-k)$$

$$u(t - t_o) = 0 \text{ for } t < t_o$$
$$= 1 \text{ for } t \geqslant t_o$$

$$g(t) = 1 - u(t-1) + u(t-2) - \cdots$$

$$\mathcal{L}[g(t)] = G(s) = \frac{1}{s} - \int_1^{\infty} e^{-s}\,dt + \int_1^{\infty} e^{-2s}\,dt - \cdots$$

$$= \frac{1}{s} - \frac{1}{s}(e^{-s}) + \frac{1}{s}(e^{-2s}) - \cdots$$

$$= \frac{1}{s}(1 - e^{-s} + e^{-2s} - \cdots) = \frac{1}{s(1 + e^{-s})}$$

in closed-form, with T = 1 (using binomial expansion)

$$G(z) = \frac{z}{(z-1)(1+z^{-1})}$$

3.3.6 Pulse-Transfer-Function of Sawtooth Ramp Waveform. The sawtooth ramp waveform is another waveform that is common in pulse control systems:

$$g(\omega t) = A\left[\frac{1}{2} - \frac{1}{\pi}\left(\sin \omega t + \frac{1}{2}\sin 2\omega t + \frac{1}{3}\sin 3\omega t + \cdots\right)\right]$$

(Fourier series)

$$g(t) = t - \sum_{k=1}^{\infty} u(t-k)$$

$$= t - [u(t-1) + u(t-2) \cdots]$$

where the transform for the ramp, $t \Rightarrow \frac{1}{s^2}$

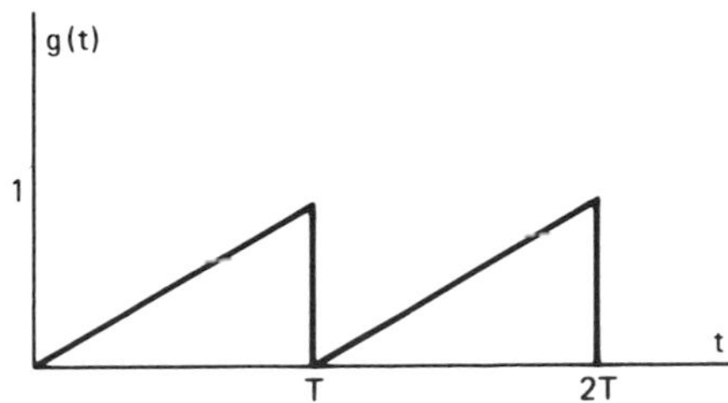

The series is $\mathcal{L}$-transformable since

$$\lim_{T > \infty} \int_o^T [g(t)]e^{-st}\,dt < \infty \text{ for } s > 0$$

$$\mathcal{L}[g(t)] = G(s) = \frac{1}{s^2} - \int_1^{\infty} e^{-s}\, dt - \int_1^{\infty} e^{-2s}\, dt - \cdots$$

$$= \frac{1}{s^2} - \frac{1}{s}(e^{-s} + e^{-2s} + \cdots) = \frac{1}{s^2} - \frac{e^{-s}}{s}(1 + e^{-s} + e^{-2s} + \cdots)$$

$$G(s) = \frac{1}{s^2} - \frac{e^{-s}}{s(1 - e^{-s})}$$

From the z-transforms table,

$$G(z) = \frac{Tz}{(z-1)^2} - \frac{z}{z-1} \cdot \frac{z-1}{1-z^{-1}} = \frac{Tz}{(z-1)^2} - \frac{z}{(z-1)^2}$$

$$= \frac{z(T-1)}{(z-1)^2}$$

3.3.7 Pulse-Transfer-Function of Periodic Square-Wave.

$$g(t) = u(t) - 2u(t-1) + 2u(t-2) - 2u(t-3) + \cdots$$

$$G(s) = \frac{1}{s}\left(\frac{1 - 2e^{-s}}{1 + e^{-s}}\right) = \frac{1}{s}\left(\frac{1 - e^{-s}}{1 + e^{-s}}\right) \cdots$$

$$\text{since } G(s) = \frac{1}{s}\,[1 - 2e^{-s}(1 - e^{-s} + e^{-2s} - \cdots)] = \frac{1}{s}\tanh\frac{s}{2}$$

$$G(z) = \frac{z}{(z-1)}\left(\frac{z-1}{z+1}\right) = \frac{z}{z+1}$$

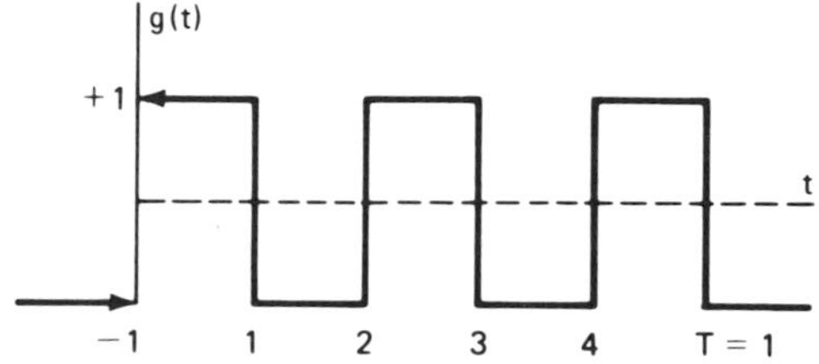

Fourier series:

$$g(\omega t) = \frac{4}{\pi}\left(\cos \omega t - \frac{\cos 3\omega t}{3} + \frac{\cos 5\omega t}{5} - \cdots\right)$$

3.3.8 Pulse-Transfer-Function of Trapezoid Pulse-Train. The trapezoid pulse-train is represented in time-domain by the following expression:

$$g(t) = \sum_{k=0,3,6}^{\infty} (t-1)u(t-k) - (t - \overline{k+1})u(t - \overline{k+1}) - u(t - \overline{k+2})$$

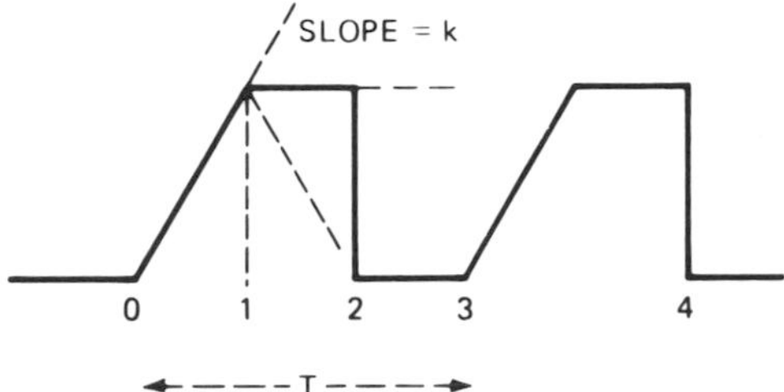

Taking Laplace transforms,

$$G(s) = \sum_{k=0,3,6\ldots}^{\infty} \left[\frac{e^{-ks}}{s^2} - \frac{e^{-(k+1)s}}{s^2} - \frac{e^{-(k+2)s}}{s} \right]$$

From z-transform tables,

$$G(s) = \sum_{k=0,3,6\ldots}^{\infty} \frac{Tz \cdot z^{-k}}{(z-1)^2} - \frac{Tz \cdot z^{-(k+1)}}{(z-1)^2} - \frac{z \cdot z^{-(k+2)}}{(z-1)}$$

$$G(z) = \sum_{k=0,3,6\ldots}^{\infty} \frac{Tz^{-k}}{(z-1)^2} [z - (1 + z^{-1})]$$

This ptf is not simple enough to express in a *closed-form* of the type, McLauren infinite-series for $(1 \pm z)^{-1}$. It is partly because of this reason that practically every feedback loop in the Q-CVTR may be alternatively treated as a sampled-data control system with the nonlinear element "saturation." The sampling process adds a "quantization" nonlinearity too, but, as mentioned earlier, it may be considered negligible due to the comparatively high rates of sampling encountered in the Q-CVTR.

3.3.9 Time-Response from Pulse-Transfer-Function of Digital Controller. If the overall pulse-transfer-function G(z) of one of the sampled-data feedback loops is obtained, the time-response of that loop is determined as illustrated in the following example by one of the most convenient methods—as that used for inverse $\mathcal{L}$-transforms.

$$G(z) = \frac{1}{1 - \frac{11}{16}z^{-1} + z^{-2} + \frac{1}{6}z^{-3}}$$

This may be, for example, one of the digital controllers in the system. The denominator can be factored by long-division.

$$G(z) = \frac{1}{(1 - z^{-1})(1 - \frac{5}{6}z^{-1} + \frac{1}{6}z^{-2})}$$

$$\left(1 - \frac{5}{6}z^{-1} + \frac{1}{6}z^{-2}\right) = \frac{1}{z^2}\left(z^2 - \frac{5}{6}z + \frac{1}{6}\right) = \frac{1}{z^2}\left(z - \frac{1}{3}\right)\left(z - \frac{1}{2}\right)$$

$$G(z) = \frac{z^3}{(z-1)(z-0.333)(z-0.5)}$$

Using partial fractions,

$$\frac{G(z)}{z} = \frac{z^2}{(z-1)(z-.33)(z-0.5)} = \frac{A}{z-1} + \frac{B}{z-0.5} + \frac{C}{z-.333}$$

$$A = z^2 \; \frac{z^2}{6} \cancel{(z-1)}/\cancel{(z-1)}(z-.5)(z-.333)]_{z=1} = 3$$

And, in the same way, the other residues: B = −3 and C = 1.

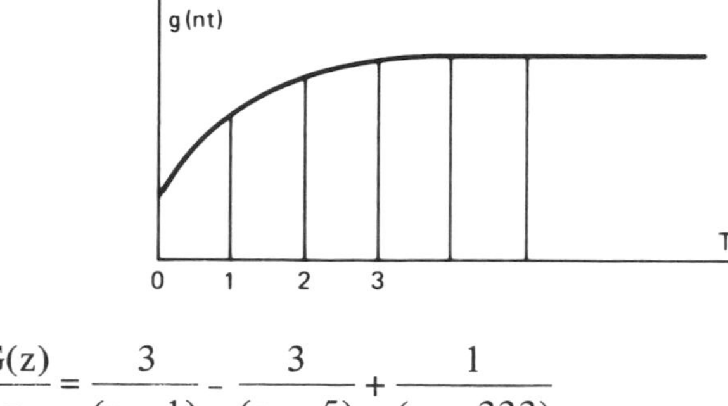

$$\frac{G(z)}{z} = \frac{3}{(z-1)} - \frac{3}{(z-.5)} + \frac{1}{(z-.333)}$$

$$G(z) = \frac{3z}{(z-1)} - \frac{3z}{(z-.5)} + \frac{z}{(z-.333)}$$

But the inverse of $G(z) = \frac{z}{(z-a)}$ is given by $g(nT) = a^n$.

$$g(nT) = [3(1)^n - 3(0.5)^n + (0.333)^n]$$

Giving integral values to n, the time-response is obtained as

$$g(0) = 1; \quad g(T) = 1.833; \quad g(2T) = 2.36; \quad g(5T) = 2.9; \quad g(100T) = 3$$

3.3.10 Physical Configuration of Digital Controller. As mentioned earlier, the delay-element is one of the most recurring pulse-transfer-functions in sampled data or digital control systems. For example, the simplest digital controller may be represented by one delay-element, two weight-setting amplitude-constants, and a simple summing "OR" logic-element:

$$\frac{C(z)}{R(z)} = D(z) = K_0 + K_1 z^{-1}$$

Less elementary digital controllers will take the following form with their corresponding pulse-transfer-functions.

Digital controllers of such elementary character with delay-elements, and more complex ones using specially shaped pulse-trains and intricate logic are very frequent in the pulse processing circuitry of the Q-CVTR. For *physical realizability*, the first weighting function in the denominator must be unity and the other coefficients can be independently controlled. (Realizability implies that the equivalent network is viable as a stable entity.) A trapezoid shaped pulse-train was explained in Section 3.3.8. It can be easily imagined how complex the overall pulse-transfer-functions become in each loop in the three interacting control systems of the Q-CVTR—the complexity increases since delay elements are not necessarily mere pure-delay. Ac-

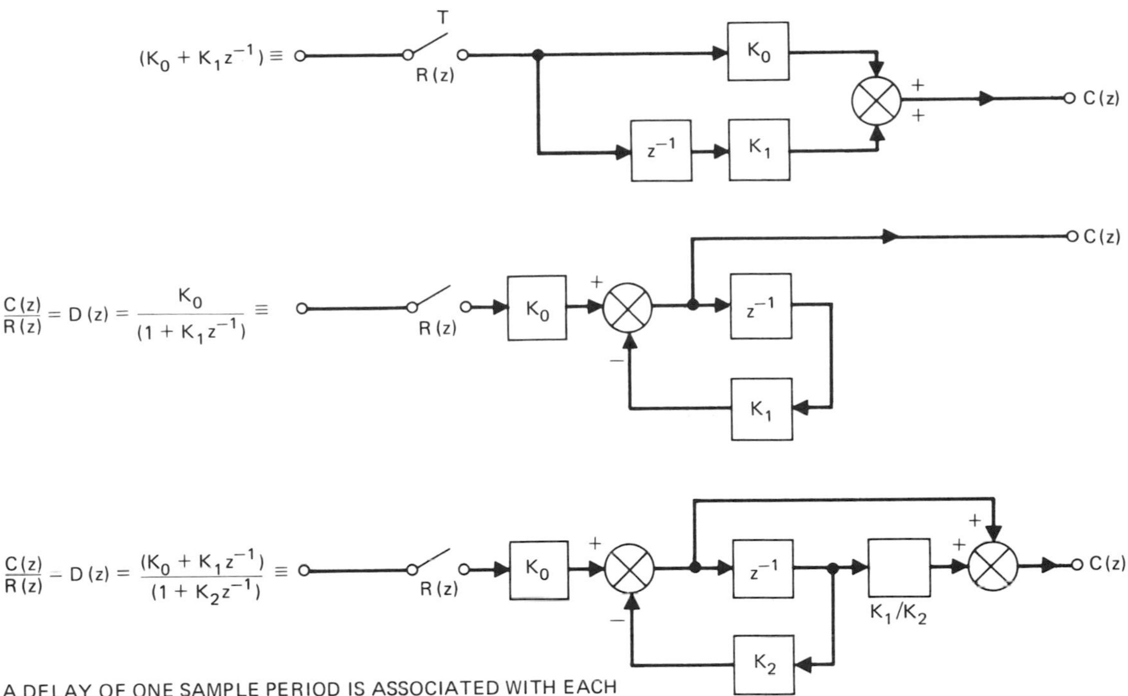

A DELAY OF ONE SAMPLE PERIOD IS ASSOCIATED WITH EACH TRANSMISSION PATH INVOLVING z^{-1}.

tive multielement delay-line networks, as used in the Q-CVTR, predominantly take the place of delay elements. Pulsed-data RC-networks are used occasionally, and they have the advantage of simplicity and economy in implementation. As remarked earlier, the various digital controllers in the reference, forward, feedback, and error-signal paths may be considered as a conglomerate of large-scale-integrated (LSI) microcontrollers. Hence, the physical realization of the various pulse-transfer-functions for the desired performance-index may be categorized as a *digital programming approach* in modern control theory. Future developments in the theory of "digital filters" may furnish a possible approach toward a theoretical study.

3.3.11 Pulse-Width Modulation (15.75 KHz, Used in Headwheel 480-Hz ac Carrier-Servo). The high repetition-rate (2T/525) pulse-width modulation (PWM) used in the 240-Hz (=T/8) headwheel servo in conjunction with the three-phase amplitude-modulated (AM) ac carrier-servo, is least susceptible to tape and system noise, since the error-data are carried exclusively by the varying pulse-width of the AFC-controlled horizontal pulse-train generated in the system. The noise content in the combined phase and velocity error-data is conveniently eliminated by limiting the random amplitude variations due to noise, before the amplitude-modulated carrier-servo becomes effective along with the compensating filter-network. The modulation or intelligence is easily recovered by the ideal motor-demodulation after converting the PWM into an AM by the simple technique of a low-pass filter (for the removal of the high-frequency 15.75-KHz pulse components) and an *ac carrier-servo*. Since PWM is a nonlinear sampling technique, standard linear z-transform techniques are not applicable. From a theoretical standpoint, either the classical Lyapounov *direct method* or the *modern state-space techniques* may be applied to this type of nonlinear sampled-data feedback control system. Nelson and Pushkin had earlier used a classical approximation technique using the z-transforms, as applied to the limit-cycle, representing the sustained low-amplitude periodic oscillations in the system due to the nonlinear element. The method of application by the first two methods (the Modern State-Transition technique and the Lyapounov Direct Method) will be illustrated with a single-loop unity feedback system. One nonlinear element will be included to act on the clamped error-signal in the second case.

3.3.12 Modern State-Transition Technique. The modern basic *unified state-variable method* of application using integrators, amplifiers and summing devices is first illustrated with a very simple sampled-data open-loop system, before the complex problem of the PWM is interpreted with an example.

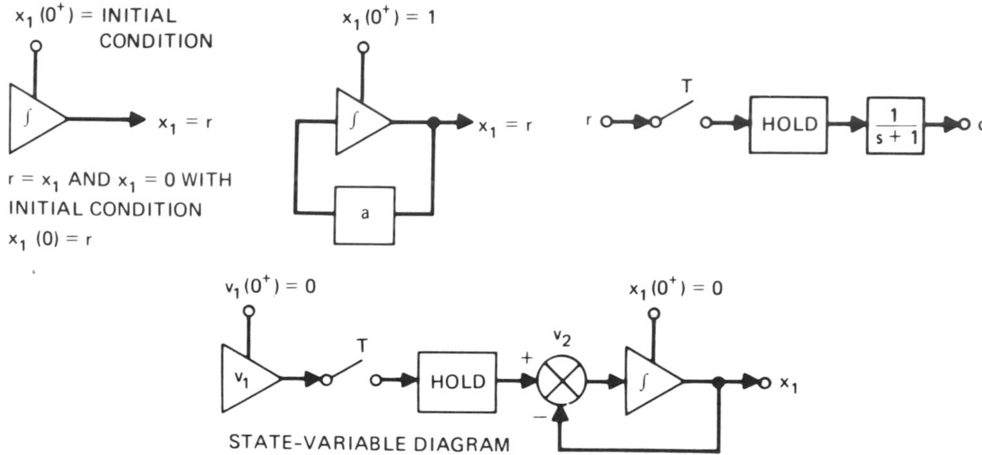

The outputs of the integrators denote the chosen state-variables, and the state-variable diagram is derived from the characteristic equation of this system. The step-function input $[r = r_0 u(t)]$ is represented by an integrator.

The ramp, $(r = r_0 + r_1 t)$, is represented by two integrators in cascade with initial conditions $x_1(0) = r_0$, $x_2(0) = r_1$, and state-variables, $\dot{x}_1 = r_1 = x_2$ and $\dot{x}_2 = 0$.

The exponential function $(r = r_0 e^{at}$ for $\geqslant 0)$ is denoted by a feedback integrator-loop. Its initial condition is $x_1(0) = r(0) = r_0$, and the state-variable is given by: $\dot{x}_1 = ar = ax_1$.

For the simple overall open-loop system in the diagram the state-variable diagram with a unit-step input takes the form shown. The initial conditions are measurable quantities. (In the design of optimal systems, it is essential that all the state-variables be accessible for measurement and observation.)

The first-order differential equations with the input, output, and the *state-variables* (*which determine the dynamic behavior of the system*) are denoted by an A-matrix. The derivative $\dot{v}$ depends on only the current state of the system (and not on its past history).

$$\begin{aligned} \dot{v}_1 &= 0 \\ \dot{v}_2 &= 0 \\ \dot{x}_1 &= v_2 - x_1 \end{aligned} \quad \equiv \overline{A} = \begin{bmatrix} 0 & 0 & 0 \\ 0 & 0 & 0 \\ 0 & 1 & -1 \end{bmatrix}$$

where overline denotes matrix.

The *state-transition equations* that specify the *initial conditions* during each transition of the state-variables at the sampling instants are then put in the following B-matrix form.

$$\begin{aligned} v_1(nT^+) &= v_1(nT) \\ v_2(nT^+) &= v_2(nT) \\ x_1(nT^+) &= x_1(nT) \end{aligned} \equiv \overline{B} = \begin{bmatrix} 1 & 0 & 0 \\ 1 & 0 & 0 \\ 0 & 0 & 1 \end{bmatrix}$$

A linear stationary sampled-data control system is described by the state-variables in linear first-order differential equations of the vector-matrix form:

$$\dot{\overline{v}}(\lambda) = \overline{A}\overline{v}(\lambda) \tag{1}$$

where

$\lambda = (t - nT), 0 < \lambda \leqslant T$

and state-vector,

$$\overline{v}(\lambda) = \begin{bmatrix} \overline{r}(\lambda) \\ \overline{c}(\lambda) \end{bmatrix}$$

The initial conditions for the state differential equation in the vector-form is given by:

$$\overline{v}(nT^+) = \overline{B}\overline{v}(nT)$$

The square-matrices $\overline{A}$ and $\overline{B}$ can be normally written from the state-variable diagram.

Taking the Laplace-transform of Eq. 1:

$$V(s) = [s\overline{I} - \overline{A}]^{-1} v(0+) \tag{2}$$

The inverse transform of Eq. 2 will give the solution to the state differential equation (Eq. 1) as:

$$\overline{v}(\lambda) = \overline{M}(\lambda)\overline{v}(0+) \tag{3}$$

$\overline{M}(\lambda)$ is termed the overall *transition-matrix*:

$$\mathcal{L}^{-1}(s\overline{I} - \overline{A})^{-1}$$

where $\mathcal{L}^{-1}$ = inverse Laplace transform.

At the sampling instant, $t = (n + 1)t$:

$$\overline{v}(\overline{n + 1T}) = \overline{M}(T)\overline{v}(nT)^+$$

$$\overline{v}(t) = \overline{M}(t - nT)\overline{B}\overline{v}(nT) \tag{4}$$

$$\overline{M}(t) = \mathcal{L}^{-1}[s\overline{I} - \overline{A}]^{-1} = \begin{bmatrix} 1 & 0 & 0 \\ 0 & 1 & 0 \\ 0 & 1 - e^{-t} & -e^{-t} \end{bmatrix} \quad \text{Since } \overline{I} = \begin{bmatrix} 1 & 0 & 0 \\ 0 & 1 & 0 \\ 0 & 0 & 1 \end{bmatrix} \text{ and}$$

$$\mathcal{L}^{-1}\left[\frac{1}{(s+1)}\right] = e^{-t} \quad \text{and} \quad \mathcal{L}^{-1}\left[\frac{1}{s(s+1)}\right] = 1 - e^{-t} \tag{5}$$

The dimension 3 X 3 depends on the number of state-variables in the system: v_1, v_2, x_1. Then, an N-matrix is defined to determine the state-variables of the system at any instant during $nT < t \leqslant (n+1)T$ by the following equation:

$$\overline{N}(t - nT) = \overline{M}(t - nT)\overline{B} \tag{6}$$

$$\overline{N}(t) = \overline{M}(t)\overline{B} = \begin{bmatrix} 1 & 0 & 0 \\ 1 & 0 & 0 \\ 1 - e^{-t} & 0 & e^{-t} \end{bmatrix}$$

$\overline{N}(T)$ is partitioned according to the dimensions of $\overline{v}$ and $\overline{x}$.

$$\text{Square-matrix } \overline{N}_1(T) = \begin{bmatrix} 1 & 0 \\ 1 & 0 \end{bmatrix}; \overline{N}_2(T) = e^{-T}; \overline{N}_3(T) = [(1 - e^{-T}) \;\; 0]$$

From the definition of the z-transform and Eqs. 4 and 5:

$$\overline{V}(z) = [1 - z^{-1}\overline{N}(T)]^{-1}\overline{v}(0) \tag{7}$$

$$\overline{v}(nT) = z^{-1}[1 - z^{-1}\overline{N}(T)]^{-1}\overline{v}(0) \tag{8}$$

Equation 7 gives the general solution to compute the system state-variables at successive sampling instants.

$$1 - z^{-1}\overline{N}_1(T) = \begin{bmatrix} (1 - z^{-1}) & 0 \\ -z^{-1} & 1 \end{bmatrix}$$

Now,

$$\overline{V}(z) = \begin{bmatrix} \overline{R}(z) \\ \overline{C}(z) \end{bmatrix} \text{ and } \overline{v}(0) = \begin{bmatrix} \overline{r} \;\; (0) \\ \overline{c} \;\; (0) \end{bmatrix}$$

Equation 7 takes the form:

$$\begin{bmatrix} R(z) \\ C(z) \end{bmatrix} = \begin{bmatrix} [1 - z^{-1}\overline{N}_1(T)]^{-1} & 0 \\ z^{-1}[1 - z^{-1}\overline{N}_2(T)]^{-1}\overline{N}_3(t)[1 - z^{-1}\overline{N}_1(T)]^{-1} & [1 - z^{-1}\overline{N}_2T]^{-1} \end{bmatrix} \begin{bmatrix} \overline{r}(0) \\ \overline{c}(0) \end{bmatrix}$$

$$C(z) = z^{-1}[1 - z^{-1}\overline{N}_2(T)]^{-1}\overline{N}_3(T)[1 - z^{-1}\overline{N}_1(T)]^{-1}\overline{c}(0) \tag{9}$$

Substituting the partitioned N-matrices in Eq. 9,

$$C(z) = \frac{z(1 - e^{-T})}{(z - 1)(z - e^{-T})} \tag{10}$$

This result is said to be obtained by the *unified approach of the state-transition analysis.* Although the method appears unduly complex for a simple problem, its utility is recognized only in general applications of nonlinear sampled-data control problems, pulse-width modulation, multirate sampled-data systems, optimal control, and time-varying systems.

3.3.13 Pulse-Width Modulation: Lyapounov's Direct Method. The classical Lyapounov Second (or Direct) Method was used by Kadota and Nelson to define the stability of a nonlinear control system with pulse-width modulation (PWM). The method is illustrated with a simple lag-type PWM example, which in fact consists of two types of nonlinearities: (1) saturation of pulse-width and (2) fixed signal-amplitude.

The error lattice-function shown is denoted by:

$$e^*(t) = \sum_{n=-\infty}^{\infty} A\{u(t - nT - \overline{mn - 1}) - u(t - \overline{n + 1}T)\}(\text{Sign})e(nT)$$

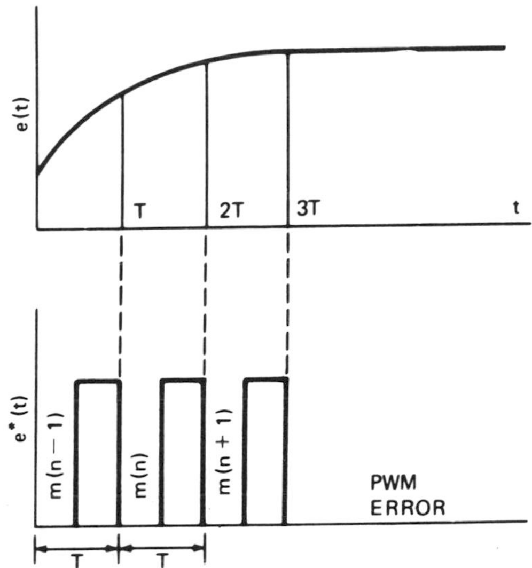

The width of the k*th* pulse is given by

$$m^1(k) = T = m(k) = a[e(kt)] \text{ for } m^1k \leqslant T \text{ and } = T \text{ for } m^1k > T$$

If the plant is denoted by $\dfrac{K}{s - 2}$ for a simple illustration of the technique, the state-variable form is

$$\dot{x} = x + b \tag{1}$$

where

$$x = c(t)$$

$$\dot{x} = \frac{dc(t)}{dt}$$

$$Ke^*(t) = b$$

The general solution of the first-order differential equation is given by:

$$x(t) = e^{2(t-t_0)}x(0) + \int_{t_0}^{t} e^{2(t-\lambda)}b \cdot d\lambda \tag{2}$$

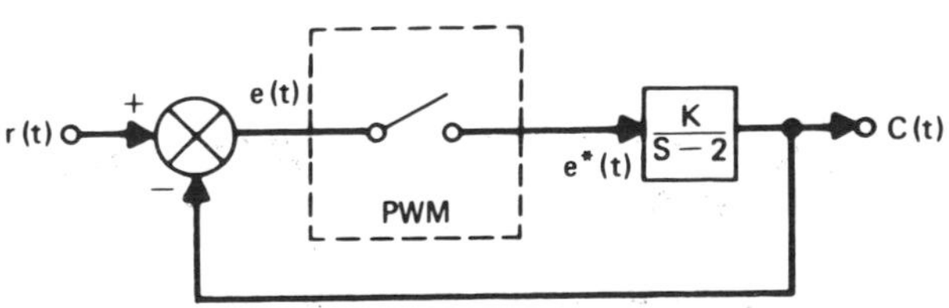

For PWM, the time element is quantized to express the above differential equation as a difference equation:

$$t = nT + m(n) \text{ for } t_0 = nT$$

$$t = (n + 1)T \text{ for } t_0 = nT + m(n)$$

Substituting these quantized time elements in the above solution (2),

$$x(nT + m) = e^{2m(n)}x(nT)$$

$$x(\overline{n + 1T}) = \{e^{2[T - m(n)]}\}x[nT + m(n)] + \int_{nT+m(n)}^{(n+1)T} e^{2(\overline{n+1T} - \lambda)}b \cdot d\lambda$$

Combining these two equations,

$$x_{n+1} = e^{2T}x_n + \frac{K}{2}(1 - e^{2m_n^1}) \operatorname{sgn}(x_n)$$

Now, a suitable Lyapounov function is chosen:

$$V(x) = x_n^2$$

$$\Delta V = x_{n+1}^2 - x_n^2 = x_n^2 \left\{ \left[e^{2T} + \frac{K(1 - e^{2m_n^1})}{2|x_n|} \right]^2 - 1 \right\}$$

The *sufficient condition for stability* according to Lyapounov's direct method is that the derivative (or first difference) must be *negative-definite*, while the function chosen is *positive-definite.*

$$\left[e^{2T} + \frac{K(1 - e^{2m_n^1})}{2|x_n|} \right]^2 - 1 < 0 \qquad \text{for all } x \neq 0.$$

The expression in brackets is monotonic during small intervals, so it can be reduced to simpler solutions, with positive constant K_1 defining the intervals as follows:

$$|e^{2T} - KK_1| < 1$$

$$\left|e^{2T} + KK_1 \frac{1 - e^{2T}}{2T}\right| < 1$$

$e^{2T} < 1$. Hence, 2 must be a negative quantity (< 0) for a positive T. That is, for stability the system transfer-function must be $\frac{K}{s + 2}$ and not $\frac{K}{s - 2}$ as assumed. In general, complex systems can be reduced to a set of first-order differential equations, and the preceding method is then applicable in each case. For less complex and ordinary nonlinear systems, the Lyapounov V-function chosen may be of a more sophisticated character like a circle, ellipse, etc. In the case of the PWM in the Q-CVTR, since a 480-Hz ac carrier modulation is simultaneously involved, the amplitude of the 15.75-KHz pulse-width modulation waveform is accompanied by the 480-Hz chopper amplitude-modulation (AM). The analysis will be therefore much more complex if AM is treated simultaneously. Hence, for analytical purposes, the AM-element can be separately treated as an ac carrier-servo.

3.3.14 AC Carrier Servo System Used in Headwheel Servo. The three-phase HAS-motor of the Q-CVTR is controlled by a three-phase *suppressed-carrier ac servo.* The

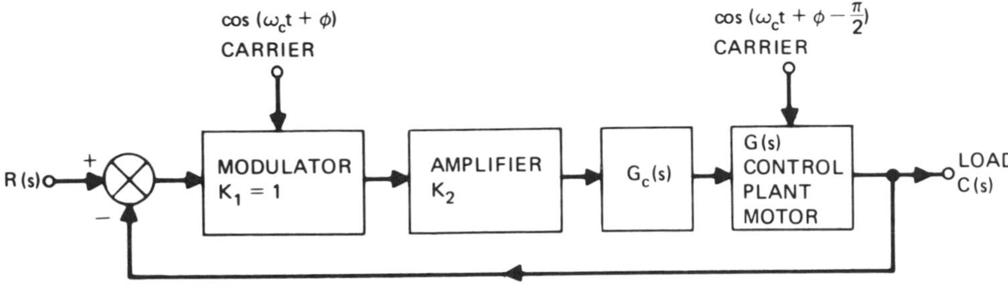

Fig. 3-10. The ac carrier servo.

carrier modulation in each phase is simultaneously accompanied by 15.75 KHz PW-modulation. Subsequent transformation of the PWM-error into an error amplitude-modulation is accomplished by means of a dual-purpose low-pass filter network, just prior to the ideal 480-Hz carrier-demodulation process by the motor itself—ideal since the motor accomplishes the function of a *harmonic-filter* simultaneously. Incidentally, the low-pass filter will act as the *ac compensation.* Lyapounov's Direct Method as applicable to a PWM nonlinear system was discussed in the preceding section.

The general procedure of analysis of an ac carrier control system is now illustrated by an exclusive ac carrier-servo example (Fig. 3-10). The motor acts as a perfect demodulator with the prescribed static and dynamic relationships between the driving voltage-envelope and the output shaft-position, provided the carrier-frequency and the magnitude remain constant. In the actual headwheel servo system, the three-phase control-plant, by way of the *rotating magnetic-field* of the stator three-phase drive, eliminates the need for the supply of a separate carrier signal at the motor (demodulation) stage. In the analysis shown for Fig. 3-10, for simplicity a two-phase servo motor (using a separate ac carrier excitation) is considered; in the three-phase case, compensation is required in each phase. The transfer-function of the motor is assumed as: $G(s) = \dfrac{7.2}{s(s + 12.5)}$. Equivalent dc transfer-function of the amplifier and the following compensation network $G_{dc}(s)$ are given. The problem is to find the closed-loop transfer-function and the transient-response to a step input.

In this particular system, since the network $G_c(s)$ is required before the demodulation (motor), it should be of the ac type. Networks of this category are usually called *class-2 compensation.* (Networks used after an exclusive demodulator, prior to motor, will be of the dc type (viz., lag, lead, lag-lead, etc.). In general an ac carrier-servo is conventionally synthesized as if it were a dc system with a separate demodulator, and then an ac network is easily derived from the dc network by means of the frequency-transformation, $s = j\omega = j(\omega_c - \omega_m)$. The formulation of the compensation is a result of the fundamental relationships in carrier system-analysis, which was briefly introduced in chapter 1.

$$G_c(s) = \frac{K_1 K_2}{2} \, \mathrm{Re}_j[G(j\omega_m + j\omega_c)e^{-j(\phi_c - \phi_m)}]$$

The corresponding relationship between the gain and the phase of an ac network and its equivalent dc network are shown. It must be noted that the theoretical implication is that the transformation is physically realizable only as a restricted case of *approximation.* The procedure involves the choice of an ac network with a gain-characteristic of the equivalent dc network, ω_c-units to the right and reflecting this characteristic about the ω_c-ordinate. It will have a phase-characteristic as derived by translating the phase-characteristic of the equivalent dc network, ω_c-units to the right and reflecting this characteristic about both the frequency-axis and the ω_c-ordinate.

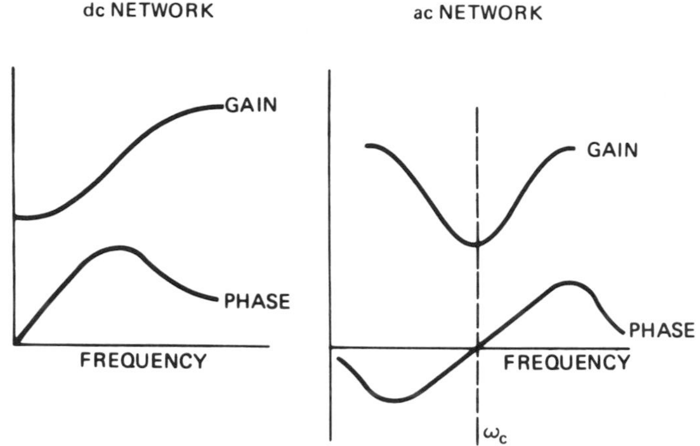

As an example, a dc lead-network and gain K are chosen in this problem as

$$G_{dc}(s) = \frac{43.5(.1s + 1)}{(.08s + 1)} \equiv \text{ac-equivalent, } G_c(s)$$

$$G_{dc}(s) = K\,\frac{k(\frac{L}{R_1}s + 1)}{(\frac{kL}{R_1}s + 1)}$$

$$G_c(j\omega) = \frac{k(j\omega T + 1)}{(kj\omega T + 1)}$$

R_2 (CHOSEN)

$K = 43.5$

$R_1 = \frac{R_2 k}{1 - k}$

$L_1 = \frac{R_1}{2}T$

CLASS-2 COMPENSATION

R_2

R_1

$L = \frac{L_1}{2}$

$C = \frac{1}{(\omega_c)^2 L}$

$$G_c(s) = \frac{LCs^2 + R_1Cs + 1}{LCs^2 + (R_1 + R_2)sC + 1}$$

If L = 1 henry and R = 20 ohms, then

$$C = 1/(377)^2 L = 1/(377)^2 \text{ farads, and } R_2 = R_1(1 - k)/k$$

where k from $G_{dc}(s)$ is found to be 1.25(=0.1s/0.08s).

$$G_c(s) = \frac{s^2 + 20s + (377)^2}{s^2 + 25s + (377)^2} \approx \frac{(s + 10 + j377)(s + 10 - j377)}{(s + 12.5 + j377)(s + 12.5 - j377)}$$

$G_c(s) = \frac{K_2}{2}$ [G(s + j377) + G(s - j377)] from ac-carrier system analysis, with amplifier gain K_2 and absence of any NL saturation, and modulator $K_1 = 1$. After substituting the ac equivalent expression for the given dc network in the preceding formula, for signal frequencies below 500 rad/sec, $G_c(s)$ is approximated to the following expression:

$$G_c(s) \approx K \frac{(s + 10)}{(s + 12.5)}$$

The time-constants are identical to those given for $G_{dc}(s)$, but the ac transformation has brought up a new gain-factor K = 54.4.

$$G_c(s) = \frac{K(s + 10)}{(s + 12.5)} = \frac{1.25 \times 43.5(s + 10)}{(s + 12.5)} = \frac{54.4(s + 10)}{(s + 12.5)}$$

With unity feedback, the closed-loop transfer function:

$$\frac{C(s)}{R(s)} = \frac{G(s)}{1 + G(s)} = \frac{391(s + 10)}{s^3 + 25s^2 + 547s + 3910}$$

For step-input,

$$C(s) = \frac{391(s + 10)}{s(s^3 + 25s^2 + 547s + 3910)}$$

By partial-fractions and residues,

$$C(s) = \frac{1}{s} - \frac{0.02}{s + 9.8} + \frac{0.53\angle 157^\circ}{s + 7.6 - j18.43} + \frac{0.53\angle -157^\circ}{s + 7.6 + j18.43}$$

The time-response or transient-response is then obtained by directly taking inverse-Laplace transforms of the above terms; the response is plotted as shown.

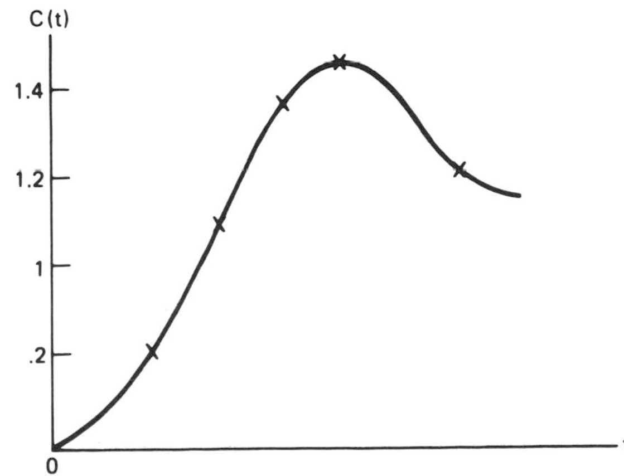

$$c(t) = 1 - 0.02e^{-9.8t} + 1.06^{-7.6t} \cos (18.43t + 157^\circ)$$

3.3.15 Effects of Carrier-Shift in ac Carrier Servo. During the advance development stages of the high-precision line-lock headwheel servo system, beat-frequency problems and stability problems based on these were encountered due to minor shifts in carrier-frequency from time to time.

The behavior of the carrier system transfer-function $G_c(j\omega)$ with drift in carrier has been investigated in terms of the drift of the poles and zeros of $G_c(s)$ by E. W. Mehelich and G. J. Murphy. As shown in Chapter 1,

$$G_c(s) = \frac{K_m K_d}{4} [G(s + j\omega_c)e^{-j(\phi_c - \phi_m)} + G(s - j\omega_c)e^{j(\phi_c - \phi_m)}]$$

The poles migrate parallel to the imaginary-axis and do affect the compensation used. According to their investigation, the effect of zero-migration is not considered serious as far as class-2 ac compensation networks are concerned. Thus, variations in carrier-frequency do result in variation in performance, due to (1) the derivation of the class-2 compensation from its designed mode of operation and (2) the effect of frequency-drift on the servo output stage. The results of Mehelich and Murphy's investigation into the variation of the closed-loop parameters, gain, and time-constant are plotted against frequency. For the class-2 compensation, with an equivalent dc transfer-function, namely, $G_{dc}(s) = \frac{k(1 + Ts)}{(1 + kTs)}$, $0 < k < 1$, the effect of the carrier frequency drift on the overall frequency transfer-function decreases as $k \longrightarrow 0$ and $G_{dc}(s) \longrightarrow k(1 + Ts)$.

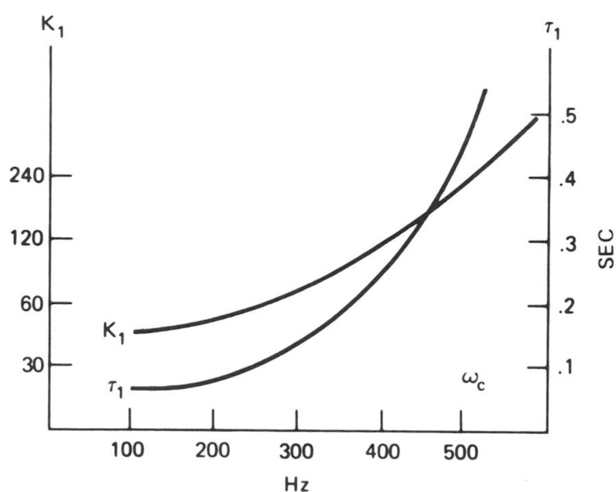

Therefore, as a solution of compromise, it is desirable to choose a pole-zero ratio as large as possible, with a bridged-T RC-network, to obtain a low sensitivity to carrier-drift, while limiting k to as small a value as possible for reduced signal/noise ratio. As a solution to the carrier-drift, a technique of adaptive control has been used with an auxiliary servo-loop, comprised of a frequency-detector, amplifier, and servo motor, for adjusting the class-2 compensation automatically. In the headwheel carrier servo, the problem has been successfully solved by a simpler means—viz., the technique of locking the phase of the carrier (and hence its frequency) to the phase-locked 240-Hz tonewheel sampling-frequency in the headwheel sampled-data control system. The ac carrier is locked in phase as a second-harmonic (480 Hz) of the phase-locked tone-wheel sampling-frequency.

3.3.16 General Comments on ac Carrier Servo System.

1. The use of ac tachometers (with rate transfer-function of differentiation, s) as feedback elements in minor loops is quite popular with ac carrier servo systems to decrease the effective time-constant of the motor for improved system stabilization.
2. While the integration of a carrier-envelope could be accomplished by T-networks in forward or feedback paths, the integration of a demodulated signal can be achieved with an inertia introduced by an air-bearing and an appropriate compensation. The effects of the nonlinearities in the driving motor are thus minimized in the case of the Q-CVTR.
3. The modulators (synchros, choppers, etc.) produce *position-modulation* on the input carrier and a *velocity modulation* on a carrier shifted in *phase-quadrature.* The position and velocity signals recovered by the demodulator process are a function of the demodulator reference-phase. This can be optimized for the null of the velocity-error.

3.3.17 Nonlinearity of Hysteresis in Headwheel Servo. The output-input nonlinear characteristic due to the hysteresis in the *asynchronous synchronous* (*HAS*) *motor* is illustrated in the accompanying diagram. The electrical hysteresis-effect occurs as a result of the inherent *flux-density versus magnetization characteristic* of the magnetic-steel used for the rotor. The hysteresis shows up as a changing fluctuation (phase-jitter) in the control signal as the corrective torque-changes operate in either direction about the PLL *set-point.* The characteristic of this nonlinear-element is a function of a *derivative-input.* The effect is considered identical to that of the *backlash* or mechanical hysteresis, which is defined as the "difference in motion between an increasing and decreasing output." (The property of mechanical backlash is further treated with an example under the vacuum-guide position control system, where the mechanical-linkages or gear-trains introduce free-play or backlash as an inherent characteristic.) A slow-speed, coarse, *on-off control system* using a relay or a con-

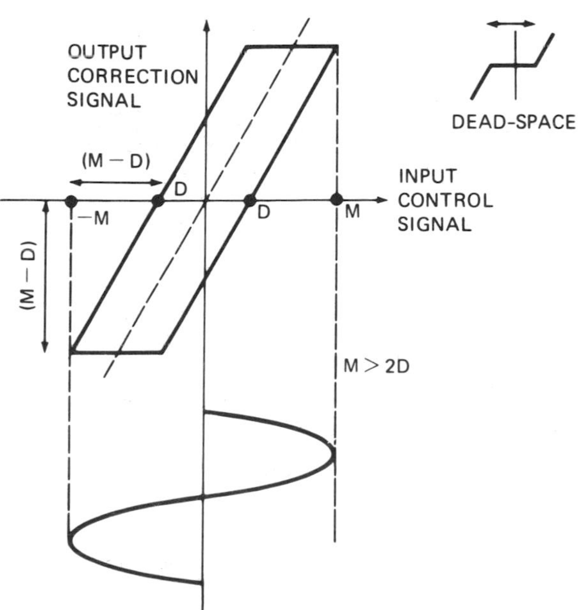

tactor also exhibits a hysteresis characteristic of a slightly different form, which is bistable in character.

The hysteresis element may have an input waveform that is different to a sinusoid. If the input is a square-wave, the hysteresis band is crossed instantly on each half-cycle, the output square-wave is in phase with the input, and the resulting transfer characteristic will then become a *dead-zone* instead of a continuous jitter of some constant value depending on the direction of the motion. That is, within a certain range of minimum error, the feedback control system is practically ineffective. The analog simulation technique used in the analysis of such systems calls for an integrator since its output, with a very rapid response and an appropriate high gain, will remain constant when its input is zero. In general, hysteresis will cause sustained oscillations of instability in an otherwise well-damped control system, and these oscillations will vanish (with a minute residual-error or a minute oscillatory transient-response) only if the linear-gain of the loop is low enough to keep, for example, the *Nyquist plot* and the $-1/G_d$ *describing-function curve* without intersecting each other. Where high gains are encountered, in practice a *90° phase-margin* is allowed to prevent low-level oscillations by means of a minor feedback-loop around the hysteresis (that is, a loop parallel to the rest of the system).

The hysteresis characteristic of the motor causes a phase-lag that is a function of the input amplitude, but it is independent of the frequency. The combined effect of hysteresis and saturation is a reduction in gain at very small and very large input amplitudes, as compared to that at intermediate amplitudes. The nonlinear differential equations involved are of considerable complexity, and they are difficult to solve.

As stated in chapter 1, the describing-function approach is the most common technique applicable to a nonlinear element of the character of hysteresis. Unlike the conventional phase-plane technique, which is applicable to a nonlinear second-order system, the describing function technique is applicable to higher-order systems. If the input is given by a sinusoidal signal ($m = M \sin \omega t$), the output waveform is expressed in Fourier series as

$$n(\omega t) = \frac{A_0}{2} + \sum_{k=1}^{k=\infty} A_k \cos k\omega t + \sum_{k=1}^{k=\infty} B_k \sin k\omega t$$

where

$$A_k = \frac{2}{T} \int_{T/2}^{T/2} n(\omega t) \cos a(\omega t) d(\omega t), \text{ a is even}$$

$$B_k = \frac{2}{T} \int_{T/2}^{T/2} n(\omega t) \sin b(\omega t) d(\omega t), \text{ b is odd.}$$

If $n(\omega t) = -n(-\omega t)$, $A_k = 0$ for an odd function.
If $n(\omega t) = n(-\omega t)$, $B_k = 0$ for an even function.
For the describing-function, A_1 and B_1 are required for only the fundamental frequency. For hysteresis, it can be shown that:

$$N(M) = \frac{K}{M} \sqrt{A_1^2 + B_1^2}, \angle \tan^{-1} \frac{A_1}{B_1}$$

where K is the positive-slope of the hysteresis characteristic.

$$A_1 = \frac{2D}{\pi M}\left(\frac{2D}{M} - 2\right) M$$

and

$$B_1 = \frac{1}{\pi}\left[\frac{\pi}{2} - \sin^{-1}\left(\frac{2D}{M} - 1\right) + \left(\frac{2D}{M} - 1\right) \cos \sin^{-1}\left(\frac{2D}{M} - 1\right)\right] M$$

From the amplitude and phase characteristics given by A_1 and B_1, it is seen that the describing-function is a function of the ratio $\frac{D}{M}$, and phase-lag occurs at low amplitudes only. Because of this theoretical explanation, low-amplitude jitter can be said to be an inevitable property of hysteresis. The basic headwheel sampled-data control system with its phase and velocity loops exhibits a speed-regulation jitter of the order of ± 1 μs as a result of the cummulative effect of motor-hysteresis and several other nonlinear elements in the system, such as saturation, pulse-width modulation, and stiction (static-friction at zero-velocity). The phase-plane "limit-cycle" is an attribute of this phase-jitter component in nonlinear control systems.

3.3.18 Determining a Compensating Network for Best Noise-Immunity by Statistical Method: A Typical Digital Filter. The determination of a compensating network for the best noise-immunity is illustrated by means of an example for a linear approximation of the continuous part of the control system (Fig. 3-11). The problem of minimizing the integral-squared-error, subject to some *constraint*, can be stated, for example, in the following way:

The actual compensation, $G_{cl}(s)$, that minimizes the integral-squared value I_f of a signal m(t) applied to a fixed element (or a transfer-function), subject to the constraint that the integral-squared-error I_e shall not exceed a specified upper limit M, is required. The slope of the input ramp-function = Ω_i.

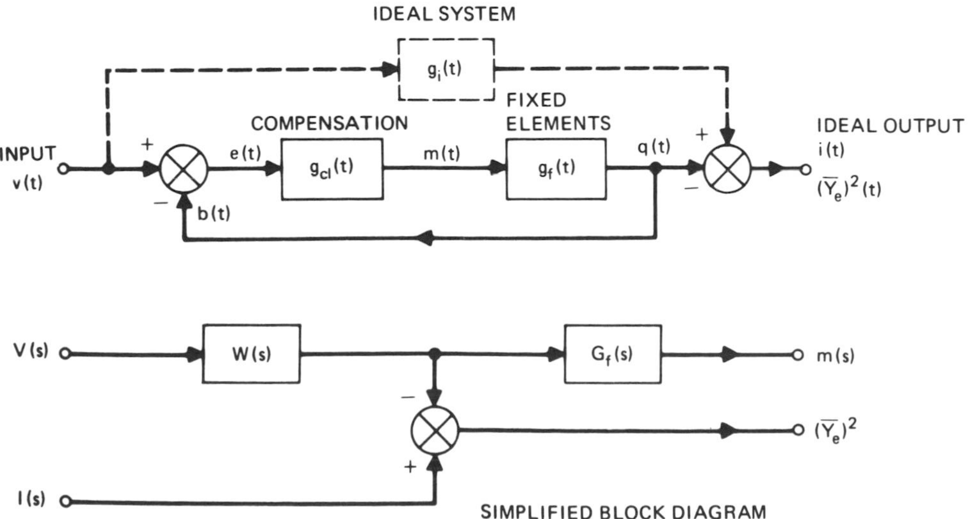

Fig. 3-11. Feedback control system, statistical representation.

Given: Fixed element $G_f(s) = \frac{K}{s^2}$; Input signal $v(t) = \Omega_i\ \delta_{-2}(t)$; Ideal output = $i(t) = v(t)$; $K = 100$; $\Omega_i = 5$ V/sec; $M = 0.025$; $H_f(s) = 1$ with unity-feedback.

$$\overline{Y}_e^2 = I_f \int_{-\infty}^{\infty} m^2(t)\, dt; \text{ and } I_e = \int_{-\infty}^{\infty} \overline{Y}_e^2(t)\, dt$$

Stochastic signals, as explained in chapter 1, are commonly characterized by *auto- and cross-correlation functions.* But this is not unique, since, in time-domain representation of the integral-squared-error, functions *auto-translation* and *cross-translation* are more convenient.

$$I_{vv}(\tau) = I_{11}(\tau) \triangleq \int_{-\infty}^{\infty} x_1(t)x_1(t+\tau)\, dt$$

$$I_{vi}(\tau) = I_{12}(\tau) \triangleq \int_{-\infty}^{\infty} x_1(t)x_2(t+\tau)\, dt$$

These functions simplify the expression of the definite-integrals of the products of the transient signals.

Then the manipulation of the transforms of these expressions in frequency-domain, on the basis of the **Weiner-Hopf criterion**, is identical to that of the *auto and cross spectral-density functions.*

In this particular problem, the classical **technique of Lagrange** (encountered in the *calculus of variations*) for minimizing a function subject to one or more constraints is applied. A synthetic function, $F = \overline{Y}_e^2 + \rho I_e$, where ρ is *the Lagrangian multiplier*, is minimized for determining the required compensation in terms of ρ.

$$v(t) = i(t) = \text{ideal output}$$

The transforms of the following translation-functions are equal.

$$I_{vi}(s) = I_{vv}(s) = I_{ii}(s) = I_{iv}(s) \tag{1}$$

$$\text{Input } v(t) = \Omega_i\ \delta_{-2}(t) \tag{2}$$

$$v(s) = \Omega/s^2$$

Now auto translation,

$$I_{vv}(s) = V(s)V(-s) = \frac{\Omega}{s^2} \times \frac{\Omega}{(-s)^2} = \frac{\Omega^2}{s^4} \tag{3}$$

From the simplified system diagram,

$$\overline{Y}_e^2 = I_f(s) = \frac{1}{2\pi j} \oint^{\infty} \left[V(s)W(s)\frac{1}{G_f(s)}\right]\left[V(-s)W(-s)\frac{1}{G_f(-s)}\right] ds \tag{4}$$

ρI_e in function $F = [\overline{Y}_e^2 + \rho I_e]$ is obtained as follows:

$$\rho I_e = \frac{\rho}{2\pi j} \oint I_{yy}(s)\, ds \tag{5}$$

where $I_{yy}(s)$ is the inverse of the auto-translation function of the error signal

$$= \frac{\rho}{2\pi j} \oint [W(s)W(-s)I_{vv}(s) - W(-s)I_{vi}(s) - W(s)I_{iv}(s) + I_{ii}(s)] \, ds$$

$$= \frac{\rho}{2\pi j} \oint [W(s)W(-s) - W(-s) - W(s) + 1] I_{vv}(s) \, ds \qquad \text{(from Eq. 1)}$$

from Eq. 3,

$$\rho I_e(s) = \frac{\rho}{2\pi j} \oint V(s)[1 - W(s)] \cdot V(-s)[1 - W(-s)] \, ds \tag{6}$$

Since $F(s) = I_f(s) + \rho I_e(s)$, the equivalent Weiner-Hopf equation in the frequency-domain is obtained for the system, by determining the coefficient of $W(-s)$ in $F(s)$. Taking the coefficient of $W(-s)$ is equivalent to taking the time-domain integral. The concept involved is that the time-function, which is zero for negative-time, will have the poles in the left half-plane.

It can be shown that the coefficient of $W(-s)$ is obtainable from the expression:

$$\psi(s)\,\Delta(s) - \Gamma(s) \tag{7}$$

This is the transform of the **modified Weiner-Hopf equation** to make it solvable in frequency-domain, where

$$\psi(s) = W(s);\ \Delta(s) = I_{vv}(s)\left[\frac{1}{G_f(s)G_f(-s)} + \rho\right];\ \Gamma(s) = I_{vv}(s)\rho$$

from Eq. 4 and F(s)   from Eq. 6

$\psi(s)$ corresponds to the system weighting function

$\Delta(s)$ and $\Gamma(s)$ correspond to the auto- and cross-correlation (or translation) functions, respectively.

Now, the *explicit* solution-formula for the Weiner-Hopf equation is obtained as follows for the transfer-function of the compensation $G_c(s)$:

If $G_c(s) = W(s)$ that minimizes $F(s)$ without constraint:

Then,

$$G_c(s) = \frac{\left.\dfrac{\Gamma(s)}{\Delta^-(s)}\right|_+}{\Delta^+(s)} \tag{7}$$

$$\Gamma(s) = I_{vv}(s)\rho = V(s)V(-s)\rho = \frac{\rho\Omega^2}{s^2(-s)^2} \tag{8}$$

$$\Delta(s) = I_{vv}(s)\left[\frac{1}{G_f(s)G_f(-s)} + \rho\right] \text{ where } I_{vv}(s) = V(s)V(-s)$$

$$= \frac{\Omega^2}{s^2(-s^2)}\left[\frac{s^2(-s)^2}{K^2} + \rho\right] = \frac{\Omega^2}{K^2}\,\frac{(s^4 + K^2\rho)}{s^2(-s)^2} \tag{9}$$

For convenience in algebraic manipulation, let $K^2\rho = 4a^4$.

$$\Delta(s) = \frac{\Omega^2(s^4 + 4a^4)}{K^2(s^2)(-s)^2} = \frac{\Omega^2(\overline{s+a}+ja)(\overline{s+a}-ja)(\overline{s-a}+ja)(\overline{s-a}-ja)}{K^2(s^2)(-s)^2}$$

To determine the poles and zeros in the left half-plane: $\Delta(s) = \Delta^{+}(s)\,\Delta^{-}(s)$.

$$\Delta^{+}(s) = \frac{\overline{(s+a+ja)(s+a-ja)}}{s^2} \text{ and } \Delta^{-}(s) = \frac{\Omega^2}{K^2}\,\frac{\overline{(s-a+ja)(s-a-ja)}}{(-s)^2}$$

$$\frac{\Gamma(s)}{\Delta^{-}(s)} = \frac{\Omega^2 \rho}{s^2 \cancel{(-s)^2}} \cdot \frac{K^2}{\Omega^2} \cdot \frac{\cancel{(-s)^2}}{(\overline{s-a+ja)(s-a}-ja)} = \frac{4a^4}{s^2(s^2+2a^2-2as)}$$

$$= \frac{C_1}{s^2} + \frac{C_2}{s} + \frac{C^3 s + C_4}{(s^2+2a^2-2as)}, \text{ using partial fractions.}$$

There is no need to find C_3 and C_4, since the corresponding poles are located in the right half-plane.

$$C_1 = \left.\frac{\cancel{s}(4a^4)}{\cancel{s}(s^2+2a^2-2as)}\right|_{s=0} = 2a^2$$

$$C_2 = \frac{d}{ds}\left[\frac{4a^4}{(s^2+2a^2-2as)}\right]_{s=0} = \left.\frac{-(2s-2a)4a^4}{(s^2+2a^2-2as)^2}\right|_{s=0} = 2a$$

$$G_c(s) = W(s) = \left.\frac{\Gamma(s)}{\Delta^{-}(s)}\right|_{+} \Big/ \Delta^{+}(s) = \left[\frac{2a^2}{s^2} + \frac{2a}{s}\right] \Big/ \left[\frac{s^2+2a^2+2as}{s^2}\right]$$

$$= \frac{2a(s+a)}{(s^2+2as+2a^2)} \tag{10}$$

where $a = (K^2 \rho/4)^{1/4}$

Thus, the compensation network $G_c(s)$, in terms of the Lagrange multiplier, which minimizes the integral-squared-error without the consideration of the constraint, is obtained. But the actual compensation, $G_{c1}(s)$, used with the necessary ρ (or a) must satisfy the constraint that the minimum integral-squared-error $\overline{Y}_e^2$ should have an upper limit M (i.e., $I_e(t) \leqslant M$). For this purpose, Eq. 4 is solved first to obtain $I_f(t)$. Then, $I_e(t)$ for the requisite constraint can be derived.

$$I_f(s) = \frac{1}{2\pi j}\oint_{-\infty}^{\infty}\left[V(s)W(s)\frac{1}{G_f(s)}\right]\left[V(-s)W(-s)\frac{1}{G_f(-s)}\right] ds \tag{4}$$

where $W(s) = G_c(s)$

$$= \frac{1}{2\pi j}\oint \frac{\Omega^2}{K^2}\,\frac{\cancel{(-s)^2}\cancel{(-s^2)}}{\cancel{(-s)^2}\cancel{(-s^2)}}\,\frac{2a(s+a)}{(s^2+2as+2a^2)}\,W(-s)\,ds$$

$$= \frac{\Omega^2}{K^2}\,\frac{1}{2\pi j}\oint \frac{[2a(s+a)]\,[2a(-s+a)]\,ds}{(s^2+2as+2a^2)\quad \text{Denominator }(-s)} \cdots \text{ ie, Numerator} = -4a^2s^2 + 4a^4$$

$$I_f(t) = \oint_0^{\infty} e^2 dt = \frac{1}{2\pi j}\oint \frac{b_1\lambda^{2(n-1)} + b_2\lambda^{2(n-2)}}{a_0\lambda^n + a_1\lambda^{n-1} + a_2\lambda^{n-2} + \cdots}\,d\lambda$$

$$I_f(t) = \frac{\Omega^2}{K^2}\left[\frac{(-1)^{n+1}}{2a_0}\,\frac{\begin{vmatrix} b_1 & a_0 \\ b_2 & a_2 \end{vmatrix}}{\begin{vmatrix} a_1 & a_0 \\ a_3 & a_2 \end{vmatrix}}\right] = \frac{-\Omega^2}{2K^2}\,\frac{\begin{vmatrix} -4a^2 & 1 \\ 4a^4 & 2a^2 \end{vmatrix}}{\begin{vmatrix} 2a & 1 \\ 0 & 2a^2 \end{vmatrix}}$$

$$I_f(t) = \frac{-\Omega^2}{2K^2} \cdot \frac{3a}{2} \tag{11}$$

where Ω and K are given constants. That is, the smaller the value of *a* (which is proportional to $\rho^{1/4}$), the smaller the minimum-squared-error, I_e, which is subject to the given constraint, M(= .025 $m^2 \cdot$ sec). Now, Eq. 6 for $I_e(s)$ enables the determination of $I_e(t)$.

$$I_e(s) = \frac{1}{2\pi j} \oint [1 - W(s)]\,[1 - W(-s)]\,V(s)V(-s)\,ds \ldots \tag{6}$$

where $W(s) = G_c(s)$

$$W(s) = \frac{2a(s + a)}{s^2 + 2as + 2a^2}$$

$$[1 - W(s)] = \frac{s^2}{(s^2 + 2as + 2a^2)}$$

$$I_e(s) = \frac{1}{2\pi j} \oint \frac{\Omega^2 (s^2)(-s)^2\,ds}{[(s^2)(s^2 + 2as + 2a^2)]\,[(-s)^2(\overline{-s}^2 + 2a \cdot \overline{-s} + 2a^2)]}$$

$\longleftarrow$ Denominator $(-s)$ $\longrightarrow$

$$I_e(t) = \frac{\Omega^2(-1)^3}{2} \left| \begin{matrix} -\frac{1}{2a^2} & 1 \\ 0 & 2a^2 \end{matrix} \right| \bigg/ \left| \begin{matrix} 2a & 1 \\ 0 & 2a^2 \end{matrix} \right| = \frac{\Omega^2}{8a^3} \tag{12}$$

Since $I_e(t) \leqslant M(= .025)$,

$$\frac{\Omega^2}{8a^3} \leqslant 0.025$$

$$a^3 \geqslant \frac{5 \times 5}{8 \times .025} \qquad a = 5 \tag{13}$$

According to Eq. 11, it is a requisite condition that *a* should be the minimum possible quantity for the minimum-squared-error. Therefore, in the final result (Eq. 13) for *a*, the "greater than" sign is not permissible. Thus, *with the constraint*, the compensation network is given by the expression:

$$G_{c1}(s) = \frac{W(s)}{1 - W(s)} \frac{1}{G_f(s)} \qquad \text{from the simplified diagram.}$$

where $W(s) = G_c(s)$

$$G_{c1}(s) = \frac{2a(s + a)}{\cancel{(s^2 + 2as + 2a^2)}} \cdot \frac{\cancel{(s^2 + 2as + 2a^2)}}{\cancel{s^2}} \cdot \frac{\cancel{s^2}}{K}$$

$$\text{Actual compensation required} = \frac{2 \times 5(s + 5)}{100} = \underline{0.1(s + 5)}:$$

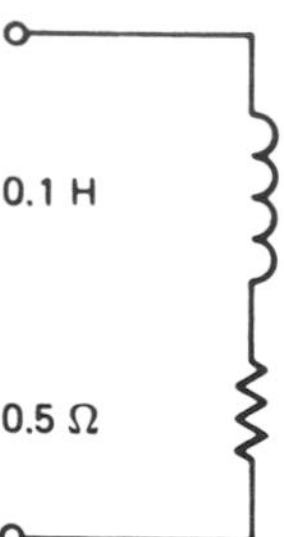

3.3.19 Adaptive (Self-Optimizing) Control System in Headwheel Servo for the Reproduction of Color from Videotape. One of the predominant random variables likely to disturb the Q-CVTR control systems is the phase of the 60-Hz system vertical information in the reference sync of the television system while a color or monochrome program is recorded or reproduced from the magnetic tape. The line 15.75-kHz information is in turn phased to 60 Hz. The studio sync generator, normally locked to the color frequency-standard 3579545 ± 5 Hz in the case of color recordings, is at times down-counted and locked to the power supply frequency during monochrome recordings; the sources of the recordings on a single network-tape may even be different regions. The power supply frequency is of course a random variable, and the headwheel control system under these conditions must be capable of adapting (or self-optimizing) itself to the changing statistics of the reference variables—that is, a changing environment.

The headwheel multirate sampled-data control system, with its third-order adaptive control and the carefully optimized time-constants of the compensating networks in the successive loops, is designed for automatically optimizing the phase of each preceding loop in the multirate system, for the final reproduction of color or monochrome signals from the tape at the requisite performance-index of a high order. The color automatic timing-control (CATC) optimizes, via the fine tape-horizontal-alignment loop (FTHA), the phasing at the monochrome automatic timing-control (MATC) stage; the latter optimizes via a tape horizontal-alignment loop (THA) that at the line-lock stage, which in turn optimizes that at the combined phase and velocity stages, all acting in sequence with appropriate time-constants. At start, an adaptive tape-vertical-alignment (TVA) loop momentarily allows the phasing of the line-lock system to enable the cascaded adaptive operational hookup during the reproduction of color or monochrome signals from the videotape. The *adaptive-controller* makes the measurements on an open-loop basis by means of an automatic-delay-line, an error-detector, and a differential-amplifier; computes the changes to be made in an analog fashion; and automatically carries out the necessary time-base phasing adjustments on the main headwheel servo system.

According to theoretical findings in *adaptive and optimal control* (of modern control theory), an adaptive control-loop, with a rate of adjustment proportional to the *measured-error in the performance-index*, is a monotonic function of the adjustable parameter. The *high-precision phase* of the 3.579545-MHz color-subcarrier is the performance-index, and the horizontal and vertical phasing are the adjustable parameters in the case of the Q-CVTR. The loops concerned are not entirely independent but frequently interact with interdependent variables.

As a case of parallel theoretical development in modern optimal control, the following analytical conclusions of an *optimum regulator system* in *dynamic-*

programming provide a semblance of applicability to the adaptive-control system under consideration, assuming that the control process is linear.*

The analytical statement of the optimum control signals is termed the *control-law*, and it is determined by the basic application of the "dynamic-programming techniques" to the optimum design of the control processes with respect to well-defined *integral performance criteria.*

For a control process characterized by the first-order differential equation,

$$\dot{\bar{x}}(t) = g(\bar{x}, \bar{m}) \text{ and } x(t_0) = x_0$$

where x, the state-variable, and m, the control signal, are scalar functions of t.

For a finite control process, the optimum-control signal m, which *minimizes the integral criterion function*, given by $I(m) = \int_{t_0}^{t_1} F(x, m)\, dt$, is obtained from the *principle of optimality* as the solution of two partial differential equations (Eqs. 2 and 3) derived from

$$\frac{-\partial f}{\partial t} = \text{minimum for all m} \left[F(x, m) + g(x, m) \frac{\partial f}{\partial x}\right] \tag{1}$$

$$\text{Function } P(x, m) \triangleq \frac{\partial f}{\partial x} = -\frac{\partial F}{\partial m} \bigg/ \frac{\partial g}{\partial m} \tag{2}$$

$$\text{Function } Q(x, m) \triangleq \frac{\partial f}{\partial t} = -F(x, m) + g(x, m) \frac{\partial F}{\partial m} \bigg/ \frac{\partial g}{\partial m} \tag{3}$$

The *principle of optimality* is stated as an optimal policy having the property that, whatever the initial state and the initial decision, the remaining decision must form an optimal control strategy with respect to the state resulting from the first decision.

In the case of a time-variant control process (like the Q-CVTR),

$$\dot{x}(t) = g(x, m, t) \text{ and } I(m) = \int_{t_0}^{t_1} F(x, m, t)\, dt \tag{4}$$

The optimal-control signal m is obtained as a solution of the partial differential equation:

$$\frac{\partial P}{\partial t} + g\frac{\partial P}{\partial x} + \frac{\partial P}{\partial m}\frac{\partial m}{\partial t} = \frac{\partial Q}{\partial x} + \frac{\partial Q}{\partial m}\frac{\partial m}{\partial x} \tag{5}$$

In the case of the multivariable systems like the Q-CVTR, the state-variable x and m will be replaced by the corresponding matrices. In actual practice, the control signals or state-variables are subject to (1) constraints such as amplitude-saturation and (2) integral constraint due to power supply limitations. The latter is handled by the method of Lagrange multiplier, as shown in the example illustrating the statistical approach. Since the analytical solution of the preceding partial differential equations is difficult, numerical methods will provide the solutions.

The analytical expressions of the optimal control, however, can be determined by optimizing an assumed linear control process with respect to a quadratic criterion function.

$$\dot{\bar{x}}(t) = \bar{A}(t)\bar{x}(t) + \bar{D}(t)m(t) \ldots \bar{x}(0)\big|_{t=0} = \bar{x}_0 \tag{6}$$

*Equations 1 through 15 from *Modern Control Theory* by Julius T. Tou. Copyright © 1964 by McGraw-Hill Book Company. Used with permission of McGraw-Hill Book Company.

$\overline{A}$ and $\overline{D}$ are coefficient and driving matrices, respectively, and $\overline{x}$ is an n-vector column-matrix representing the state of the process. The control problem is concerned with the determination of the optimal-control signal m(t), which minimizes an integral criterion-function:

$$I(m) = \int_0^T F(\overline{x}, m, t)\, dt \tag{7}$$

where F is a quadratic function,

$$F(\overline{x}, m, t) = \overline{x}'\overline{Q}(t)\overline{x} + \lambda m^2$$

$\overline{Q}(t)$ is an nxn symmetrical matrix
$\overline{x}'$ is the transpose
(λm^2) with constant λ is used to avoid any nonlinear saturation of the control signal.

If the minimum of the integral is defined by

$$f(\overline{x}, t) = \min_m . \int_t^T F(\overline{x}, m, t)\, dt$$

the condition under which the minimum of performance-index exists is obtained from the principle of optimality as the functional equation:

$$\frac{-\partial f}{\partial t} = \min_m . \left[\overline{x}'\overline{Q}(t)\overline{x} + \lambda m^2 + \sum_{j=1}^{n} \dot{x}_j \frac{\partial f}{\partial x_j}\right] \tag{8}$$

where x_j is the *jth* component of the state n-vector $\overline{x}$, and

$$\dot{x}_j(t) = \sum_{k=1}^{n} a_{jk}(t)x_k(t) + d_j(t)m(t) \tag{9}$$

with the elements $\overline{A}$ and $\overline{D}$, respectively. From Eq. 8, the minimizing control signal m_0 is obtained by equating the derivative with respect to m to zero.

$$m_0 = -\frac{1}{2\lambda} \sum_{j=1}^{n} d_j(t) \frac{\partial f}{\partial x_j} \tag{10}$$

As a solution of these three equations for $f(\overline{x}, t)$, the optimal control signal is obtained in the linear form:

$$m_0(t) = \sum_{i=1}^{n} \beta_i(t)x_i(t) \tag{11}$$

This is the *control law*, and it can be put into the configuration of an *optimal (or adaptive) regulator system* as follows:

In the system configuration illustrated in Fig. 3-12, the number of control loops gives the order of the control process. In this respect, the adaptive control system in the Q-CVTR is of order 3. The feedback signals are the measurable state-variables. With the heuristic measure-and-optimize techniques used in the case of the Q-CVTR, these measurable variables take a predominant role in design. The feedback coefficient in each loop is time-variant, but independent of the state-variables, since the

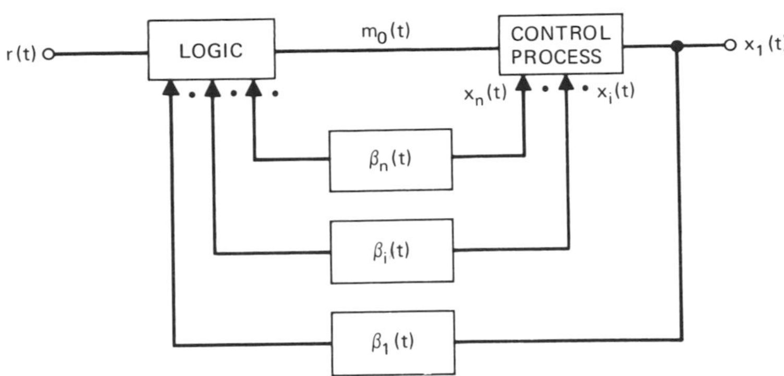

Fig. 3-12. Optimum-regulator. (From *Modern Control Theory* by Julius T. Tou. Copyright © 1964 by McGraw-Hill Book Company. Used with permission of McGraw-Hill Book Company.)

system is optimized with respect to a quadratic criterion to avoid saturation effects as far as possible.

The configuration in Fig. 3-12 becomes an *optimal control system* by including the design of an error-index, which measures the deviation of the state-variables from the desired values. This is made possible by including the "dashed" reference signal r(t). This is the closest simulation to the adaptive control system in the Q-CVTR. In the theoretical case, the optimum-control law with n time-varying feedback loops is given by:

$$m_0(t) = r(t) + \sum_{i=1}^{n} \beta_i(t)x_i(t) \tag{12}$$

where

$$r(t) = \frac{-1}{\lambda} \sum_{j=1}^{n} d_j(t)b_j(t) \tag{13}$$

$$\beta_i(t) = \frac{-1}{\lambda} \sum_{j=1}^{n} d_j(t)b_{ij}(t) \tag{14}$$

$$f(\bar{x}, t) = b(t) + 2 \sum_{j=1}^{n} b_j(t)x_j(t) = \sum_{i=1}^{n} \sum_{j=1}^{n} b_{ij}(t)x_i(t)x_j(t) \tag{15}$$

Equation 15 leads to $\left[1 + n + \frac{n(n+1)}{2}\right]$ independent ordinary first-order differential equations called the *Riccation equations*, defining the parameters b(t), $b_j(t)$, and $b_{ij}(t)$. These equations with the b parameters are numerically solved backward in time on a digital computer to obtain the optimum control signal $m_0(t)$.

3.3.20 Application of Conventional Frequency-Response Technique to Linear Approximation of Loops in Complex Sampled-Data Feedback Control System. As a contrast to the mathematically sophisticated modern optimal techniques, which are used in conjunction with the analog-simulation of the nonlinear elements and subsequent syntheses with the measured variables and the digital computation, an example of one of the simpler conventional frequency-response techniques (the *Nyquist criterion*, for example) is presented to illustrate its method of application to a linear approximation of a sampled-data feedback control system. For the sake of illustration, a typical feed-forward pulse-transfer-function is chosen at random to present the procedure for determining the stability of the closed-loop system by analyzing the open-loop pulse-transfer-function. In the case of the sampled-data control systems, there is less correlation between the phase-gain margins and the transient-response, unlike that in the case of the continuous systems; thus, frequency-response techniques are rarely extended to determine the closed-loop transient-response. (As shown earlier, the *power-series expansion of C(z) for a step input*, $R(z) = z/(z - 1)$, is a general procedure for determining the *transient-response* of linear sampled-data control systems. This is illustrated in Section 3.2.5.)

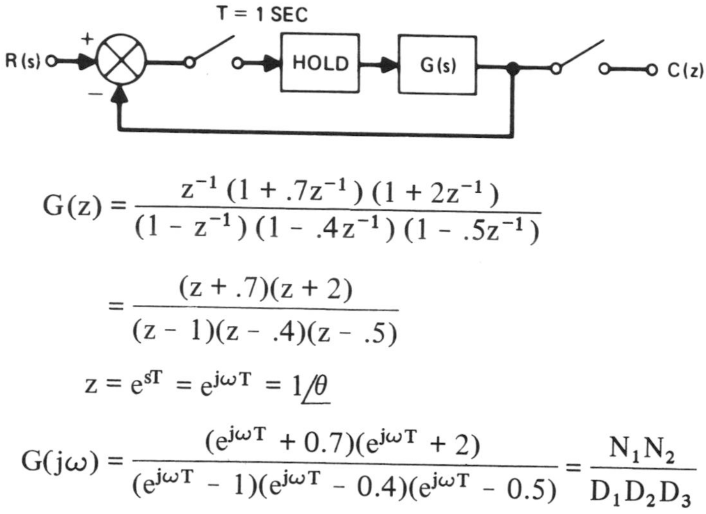

$$G(z) = \frac{z^{-1}(1 + .7z^{-1})(1 + 2z^{-1})}{(1 - z^{-1})(1 - .4z^{-1})(1 - .5z^{-1})}$$

$$= \frac{(z + .7)(z + 2)}{(z - 1)(z - .4)(z - .5)}$$

$$z = e^{sT} = e^{j\omega T} = 1\underline{/\theta}$$

$$G(j\omega) = \frac{(e^{j\omega T} + 0.7)(e^{j\omega T} + 2)}{(e^{j\omega T} - 1)(e^{j\omega T} - 0.4)(e^{j\omega T} - 0.5)} = \frac{N_1 N_2}{D_1 D_2 D_3}$$

For the pole at $z = 1$, consider an indentation to enclose the pole. (δ is infinitesimally small.)

$$G \approx \frac{1.7 \times 3}{(z - 1)(.6 \times .5)} = \frac{17}{(z - 1)}$$

Table 3-1. Nyquist Parameters.

θ	N_1	N_2	D_1	D_2	D_3	$\lvert G \rvert$,	$\underline{/G}$
0°	1.7 0°	3 0°	0 0°	0.6 0°	0.5 0°	∞ 0°	magnitude ↓ phase
45°	1.2 126.5°	2.8 14.5°	.765 160°	.76 66.8°	.73 74°	7.98 −258.8°	
90°	1.22 54.5°	2.25 29.3°	1.414 135°	1.07 111.8°	1.12 116.6°	1.62 −282.4°	
135°	.7 90.5°	1.42 29.3°	1.85 157.7°	1.3 147.6°	1.4 149.7°	.298 −335.7°	
180°	−.3 0°	1 0°	−2 0°	−1.4 0°	−1.5 0°	.072 0°	

Example: $\theta = 45°$: $N_1 = (e^{j\omega T} + .7) = 1.407 + j0.707 = 1.2, 126.5°$

where, $e^{j\omega T} = (\cos 45 + j \sin 45)$

$\cos 135° = \cos(90 + 45) = \sin 45$;

$\sin 135° = \sin(90 + 45) = \cos 45$

Let $(z - 1) = \delta e^{j\alpha}$. (As α increases from $\frac{-\pi}{2}$ to $\frac{\pi}{2}$, G is evaluated.)

α	$(z - 1)$	$G \approx \frac{17}{z-1} \approx \infty$ as $\delta \longrightarrow 0$
$-90°$	$\delta\underline{/-90°}$	$\approx \infty, +90°$
$-45°$	$\delta\underline{/-45°}$	$\infty, +45°$
$0°$	$\delta\underline{/-0°}$	$\infty, 0°$
$+90°$	$\delta\underline{/+90°}$	$\infty, -90°$

The polar Nyquist-plot (z-transform locus), with the magnitude and phase values evaluated in the preceding two tables, takes the form illustrated in Fig. 3-13. The above procedure involves

1. derivation of the pulse-transfer-function from the system block-diagram
2. plotting of the z-transform locus of the open-loop pulse-transfer-function for z traversing once around the unit-circle of the z-plane
3. determination from the pole-plot the number of counter-clockwise encirclements, for positive N, of the z-transform locus about the critical point $(-1 + j0)$
4. if the open-loop pulse-transfer-function contains no pole outside the unit-circle (open-loop stable), the system is stable for N = 0 only, and
5. if the system is open-loop unstable with poles outside the unit-circle, the system is stable for N = n - p, where n is the number of zeros of the characteristic equation, and p is the number of poles of G(z) inside the unit-circle, including that at z = 1.

In the case of the present example, the condition (4) applies, and since N = 1 from the z-transform locus, the system is *unstable.* The result can be confirmed by the Routh-Hurwitz criterion as follows: The characteristic equation, $1 + GH(z)|_{H=1} = 0$.

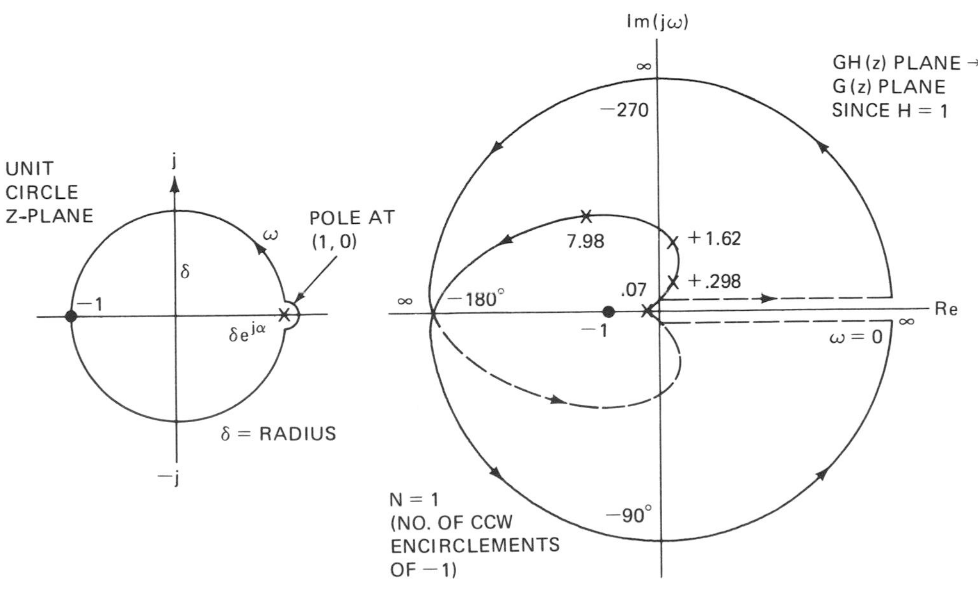

Fig. 3-13. Nyquist plot.

$$1 + \frac{(z + .7)(z + .2)}{(z - 1)(z - .4)(z - .5)} = (z - 1)(z - .4)(z - .5) + (z + .7)(z + .2) = 0$$

$$z^3 - .9z^2 + 3.8z + 1.2 = 0$$

Using the bilinear-transformation from the z- to w-plane with $z = \dfrac{1 + w}{1 - w}$,

$$\left(\frac{1 + w}{1 - w}\right)^3 - 0.9\left(\frac{1 + w}{1 - w}\right)^2 + 3.8\left(\frac{1 + w}{1 - w}\right) + 1.2 = 0$$

$$\underset{a_0}{4.5w^3} + \underset{a_1}{3.7w^2} - \underset{a_2}{5.3w} + \underset{a_3}{5.1} = 0$$

$$b_1 = \frac{a_1 a_2 - a_0 a_3}{a_1} = -11.5$$

$$c_1 = \frac{b_1 a_3 - a_1 b_3}{b_1} = 5.1$$

a_0	a_2
4.5	-5.3
a_1	a_3
3.7	5.1
b_1	b_3
-11.5	0
c_1	
5.1	

The two changes of sign in the first-column of the array place two poles in the right-half of the w-plane. Hence, the instability of the system, chosen at random as an example, is confirmed. A survey of the concepts underlying these techniques in continuous and sampled-data control systems was presented in Chapter 1.

3.4 GENERAL THEORETICAL CONCEPTS APPLICABLE TO CAPSTAN SERVO

3.4.1 Capstan Sampled-Data Feedback Control System and Its Digital Controllers. Most of the major control system features in the capstan sampled-data feedback control system are schematically presented in the comprehensive block-diagram of Fig. 3-23. The actual system electronics and the operational requirements are briefly explained in Chapter 4.

It will be seen that some of the prominent features in the capstan servo, such as (1) pulse shapers, (2) a series of digital controllers, (3) delay-elements with associated lag, (4) multirate sampling, (5) the nonlinearity of saturation, (6) the nonlinearity of the two-phase hysteresis asynchronous synchronous (HAS) motor, and other incidental nonlinear elements, are similar to those encountered in the headwheel sampled-data feedback control system. In this case, however, an ac carrier-servo with phase-modulation (in the place of PWM) and an additional digital controller in the carrier portion of the servo are exceptions. Only three rates of sampling are involved—namely, 240, 30, and 60 Hz/sec—compared to five in the case of the headwheel control system. The 30-Hz switch-lock sampled-data control system is, however, used more as a facility for the video special-effects on "pix-lock" in the headwheel control system, rather than for any stability reasons. It is theoretically not permissible to treat the aspects of analysis and synthesis of the capstan-servo as a conventional independent feedback control system since

1. the capstan sampled-data feedback control system is *intricately interacting* with the headwheel sampled-data feedback control system by way of common

output (or feedback) and input variables (in conjunction with the influential involvement of the nonlinear elements in the videotape-magnetics, the FM signal-processing circuits, and the sync digital-controller circuits)

2. the capstan and headwheel servo systems must operate in mutual synchronism, from the PLL pull-in point of view, for the precise tracking of the headwheel against the transverse FM video-tracks on the moving tape by means of the 240/30-Hz capstan control-track signal, for the best speed-regulation of the headwheel by means of the tonewheel signal.

A simple design or operational problem in the highly complex headwheel servo system will interact on the capstan-servo to throw it out-of-phase, and render the complete mechanism of the tape signal playback procedure unstable with the resultant breakup of the reproduced picture; and so it is with the headwheel servo on the incidence of some design or operational problem in the capstan-servo. Even the third sampled-data vacuum-guide position feedback control system, in its automatic mode, will interact with these two feedback control systems and make the situation doubly worse (although the safety of the quadruplex headwheel against possible destructive vacuum-guide pressure is assured by either an automatic release of the guide solenoid-pressure or a transfer of the guide to a safe position of pressure in the manual mode). It has been explained with an example under the headwheel control system how complex and unwieldy the closed-loop pulse-transfer-functions become in the case of two interacting control systems with only one feedback-loop in each.

In view of these intricate interacting features, the independent syntheses of the digital controllers in the capstan-servo according to some available, familiar theoretical technique is impractical. The successful implementation of these interacting pulse control systems by the heuristic measure-and-optimize techniques in the laboratory is an accomplished fact, and it is perhaps conceivable that modern techniques of analysis (for step- or ramp-response) using (1) an analog simulation of the nonlinear elements and (2) the subsequent computation of data on a digital computer may be attempted as a study at the present stage for the analyses and syntheses of the digital controllers in the case of the capstan-servo. Thus, possible avenues of improvement may be explored as a matter of theoretical interest (and not for any commercial assurance purpose), assuming that the other control systems are stable to the desired degree of performance-index.

The application of the theoretical techniques to (1) the multirate sampled-data feedback control, (2) the ac carrier-servo used for the hysteresis asynchronous synchronous (HAS) motor, (3) the effect of the loop delay-elements, and (4) the basic digital control systems, were explained with simple illustrative examples in the previous section. Since the digital controllers make an essential part of the capstan sampled-data feedback control system, the application and syntheses of the digital controllers in a sampled-data control system for (1) *minimal prototype response* and (2) *ripple-free response* are explained in the present chapter with simple mathematical models. A reference to the so-called staleness factor is inevitable in these applications; it is therefore included for explanatory purposes. The *extension* of the widely accepted complex s-plane *root-locus technique to the z-plane* in the case of a sampled-data control system is explained in the following section; a simple example is presented to illustrate the quintessence of this popular technique.

3.4.2 Stabilization of Capstan Sampled-Data Feedback Control System with a Digital-Controller Compensation. In practice, stabilizing networks used for analog systems can be included to provide "hold and stability" compensation for sampled-data systems, as is done occasionally in the control systems of the Q-CVTR, but tedious complications arise in the actual analyses of the approximated linear control systems in such cases due to the following reason. In case 1,

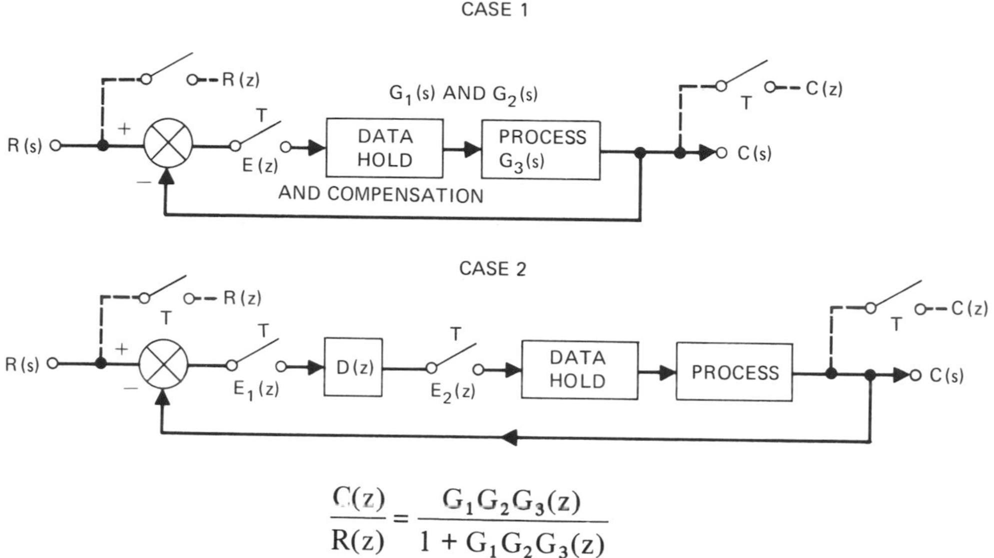

$$\frac{C(z)}{R(z)} = \frac{G_1G_2G_3(z)}{1 + G_1G_2G_3(z)}$$

In frequency-domain,

$$\frac{G(j\omega)}{R^*(j\omega)} = \frac{G_1(j\omega)G_2(j\omega)G_3(j\omega)}{1 + (G_1G_2G_3)^*(j\omega)}$$

where $G_1G_2G_3{}^*(j\omega) = \frac{1}{T} \sum_{-\infty}^{\infty} G_1(j\omega + jn\omega_s)G_2(j\omega + jn\omega_s)G_3(j\omega + jn\omega_s)$, where ω_s = sampling frequency.

To analyze by conventional Bode or Nyquist techniques, the function $(G_1G_2G_3)^*(j\omega)$ of the characteristic equation introduces the complication of the *harmonics of the sampling frequency* against the simple addition of logarithms or vectors; this manipulation is essential for examining the effect of a change in $G_2(j\omega)$ of compensation on the overall Bode or Nyquist plot. Different compensating networks require a multiplicity of overall frequency-plots to examine their effect on the system stability, which is a very tedious, difficult procedure for the measure-and-optimize design. In view of this difficulty, the stabilization of a sampled-data control system is more commonly achieved by the digital-controller shown in Case 2, which is basically a computation of an output number sequence $E_2(z)$ in a linear relation to an input number sequence $E_1(z)$ for obtaining the desired system response. Incidentally, the digital-controller can contain both the active elements and passive networks. These are normally chosen for the desired response.

$$\frac{C(z)}{R(z)} = K(z) = \frac{D(z)G(z)}{1 + D(z)G(z)} \text{ or } D(z) = \frac{1}{G(z)}\left[\frac{K(z)}{1 - K(z)}\right]$$

The following illustration explains the technique of designing an overall physically realizable and stable pulse-transfer-function K(z) of a closed-loop sampled-data control system, containing a digital-controller, for *zero steady-state error* (at the sampling instants) in response to a unit-ramp input. The *minimal-prototype response technique* enables a relatively fast settling-time as far as the transient-response is concerned. (Note: *digital controller* may be interpreted as *digital filter*.)

The importance of the digital-controller lies in its effect of cancelling any undesirable poles and zeros of the uncompensated system G(z) and replacing them with acceptable poles and zeros to obtain the desired system response. For the desired overall K(z), a unity feedback system is assumed. (For other values of feedback, the results can be extrapolated.) In general, a finite settling-time is acquired by the technique of increasing the bandwidth, while the sampling-process simultaneously tends to impose a limitation on the bandwidth.

$$\text{Assume } G(z) = \frac{(1 + 1.5z^{-1})(1 + .2z^{-1})}{(1 - z^{-1})(1 - .3z^{-1})^2} = \frac{(z + 1.5)(z + .2)z}{(z - 1)(z - .3)^2}$$

G(z) contains zeros at -1.5, $-.2$, and origin; a simple-pole at 1; and a double-pole at 0.3. To obtain a finite settling-time, the minimal-prototype response function K(z) must contain only a numerator polynomial in z^{-1} of the form: $K(z) = (a_1 z^{-1} + a_2 z^{-1} + \cdots)$. It must also contain as its zeros all the zeros of G(z) that lie on or outside the unit-circle in the z-plane.

Let

$$K(z) = (1 + 1.5z^{-1})(a_1 z_1^{-1} + \cdots) \tag{1}$$

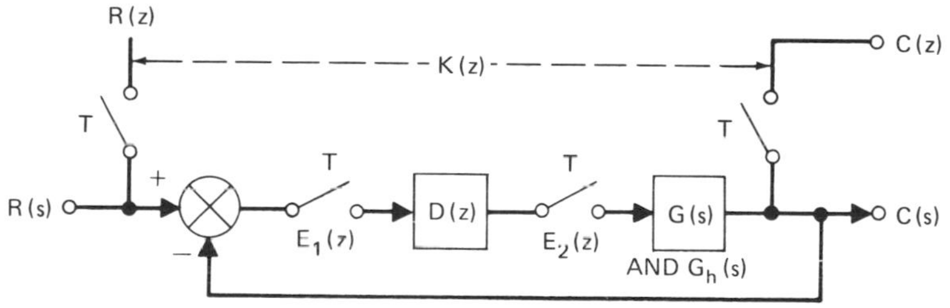

For minimum settling-time, as few terms as possible are included, consistent with other requirements. For zero steady-state error, the final value theorem is applied to find the requisite condition.

$$\lim e_1(\infty) = \lim_{z \to 1} \{(1 - z^{-1})R(z)[1 - K(z)]\}$$

where $R(z) = [Tz^{-1}/(1 - z^{-1})^2]$ for unit-ramp.

$$e_1(\infty)\big|_{s.s.} = 0 \text{ if } [1 - K(z)] = (1 - z^{-1})^2 F(z)$$

where

F(z) = 1, for a minimal prototype response function
s.s. = steady-state.

$$[1 - K(z)] = (1 - z^{-1})^2$$

An additional requirement is placed on this to satisfy the requirements: (1) the previous condition of a zero outside the unit-circle in the z-plane, and (2) a zero-error response to an input ramp-function

$$[1 - K(z)] = (1 - z^{-1})^2(1 + b_1 z^{-1} + b_2 z^{-2} + \cdots) \tag{2}$$

Thus, a slightly longer minimum finite-settling-time response will result according to the above modified equation with the constants b_1, etc. The constants a_1, a_2, etc. in Eq. 1 and b_1, b_2, etc. in Eq. 2 are a measure of the output at the sampling instants T_1, T_2, etc. due to the input at time $t = 0$. From experience, constants $b_2, b_3, \ldots$ are assumed zero to obtain a solution. (Otherwise there will be more unknowns than the number of equations.) These constants are then obtained by substituting K(z) into the expression for [1 - K(z)], and equating the coefficients of like powers in z^{-1}. This results in the following simultaneous equations (Eq. 4) relating to the coefficients.

From Eqs. 1 and 2,

$$[1 - (1 + 1.5z^{-1})(a_1 z^{-1} + a_2 z^{-2})] = [(1 - z^{-1})^2(1 + b_1 z^{-1})] \tag{3}$$

$$a_1 z^{-1} + (1.5a_1 + a_2)z^{-2} + (1.5a_2)z^{-3} + \cdots = (2 - b_1)z^{-1} + (2b_1 - 1)z^{-2} + (-b_1)z^{-3}$$

$(2 - b_1) = a_1$; $(2b_1 - 1) = (1.5a_1 + a_2)$; and $-b_1 = 1.5a_2$.

$$\left.\begin{aligned} a_1 - 1.5a_2 &= 2 \\ 1.5a_1 + 4a_2 &= -1 \end{aligned}\right\} \tag{4}$$

Then,

$$a_1 = 1.04$$
$$a_2 = -0.64$$
$$b_1 = -1.5a_2 = 0.96$$

From:

$$K(z) = a_1 z^{-1} + (1.5a_1 + a_2)z^{-2} + (1.5a_2)z^{-3}$$

$$K(z) = 1.04z^{-1} + 0.92z^{-2} - 0.96z^{-3} \tag{5}$$

Since R(z) = a unit-ramp with T assumed as 1 sec,

$$R(z) = \frac{z^{-1}}{(1 - z^{-1})^2}$$

$$C(z) = R(z)K(z) = \frac{z^{-1}(1.04z^{-1} + .92z^{-2} - .96z^{-3})}{(1 - z^{-1})^2} \tag{6}$$

The output response is obtained by taking the inverse of Eq. 6. By long-division power-series expansion.

$$C(z) = \frac{1.04z^{-2} + .92z^{-3} - .96z^{-4}}{1 - 2z^{-1} + z^{-2}}$$

$$\begin{array}{rl}
 & z^0 \quad z^{-1} \quad z^{-2} \qquad z^{-3} \qquad z^{-4} \\
 & 0 \quad\; 0 \quad +1.04 \quad\; +3 \qquad +4 \\ \cline{2-2}
(1 - 2z^{-1} + z^{-2}) & \big) \; 1.04z^{-2} + .92z^{-3} - .96z^{-4} \\
 & \underline{1.04z^{-2} - 2.08z^{-3} + 1.04z^{-4}} \\
 & 3z^{-3} - 2z^{-4} \\
 & \underline{3z^{-3} - 6z^{-4} + 3z^{-5}} \\
 & 4z^{-4} - 3z^{-5} \\
 & \underline{4z^{-4} - 8z^{-5} + 4z^{-6}} \\
 & 5z^{-5} - 4z^{-6} \ldots
\end{array}$$

$$c(t)\big|_{t=0} = 0;\ c(t)\big|_{t=1} = 0;\ c(t)\big|_{t=2} = 1.04;\ c(t)\big|_{t=3} = 3;\ c(t)\big|_{t=4} = 4$$

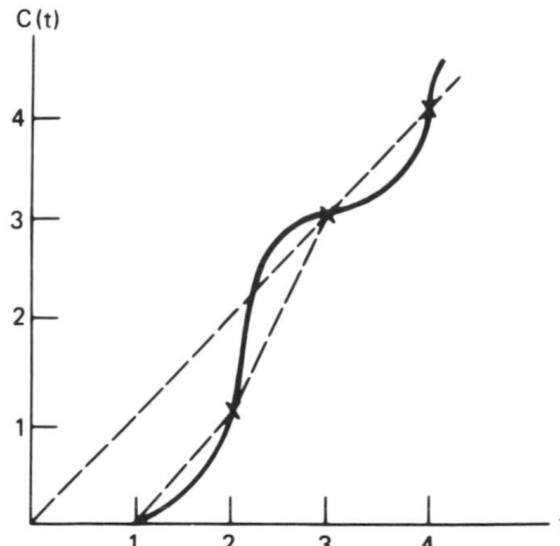

The continuous output c(t) will be seen to ripple about the final value after the system has settled to the zero steady-state error at the sampling instants only. With a step-input,

$$C(z) = (1.04z^{-1} + 0.92z^{-2} - 0.96z^{-3}) \frac{1}{(1 - z^{-1})}$$

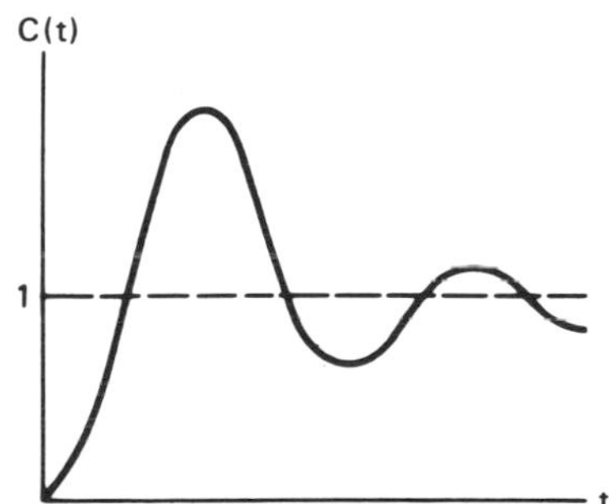

The response c(t) corresponding to the illustration can be seen to give an overshoot of as much as 200%, but it is the fastest system as far as the settling in minimum number of sampling instants is concerned. (However, if the K(z) is specifically designed for a step-input, instead of a ramp, the overshoot will be minimized.) The digital controller, which enables the physical realizability of the overall system pulse-transfer-function K(z), is then obtained from the basic relation:

$$D(z) = \frac{1}{G(z)} \frac{K(z)}{1 - K(z)}$$

where K(z) is given by Eq. 5.

3.4.3 Ripple-Free Sampled-Data Feedback Control System. One of the important requirements of the capstan sampled-data control system is that it should be ripple-free to enable a high-speed phase lockup of the synchronized interacting capstan-headwheel control system-complex. The example explained next will illustrate how the capstan control system can be designed with a special digital-controller D(z) for a ripple-free characteristic of the overall pulse-transfer-function K(z), so that the headwheel can maintain precise tracking against the video-tracks on the moving videotape right from the instant of start. With the following design technique, the output will be ripple-free, and the system can respond to a ramp-input with zero steady-state error.

For a ripple-free design with a transient of the shortest possible finite-duration, and no steady-state ripple with the step and ramp inputs, the following basic rules apply:

1. The ripple-free design is a mere *extension of the minimal-prototype response* technique explained in the previous example.
2. The feed-forward transfer function must, however, be capable of generating a continuous output function that is identical to the input function.
3. The overall K(z) must contain in this case *all the zeros* of the plant pulse-transfer-function K(z), and not merely the zeros of G(z) that lie outside the unit-circle in the z-plane.

Assume for this example:

$$G(z) = \frac{(z + .2)}{(z - 1)(z - .3)(z - .4)}$$

$$G(z) = \frac{(1 + .2z^{-1})(z^{-1})(z^{-1})}{(1 - z)(1 - .3z^{-1})(1 - .4z^{-1})}$$

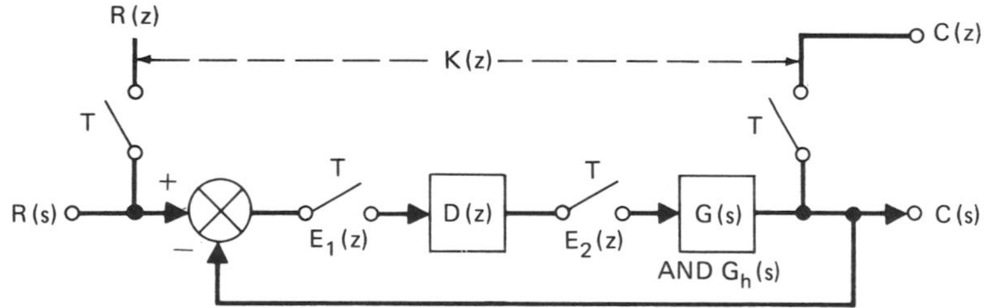

The feed-forward path with *zero-order* hold is capable of generating a continuous step or ramp function. The zeros of G(z) are located at − 0.2 and ∞ (double-order); for a ripple-free system, K(z) must contain all those zeros (not merely those outside the unit-circle).

$$K(z) = z^{-1} \cdot z^{-1}(1 + .2z^{-1})(a_0 + a_1 z^{-1}) \tag{1}$$

The second specification on k(z) is that the system be capable of following a ramp-input with zero steady-state error, in accordance with the requirement given by

$$1 - K(z) = (1 - z^{-1})^m F(z)$$

where F(z) = 1 and m = 2 for the z-transform of the ramp inputs R(z).

As in the previous example,

$$[1 - K(z)] = (1 - z^{-1})^2(1 + b_1 z^{-1} + b_2 z^{-2}) \tag{2}$$

However, an extra coefficient b_2 is included in this case to equalize the number of the unknowns and the equations. Solving for K(z) from Eqs. 1 and 2, we obtain a number of simultaneous equations with the unknowns a_0, a_1, b_1, and b_2.

From Eq. 2,

$$K(z) = (2 - b_1)z^{-1} + (2b_1 - b_2 - 1)z^{-2} + (2b_2 - b_1)z^{-3} - b_2 z^{-4} \tag{3}$$

From Eq. 1,

$$K(z) = z^{-2}(1 + .2z^{-1})(a_0 + a_1 z^{-1}) = z^{-2}a_0 + z^{-3}(.2a_0 + a_1) + z^{-4}(.2a_1) \tag{4}$$

$$\begin{aligned} 2 - b_1 &= 0 & b_1 &= 2 \\ 2b_1 - b_2 - 1 &= a_0 & b_2 &= 0.36 \\ 2b_2 - b_1 &= .2a_0 + a_1 & a_0 &= 2.64 \\ -b_2 &= .2a_1 & a_1 &= -1.8 \end{aligned}$$

From Eq. 4,

$$K(z) = 2.64z^{-2} - 1.28z^{-3} - 0.36z^{-4} \tag{5}$$

Thus, the ripple-free design contains terms in z^{-1} up to the fourth power, while, in the minimal-protype, terms in z^{-1} up to the third power only were considered. That is, the design results in a slightly longer settling-time, by one sampling-interval, but with minimized ripple, and zero steady-state error. The digital-controller is then derived from the basic equation with G(z) and K(z). As in the previous example,

$$C(z) = K(z)R(z)$$

and

$$D(z) = \frac{1}{G(z)} \frac{K(z)}{1 - K(z)}$$

$$D(z) = \frac{\cancel{(1 - z^{-1})}(1 - 0.3z^{-1})(1 - 0.4z^{-1})}{\cancel{z^{-2}}\cancel{(1 + .2z^{-1})}} \cdot \frac{\cancel{z^{-2}}\cancel{(1 + .2z^{-1})}(2.64 - 1.8z^{-1})}{(1 - z^{-1})\cancel{^{2}}(1 + 0.2z^{-1} + 0.36z^{-2})}$$

$$D(z) = \frac{(1 - 0.32z^{-1})(1 - 0.4z^{-1})(2.64 - 1.8z^{-1})}{(1 - z^{-1})(1 + 2z^{-1} + 0.36z^{-2})}$$

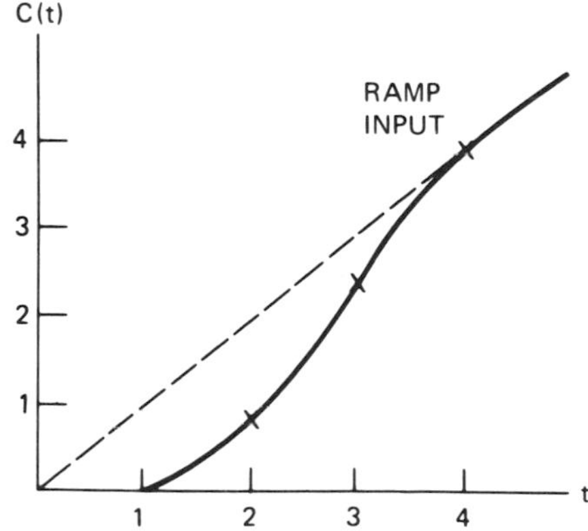

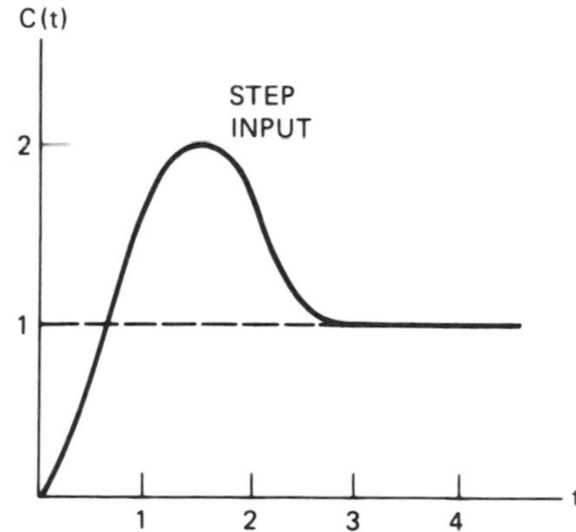

The preceding digital-controller is designed with time-delay elements to implement the overall system pulse-transfer-function K(z) with the desired characteristics.

The digital-controllers under consideration can be conveniently simulated by analog computer circuitry. A simple example will illustrate the technique.

$$D(z) = \frac{E_2(z)}{E_1(z)} = \frac{2.4 - 1.2z^{-1}}{1 + .6z^{-1}}$$

$$(1 + 0.6z^{-1})E_2(z) = (2.4 - 1.2z^{-1})E_1(z) \tag{1}$$

The difference equation for this will take the following form:

$$e_2(n) = -0.6e_2(n-1) + 2.4e_1(n) - 1.2e_1(n-1)$$

This is the simple correlation between the difference equation and the z-transform in sampled-data control systems. The digital-controller is simulated directly as follows (Fig. 3-14).

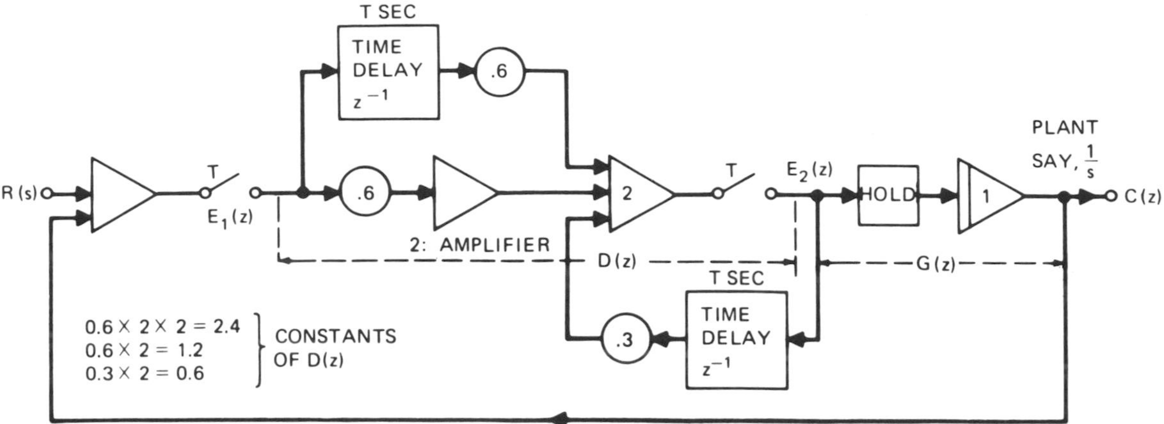

Fig. 3-14. Analog simulation of a digital controller.

The minimal-prototype and ripple-free techniques can be extended to the case of saturation so that the command sequence applied to the plant—namely, $e_2(t)$—does not exceed a specified maximum limit. This limitation will normally extend the minimal value of the settling-time, and make the digital-controller more complex. So, these designs become extremely complex in the case of the various loops encountered in the capstan and other sampled-data control systems in the Q-CVTR.

The *severe overshoot* of the sampled-data control system to the *inputs of lower order* such as the unit-step, where the system is specifically designed for a higher-order input like the ramp, can be overcome by including a term called the *staleness factor* in the system transfer-function (see step-response of the minimal-prototype and ripple-free design, p. 159).

The system ptf with the staleness factor c:

$$K_s(z) = \frac{K(z)}{(1 - cz^{-1})^n}$$

where n is positive (unity will do). The factor c can be chosen by analytical optimizing procedures or measure-and-optimize techniques in the laboratory if a ptf is available. Maximum damping is produced as c tends to unity.

3.4.4 Application of Root-Locus in Z-Plane.* The transient behavior of a sampled-data control system, such as the capstan digital servo, can be determined by the nature or location of the zeros and the poles of its closed-loop pulse-transfer-function. Starting with the open-loop ptf, the locus-plot of the closed-loop poles (which make the characteristic-roots of the system) versus the system loop-gain A, as a parameter, will aid the design of the sampled-data control system for *improved transient-response* by (1) adjustment of the *loop-gain* and (2) *redistribution of the closed-loop poles and zeros* by the method of effective digital-controller compensation.

*From *Digital and Sampled Data Control Systems* by Julius T. Tou. Copyright © 1959 by McGraw-Hill Book Company. Used with permission of McGraw-Hill Book Company.

The technique is illustrated with a simple example of a second-order sampled-data control system with a zero-order hold. The open-loop pulse-transfer functions GH(z) with finite number of poles and zeros is determined, and then the plot of the root-locus diagram in the z-plane is made in accordance with the very rules that govern the root-locus plots of continuous systems (as explained in Chapter 1).

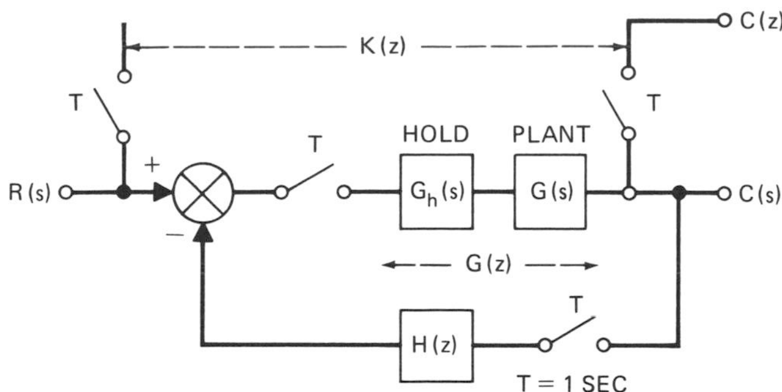

The closed-loop pulse-transfer-function:

$$K(Z) = \frac{AG(z)}{1 + AGH(z)}$$

Let $G(s) = \dfrac{A}{s(s+1)}$; H(s) = 1; T = 1 sec; GH(Z) = G(Z): A = gain; system time = constant T_m = T/K, k being a constant.

$$G_h(s) = \frac{1 - e^{-sT}}{s} = \frac{1 - e^{-s}}{s}$$

$$G(z) = Z[G_h(s)G(s)]$$

$$A_0G(z) = -1 = e^{j(180° \pm n360°)} \text{ and } A_0|G(z)| = 1$$

Characteristic equation: $1 + A_0GH(z) = 1 + A_0G(z) = 0$

where $A_0 = aTA = A$

if $p_1 = e^{-k}$, $z_1 = -\dfrac{1 - (1+k)e^{-k}}{k - 1 + e^{-k}}$, and $a = \dfrac{k - 1 + e^{-k}}{k}$

Now G(z) can be put in the form:

$$G(z) = \frac{A \prod_{k=1}^{m} (z - z_k)}{\prod_{k=1}^{n} (z - p_k)}$$

where $n \geqslant m$. From the z-transform tables,

$$GH(z) = G(z) = \left[\frac{A}{(z-1)} - \frac{A(1 - e^{-1})}{(z - e^{-1})}\right]$$

$$G(z) = \frac{0.368A(z + 0.72)}{(z - 1)(z - 0.368)}$$

The root-locus equation is obtained in the z-plane by letting $\emptyset$, the phase-angle of G(z), equal to $(\pi \pm n360°)$. With this assumption, the plot of the phase-angle equation, with the gain-factor A, as a parameter, makes one part of the root-locus for the pulsed-data system in the z-plane. It can be shown that this root-locus is represented by a circle given by the equation:

$$y^2 + (x - z_1)^2 = (z - p_1)(z_1 - 1)$$

$$[y^2 + (x + .72)^2] = [(0.72 + e^{-1})(0.72 + 1)] = 1.88 = (1.368)^2$$

In the above equation, $z_1(= -0.72)$ is the zero, and $p_1(= e^{-1} = 0.368)$ is the pole of G(z). The center of the circle is situated at the open-loop zero $z = z_1$—that is, $(z_1, 0) = (-0.72, 0)$. And, radius $= [(z_1 - p_1)(z_1 - 1)]^{1/2} = 1.368$. The graphical construction of the root-locus shown in Fig. 3-15 according to the basic rules gives an identical result. The root-locus starts from the open-loop poles at $z = 1$ and $z = p_1 = e^{-1} = 0.368$. At these locations, gain A = 0. As A is increased, the two branches of the root-locus move along the real-axis and meet at R, the breakaway point, where the two characteristic-roots of the system are equal to one another (Fig. 3-15).

In this particular example, when T is small, the roots are complex for maximum-allowable-gain. When T is large in a low-rate sampled-data system, the roots become real. The root-locus circle falls within the unit-circle. In between these values of T, the system can become unstable for a certain value of T.

From the characteristic equation,

$$1 + \frac{.368A(z + .72)}{(z - 1)(1 - .368)} = 0$$

$$A = \frac{1}{.368}\left[\frac{(z - 1)(z - .368)}{(z + .72)}\right]$$

For A = 1, the two roots are given by $(0.5 \pm j\sqrt{0.38})$. These roots for A = 1 can be obtained from the root-locus plot also, by constructing (1) the various constant-phase loci and (2) the root-locus, which is obtained by determining the points which give a vector sum of ±180° when the phase-angle loci of the two components of the open-loop pulse-transfer-function GH(z), as represented by straight-lines and circles, are superimposed upon each other. For absolute stability, all the poles of the overall closed-loop pulse-transfer-function K(z) must lie within the unit-circle. The gain at

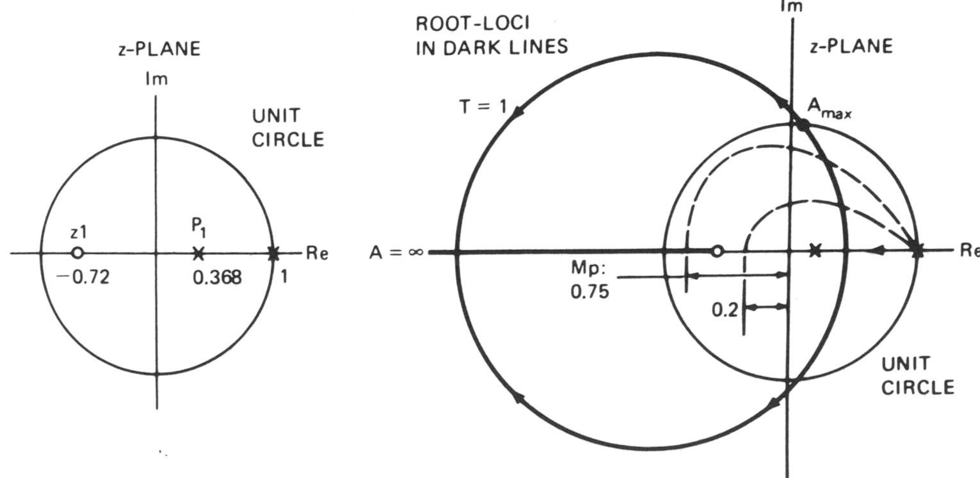

Fig. 3-15. Root-locus plot. (From *Digital and Sampled-Data Control Systems* by Julius T. Tou. Copyright © 1959 by the McGraw-Hill Book Company. Used with permission of McGraw-Hill Book Company.)

which the instability occurs is the system-gain at which the root-locus intersects the unit-circle in the z-plane. Hence, this particular value of the gain represents the maximum-allowable-gain for the stability of the sampled-data control system.

The points of intersection of the unit-circle and the root-locus circle are given by:

$$(x, y) = \left[\frac{(1 - p_1) + z_1(1 + p_1)}{2z_1}, \frac{\pm [(1 - p_1)(z_1 - 1)(1 - p_1 + 3z_1 + z_1 p_1)]^{1/2}}{2z_1}\right]$$

$$= (0.24 \pm j0.97)$$

The maximum allowable gain:

$$A_{max} = \frac{1 - p_1}{aT|z_1|} = \frac{(1 - p_1)}{.368|z_1|} = \frac{1 - .368}{(.368)(.72)} = 2.43$$

It can be shown that in the case of this second-order system, the above expression for A_{max} holds good for:

$$z_1 \leqslant \frac{-(1 - p_1)}{(3 + p_1)}$$

Beyond this value of the zero z_1, the circle of the root-locus cannot intersect the unit-circle. In general, the maximum-allowable-gain is obtained when the following condition is satisfied regarding the location of the pole p_1 and zero z_1:

$$p_1 = \frac{1 + 3z_1}{1 - z_1}$$

The relative stability of this system can be improved by reducing T, that is, for higher rates of sampling, as indicated by the following expression for the maximum-allowable-gain of this second-order sampled-data control system:

$$A_{max} = \frac{k}{T\left[1 - \dfrac{ke^{-k}}{1 - e^{-k}}\right]}$$

where $\frac{k}{T} = T_m$, the time-constant of the motor. Maximum-allowable-gain charts are available showing A_{max} plotted against T_m, with ratio $k(=T_m/T)$ as the variable.

In this example, the system can be improved for higher maximum-allowable-gain by using in front of the digital-hold, a digital-controller or compensator (as used in the case of the interacting and complex capstan-servo). As an example, a simple digitable-controller will be of the form $D(z) = A_1\left[\frac{z - a}{z - b}\right]$ for a modified plot of the root-locus.

The characteristic-roots suitable for meeting the step-response specifications can be chosen if the constant-overshoot loci (the damping-loci) are plotted for the sampled-data control system for the desired value of a damping-factor. Two examples of the damping-loci for two values of M_p are shown in the root-locus diagram. Where the root-locus circle meets the damping-loci, the required gain-constant for the desired value of damping is obtained. In general, as the order of the sampled-data control systems is increased for the root-locus plots, the roots invariably tend to move outside the unit-circle when the gain is increased, showing the necessity of maintaining by experiment and computation a rigid control on the several parameters involved in the pulse-transfer-function of a complex higher-order sampled-data control system such as the capstan servo in the Q-CVTR.

3.5 GENERAL THEORETICAL CONCEPTS APPLICABLE TO VACUUM-GUIDE SERVO

3.5.1 Vacuum-Guide Position Sampled-Data Feedback Control System. The major features of the vacuum-guide sampled-data position feedback control system are shown in the schematic block-diagram of Fig. 3-24. It will be seen that, for all practical purposes, the control system is directly interacting with the headwheel servo system, as far as the reference-input and feedback-output variables are concerned. Since the headwheel-servo, in turn, is intricately interacting with the capstan-servo, the capstan-servo does have an indirect effect on the accuracy of the guide-position. Moreover, as the tape movement is controlled by the capstan, and the moving tape is guided by the vacuum-guide assembly against the headwheel at a certain pressure, there is a semblance of direct mechanical link between the capstan-mechanics and the vacuum-guide. Thus, the position of the guide-assembly is influenced by the mutual interaction of the other two sampled-data control systems in an intricate manner. In fact, it has been found that possible mechanical-linkage problems associated with the capstan can cause minute vacuum-guide position-errors, when the guide servo is especially operated in the *manual-mode.* Minute errors of this character during the tape-playback are instantaneously corrected by the vacuum-guide servo, as it normally operates in the *automatic-mode*; these minute vacuum-guide pressure variations can be present in some tape recordings since the guide must operate in the manual-mode during the recording process at a supposedly constant predetermined pressure.

As far as this particular control system is concerned, four sampling rates are involved, which comprise the salient features in the vacuum-guide servo: (1) the 15.75-kHz sampled-data automatic frequency control (AFC) loop with a saturation NL-element, and a suitable compensation network, (2) the multirate 240/960-Hz sampled-data automatic phase control (APC) loop with a saturation NL-element, and a suitable compensation network, (3) the 960-Hz sampled-data vacuum-guide position feedback control system with a digital-controller, a saturation NL-element, and a dc compensation network, and (4) the 60-Hz suppressed-carrier amplitude-modulated ac carrier-servo with an ac compensation network and the active servo amplifier circuits. The compensation networks used in the associated 15.75-kHz AFC-loop and the 960-Hz APC-loop are not necessarily optimized from the point of view of the vacuum-guide servo in particular, since these loops are primarily expected to meet the requirements of the line-lock 15.75-kHz PLL connected with the headwheel-servo, and the FM-tape-video switching system. So, the vacuum-guide servo must rely itself on its signal-processing circuits, digital-controller, and its exclusive compensation networks for the optimum performance, irrespective of possible interference or incompatibility from any other compensation in the overall interacting control system-complex. Besides these major features, two elements of backlash (mechanical hysteresis) appear in the ac carrier-servo of the control plant when the vacuum-guide operates in the automatic-mode. The number of the nonlinear backlash-elements becomes three in the case of the subsidiary alternative manual feedback loop for the ac carrier-servo.

In view of the cumulative complexity of the several nonlinear-elements in the guide sampled-data position feedback control system, the servo is developed and designed, as explained at the outset, on the basis of the practical simpler *heuristic measure-and-optimize* technique. To obtain a uniform error-free performance for the edited picture reproduced from the videotape, the following potentiometer controls are used, all *on a preset basis.*

1. *a sensitivity control* for the dc error-signal of the 960-Hz *automatic-loop* at the output of the data-hold
2. *a sensitivity control* for the *manual ac carrier-servo-loop* to meet possible frictional variations in the vacuum-guide mechanism from machine to machine
3. *a zero-setting* for balancing the positive and negative excursions of the pressure-related *quantization* error-signal with respect to the sampling process.

The performance-index aimed at is $\pm 0.02 \mu s$ at 960 Hz/sec. In view of the theoretical implication of the several nonlinear-elements involved (including backlash), a small amount of phase-jitter corresponding to the phase-plane "limit-cycle" is inevitable. Thus the sensitivity of the automatic-loop in the overall feedback control system is optimized, with adequate gain-margin, so that (1) a *dead-zone* equal to ± 0.02 μs is the effective operational performance-index and (2) the system is sensitive enough that it immediately responds to error-signals above this figure and settles the sampled-data control system within the "dead-zone" of tolerance. The stated tolerance of error due to the 960-Hz quadruplex "skew" distortion (as a result of the recording pressure-variations) is not detectable in a picture reproduced on a 17-in. video monitor. For reproduction of color, the monochrome and color automatic-timing-correction systems in the headwheel control system will take over, in conjunction with the headwheel line-lock system, to minimize the stated tolerance of the horizontal time-displacement errors within the final headwheel performance-index of 0.005 μs, corresponding to a phase-tolerance of approximately $\pm 2°$ at the color subcarrier-frequency of 3.58 MHz. The vacuum-guide servo must perform its task to its specification for the best results in these sync time-base ATC systems.

3.5.2 Transfer-Function of Two-Phase Induction Motor Used for Vacuum-Guide Position Control System. The drive motor (or control plant) is a two-phase low rotor-inertia two-pole ac induction servo-motor, comprising of a stator with two windings, physically situated 90° apart. The direct power line drives one phase as reference, and the control signal from the 60-Hz carrier-servo drives the other phase to develop the desired torque in the presence of the vacuum-guide skew-distortion errors. The time-quadrature of these voltages is obtained in this system by the 90° phase-shift of the control signal via the servo amplifier itself.

The time-varying flux produced by the reference voltage, $e_c(t)$, represented by $\cos\left(\omega_c t + \frac{\pi}{2}\right)$ is given by the expression:

$$\phi_c(t) = K \int e_c(t)\, dt = \phi_0 \cos \omega_c t$$

where ϕ_0 is a constant.

In the same way, the flux produced by the control signal, $e_m(t)$, represented by $E_m(\cos \omega_c t)$ is given by the expression:

$$\phi_m(t) = \phi_1 \sin \omega_c t$$

These fluxes can be represented by space-vectors that vary in magnitude with time.

$$\text{At } \omega_c t = \frac{\pi}{2}, \phi_m = \phi_1 \text{ and } \phi_c = 0$$

Then, the resultant-flux in the two-phase servo-motor rotates at the synchronous speed, and the polarity of the applied control-signal with respect to the reference-signal determines the direction of rotation of the motor-shaft as a result of the induced currents in the rotor circuit due to the rotating flux. For a constant control-voltage excitation from the ac carrier-servo, the shaft rotates at approximately a constant uniform speed, over a useful portion of the normal voltage-versus-speed characteristic. The torque-speed characteristics of the induction motor are not necessarily linear, but they can be approximately linearized in practice with increased rotor-resistance. If the motor operation is centered at a point in a nonlinear region, for small displacements of the motor parameters shown, the linearization of the characteristics can be assumed, and a corresponding transfer-function can be derived.

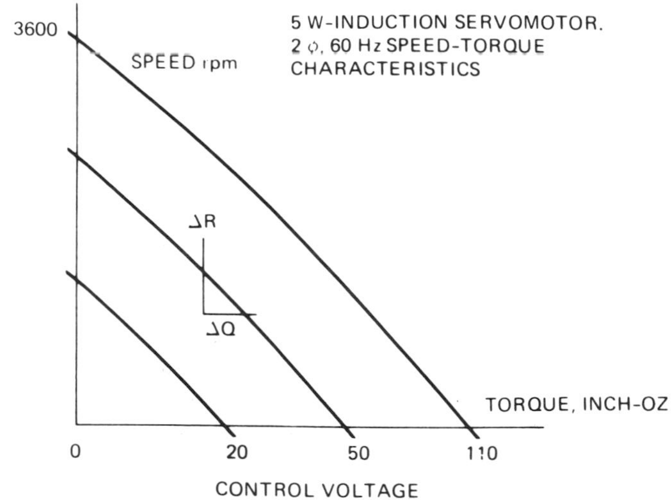

If Q = torque in in.-oz, and R = speed in rpm,

$$R = Q(k_1 + k_2)$$

where

k_1 = slope = $\dfrac{\Delta R}{\Delta Q}$

k_2 = intersection of the torque-speed curve with Q = 0.

If J = motor-driven inertia-load,

$$m = -\frac{1}{k_1}, \text{ and } k = \frac{k_2}{k_1} = \frac{\text{stall-torque}}{\text{rated control voltage}}$$

and $K = \dfrac{k}{m}$, and the time-constant of the motor, $\tau = J/m$. The parameters k and m are determined from the speed-torque characteristics supplied by the manufacturer. The parameter k_1 and, hence, m are best determined near the zero-speed operating-point of the system at the origin, where Q and speed are zero. Position-control motors, with normal operation about the zero-speed region, are usually designed to provide maximum-torque at stall. For low control-voltage excitation, the torque-speed curves are approximately linear. When the full reference-voltage is applied to the fixed-phase and the ac servo-control voltage is applied to the control-phase, the stall-torque increases uniformly as the control-voltage is increased. Then, the motor produces (1) a damping-torque that is proportional to the velocity and (2) a regular torque proportional to the control-voltage. The basic time-constant for 400-Hz servo motors is 10 to 30 msec, but it is much less for the 60-Hz motor used for the vacuum-guide servo. A typical set of values for the two-phase induction motor are given below to indicate clearly how the plant transfer-function is arrived at.

$$\text{Stall-torque} = Q_0 = 0.42 \text{ in.-oz}$$

$$\text{No-load speed} = R_0 = 3600 \text{ rpm} = \frac{3600 \times 2\pi}{60} = 378 \text{ rad/sec}$$

$$\text{Rated control voltage} = 115\text{V, } 50 \text{ Hz}$$

$$k = 0.42/115 = 3.65 \times 10^{-3} \text{ in.-oz/V}$$

As measured from the 20-V control-voltage torque-speed characteristic,

$$m = -\Delta Q/\Delta R = \frac{1}{2}\left(\frac{0.42}{378}\right) = 0.555 \times 10^{-3} \text{ in.-oz/sec}$$

For an inertia-load equivalent to the inertia of the motor, if rotor-inertia is given by 0.9 gram/cm^2,

$$J = \frac{2 \times .9}{980} \times 0.39 = 2.51 \times 10^{-5} \text{ in.-oz./sec}^2$$

$$\text{Time-constant } \tau = J/m = \frac{2.51 \times 10^{-5}}{0.555 \times 10^{-3}} = 0.045 \text{ sec}$$

$$\text{Motor constant } K = k/m = \frac{3.65 \times 10^{-3}}{0.555 \times 10^{-3}} = 6.7 \text{ V/sec}$$

$$\text{Motor transfer-function, } G(s) = \frac{\theta k}{s(js + m)} = \frac{K}{s(s\tau + 1)}$$

$$G(s) = \frac{\text{output-angle of motor-shaft}}{E_{in}} = \frac{6.7}{s(0.045s + 1)}$$

3.5.3 Popular Bode Technique. If the feedback control system is assumed as non-interacting, linear, and continuous, the following principles of design by the Bode technique apply.

The consideration of the servomotor's *corner-frequency* is very important, because it is the lowest corner-frequency in the system. In a position servo, it is followed by a $-180°$ phase-shift, which of course leads to the condition of instability in a closed-loop feedback control system if the gain-margin is not adequate. The illustration clarifies how the dB-ratio varies with the frequency for the two-phase induction motor. The voltage modulation of the control-phase is the input, while the resulting mechanical angular rotation of the shaft is the output. The low corner-frequency ω_1 is determined by (1) inertia, (2) rotor resistance, (3) friction, and (4) the torque developed. At still lower frequencies, the motor behaves like a simple integrator with a $-90°$ phase-shift. (In a high bandwidth application, a motor with a second, higher corner-frequency and hence a second time-constant is preferred to give an additional 90° phase-lag.)

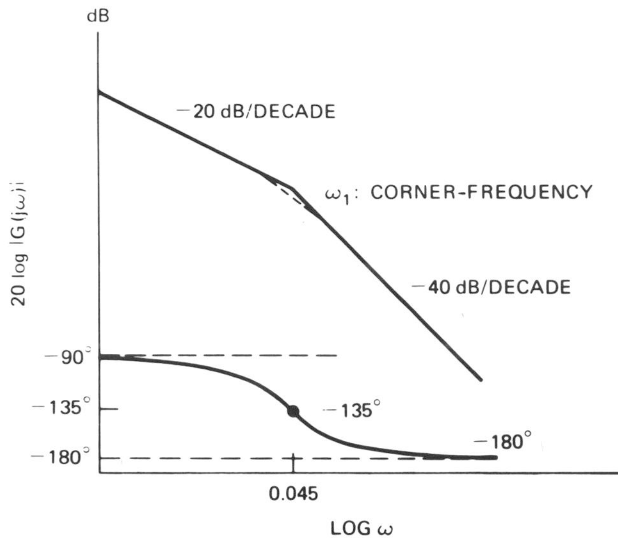

The low corner-frequency values range from 0.01 to 50 Hz with an average around 1 Hz; the higher the frequency, the less the torque. The design problem is one of working around this corner with reference to the whole system. An ac tachometer—as a differentiator for diminishing motor integration by one level—is one conventional technique to improve bandwidth. (An ac tachometer is merely a two-phase induction motor so designed that when one phase of the stator is energized from a constant-amplitude ac voltage-source, the other stator-phase produces a voltage at an amplitude that is proportional to the speed with which the tachometer is driven by the mechanical-coupling to the power-element of an angular-position control system.) The most common technique centers around the design of a compensation-network in the form of a proportional-integral RC-lag equalizer, to provide the simple means of introducing the corners ω_2 and ω_3, shown in the illustration, so as to offset the effects of the other corners. ω_3 is the *system crossover frequency*, and it is defined as the upper limit of the servo-response after the loop is closed. If ω_1 is the *natural-frequency* of the servo system, the speed of the servo-response is given by $1/\omega_1$. Corner ω_2 lowers the main loop-gain (20-dB per decade) as the frequency approaches corner ω_1, while corner ω_3 keeps the gain up at unity-gain, so that the open-loop phase-angle will not drop too soon to $-180°$. In general, the corner frequencies are adjusted as follows. ω_0 should be held high by the gain of the amplifier to minimize the error. Corners ω_2 and ω_3 should be heuristically adjusted to restrict the gain at frequencies below ω_c. Then ω_c will be a compromise between the maximum bandpass and low noise. Corner ω_1 should be kept as high as possible for rapid response. The last corner ω_4 should occur between -10 to -20 dB to give a phase-margin of 30° to 60° for the stability of the position feedback control system. The closed-loop response to a step-input is then optimized by adjusting the gain of the servo amplifier and the RC-parameters of the compensating equalizer network. An overdamped sluggish response is corrected by increasing the

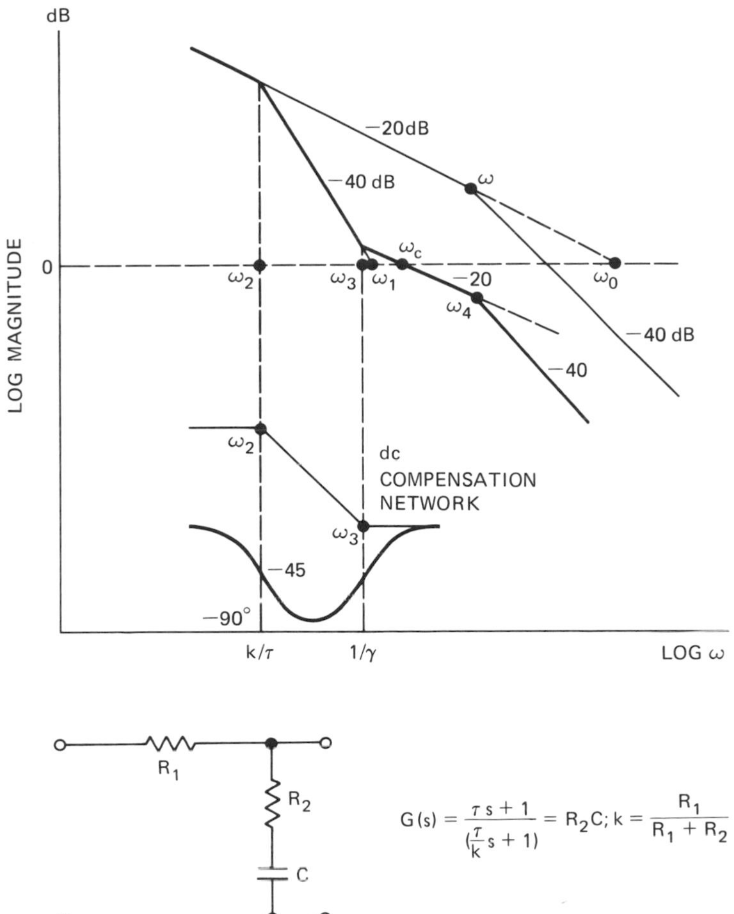

gain or decreasing the corner-frequency ω_4. An underdamped response with excessive overshoot is corrected by decreasing the gain or increasing the same corner-frequency ω_4. (See the above illustration.)

3.5.4 Application of Bode Technique to the Sampled-Data Feedback Control System. The bilinear transformation, $z = (1 + w)/(1 - w)$, maps the unit-circle in the z-plane into the entire left-half of the w-plane, and the polynomial-in-z will then become a ratio of two polynomials-in-w of the same order. Thus, the *mapping-function* will convert a multivalued transcendental function-in-s of the characteristic equation, $1 + GH^*(s) = 0$, into a single-valued polynomial-in-w. In the w-domain, the absence of the zero of the polynomial function in the right-half plane gives the condition for the stability of the sampled-data feedback control system. Since the transformation of the open-loop transfer-function into the w-plane is one of the *nonminimum-phase* category, the application of the popular Bode technique is permissible in sampled-data control systems. The following simple example illustrates the application of the Bode technique. The method enables the performance-criteria, viz., *the gain-margin and the phase-margin* establish the degree of stability of the sampled-data feedback control system.

$$\text{Assume: } GH^*(s) = GH(z) = G(z) = \frac{0.368z + 0.264}{(z - 1)(z - .368)}$$

This is the sampled z-transform of the open-loop transfer function GH(s).

$$G(w) = \frac{0.368\left(\dfrac{1+w}{1-w}\right) + 0.264}{\left(\dfrac{1+w}{1-w} - 1\right)\left(\dfrac{1+w}{1-w} - 0.368\right)}$$

$$= \frac{0.5(1-w)\left(1 + \dfrac{w}{6.08}\right)}{w\left(1 + \dfrac{w}{0.462}\right)}$$

Now the asymptotic gain and phase diagrams are plotted on a semilog paper in terms of a *fictitious frequency* v, in order to obtain the performance-criteria gain- and phase-margins (Fig. 3-16).

$$G(jv) = \frac{0.5(1-jv)\left(1 + \dfrac{jv}{6.08}\right)}{jv\left(1 + \dfrac{jv}{0.462}\right)}$$

From the following gain and phase Bode plots, it is seen that the system is reliably stable with a gain-margin of 8.5 dB and a phase-margin of 30°. With the Nichols chart, using the relationship of the log-gain versus phase-angle from the open-loop response, the closed-loop response, the resonant-peak M_p, and the fictitious resonant-frequency can be found as in the continuous systems.

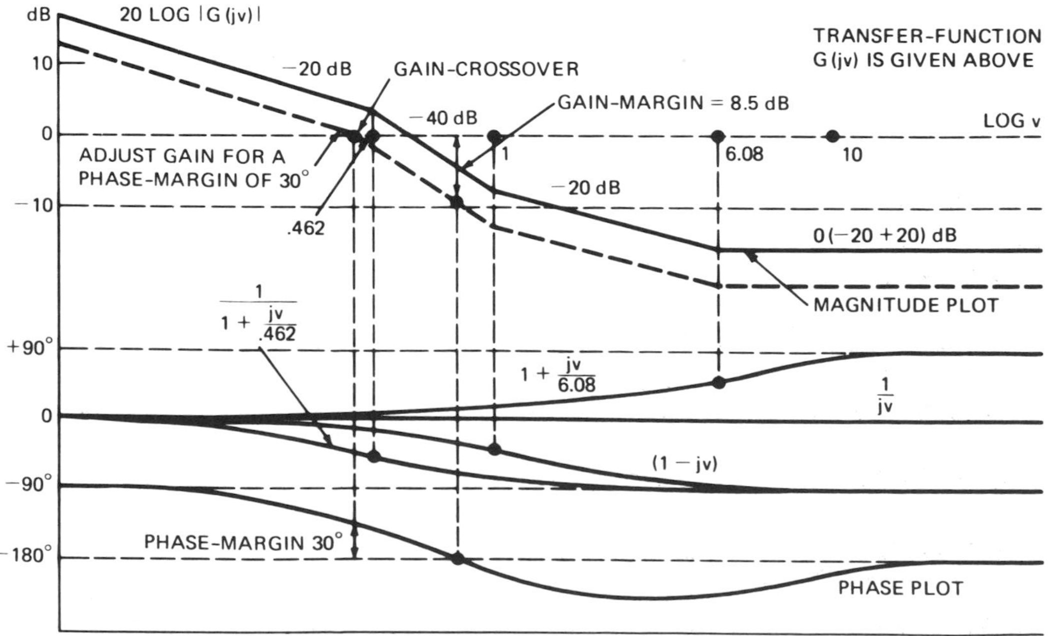

Fig. 3-16. The Bode-plot technique in sampled-data control system.

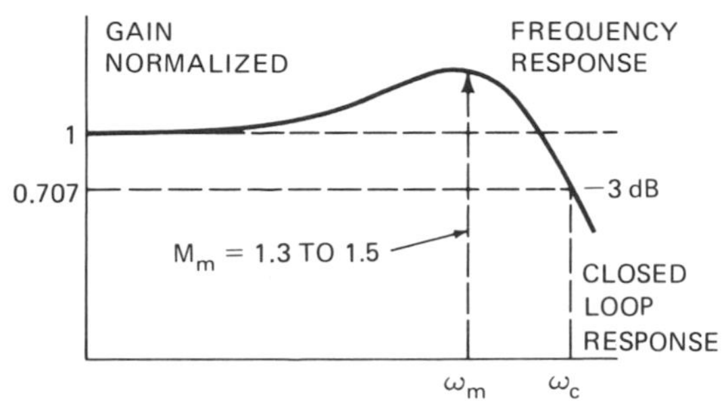

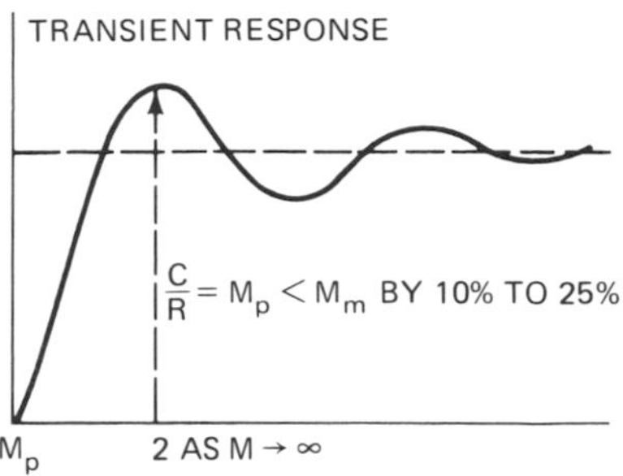

The *closed-loop response* can be obtained from the gain-phase plot of the *open-loop GH(jv) transfer-function* by superimposing and inspecting the points of intersection of this gain-phase characteristic on the curves of the Nichols chart in the *complex w-plane*, which correspond in rectangular-coordinates to the *loci of the constant-magnitude M-circles and the constant-phase α-circles in a closed-loop feedback control system.* When the Bode plot is extended to the Nichols chart, the GH(jv)-plot is modified by means of the 1/H(jv)-plot since the M and α-loci curves in the *Nichols chart* are based on *unity* feedback.

3.5.5 Automatic Frequency Control in Sampled-Data Feedback Control System. The automatic frequency control (AFC) is one of the most common facilities in the pulse, subcarrier, and FM-AM carrier subsystems of color television in several functions. In a vacuum-guide sampled-data feedback control system in the videotape recorder, it is seen from Fig. 3-24 that both the reference 960-Hz sampling-pulse information and the output feedback signal with the 960-Hz skew-distortion information are obtained from separate prior AFC-loops. Thus, the error and the reference signals applied to the automatic vacuum-guide sampled-data control system are subject to the influence of the compensation networks already used in the two immediate AFC-loops involved. It may be noted that, in general, when the required degree of accuracy in phasing is stringent in the case of these pulse-frequency or carrier circuits, the terms *automatic frequency control* (*AFC*) and *automatic phase control* (*APC*) become synonymous for all practical purposes. The basic function of the circuits is merely the local regeneration of a fresh, noise-free pulse-train (or carrier), the frequency and phase of which are automatically controlled with high-precision to track those of the incoming local television synchronizing signals; the reproduced tape synchronizing-signal is interspersed with noise and other distortion components. Both the television synchronizing pulse-train and the control-track signal belong to this category, since they are recovered from the magnetic videotape; the former in particular is separated from the composite-video via the FM record-playback system, which constitutes of several nonlinear-elements, moire, and noise-content. The pulse-trains are simultaneously contaminated with the effects of the phase-jitter tolerances in the nonlinear sampled-data control systems in the videotape recorder.

In the videotape recorder, the bandwidth and compensation networks used in the AFC-circuits are determined on a heuristic measure-and-optimize basis, as a compromise between such conflicting factors as (1) fast-response versus sluggish-response in one or more subsequent subsystems and (2) freedom from high-frequency noise components versus low-frequency "hum" and distortion components in subsequent circuits. The bandwidth and compensation best for one subsequent system is not nec-

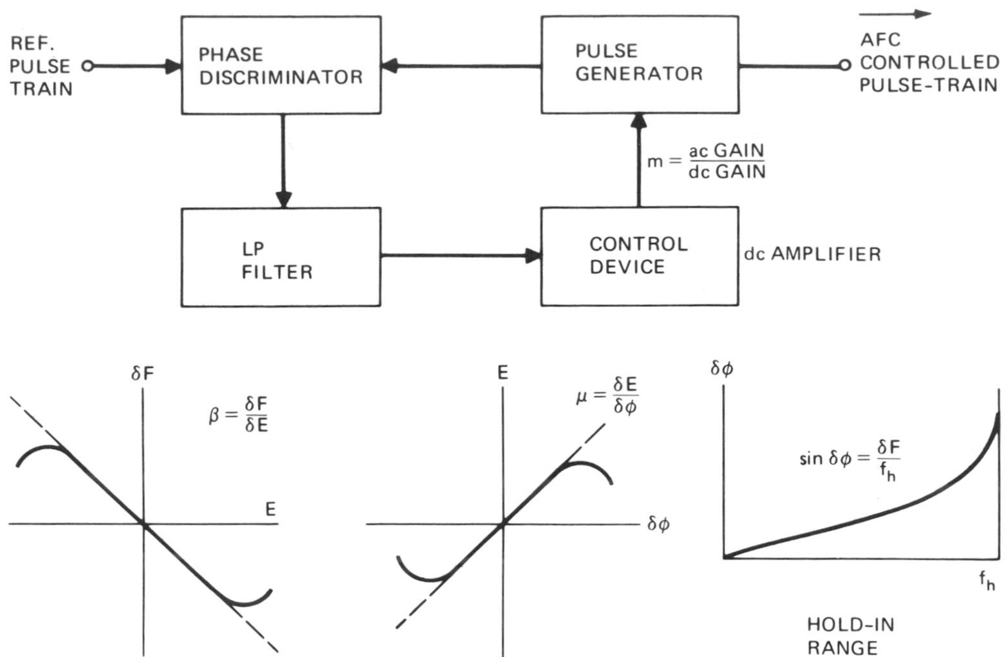

Fig. 3-17. Basic APC loop.

essarily the most suitable (if it is not disturbing) to another system in the multiloop chain. Hence, the selection of the compensation networks of both dc and ac categories is a result of mainly a measure-and-optimize approach, for the best possible performance-index under actual, statistically normal operating conditions (that is, in the absence of interacting system-problems of an unpredictable character from elsewhere in the color videotape recorder).

The basic concepts of the AFC/APC take the following form in brief (Fig. 3-17). The APC sampled-data feedback control system gives greater phase accuracy for a given noise bandwidth and detuning. The parameter m, indicating the ratio of the ac to dc transmission through the integrating low-pass filter with a limited transmission gain-factor, Q, determines the compensation. In the AFC case, the pulse-generator may be an astable multivibrator, the frequency of which is precisely controlled by an external transistor operating as a *reactance or voltage control* device. The loop characteristics are determined by (1) the sensitivity β of the control device, indicating the control-voltage E versus the frequency-shift ∂F in the pulse-generator, and (2) the gain-constant μ of the synchronous phase-discriminator, indicating the phase-difference $\partial\phi$ between the incoming reference pulse-train and the local generator versus the control-voltage E, and (3) the *static hold-in characteristic* of the APC-loop, as given by the static phase-error $\partial\phi$ produced to obtain the control-voltage E for a frequency-shift of ∂F in the pulse-generator. The *dc loop-gain* $|\mu\beta|$ is numerically equal to the *frequency holding-range* f_h, and the hold-in characteristic is obtained from the μ and β characteristics.

By increasing the dc loop-gain and hence the frequency holding-range f_h, the static phase-shift is held minimum, independent of the noise-integrator properties of the loop. The noise-bandwidth for the usual forms of integrators should be less than 100 Hz for equivalent performance with a high-Q filter. If T is the transient time-constant of the loop when the transfer characteristic of the feedback-loop filter is unity, the holding-range f_h corresponds to $\frac{1}{2}\pi T$.

How the system actually pulls into phase is very complex to explain, since the process is determined by the solution of the nonlinear portion of the dc loop-gain characteristic. To minimize the important pull-in time to the order of 0.1 to 0.5 sec, the static phase-error must be minimized. To start with, the AFC must have the ability for the best pull-in to recognize a frequency-difference in the reference pulse-train and noise. The capacitor, integrating the error-information in the low-pass filter, permits the dc gain to exceed the ac gain, and makes it possible for a rapid pull-in performance. Except near the limit of the *pull-in range*, the *pull-in time*, and the noise-bandwidth are closely interrelated with frequency-detuning effects in a nonlinear fashion. The frequency pull-in characteristic of the standard AFC-loop indicates that within a range roughly two-thirds that of the noise-bandwidth, the pull-in is effectively instantaneous.

A typical filter used in the AFC feedback loop, and its *dynamic characteristic* take the following form:

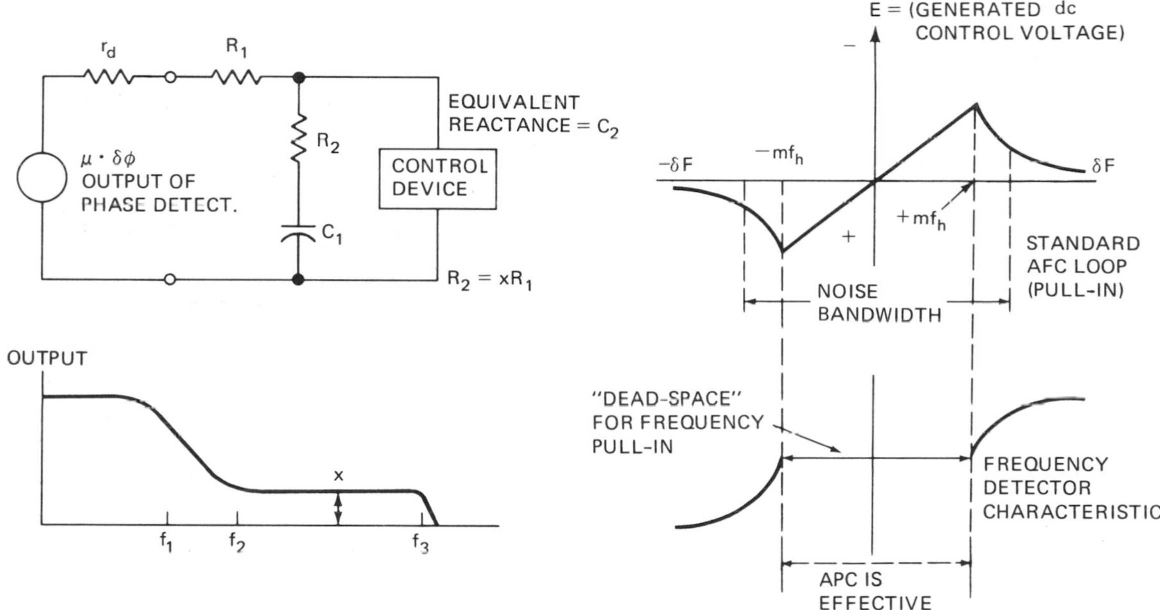

The ratio of ac-gain/dc-gain through the network = m = $\dfrac{x}{1+x}$. Then, the *loop noise-bandwidth* is given by

$$\frac{\pi}{2}(mf_h + f_2) = \pi\left(mf_h + \frac{1}{2\pi xT}\right)$$

where

$f_h = |\mu \cdot \beta|$ is the dc loop-gain in Hz/s
mf_h is the ac loop-gain in Hz/s and $xC_1R_1 = xT$

These *loop-gains have the dimension of cycle/sec* as a result of the definitions given to μ and β. If k, the *damping-constant*, determines the shape of the pass-band of the filter at the cutoff frequency, it is given by the expression:

$$k \approx \frac{\pi}{2}(mf_h \cdot xT)$$

For the preceding network, it can be shown that

$$f_1 = \frac{\pi}{2}(r_d + R_1)C_1 \text{ where } C_1 >> C_2 \text{ and } m = \frac{R_2}{(R_1 + r_d)}$$

$$f_2 = \frac{\pi}{2}\left(\frac{R_2}{C_1}\right) \text{ and } R_1 >> R_2$$

$$f_3 = \frac{\pi}{2}(R_2C_2)$$

The optimum value of k is adjusted between 0.5 and 1 for the best pull-in performance of the AFC-loop. The term *noise-immunity* is sometimes used and is defined by the expression: $I_n = 1/(0.31\ mk)$.

The actual compensation networks used for the AFC loops in this system do not follow the general design trend presented here, since they are specially optimized to meet the response requirements of primarily the high-precision line-lock control subsystem in the headwheel sampled-data control system.

By means of an auxiliary frequency-difference detector with the characteristic shown, the function of the AFC in the absence of the external reference-synchronizing signal and the function of the high-gain automatic phase control in the presence of the external reference-synchronizing signal can be automatically separated; that is, the AFC frequency-detector becomes automatically ineffective in the presence of the external sync by virtue of its special pull-in characteristic. Several variations in AFC/APC circuits are encountered in color television, and there are several papers in the literature on the subject, but none specifically based on the application of the z-transform techniques. However, Tsypkin (1964), in Russian literature, has treated the subject of automatic frequency control along the sampled-data z-transform approach. He considered the integrating circuits as ideal and non-ideal and approximated the response of the phase discriminator, the local pulse (or carrier) generator, and the amplifier to algebraic equations, instead of differential equations (by ignoring the delay-elements and the time-constants). Even with such gross simplification, the formulation of the results of analysis is rather complex for direct application in circuit design.

3.5.6 Backlash Nonlinear Elements in the Vacuum-Guide Servo. Free-play and mechanical-hysteresis are other defining terms for backlash. A servo-mechanism like the vacuum-guide position-servo with backlash in the gear-trains (or mechanical linkage) produces a continual chattering or hunting since the linkage is located inside the feedback loop. The backlash nonlinearity is frequenctly converted into a dead-zone or dead-space nonlinearity by using spring-loading, so as to prevent a constant-wear on the gear-train and hence a worsening situation with respect to the limited free-play. The dead-zone merely indicates a final tolerance in the position-error in an otherwise stable servo-mechanism. The mechanical spring-loading device was actually used during earlier stages of development, and it was naturally discarded in the later versions of the Q-CVTR, which had refined electronics and an improved guide-mechanism. The final residual guide-error due to the backlash element is restricted to the minimum dead-zone specification of ±0.02-mil position-error; and this corresponds to an undetectable ±0.02 μs time-displacement *skew-distortion* error in an average-size television receiver.

As far as the feedback position-control system is concerned, since both backlash and dead-zone input-output characteristics represent different forms of nonlinearity, they can be simulated for analytical purposes on an analog computer—for example, the differential analyzer shown in Fig. 3-18.

The backlash is similar to the dead-zone with the difference that the output e_0 is not zero during the dead-space but of some constant value, depending upon the direction of motion of the driver. The analog-simulation of backlash calls for an integrator because its output will remain constant when its input is zero. Since the output of the integrator represents the displacement of the follower, it must have a very rapid response and hence a large gain.

A servo-mechanism such as the vacuum-guide position-control system, containing one or more backlash nonlinear elements, can be stabilized by one of the three conventional techniques described below. In the guide position-servo of the Q-CVTR, actually (1) a sensitivity-adjustment, a low-pass noise-immunity filter, and a digital-controller compensation for the interacting control system in the automatic feedback loop, and (2) a dc compensation after the error-sampling, and (3) an ac compensation network in the ac carrier-servo, are all effectively used for reliable operation by the heuristic measure-and-optimize design technique. The desired optimum specification of the final position-error is met unfailingly within ±0.02 mil of vacuum-

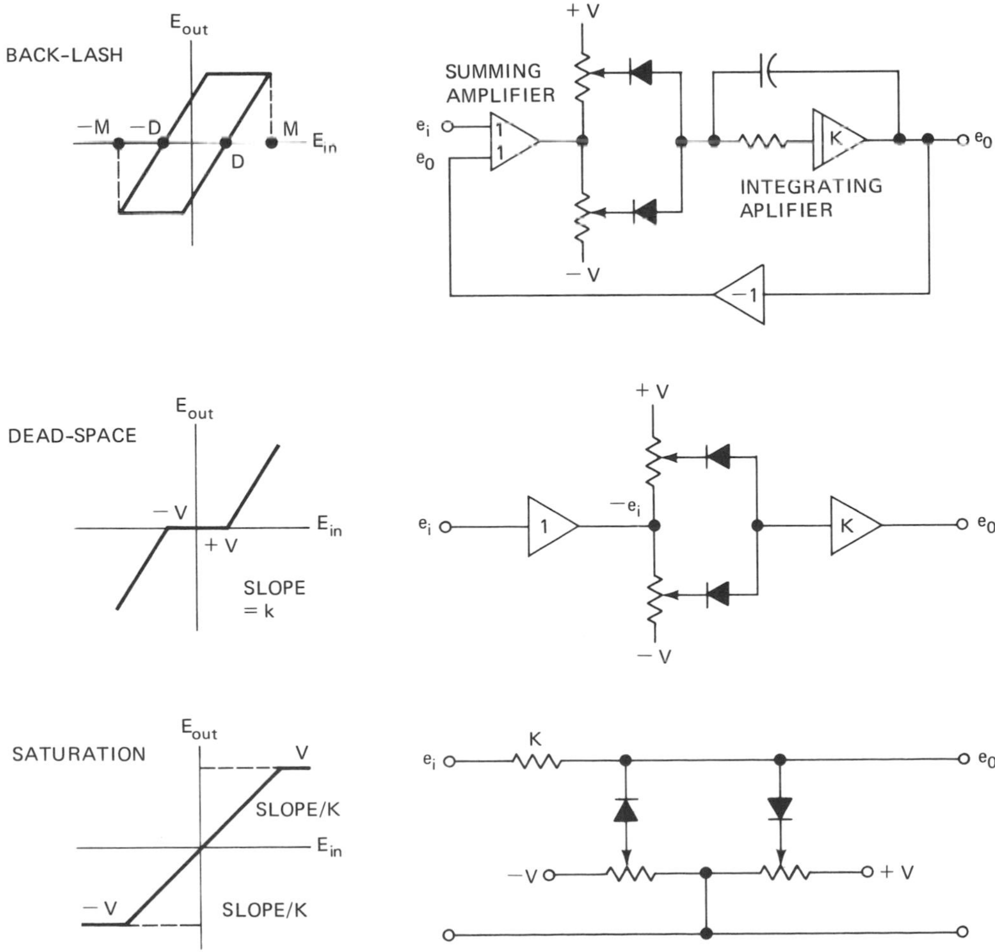

Fig. 3-18. Analog-simulation of backlash, dead-zone, and saturation nonlinearities.

guide pressure to enable automatic headwheel penetration against the moving videotape for the reproduction of black-and-white or color picture signals from the recorded videotape.

The following analysis presents some parallel theoretical concepts:

1. The stabilization of the backlash nonlinearity in a position control system by *optimum gain adjustment:*

The describing-function $G_D(M, \omega)$ of the nonlinear element is considered for determining the stability of the closed-loop control system.

$$\left.\frac{C(s)}{R(s)}\right|_{s=j\omega} = \frac{G_D(M, \omega)G(j\omega)}{1 + G_D(M, \omega)G(j\omega)}$$

where $G(j\omega)$ corresponds to the plant transfer-function. The zeros of the characteristic equation determine the stability.

From the characteristic equation,

$$G(j\omega) = \frac{-1}{G_D(M, \omega)}$$

If a combination of amplitude M and frequency ω can be found that satisfies the the preceding equation, the system can have sustained-oscillation with resulting instability. The gain-phase plots, derived from the Bode magnitude and phase readings, lend well to determine the stability of the overall control system from the describing-function, by using two separate sets of loci corresponding to $G(j\omega)$ and $-1/G_D(M, \omega)$ on the same graph. The plot of the describing-function will commonly take the form of a family of curves for different input magnitudes M and frequencies ω. The intersection of the two separate loci, corresponding to $-1/G_D(M, \omega)$ and $G(j\omega)$, represents the magnitude and frequency of the sustained-oscillation, and the absence of the intersection represents a stable system with some finite error. The relative distance between the two at an otherwise potential intersection is a criterion of the relative stability and an approximate disposition of the transient-response.

Reduction of gain is the most common technique employed to effect the separation of the two loci under consideration. Independent sensitivity-adjustments both before and after the chopper-modulator proved to be a successful technique in the case of the vacuum-guide servo with several backlash elements.

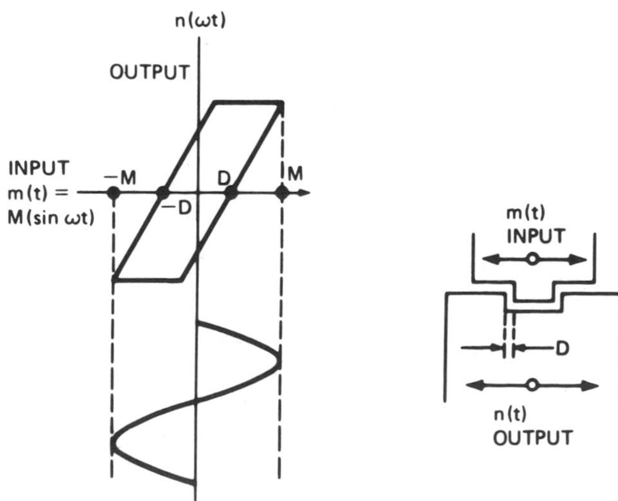

If K represents the overall system gain, and $G^1(j\omega)$ represents the transfer-function and hence the poles and zeros of the system alone, then the characteristic equation takes the form:

$$1 + G_D(M, \omega)G(j\omega) = 0$$

where $G(j\omega) = KG^1(j\omega)$

Now,

$$G(j\omega) = -1/G_D(M, \omega)$$

$$G^1(j\omega) = -1/KG_D(M, \omega)$$

As an example, assuming the system as an approximated continuous system,

$$\text{Let } G(j\omega) = \frac{1.5}{j\omega(1 + j\omega)^2} \text{ where } K = 1.5.$$

For backlash, the *describing-function* is given by:

$$-1/G_D(M, \omega) = -(1/M\sqrt{A_1^2 + B_1^2}), \underline{/\tan^{-1}} \frac{A_1}{B_1}$$

where

$$A_1 = \frac{2}{\pi}\frac{D}{M}\left(\frac{2D}{M} - 2\right)M$$

$$B_1 = \frac{1}{\pi}\left[\frac{\pi}{2} - \sin^{-1}\left(\frac{2D}{M} - 1\right) + \left(\frac{2D}{M} - 1\right)\cos\ \sin^{-1}\left(\frac{2D}{M} - 1\right)\right]M$$

Now, the gain-phase characteristic $-1/KG_D(M, \omega)$ is plotted with gain K = 1.5 and several values of D/M (from 0.9 to 0.1). The gain-phase characteristic $G^1(j\omega)$ is also plotted on the same graph as shown (Fig. 3-19). It results in a stable limit-cycle (sustained-oscillation of a limited amplitude) at the intersection where the amplitude of oscillation M is *maximum*. The system is unstable between the points of intersection. In general, if the locus of all the values of $-1/KG_D$ is plotted, then the system will be stable only if $-1/KG_D$ locus does not intersect $G^1(j\omega)$ locus, and $-1/KG_D$ locus lies entirely on the left-hand side of the $G^1(j\omega)$ locus when the latter is traversed in the direction of the decreasing frequency.

Now, in Fig. 3-19, when the gain K = 1.3, the intersection of the two loci under consideration is just avoided to effect a stable control system with minimum residual error. Since a gain-margin is desirable, K is made unity in the above case to obtain the modified describing-function (by reduction of gain) well above the $G^1(j\omega)$-locus, as shown by the dotted characteristic. It may be noted that a trade-in between the gain-margin and the residual-error is an inevitable consequence of the backlash nonlinearity in a position-control system such as the vacuum-guide servo, and the actual amount of free-play in the gear-train is the ultimate determining-factor for minimum steady-state residual-error of a dead-space character.

2. *The phase-lead compensation* is the second conventional technique for stabiliz-

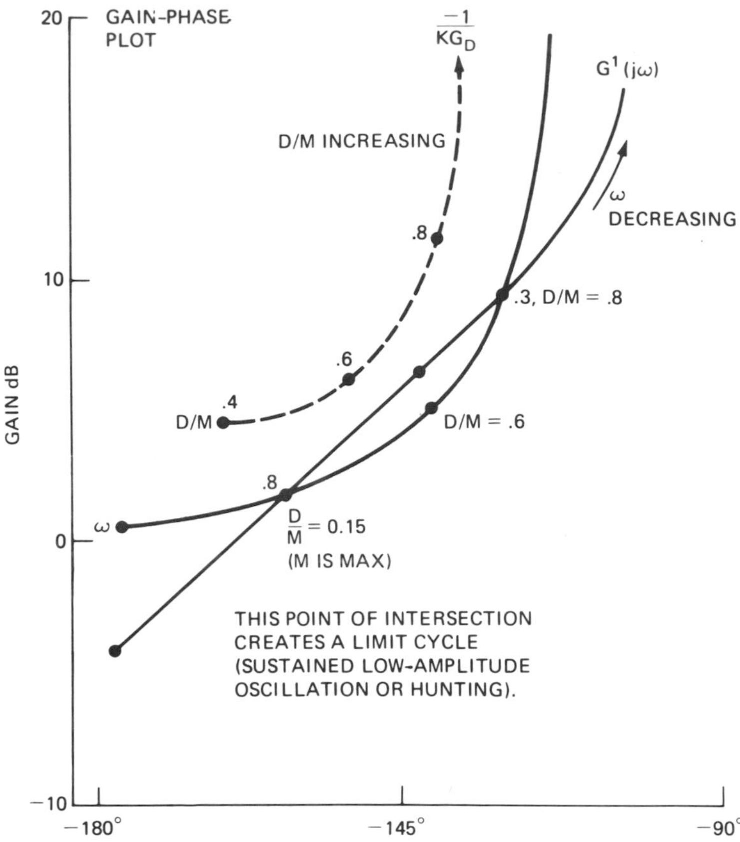

Fig. 3-19. Describing-function technique for the backlash nonlinearity.

ing a feedback position-control system with backlash. In the preceding example, the compensated system transfer-function will take the following form:

$$G_c(j\omega) = \frac{1.5}{j\omega(1 + j\omega)^2}\left(\frac{1 + j\omega T_1}{1 + j\omega T_2}\right)$$

where $T_1 > T_2$ for a phase-lead network with the characteristics shown.

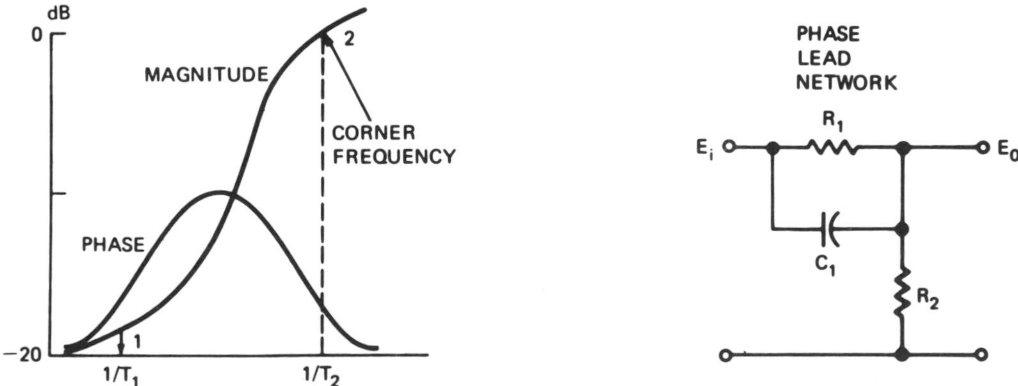

With the *lead-network*, the phase-angle of the output leads that of the input, and the magnitude of the output increases with the frequency of the sinusoidal excitation. (It is generally used for type 1 and 2 systems.) The demerit of the phase-lead compensation is that more gain is required with the implied dc attenuation. Noise-immunity is reduced due to the boost in high-frequency response. There is also a tendency toward saturation in the presence of transients.

In the case of the backlash nonlinearity, the system is stabilized by choosing a suitable set of values for T_1 and T_2. If $T_1 = 0.8$ and $T_2 = 0.4$ for the preceding case, the intersection of the two loci for $G_c(j\omega)$ and $-1/G_C(M, \omega)$ is safely prevented, as in the case of the reduction-of-gain technique.

3. *Velocity or rate-feedback* is the third conventional technique for stabilizing the position-control system with backlash. The tachometer feedback compensation, along with the unity-feedback, gives the combined transfer-function $[1 + K(j\omega)]$ for the feedback-path $H(j\omega)$. Now the characteristic equation takes the form:

$$1 + KG_D(M, \omega)G(j\omega)H(j\omega) = 0$$

$$G(j\omega)H(j\omega) = \frac{1.5}{j\omega(1 + j\omega)^2}[1 + K^1 j\omega]$$

The choice of the value of K^1 determines the elimination of the point of intersection of the two loci, $-1/G_D(M, \omega)$ and $G(j\omega)H(j\omega)$, as in the two previous techniques.

Since superposition of signals is not valid for nonlinear systems, and since the describing-function technique is an amplitude-sensitive approach, the solution arrived at with the describing-function cannot be extended to determine low-amplitude transient-response. It is, however, a valid conclusion that the transient-response should improve when the gain-phase loci of the describing-function and the remaining system transfer-function lie farther apart. (There is a practical brute-force technique called *dither* for maintaining the output of a position-control system having a backlash at its correct average-value, by applying oscillating input signals (50 to 100 Hz) with frequencies higher than the system-bandwidth around 5 Hz. The method is not usually recommended for control systems such as those in the Q-CVTR that operate continuously, since the increased wear on the mechanics of the system is a serious disadvantage.)

4. *Modified Routh-Hurwitz approach for the determination of stability.* The describing-function of a nonlinear-element, $G_D(|M|\omega)$ depends on the frequency and the amplitude of the input signal. If the describing-function is assumed as a real-number (when hysteresis is absent and dead-space is effective), and as a function of a specific amplitude, such as $G_D(|M|)$, then in the system example under consideration, as a condition of absolute stability of a quasi-linear system, the characteristic equation $1 + G_D(M, \omega)G(j\omega) = 0$ will take the following form:

$$(j\omega)^3 + 2(j\omega)^2 + (j\omega) + 1.5G_D(M_1) = 0$$

$$\text{if } G(j\omega) = \frac{1.5}{j\omega(1 + j\omega)^2}$$

The Routh-Hurwitz criterion is applicable in this quasi-linear case since the number chosen (the absolute magnitude) for the describing-function is real and *not* complex. (When D = 0 in the expression of the describing-function, it becomes a real-number.)

The Routh-Hurwitz array is given by:

$$\begin{vmatrix} 1 & 1 \\ 2 & 1.5G_D(|M|) \\ 1 - .75G_D(|M|) & \\ 1.5G_D(|M|) & \end{vmatrix}$$

For stability, all the terms in the first column must have the same sign for the location of the poles of the quasi-linear system in the left-half plane.

$$1 - 0.75G_D(|M|) > 0$$

$$G_D(|M|) < 1.25$$

Therefore, for stability, the describing-function of the backlash nonlinearity must be less than the absolute magnitude of 1.25.

As a matter of passing interest, the amplitude-sensitive describing-function characteristics for (1) backlash, (2) dead-zone, and (3) saturation are indicated below:

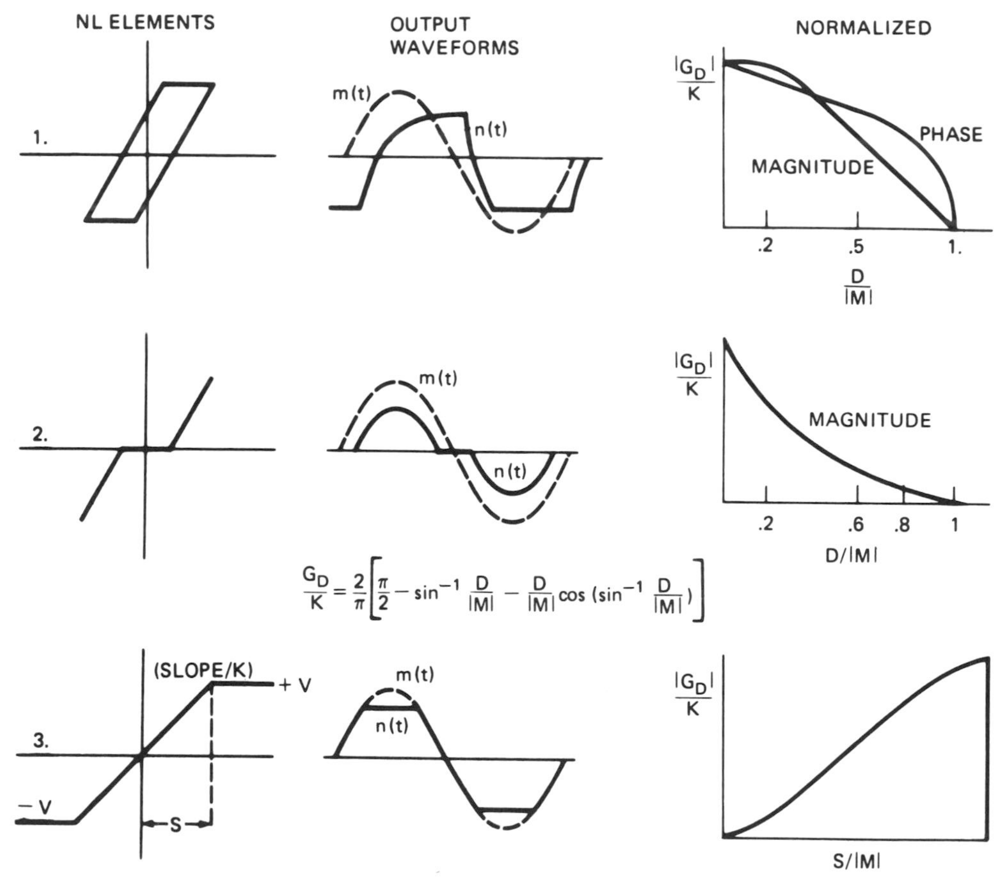

DESCRIBING FUNCTIONS

3.5.7 Saturation Nonlinearity. It is clear from the schematic block-diagrams of the sampled-data control systems in the Q-CVTR, including the vacuum-guide position-servo, that saturation as a nonlinear element is one of frequent incidence, since error-sampling is exclusively performed employing a reference trapezoid pulse-waveform. Specialized approaches like classical phase-plane limit-cycle analysis, describing-function technique, Lyapounov's Direct Method, Popov's stability criterion, and modern state-transition technique with matrix manipulations of the state-variables are available to treat this case of nonlinearity. Since modern state-transition techniques provide more or less a generalized approach for analysis and synthesis of digital control systems, this technique will be illustrated for a simplified version of the vacuum-guide position-control system with only one saturation nonlinear-element. (The complex interacting vacuum-guide position-control system has actually three saturation nonlinearity-elements in three directly associated feedback loops, and two backlash hysteresis elements due to the relevant gear-trains in the guide-mechanism.)

Before attempting the case of the more complex sampled-data system, a simple example of the saturation of an amplifier in a continuous feedback system is presented first to illustrate the *piecewise-linear approximation technique.*

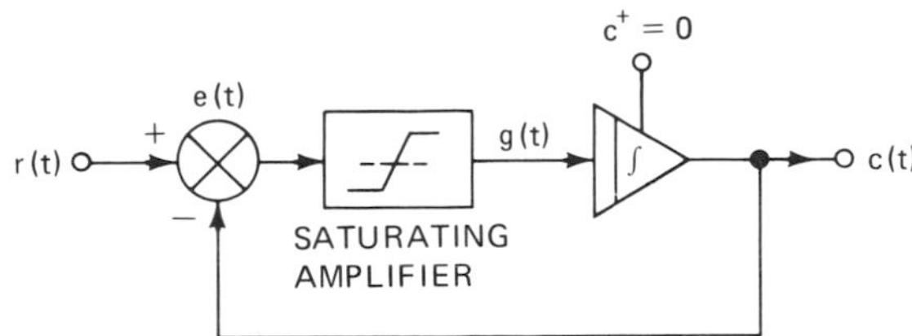

Assume the input voltage range as ±1 V. If the element saturates at ±5 V, it is seen that two linear operating regions are available. These individual linear characteristics are considered in a piecewise-linear fashion to obtain an approximate composite-response of the overall nonlinear system.

Before the nonlinear-element is effective:

$$e(t) = r(t) - c(t) \tag{1}$$

$$g(t) = 5e(t) \tag{2}$$

$$c(t) = \int g(t)\, dt \tag{3}$$

After saturation, only Eq. 2 undergoes modification.

$$g(t) = \pm 5 \text{ for } e(t) > 1 \text{ and } < 1, \text{ respectively.} \tag{4}$$

Assume zero initial-condition. With a step-input of 10 V, the output during the linear and saturated portions of the characteristic is given by equations:

$$\text{linear } c(t) = \int_0^t 5\, dt = 5t \tag{5}$$

$$\text{saturated } c(t) = \int_{t_1}^t (10 - c)\, dt \tag{6}$$

where t_1 denotes the instant of saturation in the waveform.

Differentiating,

$$\frac{dc(t)}{dt} + c = 10$$

Hence, the solution

$$c(t) = 10 - e^{-(t-1.8)}$$

when $e^\circ = 1$, $c = (10 - 1) = 9$, and t_1 is 1.8 sec.

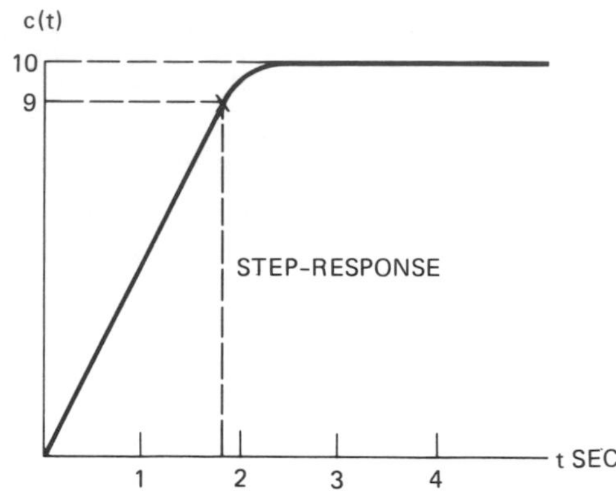

The final and initial values of the linear and saturated regions, respectively, are equal, since the boundary conditions between the segments are continuous. At the boundary value 1.8 sec:

$$c(t) = c(1.8) = 10 - e^{-(1.8-1.8)} = 9 = 5t = 5(1.8).$$

Hence, the composite solution obtained by piecewise-linear analysis is expressed by the following equations and plotted.

$$\text{Linear } c(t) = 5t \text{ for } 0 \leqslant t \leqslant 1.8$$

$$\text{Saturated } c(t) = 10 - e^{-(t-1.8)} \text{ for } t \geqslant 1.8$$

3.5.8 Application of Modern State-Transition Technique to Saturation (Variable-Gain Concept). Assume the transfer-function $G(s) = 1/s(s+1)$ with a sampling period T of 1 sec. The problem is one of combined analysis and synthesis of a pulse-transfer-function in the presence of saturation (±1 V) for a suitable digital-controller D(z), so that a 1.5 V step-function response will settle down in, say, four sampling-periods with zero initial conditions. It can be shown in general by this very method that in the absence of the saturation nonlinear-element, the sampled-data feedback control system with a step-function input could be made to settle down in less number of sampling periods. The effect of the nonlinear-element is thus an elongation of the settling-time for steady-state dead-beat performance.

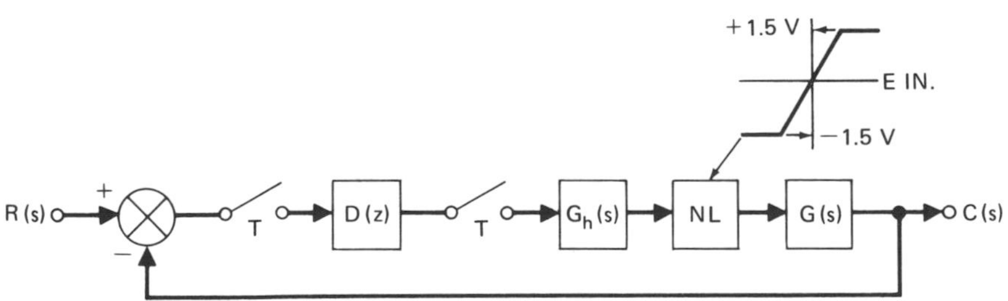

The technique of synthesis is carried out along the following lines:

1. The nonlinear-element is represented by a *variable-gain element* K_n, in lieu of the requisite digital-controller plus the nonlinear-element (Fig. 3-20). K_n will have

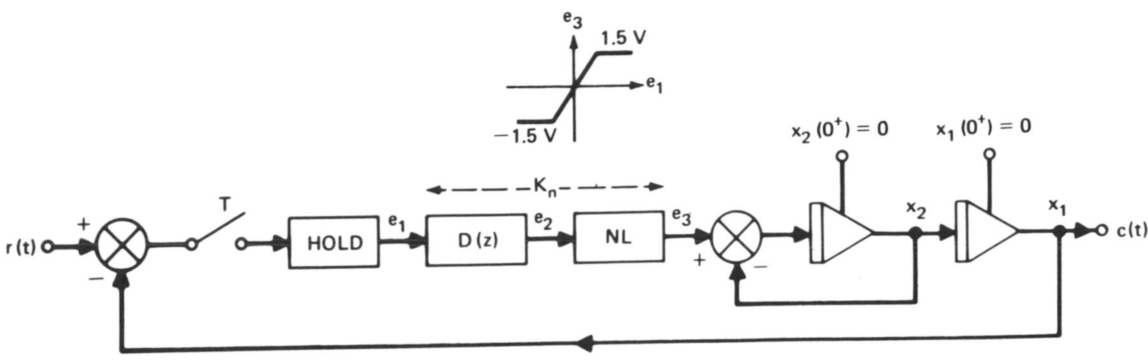

Fig. 3-20. State-variable diagram, saturation.

different values during different sampling periods and is dependent upon the nonlinear characteristic. At any sampling instant, the input e_1 and the output e_3 of this combination are interrelated in the following state-variable diagram by the expression:

$$e_3(nT^+) = K_n e_1(nT^+) \tag{1}$$

at any sampling-instant $t = nT^+$, where K_n is the gain-constant of the combination during the $(n + 1)th$ sampling-instant. The transition-matrix $\overline{M}$ and the $\overline{B}$ matrix of the system, as explained earlier, are obtained by inspection of the state-variable diagram.

2. The input signal e_1 is then found from $\overline{M}$ and $\overline{B}$ by the use of the relations:

$$\overline{v}(nT^+) = \overline{B}\overline{v}(nT)$$
$$\overline{v}(n+1)T = \overline{M}(K)\overline{v}(nT^+)$$

{The symbol of the matrix is denoted by the horizontal-stroke above.

3. The gain-constants of K_n and the combination-output e_3 are then determined from the above results for analytically evaluating the input signal e_2 to the nonlinear-element in the form:

$$E_2(z) = a_0 + a_1 z^{-1} + a_2 z^{-2} + \cdots + a_k z^{-k}$$

where the coefficients a_i are graphically determined from the nonlinear characteristic.

The input and output signals e_1 and e_3 to the *D(z), nonlinear* combination assume the following form:

$$E_1(z) = e_1(0^+) + e_1(T^+)z^{-1} + \cdots + e_1(kT^+)z^{-k}$$
$$E_3(z) = K_0 e_1(0^+) + K_1 e_1(T^+)z^{-1} + \cdots + K_k e_1(kT^+)z^{-k}$$

where $K_0, K_1 \ldots$ are gain-constants K_n.

4. The pulse-transfer-function of the digital-controller is then synthesized for minimum settling-time from the final expression:

$$D(z) = \sum_{i=0}^{k} a_i z^{-i} \Big/ \sum_{i=0}^{k} e(iT^+)z^{-i}$$

(This procedure is a generalized approach to synthesize sampled-data control systems containing different types of nonlinear elements. The technique is applicable to multirate, variable-rate, and time-varying sampled-data control systems too.)

The square-matrices $\overline{A}$ and $\overline{B}$ are first written by examining the state-variable diagram and by using the basic state-variable vector-matrix equation, $\dot{\overline{v}}(\lambda) = \overline{A}\overline{v}(\lambda)$. It represents the first-order differential equation with the input, the output, and the state-variables that determine the dynamic behavior of the system.

$$r(t) = \text{unit-step} = 1.5 \text{ V}$$

$$\dot{r} = 0; \dot{x}_1 = x_2; \dot{x}_2 = e_3 - x_2; \dot{e}_1 = 0.$$

At any sampling-instant,

$$t = nT^+: e_3(nT^+) = K_n e_1(nT^+)$$

where $\overline{v}$ = column-matrix $[r_1 x_1 x_2 e_1]$, and the square-matrix represents $\overline{A}$.

$$\begin{bmatrix} \dot{r} \\ \dot{x}_1 \\ \dot{x}_2 \\ \dot{e}_1 \end{bmatrix} = \begin{bmatrix} 0 & 0 & 0 & 0 \\ 0 & 0 & 1 & 0 \\ 0 & 0 & -1 & K_n \\ 0 & 0 & 0 & 0 \end{bmatrix} \begin{bmatrix} r_1 \\ x_1 \\ x_2 \\ e_1 \end{bmatrix}$$

The state-transition equations are given by

$$r(nT^+) = r(nT); x_1(nT^+) = x_1(nT); x_2(nT^+) = x_2(nT)$$

and

$$e_1(nT^+) = r(nT) - x_1(nT)$$

Now, the square-matrix $\overline{B}$ is defined by the state-transition equations:

$$\overline{v}(nT^+) = \overline{B}\overline{v}(nT)$$

$$\begin{bmatrix} r(nT^+) \\ x_1(nT^+) \\ x_2(nT^+) \\ e_1(nT^+) \end{bmatrix} = \begin{bmatrix} 1 & 0 & 0 & 0 \\ 0 & 1 & 0 & 0 \\ 0 & 0 & 1 & 0 \\ 1 & -1 & 0 & 0 \end{bmatrix} \begin{bmatrix} r(nT) \\ x_1(nT) \\ x_2(nT) \\ e_1(nT) \end{bmatrix}$$

where with a step input of 1.5 V

the initial state vector $\overline{v}(0)|_{m=0}$ = column matrix $[1.5 \quad 0 \quad 0 \quad 0]$

the square-matrix represents $\overline{B}$. The overall transition-matrix $\overline{M}(\lambda)$ is given by:

$$\overline{M}(\lambda) = \mathcal{L}^{-1}[s\overline{I} - \overline{A}]^{-1} = \mathcal{L}^{-1} \begin{bmatrix} s & 0 & 0 & 0 \\ 0 & s & -1 & 0 \\ 0 & 0 & s+1 & -K_n \\ 0 & 0 & 0 & s \end{bmatrix}^{-1}$$

$$\overline{M}(\lambda) = \mathcal{L}^{-1} \left\{ \frac{1}{s^3(s+1)} \begin{bmatrix} s^2(s+1) & 0 & 0 & 0 \\ 0 & s^2(s+1) & s^2 & sK_n \\ 0 & 0 & s & K_n s^2 \\ 0 & 0 & 0 & s^2(s+1) \end{bmatrix} \right\}$$

$$\overline{M}(\lambda) = \mathcal{L}^{-1}\begin{bmatrix} \frac{1}{s} & 0 & 0 & 0 \\ 0 & \frac{1}{s} & \frac{1}{s(s+1)} & \frac{K_n}{s^2(s+1)} \\ 0 & 0 & \frac{1}{s+1} & \frac{K_n}{s(s+1)} \\ 0 & 0 & 0 & \frac{1}{s} \end{bmatrix} = \begin{bmatrix} 1 & 0 & 0 & 0 \\ 0 & 1 & 1 - e^{-\lambda} & (\lambda - 1 + e^{-\lambda})K_n \\ 0 & 0 & e^{-\lambda} & (1 - e^{-\lambda})K_n \\ 0 & 0 & 0 & 1 \end{bmatrix}$$

Thus, the above transition-matrix $\overline{M}(\lambda)$ is expressed as a function of the variable-gain K_n. It has different values at different sampling instants.

The state-transition equations for the basic control system are given by:

$$\overline{v}(nT^+) = \overline{B}\overline{v}(nT) \tag{1}$$

$$\overline{v}(\overline{n+1\,T}) = \overline{M}(T)\overline{v}(nT^+) = \overline{M}(T)\overline{B}\overline{v}(nT) \tag{2}$$

For $n = 0$, $\overline{v}(0^+) = B\overline{v}(0)$ where $\overline{v}(0)$ is the given initial state-vector.

At the same time $t = T$,

$$\overline{v}(T) = \overline{M}_0(T)B\overline{v}(0) = \overline{M}_0(T)\overline{v}(0^+) \tag{3}$$

Since the transition-matrix $\overline{M}_0(T)$ is a function of the gain-constant K_0 of the variable-gain element—D(z), nonlinear combination—during the first sampling-instant, the state-vector $\overline{v}(T)$ at $t = T$ is also a function of K_0.

In the above example, for **n = 0**,

$$\overline{v}(0^+) = \overline{B}\overline{v}(0) = \begin{bmatrix} 1 & 0 & 0 & 0 \\ 0 & 1 & 0 & 0 \\ 0 & 0 & 1 & 0 \\ 1 & -1 & 0 & 0 \end{bmatrix}\begin{bmatrix} 1.5 \\ 0 \\ 0 \\ 0 \end{bmatrix} = \begin{bmatrix} 1.5 \\ 0 \\ 0 \\ 1.5 \end{bmatrix}$$

Since $T = 1$ sec,

$$\overline{M}(\lambda) = \overline{M}_n(T) = \overline{M}(K_n) = \begin{bmatrix} 1 & 0 & 0 & 0 \\ 0 & 1 & 0.63 & 0.37K_n \\ 0 & 0 & 0.37 & 0.63K_n \\ 0 & 0 & 0 & 1 \end{bmatrix}$$

Now, from Eq. 3,

$$\overline{v}(T) = \overline{M}(K_0)\overline{v}(0^+) \tag{4}$$

since $\overline{M}_0(T)$ is a function of K_0. And $e(0^+) = r(0^+) = 1.5$, since both $\overline{B}$ and $\overline{v}(0)$ are known and $\overline{v}(0^+)$ is defined. Since $e(0^+)$ exceeds the saturation limit (± 1V) of the nonlinear element, K_0 should be considered as the maximum allowable value of the variable-gain K_n during the first sampling-period.

$$K_0 = \frac{e_3(0^+)}{r(0^+)} = \frac{1}{1.5} = 0.67$$

since saturation is effective at ±1 V. During the first sampling-period, the transition-matrix becomes

$$\overline{M}(K_0) = \begin{bmatrix} 1 & 0 & 0 & 0 \\ 0 & 1 & 0.63 & 0.24 \\ 0 & 0 & 0.37 & 0.41 \\ 0 & 0 & 0 & 1 \end{bmatrix} \qquad \begin{matrix} 0.37 \times .67 = .24 \\ 0.63 \times .67 = .41 \end{matrix}$$

From Eq. 4, the state-vector at $t = T: \overline{v}(T) = \overline{M}(K_0)\overline{v}(0^+)$

$$\overline{v}(T) = \begin{bmatrix} 1 & 0 & 0 & 0 \\ 0 & 1 & 0.63 & 0.24 \\ 0 & 0 & 0.37 & 0.41 \\ 0 & 0 & 0 & 1 \end{bmatrix} \begin{bmatrix} 1.5 \\ 0 \\ 0 \\ 1.5 \end{bmatrix} = \begin{bmatrix} 1.5 \\ 0.36 \\ 0.62 \\ 1.5 \end{bmatrix}$$

From Eq. 1,

$$\overline{v}(T^+) = \overline{B}\overline{v}(T) = \begin{bmatrix} 1 & 0 & 0 & 0 \\ 0 & 1 & 0 & 0 \\ 0 & 0 & 1 & 0 \\ 1 & -1 & 0 & 0 \end{bmatrix} \begin{bmatrix} 1.5 \\ 0.36 \\ 0.62 \\ 1.50 \end{bmatrix} = \begin{bmatrix} 1.5 \\ 0.36 \\ 0.62 \\ 1.14 \end{bmatrix}$$

$$e_1(T^+) = 1.14$$

But the value of $e_1(T^+)$ is higher than the saturation limit of 1 V:

$$e_3(T^+) = 1 \text{ and } K_1 = \frac{e_3(T^+)}{r(T^+)} = \frac{1}{1.14} = 0.88$$

During the second sampling-period, the transition-matrix becomes

$$\overline{M}(K_1) = \begin{bmatrix} 1 & 0 & 0 & 0 \\ 0 & 1 & 0.63 & 0.32 \\ 0 & 0 & 0.37 & 0.55 \\ 0 & 0 & 0 & 1 \end{bmatrix} \qquad \begin{matrix} 0.37 \times 0.88 = 0.32 \\ 0.63 \times 0.88 = 0.55 \end{matrix}$$

n = 1 for the second sampling-period from Eq. 2

$$\overline{v}(2T) = \overline{M}(K_1)\overline{v}(T^+) = \begin{bmatrix} 1 & 0 & 0 & 0 & 1.5 \\ 0 & 1 & 0.63 & 0.32 & 0.36 \\ 0 & 0 & 0.37 & 0.55 & 0.62 \\ 0 & 0 & 0 & 1 & 1.14 \end{bmatrix} = \begin{bmatrix} 1.5 \\ 0.36 + 0.39 + 0.37 \\ 0.23 + 0.63 \\ 1.14 \end{bmatrix}$$

$$\overline{v}(2T^+) = \overline{B}\overline{v}(2T) = \begin{bmatrix} 1 & 0 & 0 & 0 \\ 0 & 1 & 0 & 0 \\ 0 & 0 & 1 & 0 \\ 1 & -1 & 0 & 0 \end{bmatrix} \begin{bmatrix} 1.50 \\ 1.12 \\ 0.86 \\ 1.14 \end{bmatrix} = \begin{bmatrix} 1.50 \\ 1.12 \\ 0.86 \\ 0.38 \end{bmatrix}$$

$$e_1(2T^+) = 0.38$$

Since $e_1(2T^+)$ is less than the saturation limit, the determination of gain-constant K_2 is not repeated as in the previous cases; it calls for other conditions toward dead-beat performance (zero steady-state error).

For n = 2:

$$\overline{v}(3T) = \overline{M}(K_2)\overline{v}(2T^+) = \begin{bmatrix} 1 & 0 & 0 & 0 \\ 0 & 1 & 0.63 & 0.37K_2 \\ 0 & 0 & 0.37 & 0.63K_2 \\ 0 & 0 & 0 & 1 \end{bmatrix} \begin{bmatrix} 1.50 \\ 1.12 \\ 0.86 \\ 0.38 \end{bmatrix} = \begin{bmatrix} 1.50 \\ 1.66 + 0.14K_2 \\ 0.32 + 0.24K_2 \\ 0.38 \end{bmatrix}$$

$$\overline{v}(3T^+) = \overline{B}\overline{v}(3T) = \begin{bmatrix} 1 & 0 & 0 & 0 \\ 0 & 1 & 0 & 0 \\ 0 & 0 & 1 & 0 \\ 1 & -1 & 0 & 0 \end{bmatrix} \begin{bmatrix} 1.50 \\ 1.66 + 0.14K_2 \\ 0.32 + 0.24K_2 \\ 0.38 \end{bmatrix} = \begin{bmatrix} 1.50 \\ 1.66 + 0.14K_2 \\ 0.32 + 0.24K_2 \\ -0.16 - 0.14K_2 \end{bmatrix}$$

For n = 3:

$$\overline{v}(4T) = \overline{M}(K_3)\overline{v}(3T^+) = \begin{bmatrix} 1 & 0 & 0 & 0 \\ 0 & 1 & 0.63 & 0.37K_3 \\ 0 & 0 & 0.37 & 0.63K_3 \\ 0 & 0 & 0 & 1 \end{bmatrix} \begin{bmatrix} 1.50 \\ 1.66 + 0.14K_2 \\ 0.32 + 0.24K_2 \\ -0.16 - 0.14K_2 \end{bmatrix}$$

$$= \begin{bmatrix} 1.5 \\ (1.66 + 0.14K_2) + (0.2 + 0.15K_2) + (0.36K_3 - 0.05K_2K_3) \\ (0.12 + 0.09K_2) + (0.6K_3 - 0.09K_2K_3) \\ 0.16 \quad 0.14K_2 \end{bmatrix} = \begin{bmatrix} r(nT) \\ x_1(nT) \\ x_2(nT) \\ c_1(nT) \end{bmatrix}$$

Then the system steady-state error is zero for $t \geqslant nT$, if

$$x_1(nT) = r(nT) \text{ and } x_2(nT) = 0$$

where $x_1(nT)$ and $x_2(nT)$ are functions of the successive constants of the variable-gain K_n. Hence, for dead-beat performance, the successive values of K_2 and K_3 are obtained by solving the equations:

$$1.86 + 0.29K_2 + 0.36K_3(1 - 0.15K_2) = 1.5 \tag{5}$$

$$0.12 + 0.09K_2 + 0.6K_3(1 - 0.15K_2) = 0 \tag{6}$$

$$0.18K_2 + 0.22K_3(1 - 0.15K_2) = -0.22$$

$$0.03K_2 + 0.22K_3(1 - 0.15K_2) = -0.04$$

$$K_2 = \frac{-0.18}{0.15} = -1.2 \text{ and } K_3 = -0.025$$

$$e_1(3T^+) \text{ from } \overline{v}(3T^+) = -0.16 - 0.14K_2 = 0.01$$

Hence,

$$E_1(z) = e_1(0^+) + e_1(T^+)z^{-1} + e_1(2T^+)z^{-1} + e_1(3T^+)z^{-1}$$
$$= 1.5 + 1.14z^{-1} + 0.38z^{-2} + 0.01z^{-3}$$

and

$$E_3(z) = K_0e_1(0^+) + K_1e_1(T^+)z^{-1} + K_2e_1(2T^+)z^{-1} + K_3e_1(3T^+)z^{-1}$$
$$= 0.67(1.5) + 0.88(1.14)z^{-1} - 1.2(0.38)z^{-2} - 0.025(0.01)z^{-3}$$
$$= 1 + z^{-1} - 0.46z^{-2} - 0.00025z^{-3}$$

Since the saturation characteristic is linear between the two limits, $E_3(z) = E_2(z) =$ z-transform for the input signal to the nonlinear-element. Now

$$D(z) = \frac{\sum_{j=0}^{k} a_j z^{-j}}{\sum_{j=0}^{k} e(jT^+)z^{-j}}$$

$$D(z) = \frac{1 + z^{-1} - 0.46z^{-2} - 0.00025z^{-3}}{1.5 + 1.14z^{-1} + 0.38z^{-2} + 0.01z^{-3}}$$

Hence, the desired pulse-transfer-function compensation

$$= \frac{0.66(1 + z^{-1} - 0.46z^{-2} - 0.00025z^{-3})}{(1 + 0.76z^{-1} + 0.25z^{-2} + 0.007z^{-3})}$$

The step-function response of Fig. 3-21 is plotted as explained in Section 3.3.3.

It is clear from the digital-controller compensation synthesized that the output response settles down to dead-beat performance in four sampling periods with the saturation nonlinear-element. (It can be shown by a similar procedure that the response of an otherwise linear system, without saturation, would have settled down as shown in two sampling-periods.) If the nonlinear-element is not linear between the saturation limits, the numerator $E_3(z)$ can be modified from the nonlinear characteristic graphically.

Thus, the *modern* control theory with *state-transition* (state-space) technique does provide a *generalized analytical approach* to simple linear and nonlinear control systems.

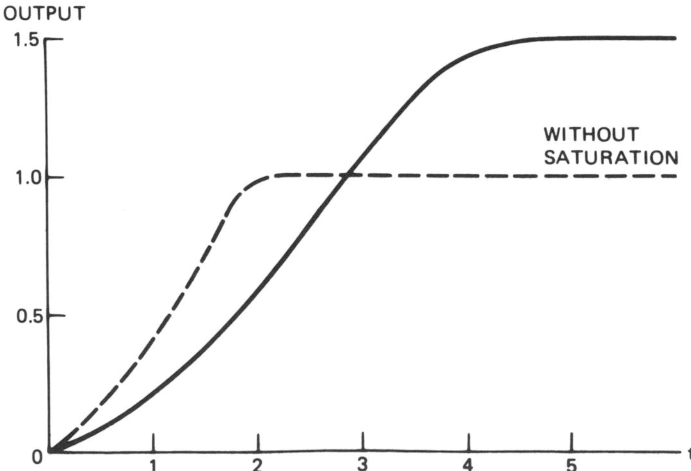

Fig. 3-21. Step-function response with D(z) compensation.

3.5.9 Principle of Invariance or Invariant-Embedding. The brief theoretical interpretation of the interacting multidimensional control processes in the Q-CVTR will be concluded with a short note on the principle of invariant-embedding. This is the latest state-space optimal technique for reducing a multidimensional interconnected control system-complex, like that of the Q-CVTR, to a set of isolated one-dimensional control processes. In this connection, there is a reference in Russian literature to the principle of invariance. The procedure is akin to the coordinate-transformation known in the American literature as the *invariant-embedding* procedure. In optimal control, the principle is applied to enlarge the single dimension of a state-vector by adding a new coordinate, so as to allow maximization or minimization with respect to the new coordinate, according to Pontryagin's maximum principle. Modern *dynamic programming* theory, used for solving multidecision or multidimensional problems, is also based on the principle of invariant-embedding. (A brief theoretical introduction to Pontryagin's maximum principle and the dynamic-programming technique was given in Chapter 1.

By using the invariant-embedding concept, a highly complex unsolvable problem is "embedded" into a class of simpler solvable problems for obtaining the solution. That is, the multidimensional optimization process is replaced by the problem of solving a sequence of single-dimensional processes. The principle of dynamic programming is founded on an *optimal policy* with the following meaning. Whatever the initial state and the initial decision, the *remaining decision or process must form an optimal control strategy* with respect to the state resulting from the first decision or process. This optimal policy is based on the concept of invariant-embedding, the powerful mathematical tool that would eventually provide a method to study and develop theoretical solutions to a multidimensional interacting complex digital control problem like that of the Q-CVTR.

3.5.10 Complexity of Digital Control Systems in Q-CVTR. The theoretical configuration and the interpretation of the three complex interacting sampled-data feedback control systems used in the quadruplex color videotape recorder are illustrated in Figs. 3-22 through 3-24 in order to provide a channel of communication between theory and practice in the case of a highly complex digital control system.

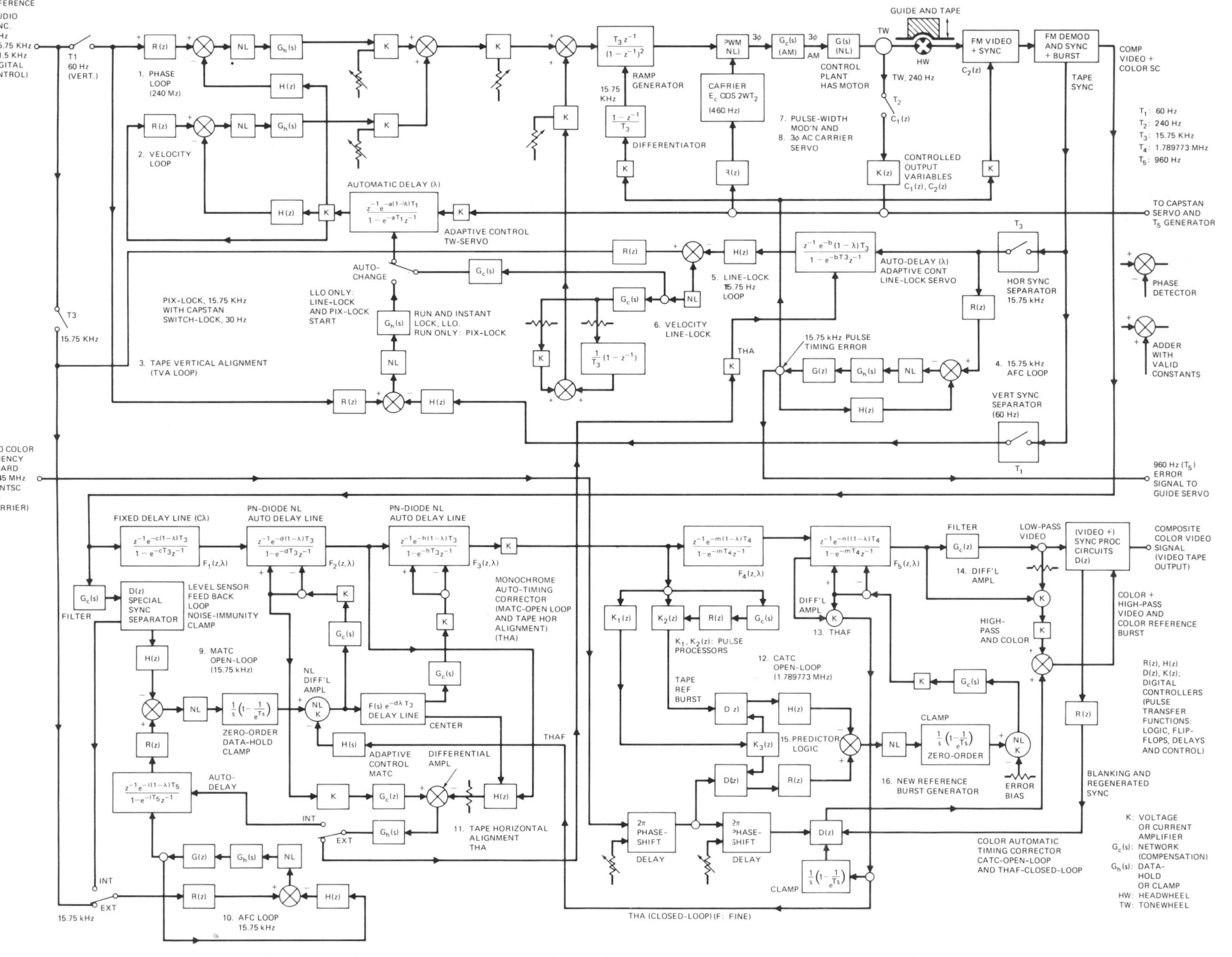

Fig. 3-22. Theoretical representation of head-wheel, sampled-data, feedback control system. (Multirate, multivariable, adaptive, interacting nonlinear control system in broadcast quadruplex color videotape recorder.)

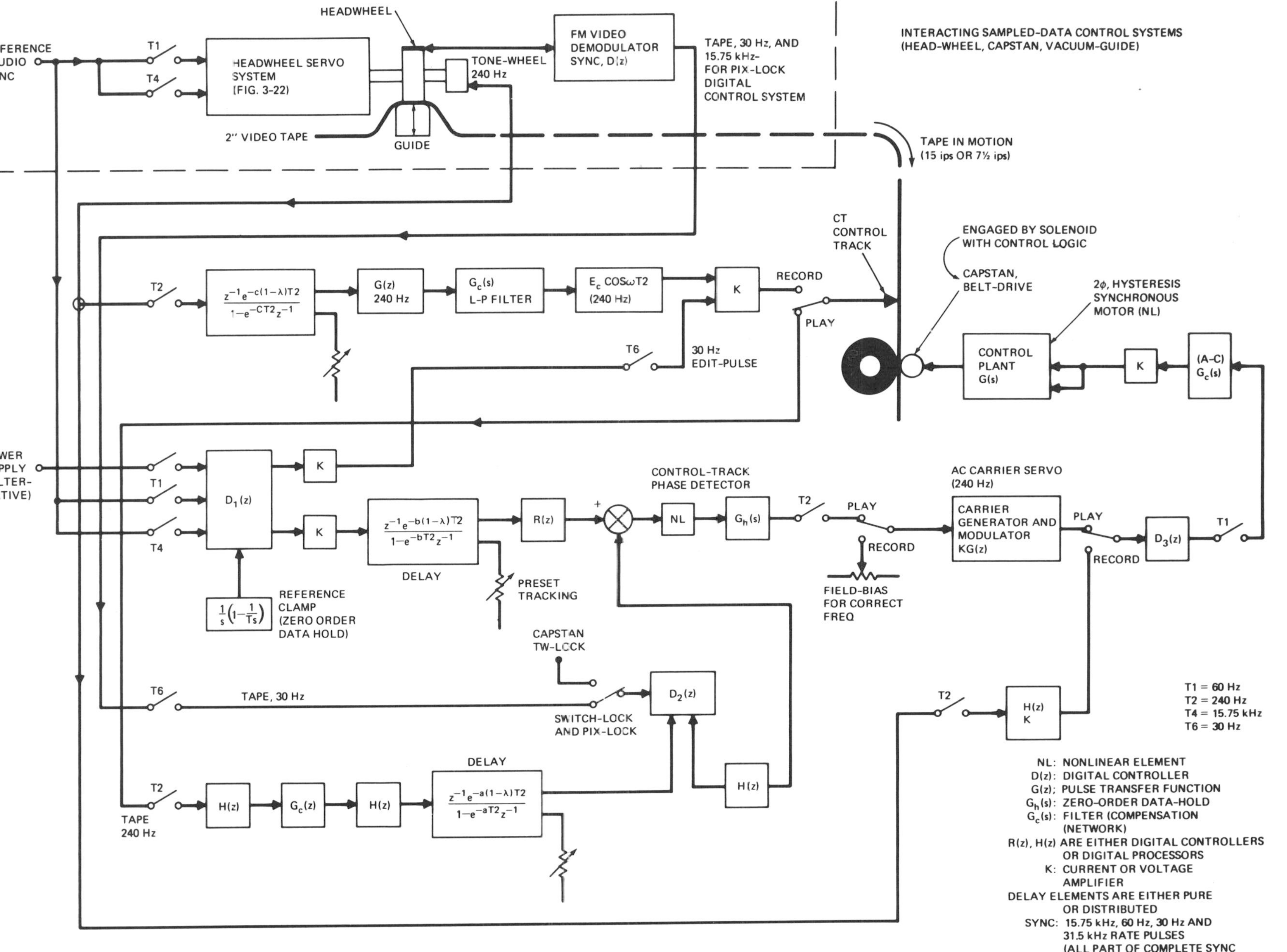

Fig. 3-23. Theoretical representation of capstan sampled-data feedback control system. (240-Hz capstan servo and 30-Hz capstan switch-lock servo for pix-lock of Q-CVTR.)

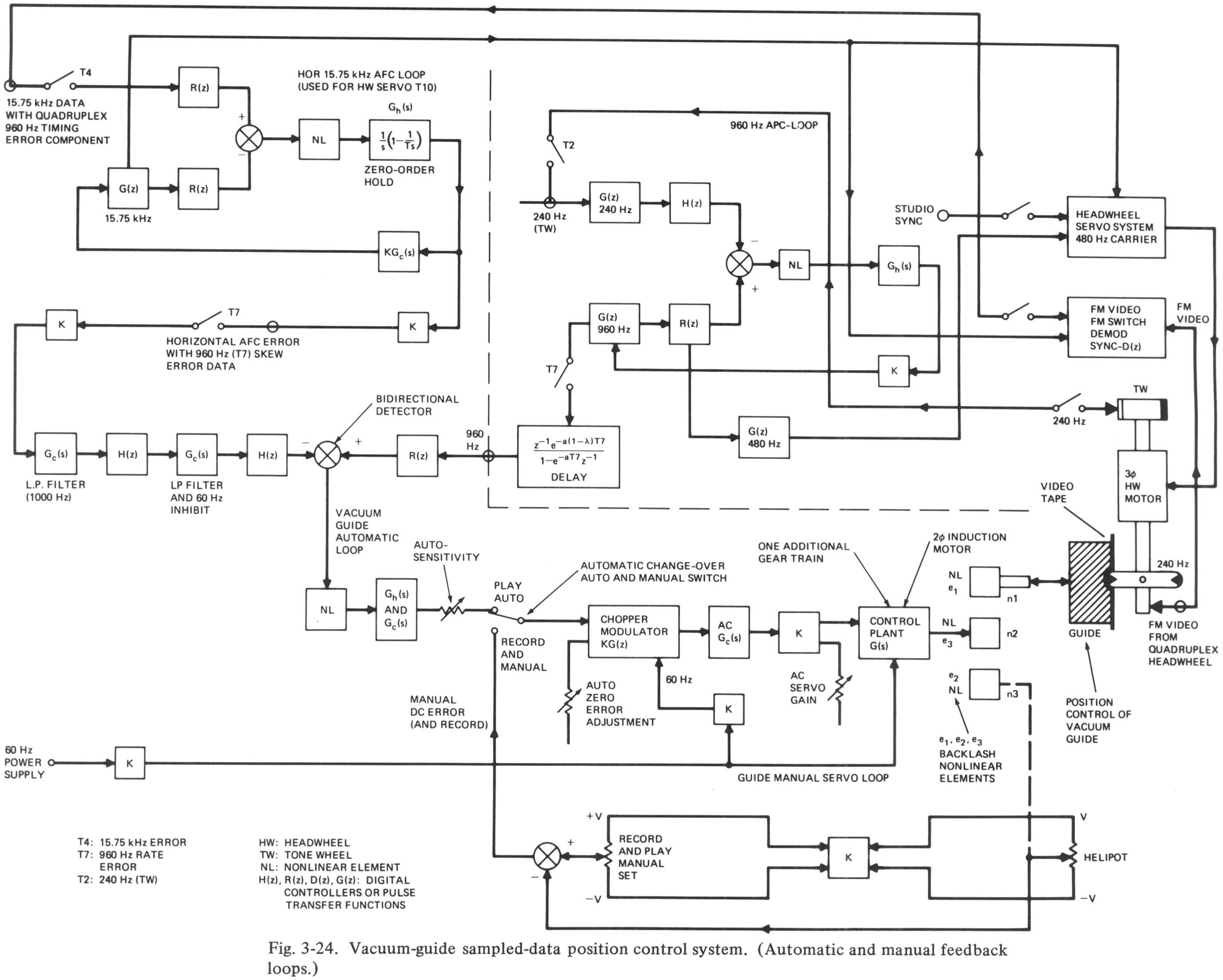

Fig. 3-24. Vacuum-guide sampled-data position control system. (Automatic and manual feedback loops.)

4 Development and Design of a Complex Digital Control System in Commercial Practice (Broadcast-Quality Quadruplex Color Videotape Recorder)

The following presentation includes the system electronics and the technology of magnetic-tape as required in a conventional transmission and receiving system of color television. The complex interacting sampled-data or digital control systems of a quadruplex color videotape recorder (Q-CVTR) used in high-quality broadcast color television throughout the world are chosen for the exposition of this subject. In actuality, the overall digital feedback control system is comprised of several phase-lock loops (PLLs) at various rates of sampled-data. In order to cover the latest status of videotape in general, the principles of home videotape recording are briefly covered in this chapter.

4.1 INTEGRATION OF Q-CVTR AS MAJOR TECHNICAL FACILITY IN UP-TO-DATE BLACK-AND-WHITE AND COLOR TELEVISION SYSTEMS

The quadruplex color videotape recorder is a high-quality broadcast magnetic-tape recording and reproducing medium for standard monochrome (black-and-white) and color television programs, and the associated audio information. The system is designated as *quadruplex* since basically the unique system-technique makes use of four magnetic heads *in quadrature* on the periphery of a rotating headwheel as the nucleus for handling the frequency-modulated picture and color information and the television synchronizing pulse-data.

A program of up to a maximum of 96 min. is recorded on a single 14-in. reel, containing 7200-ft-long, 2-in. wide, mylar-polyester base, 1.4-mil-thick oxide-coated videotape. This can be immediately reproduced after a fast rewind time of 4 min. It can be played back at least a hundred times without noticeable deterioration in the quality of the color picture. Rerecordings from tape to tape, called *dubs*, are generally satisfactory up to at least a third generation; much progress is evident in this respect in a modern Q-CVTR that meets more rigorous specifications in respect to signal-to-noise ratio, FM intermodulation products, transient-response in terms of a k-factor (which is measured by means of a special test technique), and group- or envelope-delay distortion.

In view of the high head-to-tape velocity, the high tension, and the comparatively large penetration (or pressure) exerted on the tape, the videotape is manufactured to meet rigorous specifications regarding physical and magnetic characteristics. The tape employs synthetic gamma-ferric-oxide coating of 0.4-mil thickness, having needlelike particles less than 1 micron and oriented *transversely* (and *not longitudinally* as in the low-frequency audio recording case). The dark reddish-brown high-output oxide particles are uniformly suspended in a heat-resistant binder system and treated with a dry lubricant to reduce static and dynamic friction to a minimum. With the head-to-tape motion, the flux-lines from one pole-piece enter the tape as a fringing-field at the gap (where the flux-density is lower than in the core), and return to the core at the other pole-piece. The oxide magnetic-domains on the tape, acting like micromagnets, align themselves along the flux-lines to make the magnetic pattern, according to the density and direction of the flux-lines at the trailing-edge of the gap, as the tape moves on the rotating magnetic-heads (Fig. 4-1a and b). For television-tape, extreme smoothness of coating is essential for the best figure of signal-to-noise (S/N) ratio, especially at the higher FM-frequencies, since the FM-carrier in a *wide-band* carries the recorded information.

The original kinescope video recording on film is not an ideal signal processing method due to (1) the inherent demerit of nonlinear distortion in the gray-scale (or the brightness-level) of the image, (2) problems of shutter-flicker, and (3) the time interval involved in film processing. Therefore, various attempts to *multiplex* the broadband picture information at high tape-speeds, even up to 300 in./sec, were made right from the early stages of television broadcasting, but they did not materialize in a practical way because of the high tape-speeds involved and hence the enormous quantity of tape required. Regular quadruplex videotape recorders made a welcome breakthrough into the world market around 1958, as produced by only two research-and-development manufacturers, Ampex and RCA. In 1961, RCA came up with the most versatile and practical high-quality product in this field, the transistorized Q-CVTR in a single console (as compared to about half-a-dozen racks of the earlier vacuum-tube version equipment). Standards for video recording are laid down by the Society of Motion Picture and Television Engineers (SMPTE) for the reproduction and the interchangeability of tapes recorded on other Q-CVTRs in the field. Tape is versatile and indispensable for delayed telecasts in preplanned (or automated) international or long-distance coast-to-coast network production-schedules in television programming. It is equally indispensable for television rehearsals and subsequently edited production *repeat-sequence* of action. A very high percentage of the regular television features, broadcast at the present time in both color and monochrome, are played back from tape during the telecast—it is so throughout the world, even in the case, at times, of world-wide digital television via the synchronous satellites.

There are many applications of the videotape-medium besides broadcasting. Military and governmental agencies, aviation radar, hospitals, universities, schools with *closed-circuit television* educational facilities, and several industrial, travel,

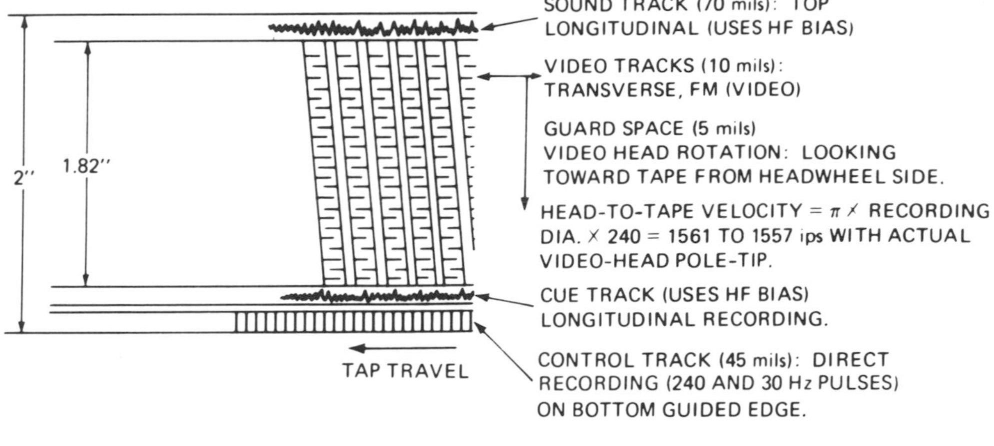

Fig. 4-1. *a*, Recorded data on Q-CVTR videotape for monochrome and color television signals. *b*, Fringing magnetic flux in tape recording.

and banking firms are some of the more recent users. In view of the diverse quality requirements, especially for less rigorous applications in closed-circuit television, simpler, single and double rotating-head, slant-track (or *helical-scan*) monochrome video recording versions for limited usage are available from several manufacturers. There is even a special purpose dual-channel wide-band octaplex video recorder. In addition, practical, low-priced ($1000) home video-cassette television recorders with simpler basic features for the vast consumer market are presently available. Even in the standard compatible quadruplex version, simpler solid-state integrated-circuit medium-priced machines are also available for high-quality closed-circuit applications. (The prices of the Q-CVTRs currently vary from $20,000 to $50,000, depending upon the user's requirements.) The high-quality Q-CVTR, with its highly-complex but reliable sampled-data feedback control systems, has come to stay as a world standard in meeting the SMPTE standards and the regulations of the Electronics Industries Association (EIA) and of the Federal Communications Commission (FCC) in the United States.

4.1.1 Television Video and Synchronizing Information for Monochrome and Color Transmission. The video spectrum required for the transmission of a picture with appreciable detail or resolution is very large, as compared to other channels, of communication. If the maximum frequency-component in the video signal is f, the

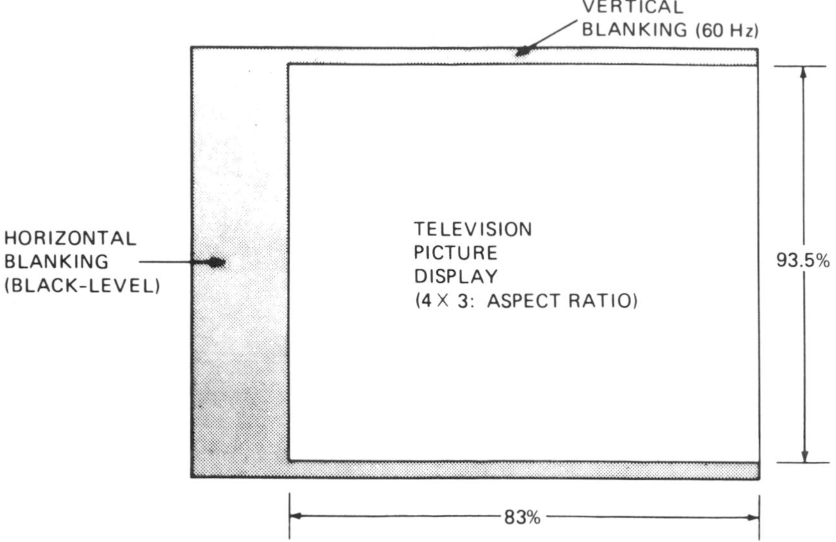

Fig. 4-2. Picture and blanking areas (television display).

number of picture-elements generated per second is 2f for the white (positive-going) and black (negative-going) halves in a cycle.

Then the number of picture elements per line is $\frac{2fh}{nF}$, if n is the number of lines in the picture-frame, F is the number of frames per second, and h is the fraction of a line-interval actually used for picture information (Fig. 4-2).

If A is the aspect-ratio (width/height), the horizontal resolution $H = fh/nFA$. If v is the fraction of the vertical-frame actually used for the picture information, and K is a constant, the vertical resolution $V = Kvn$. Ratio of resolution,

$$m = H/V; f = \tfrac{1}{2} Kmn^2 \cdot FA \cdot v/h,$$

where

$K = 0.7$,
$F = 30$ frames/sec,
$A = \frac{4}{3}$,
$v/h = 93.5/83$

Hence, *video bandwidth*,

$$f = 15.8\ mn^2 = 4\text{MHz/s, if } n = 525 \text{ lines and } m = 0.92.$$

Since 30 picture-frames/sec will produce noticeable flicker for visual perception, *interlace-scanning* is used with 60 fields/sec, but using half the number of lines/sec (i.e., 262.5 lines/field) to maintain exactly the same 4 MHz/sec bandwidth. So the television receiver or monitor actually scans 262.5 lines, 60 times/sec, and its design calls for the *interlace* or alternating of the horizontal lines belonging to two consecutive odd- and even-number fields, so that the lines of each field appear equidistant in the picture-frame displayed at the rate of 30 frames/sec.

The *vertical resolution* depends on good interlace, and it is, in a system, independent of actual bandwidth; it is approximately 340 lines, considering a vertical blanking interval, and a factor for imperfections in interlace. The horizontal resolution, determining the definition of the vertical lines in an image, is limited by the system

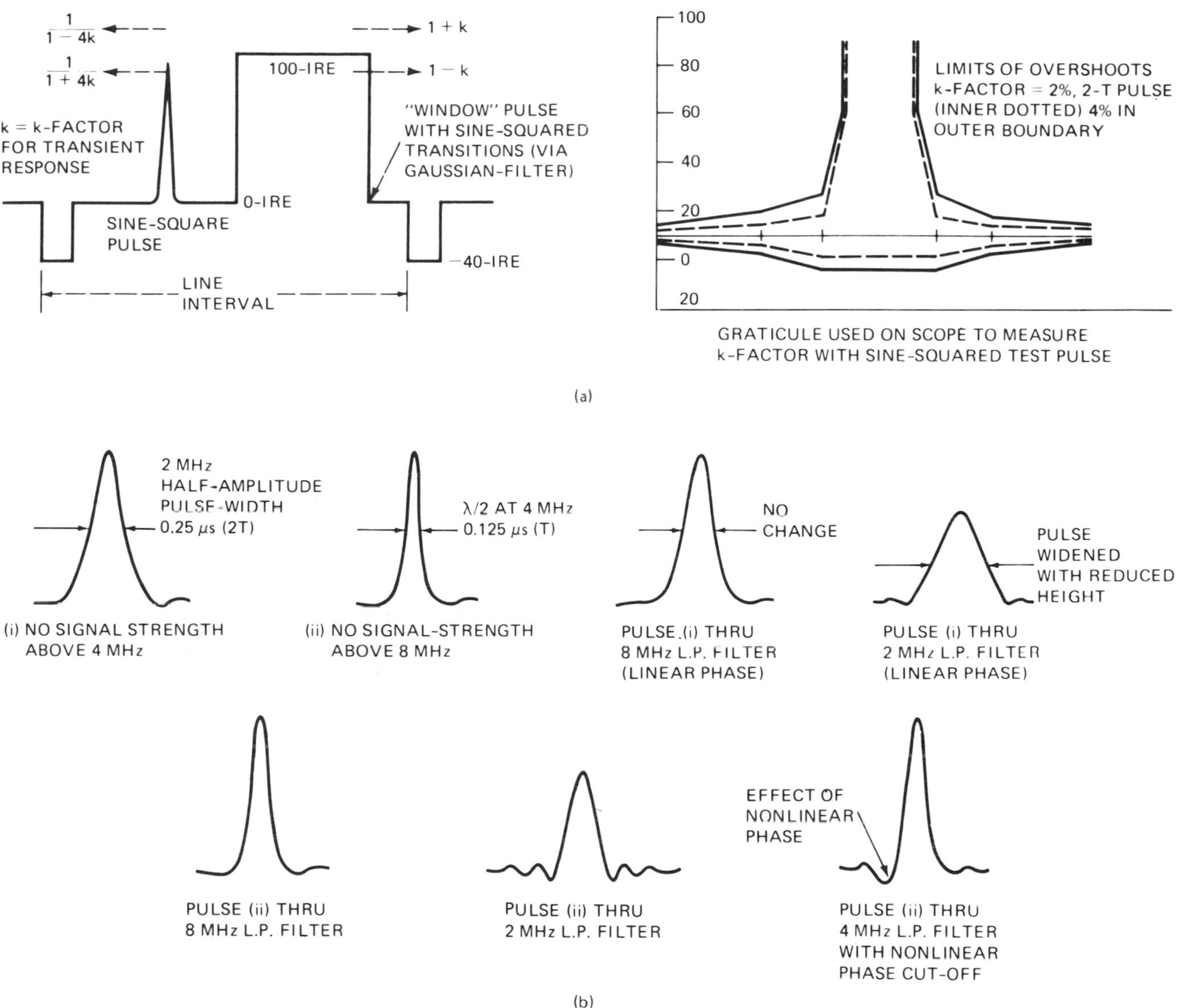

Fig. 4-3a. Hypothetical pulse-and-bar video test signal. Pulse increases in amplitude for excessive high-frequency gain; decreases in amplitude for excessive low-frequency gain.

Fig. 4-3b. Distortion of sine-squared test pulse through video system.

bandwidth, provided the scanning-beam of the cathode-ray tube (CRT picture tube) is properly focused with a minimum of astigmatism. The *horizontal resolution* is approximately 360 lines for a 4.5-MHz/sec video-spectrum.

The product of the *bandwidth* and the actual *pulse rise-time* gives a parameter indicating *transient-response.* It is determined by (1) the residual margin of phase-distortion and (2) the magnitude of high-frequency compensation in a video system. A horizontal resolution of 360 lines corresponds to a video pulse rise-time of slightly less than 0.08 μsec. The conventional *k-factor* of a high-quality video system is normally aimed at a specification of 2%; and, for the purpose of measurement, a hypothetical *pulse-and-bar test signal* (Fig. 4-3a) is devised for checking the transient-response and the approximate *envelope-delay-distortion* at the high-frequency end. Performance limits are given in terms of the limiting-factor k, for which numerical

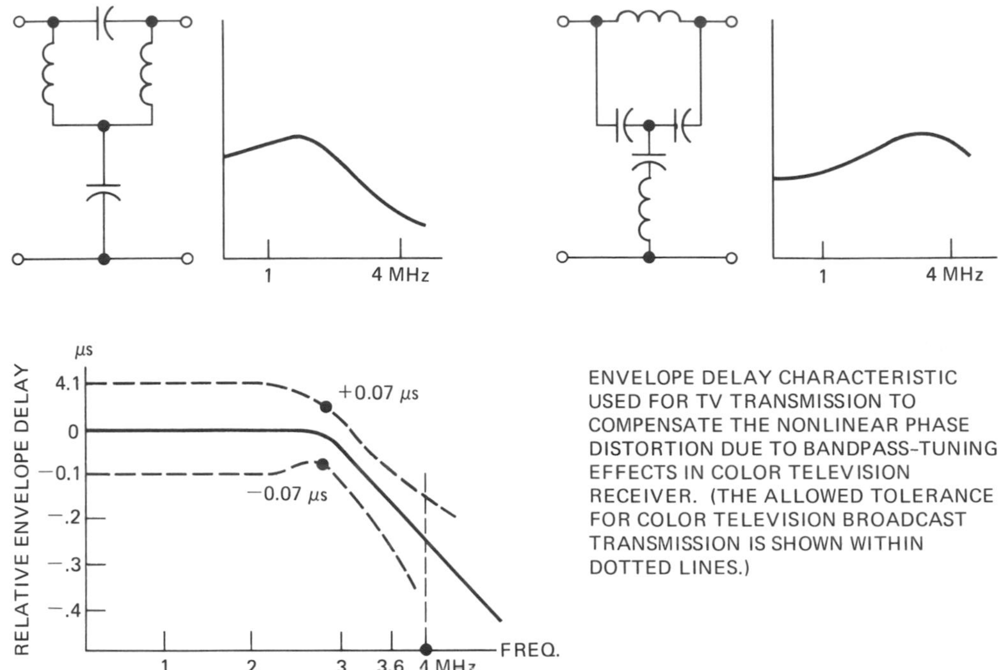
1
4 MHz
1
4 MHz
μs
4.1
0
−0.1
−.2
−.3
−.4
+0.07 μs
−0.07 μs
1
2
3
3.6
4 MHz
FREQ.
RELATIVE ENVELOPE DELAY
ENVELOPE DELAY CHARACTERISTIC USED FOR TV TRANSMISSION TO COMPENSATE THE NONLINEAR PHASE DISTORTION DUE TO BANDPASS-TUNING EFFECTS IN COLOR TELEVISION RECEIVER. (THE ALLOWED TOLERANCE FOR COLOR TELEVISION BROADCAST TRANSMISSION IS SHOWN WITHIN DOTTED LINES.)

Fig. 4-4. Two typical video equalizer network sections used for envelope (group) delay compensation.

values are assigned in individual specifications for the various parts of the tape record/playback system, as it is done, for instance, in television microwave-links and in studio equipment. A rating k-factor in percentage, is then assigned to the system in terms of the relative-amplitude of the reflection or overshoots/undershoots produced as a result of phase-distortion. Either the *sine-squared* T-pulse, with a *half-amplitude-duration* of 0.125 μsec (half-period at 4 MHz/sec), or the sine-squared 2T-pulse, with a half-amplitude-duration of 0.25 μsec is recorded on the tape along with a "smoothed" linear low-pass limited 40-μsec-wide *window/bar* test waveform.

The pulse is an ideal filtered impulse, while the bar is shaped to have integrated *sine-squared transitions* with a rise-time of 0.25 μsec. For acceptance, the ratio of the amplitude of the bar-response to the amplitude of the 2T-pulse response should fall within the limits of 1 ± 4k. The waveforms of the sine-squared test pulse shown in Fig. 4.3b explain the method of system evaluation by this technique. In practice, one can read the k-factor directly on a specially calibrated graticule over the CRT oscilloscope-display of the sine-squared test pulse at the output of the system on investigation. Special test tapes, made available with the sine-squared test pulse on a stair-step waveform depicting the various levels of brightness, are not suitable at this time for the k-factor measurement.

The system transient-response, as assessed by the sine-squared test pulse technique, can be modified, to meet the desired specification, by including specially designed *envelope or group-delay compensating networks* to provide *phase correction.* Correct delay-compensation will result in an effective linear-phase system, which is an essential requirement in the Q-CVTR record/playback system. The envelope-delay equalizer is essentially a bridged-T network composed entirely of reactive elements. The type of network sections commonly used for this purpose are shown in Fig. 4-4. Each section of the network is a passive all-pass constant-resistance unit, having a flat amplitude-response with frequency and a negligible dissipation. The *all-pass network* is by definition the system transfer-function, whose poles are situated in the left-half complex-plane, and whose zeros are situated in the right-half complex-plane as mirror-images of the poles about the imaginary-axis. Lattice and bridge networks of this category are *nonminimum phase* in character. (A minimum-phase network, on the other hand, will have no zeros in the right half-plane.)

The dc-component in a video signal is inserted by means of a *line-to-line clamp*, which charges or discharges a coupling-capacitor to a dc-reference; this dc-reference commonly represents the blanking-level of the video signal to assure a common black-level reference for the whole picture. Black-level inadequacies on the part of the video system are specified in terms of *tilt*, and it is limited to less than 2% according to specification.

The television horizontal scanning-rate = 262.5 X 60 = 15.75 kHz/sec. This gives a duration of 63.35 μsec to a horizontal line-interval (American Standards). The television vertical-interval of a field (1/60 sec) works to 16.67 msec (Fig. 4-5a–c).

The half-line discrepancy between the *odd and even-number fields* is taken advantage of in deriving the 30-Hz/sec frame-pulse from the synchronizing pulse-train (sync), consisting of the horizontal 15.75-kHz/sec pulses and the vertical 60-Hz/sec pulses. The *equalizing pulses* preceding and following the vertical sync-pulses facilitate a satisfactory interlace of the lines in the two alternate fields of a picture-frame in a television receiver or monitor. The *vertical-blanking* of approximately 19 lines affords time for the *vertical flyback* of the synchronized 60-Hz/sec saw-tooth vertical-scan waveform generated in the television/monitor/receiver; the flyback takes place at the *black blanking-level. Retrace* and *flyback* are synonymous.

In the case of the NTSC color television signal, a *color synchronizing burst* is inserted in addition on the *back-porch* during the horizontal blanking-interval succeeding the horizontal synchronous pulse; it is then transmitted along with the *color-subcarrier* and the *monochrome* information in the picture in order to provide a reference for the *color-phase-lock* in a digital phase-lock loop (PLL) sampled-data control system in the color television receiver. Necessary precaution is taken to gate out the color synchronizing-burst during the vertical sync-interval in order to eliminate possible interference to the correct operation of the vertical-hold in receivers, whether color or black-and-white, in the compatible NTSC color television system. By the way, a color multiplexing device called the *colorplexer* or *color encoder/multiplexer* processes the three primary color signals (green, red, and blue) from a three- or four-channel, regular or film color television camera to transmit the *compatible* NTSC color television signal for reception in both color and ordinary black-and-white television receivers. The color or chroma, in accordance with the specification-figures in Table 4-1, is added to the monochrome component (which is derived from either a separate monochrome camera-tube or from the three primary color components by matrixing) in the form of two individual quadrature-components of color-subcarrier at 3.579545 MHz/sec. I, *the broad-band In-phase component* and Q, *the narrow-band Quadrature component*, are each modulated in (1) *amplitude for the saturation* of the pertinent color in the spectrum of the *color-triangle* and (2) *phase for the hue* of the corresponding color component. The Q-CVTR tape-medium is actually concerned with the recording of the multiplexed NTSC color signal in the FM (frequency-modulated) format.

True color signal transmission, in particular, demands extremely rigorous specifications on the part of the intervening FM, AM, tape, and video transmission equip-

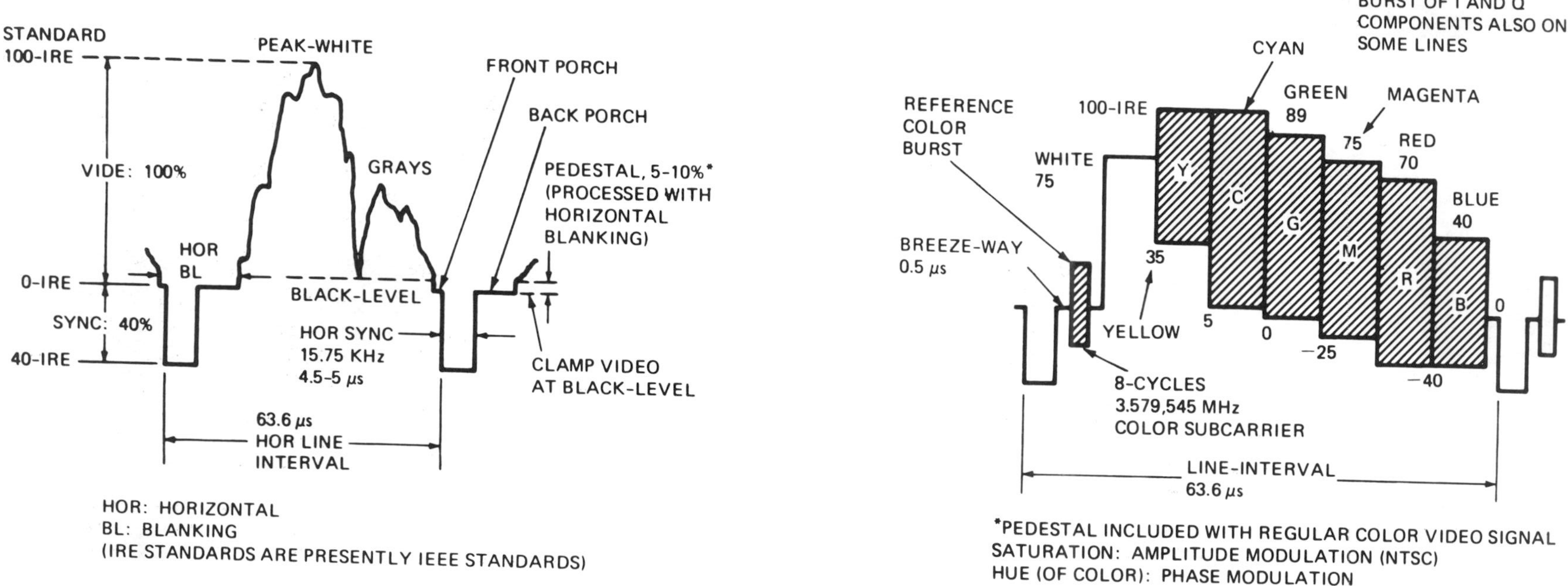

Fig. 4-5a. Line-interval of a monochrome (black-and-white) composite video signal used in television broadcasting.

Fig. 4-5b. Line-interval of NTSC color video signal (color-bar test signal). Pedestal included with regular color video signal. Saturation: amplitude modulation (NTSC). Hue (of color): phase modulation.

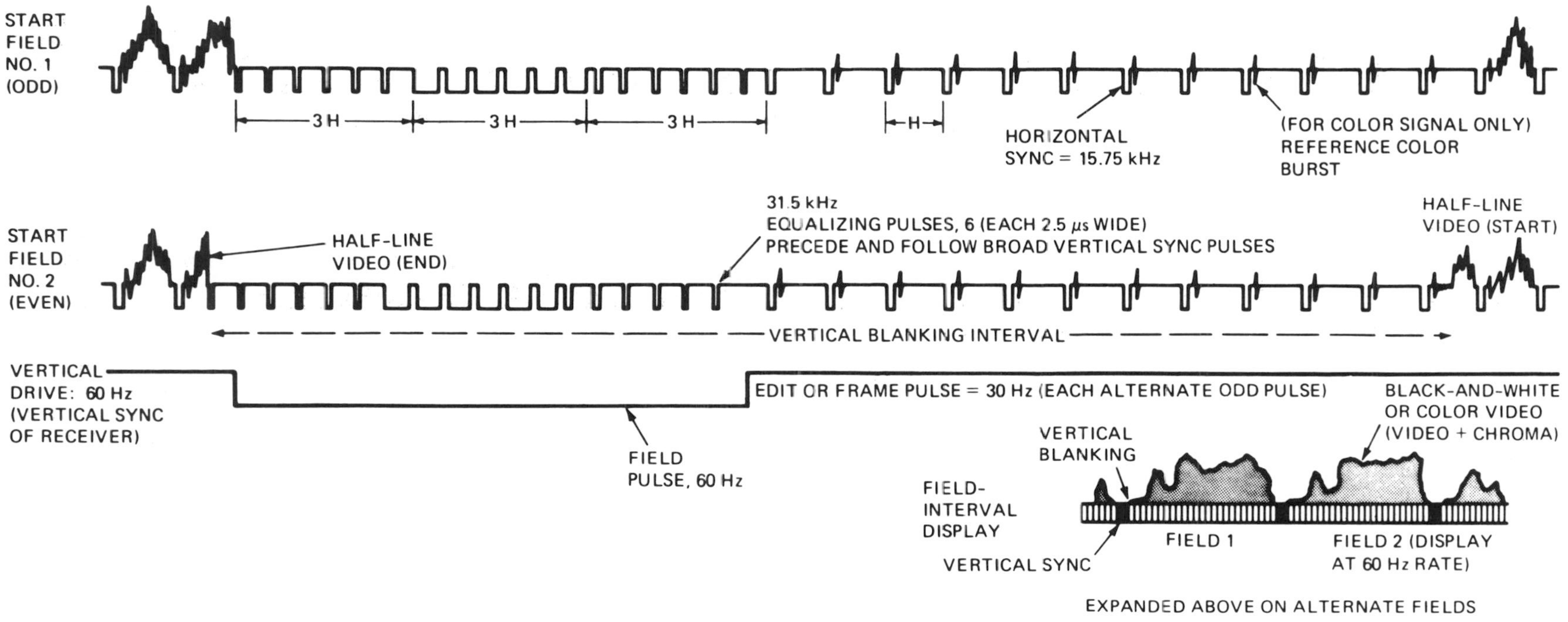

(c)

Fig. 4-5c. The vertical intervals with vertical sync and equalizing pulses in the alternate odd and even fields of a composite video signal (NTSC color signal).

Table 4-1. NTSC *Colorimetry:* Matrixing of the Color Spectrum for the Most Pleasing Visual Appreciation.

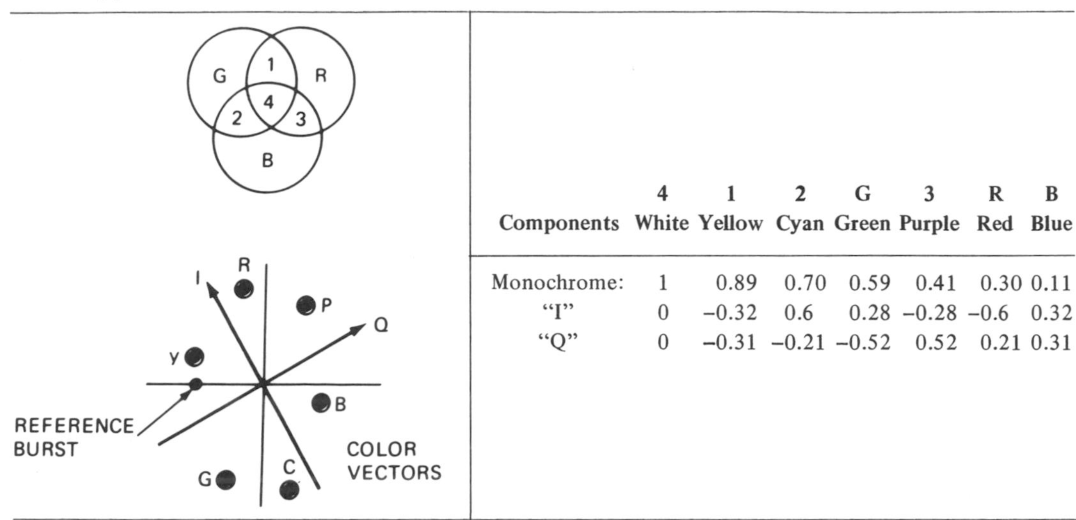

Components	4 White	1 Yellow	2 Cyan	G Green	3 Purple	R Red	B Blue
Monochrome:	1	0.89	0.70	0.59	0.41	0.30	0.11
"I"	0	−0.32	0.6	0.28	−0.28	−0.6	0.32
"Q"	0	−0.31	−0.21	−0.52	0.52	0.21	0.31

ment between the color television camera and the color television receiver, in respect to (1) the transient-response and the implied *envelope-delay distortion* and (2) the *differential-gain* and the *differential-phase* at the color subcarrier-frequency 3.58 MHz/sec. The last two parameters indicate the amplitude and phase linearity, respectively, of the color video information on the entire gray-scale.

The divisible interrelationship between the color subcarrier-frequency and the vertical- and horizontal-scan frequencies, as maintained by the station color frequency-standard and the sync pulse generating equipment, is shown in Table 4-2. The European broadcasting system, with the coordinating authority of the Consultative Committee, International Radio (CCIR), have adopted a different color subcarrier-frequency for the European line-standards used in PAL (phase-alternation by line) color television system. In principle, the subcarrier-frequency is chosen as an exact *odd-multiple of one-half the line-rate*, to achieve *dot-interlace*, with *spectrum-interleaving* of *luminance* (brightness) and subcarrier information. One Q-CVTR must be capable of handling the scan and color frequency standards of both the United States and Europe, with due adherence to the respective specifications regarding S/N ratio and envelope-delay tolerance. The various *sampling rates* referred to herein are therefore subject to slight alteration in the case of the present-day European television-scan systems, viz., 405 line/50 Hz/sec (United Kingdom), 819/50 (France), and 625/50 (United Kingdom and rest of Western Europe) as compared to the American 525/60 television-scan rates.

The video spectrums used for the transmission and reception of the monochrome and NTSC color television signals are shown in Fig. 4-6a and b.

During the last few years, the interest in color television has escalated at a fast pace throughout the world. In the United States, the development of the videotape recorder in color recording and reproduction has now progressed to such a remarkable extent that, with experienced operating personnel or automatic techniques, videotape color program material can be presented as indistinguishable from the original live camera signal. At present, Europe, Japan, and Canada are the other countries that have regular color broadcasts. At a time when satellite worldwide color television broadcasting is feasible, Eastern Europe, France, and the Soviet

Table 4-2. Color Standards, United States and CCIR.

NTSC (United States)	CCIR (Europe)
1. Subcarrier frequency: 3.579545 MHz/sec	4.4296875 MHz/sec (NTSC = original) 4.43361875 MHz (NEW: PAL and SECAM)
2. Video spectrum, effective: 4.5 MHz/sec	5.5 MHz/sec
3. Horizontal frequency $= \frac{4.5 \times 10^6}{286}$ $= 15.73426$ kHz/sec and $15.73426 = \frac{3.579545 \times 2 \times 10^3}{455}$ and $455 = 13 \times 7 \times 5$ (for division) (The horizontal frequency is slightly altered from the monochrome 15.75 kHz/sec rate, to make it the 286th harmonic of the vision and sound carrier separation, namely, 4.5 MHz/sec.)	$\frac{5.5 \times 10^6}{352}$ $= 15.625$ kHz/sec and $15.625 = \frac{4.4296875 \times 2 \times 10^3}{567}$ and $567 = 7 \times 9 \times 9$ (for division) 5.5 MHz/sec = Vision and sound carrier separation.
4. Field frequency $= \frac{15.73426 \times 2 \times 10^3}{525}$ $= 59.94$ Hz/sec For monochrome, it is 60 Hz/sec. $525 = 7 \times 5 \times 5 \times 3$ (for division)	$\frac{15.625 \times 2 \times 10^3}{625}$ $= 50$ Hz/sec (same as monochrome) The European television receiver/monitor/CRT requires a heavier DC filter-pack in view of the lower 50-Hz/sec power supply. $625 = 5 \times 5 \times 5 \times 5$ (for division) The color receiver requires a single-line delay-line in "PAL" color system.

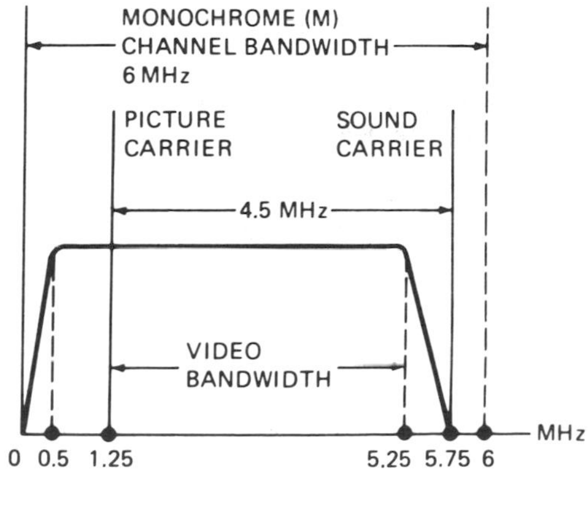

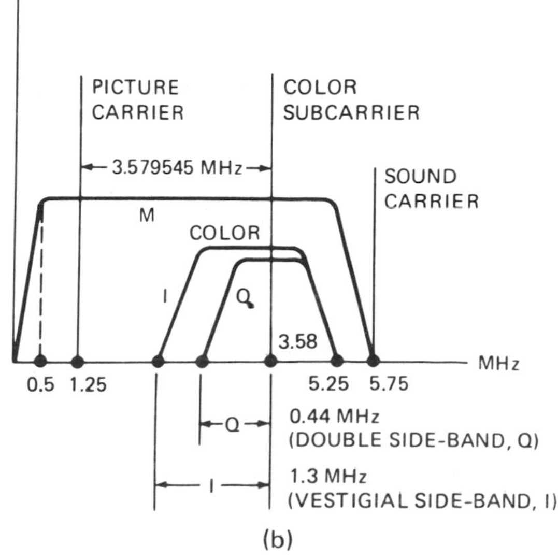

Fig. 4-6. The video spectrum used for NTSC compatible television transmission and reception in the United States for monochrome and color. *a*, Monochrome television signal spectrum (M). *b*, Color television signal spectrum (NTSC).

Union have opted for a French/Russian exclusive SECAM color television system, requiring a higher line-frequency rate, 819 line/sec, and a costlier and broader video bandwidth. SECAM (*Sequentiale à Memoire Couleur*) is a French variation of a sequential FM/AM color multiplexing system. The rest of Western Europe has adopted the German PAL system (Phase Alternation by Line), which is a mere variant of the NTSC simultaneous compatible color system. In the PAL system, probable error susceptibility in maintenance to any distortion of the character of phase-errors in long-distance color television transmission equipment is overcome by averaging out these errors on alternate lines by including a single-line delay line in color television receivers.

In television studios, to most of the equipment including videotape recorders, four common pulse-train signals are distributed according to the needs from (1) the station crystal-controlled, (2) color frequency-standard controlled, or (3) *gen-locked* sync pulse generator (if it is phase-locked to an incoming external composite video signal). They are designated as (1) the horizontal-drive at 15.75 kHz/sec, (2) the vertical-drive at 60 Hz/sec, (3) blanking, consisting of the wider horizontal-rate and vertical-rate blanking pulses, and (4) complete-sync, consisting of the horizontal 15.75-kHz/sec synchronizing pulses and the vertical 60-Hz/sec synchronizing pulses with the associated equalizing pulses. In color television studios, the color or chroma-subcarrier and the burst-flag pulse-train (for placement of synchronizing reference-burst on color signals) are also distributed. The Q-CVTR requires the station complete-sync for the record and playback of monochrome signals, and the 3.58 MHz/sec subcarrier also for color videotapes. For recording test purposes, the Q-CVTR also requires a composite color-bar signal (Fig. 4-5a), consisting of sync, blanking, the three primary colors (red, green, and blue), and the three secondary colors, (cyan, purple, and yellow), and black, white, I, and Q color components. A special instrument, a vectorscope, is also required to indicate the saturation and hue of each of these color components in the NTSC color television signal (Table 4-1).

4.2 QUADRUPLEX VIDEO RECORDING TECHNIQUE AND PRINCIPLES

4.2.1 Videotape and Bandwidth Requirements. A tape-speed of at least $7\frac{1}{2}$ inches per second (ips) relative to the magnetic-head is essential for audio recording of a 15-kHz/sec bandwidth, using a magnetic playback head-gap of nonmagnetic material of feasible dimensions (0.05 to 0.5 mil). This gap must be smaller than the shortest recorded wavelength. One-half mil corresponds to half-a-wavelength at 15.75 kHz/sec. The recorded wavelength must be twice as long as the gap-width for the necessary high-frequency preemphasis, good reproduction, and S/N ratio. Preemphasis is required to allow the compensation of losses in the magnetic-core structure. As a comparison, for a satisfactory S/N ratio of 42 dB at 4-MHz/sec video bandwidth, a relative head-to-tape speed of at least 400 ips would be necessary, if information could be packed up to, for example, 10 kHz/sec/in. at the above speed. In the case of the longitudinal-track audio recording, approximately 2 kHz/sec per inch (15 kHz/sec at $7\frac{1}{2}$ ips) of tape-speed is allowed.

In the broadcast quadruplex color videotape recorder, the required wide-band frequency-modulated (FM) video recording is likewise accomplished by means of 2-in.-wide magnetic videotape at a low tape-speed of 15 ips, and a head-to-tape velocity of the order of 1500 ips, by actually using fine narrow tracks of information, 10-mil wide with 5-mil spacing. The FM-video tracks are transversely recorded across the tape by a rotating headwheel (HW) with four magnetic-heads, which are spaced along the periphery in quadrature, tolerance-wise, as accurately as possible. For extended life and the requisite high head-to-tape pressure in the case of video recording, the projecting pole-tip material is a special upgraded aluminum-iron-silicon alloy. Alphecan is a successful example with an average-life of 500 hr, and there is evidence of future progress in this respect. Associated audio information is recorded longitudinally along the top edge of the tape at the conventional 15 ips, while a 240-Hz/sec control-track signal is recorded along the bottom edge of the tape to indicate the precise position or timing of the rotating HW, relative to the recorded FM-video information. The width of the tape is greater than the 90° arc on the headwheel, to allow a slight overlap of information between the end of each transverse-track and the start of the next. This specific procedure permits the reassembling of the FM-video, which is recorded in a series of transverse tracks, into a smooth continuous FM signal output. A recorded-track accommodates 18.4 television horizontal-lines of 63.5-μsec duration each, thus permitting a two-field (one frame/picture) FM-data of 32 tracks within half an inch of tape. The signal-overlap is eliminated during playback by electronic switching to allow actually an average of slightly above 16 line/track (Fig. 4-7a).

Information could be packed in finer tracks of 5-mil width and 2.5-mil spacing by employing an appropriate magnetic-head design on the headwheel, thus making it possible to economize on tape by half, by moving the tape at $7\frac{1}{2}$ ips for most applications–at half the standard 15-ips tape-speed. The Q-CVTRs have a direct pushbutton switching facility for these two tape-speeds, depending on the actual headwheel assembly used. For interchangeability, the wide-track 15-ips headwheel can, however, be used for $7\frac{1}{2}$-ips recordings and playbacks without any inconvenience.

4.2.2 FM-Video. Since the transfer characteristic of the magnetic medium is nonlinear (like the movie film) with respect to the brightness scale, a special FM technique is developed for recording the video spectrum from about 10 Hz to 4 MHz, a gamut of 18 octaves. The FM technique, as an advantage, allows the recording of signals at a constant level, high enough to saturate the tape; FM also permits, during playback amplitude-limiting, for the attenuation of extraneous noise and improvement of S/N ratio. In the case of the longitudinal audio-tracks, high-frequency bias solves the nonlinearity problem under reference; during recording, the ac flux due to the fringing-field about the gap is left symmetrical about a positive or negative flux-value (with respect to the origin, in the plot of the flux-density versus magnetization). The flux under consideration is thus linearly determined by the recording current. For the high degree of linearity required in the case of the video and chroma at all brightness levels, FM solves the problem since the symmetrical amplitude-nonlinearity, produced by the characteristic of the magnetic-tape, does not alter the timing between successive crossings of the zero-axis. The preservation of these timing relations–or, in other words, the correct wavelengths–permits the balanced demodulation of the FM video information without distortion. It has been established that, in the case of the monochrome video, for a satisfactory S/N ratio, the requisite 4-MHz/sec bandwidth, as dictated by the rise-time requirements, necessitates a peak frequency-deviation of 2.5 MHz/sec as the 100% modulation reference.

If, in order to prevent the nonlinearity caused by the magnetic medium, the FM

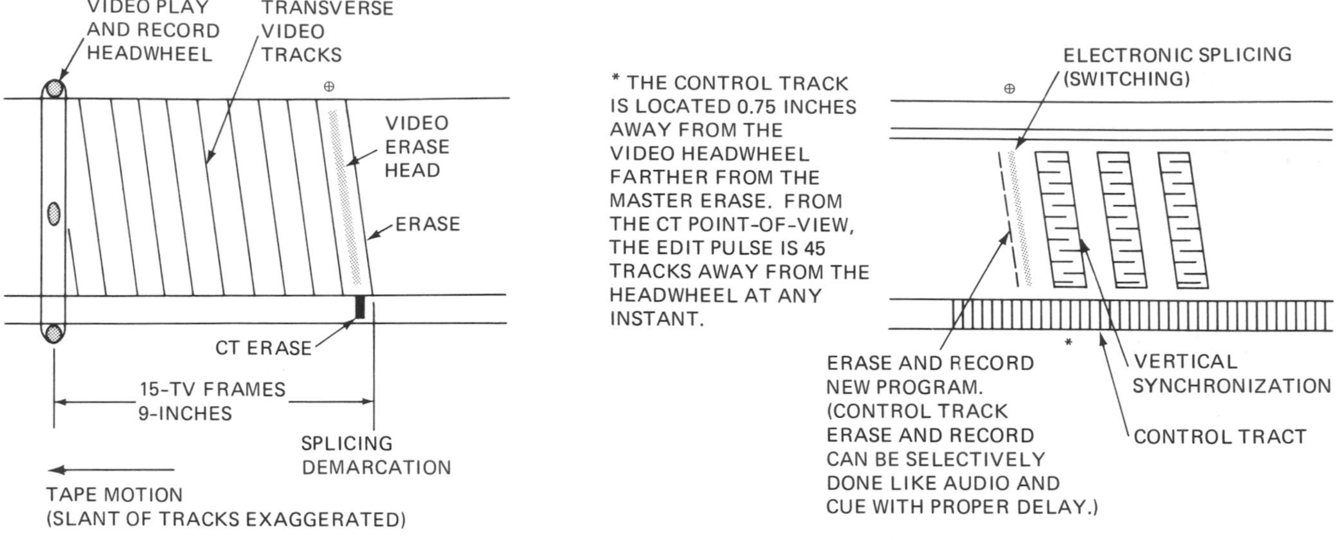

(a)

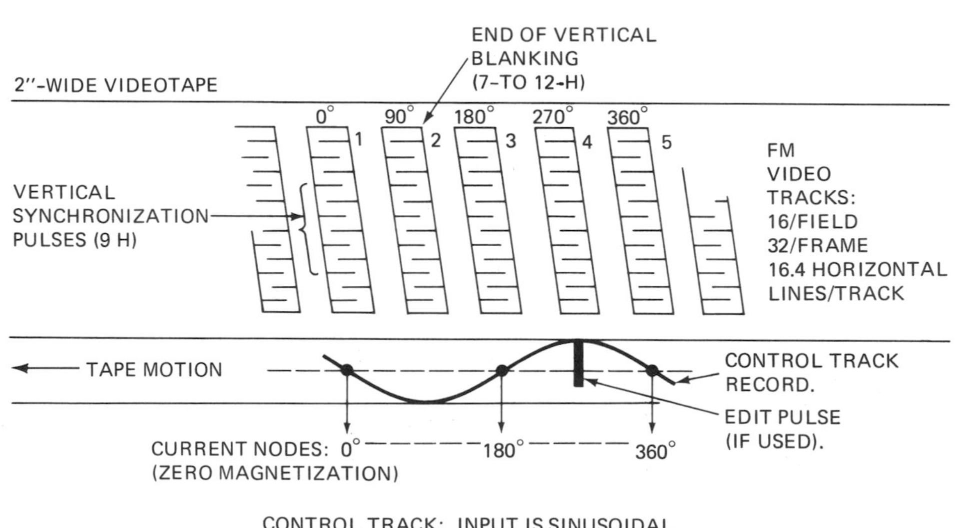

(b)

Fig. 4-7. *a*, Electronic splicing; line of demarcation for switching system (from playback to record) (*Courtesy of RCA*). *b*, Control-track timing with respect to the four FM-videotrack per head-wheel revolution.

carrier is chosen as 5 MHz/sec, with the picture blanking-level clamped at the FM carrier, then a standard composite-video input signal (0.3-V sync and 0.7-V video) produces the frequency-deviation as shown in Fig. 4.8a. For a 4-MHz/sec bandwidth, the lowest sideband will occur at 1 MHz/sec. For frequency-deviation less than one-half the modulation frequency, corresponding to the 0.5-modulation-index, the sideband energies approach the amplitude-modulation properties, the sideband energies above a single pair being negligible—this is confirmed by theory.

The total FM-range of 7 MHz/sec in plot shown in Fig. 4-8a thus converts the 18-octave video signal to a video-spectrum that is less than four octaves, as a practical solution to the videotape recording process in a linear fashion. For direct longitudinal magnetic recording, on the other hand, 10 octaves is about the practical limit.

The frequency modulation of the composite video signal is accomplished by mixing two oscillator signals, one at 46 MHz/sec, proportionately deviated as a function of the amplitude of the applied video signal, and the other at the master oscillator frequency, fixed at 51 MHz/sec. If automatic frequency control (AFC) is used for this purpose, small frequency corrections will be applied to the master oscillator by a frequency discriminator and an AFC-loop in order to hold the output centered at the heterodyne-carrier 5 MHz/sec.

In the case of color, a smaller FM-deviation of 1 MHz/sec was chosen to start with (about an FM-carrier of 5.79 MHz/sec), so as to minimize the background beat-patterns (otherwise called *moire*) due to the heterodyne effects of (1) the color processing subcarrier, (2) the intermodulation products produced in the FM amplification system and the crystal mixers, and (3) the *folded* side-bands produced due to the third-harmonic content in limiters and demodulators, in the original double-heterodyne color processing technique. This technique of reproduction of color from tape by phase-error compensation is presently obsolete. (This original *narrow-band* FM and double-heterodyne approach to taping color involved some compromises regarding (a) overall frequency-response, (b) S/N ratio, (c) excessive differential-gain and differential-phase of color information, as a result of the nonlinearities produced by the narrow-band FM-deviation of 1 MHz/sec, and (d) time-base low-frequency jitter, since no precise phasing was possible with respect to the headwheel servo system and its phase-jitter content.) The present-day color reproduction from tape, by means of a color (digital/sampled-data) automatic timing-correction system (CATC), which involves extremely accurate phasing over and above the high-precision line-lock HW feedback control system, solves these problems. Therefore, the present high-band technique with a wide FM-deviation from 7.06 to 10 MHz/sec with the carrier at 7.9 MHz/sec, eliminates the original need for restricting the FM-deviation for color. The *wide-band technique* achieves the theoretically possible improved S/N ratio at minimal differential-gain and differential-phase. In practice, for compatibility to play back any tape, a switching facility is provided in a modern Q-CVTR for both wide- and narrow-band FM-deviation.

With the latest recording technique, using

1. a common wide-band FM-deviation for both monochrome and color
2. optimized low-impedance magnetic-head design in respect to S/N ratio and head-resonance
3. appropriate and precise video pre- and post-emphasis for recording and playback, respectively
4. suitable nonminimum-phase envelope-delay equalizers
5. linear-phase filter networks
6. optimum phase equalization of the FM recording signal as if it were a phase-modulation

the Q-CVTR gives excellent results in both monochrome and color in respect to not only S/N ratio (around 45 dB), but also undetectable Moire and intermodulation products. The improvement in transient-response results in fine k-rating of the order of 2%. In short, as compared to narrow-band FM, theory indicates that wide-band

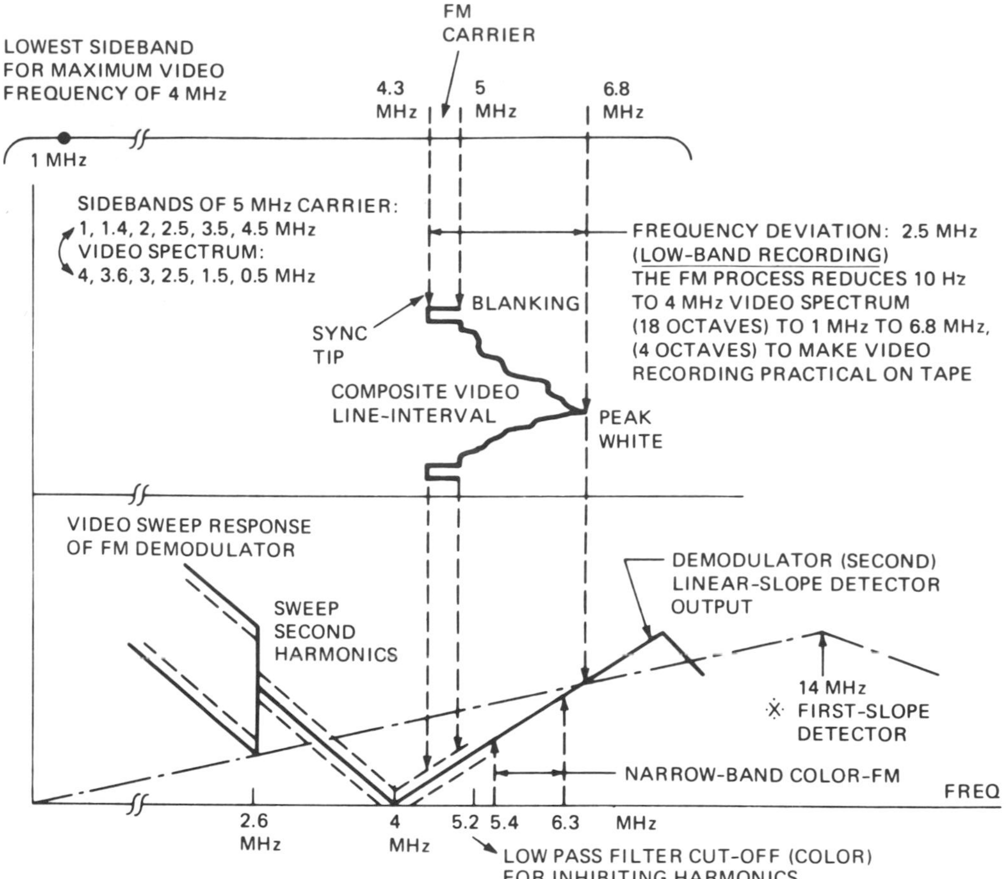

※ THE SLOPE USED IN ORIGINAL Q-CVTR IS SECOND-SLOPE AS SHOWN ABOVE.
THE PRESENT HIGH-BAND HIGH-QUALITY Q-CVTR USES FIRST SLOPE (—— - ——)
FOR BOTH LOW AND HIGH BANDS.

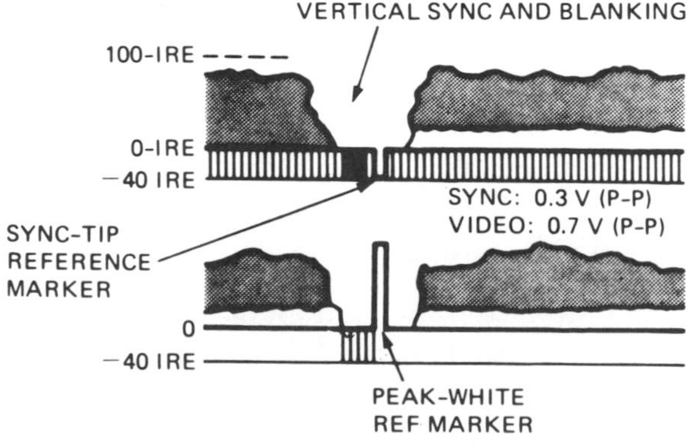

WITH INCORRECT FM-DEVIATION SETTING, THE MARKER REFERENCE PULSE AMPLITUDE WILL LIE IN-BETWEEN

Fig. 4-8a. Frequency modulation FM-versus-video and sideband energy distribution and prerecord calibration for FM-deviation. (*Courtesy of RCA*)

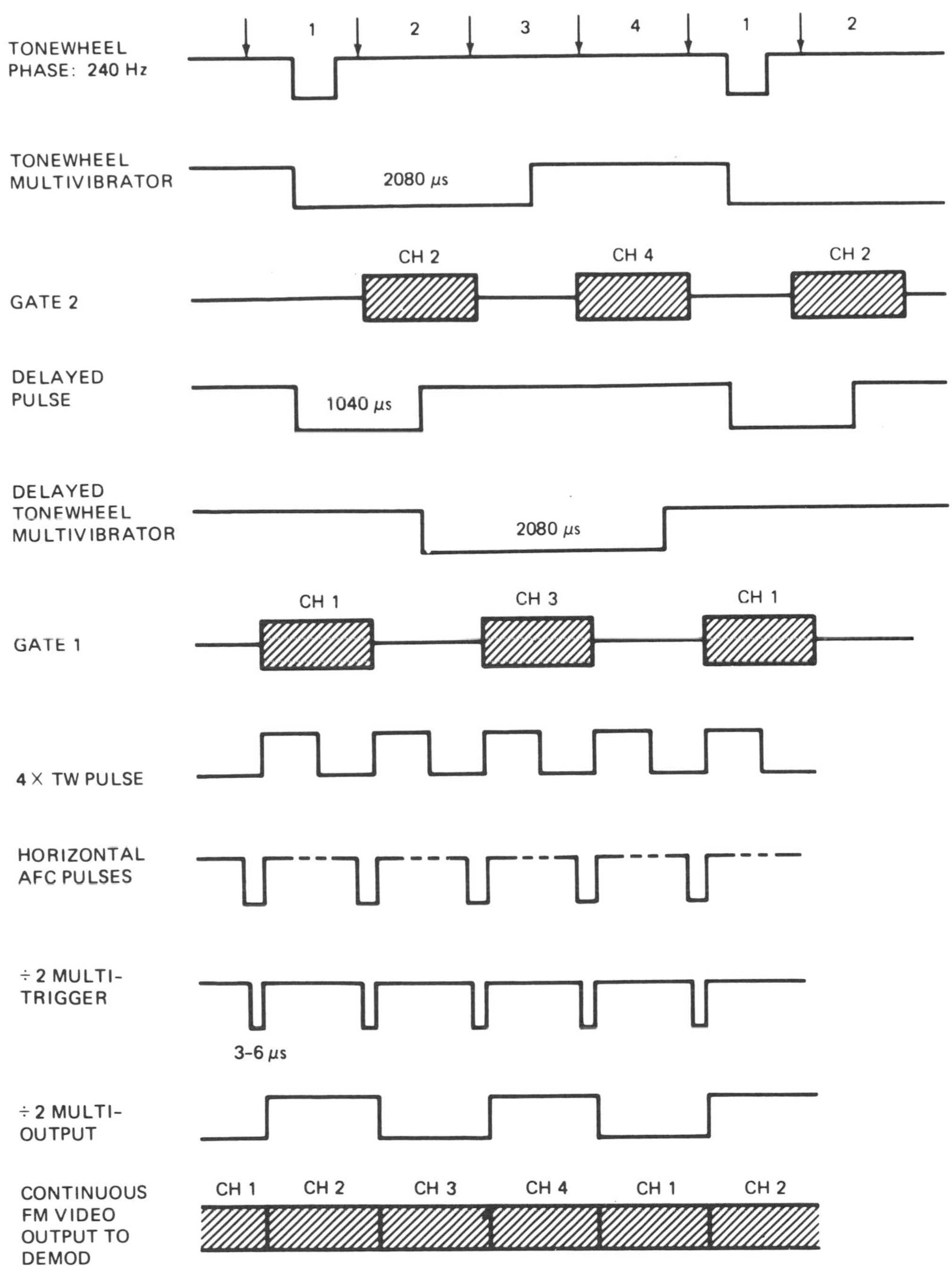

Fig. 4 8b. FM switcher waveforms to reproduce the four video-head signals, along with video-overlap content.

FM affords best S/N ratio for a larger range of input carrier levels, and a greater amount of allowable FM-limiting.

In FM that uses wide frequency-deviation, the sideband spectrum contains terms of not only Bessel functions, but also error functions; since the computation of these terms along with the imminent traces of spurious crosstalk in the signal processing system is cumbersome for actual design purposes, special-purpose test equipment is required for the design of the FM system in the recorder to measure and minimize the intermodulation products. The Moire caused by these products is rendered unnoticeable by minimizing the signal-processing nonlinearities.

4.3 CONTROL AND EDITING

4.3.1 Logic and Control System in Q-CVTR. The various operational functions of the machine, such as "wind" (forward or reverse, fast and slow), "standby for play," "play," "set up for record," "master record," separate "audio and cue record," system analog and digital data-protection features, two-speed selection, electronic splicing, logic and control switching circuits, television-scan selecting facilities, monochrome or color, etc., are all controlled by means of a multiplexer/digital-controller that consists of flip-flop memories, inhibit and interlock circuits, and semiconductor diode logic-matrix. Thus, momentary push-button switching with light indicators allows, according to requirement, the operation of the capstan motor, the tape pinch-roller actuated by a rotary-solenoid, headwheel motor, tape-reel motors, tape vacuum-guide solenoid, reel-brake solenoids, record group-transfer, playback group-transfer, direct mod-demod (modem) signal-processing operation, etc. In the "wind" mode, the tape is pulled by the take-up reel while the supply-reel provides tension by pulling in the opposite direction. The supply-reel motor, take-up reel motor, and capstan motor simultaneously control the tape-motion; during record and playback, however, the tape-speed is primarily governed by the capstan alone, while the tape-reels merely provide a constant torque. The self-checking *fault indicator* system monitors the proper functioning of all the servos, as well as the important signals (erase signal, control-track signal, head currents, etc.) when the machine is on record-mode, thus assuring the operator that all circuit functions are normal and "go" during the recording of the television program.

4.3.2 Video Recording Mode. The incoming composite-video signal is amplified, preemphasized, and modulated to a double-sideband FM signal, extending from 4.28 to 12 MHz, depending upon the FM-deviation used. Special-purpose devices, designated FM standards, and FM references provide linear-phase low-pass filtering and preemphasis (high-frequency boost) to the video signal being recorded, and post-emphasis (corresponding high-frequency roll-off) for the video signal being played back, depending on (1) the television scan standards, (2) monochrome or color, and (3) high- or low-band recording. The exact frequency-deviation of the FM signal, corresponding to the modulating composite-video input, is actually calibrated by means of a set of crystal-controlled reference frequencies for sync-tip, blanking, and peak-white levels in the video signal, as shown in Fig. 4-8a.

Four individual FM-record amplifiers, via respective adjustable delay-lines (for compensating the quadrature timing-errors of the four magnetic-heads of the actual headwheel assembly in use), individual magnetic head-resonance phase-equalizer networks, and transfer-relay contacts in the preamplifier, apply the FM-record signal at the appropriate saturating head-current level to the four video magnetic-heads, through the slip-ring brush assembly or a rotary video transformer on the shaft of the headwheel motor.

The precision of manufacturing tolerances of the four (headwheel) quadrature-errors has presently advanced far enough (within 6 nsec) that record/playback delay-lines, in individual FM channels, are no longer essential in the latest Q-CVTR with a built-in monochrome automatic timing control (MATC).

The videotape from the supply-reel proceeds first to the master-erase head for the erasure of the previous video information, if any. The erasure of the control-track signal on the tape is selectively controlled for electronic splicing purposes. The erasure is done by subjecting the tape to a saturating alternating field (at 87 kHz/sec) and then reducing the magnitude of the field to zero as the tape moves on toward the headwheel disk. The master-erase head, with its comparatively wide gap, allows the flux to reverse itself at 87 kHz/sec as the tape moves across the gap. An erase-current on the order of about 4 A overrides the recorded signal, and the negative and positive half-cycles cancel each other across the gap to effect the erasure. The vacuum-guide assembly establishes suction so as to curve the tape transversely and fit the headwheel so that the heads penetrate the tape to the precise nominal pressure that is manually preset by means of the guide manual-servo system. A tonewheel slot on the headwheel-shaft causes, for the use of the servo systems, the generation of a 240-Hz/sec tonewheel signal in a pickup coil, which is mounted adjacent to the rotating rim of the tonewheel. The slot on the disk rotating at 240 rps provides a variable reluctance. The generation of the 240-Hz/sec tonewheel synchronizing control signal by means of this transducer makes the tonewheel device essentially basic to the associated sampled-data feedback control systems.

The tape then passes under the control-track magnetic-head to record a freshly processed 240-Hz/sec sinusoidal signal (along with a supplementary 30-Hz/sec frame-pulse), as a control-track for the subsequent use of the servo systems during the tape-playback. Figure 4-7a illustrates (1) the pattern of the control track signal and (2) the timing relationship of the control-track with respect to the transverse video tracks.

After passing by three more tape-lacing posts, each containing two magnetic-heads (one for the erasure and recording of the audio and cue tracks, the audio-heads being located 9 in. ahead of the corresponding video-tracks, and the other for the simultaneous monitoring of the *playback* audio and control-track signals), the tape passes onto the capstan spindle and pinch-roller, the take-up tension arm, and finally the take-up reel. In the "record" mode, the headwheel 240-Hz/sec tonewheel servo is phase-locked by means of its feedback control system. The tape capstan-servo is simultaneously synchronized to headwheel-servo by way of the tonewheel pulse, while the vacuum-guide servo is operated as a regular manually-preset closed-loop position-control system using a bridge null-setting.

4.3.3 Video Playback Mode. The video, audio, cue, and control-track magnetic-heads function as "playback" heads in this mode, with the master-erase, audio, and cue erase-heads deenergized. The headwheel 240-Hz/sec (tonewheel) servo, with its velocity and phase-lock loops, becomes automatically effective. The capstan 240-Hz/sec servo is simultaneously phase-locked along with the 30-Hz/sec capstan switch-lock servo. The headwheel 15.75-kHz/sec line-lock servo system then goes into operation to "pix-lock" the picture information from the tape at the accuracy needed for synchronizing it to the local television camera signals. Thereafter, the open-loop monochrome automatic timing correction system (MATC) and the color automatic timing correction system (CATC) become operative in this order for the accurate reproduction of color from videotape. Where color is not involved, MATC alone will be effective (1) in extending picture-stability concerning phase-jitter beyond that possible with normal 15.75-kHz/sec pix-lock, and (2) in eliminating minute horizontal time-phase errors due to head-quadrature tolerance and scallops, and (3) the minute residual guide skew-error below 0.02 μsec. All this corrective action takes place provided the FM-video information on the tape is appropriately signal-processed for the extraction of the composite-video signal and its synchronizing information from the videotape. If either one of the headwheel or the capstan servo systems fails for any reason, the picture breaks up, and the headwheel servo indicator lights up. The picture breaks up in either case.

In the playback mode, as soon as the headwheel velocity-loop and the capstan-servo are correctly phased, the video-tracks are read by the four video magnetic-heads sequentially; the FM signals are fed to the individual FM playback amplifiers via the slip-rings or the rotary transformer, the record/play transfer-relay contacts, and the four-channel preamplifier. Four adjustable delay-line amplifiers (in the absence of the MATC accessory) then correct the quadrature-errors according to any preset manual adjustment for the specific headwheel assembly in use; it is of course adjusted with the help of the picture on a video monitor. Gain and FM equalization presets compensate for the head-resonance and other variations in the response of the individual video-heads. The equalization employed for the aperture compensation of the high-frequency gap losses must be linear-phase to avoid envelope-delay distortion in the demodulated video. To achieve uniform playback video signals while splicing different program material recorded with different headwheel assemblies on a single tape, it is essential that all the video recordings on that tape be orginally made with the same FM carrier-deviation. The pix-lock and MATC accessories will then enable a fine playback as a *real-time* camera signal.

During *playback*, heads 1 and 3 do not read FM at the same time because of the quadrature head-spacing; in view of the blank spacing, the two signals can share a single channel to begin with. So it is with heads 2 and 4. Since an overlap of approximately three television horizontal-lines is allowed between the data on two adjacent tracks, it is the function of the FM-switcher to gate the correct FM-video signal during the overlap-interval by means of phase-locked 960-Hz/sec gating-pulse, which identifies the overlap interval in conjunction with the 15.75-kHz/sec horizontal sync-pulses from the processed composite-video signal (Fig. 4.-8b). The combined continuous FM signal from the four video-heads is then equalized and limited by balanced amplification against the zero-crossover, to a gain-equivalent of about 55 dB with respect to a 1-V peak-to-peak FM signal. The limiting minimizes not only an incidence of amplitude-modulation, but also losses of the picture-data from the tape (seen normally as white dots and streaks on the picture) because of the so-called occasional *dropout* phenomenon in magnetic-tapes.

The dropout usually occurs due to the basic oxide-coating imperfections in the manufacturing process, but in the Q-CVTR their rate of occurrence or intensity can be enhanced by such factors as (1) the nature of a scene (of the type of a dark background), (2) the head-to-tape contact pressure due to improper penetration, (3) inadequate limiter-gain, and (4) improper transient-response of the FM playback system. The dropouts due to these system imperfections can be eliminated by careful design and operation; for tape imperfections, an *electronic dropout compensator* accessory can automatically level down these dropouts either to the less imperceptible gray-level, or alternatively, by the insertion of the preceding picture-data on a line during the dropout interval by employing a video delay-line technique.

The balanced FM demodulator (a discriminator of the form of a diode-bridge) follows the limiter to extract or separate the video information from the FM signal. For example, with the original deviation used for monochrome and color, the *single-line second-slope of the frequency discriminator* extends from 4 to 8 MHz/sec, as shown in Fig. 4-8a. The residual of the balanced second-harmonic at left in the figure is actually cut off at 5.2 MHz/sec; therefore, a linear-phase low-pass filter, with a cutoff at 4.5 MHz/sec, is employed to eliminate the second harmonic residue. The figure illustrates the *first-slope detection* used with the present wideband-FM system also. The FM demodulator output can be directly switched adjacent to the modulator output on an oscilloscope to monitor the signal being recorded at any time in a *back-to-back mode*. A switching facility is also provided for preferably making *RF copies* of prerecorded tapes by directly connecting the FM output of a secondary playback machine to the FM recording set-up in the recording machine, in order to avoid, during these so-called *RF dubs*, error tolerances in the mod-demod and video processing circuits in the two machines concerned.

After the requisite postemphasis (as a compensation to the preemphasis used during recording), depending on whether the signal is monochrome, color, high-band, or low-band, a video signal-processing amplifier takes over the composite-video signal and:

1. inserts the dc component of the video signal by means of a black-level clamping circuit
2. removes any FM switching-transients extending above the peak-white and below the black-level
3. adds a freshly regenerated video-pedestal of blanking and the synchronizing horizontal and vertical pulse-waveform (which is phase-locked to the videotape-sync pulses by means of an automatic frequency control-loop, functioning as an all-electronic sampled-data feedback control system)
4. functions as a tape video output distribution amplifier to the picture monitor, the local television broadcast transmitter (VHF or UHF), and the long-distance microwave links or synchronous satellite terminals.

The regenerated complete-sync is practically free from all forms of noise and disturbances that might arise from the tape-playback, especially in the case of the *multigeneration dubs*; this procedure partially assures the reliability of the various sampled-data feedback control systems employed in the machine.

When MATC and CATC accessories are used for the reproduction of color from the tape, the demodulator composite-video output is directly switched over to these two automatic electronic video delay-control systems in sequence, and finally applied to the above signal-processing amplifier. The CATC system, incidentally, separates chroma and reference-burst in the color signal, clamps the chroma signal and removes any disturbances during the horizontal-blanking interval, and regenerates a fresh reference color-burst at 3.58 MHz/sec on the back-porch of the composite-video. With the color-ATC, the new reference-burst and the chroma are phase-locked to the station color-subcarrier frequency-standard in order to facilitate *special-effects* between the tape color signal and the other color camera signals originating on the station complete-sync. The horizontal- and vertical-synchronizing pulses of the studio are either simultaneously phase-locked or switched over to the horizontal and vertical tape-synchronizing pulses regenerated in the video-processing amplifier.

4.3.4 Electronic Splicing. Prior to the development of the electronic splicing accessory to the Q-CVTR, direct mechanical splicing, with the aid of an *optical tape-splicer*, was the only available method. At the approximate spot, where an "add-on" or an "insert" splice of new tape program material is required, the video-tracks on the tape are developed with a liquid-solution of iron-oxide (the commercial solvent is called *visimag*) to locate on the then-visible *magnetic-pattern* the nearest 30-Hz/sec edit (frame) pulse with a 40X microscope. The tape is cut along the *guard-spacing* just after the track containing the *edit-pulse*, and a special splicing tape is dispensed in position to form a clean, solid, square, butt-splice, in order to eliminate *picture rollover* on a video monitor or receiver and/or temporary picture breakup following a splice.

The present tape-editing technique using electronic splicing is quick and versatile. With its automatic facilities for adding in continuation or inserting at any spot, as required, new program material or commercials from a live or film camera or from another tape reproduction, hardly any trace of disturbance, such as that which occurs in mechanical splicing, is noticeable. With the availability of this facility, the broadcast television stations have the unique advantage of directly putting "on air" an edited television tape-program that has already gone through

1. the usual studio switching requirements
2. the special effects of different signals from different cameras, films, other videotapes and even "retakes" of scenes.

The tape recorder used for editing can be started and stopped at random between scenes or for *animation effects.* For best results, the machine used for recording the original program is used for editing too. The electronic splicing accessory

1. modifies the switching logic of the standard Q-CVTR with splice logic, splice control, and splice timing circuitry
2. requires an alternative master-erase head, with a separate *selective-erase head* for the control-track, as required for the "insert" of the electronic splicing
3. enables an undetectable electronic splice, either during the guard-spacing after vertical-sync pulses, or during a four-track overlap (with the FM saturation process aiding the new recording during the overlap interval) in the case of the program material from certain tapes, that involve problems relating to the control-track due to the phasing discrepancies and the erase-delay timing-variations of different headwheel assemblies.

Since the master-erase head is effective for recording only, it is automatically isolated off the moving tape by a *tape-lifter* during the playback in order to extend the tape life.

The actual splicing manipulation requires that the tape to be edited is simply played back with the picture on the monitor; at the appropriate moment, the record button is pressed so that the erase current is fed to the master-erase head, and the FM record-current is fed to the headwheel for recording the new scene. To prevent

1. the servo system-disturbances caused by the playback-to-record switching manipulation
2. the difference in phasing of the reproduced tape signals from the previous recording on that tape, and the incoming video signal for the proposed new recording
3. the undesired overlap of recordings due to the distance between the locations of the video-record and master-erase heads, and hence the resulting disturbance to sync-processing and servo systems (due to the conflicting pulse-timings)

the electronic splicing accessory, respectively, provides for the following features:

1. the servo systems make use of the same sources of headwheel and capstan drive signals on switch-over to the *record* mode as were used during the *playback* mode (while the headwheel servo remains phase-locked to a free-running 60-Hz/sec oscillator)
2. the high-precision phasing of the *playback* signal to the new incoming composite-video signal (and hence the station-sync) by means of the headwheel pixlock servo system
3. the exact determination of the splicing-transition by means of a coincidence pulse-gate, which would detect the first vertical-sync pulse after a cue signal (Fig. 4-7b).

The recognition of the abovementioned vertical-pulse would allow the master-erase head to commence the erasure of the previous recording. After a precise duration of 15 frames (for the time of the tape-travel from the master-erase head to the recording headwheel), the FM record-current is turned onto the headwheel magnetic-heads so that the new recording would commence exactly at the demarcation of the splicing-transition where the erasure had earlier started.

It is seen from these basic characteristics of electronic splicing, that videotape-editing would involve accurately-timed switching- and control-flags for not only the master-erasure and record head-currents of the FM-video, audio, and control-track, but also the three sampled-data feedback control systems for headwheel, capstan, and vacuum-guide. For a smooth undetectable transition from the original to the new program material at the splice demarcation, the automatic vacuum-guide servo as well is naturally expected to enable the new recording to be made at the corresponding *playback* pressure. For example, a 1-hr television feature program on tape, assembled according to the preceding flexible editing technique in just a few hours by a skilled video engineer (under the direction of a television program director), would make as fine a *feature presentation* as a movie shot and edited on film in several days or weeks. As a result, videotape facilitates an economic presentation of feature programs.

4.3.5 Electronic Editing. Tape editing accomplished with the electronic splicer requires, on the part of the video engineer, a considerable amount of skill and practice related to synchronization or timings of sound, video, and special effects, if a program equivalent to *animated trick-photography* on film is desired. So the *electronic editor* accessory, as a complement to the *electronic splicer*, is the next available facility for achieving

1. foolproof animated picture information
2. splice-preview before it is actually accomplished on tape
3. operationally simplified automatic tape editing, correct to a single frame at 30 Hz/sec
4. synchronization of the audio and video splicing on a frame-by-frame basis
5. a first-generation master-edited tape with special effects.

At the right instant, the electronic editor will be able to start a second Q-CVTR or a television film camera automatically for recording on a master-tape a previously previewed playback signal at the end of any *chosen frame*.

The principle of operation is based on inserting a "marker" *tone-burst* on the available cue audio-track of the videotape by means of a special "advance-cue" record/erase magnetic-head on the tape transport. The markers, established on a rehearsal basis, enable the electronic switching of the recording and monitoring circuits with appropriate logic and control circuitry. Every phase of the editing procedure may be *previewed* several times without any effect on the master-tape, while it is produced in its final format for a television broadcast as an all-in-all first-generation recording.

4.4 HEADWHEEL SERVO SYSTEM

4.4.1 Introduction. The motor, which rotates the headwheel carrying the four magnetic-heads in quadrature on its periphery against the moving tape, is one of synchronous hysteresis type, originally designed for 300-Hz/sec synchronous operation. However, it is not operated synchronously; three-phase power supply at 480 Hz/sec is applied in an ac carrier-servo system, and speed is held back for an asynchronous mode of operation at 240 rps by the headwheel sampled-data digital feedback control system. It is determined that this asynchronous operation of the synchronous motor (HAS motor), as a hybrid synchronous-induction motor, gives a suitable acceleration characteristic for establishing the phase-lock in the minimum possible time-interval, under different operating modes. The servo system actually delivers just enough power to exactly balance the load at the speed of 240 rps.

The headwheel, capstan, and guide interacting servo systems work in coordination to establish the precise timings required for video recording and reproduction, and the requisite *basic* feedback signal for these three sampled-data control systems is provided by means of a tonewheel mounted at the other end of the headwheel motor-shaft. As the motor runs, a notch on the tonewheel, at the instant of passing through the field of a fixed reluctance-type pickup coil situated at its rim, generates one narrow output spike per revolution, to present a periodic indication of both the speed and the angular-position of the headwheel. The generation of the 240-Hz/sec signal thus makes the tonewheel a *transducer for developing the basic analog-to-digital relationship* for the three complex digital feedback control systems used in the Q-CVTR.

The nominal speed of the headwheel motor, 240 rps, which determines the precise head-to-tape speed, is very rigorously/precisely controlled by the headwheel feedback control system, while recording and reproducing the composite-video information on the slowly moving tape. The schematic block-diagram shown in Fig. 4-9 reveals the characteristic multirate multiloop features of the headwheel servo, depicted by

1. the low sampling-rate 240-Hz/sec velocity feedback-loop
2. the 240-Hz/sec phase feedback-loop
3. the high sampling-rate 15.75-kHz/sec velocity and phase feedback-loops, otherwise known as the "line-lock" system for high-precision CRT time-base stability.

In the *record mode*, the headwheel-speed of 240 rps is controlled at exactly four times the vertical scanning-frequency of 60 Hz/sec; this reference is generated by frequency division from the crystal oscillator of the station sync pulse generator or the temperature-controlled color-subcarrier crystal-oscillator in color telecasts. The actual headwheel-speed is, however, primarily controlled by way of closed-loop

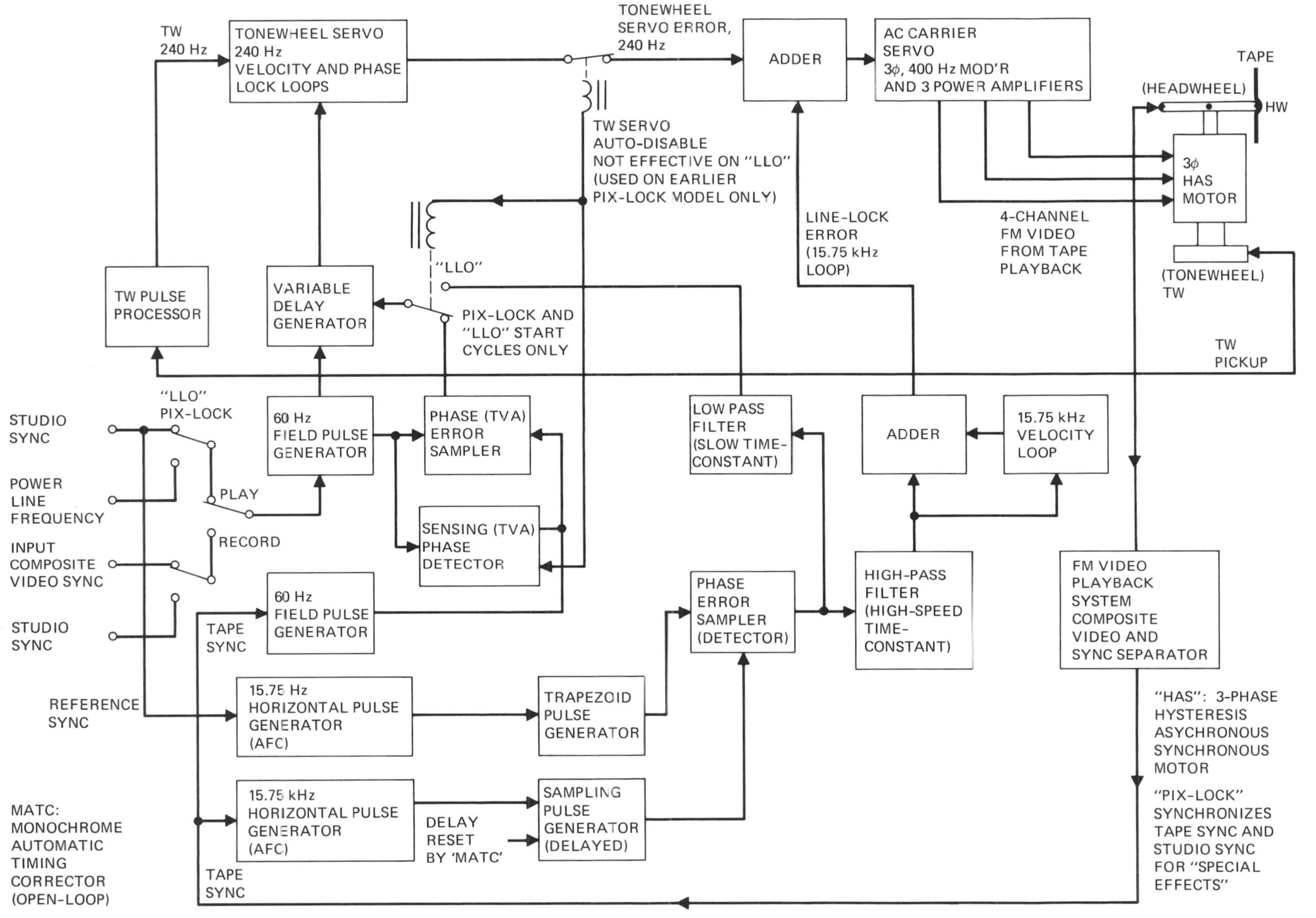

Fig. 4-9. Simplified block-diagram: headwheel sampled-data control system in pix-lock and line-lock-only modes of the Q-CVTR tape playback system. (*Courtesy of RCA*)

headwheel control system, in which an error-detector compares the frequency and phase of the signal produced by the tonewheel against a 60-Hz/sec reference-pulse at the television vertical field-rate; the periodic pulse-train is derived from the very composite-video signal being recorded on the tape. A reference-pulse generator in the Q-CVTR system automatically provides the requisite 60-Hz/sec reference field-pulse, as well as a 30-Hz/sec frame-pulse from the composite-video signal during the *record* mode, and from the station sync waveform during the *playback* mode. This procedure assures that each television-field is recorded in exactly *16 complete transverse-tracks* on the tape, since each revolution of the headwheel motor at 240 rps reproduces four magnetic-tracks with the four FM-video magnetic-heads on the moving tape (240 X 4 = 960 track/sec and 960/60 = 16 track/field). At the correct speed of the headwheel, the control-pulse processed off the tonewheel signal bears the correct phase-relationship to the reference-pulse and produces no effective control-error signal. The control system-stability then remains in equilibrium. Since the tonewheel maintains a fixed angular-position with respect to the headwheel on the same shaft, the vertical sync pulses at 60 Hz/sec are invariably recorded by magnetic-head No. 1 at a precise predetermined spot halfway through the scanning interval. The instant the headwheel-servo is thus phase-locked, the 240-Hz/sec tonewheel control-signal serves as the primary-reference to the 240-to-60-Hz/sec counter, for the simultaneous (actually subsequent) phasing of the capstan-servo system, to run the tape at exactly 15 ips from the supply to the take-up reel. In short, the tonewheel control-pulse synchronizes the capstan motor to the headwheel motor.

At the same time, to provide a continuous record of the angular-position and speed of the headwheel motor during the recording of a video signal, the 240-Hz/sec tonewheel-control-pulse is reshaped to a sinusoidal waveform and recorded as a control-track signal along the guided-edge of the moving tape by means of a separate control-track magnetic-head (see Fig. 4-10). The latter is located adjacent to the vacuum-guide of the headwheel assembly.

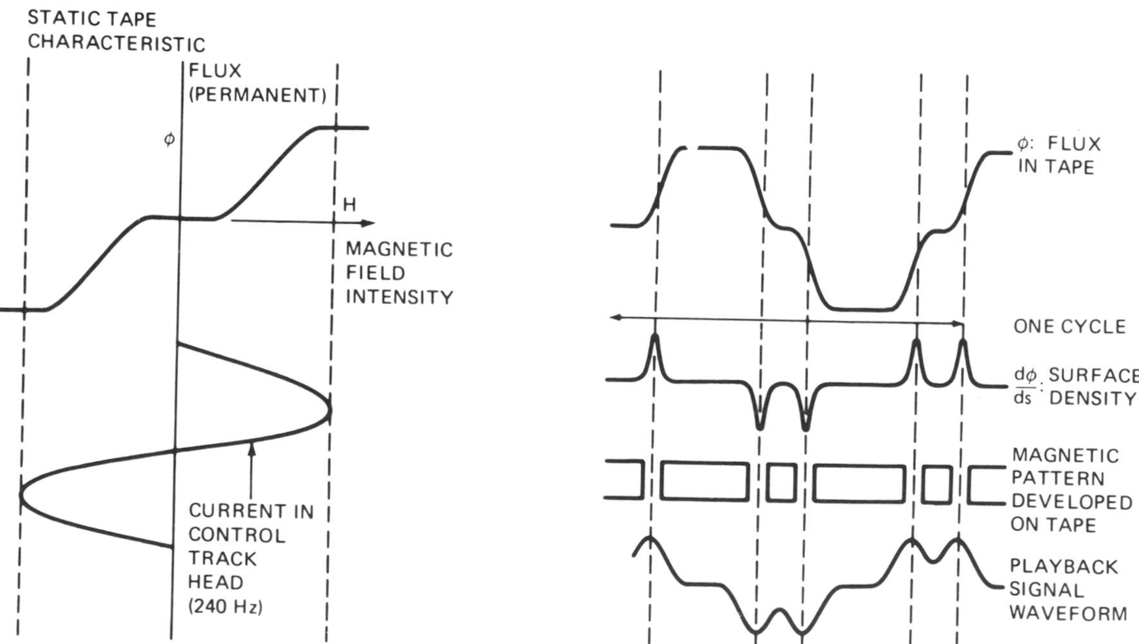

Fig. 4-10. Control-track signal in record and playback modes. Direct recording (nonlinear process).

The control-track corresponds to the sprocket holes in motion-picture film since, during playback, the control-track signal enables the capstan-servo to control the motion of the videotape at the exact original recording speed of that particular program. A frame-pulse at 30 Hz/sec, extracted from the synchronizing pulse-waveform of the recorded composite-video signal, is simultaneously recorded by adding it as a complement to the control-track sinusoidal signal in order to (1) aid in tape-editing and (2) provide a frame *switch-lock* facility in some versions of the Q-CVTR machines. The headwheel closed-loop servo will therefore necessarily maintain the relative phase-relationship of the FM television signal recorded on the transverse-tracks with respect to the above frame-pulse of the control-track. In other words, servo feedback control-stability relative to the scanning-frequencies is thus maintained to prevent the slow phase-drift that is otherwise possible.

In the *playback* mode, the headwheel speed is generally held close enough to allow (1) the scanning frequency-tolerance in video monitors and television receivers, and (2) exact split-second time schedules used in the editing of tape programming. In this case, the 60-Hz/sec reference signal, which is compared with the tonewheel control signal for feedback control, is derived from the studio sync generator or from the power supply as an alternative, in the event that a sync generator is not available (as in the case of some simplified versions of the quadruplex playback machines used away from the television studio).

It is seen from the above general operational requirements of a typical Q-CVTR that the headwheel control system (along with the other two control systems) necessarily requires timewise correctly processed pulse waveforms as individually identified in the block-diagram of Fig. 4-11. These waveforms are processed with the appropriate digital logic and pulse forming circuitry in three separate subsystems, generally called *reference generator, tonewheel pulse processor,* and *tape sync processor.*

4.4.2 Headwheel Velocity and Phase 240-Hz/sec Digital Feedback Control Systems (Basic Headwheel-Tonewheel Servo). A simplified block-diagram of the headwheel-tonewheel servo that includes both the velocity- and phase-feedback loops is shown in Fig. 4-12.

The headwheel-tonewheel servo system primarily performs two functions:

1. the phase of the tonewheel-pulse is compared to that of the 60-Hz/sec fixed reference-pulse obtained from the reference-pulse generator to derive the exact phase information needed for the control or regulation of the headwheel motor-speed
2. the period of the tonewheel-pulse is compared to a fixed "ball-park" auxiliary time-interval, as a preparatory "coarse-servo" system for the subsequent *fine-phase* control-loop just mentioned.

These are called *phase-* and *velocity-loops*, respectively. The error signals at the output of the *sample-and-hold* networks, following the phase error-sampling are matrixed for addition in such a manner that the velocity-error signal would predominate when large speed-errors occur, and the phase-error signal would be automatically effective when comparatively small errors occur.

Close control is achieved through the phase-error detector, when a trapezoidal waveform derived from the 240-Hz/sec tonewheel signal is sampled by pulses derived from the 60-Hz/sec reference source. A trapezoid waveform in the place of the

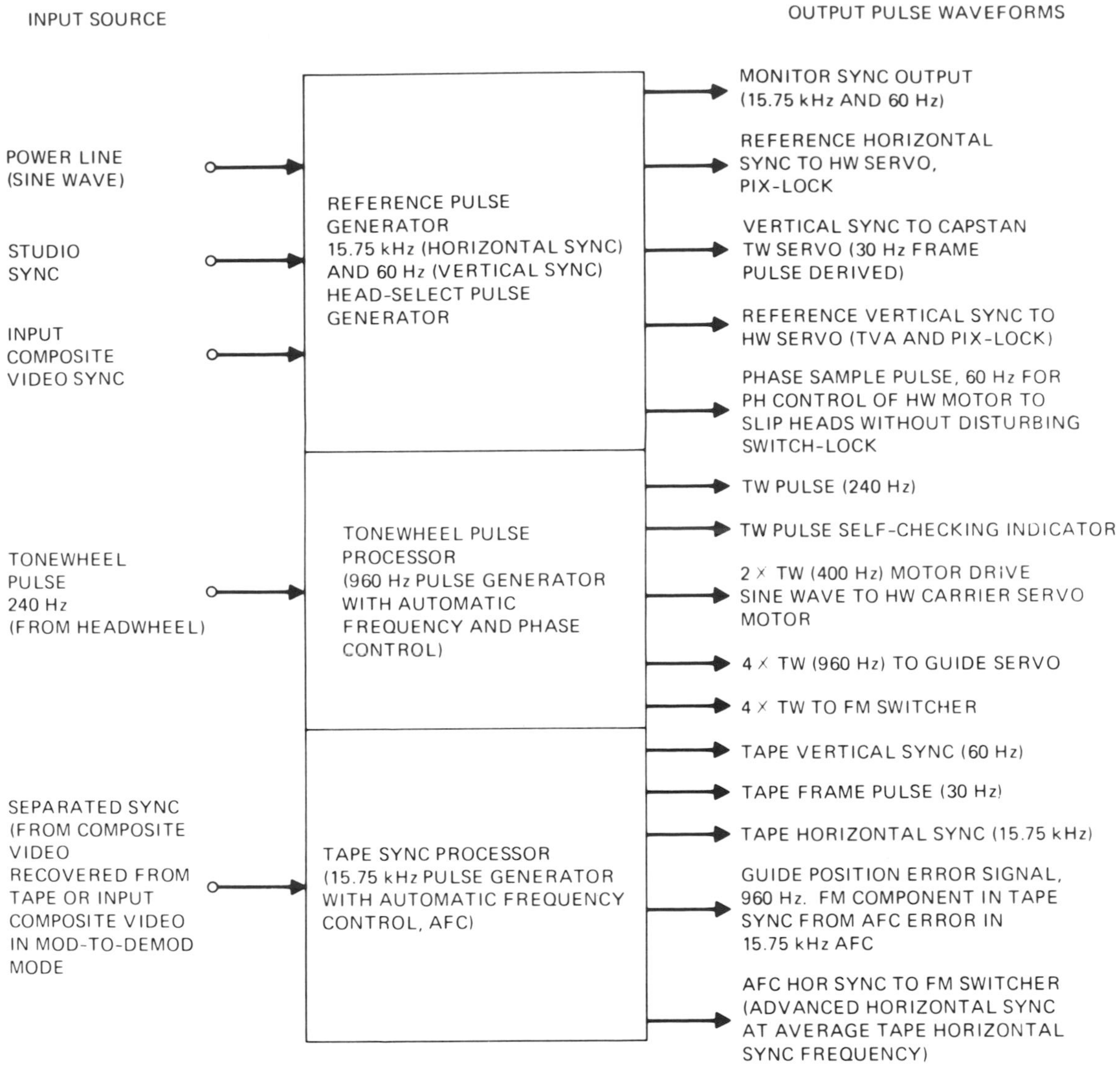

Fig. 4-11. Auxiliary pulse processing devices required for the sampled-data control systems in the Q-CVTR.

more common sawtooth (ramp) waveform is preferred for this purpose to obtain a steep slope and hence a narrow lock-in range of the order of 100 μsec for the digital phase-lock (PLL). This technique allows sufficient accuracy for the general black-and-white signal operation of the Q-CVTR with a relative PLL tape-to-sync jitter of the order of 1μsec during playback. This form of jitter is of course not seen in television receivers because the associated synchronizing information and the picture seen have the same low-frequency jitter components, and the television vertical and horizontal scans closely follow the playback-timing of the synchronizing information only. The extent of the recorded jitter can be examined, however, on the picture-monitor of the Q-CVTR by superimposing the playback tonewheel pulses on the picture, when the monitor is synchronized to the (internal) sync of the composite-video. This jitter actually corresponds to the stable inner limit-cycle in the phase-plane, as a characteristic of the "nonasymptotic" stability of a nonlinear system (with saturation elements, hysteresis, etc.).

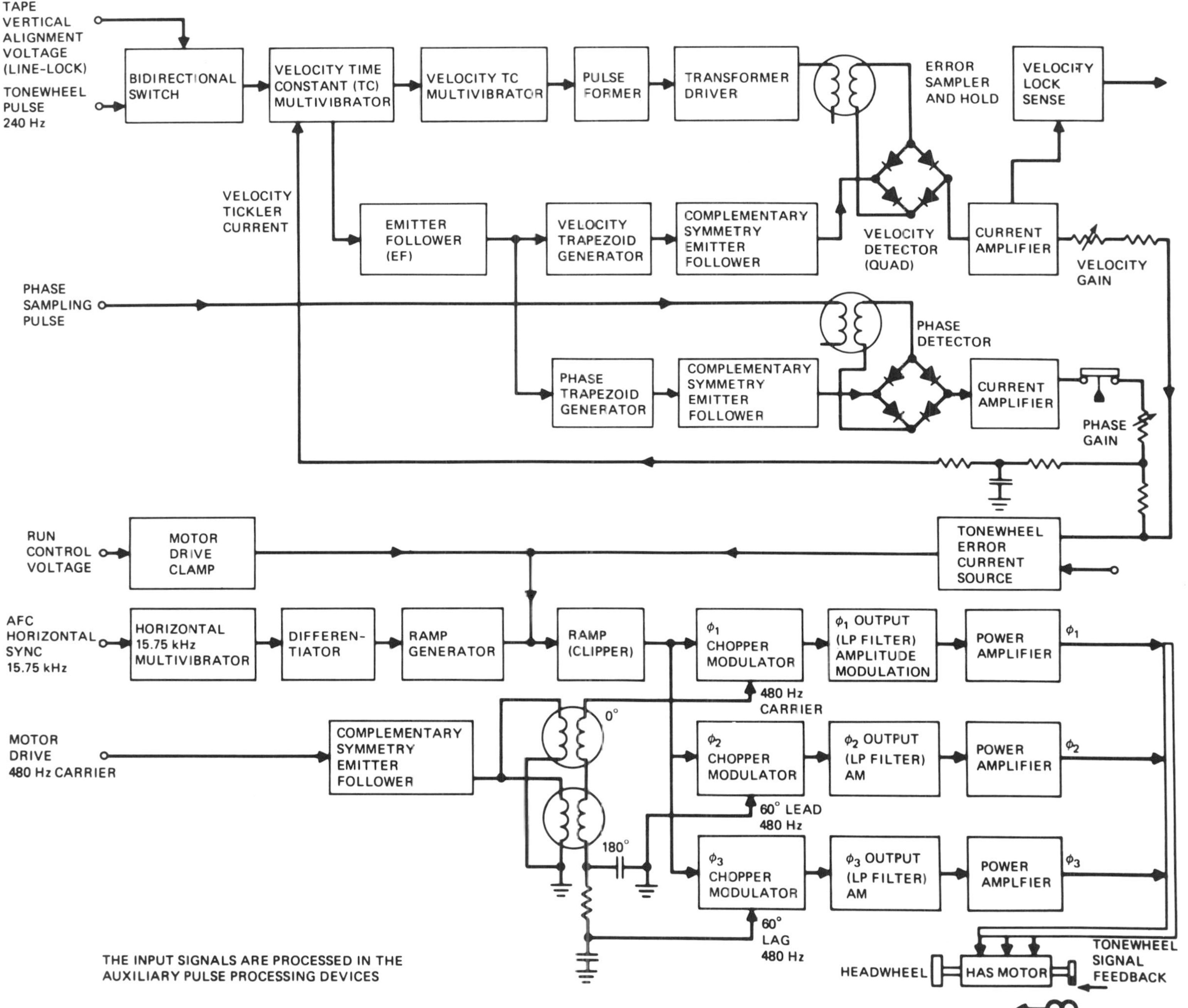

Fig. 4-12. Schematic block diagram. Headwheel combined velocity-and-phase sampled-data feedback control system (tonewheel servo). (*Courtesy of RCA*)

The wide-range velocity-error detector is essential to reduce the time required for the headwheel-servo to pull into the normal control region of stability

1. when the machine is started
2. when the reproduction is momentarily interrupted, for example, by partial loss of signal from the tape (due to a bad splice or *drop-out*) or by a potential high-intensity system-disturbance of external origin.

The system is immune to noise disturbances of the usual order. The velocity-error detector operates by measuring the period of the tonewheel pulses, and, systemwise, this is an elementary form of a *digital adaptive-control* system, since as a result, the servo system-parameters automatically change for producing a fine control of the motor speed. After the velocity pull-in, the effect of the "coarse" velocity-error signal is, circuitwise, automatically minimized (to prevent possible conflicting data to the subsequent ac carrier-servo system) by feeding a small portion of the phase-error signal into the velocity timing circuits. This *minor-loop* effectively modulates the trapezoidal slope in such a way that the velocity-error signal will sample at an approximately constant dc level, while the phase-error signal exercises the actual control via the following 480-Hz/sec carrier-servo system.

The phase and velocity loops may be theoretically interpreted also as a *proportional-plus-derivative* control:

$$m = Ae + AD\frac{de}{dt}$$

Hence, transfer function,

$$G(s) = \frac{m}{e} = A(1 + sT)$$

where m is the change of output-speed, e is the error from the set-point, A is the *proportional- or phase-sensitivity*, and T is the *derivative-time.* The derivative-action causes a change in output-speed, which is proportional to the time-rate of change of error from the set-point. Hence, the *derivative-control can be designated as a velocity feedback control.* The preset derivative action causes an initial jump in output-speed, which is proportional to the derivative-time T. In short, the velocity loop, with its relatively large pull-in range, acts as an automatic *reset* for large drifts in the error of the phase-lock loop (Fig. 4-13).

The **derivative-controller** behaves like a *linear lead network* with a zero-frequency gain of A. At the derivative-corner, the gain is 1.414 and the phase 45°. Increasing-T moves corner to the left and vice versa. The second effect of the derivative-response is to *reduce the recovery-time* following a load-change by increasing the operating frequency. Hence, derivative action in effect permits use of proportionally higher phase-sensitivity.

Referring to the sampling pulse-waveforms and error set-points, related to the phase and velocity-feedback control loops in Fig. 4-14, it may be noted that the interval "a" between the sampling-pulse and the leading-edge of the preceding tonewheel-pulse is determined by the electrical time-constants only, and is independent of the actual tonewheel speed. Interval "b" between the successive mid-points on the trapezoidal slope depends on the speed of the tonewheel and equals the sampling pulse-period only at 240 Hz/sec. Since the effective velocity-error signal develops as soon as the tonewheel pulses appear at start, the pull-in range of

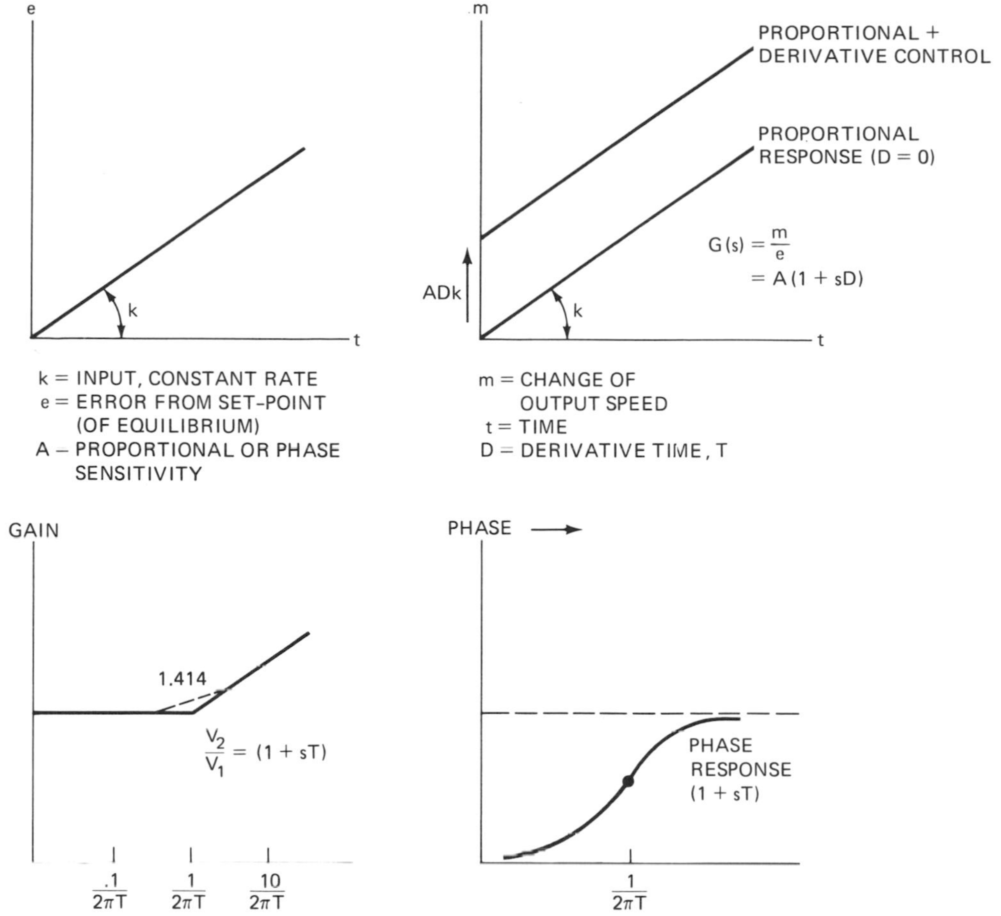

Fig. 4-13. Derivative control (effect of velocity loop).

the velocity-loop is correspondingly much larger. With the head-wheel running at an incorrect speed, the relative phases of the sampling-pulses and the trapezoid are slightly different. For example, if the headwheel is much faster, the trapezoid-period is shortened slightly to cause the set-point in detector to move below the optimum level and produce an error within the pull-in range of the phase-lock loop. The detector, therefore, effectively measures the period between the successive tonewheel pulses (the period/time-delay of the velocity-loop being 4166 μsec at 240 Hz/sec), and modifies the headwheel-servo system-characteristic by automatically enabling the appropriate loop effective. The technique thus introduces a basic format of a *digital adaptive feedback control system.*

Since, in the *record* mode, it is essential that the headwheel be locked to the 60-Hz/sec field-pulse, associated with the input composite-video signal being recorded in the FM format, a self-checking system, by means of the logic interrelating the various pulse-timings, senses the missing error set-point during the 100-μsec trapezoidal slope to automatically illuminate the headwheel warning-indicator light, in the event the headwheel-servo is out of phase-lock. (As mentioned earlier, a self-checking subsystem of this character for the headwheel, capstan, and guide control systems, and for the more important functional signals provides an assurance to the program recording-aspects of the studio operations. This subsystem is designated as *Indicator.*)

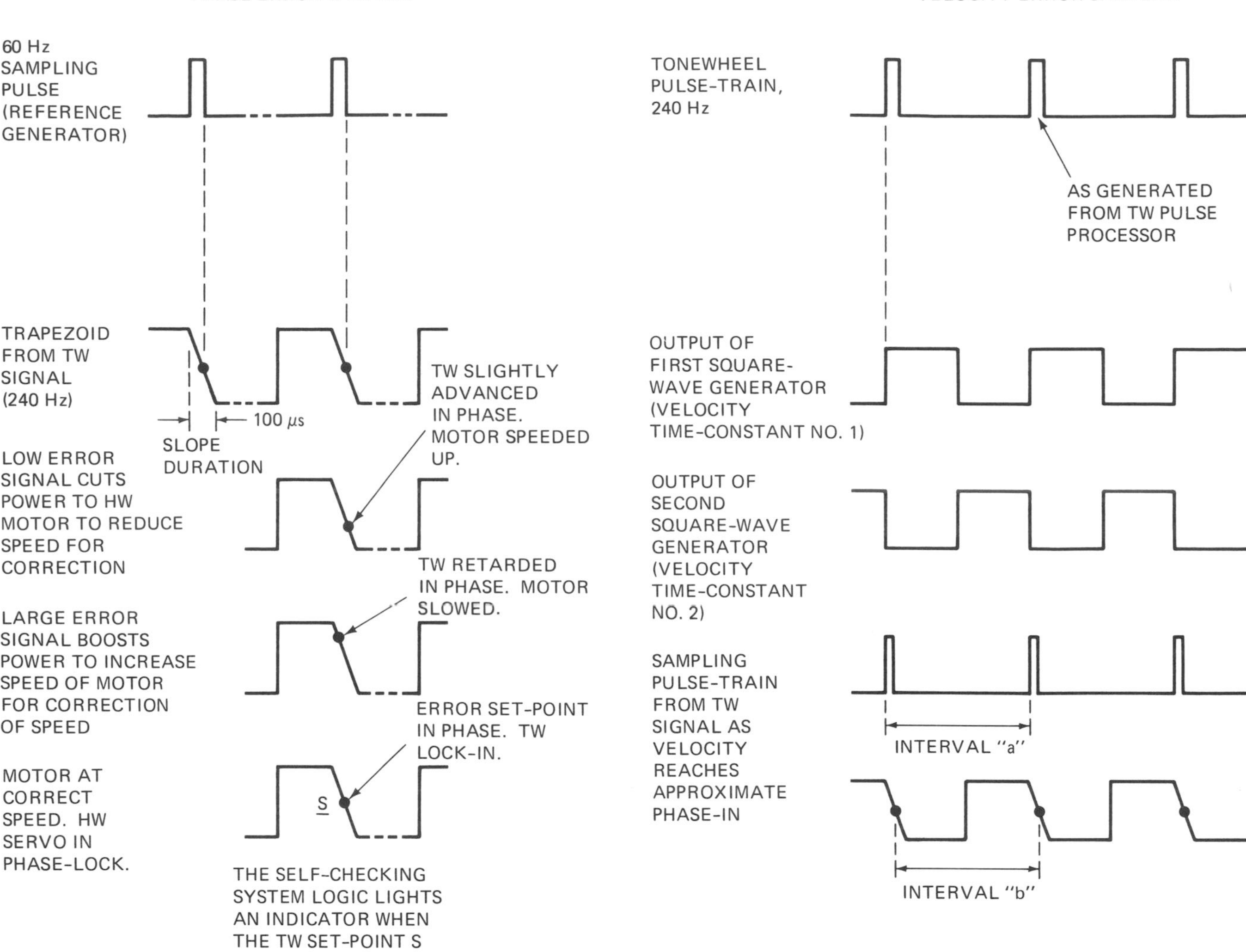

Fig. 4-14. Sampling pulse waveforms and error set-points; headwheel velocity and phase feedback loops (tonewheel 240-Hz servo).

The error signal from the tonewheel-servo is pulse-width modulated in effect at the line-frequency rate (15.75 Hz/sec) to enable the headwheel-servo to provide the fine accuracy required in precision *pix-lock* mode. The pulse-width modulated error signal is for this purpose applied to an ac carrier-servo system. The carrier 480 Hz/sec of this servo is phase-locked to the 240-Hz/sec tonewheel control signal to eliminate any disturbance due to possible undesirable beats in the carrier subsystem. Since a three-phase hysteresis asynchronous synchronous HAS motor, driven by three individual power amplifiers and their Y-delta connected output transformers, is used for driving the headwheel, the carrier is split into three carrier components at a phase-shift of 120° from one another. Then, the *pulse-width modulated error signal amplitude modulates* each one of the three carrier drives in a bilateral *transistor-chopper circuit*, the magnitude of the pulse-width determining the amplitude of the 480-Hz/sec carrier of each phase at any instant. (This specific type of transistor provides balanced operation on both positive and negative half-cycles, with both the collector and the emitter operating as the collector in turn.) When the line frequency

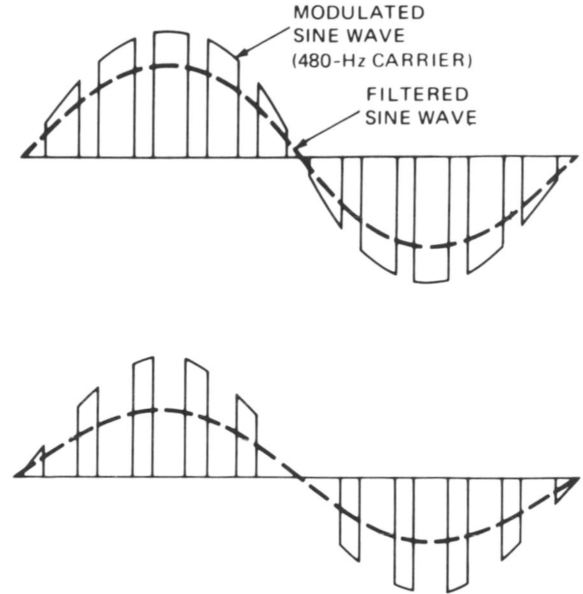

Fig. 4-15. Pulse-width versus amplitude modulation in the 480-Hz ac carrier servo of headwheel servo.

component is filtered out, as shown in Fig. 4-15, the driving power of the amplitude-modulated carrier approximately depends on the error information at any particular instant, since the headwheel servo as a whole is primarily a nonlinear feedback control system. The hysteresis hybrid-synchro-induction HAS motor itself functions as *an ideal demodulator* in this ac carrier setup.

4.4.3 Headwheel 15.75-kHz/sec Line-Lock Sampled-Data Feedback Control System: *Pix-Lock* and *Line-Lock Only* Modes of Operation. The horizontal 15.75-kHz/sec high-precision line-lock headwheel control system operates in conjunction with the frame 30-Hz/sec capstan *switch-lock* servo system, to provide a Q-CVTR facility termed *pix-lock*, for special television programming techniques of the category of *special-effects* between the videotape signals and the live or film monochrome/color television camera signals. The special-effects, for example, may involve picture superimpositions or partial electronic-insertions for the simultaneous broadcast of a part of the picture-frame from the tape-reproduced picture and the remaining part from the live or film camera signal. The line-lock and switch-lock digital control systems, in short, synchronize the Q-CVTR playback signals to the horizontal and vertical scanning frequencies used for the other camera signals in the studio. The relative horizontal picture-jitter (theoretically equivalent to the limit-cycle in the phase-plane for a nonlinear feedback control system) effective in the case of the previous 240-Hz/sec tonewheel servo system may reach a maximum up to 1 μsec (corresponding to an equivalent horizontal picture-jitter of 0.25-in. on a 17-in. monitor) with respect to a live camera picture, when the latter is simultaneously displayed on the same video monitor. (The above unnoticeable relative-jitter is of course acceptable for the display of exclusive tape signals.) The high-precision headwheel control system minimizes the 1-μsec jitter to an extent of ±0.1 μsec (or ±0.07 μsec in the more sophisticated air-bearing headwheel assembly); the precision of tape picture-stability thus realized is not objectionable for camera switching purposes and for simultaneous display with a live studio camera signal when the broadcast signal is observed on an ordinary *black-and-white* television receiver.

The Q-CVTR normally takes less than 2 or 3 sec to resynchronize and lock the tape (color) picture if the system is subjected to a high-intensity low-frequency disturbance. Under these conditions, the lock-in of the 15.75-kHz/sec headwheel line-lock control system would be fairly instantaneous (of the order of 0.1 sec) but the low-rate 30-Hz/sec capstan switch-lock digital feedback control system would take up most of the effective 2- to 3-sec interval. So, in the event the line-lock alone is sufficient (for general studio switching purposes without the *switch-lock* facility), the headwheel 15.75-kHz/sec high pulse-rate digital feedback control system by itself is more effective for instantaneous stabilization of the picture under the above condition. Provision is, therefore, made in a modern Q-CVTR for this *line-lock only* mode, especially effective in the case of a preedited tape color-reproduction, since a low-frequency disturbance or dropout would momentarily roll the complete picture vertically for an instant instead of breaking up the entire picture or color for 2 or 3 sec. The former mechanically spliced videotape is another instance where this feature is quite effective in minimizing the effect of a disturbance to a mere blink, in the event of an error in the insertion of the vertical-sync at the splice-transition.

The principles involved in the pix-lock control system can be explained as follows with the aid of the schematic block-diagram shown in Fig. 4-16. On start, the headwheel/tonewheel control-loop is immediately effective, and, as the headwheel motor-speed approaches the exact 240 rps, the error voltage is gradually reduced, and less power is delivered to the motor via the modulator in the 480-Hz/sec carrier-servo. The PLL error- and drive-signals respond to minor error voltages very rapidly with the effective time-constants in the system. The frequency chosen for the carrier, by doubling the 240-Hz/sec tonewheel frequency, allows smooth and fast error correction without interference from any beat-frequency components. No sooner do the 240-Hz/sec headwheel- and capstan-servos lock up, then the tape composite-video signal becomes available from the FM system to provide sync and its associated 30-Hz/sec frame-pulse. The latter acts as a reset on a 240/30-Hz/sec binary frequency-divider system and initiates a *slip head-tracks* procedure on the part of the 240-Hz/sec capstan servo to enable the capstan 30-Hz/sec *switch-lock* mode of operation. Since vertical timing errors often occur in television systems due to the coarse insertion of the recorded vertical-sync pulses in composite video, adaptive *tape-vertical-alignment* (TVA) is essential for the automatic transfer of the headwheel 240-Hz/sec tonewheel control to the high-precision headwheel 15.75-kHz/sec line-lock control subsystem.

The TVA is a high-gain variable-delay digital electronic feedback control system, and the system location of this complementary adaptive feedback loop is shown in the line-lock servo block diagram of Fig. 4-16. It compares the phase (timing) of the 30-Hz/sec frame-pulse derived from the tape-sync with that of the local sync to provide a dc error-voltage at the output of the hold and compensation networks. This error-voltage in turn controls a variable-delay pulse generator, inserted in the path of the 60-Hz/sec reference-pulse derived from the local studio complete sync, and thus enables the control of the angular-momentum of the headwheel via the headwheel 240-Hz/sec tonewheel control subsystem. The phase-shift of the tape playback-signal is hence automatically corrected for the timing discrepancies in the insertion of the vertical-sync pulses on the recorded video-tracks. The TVA with its high sensitivity adaptively corrects the horizontal timing of the reproduced

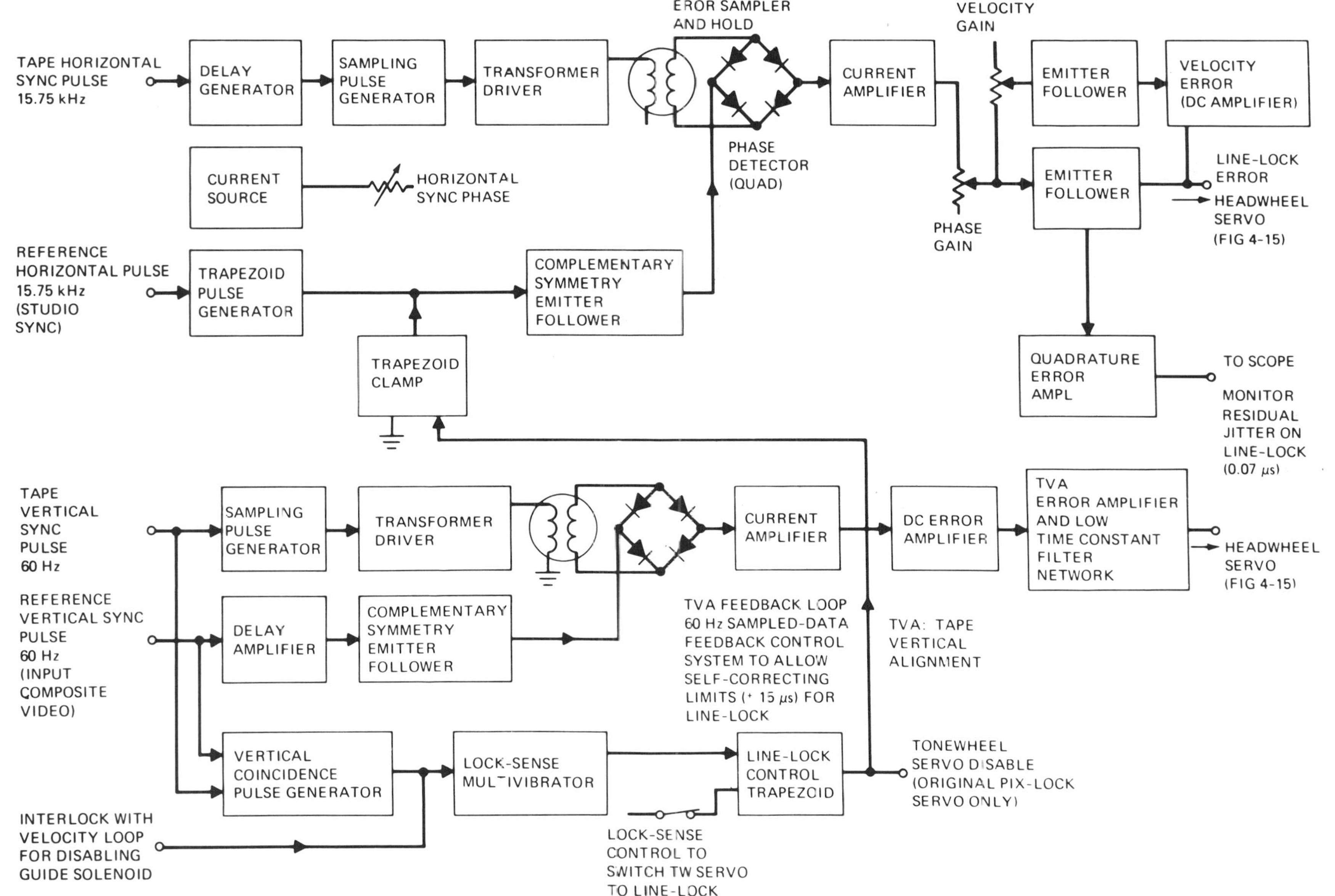

Fig. 4-16. Line-lock 15.75-KHz sampled-data feedback control system for pix-lock operation of Q-CVTR. System works as part of headwheel servo. (Pix-lock operates in conjunction with switch-lock of capstan servo.) (*Courtesy of RCA*)

picture to within plus-or-minus a few microseconds from the exact position required for pix-lock. The residual error cannot be corrected at this stage, since the loop-gain is not adequate at the low TVA sampling-rate of 60 Hz/sec. This slow-acting vertical-rate PLL, with a large time-constant, does not normally interfere with the head-wheel 240-Hz/sec PLL at the start-up and the lock-in of the headwheel control system, but it develops an error-signal only if the headwheel-speed is correct enough for the close proximity of the tape and reference 60-Hz/sec pulses. The instant the TVA-loop becomes effective, the TVA error-voltage, while acting on the vertical correction, simultaneously operates a relay to switch over the headwheel carrier-servo drive system from the 240-Hz/sec tonewheel feedback-loop to the high sampling-rate 15.75-kHz/sec feedback-loop. The sampled error information is then presented at a sufficiently high sampling rate, and the line-lock control subsystem builds up its high sensitivity to respond to minute timing errors of the order of ±0.1 μsec very rapidly. With the incidence of a heavy disturbance on a rare occasion or a long-duration picture dropout (most tape *dropouts* usually cause an instantaneous white-streak, and hence they are not effective in causing noticeable disturbance to the control subsystems), the system automatically goes through the complete lock-up cycle for pix-lock in less than 3 sec—namely, the velocity-loop to the phase-loop, then the TVA-loop, and finally to the line-lock loop plus the capstan switch-lock.

In the high-precision line-lock loop, the clean horizontal-drive pulses at the 15.75 kHz/sec as derived from the tape sync and the local sync are sampled and compared in a *phase comparator*; the dc error-signal, with the appropriate hold and compensation network, is applied to the ac carrier-servo along two signal paths of adjustable gain, one direct through a high input-impedance emitter-follower, and the other by way of a velocity amplifying device. The latter employs a differentiating network in conjunction with a common-base transistor amplifier. Only changes in the dc error-signal magnitude are involved in this minor-loop in view of the differentiating network used, and these changes therefore pertain to minute headwheel velocity-errors. A low-pass filter network, which functions as a *simple phase-lag network*, is included in this high-gain loop to prevent possible overloading of the following carrier-servo modulation by the heavier transients in the system, due to switching, noise, etc. For monitoring purposes, the minute dc error-signal, corresponding to the timing errors of the order of ±0.1 μsec, is differentiated, amplified, and ac-coupled to the calibrated CRT monitor to enable the time-error readings in terms of a voltage-scale. The readings on this special scale make the "performance-index" for this modern digital control system.

In the case of the alternative *line-lock only* (LLO) mode, the line-lock system is modified as shown in the simplified block-diagram in Fig. 4-9, for the *LLO* versus *pix-lock* modes of operation.

With this final *LLO* system development, the automatic *TVA* relay switch-over between tonewheel (TW) and pix-lock/LLO modes is further improved to automatic effectiveness in, circuitwise, permanently combined 240-Hz/sec TW and 15.75-kHz/sec line-lock error signals. That is, as an example, the operation of the velocity- and phase-loops in the headwheel 240-Hz/sec control subsystem is presently effective without any actual relay operation. The HW-LLO mode, without capstan switch-lock, operates, in principle, along two signal paths, using error signals of two different bandwidths, as far as the 15.75-kHz/sec error-sampler is concerned. The *slow-acting low-pass signal path* automatically modifies the amount of 60-Hz/sec reference pulse-delay in the regular TW phase-loop, while the *fast-acting high-pass signal-path* alone contributes to the regular line-lock phase- and velocity-loops. As a major difference, in the case of the pix-lock mode, the reference pulse-delay in the TW phase-loop is controlled by the error from the exclusive 60-Hz/sec TVA-loop, while in the case of the LLO, a low-frequency component from the 15.75-kHz/sec line-lock acts in conjunction with the effect of the TVA-loop.

The novel pulse-sampling method used for the 60-Hz/sec TVA-PLL, as required for the line-lock, is interesting in that it presently follows the common practice of *phase comparison in pulse techniques.* The tape- and reference-pulses will be appropriately aligned by the TVA-loop to prepare the *transfer* of TW-servo to the line-lock servo. Usually, the reference-pulse generates the requisite ramp or trapezoid waveform; the pulse to be controlled, phasewise, generates the narrow pulses required for sampling the sloping-waveform about the set-point, so that the dc error-voltage output of the "hold" network in the digital feedback control system is then a function of the sampling pulse-timing with respect to the set-point on the sloping-waveform. In the TVA-loop, on the other hand, the reference pulse generates a rectangular-waveform, whereas the pulse to be controlled, phasewise, generates a specially designated bracket-pulse as shown in Fig. 4-17.

In both cases, the digital feedback control systems will lock-in to the timing where the identical dc error-voltage output, in either case, is equivalent to the voltage at the set-point of the reference-trapezoid. In the bracket-pulse control subsystem, the dc error-voltage output will be actually a resultant of the upper and lower levels of the rectangular waveform during the pulse-sampling time-interval. The leading-edge of the sampling bracket-pulse precedes the phased leading-edge of the rectangular reference waveform, while in the trapezoid the leading-edge of the narrow sampling-pulse follows the optimally timed start of the trapezoid-slope. However, in either case, a delay equivalent to a half-bracket pulse-width or half the slope-width, respectively, is required in one of the two signal paths to align the reference-pulse and the sampling waveform for the correct lock-in of the combined PLL.

4.4.4 Monochrome Automatic Timing Corrector. The monochrome automatic timing corrector (MATC) is an all-electronic open-loop 15.75-kHz/sec digital control system, mainly independent of the controlled-variable (that is, the control of the motor), although it does indirectly contribute a fine correcting influence on the closed-loop line-lock feedback control system on an adaptive-control basis. The MATC operates electronically on the reproduced composite monochrome- or color-video, with variable time-delay as the control-parameter. The fine relative phase-jitter tolerance of ± 0.07 μsec is perhaps the ultimate precision as far as the headwheel digital feedback control system is concerned (as obtained by the line-lock control system with an air-bearing headwheel motor assembly). The MATC goes a step further. This digital control system for the monochrome composite-video signal reproduced from the tape enables a maximum reduction-factor of 25 to 1 for the input horizontal time-base errors extending to a cumulative range of ± 1 μsec. The device automatically minimizes the headwheel-servo jitter-tolerance and the time-displacement errors of the quadruplex magnetic-head system, namely, the *geometric distortion of the horizontal time-base* to an accuracy of ± 0.02 μsec, a performance-index that is more than *sufficient as far as the detectable time-base stability* of a monochrome CRT picture-display is concerned (especially in the case of the special-

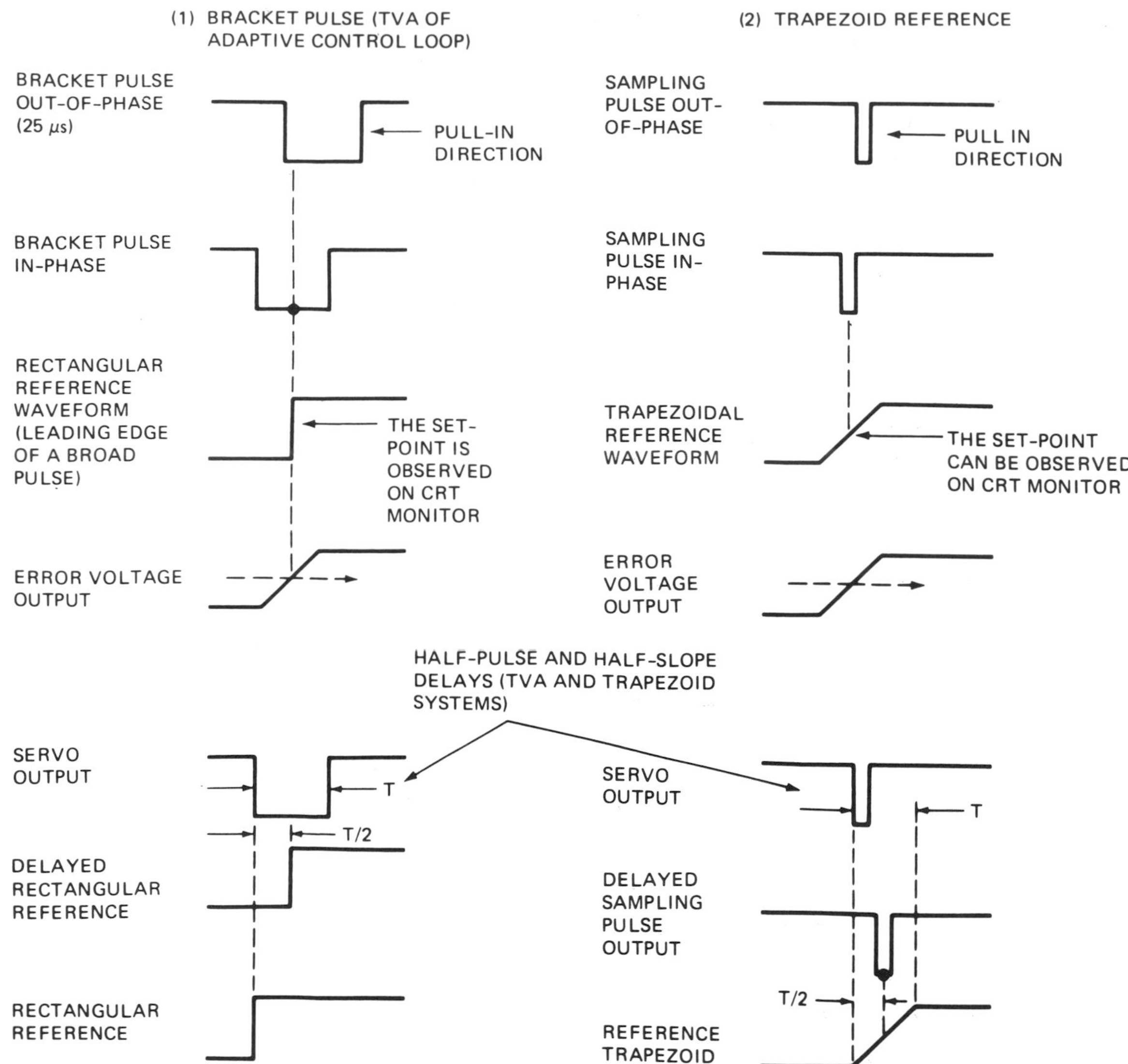

Fig. 4-17. Sampled-data control with (1) bracket pulse (tape vertical alignment) and (2) trapezoid reference. (*Courtesy of RCA*)

effects between the tape and television camera signals). Such geometric distortion is produced in a Q-CVTR by any one of the following four problems:

1. minimal jogs or skew-distortion caused by mechanical misalignment of the vacuum-guide in a direction parallel to the headwheel assembly, when an automatic guide servo for this purpose is not accurate enough for any reason
2. scallops or bows caused by mechanical misalignment of the guide in a direction perpendicular to the headwheel panel
3. vertical steps or repeated horizontal-displacements in groups of 16 CRT picture-lines, as caused by the quadrature-errors due to the headwheel manufacturing-tolerances
4. system low-frequency hum pickup.

Prior to the advent of the MATC system, these problems required special maintenance and critical manual adjustments by operating personnel. Also, for the extremely high horizontal time-base accuracy (of the order of 0.005 μsec) required for the reproduction of true color from the videotape, the MATC in conjunction with the pix-lock (or the LLO) is an essential preparatory stage to the color automatic timing corrector (CATC) used in a modern Q-CVTR. The color performance-index of 5-nsec error corresponds to ±3° phase-tolerance at 3.58 MHz/sec.

The operation of the MATC is based on the principle of an "automatic delay line" (*electronically variable delay-line*), which employs inductor coils and reverse-biased PN silicon-capacitor diodes in the place of the lumped-constant fixed capacitors. At a semiconductor PN-junction, the depletion or transition region widens with reverse-potential, and this naturally decreases junction-capacitance as if the thickness of a dielectric is varied. The silicon semiconductor device that uses this phenomena is called a *varicap* or *silicon-capacitor*, and a typical variation is 120 to 22 pF for a change in reverse-potential of the order of 0.1 to 25 V. The dc error information, obtained from a phase comparator at the horizontal 15.75-kHz/sec sampling-rate, controls the variable-capacitance of the PN-diodes or varicaps in the automatic delay-line (ADL), to produce an automatic timing-error compensation, on an open-loop basis, in a video delay amplifier containing a built-in fixed delay-line; it is designed for the video-passband on a linear basis. The simplified system is illustrated in the schematic block-diagram shown in Fig. 4-18. A typical silicon-alloy varicap (for example, TRW-4101-1) used in the ADL may have a capacity variation such as 180 pF ±25% for an error-range of 0.3 to 2.8 V.

In addition, the MATC control system requires an adaptive-control facility. It is necessary that the control of the ADL within the specified error-limits of ±1 μsec be maintained on a line-by-line basis automatically by a "tape horizontal alignment" (THA) digital feedback control device, somewhat similar in principle to the TVA-loop used in the line-lock control system.

The line-by-line timing correction employed in the MATC requires that this accessory be primarily an open-loop control subsystem, because obviously a closed-loop sampled-data system cannot respond fully in one sample-interval. The timing errors, produced by the quadrature magnetic-head switching at a predetermined timing of the horizontal sync-pulse, are measured by deriving the error from the trailing-edge of the tape-sync, so that the proper delay-correction can be made for the next horizontal line. In the actual circuit, the ADL is split into two separate halves to reduce by half the duration of the transient produced by the application of the timing-error signal (due to the changes in delay), as the trailing-edge of the sync appears at the input of the ADL. The error signals applied to the two halves are therefore compensated by means of a fixed delay-line. In addition, to prevent the transient extending into the video-passband, a fixed delay-line of a duration of 2.5 μsec precedes the two ADL halves, each providing a maximum delay of 1.5 ± 0.25 μsec. The transient referred to should not interfere with either the leading-edge of the sync (which is used by the receivers as horizontal scan-trigger), or the back-porch color reference-burst required by the color television receivers. The transient is naturally caused by the reflections produced by the unavoidable *mismatches* within the length of the delay line due to the stray-capacity effects. Hence, the delay-modulation obtained with a delay-line of a given length is limited, although the error-drive amplifiers used for this purpose present negligibly low-impedance over the video-passband.

Each ADL in effect uses two sets of silicon-capacitors, as shown in Fig. 4-19, to cancel the component due to the varying-bias from the output signal. To meet this

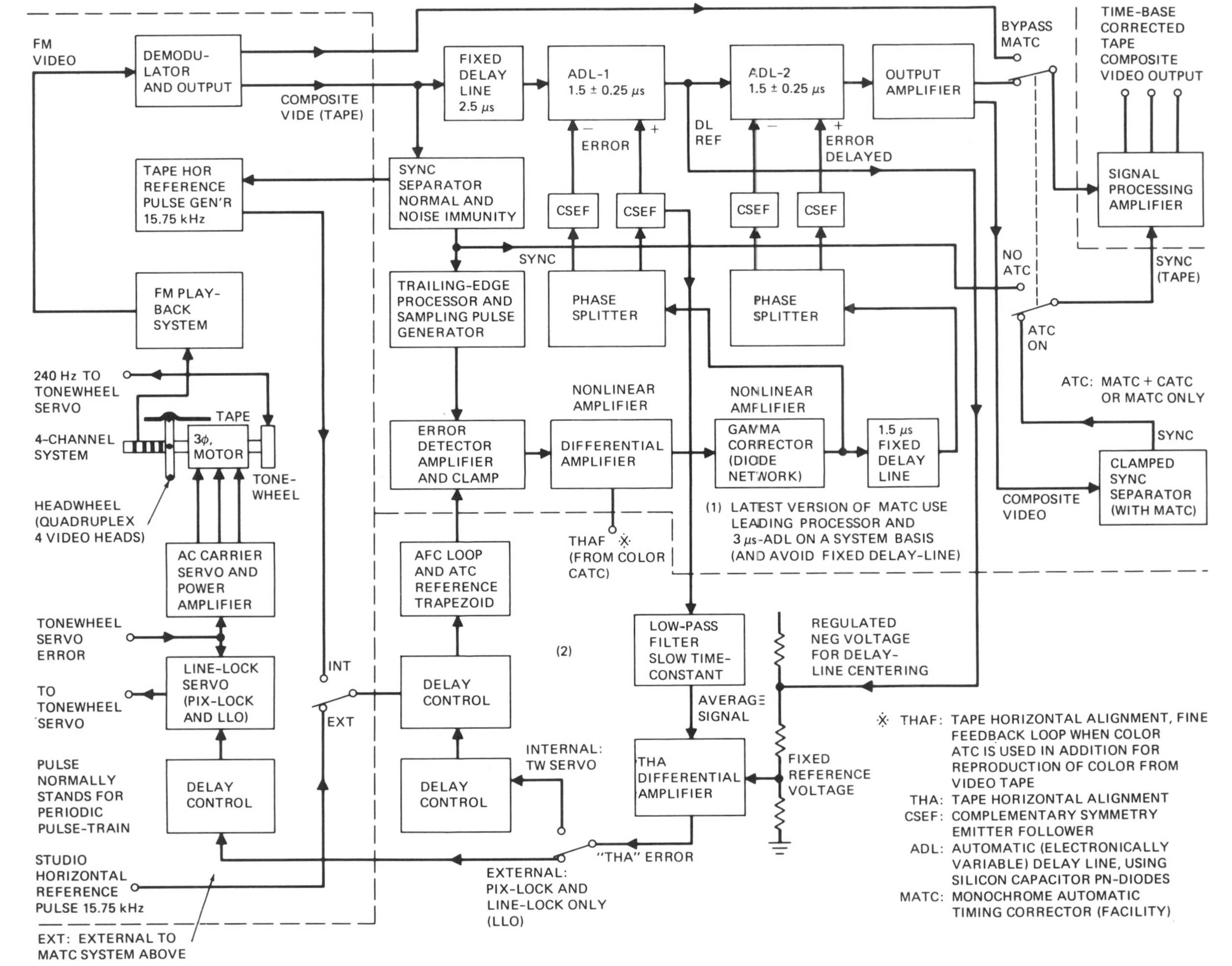

Fig. 4-18. Open-loop monochrome automatic timing corrector, sampled-data control system, and associated tape horizontal alignment feedback control loop of headwheel line-lock control system. See Fig. 4-21 for timings of waveforms in internal (tonewheel servo) and external (pix-lock) modes. (*Courtesy of RCA*)

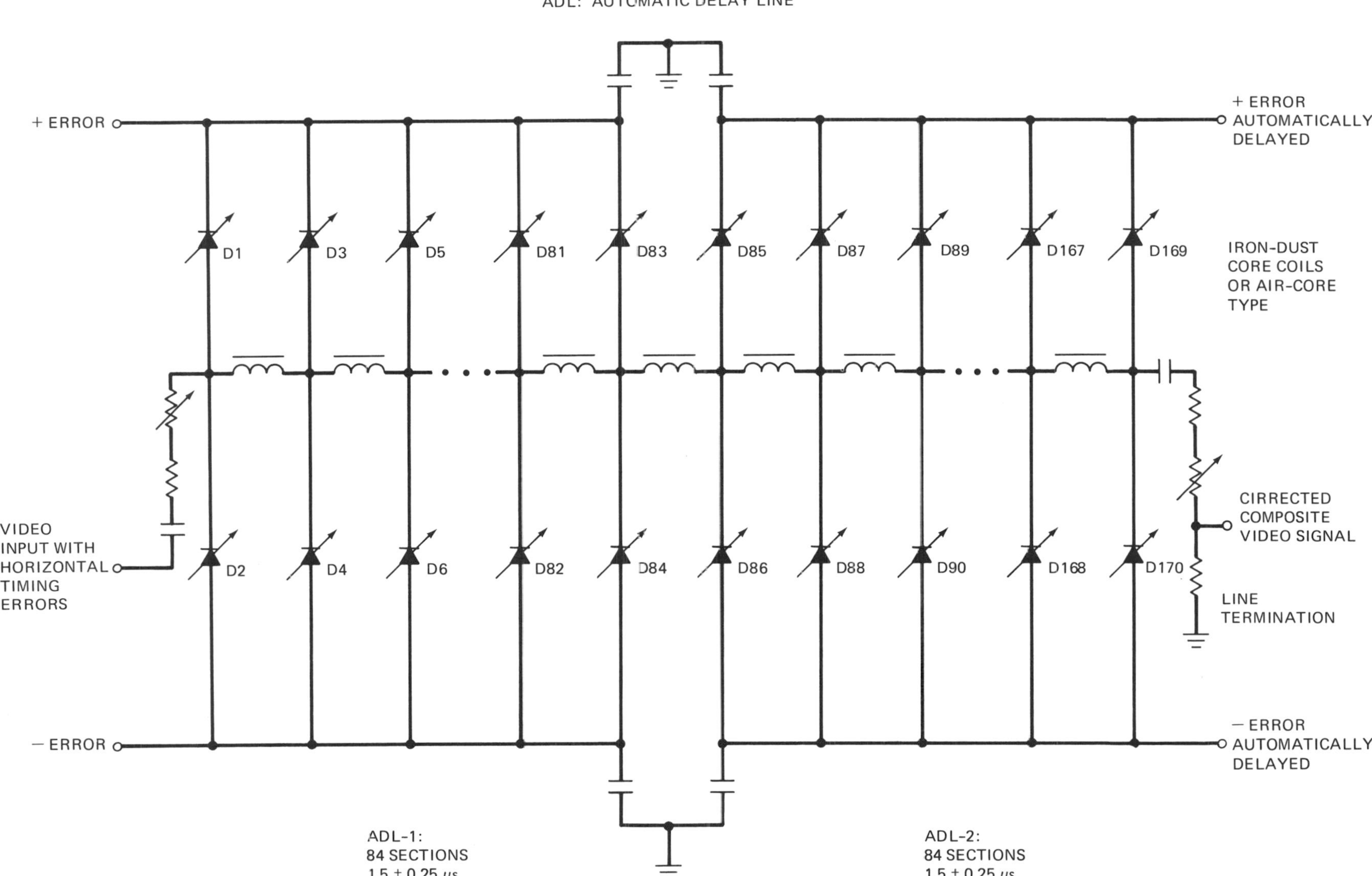

Fig. 4-19. Typical automatic delay line or electronically variable delay line, used in the all-electronic, sampled-data control system, open-loop MATC.

requirement and the cancellation of differential-phase, the dc levels of the phase-split error information, applied on either diode-bus in a push-pull manner, are kept balanced equally above and below the potential applied to the center-bus, to which the composite-video signal is applied for the time-base error correction.

The error-detector, which compares the phase of the relevant 15.75-kHz/sec sampling and trapezoid pulse-waveforms, produces an output voltage that changes linearly with the phase-difference between these two pulse signals. The varicap is a highly nonlinear device, and since the delay varies as the square-root of the capacitance, the ADL operates in principle in a nonlinear fashion, voltage versus time-delay. Therefore, the error-voltage from the phase-comparator is applied to a nonlinear amplifier, that is carefully counter-matched against the transfer-characteristic of the ADL. A linear differential-amplifier, using a variable *gamma-corrector* for the desired nonlinear transfer-characteristic, makes a suitable *nonlinear amplifier* for this purpose. The error signal, reprocessed in this manner, is phase-split into two sets of signals, in conjunction with a fixed delay-line (1.5 μsec), to extrapolate the two halves of the ADL for the automatic line-by-line timing-error corrections. It is, in short, a time-base error-measuring device. The THA, by means of a differential-amplifier, measures the timing-errors between the adjacent CRT horizontal-lines at the beginning of each line and applies correction for the duration of that particular line, thus automatically extrapolating the dc level of the error incursions due to the horizontal line-by-line time-displacements about the electrical-center of the ADL. As seen from the schematic block-diagram of the THA-loop, an ATC error-controlled nonlinear amplifier in the THA feedback-loop, via the headwheel line-lock control subsystem, produces an average-signal about the center of the ADL correction-range in a symmetrical manner.

Since the MATC operates in conjunction with the substantially more refined color automatic timing-corrector (CATC) to reproduce true color from videotape, it is interesting to note that the "coarse" THA feedback-control, associated with the MATC accessory needs readjustment instant-to-instant (by a fine automatic-reset). This is achieved by the control of the differential-amplifier that drives the nonlinear gamma-network, by means of the "fine-THA" feedback-control subsystem (associated with the CATC system) for improved reliability of the color phase-lock. The fine-THA control from the CATC is normally slow-acting like that in the MATC control accessory. It is, however, rendered fast-acting to effectively reduce phase-jitter when color FM-videotapes recorded on a Q-CVTR, employing the earlier heterodyne color technique, are played back on a modern Q-CVTR equipped with line-lock, MATC, and CATC for color reproduction. With the earlier heterodyne color technique, the color-subcarrier at 3.58 MHz/sec will be incompatible with respect to the sampling-rates used in the headwheel control subsystem—whereas the color-subcarrier is phase-locked in the present case. It may be noted that the fine and coarse closed-loop THA nonlinear feedback-control circuits operate sequentially with a restricted bandwidth as minor supplementary feedback-loops on the primary 15.75-kHz/sec line-lock digital feedback control subsystem to enable *high-precision speed-regulation* of the headwheel motor for undetectable low-frequency phase-jitter. In general, the PLL compensation at the "hold" network associated with the respective dc error-signal is designed on a measure-and-optimize basis in this highly complex system. In practice, rigorous computation is extremely time-consuming for such complex commercial products. The compensation used determines a suitable speed of response with maximum immunity to noise; in other words, the effective bandwidth of the feedback-loop concerned, as compared to that of the main control subsystem, is determined heuristically.

The MATC can function as a corrector of geometric-distortion in two different modes of the operation of the Q-CVTR, with and without "pix-lock" in external and internal modes, respectively. The internal tonewheel-mode of the headwheel control subsystem (with or without the capstan "switch-lock" accessory) automatically controls the MATC facility so that the reference-synchronizer used for the open-loop MATC is timed by deriving an average-timing from a long time-constant automatic frequency control-loop (AFC) that is locked to the tape sync pulses (Fig. 4-20).

The *AFC phase-lock loop* compares the phase of the sampling-pulse derived from the leading-edge of the tape sync pulse, and the phase of the trapezoid-waveform derived from the studio reference sync pulse. The dc-error-signal obtained from this phase-comparator (a diode-quad) controls the timing of the reference-trapezoid used in the main ATC phase-comparator. (The *diode-quad* used for the phase-comparison as an error-sampler operates like a switch that is normally open and controlled by the narrow sampling pulse. This diode-switch closes for the duration of the sampling-pulse, and during that instant applies the voltage at the set-point on the slope of the trapezoid to an RC-network with the desired time-constant to "hold" or "smooth-out" the rapid phase or timing-variations. The effective dc error-voltage at the output of the quad provides a measure of the average-time between the start of the AFC trapezoid and the leading-edge of the sampling-pulse.) The AFC-loop, as an all-electronic digital control subsystem, prevents the slow drifts in the ATC reference-trapezoid from causing the error-signal to exceed the correction-range of the variable delay-line. *In the internal mode*, the THA current in the THA-loop has a supplementary delay-control on the timing of the tape horizontal-pulses used for the AFC-loop. *In the external pix-lock (or LLO) mode*, the AFC is locked to the local studio-sync as required in the case of the pix-lock control subsystem. Also, the THA control-voltage is applied via a slow-acting large time-constant "hold" network to reset the average-timing of the reference-pulses used for the phase-comparator in the closed-loop line-lock digital control subsystem. Thus, the effective low-frequency pix-lock control-jitter is further reduced from ± 0.07 to ± 0.02 μsec, as a result of the combined effect of (1) the slow-acting closed-loop THA control on the headwheel motor (via line-lock) and (2) the line-by-line open-loop MATC control. The appropriate timing relations pertaining to the MATC internal and external modes are shown in Fig. 4-21a and b.

4.4.5 Color Automatic Timing-Corrector. The color automatic timing-corrector (CATC) is, like the MATC, an open-loop all-electronic sampled-data control subsystem, exclusively used for the reproduction of color from videotape, with the MATC operating as a base accessory.

The inherent quadruplex picture-segmentation error-tolerances of the order of 0.02 μsec and the pix-lock time-base phase-jitter of the order of 0.1 μsec, inherent in a monochrome Q-CVTR—using (1) four magnetic-heads on the periphery of a headwheel and (2) three independent but closely interacting nonlinear digital feedback control subsystems—are undetectable in the case of the monochrome television signals. However, for color reproduction, timing variations such as 0.02 and 0.1 μsec correspond to 27° and 130°, respectively, at 3.58 MHz/sec, the color-subcarrier used in color television. Standard color specifications emphasize a requirement of

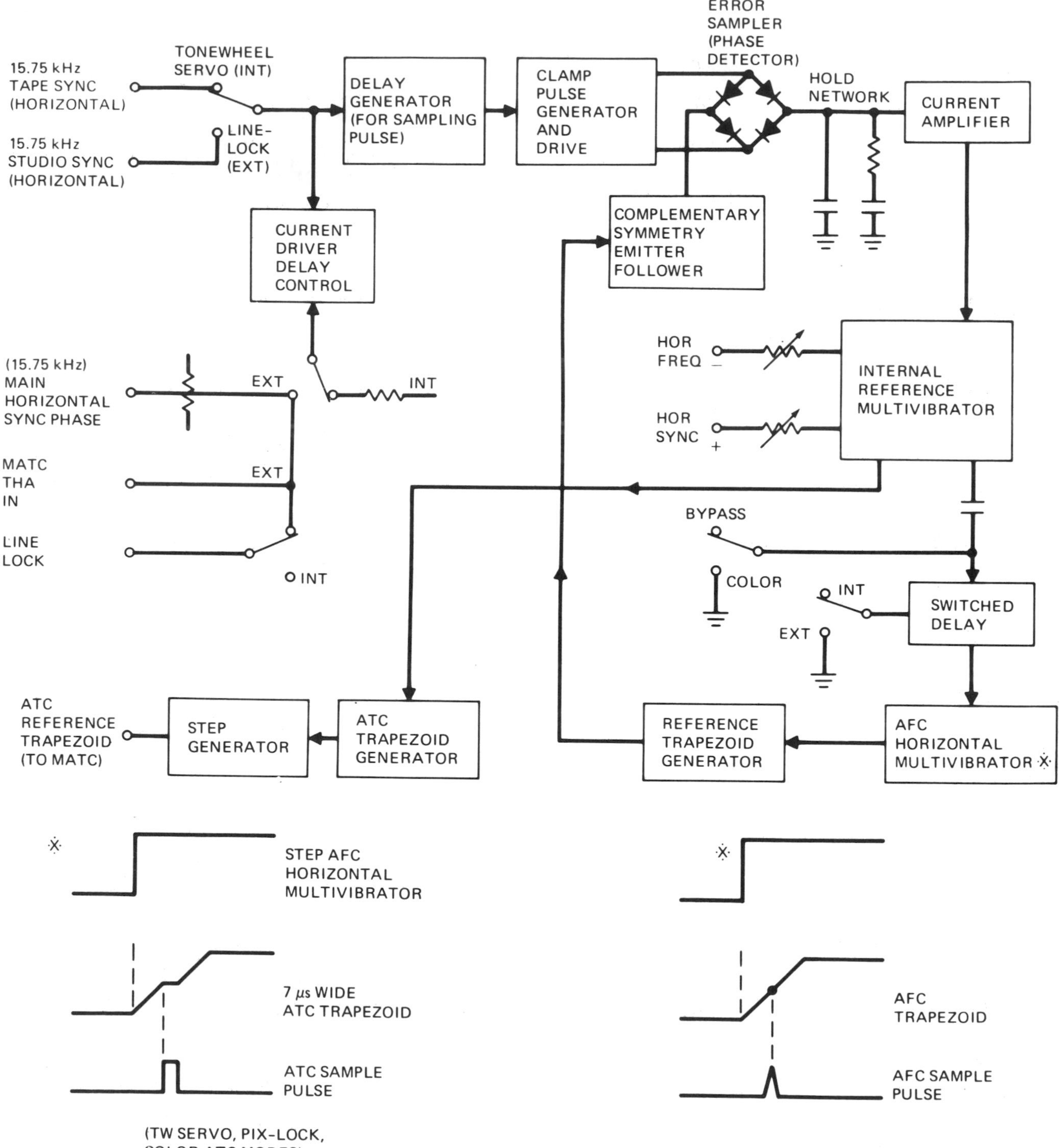

Fig. 4-20. Automatic frequency control loop used for the open-loop MATC, in conjunction with tape horizontal alignment feedback loop. (*Courtesy of RCA*)

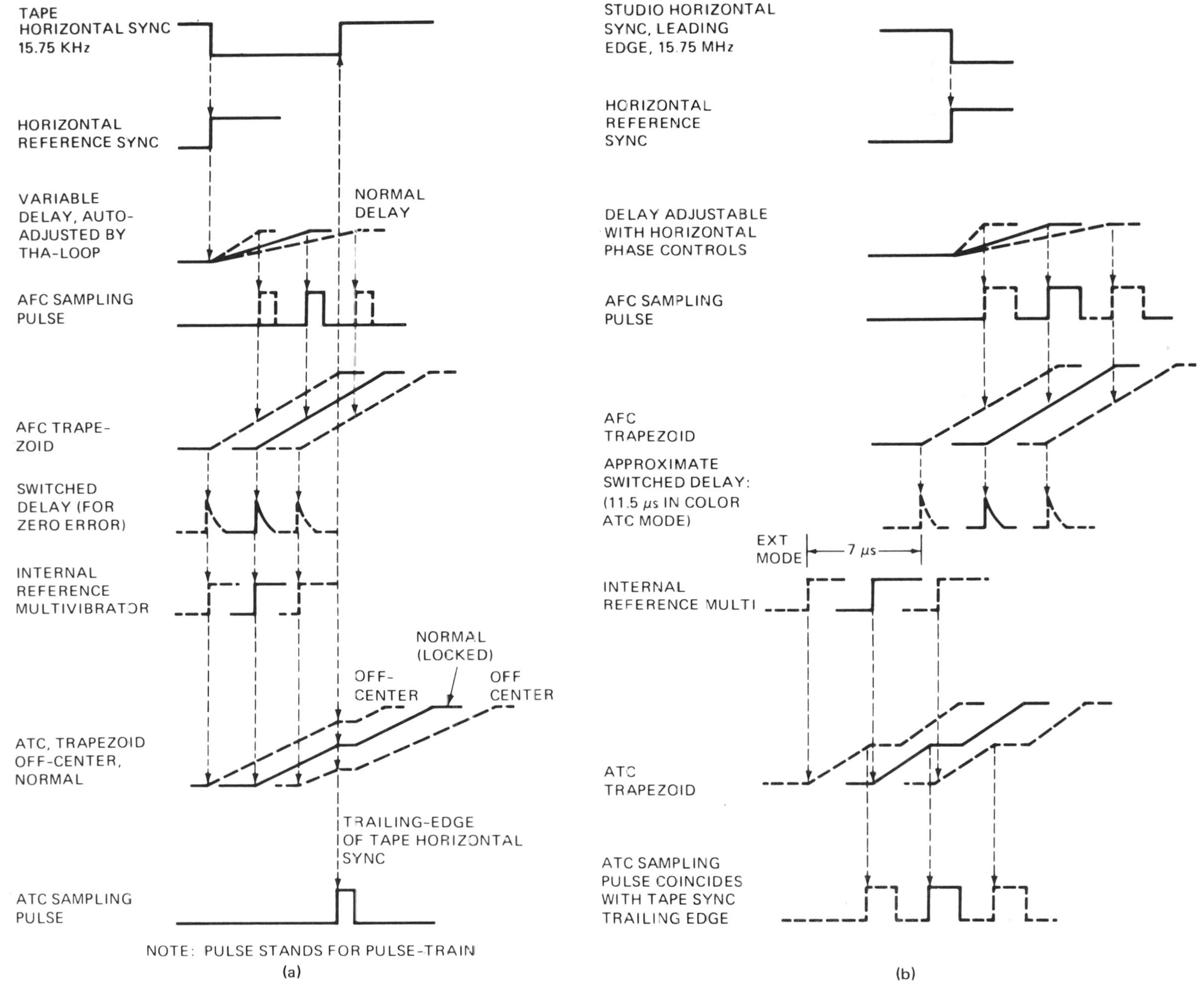

Fig. 4-21a. ATC waveform timings. Internal mode, play (tonewheel 240-Hz servo). (*Courtesy of RCA*)
Fig. 4-21b. ATC waveform timings in external pix-lock mode and color ATC mode.

less than ±2.5° tolerance for true color-reproduction. Hence, the MATC and the CATC subsystems are sequentially used as essential accessories in the Q-CVTR used for the reproduction of color from videotape. The required ±2.5° phase-tolerance approximately corresponds to a time-base stability of 4 nsec.

In the case of the MATC, one sample per CRT line-interval is provided by the sync pulse-train at 15.75 kHz/sec, and, since a relatively wide bandwidth is involved, the quantization and random noise of the tape reproduction around the threshold-visibility level in the monochrome represents a noticeable phase-jitter in the color reproduction from videotape. The CATC digital control system, using the higher sampling-rate available in the form of eight continuous cycles of reference color-burst at 3.58 MHz/sec at the beginning of each CRT line-interval during the horizontal-blanking interval, and possible use of a relatively narrow bandwidth and hence faster-response, further compresses the phase-jitter due to the abovementioned random noise to an unnoticeable extent. The desired fine time-base correction is therefore conveniently accomplished by comparing the phase of the color-subcarrier reference-burst with a fixed reference such as the studio color frequency-standard (3.579545 MHz/sec). Systemwise, the 15.75-kHz/sec sync pulse-train is phase-locked to the subcarrier, but there is bound to be, even with the use of the MATC digital control system, some residual phase-jitter of the order of a few nanoseconds. This phase-jitter takes the form of a time-modulation of the sync-pulse with respect to the subcarrier, as a result of the tolerances involved in practical analog and digital pulse circuitry. The CATC digital control process, therefore, actually corresponds to a further compression (on an open-loop basis) of the phase-space limit-cycle of an already extremely small dimension in a nonlinear control system-complex. Thus, for color reproduction, by cascading the MATC and the CATC digital control subsystems, the final phase-drift at 3.58 MHz/sec can be reduced to a negligible extent of the order of 4 nsec, by another reduction-factor of 25-to-1 over and above the MATC correction-factor. The accuracy at this final stage corresponds to an overall speed-regulation of 0.0017% with respect to 240 rps as far as the whole headwheel control subsystem is concerned. The CATC, therefore, employs a corresponding automatic delay-line (ADL) type open-loop digital control system, at a 3.58 MHz/sec sampling-rate; an associated *fine-THA* feedback control-loop acts as a backup to the immediately preceding MATC digital control subsystem of the headwheel-servo to provide an additional adaptive-control service. The error-detector this time utilizes the 3.58-MHz/sec color reference-burst, instead of the horizontal sync pulse. The simplified schematic block-diagram of the CATC digital control system is shown in Fig. 4-22. (Error-sampling is actually performed at half the carrier rate.)

In the case of the overall CATC-PLL, a delay control-range of 360° or 0.28 μsec at 3.58 MHz/sec is all that is mandatory; hence, a comparatively short ADL is adequate, unlike the dual-section ADL in the monochrome case, to cover the delay-change within the reference-burst of the line-blanking interval. Otherwise, the control subsystem is identical to the MATC-PLL in respect to the major elements like the nonlinear amplifier, the pulse driving system, the fixed delay-line, and the THA feedback-loop. To achieve color reproduction that is free from phase-jitter, the *fine-THA control* from the CATC 3.58-MHz/sec digital control subsystem, as an adaptive-control device, measures the accuracy by means of the differential amplifier and the error-sampler and resets within the appropriate limits the error of the *coarse-THA feedback loop*, associated with the MATC and the line-lock 15.75-kHz/sec digital control subsystems.

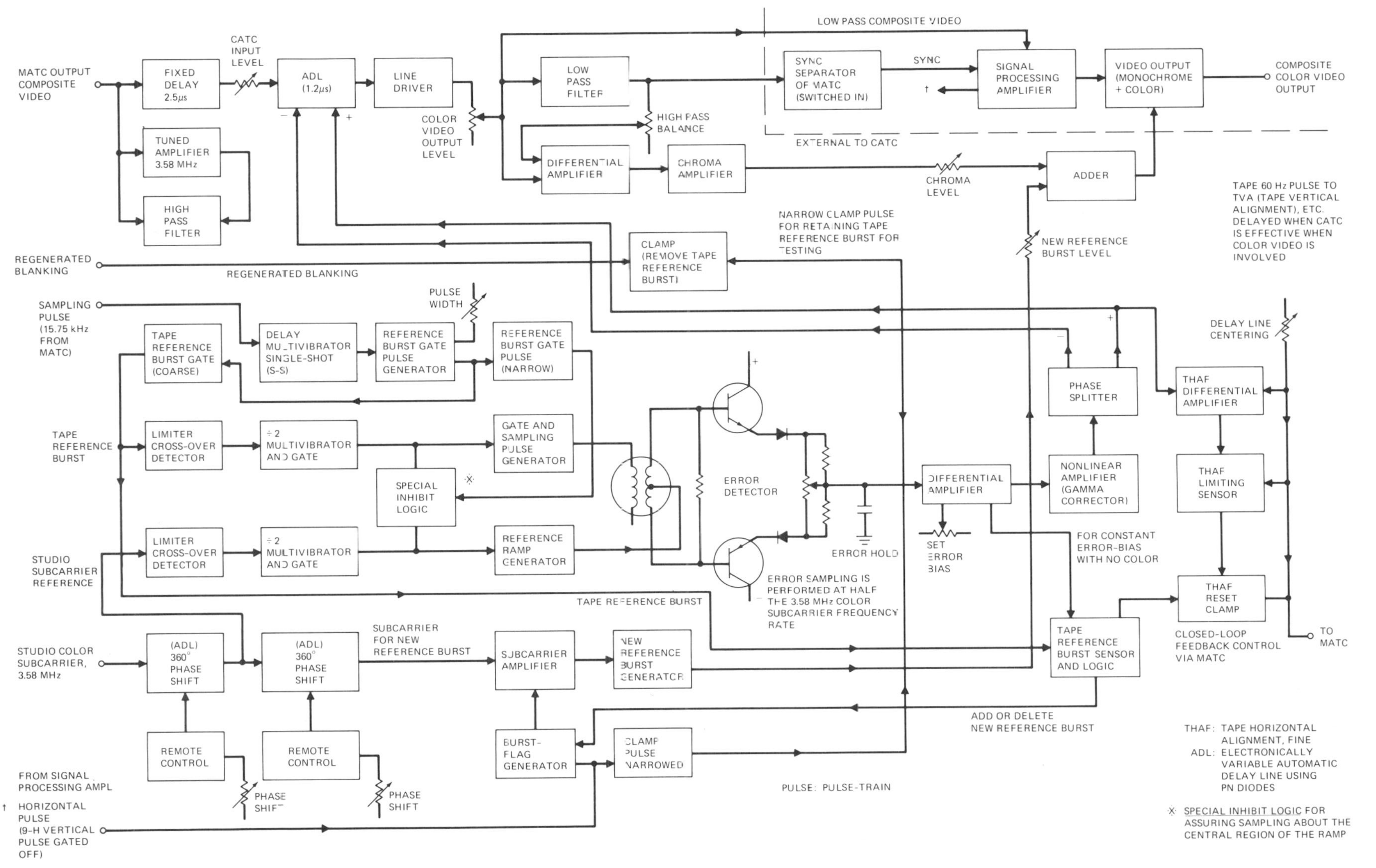

Fig. 4-22. Open-loop color automatic timing corrector system (CATC) and the fine tape horizontal alignment (THAF) feedback loop to MATC. *(Courtesy of RCA)*

The phase error-detector in this case should handle a linear delay-range of 360° at 3.58 MHz/sec, and since the full 360° range is not practical for an effective color phase-discriminator, it is operated to cover the full range within 180° by using the narrow sampling pulses derived from the tape color reference-burst, and the conventional ramp waveform formed from the local color-subcarrier signal, after the necessary division-by-two by means of their respective binary-counters. Special inhibit-logic in the burst-channel restricts the sampling-interval to the center-half of the ramp waveform to assure the phase-lock of the color information in the video signal, and the automatic digital interloop transfer in the event of a heavy disturbance or a dropout of long duration. It is mandatory that the open-loop ADL digital control subsystem correct the phase-errors in the composite color-video signal, reproduced according to the error-data, on a line-to-line basis. The phase-stabilized composite color-video signal, thus retrieved from the tape, is further reprocessed in two separate channels, one low-pass and the other high-pass. These processing-amplifier circuits

1. clean up blanking and sync
2. insert a *freshly regenerated reference-burst* on the back-porch, with a provision for the "fine" phase-control of the reference-burst with respect to the retrieved chroma
3. individually *dc-clamp* the monochrome and chroma-channel components about the black-level, subject to a provision of the control of the *pedestal-level* to under 10%
4. provide a *coarse-phase delay-line* cable-compensation for matching the phase of the tape color signal to that of the other color signals in the studio system
5. distribute the combined composite color-video signal thus reproduced from the tape with essentially no restraints in the correct bandwidth or color response.

4.5 CAPSTAN DIGITAL FEEDBACK CONTROL SYSTEM

4.5.1 Introduction. The capstan digital feedback control system maintains the longitudinal tape-speed at 15 ips (or $7\frac{1}{2}$ ips, if required) by means of a capstan-spindle on the shaft of the capstan two-phase hysteresis synchronous motor, while the tape is gently moved against the friction of a pinch-roller. The tape is run transverse to the rotating-headwheel at the appropriate pressure, as impressed by the set-position of a vacuum-guide servo. Either friction-coupling or an improved belt-drive system are used for the operation of the capstan motor. The interacting capstan digital control system synchronizes the phasing of a recorded *longitudinal control-track (CT) signal* to the transverse video-tracks, while they are recorded by the four magnetic-heads of the headwheel assembly. The timing-reference is provided by the 60-Hz/sec field-rate vertical reference; the latter is in turn derived from the studio complete sync or the complete sync separated from the input composite-video signal or the 60-Hz/sec power-line frequency. This figure is 50 Hz/sec for the European television standards.

A continuous record of the headwheel motor-speed is provided by means of a separate magnetic-head along the guided-edge of the tape as the control-track. The actual 240-Hz/sec tonewheel signal, generated by the headwheel motor-assembly is reprocessed and used for the control-track signal. In the record mode, the capstan-motor can be driven by the 60-Hz/sec power at the power-line frequency, but, in practice, for the best playback tracking results, the motor is phase-locked to the controlled tonewheel 240-Hz/sec signal by means of a PLL at the 60-Hz/sec rate by using an oscillator and a closed-loop digital feedback control system.

(It may be noted that the exact rates of sampling in the Q-CVTR, for the interacting capstan, headwheel, and vacuum-guide servo digital control systems, depend on the television scan-line standards used in the country of usage around the world. The sampling rates indicated throughout this work apply to the domestic standards in the United States and elsewhere. Special digital equipment for international scan-conversion is available in some major broadcasting studios.)

To recover the video signal from the tape, the speed of the tape is tightly controlled by comparing the phase of the signal read from the control-track and a corresponding reference-signal derived from the studio complete-sync, as the headwheel scans the moving-tape along its transverse video-tracks with the aid of the digital information picked up by the tonewheel of the headwheel-assembly. This procedure insures that exactly four video-tracks are pulled past the headwheel-assembly during each rotation of the headwheel. If the phase-tracking is not correctly maintained by the capstan-servo, the relative position of the heads with respect to the tracks will slowly drift along the length of the tape to the guard-spacing between two video-tracks (instead of exactly remaining over the tracks), and thus produce a gradual tear-off in the picture.

The physical orientation of the notch on the tonewheel, which is mounted on the headwheel motor-shaft, with respect to one of the four quadrature magnetic-heads on the periphery of the headwheel, determines which of the four heads will record the 60-Hz/sec reference vertical-sync in the composite video-signal being recorded. The *video-head No. 1* can be arbitrarily chosen to record *vertical-sync*, and it will then be approximately situated at the center of the tape, when the notch is directly over the tonewheel magnetic-head.

The failure of any feedback-loop in either the capstan-servo or the headwheel-servo will cause complete breakup of the picture-synchronization and hence the picture-display on the monitor of the Q-CVTR. Guide position-servo failure causes only skew-distortion of the verticals in the picture; if the guide-pressure is too low or too high, both the headwheel-servo and capstan-servo fail. During playback, if the picture breaks up, the system self-checking device in the Q-CVTR, with its built-in logic, will instantaneously give a red-light indication if it is the headwheel-servo in particular that is out of phase-lock. The *self-checking system* is otherwise restricted to servo and signal fault-indication in *record-mode only*, since it is impractical to monitor the recorded tape-signal while recording is in progress in a normal single-headwheel Q-CVTR system. (It is feasible to couple the record-system of one Q-CVTR with the playback-system of an adjacent Q-CVTR by means of the tape, and simultaneously and continually monitor the recorded video signal on the second machine, thus entailing no delay between record and play.) While recording is proceeding normally on a Q-CVTR, the complete FM signal-processing system alone is simultaneously monitored on a mod-demod (modem) basis, while the fault indicator system provides a simultaneous check on the complete *recording* system.

In the capstan-servo, the 30-Hz/sec television frame-pulse needs a special mention (as stated earlier) since

1. as a so-called *edit pulse*, it is used as a *frame-marker* during the tape-editing procedures in making mechanical or electronic splicing, for vertical roll-free picture-transition during playback

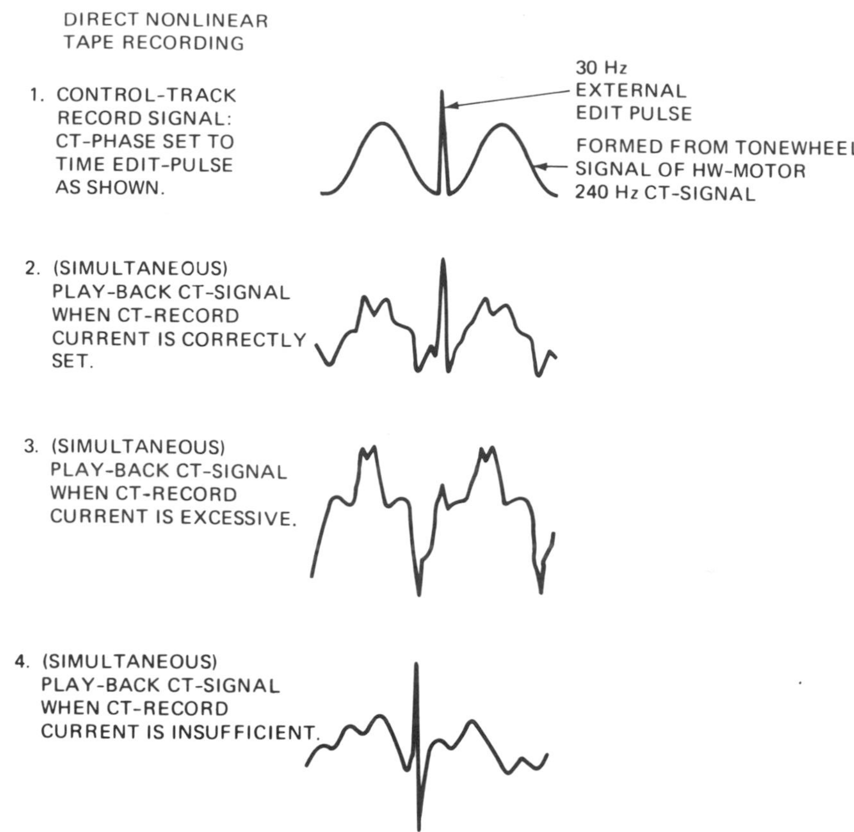
DIRECT NONLINEAR
TAPE RECORDING
1. CONTROL-TRACK
RECORD SIGNAL:
CT-PHASE SET TO
TIME EDIT-PULSE
AS SHOWN.
30 Hz
EXTERNAL
EDIT PULSE
FORMED FROM TONEWHEEL
SIGNAL OF HW-MOTOR
240 Hz CT-SIGNAL
2. (SIMULTANEOUS)
PLAY-BACK CT-SIGNAL
WHEN CT-RECORD
CURRENT IS CORRECTLY
SET.
3. (SIMULTANEOUS)
PLAY-BACK CT-SIGNAL
WHEN CT-RECORD
CURRENT IS EXCESSIVE.
4. (SIMULTANEOUS)
PLAY-BACK CT-SIGNAL
WHEN CT-RECORD
CURRENT IS INSUFFICIENT.

Fig. 4-23. Direct nonlinear tape recording of control-track signal during record.

2. as a tape frame-pulse, it is (after extraction from the tape control-track signal) used for the 30-Hz/sec capstan *switch-lock* control subsystem in some earlier Q-CVTR machines for separate *switch-lock* and *pix-lock* facilities.

The frame-pulse, obtained by comparing the studio horizontal 15.75-kHz/sec and vertical 60-Hz/sec signals in coincidence-gate logic, is superimposed, as shown in Fig. 4-23, on the 240-Hz/sec control-track signal and directly recorded (in a nonlinear fashion) along the guided-edge of the tape to keep an accurate timing-record of the capstan motor-speed during recording. The timing of the edit-pulse with respect to the control-track signal, as shown in Fig. 4-23, can be readily adjusted while recording. This figure shows for guidance the waveforms that result with excessive, insufficient, and correct amounts of recording currents in this direct nonlinear magnetic recording medium.

The control-track magnetic-head is located in a typical Q-CVTR on the headwheel assembly itself. Since the sinusoidal control-track signal is actually processed from the tonewheel signal, it keeps an accurate record of the headwheel motor-speed too. It is timed by a variable-delay circuit to bear the appropriate phase relationship to the four video-heads and the control-track frame-pulse. A simultaneous playback-head is provided to monitor this important control-track signal, and allow the adjustment of its correct recording level, if necessary.

4.5.2 Capstan 240-Hz/sec Tonewheel Digital Feedback Control System. During recording, the exact timing-relationship between the tape-speed and the quadruplex magnetic-head scanning is achieved by synchronizing the phase of the capstan motor-speed to the headwheel motor-speed. A simplified schematic block-diagram of the capstan tonewheel-servo in record and playback modes is shown in Fig. 4-24a and b.

The phase synchronization between the capstan and the headwheel motor-speeds is accomplished by starting with the tonewheel pulse and using it to trigger a sequence of binary-counters for a division of 4, from 240 to 60 Hz/sec, to obtain the two 60-Hz/sec signals in quadrature that drive the capstan two-phase hysteresis synchronous motor via two power amplifiers. With the latest capstan two-speed asynchronous hysteresis motor for the 15- and $7\frac{1}{2}$-ips tape-speeds, one power-drive will do by using a quadrature-capacitor right at the motor-winding. The sinusoidal signals in quadrature are generated by low-pass filtering two 60-Hz/sec square-wave signals. These in turn are formed by starting with either a 240-Hz/sec oscillator signal (for playback) or the tone-wheel signal (for record), via two signal paths, one a master-binary signal path and the other a slave-binary signal path. Each signal is triggered by the respective 240-Hz/sec pulse-waveform in opposite phase. To insure motor rotation in the correct direction, appropriate diode gating-logic is employed to control the phase of the slave-binary so that its output will always quadrature-lead the master-binary signal output. Since a free-running 240-Hz/sec oscillator is available for the control of the capstan-servo during playback, and since the headwheel requires only a short time-interval to phase-lock the corresponding velocity- and phase-loops for the availability of the correct tonewheel pulse, the capstan binary-counters are momentarily driven at the start by the above free-running 240-Hz/sec pulse input instead of the tonewheel pulse. The technique provides a protection against possible heavy currents in the power drive amplifiers and the motor due to way-out improper count of the tonewheel pulse as the headwheel motor starts to run. No sooner the headwheel is phase-locked than the capstan-speed is automatically locked to the headwheel-speed by the automatic changeover of this control to the tonewheel pulse. Thus, the control-oscillator and the feedback-loop are isolated from the capstan-servo in the record-mode. The procedure adopted above for the synchronization of the headwheel- and capstan-speed with this precaution insures that the video-tracks imprinted on the tape follow the standard geometric-pattern for the interchangeable reproduction of tapes on a standard Q-CVTR of any make.

In the playback-mode, the free-running 240-Hz/sec local oscillator (a square-wave multivibrator) takes over the control of the capstan tonewheel servo by way of a cascade of three binary-counters for a division of 8. The free-running oscillator-frequency is, however, tightly controlled by means of the 240-Hz/sec sampled-data feedback control system, if the tape tends to move faster or slower. The tape-speed is corrected, as the error signal, produced by the conventional phase comparison of the sampling-pulse (derived from the control-track signal) and the 30-Hz/sec reference trapezoid-waveform (derived from the 60-Hz/sec vertical-sync), modifies the local oscillator-frequency as required for correct capstan-speed. The control-track signal is picked up from the tape by means of the record/playback control-track head, amplified, limited, delayed by a variable amount, and shaped into a clean 240-Hz/sec pulse-train in the auxiliary pulse-forming circuits of the capstan servo system. The horizontal- and vertical-sync pulses from the studio sync pulse generator are combined in a special coincidence-gate logic circuit to produce the 30-Hz/sec frame-pulse; shaped as a trapezoid-waveform, this makes the reference for the

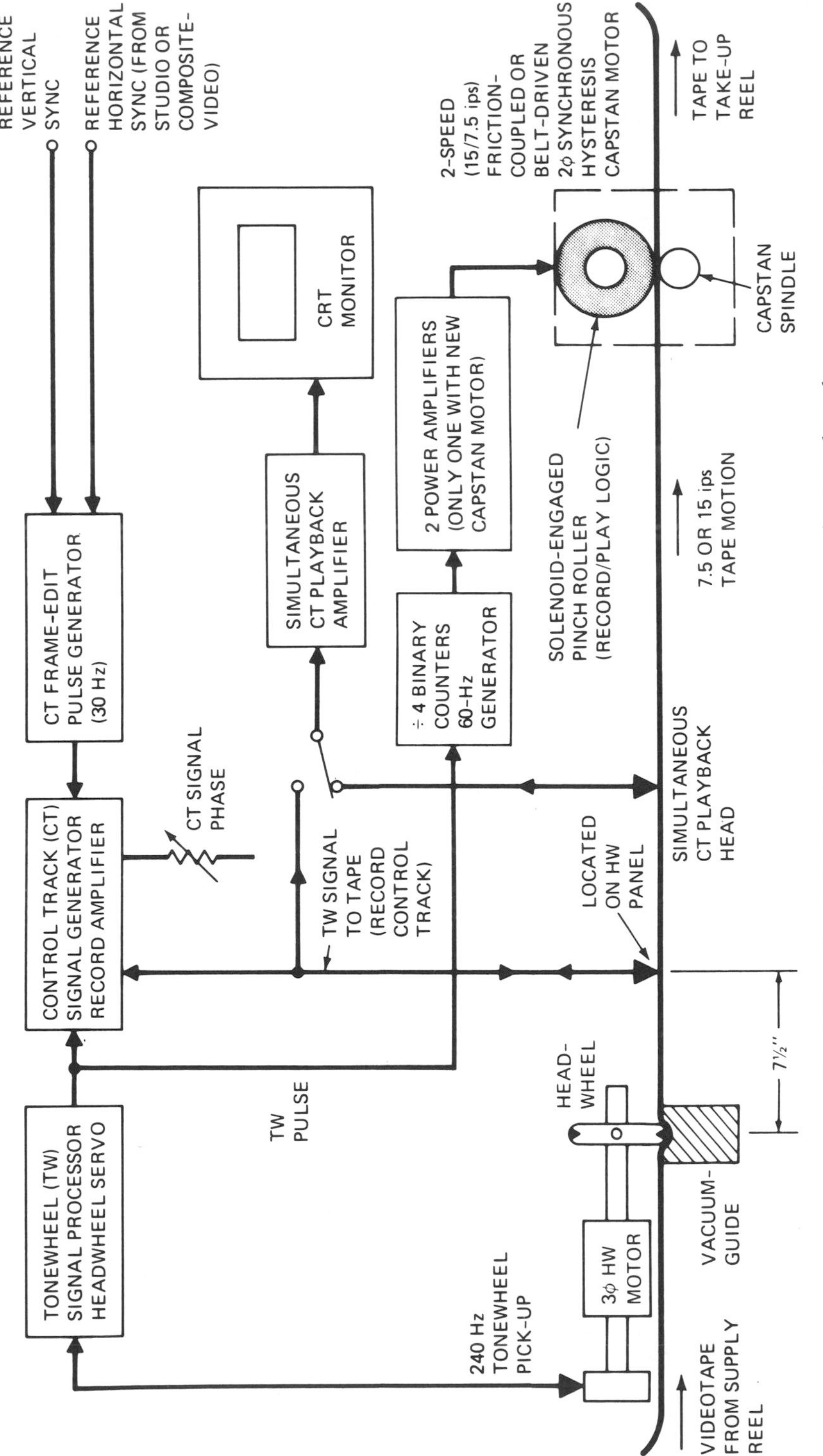

Fig. 4-24a. Simplified block-diagram capstan servo in record mode.

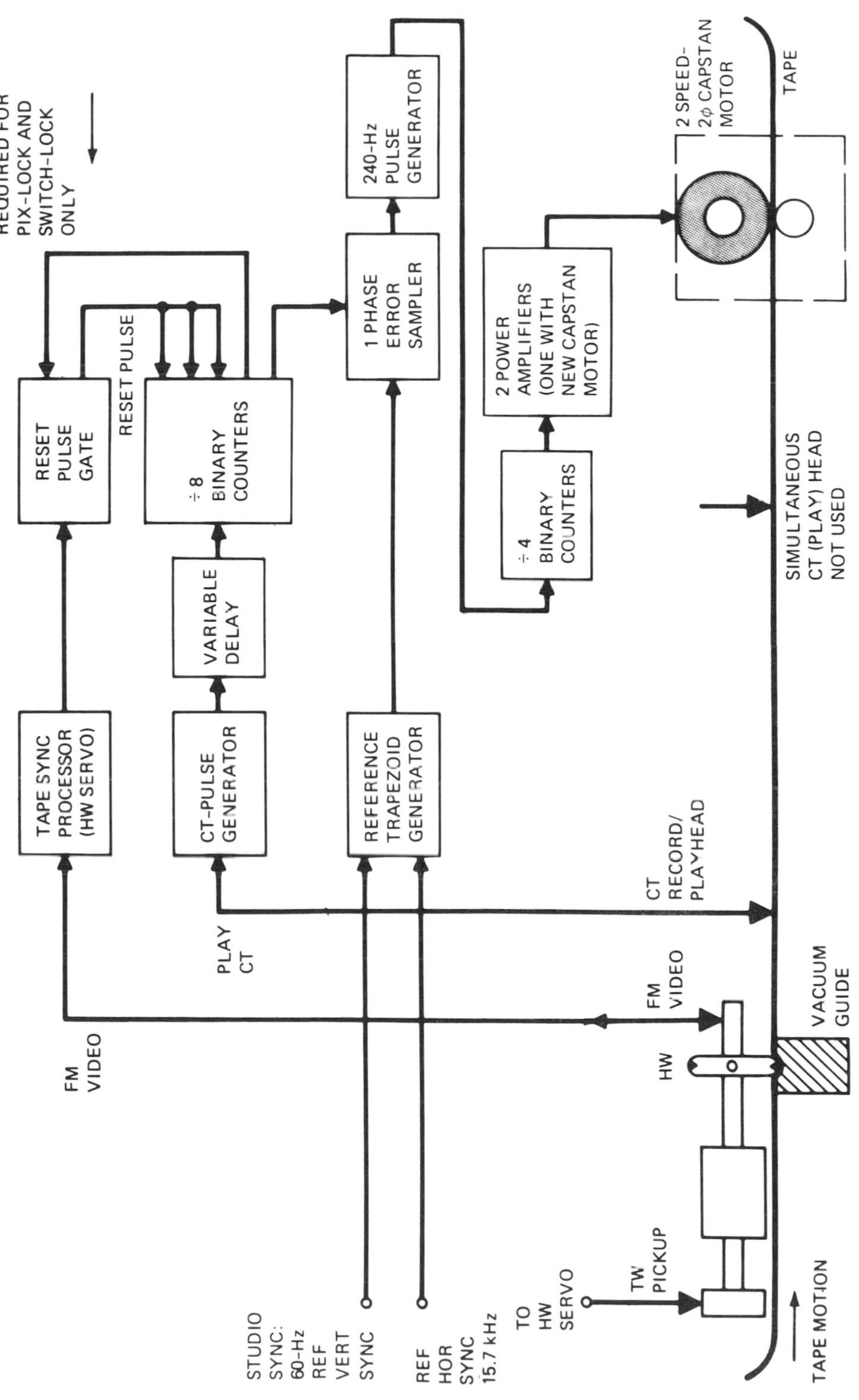

Fig. 4-24b. Simplified block-diagram, capstan servo, playback mode.

above phase comparison. A trapezoid, which is actually a slice of the more common ramp waveform, gives a steep-slope, and hence a tight lock-in range (as in the case of the headwheel control-loops) for the stability of the digital feedback control system to assure sufficient accuracy and fast response. (The trapezoid incidentally produces the desired nonlinearity for best operation, namely, saturation in practically every loop.) If the motor-speed has any tendency to increase or decrease, first the frequency and then the phase of the control-track pulse-train follows this change, and the sampled-error signal, after the hold, causes the speed of the controlled-plant (capstan motor) to decrease or increase, respectively, until the effective count-down 30-Hz/sec control-track pulse coincides in phase with the stable 30-Hz/sec timing reference-pulse from the studio. This is indicated by the stability of the sampling set-point close to the center of the reference trapezoid-slope on the waveform monitor. Then the capstan motor-speed is phase-locked to the studio reference-frequency for stable, vertically roll-free, picture-reproduction from tape. (If a studio sync pulse generator is not available, the Q-CVTR can still play back a tape with a 60-Hz/sec reference-signal derived from the power-line.)

As part of the electronics of the capstan-servo, a variable-delay multivibrator is included in the signal path of the processed control-track pulse-train to provide a precise manual control facility for the phase-adjustment of the capstan over a wide range. This control-track phase-control enables the precise centering of the rotating magnetic-heads over the recorded video-tracks on the moving tape, for the maximum output of the FM-switcher with minimum of noise, as indicated on the waveform monitor, while the vacuum-guide servo maintains the correct pressure between the tape and the headwheel.

In the capstan tonewheel-mode, the control-track phase-control can be used also to *slip tracks* so that the video head No. 1 will playback information from the video-track containing the vertical-sync under normal operating conditions, or for that matter, any of the other three tracks, if so desired for any special purpose. This alternative is not preferred normally, since the quadrature-delay errors of the four magnetic-heads in the FM playback system need readjustment all over again, if the MATC facility is not available. The capstan tonewheel servo can be operated for test purposes as an open-loop control system by replacing the sampled error-signal to the local oscillator by means of manually adjusted dc control-current.

4.5.3 Capstan 30-Hz/sec Switch-Lock Digital Feedback Control System. With the capstan and headwheel 240-Hz/sec tonewheel digital feedback control systems, the reliable reproduction of the tape composite-video signal does not directly insure the exact synchronization of the tape composite-video to the syncs available in the studio—either horizontally or vertically. Under these circumstances, if a tape-playback signal is switched "on air" directly after a camera-signal (processed with the studio-sync), a vertical rollover of the picture is imminent in television receivers, since at the instant of the picture transition, the effective 30-Hz/sec frame-syncs of the two picture-signals are out of phase. The situation is identical when a studio switches "on air" a remote composite video-signal directly after a studio camera-signal. A PLL *genlock* electronic digital feedback control system at the studio can synchronize the local camera-signals to the remote-signal to accomplish a smooth transition and subsequent special-effects or montage between the remote and local camera signals. This genlock is of course ineffective in the case of the tape-recorder because the phase-synchronization probelm is complicated by the phase-jitter components of the 240-Hz/sec tonewheel servo systems.

In the capstan tonewheel-mode, the phasing of the 240-Hz/sec control-track pulse with reference to the 30-Hz/sec trapezoid-waveform (derived from the studio sync) is such that the probability of synchronization at the beginning of the binary-count is only 1 in 8. Therefore, there is a substantial need for tape vertical-synchronization, as far as the television receivers or monitors are concerned. The capstan 30-Hz *switch-lock* digital feedback control system will specifically meet this requirement. It may be noted in this context that the *switch-lock* mode of capstan-servo is automatically effective when a Q-CVTR in the *pix-lock* mode provides fully synchronized tape-signals for special-effects at the studio. It is not effective in the case of the *LLO* mode, since the high-speed line-lock alone will be in use in this particular mode for a special purpose.

During the earlier stages of videotape recorders, the 30-Hz/sec edit frame-pulse, retrieved from the control-track signal, was the obvious means for operating the capstan-servo in the *switch-lock* mode by phase comparison with the frame-sync derived from the studio-sync. The edit-pulse, presently used for splicing purposes, is no longer necessary for the switch-lock facility. The switch-lock is presently accomplished by means of a "reset" pulse derived from the tape-signal with appropriate logic. When the machine is started to playback a videotape, the headwheel and capstan 240-Hz/sec tonewheel servo systems enable the reproduction of a stable composite-video picture signal. At the next instant, the separated tape vertical- and horizontal-syncs are compared in coincidence-gate logic to produce the tape frame-trigger. The 30-Hz/sec "reset" pulse formed out of this frame-trigger *resets* the divide-by-eight binary-counters, and causes the capstan-servo to shift tracks until the frame sync-alignment (or in other words, a "coarse" vertical alignment) is effected. With this switch-lock operation, video head-1 picks up the information from the track that contains the vertical sync (track-1). A schematic block-diagram of the switch-lock mode of capstan-servo is shown in Fig. 4-25a and b, along with the effective digital waveforms during the transition to the switch-lock mode of operation.

The coarse vertical-alignment is all that is achieved by the capstan *switch-lock* servo, so as to prevent the picture rollover in the television receivers as a result of a studio camera-to-tape switch-over and vice-versa. The control-track phase-control used in the capstan tonewheel-servo is equally effective in the *switch-lock* mode for the precise centering of the rotary video-heads over the recorded tracks. For the *pix-lock* control, however, more precise vertical-framing must be established as far as the horizontal line-frequency rate 15.75 kHz/sec is concerned. The *switch-lock* and tonewheel control subsystems do not recognize any errors in the alignment pertaining to the exact placement of the vertical-sync on the recorded track. The complementary TVA *adaptive* feedback control system, earlier described under the *pix-lock* servo, establishes the necessary *fine vertical-alignment* for the complete tape-to-studio signal synchronization after the successful incorporation of the PLL at the line-rate. The TVA thus eliminates the need for a *manual* control-track phase-control setting at start.

4.5.4 Two-Speed Capstan Servo. A two-speed (1800/900 rpm) fractional hp (1/200 to 1/100) hysteresis-asynchronous type ac motor is presently available with two pairs of windings. It has a built-in provision to *switch it over to half-the-speed* by externally changing one servo-controlled power-drive output to two alternative pairs of quadrature-windings associated with two alternative poles (8 and 4 for the tape-speeds of $7\frac{1}{2}$ ips and 15 ips, respectively). That is, the PLL power control-supply itself is connected via a single power amplifier to the two effective windings of the

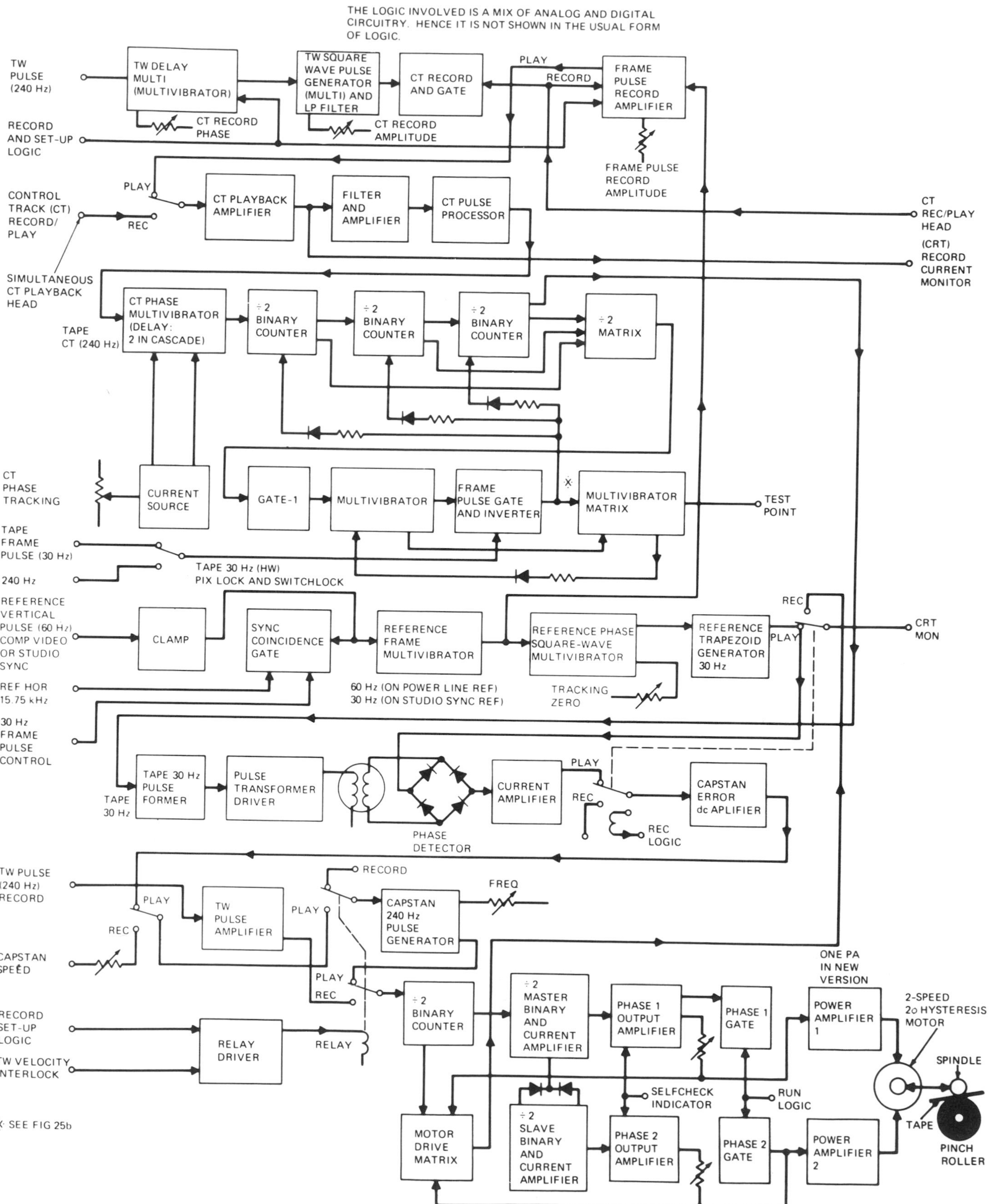

Fig. 4-25a. Combined tonewheel (240 Hz) and switch-lock (30 Hz), sampled-data feedback control system for capstan. (See Fig. 4-24b for difference between tonewheel and switch-lock playback modes.) (*Courtesy of RCA*)

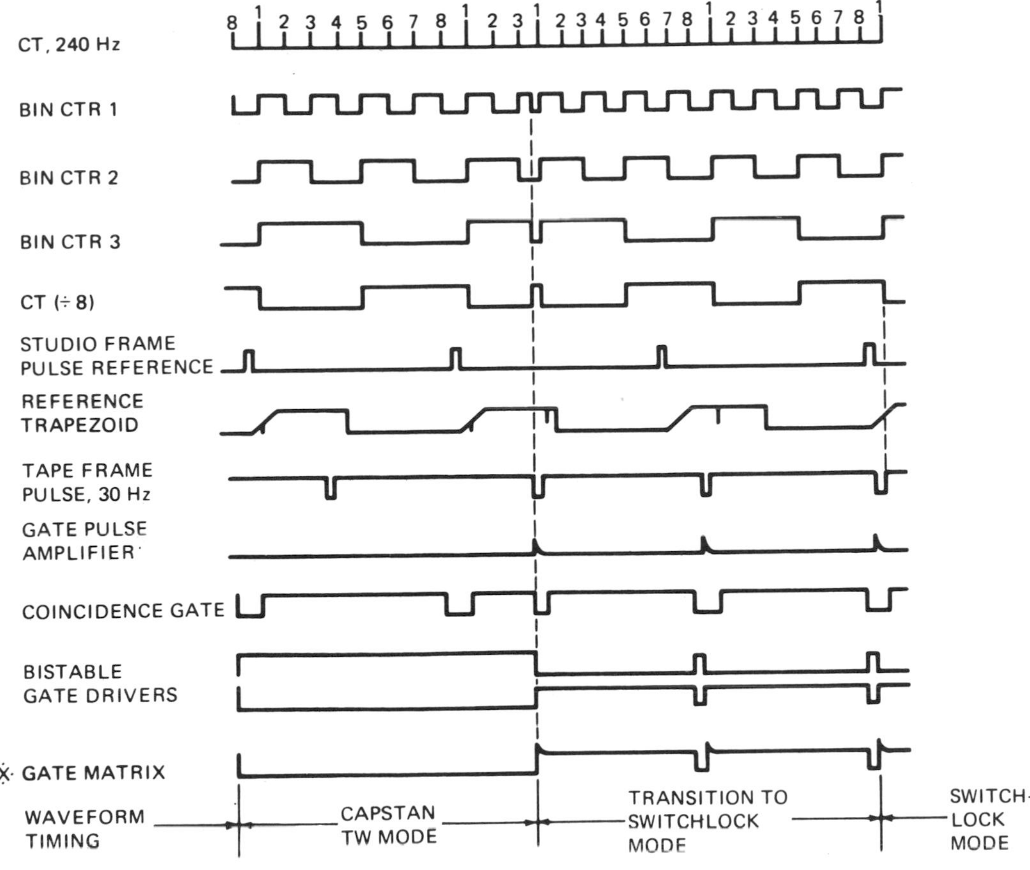

Fig. 4-25b. Pulse gating process during transmission from tonewheel to switch-lock mode of capstan servo operation. (*Courtesy of RCA*)

two-phase asynchronous synchronous HAS motor at the desired speed. One of the two windings receive the control-power by way of a quadrature phase-shift capacitor. The capstan-spindle is either friction-coupled to the motor-shaft, or preferably belt-driven as in the latest machines for higher reliability in respect to traction.

The low-speed capstan digital feedback control subsystem is identical to that used for the 15-ips tape-speed, except for minor modifications of "hold" and RC-equalizer compensation networks at the output of the phase error-sampler. The capstan-motor is also operated like the headwheel motor as a hybrid induction-synchronous ac motor. The headwheel servo system operates identically at both the tape-speeds, using the appropriate 10- or 5-mil FM video-track (15 or $7\frac{1}{2}$ ips, respectively) plug-in headwheel assembly. A modern Q-CVTR must have this two-speed reel-interchange facility for doubling the record/playback times on standard videotape-reels at half the standard 15-ips tape-speed. (From the overall-system point of view, the two-speed technique merely entails the automatic switch-over of minor changes to audio equalization only.)

4.6 VACUUM-GUIDE SERVO

4.6.1 Introduction. The vacuum-guide 960-Hz/sec digital feedback position-control system serves the important function of automatically guiding the moving-tape into

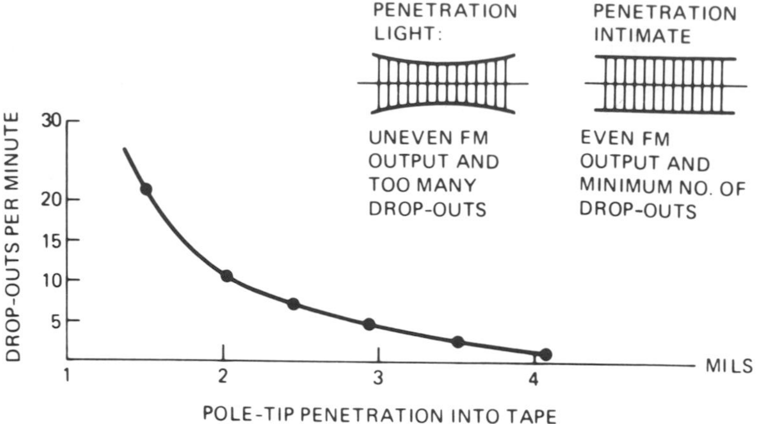

Fig. 4-26a. Typical rate of dropouts with head penetration. Insufficient guide pressure or worn head-wheel pole-tips must be avoided.

16-LINES PER FIELD

TWO PARALLEL VERTICALS IN THE PICTURE

JOGS (OR SKEW DISTORTION) OF VERTICAL BARS IN PICTURE, DUE TO INSUFFICIENT HEAD PENETRATION DURING PLAYBACK.

JOGS OF VERTICAL BARS IN PICTURE: HEAD PENETRATION EXCESSIVE.

THE ATMOSPHERIC PRESSURE HOLDS THE TAPE AGAINST THE GUIDE

HEADWHEEL (HW) AND POLE-TIP (ALFECON) OF 1.5 TO 3.5 MILS HW: 240 REV'S PER SEC

TAPE

5 MILS OF GUIDE MOVEMENT

0.1″

VACUUM

AIR-TAKE

GUIDE-POSITION CALIBRATED WITH RESPECT TO THE HEADWHEEL TO DETERMINE VIDEO HEAD PENETRATION INTO THE TAPE.

GUIDE

HEAD (HW) PENETRATION INTO TAPE AGAINST GUIDE

X X ← HW AXIS

GUIDE CENTER

HEAD-WHEEL (HW)

VIDEO TAPE, 2 IN. WIDE

HW RADIUS = 1.032 IN.
GUIDE RADIUS = 1.0334 IN.
(−0.0005 IN.) TOLERANCE

PARALLEL MISALIGNMENT AS ABOVE (HW AND GUIDE CENTERS NOT COINCIDENT) PRODUCES JOGS OR TIMING DISCONTINUITIES.

EQUALLY SPACED SYNC ON TAPE

HW ROTATION

POLE-TAP PRESSURE (ONE OF FOUR)

PULSES FARTHER APART. HENCE TIMING DISCONTINUITIES DEPEND ON DEPTH OF PENETRATION OF HW DURING PLAYBACK, DIFFERING THAT DURING RECORD.

A HORIZONTAL SYNC, 15.75 kHz HEAD NO. 1

B HEAD NO. 2

C AVERAGE FREQUENCY AT AFC OUTPUT (TAPE SYNC PROCESS)

EXPANDED TIME SCALE

A

B

C

TIMING COINCIDES WITH THE AVERAGE NEAR THE CENTER OF HEAD-SCAN INTERVAL

AVERAGE HORIZONTAL FREQUENCY 15.75 kHz

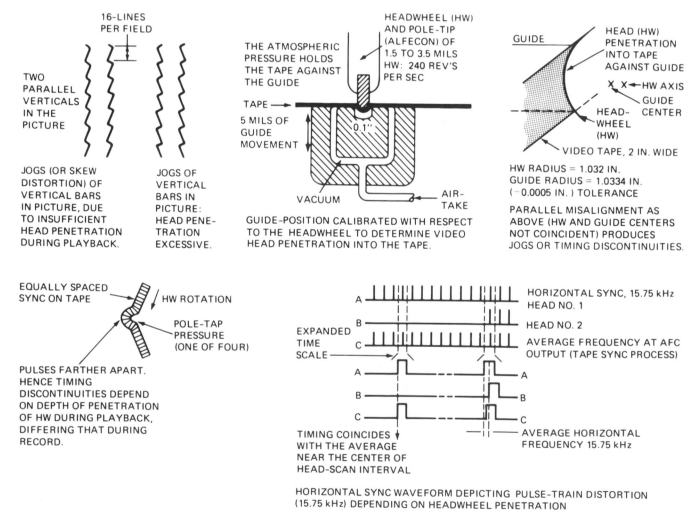

HORIZONTAL SYNC WAVEFORM DEPICTING PULSE-TRAIN DISTORTION (15.75 kHz) DEPENDING ON HEADWHEEL PENETRATION

Fig. 4-26b. Head-penetration and the effect of parallel misalignment. (*Courtesy of RCA*)

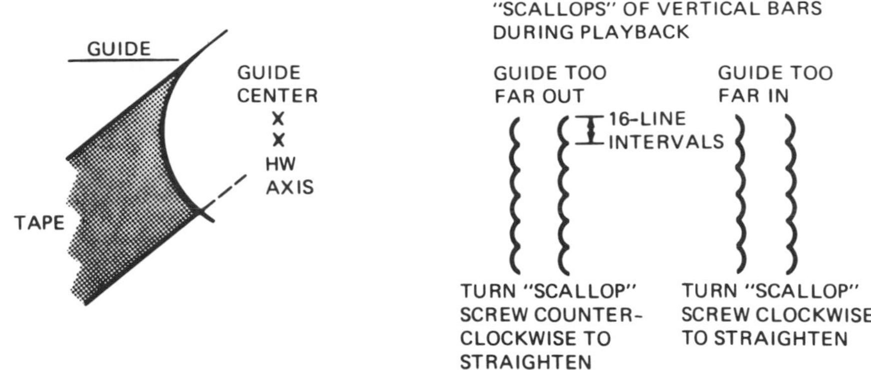

Fig. 4-26c. Scallops due to perpendicular misalignment of guide-center and headwheel axis.

intimate contact with the rotating headwheel *at the recorded pressure* precisely. Intimate head-to-tape penetration is mandatory for recording the FM video signal at particularly the higher frequencies. Picture dropouts, appearing as white-specks on the picture-display, are usually due to insufficient recording pressure, and short rotary-video-head-life may sometimes be due to excess recording pressure and hence during the subsequent playbacks. The position of the tape-guide (or shoe) in respect to the headwheel, during the usual span of the magnetic-head-life (approximately 1000 hr at the present time), is so adjusted that the rotating heads project and indent the tape within a maximum range of 1.5 to 4 mils, depending on the head wear-out. The manual position-control of the vacuum-guide servo is calibrated on a scale for the pressures that correspond to an adjustable position from +2 to −3 mil. The automatic vacuum-guide digital control subsystem must be capable of correcting the guide positional-errors within the 5-mil range to an accuracy of at least ±0.02 mil or 0.4%. A typical rate of dropouts versus the head-penetration into the moving-tape is indicated in Fig. 4-26a.

As the pole-pieces of the video-heads protrude over the rim of the headwheel to about 0.9 to 3.6 mil (depending on the head-wear), the tape is stretched in a localized indentation. The stretching takes place across the tape in a *transverse* direction as the head rotates. As one of the basic factors for determining the play-back-compatibility, it is the function of the automatic vacuum-guide digital control subsystem to maintain the position of the vacuum-guide during the playback of a tape by any headwheel assembly at exactly the same indentation or pressure that was used for recording a particular bit of the program material. The program material may be edited on the tape by electronic or mechanical splicing; and normally the headwheel of a different headwheel assembly with an entirely different head-wear may have been used. At the same time, the stretching of the tape resulting from the indentation is essential

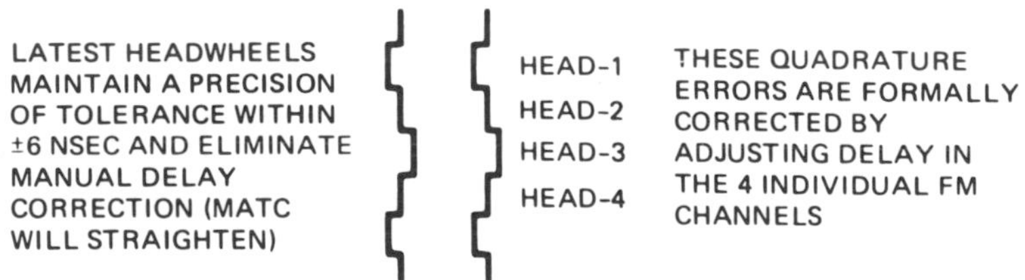

Fig. 4-26d. Quadrature placement of errors of the four heads on the periphery of the headwheel.

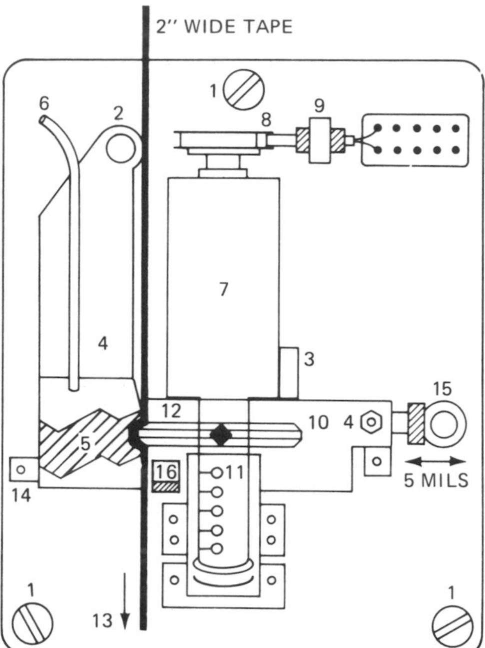

1. MOUNTING SCREWS.
2. PIVOT TO RELEASE GUIDE FROM HW.
3. GUIDE RELEASE LEVER.
4. GUIDE FRAME.
5. VACUUM-GUIDE.
6. VACUUM AIR-TAKE.
7. 3-ϕ ASYNCHRONOUS SYNCHRONOUS HYSTERESIS MOTOR (HAS) SPEED REGULATION AT 14400 RPM.
8. TONE-WHEEL WITH SLOT.
9. TONE-WHEEL PICK-UP COIL.
10. HEADWHEEL (HW) WITH 4 VIDEO (FM) MAGNETIC HEADS.
11. SLIP-RING AND BRUSH ASSEMBLY FOR FM VIDEO (SHIELDED).
12. PENETRATION OF HEAD-TIPS INTO MOVING TAPE.
13. MOVING TAPE (15 OR 7½ ips).
14. HEAD "SCALLOP" TIMING ERROR CORRECTION SCREW.
15. GUIDE CONTROL ECCENTRIC.
16. CONTROL-TRACK HEAD.

MOTOR RUN BY EITHER BALL-BEARING OR AIR-BEARINGS. FOR PIX-LOCK AIR-BEARING HW PROPOSED FOR HIGHER PRECISION.

Fig. 4-27a. Headwheel panel (plug-in assembly). Headwheel panel factory-set approximately every 500 hours. (*Courtesy of RCA*)

for minor dimensional variations resulting from (1) the wear of pole-tips, (2) the mechanical-tolerance differences in machines, and (3) the minute contraction or the expansion of the tape with temperature. Hence, when the pole-tips wear off to the minimum allowed, the head-wheel assembly shown in Fig. 4-27a must receive factory attention for magnetic-head replacement.

In record-mode, the vacuum-guide is maintained at a fixed position in respect to the tape and the headwheel, and this results in an *optimal amount of stretch* for the program recorded. In practice an optimized pressure, corresponding to a mechanically-preset arbitrary 2-mil position-setting, is chosen for the best recordings to enable the best possible reproduction at maximum SNR, as a compromise between various factors like (1) minimum dropouts, (2) desired frequency-response, and (3) mini-

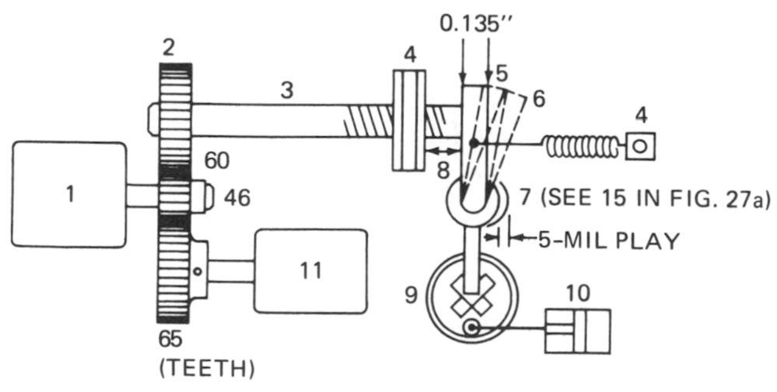

1. 2-ϕ INDUCTION MOTOR 144 RPM (INTERNAL GEAR-RATIO 10:1)
2. GEAR-TRAIN TO LEAD-SHAFT LINKAGE AND HELIPOT.
3. LEAD SCREW.
4. FRAME AND SPLIT-NUT.
5. GUIDE CONTROL ARM.
6. CONTROL ARM RELEASED.
7. GUIDE CONTROL ECCENTRIC.
8. 5-MIL MOVEMENT OF SHAFT FOR GUIDE TAPE PENETRATION RANGE.
9. ROTARY GUIDE SOLENOID.
10. DASHPOT FOR SOLENOID OPERATION TO ENGAGE VACUUM-GUIDE.
11. HELIPOT.

Fig. 4-27b. Vacuum-guide mechanical linkage. (*Courtesy of RCA*)

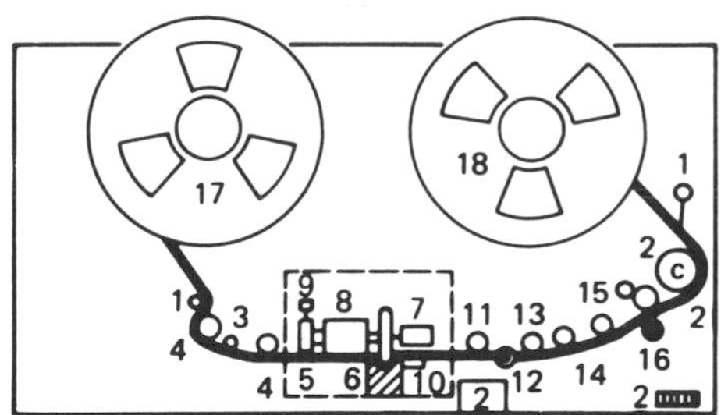

1. STABILIZING ARMS.
2. COUNTER (TAPE MOVEMENT) AND HOURS INDICATOR (HEADWHEEL).
3. MASTER AND CONTROL-TRACK ERASE HEADS.
4. AIR-GUIDES.
5. PLUG-IN HEADWHEEL PANEL ASSEMBLY.
6. VACUUM-GUIDE.
7. FM-VIDEO BRUSH ASSEMBLY.
8. HAS MOTOR AND HEADWHEEL (HW).
9. TONEWHEEL (TW) AND PICK-UP HEAD.
10. CONTROL-TRACK (CT) HEAD.
11. AUDIO AND CUE ERASE HEADS.
12. TAPE GUIDE-POST.
13. AUDIO AND CUE RECORD/PLAY HEADS.
14. SIMULTANEOUS PLAY HEADS.
15. AIR-GUIDE.
16. CAPSTAN AND PINCH-ROLLER.
17. SUPPLY REEL (AND MOTOR).
18. TAKE-UP REEL (AND MOTOR).

Fig. 4-27c. Tape transport assembly (Q-CVTR). (*Courtesy of RCA*)

mum phase-jitter of the headwheel-servo. Standard test tapes are made available to the users to present typical recordings of a comprehensive test-pattern at the optimum indentation. So, the test-pattern is played back at occasional intervals during the life of a headwheel to manually *preset the optimum recording pressure for the playback of the test-pattern* without any *skew* distortion; that vacuum-guide setting is used for any fresh video recording.

4.6.2 Timing Discontinuities in Picture. In playback-mode, if the vacuum-guide position is not correct, resulting in too much or too little a stretch of the tape, compared to what was actually used for recording the pertinent program material, the relative head-to-tape speed will be irregular, and minor timing-errors will result in the output signal and hence the CRT television picture. However, the angular timing will remain correct since the angular-velocity of the headwheel is appropriately maintained by the headwheel servo system. The time-displacement errors appear in the form of discontinuities at the instant of switching from one magnetic-head to the other, at 16-line intervals. This kind of geometric distortion in the picture, usually called the *jogs* or *skew*, is the effect of the parallel-misalignment of the guide-center and headwheel-axis, as shown in Fig. 4-26b.

If a sync pulse-train containing these timing discontinuities is applied to the phase-comparator for producing the dc error-signal in the automatic frequency control (AFC) electronic digital feedback control subsystem (as used in the auxiliary tape sync processing system of the headwheel-servo, with a *flywheel* time-constant for the "hold" network slightly greater than the 16-line head-scanning interval), a uniform pulse-train with an average-frequency approximately equal to 15.75-kHz/sec is produced by the controlled AFC multivibrator-generator; *the timing discontinuities are eliminated as far as the horizontal-sync is concerned.* The synchroguide in the television receivers, and hence their horizontal-scan deflection circuits, respond similarly. The synchroguide in the television receivers is meant for the same purpose to straighten up verticals in the picture in the presence of noise. But the guide position-errors will show up, since the timing-errors in blanking and picture information remain, and produce *jogs* or *skew-distortion*, in spite of the uniform line-scanning rate. The skew-error signal is developed by the AFC whenever the controlling tape horizontal-sync frequency varies from the average line-frequency generated by the AFC. The 960-Hz/sec dc error-signal, as derived from the AFC frequency/phase discriminator, is used to operate the vacuum-guide 960-Hz/sec digital position feedback control subsystem to automatically eliminate the skew-distortion before the picture is transmitted. The vacuum-guide servo has the facility to eliminate these errors by its manual control, as an alternative, if it is transferred to the manual-mode. The pulse-train and the skew-distortion (as indicated by *jogs with positive-slope* for insufficient guide-pressure and *jogs with negative-slope* for excessive guide-pressure with reference to the recorded program material) are illustrated in Fig. 4-26b. For example, an inaccuracy of 1-mil in the guide-position produces on a 17-in. playback monitor or television receiver about 0.25-in. distortion for every 16-line group in a field (or every 32-line group in a properly interlaced picture-frame).

In addition, there are other forms of *geometric distortion. Scallops*, in the form shown in Fig. 4-26c, are caused by *perpendicular-misalignment* of the guide-center and the headwheel-axis. These mechanical errors are corrected for any particular headwheel assembly by adjusting a preset-screw, which is available for the purpose of maintenance. However, the pulse processing circuits in the automatic vacuum-guide digital phase-lock loop should not respond to the timing errors of this parabolic category, although they occur at the same rate as the jogs. The other form of geometric-distortion commonly encountered is that due to the "quadrature-errors" shown in Fig. 4-26d, and this distortion is due to the headwheel assembly itself, since the required quadrature-placement of the four magnetic-heads is invariably inaccurate by a few seconds as a result of the mechanical-tolerances involved in manufacture. (These quadrature-errors can be normally corrected by manually adjusting a fine delay-network in each one of the four FM-recording and FM-playback channels.)

As explained, with the MATC accessory of the recorder, the preceding three forms of geometric distortion in the picture can be eliminated to an undectable extent. But, when some of these errors in conjunction with possible hum, or AFC S-distortion, etc., compound in the same direction, the accuracy obtainable from the MATC digital control subsystem may be inadequate. So, since most of the timing-error is normally contributed by the jogs, the automatic vacuum-guide servo has an important function to perform; it is more so in the case of the phasing of the color-ATC system in view of the extremely high degree of timing-accuracy needed for the clean *unbanded* uniform reproduction of color, during the head-to-head transitions in the picture-frame.

4.6.3 Vacuum-Guide Servo, Mechanical Design Tolerances. The mechanical assembly of the vacuum-guide on the headwheel panel (front and back views) is illustrated in Fig. 4-27a. The motor-shaft remains stationary without any mechanical-jitter under the zero-error conditions. It develops torque in one direction or the other only in response to a 960-Hz/sec error-signal from the 15.75-kHz/sec AFC-device of the headwheel digital control subsystem. The movement of the vacuum-guide is smoothly controlled via the appropriate mechanical-linkage consisting of three gear-trains, one inside the two-phase induction motor assembly, the second for the helipot of the manual control-loop, and the third for the lead-screw controlling the position of the vacuum-guide via an eccentric.

If the tape-guide and the headwheel remain concentric, the incremental difference of the track-lengths subtended by the 90° angular-distance between two successive magnetic-heads is equal to $\pi/2$ multiplied by the change in guide-position. Hence, for 1-mil change in guide-position, the track-length between two magnetic-heads is given by $\pi/2$ or 1.57 mil.

A change of 1.57 mil = 0.1% variation in track-length.

And one head-scan = $\frac{1}{4}$ X the period of the headwheel rotation at 240 Hz/sec = $\frac{1}{4}$ X 4167 μsec.

Thus, 0.1% variation in track-length = $\dfrac{4167 \times 0.1}{4 \times 100}$ = 1.04 μsec approximately.

Hence, the guide position-change of 1 mil = 1 μsec = $\frac{1}{4}$ in. on a 17-in. wide monitor since the duration of a line-interval corresponds to 63.5 μsec.

Now, the pitch of the lead-screw = 24 turns/in.

A 5-mil travel of guide requires 0.133 in. of lead-screw axial movement.

Hence, 0.133-in. of axial movement = 3.19 turns of lead-screw.

With a gear-ratio of 65 : 60 to the helipot (the manual servo pick-off point): 0.133-in. axial movement = $\dfrac{3.19 \times 60}{65}$ = 2.55 turns of helipot.

That is, a 5-mil guide movement is approximately equivalent to three turns of helipot.

The accuracy required of the guide servo = 0.01-in. on the 17-in. monitor.

And, 1-mil guide-error = 1μsec or 0.25-in. jog on the 17-in. monitor.

Thus, 0.01-in. jog = $\frac{1}{25}$ or 0.04 mil of guide-error.

So, the maximum error allowed for the manual or automatic guide-servo is 0.04-mil penetration, which is equivalent to a 0.04-μsec (or ±0.02 μsec) timing-error in the picture; this corresponding to 0.01-in. error-discontinuity in the verticals of a 17-in. monitor.

For the manual-servo with a 5-mil range, the error permissible for the total 5 mils = $\frac{1}{25} \times \frac{1}{5} = \frac{1}{125}$ of the total 5-mil range.

With 60 gear-steps and three full-turns for the helipot, a 5-mil guide-movement is equivalent to 180 gear-steps.

Hence, a tolerance of 0.04 mil = a tolerance of 180 X 0.04/5, or a tolerance of above 1 gear-tooth.

Thus, the guide-servo must operate to an accuracy of 1 step on the gear-train as a nonlinear subsystem, with the backlash of the two gear-trains constituting the major nonlinear element in the case of the vacuum-guide digital control subsystem.

Including the gear-train used for the helipot, the combined effect of the back-lash in three gear-trains makes a more pronounced nonlinear-element in the case of the manual guide-servo.

Thus, the vacuum-guide servo in the automatic-mode is more accurate as far as the tolerance in timing-error is concerned.

4.6.4 Vacuum-Guide Servo, Logic. The guide-servo is disabled when the velocity feedback-loop in the headwheel digital control subsystem fails to lock-in during record or playback. With the logic used for the operation of a rotary guide-solenoid, this failure disconnects power to the two-phase induction motor driving the vacuum-guide and throws the guide-solenoid control-arm off the lead-screw, and isolates the heavy vacuum-guide assembly and hence the tape from the headwheel, as a protection for the latter. The mechanism is illustrated in Fig. 4-27b and c.

When the headwheel-tracking fails due to a failure of the capstan digital control subsystem, the tape-sync becomes excessively noisy, and automatically changes over the normally automatic guide-servo to a preset optimum manual-position. Otherwise, the disturbances and noise in the sync under this condition might throw the vacuum-guide digital feedback control subsystem out of control and cause damage to the headwheel-assembly and the magnetic-heads, by jamming the vacuum-guide against the rotating headwheel.

A third operational logic feature is necessary. In the record and setup modes of operation, no tape-sync is available for the guide-servo. It is then automatically switched over to a calibrated manual, bridge-type feedback control-loop with the helipot. Since a carefully-preset vacuum-guide position-setting with a test-tape is necessary for recording at optimum pressure, an exclusive manual vacuum-guide control-loop is mandatory for the record-mode, other than that used for the manual playback-mode. So, in the record-mode, the predetermined and preset "record" guide-position will become automatically effective for recording at optimum guide-pressure.

The above logic and switching features are accomplished by conventional diode-matrixing and flip-flop circuits as indicated in the simplified block-diagram of Fig. 4-28. (All operational manipulations like master-record, audio-record, play, setup, play-standby, wind, electronic-splice switching, etc., are all done in a similar fashion, the flip-flops providing the temporary memory where needed.)

4.6.5 Calibrated Manual Servo System for Vacuum-Guide Servo. Although under normal playback conditions, the guide-servo must be self-correcting as a digital feedback position-control subsystem, one must be able to manually set or preset the guide-pressure in the record-mode, and manually correct the skew-distortion during the playback-mode, as a calibrated continuous bridge-type feedback control system in the case of test-runs and excessively noisy reproduction due to momentary mistracking of the capstan-servo, etc. A schematic block-diagram of the remote-control device for the vacuum-guide in the manual-mode is shown in Fig. 4-29.

In the manual-mode, a dc error-signal is developed by an unbalance in the bridge control-loop—which consists of a helipot geared to the motor-shaft, and either the record or the playback manual-control potentiometer. These are connected across a fixed dc potential of approximately 3 V. The gain of the ac servo amplifier must be such that the bridge is continuously brought close enough to the balance for null-output, so that the error, as a minimum tolerance, should not exceed $\frac{1}{125}$ of the fixed dc potential. That is, a minute error of this dimension in either the positive or the negative direction must deliver enough power to the control-winding of the induction motor to cause it to move just a step ahead or backward, according to the calculation figured out earlier. At the same time, it must be assured that the sensitivity of the servo is such that, from the viewpoint of tolerance, the drift-effects in the chopper (modulator) and the following transistor stages should not reach the above limit.

When the manual vacuum-guide position is reset to follow the calibration on the manual control, the dc error-signal due to the immediate unbalance of the bridge-

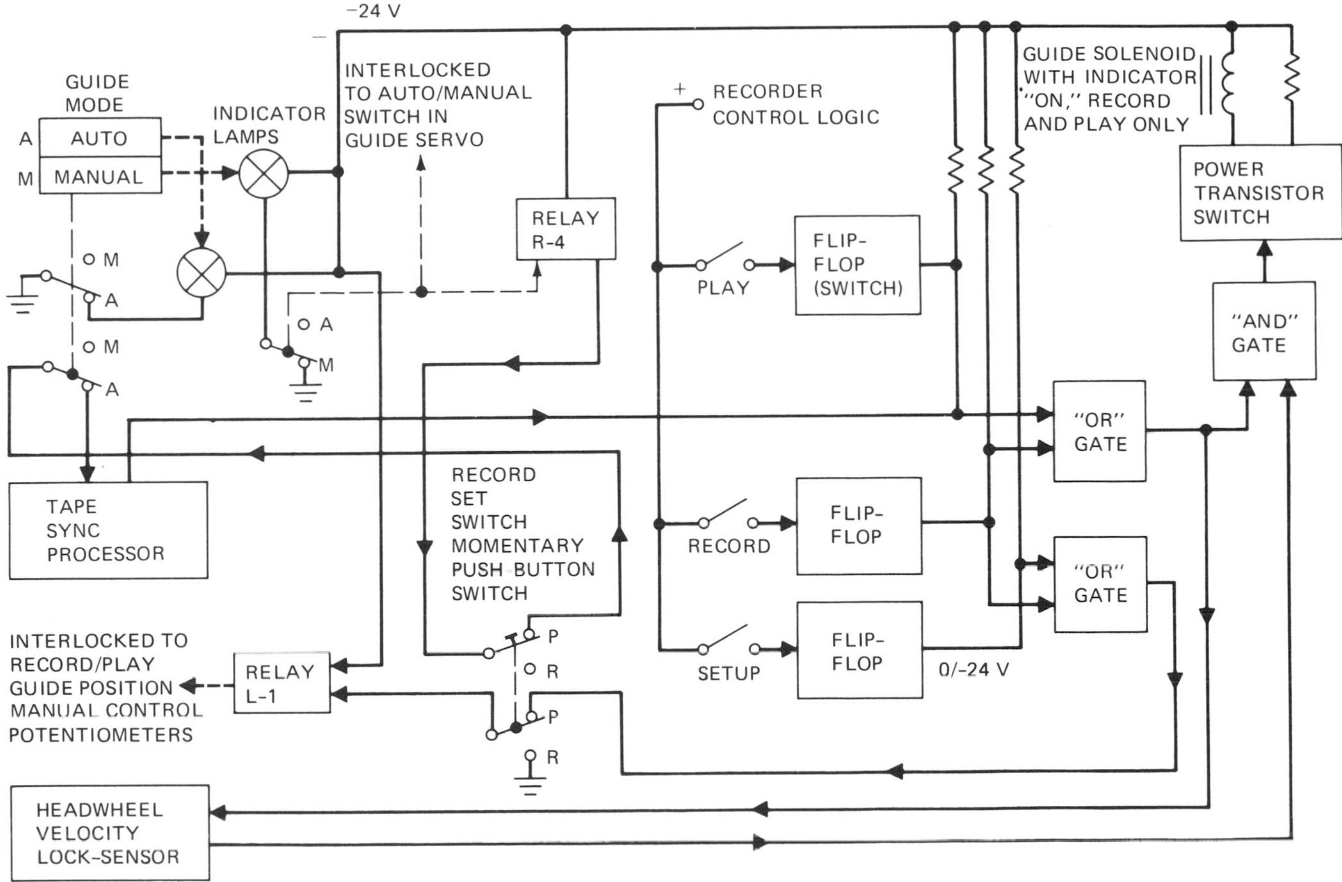

Fig. 4-28. Logic associated with vacuum-guide sampled-data control system. (*Courtesy of RCA*).

circuit is converted by the bilateral transistor-chopper to an ac signal. The chopper, in short, converts the dc error-signal from the unbalance of the bridge-circuit to an amplitude-modulation of the 60-Hz/sec ac-carrier, which reverses in phase as the polarity of the applied dc error is reversed about the zero-reference of the bridge-circuit. The ac signal is amplified to an extent of approximately 46 dB in the servo amplifier to develop the requisite sinusoidal control-voltage of about 250-V peak-to-peak for the control-winding of the two-phase induction motor. The output-drive of the push-pull class-AB transistor power amplifier (working in an avalanche or delayed-collector-conduction mode for high-sensitivity and minimum power-dissipation) is capacity-tuned, to incidentally provide the quadrature phase-shift to the control-voltage applied to the motor, while the direct ac power is applied to the reference-winding of the two-phase induction motor. The LC-tuners used in the transformer-coupled ac-carrier stages incidentally provide ac-compensation for the improved stability of the overall nonlinear vacuum-guide servo system. The motor-shaft rotates in such a direction that the error is cancelled, the sense of the shaft-rotation being determined by the quadrature-lag or -lead of the control-voltage with respect to the reference-voltage. So, as the position of the vacuum-guide is manually set, the bridge-circuit in the feedback-loop is continuously balanced to a stable equilibrium with a resultant null-signal as the final steady-state servo control-voltage output.

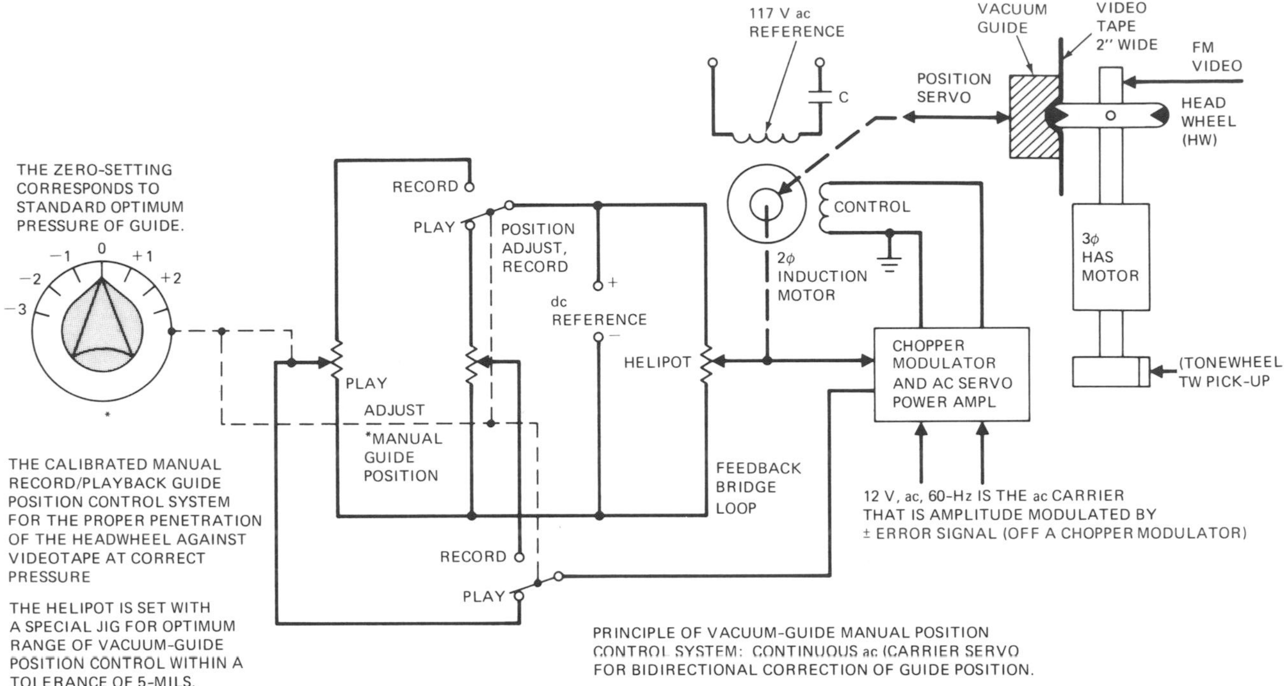

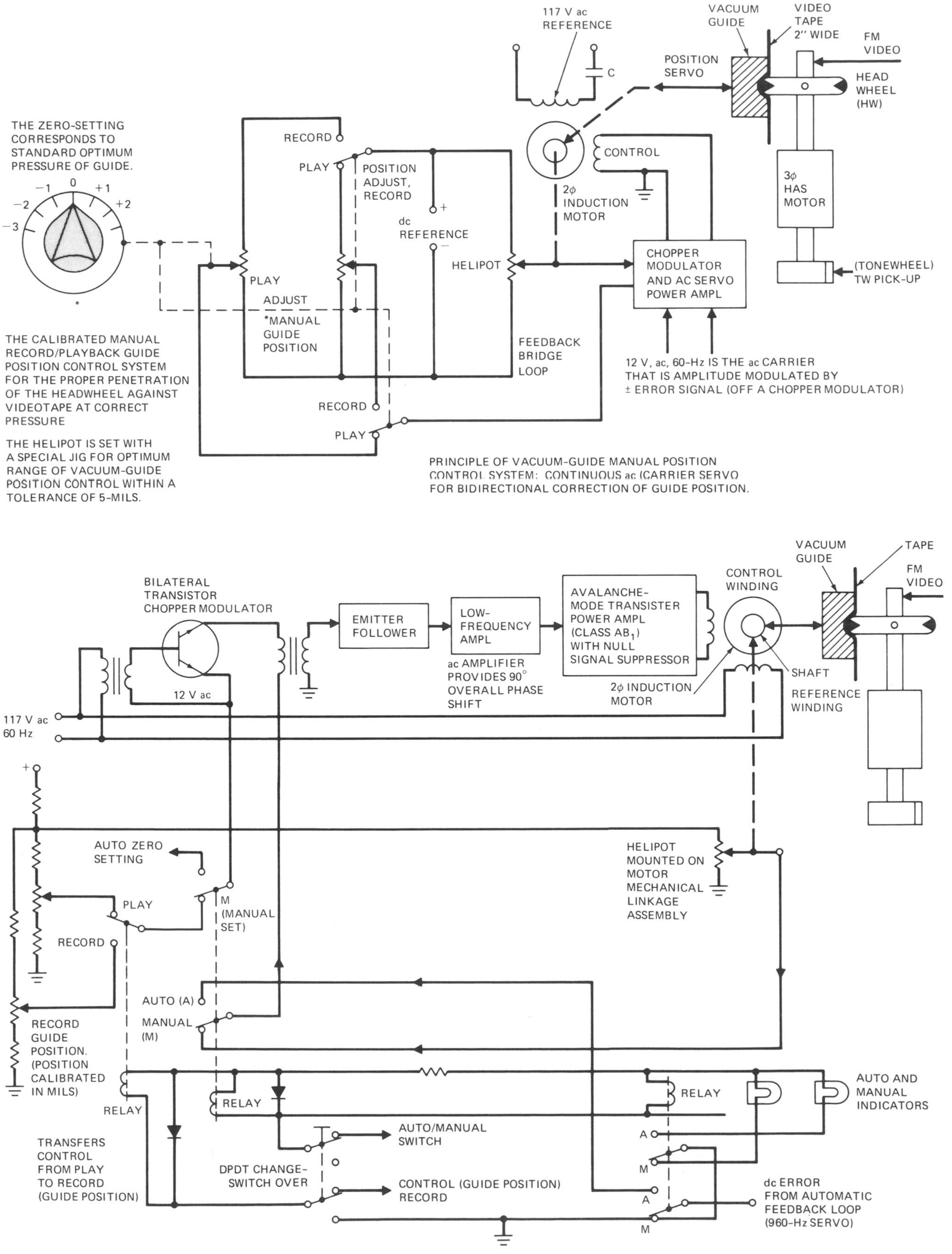

117 V ac
REFERENCE
C
VACUUM
GUIDE
VIDEO
TAPE
2" WIDE
FM
VIDEO
POSITION
SERVO
HEAD
WHEEL
(HW)
THE ZERO-SETTING
CORRESPONDS TO
STANDARD OPTIMUM
PRESSURE OF GUIDE.
−1
0
+1
−2
+2
−3
RECORD
PLAY
POSITION
ADJUST,
RECORD
CONTROL
2ϕ
INDUCTION
MOTOR
3ϕ
HAS
MOTOR
dc
REFERENCE
HELIPOT
CHOPPER
MODULATOR
AND AC SERVO
POWER AMPL
(TONEWHEEL)
TW PICK-UP
PLAY
ADJUST
*MANUAL
GUIDE
POSITION
THE CALIBRATED MANUAL
RECORD/PLAYBACK GUIDE
POSITION CONTROL SYSTEM
FOR THE PROPER PENETRATION
OF THE HEADWHEEL AGAINST
VIDEOTAPE AT CORRECT
PRESSURE
FEEDBACK
BRIDGE
LOOP
12 V, ac, 60-Hz IS THE ac CARRIER
THAT IS AMPLITUDE MODULATED BY
± ERROR SIGNAL (OFF A CHOPPER MODULATOR)
RECORD
PLAY
THE HELIPOT IS SET WITH
A SPECIAL JIG FOR OPTIMUM
RANGE OF VACUUM-GUIDE
POSITION CONTROL WITHIN A
TOLERANCE OF 5-MILS.
PRINCIPLE OF VACUUM-GUIDE MANUAL POSITION
CONTROL SYSTEM: CONTINUOUS ac (CARRIER SERVO
FOR BIDIRECTIONAL CORRECTION OF GUIDE POSITION.
VACUUM
GUIDE
TAPE
FM
VIDEO
BILATERAL
TRANSISTOR
CHOPPER MODULATOR
EMITTER
FOLLOWER
LOW-
FREQUENCY
AMPL
AVALANCHE-
MODE TRANSISTER
POWER AMPL
(CLASS AB1)
WITH NULL
SIGNAL SUPPRESSOR
CONTROL
WINDING
12 V ac
ac AMPLIFIER
PROVIDES 90°
OVERALL PHASE
SHIFT
2ϕ INDUCTION
MOTOR
SHAFT
REFERENCE
WINDING
117 V ac
60 Hz
+
AUTO ZERO
SETTING
PLAY
M
(MANUAL
SET)
RECORD
HELIPOT
MOUNTED ON
MOTOR
MECHANICAL
LINKAGE
ASSEMBLY
AUTO (A)
MANUAL
(M)
RECORD
GUIDE
POSITION.
(POSITION
CALIBRATED
IN MILS)
RELAY
RELAY
RELAY
AUTO AND
MANUAL
INDICATORS
TRANSFERS
CONTROL
FROM PLAY
TO RECORD
(GUIDE POSITION)
AUTO/MANUAL
SWITCH
DPDT CHANGE-
SWITCH OVER
CONTROL (GUIDE POSITION)
RECORD
A
M
A
M
dc ERROR
FROM AUTOMATIC
FEEDBACK LOOP
(960-Hz SERVO)

Fig. 4-29. Vacuum-guide manual ac carrier servo. (*Courtesy of RCA*)

If the servo amplifier gain is not correctly set and excessive, the mechanical backlash nonlinearity involved in the three gear-trains of the mechanical-linkage will tend to drive the manual feedback-control system into a limited-oscillation (termed *chatter* or *hunting*). The oscillation corresponds to an excessive phase-plane limit-cycle of a nonlinear system or the stable limit-cycle encountered in the gain-phase plot of the describing-function technique. The hunting problem will appear as a back-and-forth rotation of the shaft, and the null-signal will be replaced by an ac control-winding signal with *continuous phase-reversals.* On the other hand, if the gain is not adequate, the resettability of the vacuum-guide about the precise optimum-pressure, as determined by the calibration of the manual position-control setting, will be affected with *sluggishness and dead-space* as the result of another version of nonlinear element. Since the friction of the vacuum-guide mechanism and the other mechanical parts associated with the headwheel and the tape-transport system vary from machine to machine, due to the cumulative effect of the mechanical-tolerances involved on a tape-transport, a carefully adjusted preset manual gain-control is included for the best resettability of the vacuum-guide position in each machine.

The normal working-position of the lead-screw shaft is preset with reference to the helipot and the calibrated position-control potentiometer, by means of a special vacuum-guide position-alignment jig, designed for this specific purpose. This centering or alignment procedure is a basic step for the correct operation of the vacuum-guide digital control subsystem in its manual and automatic modes of operation.

4.6.6 Vacuum-Guide 960-Hz/sec Digital Feedback Position-Control System. A simplified block-diagram given in Fig. 4-30 illustrates the principle of the vacuum-guide servo operating in the automatic-mode for maintaining a reliably stable picture without skew-distortion, as the videotape presents the spliced program-material, that is invariably recorded at different vacuum-guide pressures on other machines using headwheels of varying degrees of wear-out.

As stated earlier, the vacuum-guide position error-signal is developed at the output of the horizontal-AFC frequency (phase)-discriminator in the headwheel digital control subsystem by comparing the 15.75-kHz/sec horizontal synchronizing pulse-frequency from the tape with the frequency of the voltage-controlled AFC horizontal-frequency multivibrator. The hold and compensation network used for this all-electronic 15.75-kHz/sec digital feedback frequency-control subsystem, discriminatingly leaves undisturbed the transitional magnetic-head switching 960-Hz/sec *skew* error-components due to the repeated timing-discontinuities that are produced by the changes in the vacuum-guide pressure. These minute 960-Hz/sec error-components from the AFC-loop appear in the shape of a distorted trapezoidal-waveform, of a polarity determined by either an excess or an insufficiency of vacuum-guide pressure with reference to the original vacuum-guide pressure during recording. A minute error-signal such as 10 mV peak-to-peak is converted to a 960-Hz/sec sinusoidal-waveform by passing the signal through a low-pass filter with a 3-dB cutoff at about 1000 Hz/sec. The error-components produced by the *scallops* and the quadrature-errors at the output of the low-pass filter are either negligibly minute or out-of-step with the skew-errors when they are appreciable, and hence do not affect the vacuum-guide digital subsystem in any way; the *scallops* can be easily corrected on the vacuum-guide assembly by means of an Allen-wrench during the test-run of the tape.

The minute 960-Hz/sec distorted-waveform error-signal is amplified in an "integrated-injection type of logic signal-processor" (I^2L) and clipped from peak-to-peak by means of diode-limiters, and shaped to a regular trapezoidal-waveform before application to a bidirectional diode-bridge. Due precaution is taken in the design for suppressing the low-frequency transients from the headwheel-servo system, and also potential periodic high-amplitude 60-Hz/sec error-components, that might arise out of the vertical-sync timing-errors in the television signal-processing systems used at the time of recording. The 960-Hz/sec error signal-processor acts as a kind of digital-controller in the vacuum-guide control subsystem.

The I^2L signal-processing circuit consists of two pairs of cascaded PNP-NPN transistors in this order. The transistors are tied to a common negative voltage supply as was illustrated in Fig. 2-11. The NPN transistor is operated in its forward or cutoff region to avoid saturation and hence time-delay (at low and high pulse-repetition-intervals), while the PNP transistor functions as a current-source to the NPN transistor. In the discrete PNP-NPN transistor circuit-configuration of I^2L technique, the PNP stage that, in principle, functions as the current source to the following NPN, is simultaneously improvised (by the expediency of ac-coupling) as a signal handling stage in view of the discrete nature of the transistors (unlike those used in modern large-scale integration at very low operating voltage). The germanium crystal-diodes used in this circuit serve the same high-speed switching function as the Schottky-diode complements of the latest Schottky-I^2L LSI technique.

The polarity of the trapezoidal-waveform at the input to the bridge-circuit is determined by the polarity of the vacuum-guide skew error-signal. Error-sampling is accomplished against the slope of the trapezoid-pulse at the 960-Hz/sec rate by using narrow sampling pulses of opposite polarity. The sampling-pulses are derived by phase-splitting a 960-Hz/sec pulse-train obtained from the tonewheel pulse processing system in the headwheel-servo. The 960-Hz/sec pulse-train is generated and phase-locked to the 240-Hz/sec tonewheel pulse-train used for the headwheel- and capstan-servo subsystems. Thus, the vacuum-guide servo sampling is synchronized to the other sampling subsystems in the Q-CVTR. The bidirectional diode-bridge circuit is so designed that, during the sampling process, the error-signal is clamped to the ground-potential at either the positive or the negative peaks, depending on the polarity of the vacuum-guide error. With the hold and dc compensating-network at the output of the error-sampler, a fluctuating dc output appears, and it is either positive or negative with respect to the ground. An RC lag-network and a sensitivity-control are included in the automatic-loop for improving and centering the feedback stability-conditions against possible *hunting* due to the stable limit-cycle of a nonlinear system. The excursions of the dc error-signal are limited to ± 1.5 V by peak-to-peak diode-limiters to make it a nonlinear system with saturation. So, the clipped-amplitude of the trapezoidal error-pulse determines the level of saturation in this case.

As in the manual-mode, the bilateral-transistor chopper converts the $\pm$dc error-signal into a 60-Hz/sec sinusoidal control-signal reversing in phase by 180°. The gain of the ac servo amplifier is preset to obtain the maximum output-drive and hence the maximum-torque of the two-phase induction motor at the above-specified ± 1.5-V dc error-input levels. The limiting prevents possible undesirable saturation in the following ac amplifier stages. The nonlinear design technique enables maximum speed of correction at both comparatively high- and low-level error-signals, and that is the most common situation in practice. For the minimum possible vacuum-guide error-signals encountered, the overall system-gain used in the automatic feedback-loop for developing full torque is approximately 86 dB.

Incidentally, the 2-W output, push-pull, class-AB transistor power-output stage associated with this digital position-servo amplifier employs a PNP-NPN pair in series-mode on either arm of an *Avalanche* high-sensitivity low-current power-amp. The technique enables the speedup of operation and cancellation of the residual signal in the control-winding of the motor at the fine zero control-setting under steady-state operation. It prevents potential low-level oscillation below the tolerance of error.

The sampling against the slope of the trapezoid waveform of the error-signal, with the incidental dc-clamping technique against a set-point on the slope, provides the basis for a zero-reference by means of a preset zero-setting control. The facility enables the operation of the Q-CVTR with an unnoticeable residual vacuum-guide error or skew-distortion (within a dead-space of ±0.02 mil of residual position-error on a 17-in. CRT monitor/receiver), irrespective of the origin of the tape program-material. The position of the preset zero-control at the stage of the chopper is shown in Fig. 4-30. By playing a standard test tape, the precise zero-adjustment is carefully preset in conjunction with the automatic sensitivity control whenever a new headwheel panel assembly is installed. The sensitivity of the manual ac servo is factory-preset in each Q-CVTR shipped.

The overall development and design of the nonlinear vacuum-guide digital feedback position-control subsystem, on a heuristic measure-and-optimize basis, is a result of trade-in and counterbalance between a score of conflicting factors such as

1. the speed of correction (about 0.25 sec for an error of a 0.25 in.)
2. minimum settling-time for the steady-state *zero-error* operation
3. the torque obtainable from a reliable 60-Hz/sec servo transistor power amplifier, as compact and simple as possible, without such complications as forced air-cooling and elaborate heat-sinks
4. sufficient gain-margin for stability without any trace of *hunting* due to the stable limit-cycle of a nonlinear phase-lock loop, in either case of error-polarity
5. compatibility to cope with the compensation networks and their response in the associated AFC and other feedback loops in the interacting headwheel and capstan digital feedback-control subsystems
6. machine overall-system disturbances, if any
7. the consideration of extremely minute and large error-signals
8. the noise inherently present in sync from the FM system if the capstan-tracking is marginal due to some system problem
9. beta-variations in large-production transistor samples, although aging is not a problem as in the case of the vacuum-tubes
10. incidence of possible disturbances in the vacuum-guide control-loop from the capstan-mechanics, via the tape
11. unfailing reliability of operation on a long-term basis within the extreme ranges of temperature on a large-scale production of these machines.

As far as the overall Q-CVTR performance is concerned, the vacuum-guide servo will enable, at the preset optimum-settings of the zero and sensitivity controls, the display of a picture with undetectable skew-distortion, the instant the headwheel and capstan servo subsystems lock-in. The digital PLL will correct *jogs* smoothly in a

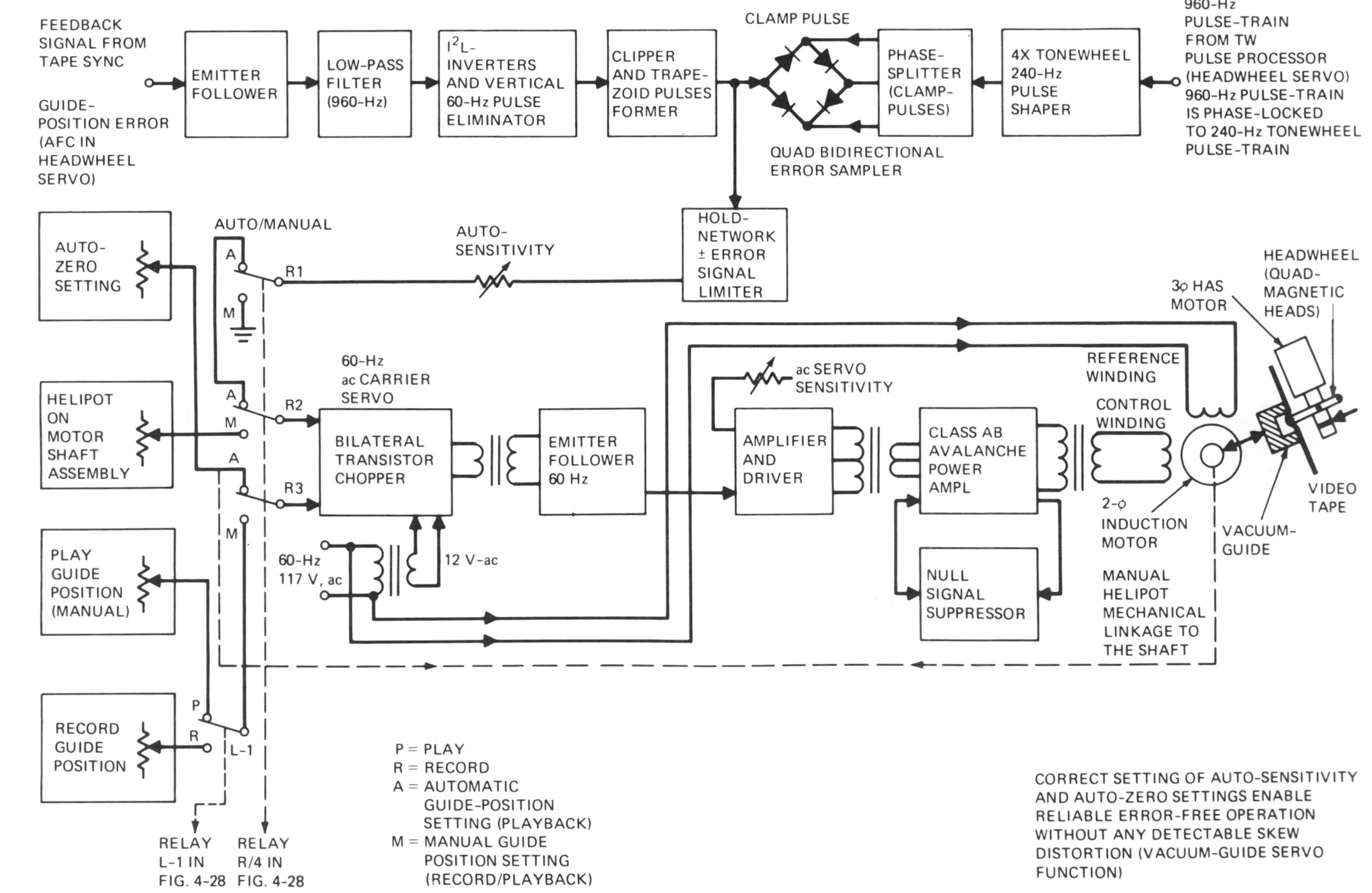

Fig. 4-30. Schematic block-diagram, vacuum-guide sampled-data feedback position control system in automatic and manual modes of operation. (*Courtesy of RCA*)

fraction of a second without noticeable overshoots, when splices recorded at different vacuum-guide pressures are encountered in the tape program-material.

4.6.7 Automatic Velocity-Error Compensator. High-quality broadcast videotape color television programming is internationally accomplished at this time with the VTRs of two manufacturers, AMPEX and RCA. The velocity-error accessory works in conjunction with the CATC of RCA (or the corresponding COLORTEC of AMPEX) to eliminate *color banding* and other velocity-errors when videotape interchange is involved for broadcast reproduction. The 16-line magnetic-head hue-banding shows up as a result of slightly-differing mechanical dimensions and tolerances, especially those associated with the height of the tape-guide assembly with repsect to the moving-tape. (The AMPEX accessory corresponding to the MATC of the RCA recorder is called AMTEC.)

The compensator incidentally allows the correction of other velocity-errors such as "once around" and engagement errors. The facility enables the best color reproduction results, since changes of color saturation within the normally undetectable bands may be eliminated by precision mechanical adjustment of the guide-height with respect to the tape.

The velocity-error compensator employs a random-access solid-state memory (RAM) to delay the digitized color reference-burst by one line-interval and make it possible for the phase-comparison of the 3.58-MHz/sec reference-burst to that of the eight-cycle reference-burst on the succeeding horizontal line. The error-signal, representing the chroma phase-difference, derives a linear-ramp, and its height is hence proportional to the phase-error. The ramp in turn automatically corrects the resampling-timing in the output digital-to-analog converter of the color video signal to cancel the effects of the accumulated velocity-errors in the horizontal line to present a true color-picture of the videotape. The device is particularly an essential videotape accessory in television studio operations for the multiple generation duplication of color television programs.

4.6.8 Typical Specification of Quadruplex Color Videotape Recorder. The following overall performance-data of a typical quadruplex color videotape recorder (Q-CVTR) gives a comprehensive impression of the basic characteristics, the high-degree of precision-reproduction expected of these recorders, and the complex interacting digital control system involved for high-quality color broadcasts. (See Figs. 4-31 and 4-32.)

Recording Medium: Oxide-coated video magnetic-tape, 1.4-mil thick, 2-in. wide. Examples: 3M Scotch videotape type 379 or 179.

Tape-Speed: 15 ips (60-Hz/sec power), $7\frac{1}{2}$ ips for special use. 39.7 cm/sec (50-Hz/sec power, overseas).

FM-Video Head-Life: (Headwheel assembly returned to factory for replacement by a new headwheel assembly.) Average of 500 hr of usage with Alfecon pole-tips. The FM video-head's magnetic *gap* = 0.9 mil.

Picture-and-Sound Separation: 18.5 frames, sound leading (United States). 14.8 frames, sound leading (overseas).

FM-Deviation, Low-Band: (for different television standards)

1. monochrome (405 lines/525 lines): 4.3 to 6.8 MHz/sec
2. color (525 lines): 5.4 to 6.3 MHz/sec
3. monochrome (625 lines/819 lines): 5 to 6.8 MHz/sec.

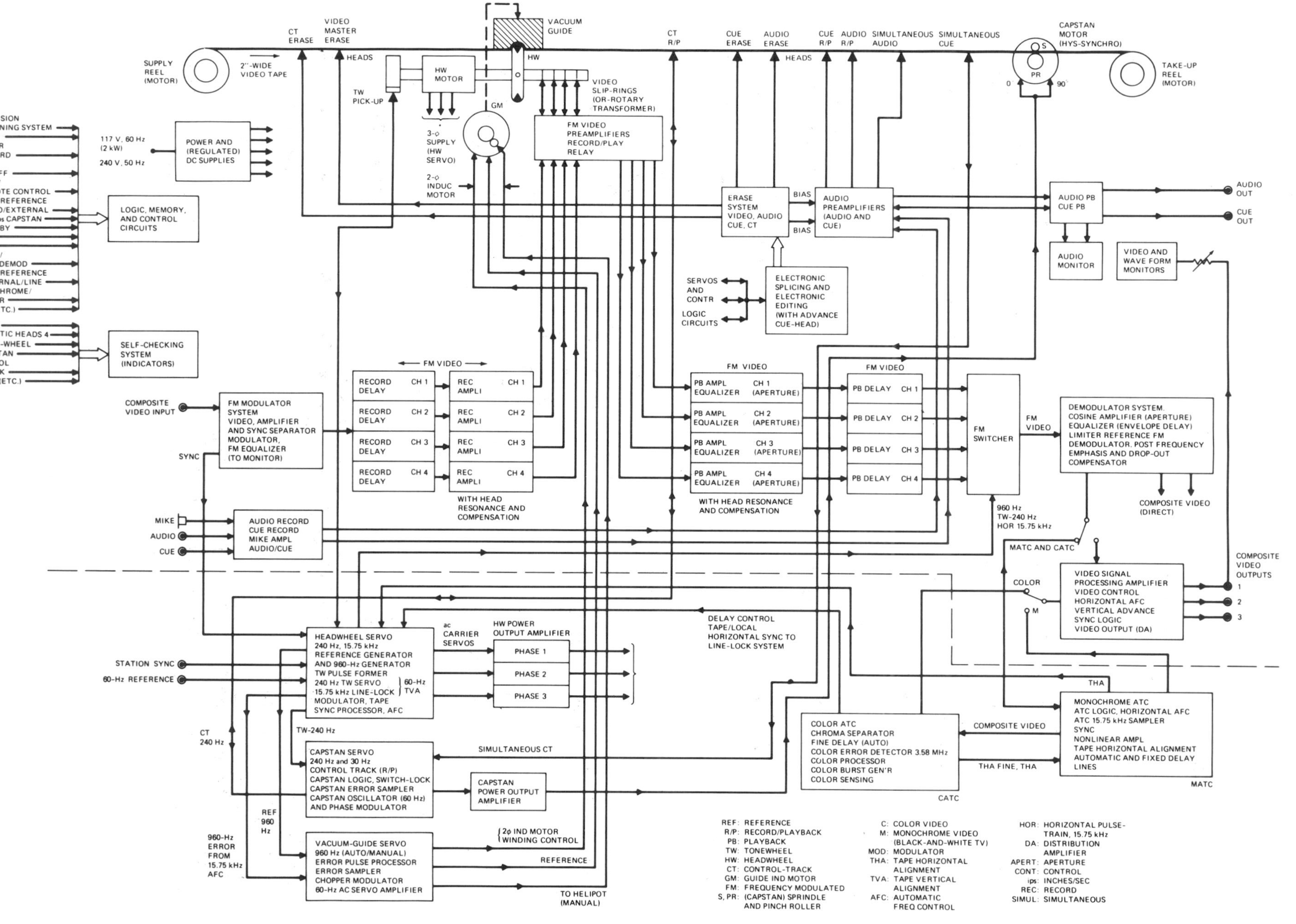

Fig. 4-31. Quadruplex color videotape recorder (Q-CVTR)—overall system block diagram. *(Courtesy of RCA)*

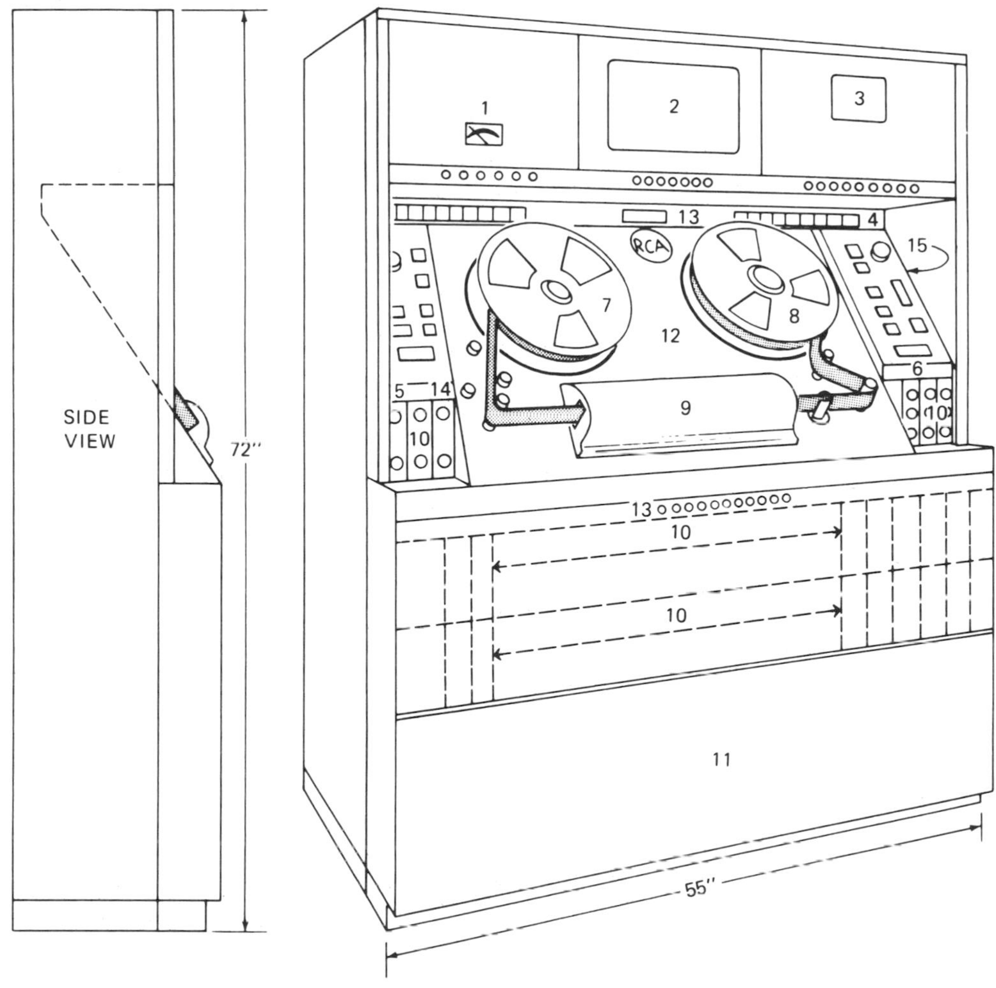

1. AUDIO MONITOR
2. VIDEO MONITOR
3. WAVEFORM MONITOR
4. SELF-CHECKING INDICATOR SYSTEM
5. RECORD CONTROL PANEL
6. PLAYBACK CONTROL PANEL
7. SUPPLY REEL
8. TAKE-UP REEL
9. QUADRUPLEX HEADWHEEL AND OTHER MAGNETIC HEADS
10. PLUG-IN SOLID-STATE CIRCUIT MODULES (SYSTEM ELECTRONICS AND SAMPLED-DATA CONTROL SYSTEMS HARDWARE) . . . COVERED.
11. POWER SUPPLIES, HEADWHEEL BLOWER, AIR-BEARING PUMP, VACUUM PUMP, FUSES, RELAYS
12. TAPE TRANSPORT ASSEMBLY
13. METER INDICATOR
14. DIGITAL LOGIC AND CONTROL
15. AIR-SOLENOID VALVE FILTER, PRESSURE GAUGE, PRESSURE REGULATOR, MUFFLER FOR NOISE

Fig. 4-32. Typical Q-CVTR (RCA TR Series). (*Courtesy of RCA*)

FM High-Band: (Monochrome and color) FM bandwidth 12 MHz/sec
1. (405 lines/525 lines): 7.06 to 10 MHz/sec
2. (625 lines/819 lines): 7.2 to 9.3 MHz/sec.

Recording Time: 14-in. reel, 96 min. (7200 ft); $12\frac{1}{2}$-in. reel, 64 min. (4800 ft); 8-in. reel, 32 min. (2400 ft).

Rewind Time: 96-min. recording: approximately 4 min. (This is all the interval that is required between a fresh recording and the first broadcast playback.) 32-min. recording: less than $1\frac{1}{2}$ min.

Starting Time:
1. from stop-mode: less than 3 sec (TW/LLO)
   less than 5 sec (Pix-lock)
2. from standby: less than 2 sec (TW/LLO)
   less than 4 sec (Pix-lock).

Stopping Time: (Record or Playback): less than 0.2 sec.

System Timing Reference:
1. Recording: Incoming composite-video or studio sync generator
2. Playback: Studio sync generator or power-line.

Phase-Jitter, Pix-lock (for special-effects or synchronization with camera signals): ±0.07 μsec, maximum (with air-bearing headwheel panel)
±0.1 μsec, maximum (with ball-bearing headwheel panel).

Horizontal Phase (*Pix-lock*): ±2-μsec range of adjustment for tape horizontal-sync, relative to station horizontal-sync. The stability of this adjustment is ±0.3 μsec for a 1-hour interval (Adaptive-Control).

Total geometric-distortion allowed for Pix-lock input: ±1 μsec. This is the total variation of the average timing-error of the 16-line sections in the picture with respect to the local sync.

Vertical-Line Displacements (at the transitions of the four FM-video headbands): 0.02 μsec or less with the channel delay-line adjustment, and the vacuum-guide *scallop* correction. Less than 0.02 μsec (undetectable) with the MATC, and 0.005 μsec with the color-ATC accessories.

Velocity-Error Correction: Uniform 0.005 μsec (without color-banding effects).

Video S/N Ratio: Low-band FM; better than 40 dB. High-band FM; 46 dB (USA) with the American NTSC 3.579545-MHz/sec color-subcarrier. 43 dB with the higher 4.43361875-MHz/sec subcarrier (overseas) of the PAL color television system.

Video Bandwidth: 405/525-line: ±1.5 dB to 5 MHz/sec; 625/819 line: ±5 dB to 5.5 MHz/sec.

Transient-Response (0.2 μsec sine-squared pulse, k-rating): 2%.

Moire: The amplitude of the largest spurious beat-frequency component: 3% at 3.58 MHz/sec and 5% at 4.43 MHz/sec (overseas).

Differential-Gain: 5%

Differential-Phase: 5°

Low-Frequency Linearity (*Tilt*): 2%

FM-Limiting in Playback: 55 dB

Audio Frequency-Response: ±2 dB from 50 Hz/sec to above 15 kHz/sec.

Audio S/N Ratio: Better than 55 dB.

Cue Frequency-Response: ±3 dB from 300 Hz/sec to 6 kHz/sec at 34-dB S/N ratio.

Ambient Temperature and Humidity: Between 35° and 110°F at 20-to-90%.

Ambient Noise-Level: Not more than 60 dB above threshold.

Typical Dimensions and Weight: 55″ X 72″ X 26″ and 1450 lb

Power Required: 110 to 130 V ac, 60 Hz/sec ±3% (United States), single-phase, 2- or 3-wire, 2 kW. 200 to 260 V ac, 50 Hz/sec ±5% (overseas, single-phase, 2kW).

Video Input-Level: Composite-video signal, color or monochrome: 0.5 to 1.4 V, peak-to-peak into 75 ohms, bridged or terminated.

Audio Line-Input: −10 to +8 VU into a 150- or 600-ohm balanced- or unbalanced-load. (Same for cue line-input.)

Sync (studio-sync essential for Pix-lock): 3 to 5 V peak-to-peak negative polarity.

Color-Subcarrier (for color playback or broadcast): 1.5 to 2.5 V peak-to-peak of color-standard; studio-sync derived by division from the color-standard, 3.58 MHz/sec (United States) and 4.43 MHz/sec (overseas).

RF-Copy Input: FM-input recording with FM-playback output of another machine: 1 V peak-to-peak of FM-video, 75-ohm sending and receiving impedances.

Video Line-Output: Standard composite monochrome or color television signal, 1 V peak-to-peak into 75-ohm sending and receiving-end impedances. Video, sync, pedestal, color reference-burst, and chroma levels adjustable as required.

Audio Line-Output: +8 VU maximum into 150/600-ohm balanced- or unbalanced-line.

RF-Copy Output: 1 V peak-to-peak FM-video output into 75-ohm sending and receiving-end impedances.

Sync Output: Standard sync, 3.5 to 5 V peak-to-peak from the tape composite-video signal.

4.6.9.1 Helican-Scan (Slant-Track) Color Videotape Recorders. Several Japanese models of helican-scan (slant-track) color videotape recorders (VTR), such as Sony, and one notable American version by the International Video Corporation (IVC) are popular at this time.

The special features of the IVC model-825 are described herein along with the specifications and the illustrations of the tape-format, alpha-wrap and "head-drum." This particular comparatively low-cost, space-saving, minivideo recorder is also used in broadcast work for rehearsals, preview, and editing decisions. Some Japanese versions are also used. The machine uses a 1-in. videotape and tape-reel having an 8-in.-diameter hub-size. A total of 2150 ft of tape plays back at 6.91 ips in 1 hr. Twelve-inch-long, widely-spaced slant-tracks of video-scan are recorded by means of a *single-crystal Ferrite video-head* with a guaranteed lift-time of a thousand hours.

The machine uses the preferred *alpha-wrap* as opposed to the previous *omega-wrap* around the head-drum to enable the successful head-to-tape contact for the complete field of the television-scan, with the exception of the vertical blanking-interval. The video information is blanked out anyway in this interval. For the reproduction of color by two *digital time-base correctors*, the recorder adopts a digital pulse-interval modulation technique (PIM) to provide a significant improvement in SNR and frequency-response. The digital technique minimizes the problems of the previous moire-effects in these machines, and allows the full utilization of the width of the tape for the digital video-tracks, without any cross-talk from the simultaneously recorded audio, control, and cue tracks at the two edges. The latest silicon medium-scale integrated circuits enable excellent carrier-frequency balance in the PIM and limiter circuitry.

An air-bearing tape support system is used for the tape traveling around the head-drum to allow a smooth tape-movement for optimum time-base stability, for the reproduction of color in NTSC/PAL/SECAM color television systems. As regards the tape format, comparatively wide guard-bands of 3.6 mil between the 6-mil-wide video-tracks prevent, with the capstan-servo used, any tracking problems of the recorded data due to minute differences in the alignment of the tape with respect to the video-head. This naturally facilitates an unfailing tape-interchangeability between the IVC machines. As the tape is threaded through its guide tape-path, interlocked electrical controls and dynamic-breaking safeguard against incorrect push-button switching. The portability of the minirecorder is of course a great advantage. Incidentally, at the end of its life, the video-head can be replaced in 1 min. A tape-metering capstan at the supply-side of the tape-path, minimizes tape-tension on the head-drum for precise speed-control of tape and stability of picture. A *still* picture can be displayed by a *stop-motion* facility.

The mechanical construction features modularized circuitry and four-motor servo-controlled design that assures fast lockup and picture-stabilization and provides fast-forward and rewind of full reel in 1 min. Push-button switching will allow remote control. The capstan-servo permits the synchronization of the signal to a composite-video signal or the 60-Hz power supply, if the built-in accessory, automatic digital time-base correctors are used for black-and-white and color.

Specifications, IVC-825 Helical-Scan Color Video Tape Recorder
Tape-Speed: 6.91 ips ± 0.15%
Tape and reel: 1-in.-wide, *alpha-wrap* tape on NAB-hub, 8-in. reel
Case-mounted: 24" × $11\frac{1}{2}$" × $13\frac{1}{2}$", 78 lb
Rock-mounted: 19" × $12\frac{1}{4}$" × $9\frac{5}{8}$" deep, 59 lb
Controls: momentary push-buttons: play, record, rewind, fast-forward, stop; knobs: on/off, tracking, tension, color-lock, video, cue, and audio record-levels (with video and audio level-indicators)
Power: 110 to 130 V ac, 60 Hz: 350-W maximum; 230-V/50-Hz operation as option
Video: Bandwidth: 30 Hz to 5 MHz; +1 dB to −3 dB
Signal-to-Noise ratio: 45 dB minimum
Differential gain: 5% maximum
Input level: 0.5 to 2.0 V, peak-to-peak, composite-video signal
Input Signal: Any line-standard, 60 fields, monochrome/NTSC color signal (custom PAL/SECAM); 75-ohm termination inside recorder
Outputs: Two outputs, one for monochrome and one for color; both adjusted for 1-V composite, positive-going video into 75-ohm line
Stability: <10 μsec on ramp, without automatic time-base correctors
Audio (two channels): Ch 1: 75 Hz to 10 kHz, ±3 dB
Ch 2: 250 Hz to 7.5 kHz, ±4 dB
Signal-to-Noise, audio: 40 dB peak-to-peak signal to rms noise
Interchannel cross-talk: −40 dB minimum
Flutter and wow: 0.2% maximum
Inputs (audio): microphone input is 0.2-mV minimum, 200-ohms nominal; line-input is −20 to +16 dBm (+4 dBm nominal), 600-ohms balanced or unbalanced
Outputs: Audio 1: Balanced or unbalanced into 600 ohms at +4 dBm
Audio 2: Unbalanced into 600 ohms at +4 dBm

4.6.10 Specifications of Typical Data Tape Recorders. The special features and specifications are listed herein for three tape recorders used in industrial instrumentation, airborne acquisition, and high-density data mass-storage applications.

1. *Industrial Instrumentation Tape Recorder (Bell & Howell Datatape Type 4020).* Low-cost, industrial, portable, IRIG-compatible, laboratory-quality recorder/reproducer . . . Seven bidirectional tape speeds . . . Automatic record transfer . . . Tape footage counter . . . Direct or FM signal electronics . . . Tape-lock servo . . . Peak-reading monitor meter . . . FM calibrator . . . Edge-track, voice recording, and tape-shuttle.

Tape-speeds: 60, 30, 15, $7\frac{1}{2}$, $3\frac{3}{4}$, $1\frac{7}{8}$, 15/16, and 15/32 ips
Tracks: 7 or 14

Reel size: $\frac{1}{2}$- or 1-in.-wide tape on $10\frac{1}{2}$-in. reels
Record method: Direct wideband–400 Hz to 1 MHz
FM extended band–dc to 80 kHz
FM wideband group 1–dc to 40 kHz
FM intermediate band–dc to 20 kHz
Digital–6000 bps (bit/sec) with optional PC8 analog-to-digital system
Size: 17 × 23 × $13\frac{1}{2}$ in., 120 lb
Power: 115/230 V, AC, 50/60/400 Hz

2. *Airborne Acquisition Tape Recorder, Type MARS-1000/2000.* Smallest multiband airborne recorder available accepts $10\frac{1}{2}$-in. tape reels . . . Choice of analog direct or FM signal electronics . . . Choice of side-mount or end-mount connectors for best installation facility . . . 14/28/42 tracks . . . Full IRIG compatibility . . . Optional remote monitor test unit.

Tape Speeds: 60, 30, 15, $7\frac{1}{2}$, $3\frac{3}{4}$, $1\frac{7}{8}$ ips
Tracks: 14 (2000); 28/42 (1000)
Reel size: 1-in.-wide tape on $10\frac{1}{2}$-in. reels
Record method: Direct wideband–400 Hz to 1 MHz
Direct intermediate band–100 Hz to 250 kHz
FM wideband, Group II–dc to 250 kHz
FM wideband, Group I–dc to 40 kHz
Environment: MIL-E-5400 (airborne defense)
Size: 12 × 16 × 5 in., 38 lb with tape
Power: 25 to 30 V dc, 115 W (nominal)

3. *High-Density Mass Data Storage (System 100).* Data rates to 100 MBPS . . . E-NRZ tape format . . . Error detection and correction (EDAC) for bit error rates less than 1 in 10^{10} . . . Seven tape speeds . . . Internal or external clock controls . . . Confidence monitor . . . Multiple transport synchronization for rapid throughput rates . . . LED indicators for rapid fault isolation . . . Accessory bit-error rate tester and circuit card tester.

Tape speeds: 120, 60, 30, 15, $7\frac{1}{2}$, $3\frac{3}{4}$, $1\frac{7}{8}$ ips
Tracks: 7, 14, 28/42. Bit-error rate (BER) = 1 in 10^{10} for 14 tracks
Data rates:

Tape Speed (ips)	*28 Tracks (MBPS)*	*14 Tracks (MBPS)*
120	100	50
60	50	25
30	25	12.5
15	12.50	6.25
$7\frac{1}{2}$	6.25	3.12
$3\frac{3}{4}$	3.12	1.56
$1\frac{7}{8}$	1.56	0.78

4.6.11 Automatic Time-Base Correctors (Monochrome and Vernier/Color Accessories). These two digital accessories are usually supplemented by (1) a dropout compensator and (2) velocity-error compensator for the best possible color reproduction from low-cost portable helical-scan tape recorders, to enable the playback of remotely recorded programs on broadcast. Actually, the time-base corrector setup can be used in the place of the original MATC and CATC automatic delay-line accessories in the case of the Q-CVTRs. Ampex TBC-800 is an example of the above complete setup of the digital time-base correctors (TBC).

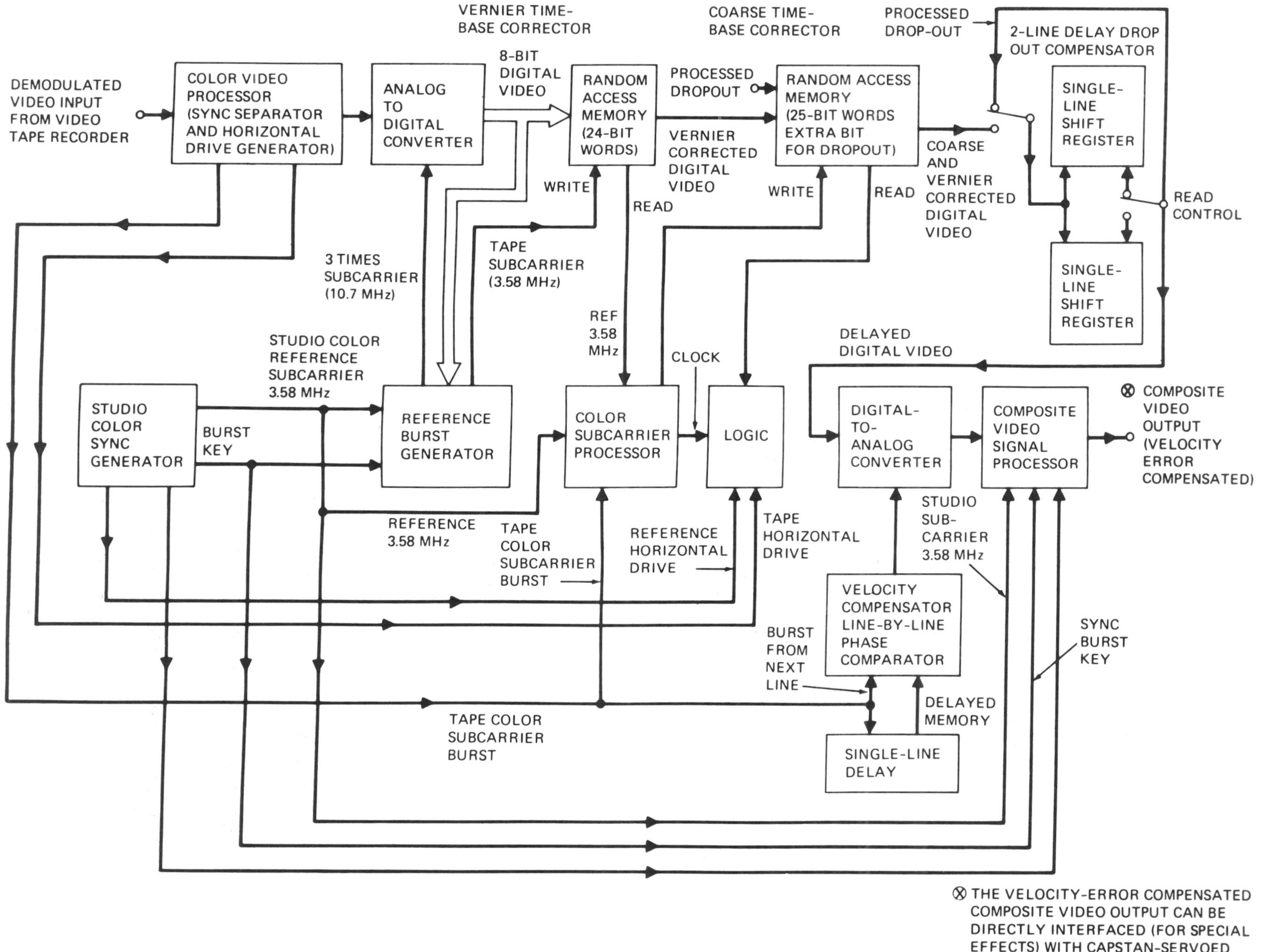

Fig. 4-33. TBC-800 digital time base corrector for videotape recorder. (Reprinted from the March 1976 *SMPTE Journal* with the permission of the Society of Motion Picture and Television Engineers, Inc. Copyright © 1976 by the Society of Motion Picture and Television Engineers, Inc., 862 Scarsdale Ave., Scarsdale, NY 10583.)

The reproduced composite-video signal is further processed in the TBC and converted from the analog-video to a coded digital-word of 8 bits (4 bits at a time). There are approximately 683 such samples per horizontal-line (including the blanking interval). The 8-bit words are generated at exactly three times the off-tape color-subcarrier rate, viz., 10.7 MHz. They are combined into 24-bit words that occur at one-third the above rate, viz., 3.58-MHz color-subcarrier rate. The data words are loaded into a random-access memory file, at a rate directly related to off-tape sync and reference-burst, which naturally include all the time-base errors formerly corrected by the MATC and CATC of the Q-CVTRs). The digital signal is then read out of the memory at a slightly different rate, which corresponds to the sync and the subcarrier of the local studio, or alternatively to a regular built-in sync and burst-generator inside the TBC. A simplified block-diagram of the TBC setup is illustrated in Fig. 4-33. Presently, the whole system can be further simplified economically by the use of a miniature (preferably I^2L) microcomputer system.

4.6.12 Home Videotape Recorders. As regards low-cost consumer products, Japan's state-sponsored research and development in private enterprise under the authority of a public-sector is highly effective in creating a large international market. Sony, Sanyo, Toshiba, Matsushita, Hitachi, Sharp, and Mitsubishi comprise the private sector, and produce products for some of the major American corporations. AKAI, VTS-110DX, 0.25-in.-wide tape, portable home-VTR, at a price that is comparable to a color television console under $1000, is briefly described herein. As an alternative to the use of a VHF converter and a television receiver, the setup may include a low-cost video camera and a portable 3-in. monitor. This typical home-VTR uses $\frac{1}{4}$-in. videotape and dual long-life single-crystal ferrite video-heads for economy. The head-drum motor is a dc servo-motor for better picture stability. Stationary full-track erase, control/audio, and side-track erase heads, and automatic audio- and video-level control facilities are included. A brushless micromotor for capstan provides constant-speed at $11\frac{1}{4}$ in./sec, to allow 20 min. of recording on an 1100-ft, 4-in. reel. Video SNR is better than 40 dB. Two rechargeable 6-V batteries allow 1 hr of continuous VTR use, at a power-consumption of 14 W.

Japan's Sony Beta-Format and Victor Video Home System (VHS), two incompatible $\frac{1}{2}$-in. wide, dual-head color-video cassette systems, are the latest products vying for *tape-economy*—which is the most dominant factor in consumer VTRs. (A Matsushita version is also available.) These machines use less tape/hour in narrow video-tracks than the previous home color VTRs, giving 2- and 4-hr playback times. The reel and head-drum mechanisms and the video-tracks of the two machines are illustrated in Figs. 4-34 and 4-35. The rotating drum of the Beta-Format VTR is a 3-assembly sandwich, with a wheel carrying two heads rotating between two stationary drums using air-friction for the tape. The lower drum has a precise ridge to guide the tape helical-lacing around the drum. The VHS version employs a rotating upper section for the head-drum. The upper section of the drum carries two video-heads attached diametrically to one another.

Luminance and chrominance signals are recorded alike in both the Beta-Format and VHS machines, but they differ in the method of eliminating the track-to-track cross-talk of the heterodyned color-subcarrier, which is recorded AM at 688 and 629 KHz/sec, respectively. The video is recorded in FM at a bandwidth of 3-to-5 MHz. The writing-speed on the Beta VTR is 6.9 meter/sec. The chroma is recorded in this machine by a *phase-inversion method by interleaving* their spectrums on ad-

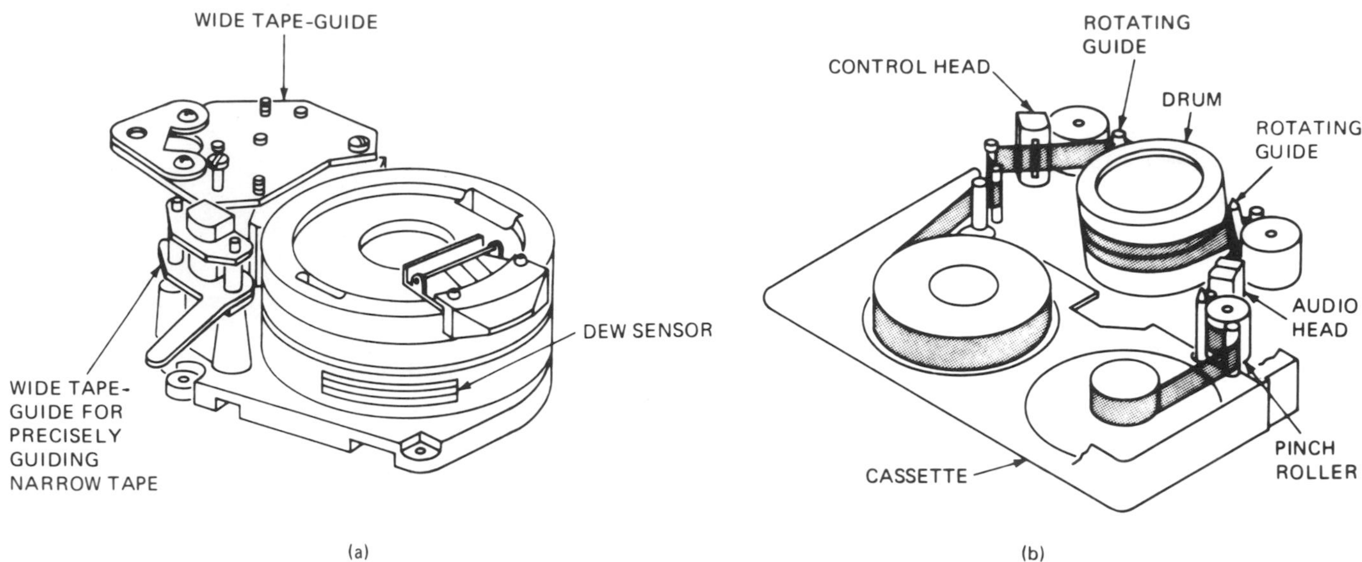

Fig. 4-34*a*. Simple loading. Victor's VHS cassette-loading configuration is new for videotape recorders. Its advantage is that it produces a somewhat simpler tape path with a correspondingly simpler arrangement of mechanical parts, so that cartridge insertion easily positions the tape. *b*, Tape-guide. Sony's wide tape-guide is designed to keep the videotape accurately in position and facilitate the use of narrow audio and control tracks. The guide also makes it possible to interchange cassettes made for Beta D Format machines. (Reprinted from *Electronics*, November 24, 1977: Copyright © McGraw-Hill, Inc., 1977.)

jacent tracks; a single-line delay-line is involved in the playback signal-processing. In the VHS version, the phase of the chroma signal (of a full television field) is *quadrature-shifted* in each successive horizontal-line, and double-limiting is used during playback. The high- and low-frequency ends of the FM signal are separated by high- and low-pass filters with fine roll-offs to minimize the noise and phase-jitter of the reference color-burst.

4.6.13 Home Video-Disks. Three video-disk systems are presently available at an approximate price of $500 to playback color programs from *motion-picture and documentary disks* into home television receivers:

1. *Philips/MCA video-disk system*, rotating at 1800 rpm/min, is scanned by a laser with a life of approximately 8000 to 10,000 hr. The disk price will vary from $2 to $10, depending on playing time. Frame repetition, slow- and fast-motion, and frame-freezing are feasible with coded frame numbers.
2. *RCA video-disk system*, rotating at 450 rpm/min., uses a gliding, capacitive, sapphire-stylus with 300 to 500 hr of playing-time. Luminance, chrominance, and audio signals are encoded in a relief-pattern that is pressed into a vinyl-disk having a metal-coating, which in turn is protected by a styrene film and a lubricating surface. Disks are priced at approximately $10 for a 1-hr. recording.
3. *TelDec (Telfunken/Decca) video-disk player* is a mechanical "skid-needle" player, with an optional system. A Zenith floppy-disk is compatible to the rigid disk of TelDec. The player is priced at $600.

All these low-cost consumer products are feasible only at the present state-of-the-art because of the successful mass-production of the MSI and the custom-LSI linear and digital integrated circuits.

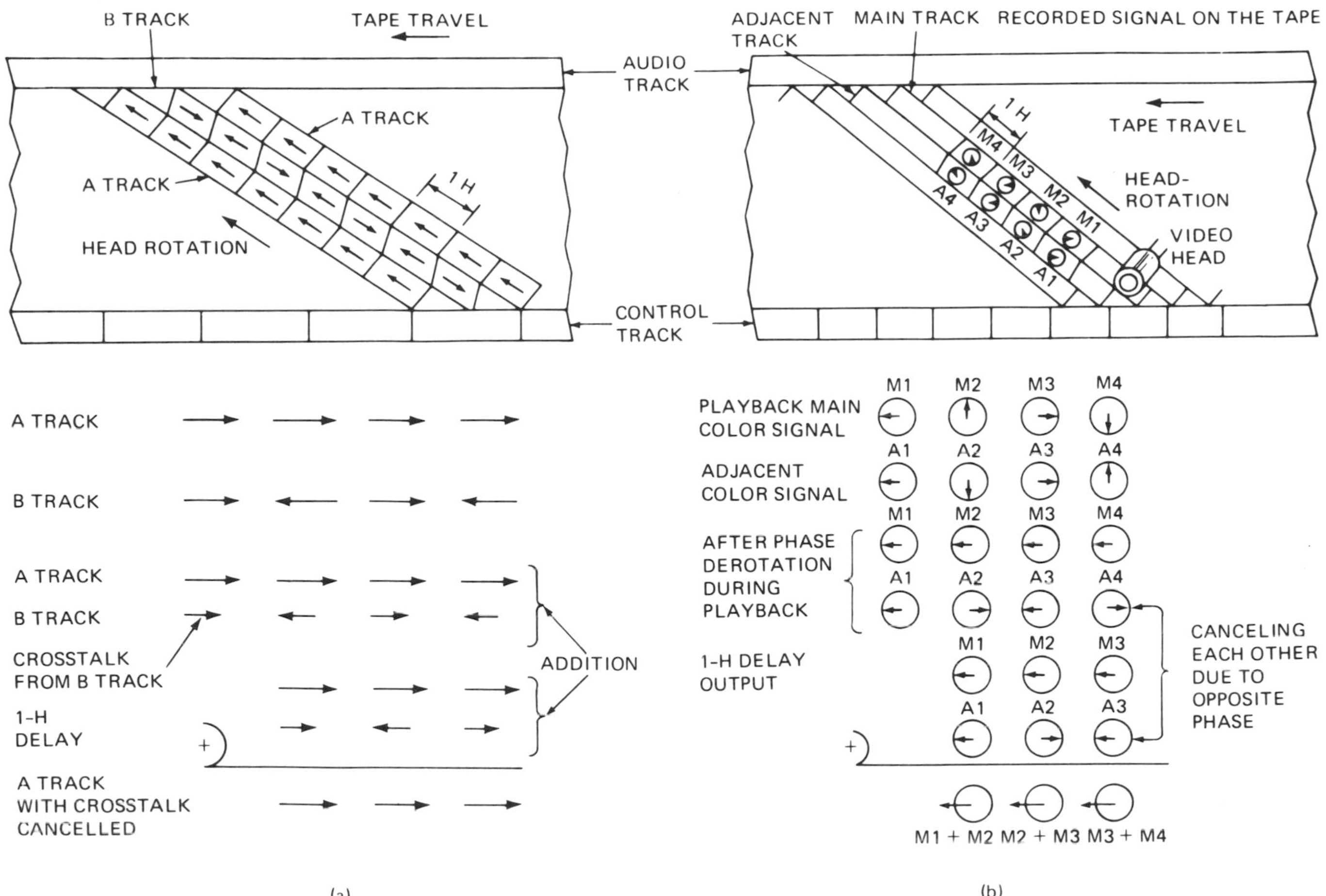

Fig. 4-35. Canceling. Crosstalk elimination techniques counterbalance phase shifts in the chroma signal. In Sony's method (a), phase of alternate horizontal lines on B track is inverted. Victor's (b) involves 90° phase shifts for each successive horizontal line. (Reprinted from *Electronics*, November 24, 1977: Copyright © McGraw-Hill, Inc., 1977.)

4.7 NOTES ON NTSC/PAL/SECAM COMPATIBLE COLOR TELEVISION SYSTEMS

The early NTSC color television studio-terminal equipment and the television receivers exhibited a few deficiencies in respect to color-stability due to the use of the power-consuming filament-heated vacuum-tubes and then the temperature-sensitive germanium transistors. With the introduction of the stable silicon bipolar and MOS semiconductor technology, the stability of color-performance has gradually improved to the present high standards, and practically all television programming in the United States is broadcast in color; in addition, the color television receiver with its superior performance is as commonplace as the black-and-white television receiver at this time. Incidentally, the recent Japanese television receiver and the low-cost videotape technology have made a fine contribution, whereas the American research-and-development electronics technology in this field remained stagnant during the early 1970s.

In the United States, the development of the color videotape recorder has progressed to such an extent that videotape and film color program material can be uniformly telecast as indistinguishable from live color-camera programs. The

Western Hemisphere and Japan have naturally settled down to the AM/PM, NTSC compatible color television system (first pioneered by the RCA Corporation) at the American 525/60 line-standards. The NTSC signal is represented by the following equations:

$$E_M = E_Y^1 + \underbrace{[E_Q^1 \sin(\omega t + 33°) + E_I^1 \cos(\omega t + 33°)]}_{\text{Luminance + chrominance}} \tag{1}$$

Below a bandwidth of 0.08 MHz, it can be represented by the equation (as applicable to the general color television receivers):

$$E_M = E_Y^1 + 0.493(E_B^1 - E_Y^1) \sin \omega t + 0.877(E_R^1 - E_Y^1) \cos \omega t \tag{2}$$

(Luminance + color difference components)

where the quadrature color-component of Eq. 1,

$$E_Q^1 = 0.41(E_B^1 - E_Y^1) + 0.48(E_R^1 - E_Y^1) \tag{3}$$

and the in-phase color-component of Eq. 2,

$$E_I^1 = -0.27(E_B^1 - E_Y^1) + 0.74(E_R^1 - E_Y^1) \tag{4}$$

and the monochrome (luminance or brightness component),

$$E_Y^1 = 0.59E_G^1 + 0.3E_R^1 + 0.11E_B^1 \tag{5}$$

(Green + red + blue) of the basic color triangle

The CCIR in Western Europe and the rest of the world use 625/50 line-standards due to their 50-Hz/sec power distribution system, and they have opted to the modified NTSC, viz., PAL (Phase Alternation by Line) color television system. However, France, the USSR, and some of the East European countries have opted for the French-modified FM/AM, SECAM (Sequentiale Couleaur a Memoire) at the 819 or 625/50 line-standards. At a time when worldwide color television broadcasting is both feasible and at times commonplace by synchronous satellites, these three standards may be somewhat inconvenient. However, all international broadcasts, and video/audio/data communications by satellites are presently conducted by digital pulse-code modulation (PCM, and frequency or time division multiplex FDM/TDM), digital computer networks, and short-hop microwave/optical and fiber-optics communications; thus, the inconvenience referred to is not a major problem for international interchange of programs. Highly reliable digital switching systems, digital video synchronizers (as replacement to the previous studio gen-lock synchronizing facilities), digital time-base correctors, and digital scan-converters supplement the previous analog/linear studio-terminal equipment to facilitate international broadcasts at this time. In the near future, linear and digital LSI (large-scale integrated) microcontrollers and microcomputer systems using solid-state memories will take over these tasks and considerably improve the situation in respect to economy, power-consumption, program-automation, and reliability at the television studios throughout the world.

4.7.1 PAL Color Television System. In the NTSC system, as a result of improper maintenance or transmission errors, or multipath interference or differential-phase problems in terminal equipment, an objectionable phase-shift in hue used to show up as drift in colors during the early 1960s. In the CCIR-PAL color television system, the phase of the chrominance information and the reference-burst on the alternate horizontal-lines is switched over by 180°, so that a positive-error under the above conditions on a horizontal-line is cancelled by the corresponding negative-error on the succeeding line. The *algebraic summation* of the two opposite phase-errors (up to a subjective extent of 70°) can actually take place by optical integration on the screen of the kinescope (CRT) and in the viewer's eye or by means of the regular one-line delay-line in a deluxe television receiver at an extra cost of about 12%. That is, PAL receivers do not require a hue-control for adjusting the color. However, color reproduction of the PAL tape-signal involves the same complexity as the NTSC signal, requiring the monochrome and color automatic time-corrector accessories. Also, the phase-errors in the NTSC signal-display will turn out as saturation-errors in the PAL signal-display.

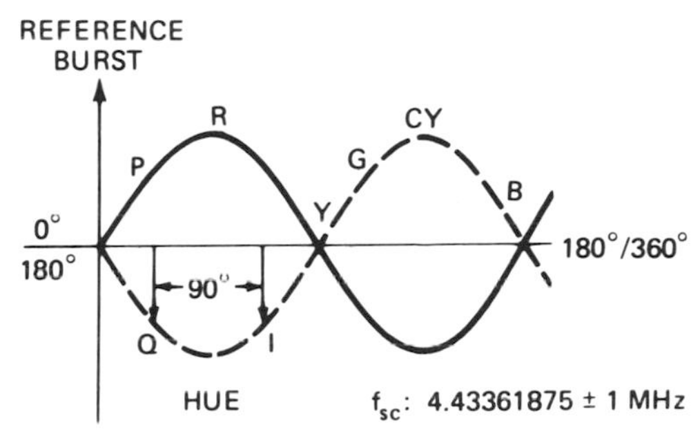

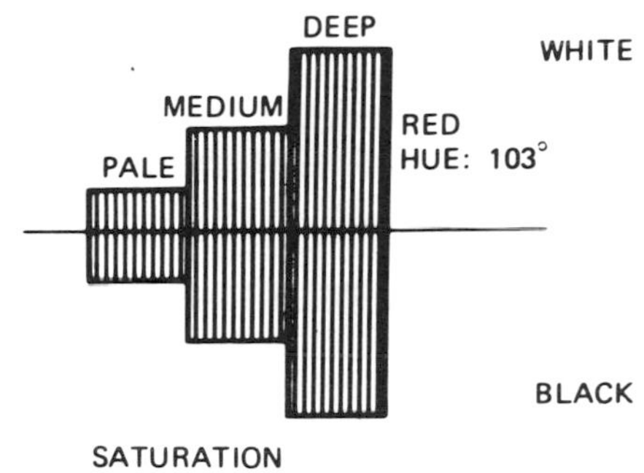

PAL: $\dfrac{f_{sc} - \frac{1}{2} f_{field}}{284 - \frac{1}{4}} = 15{,}625$ Hz/sec

10 to 12 cycles

PAL subcarrier phase-shifted 180° on every alternate horizontal line.

NTSC: Vestigial-sideband bandwidth: 1.25 MHz | **PAL**: 0.75 MHz
Main sideband (Luminance): 5.5 MHz | 5 MHz
8 cycles only: $f_{sc}(2/567)$
Video to sound spacing: 6 MHz | 5.5 MHz

The PAL television signal is basically represented by the following equation:

$$E_M = E_Y^1 + E_Q^1 \sin(\omega t + 33°) \pm E_I^1 \cos(\omega t + 33°)$$

The polarity of the $E_I^1 \cos(\omega t + 33°)$ chroma component is positive during the odd-number lines of the first and second fields, and the even-number lines of the third and fourth fields. The horizontal synchronizing reference-burst is also subject to the above condition.

4.7.2 SECAM Color Television System. Principle: The eye *detects* changes in *brightness* much easier than shift in *hue*. In the NTSC color system, full brightness (luminance) and color (chrominance) information are transmitted simultaneously. The chrominance information is much greater than what the eye can actually resolve. So, in the French SECAM color system, only the brightness signal is transmitted in full on excessive number of 819 horizontal-lines, and the FM-AM chrominance information is transmitted *on every other horizontal-line to make vertical color resolution half* that of the NTSC color. A single line-interval delay-line in the television receiver stores the chroma for reinsertion on the succeeding line for full color display.

When compared to the NTSC color-signal display at 525 horizontal-lines, it is normally difficult to detect the reduced color-resolution due to the preceding technique. The expediency therefore minimizes errors that the NTSC system could exhibit under improper maintenance conditions. Because of the above simplification in color-processing, the requirements of time-base stability are identical to those of the black-and-white picture. That is, in the case of the VTRs, a simple pix-lock control system will do (without MATC and CATC accessories) for reproduction of SECAM color. No saturation and hue controls are required in the preset home television color receivers. Standard brightness and contrast controls will do as in the black-and-white receivers. However, the deficiencies in the SECAM system will show up in the distant fringe-areas at low-level subcarrier levels. Also, the partial black-and-white receiver incompatibility is noticeable since the subcarrier cannot be automatically suppressed as in the case of the NTSC and PAL color television systems. The SECAM and PAL color television receivers generally cost about 10% higher than the NTSC color television receiver. *Note:* NTSC Television *receivers* use the simplified *color-difference signal* approach and not the I and Q components.

The SECAM color television signal can be represented by the following equation:

$$E_M = E^1 + A \cos(\omega_{SC} + E^1_C \, \Delta\omega_{SC})t,$$

where E^1_C = color-difference signal D^1_R or D^1_B

$$D^1_R = -1.9(E^1_R - E^1_Y) \text{ and } D^1_B = 1.5(E^1_B - E^1_Y)$$

$\Delta\omega_C/2\pi$ = the frequency-deviation corresponding to the unity amplification of the preemphasized color-difference signal

A, as a function of $E^1_C \, \Delta\omega_{SC}$, determines the amplification of the chroma signal. In

PURPLE RED YELLOW GREEN CYAN BLUE

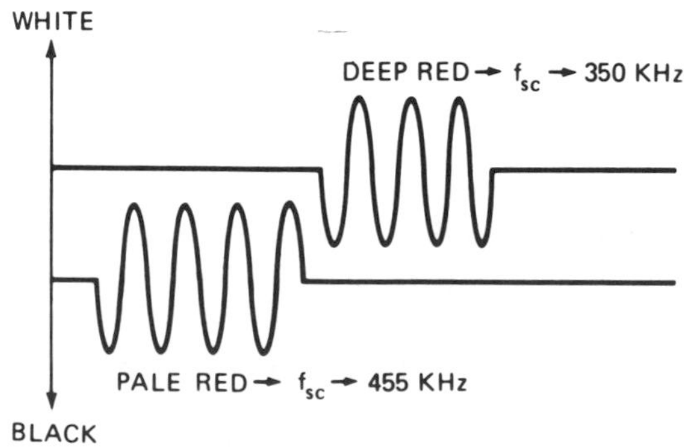

SECAM: FM Subcarrier: 4.43361875 MHz: (f_{sc})
Vestigial-sideband: 1.25 MHz
Horizontal line frequency ≈ 15,625 Hz = (f_{sc}/284)
Main sideband: 0 to 6 MHz
Video to sound spacing: 6.5 MHz
Subcarrier phase-shifted 180° every alternate line
Reference Burst: Subcarrier during six lines of each field-blanking interval. Modulated in frequency and amplitude to correspond to a standard color difference signal D'_B or D'_R. Starts 5.7 ± 0.3 μsec after the leading-edge of the line-pulse.

the absence of color, E_Y^1 max = 1; then A = 0.1 (SECAM); A = 0.115 (PAL); A = 1 (NTSC).

$$D_R^1 \text{ and } D_B^1: 1 \text{ MHz} < 2 \text{ dB down,}$$
$$1.5 \text{ MHz} > 5 \text{ dB down,}$$
$$2 \text{ MHz} < 20 \text{ dB down.}$$

4.8 VERTICAL APERTURE COMPENSATION ALIAS VIDEO CONTOUR ENHANCING

The electro-optical television pickup tubes, such as plumbicons, vidicons, image-orthicons, and CCD cameras, in general can physically reproduce sharp changes in image contours through a gradual transition in amplitude, and not as a high-speed sharp amplitude-transition, as a result of the following factors:

1. The finite size of the electron beam that scans the charge image.
2. The continuous flow of charge from the minute target elements at high potential (due to the highlights in the scene) to the adjacent elements at lower potentials.
3. The electron-beam bending at sharp picture transitions.
4. Some unavoidable loss of resolution in the associated optics.

Horizontal aperture compensation in television cameras is well known as a *peaking* facility for the higher video frequencies, and it is generally accomplished with a linear phase relationship by using an open-circuited broad-band video delay line. The signal reflected back, up along the delay line, is appropriately added to the original signal to accomplish the implied *horizontal aperture compensation.* For vertical resolution, the line structure and the vertical field-interlace of the picture frame introduce a factor of discontinuity into the vertical picture detail. Experimental evidence indicates that the vertical aperture compensation is as important as the horizontal compensation in producing a noticeable improvement in the picture sharpness by adding a definite amount of "snap" to both the black-and-white and color television pictures. With the introduction of high-quality ultrasonic delay lines (acoustic glass and elastic magneto-strictive delay lines) of one line-interval scan-time equal to 63.49 μs for a 525-line television system, vertical aperture compensation has become a practical proposition for general implementation in television broadcasting. As a result, the *contour enhancing* facility is presently a common feature in cameras and tape recorders.

4.8.1 Horizontal Video Contour-Enhancing. The horizontal video aperture compensation is a regular facility in all television cameras, and it is commonly implemented by the emitter bypass capacitor boost followed by an RC-phase equalization network. When vertical aperture compensation is provided for contour-enhancing effect, independently controlled simultaneous horizontal contour-enhancing is included to present a uniform and balanced picture-crispening effect in both horizontal and vertical resolution. Under low light levels, however, the horizontal enhancement is reduced automatically to obtain an improved signal-to-noise ratio (SNR), while the video signal is boosted in contrast; the vertical aperture compensation is not restricted under this condition since it does not contribute to the SNR-figure.

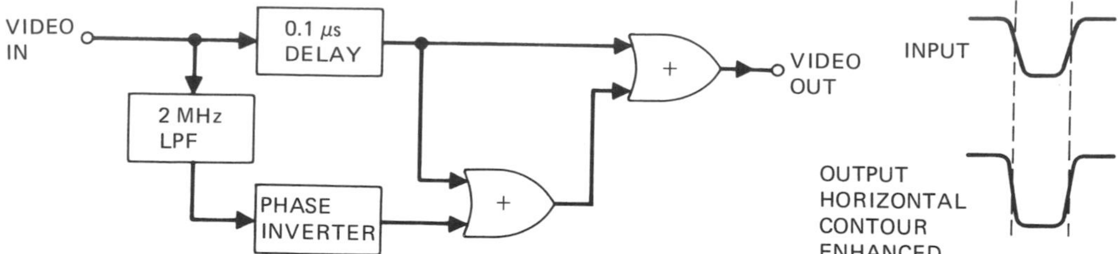

Fig. 4-36. Contour enhancing (horizontal).

Horizontal video contour-enhancing is generally accomplished by one of the two following techniques to insert the desired degree of crispness. To alleviate the SNR problem, the first technique (Fig. 4-36) uses nonlinear circuitry to decrease the apparent rise-time of an isolated video step-input transition, while the contour correction signal is restricted in bandwidth to a suitable pass-band. The overall signal-processing effect gives an illusion of larger bandwidth and hence higher horizontal resolution. Repetitive patterns representing frequencies beyond cutoff are exempt from this effect. The nonlinear element affords larger boost for larger differences in amplitude and vice versa. That is, the technique consists of merely adding to a waveform with a slow transition a second waveform representing the difference between the desired waveform and the original waveform. A delay line is included in the forward path to compensate the effective delay in the filter network.

In the second technique (Fig. 4-37), the signal is double-differentiated, and hence slightly time-delayed, to provide a correction signal. The RC equalizing networks associated with the following diodes control the duration of the added spikes in either direction of the original transition in amplitude. In both techniques, a 0.1-μs delay line compensates for the delay associated with the above signal processing.

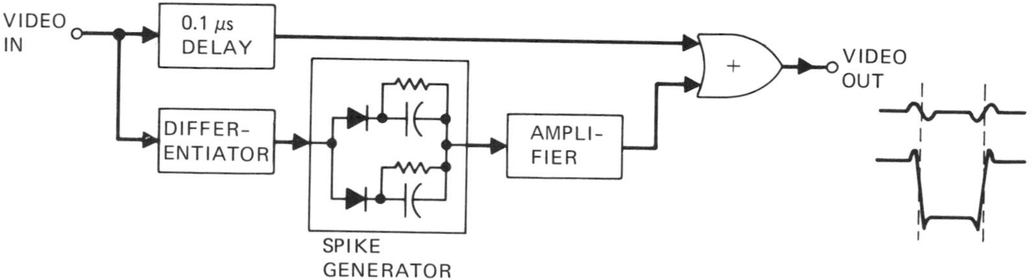

Fig. 4-37. Horizontal contour enhancing, second method.

4.8.2 Vertical Contour Enhancing (Vertical Aperture Compensation). The original technique of contour enhancing is shown in Fig. 4-38. The technique compares three successive lines (in a field) with each other and develops a correction signal; it is then added to the second line. The waveforms shown in Fig. 4-39 illustrate the way the large-area vertical contours are enhanced.

Glass acoustic delay lines, which are one line-interval long and manufactured by both Corning Glass Works and the Microsonics Corp., can be operated up to a carrier frequency of 40 MHz to accommodate the amplitude-modulated video signals of standard bandwidth. The broadcast color cameras use carriers ranging from 10 to 20 MHz. Since these acoustic delay lines do not pass the low video frequencies, the video signal amplitude modulates the carrier, which is in turn transmitted into the

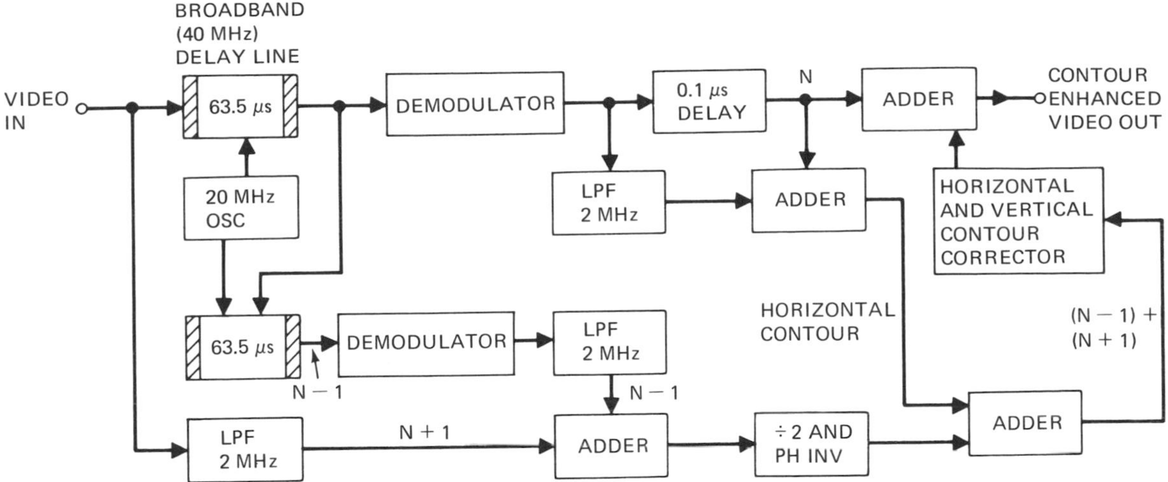

Fig. 4-38. Vertical contour enhancing.

acoustic delay line via a switchable transducer. An identical transducer at the other end feeds a high-gain radio-frequency amplifier to compensate for the excessive insertion loss of 40 to 60 dB in the delay lines. The amplified radio-frequency signal is then demodulated to reproduce the original video signal delayed by one line-interval. Loss and bandwidth characteristics are interdependent; choice of line terminations, carrier frequency, and transducer capacity determine the proper operation of this device. The best attenuation of spurious signals is achieved in medium delay-time ranges and at higher carrier frequencies.

The European PAL color-television receiver employs a one-line-interval delay line for its phase alternation on a line-to-line basis. The Telefunken VL3 glass acoustic delay line used in this application ($10) also makes a suitable choice for low-budget closed-circuit television applications. As an example of specification, the VL3 delay

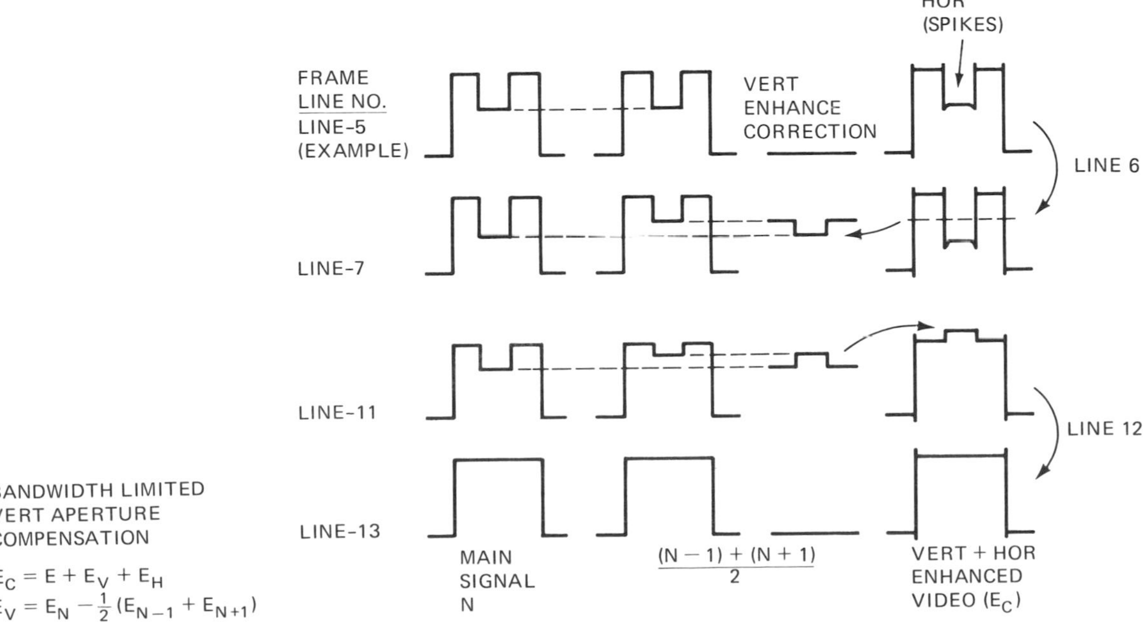

Fig. 4-39. Vertical contour enhancer waveforms.

line, with a delay of 63.94 μs (corresponding to the CCIR line frequency rate 15.625 kHz), operates into 100-ohm terminations at the European color subcarrier frequency 4.433619 MHz, and the upper and lower sidebands cover the bandwidth 3.4 to 5.2 MHz. This passband characteristic allows video contour enhancing up to a video bandwidth of 1 MHz, and that is adequate for large- and medium-area picture information observed from normal viewing distances. (Approximate dimensions of PAL delay line: 5 X 2 X 0.5 in., 0.3 lb; delay ±0.005 μs, 10° to 60°C.)

4.8.3 A New Contour-Enhancing Technique Using a Single One-Line-Interval Delay Line Instead of Two. A modulator and demodulator are essential anyway whenever an ultrasonic (acoustic glass) delay line is used. Under these circumstances, if the modulator and demodulator are replaced, respectively, by a doubly balanced modulator and a synchronous detector for handling two quadrature signals at 4.43 MHz (such as the I and Q signals in the NTSC and CCIR color encoders), one 63.5-μs delay line can function to provide the requisite two-line-interval delay as well. This is easily accomplished by recirculating the video signal, which is first transmitted once through the delay line, after a 90° phase-shift, as shown in Fig. 4-40. This technique will be cost-effective in the case of economy (black-and-white and color cameras used in closed-circuit applications). However, design precautions are necessarily involved due to the potential quadrature cross-talk problems in a combined amplitude and phase-modulation system using the 4.43-MHz subcarrier. The upper sidebands extend to 0.77 MHz above the carrier (in a vestigial fashion like the I-component in a color encoder), while the lower sidebands carry the regular 1-MHz video components.

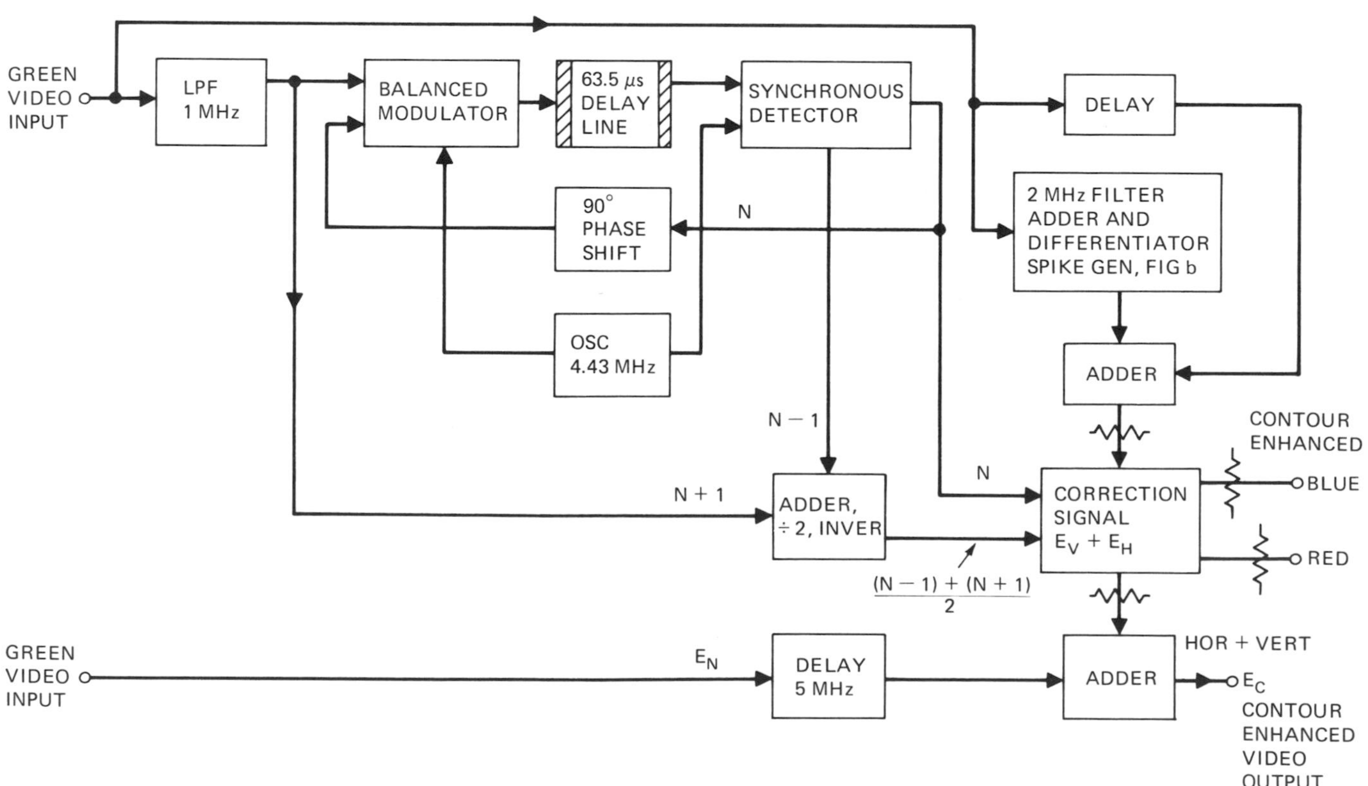

Fig. 4-40. Vertical aperture compensation, contours out-of-green (using a single delay line (63.5 μs) in the place of the two delay lines of Fig. 4-38.

4.9. SUPERHIGH-BAND VIDEO RECORDING

Signal-to-noise ratio is improved by wide-band frequency deviation in order to relieve intermodulation cross-talk problems between the interleaved luminance (monochrome) and chroma components of the NTSC color signal. The allowable extent of frequency deviation is mainly determined by the FM-video magnetic-head resonant frequency. Early Q-CVTRs had to use narrow-band (low-band) color because of the restricted low carrier-frequency due to lower FM-video resonance. The latest single-crystal ferrite videoheads with resonances above 11 MHz allow implementation of high-band and even superhigh-band video recording and thus enhance signal-to-noise ratio and quality to such an extent that the live color television camera and Q-CVTR (or a new BCN continuous-field twin "1-$\frac{1}{2}$-head," 1-in. tape-width, helical-scan color VTR) signals cannot be recognized from one another in a color television receiver. The BCN helical-scan machine uses a digital time-base corrector for reproduction of color in a so-called type-C format.

The FM-video frequency bands of the monochrome, low-band color, high-band color (SMPTE), and superhigh-band color are illustrated in the following diagram.

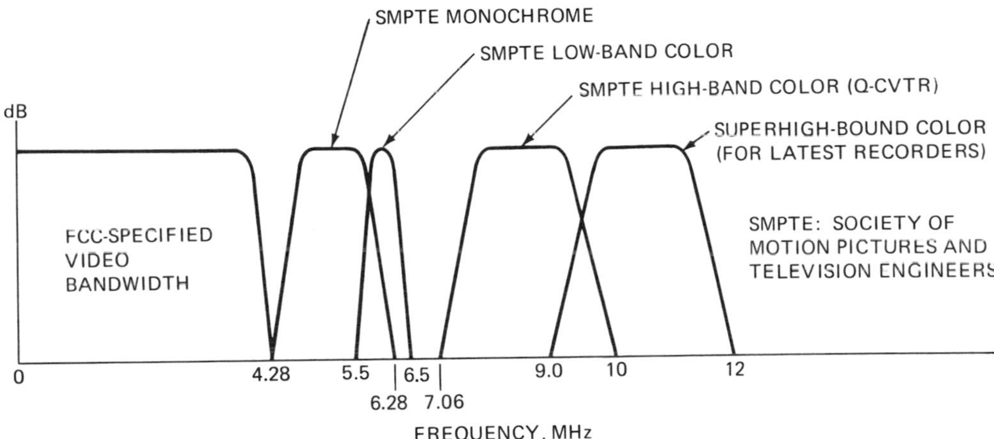

5
Modern Computer-Controlled Digital Control Systems, Microcomputers, and Digital Filters

5.1 INTRODUCTION—SIGNAL PROCESSING AND DATA-FORMATTING IN DIGITAL CONTROL SYSTEMS

In view of (1) the inconvenience and problems the control systems engineer faces in applying the multiplicity of available theory to complex sampled-data control systems in commercial practice, (2) the unexpected explosive developments in the large-scale integration of LSI microprocessors/microcontrollers and solid-state LSI memories, and (3) the ready availability of high-resolution 10- and 12-bit analog-to-digital and digital-to-analog converters, and 4- and 5-bit quantizers for phase comparators, two potentially successful and convenient approaches are naturally preferred for the development and design of complex digital control systems at this point in time.

In regard to these control systems, the obvious route is oriented in the direction of the microcontrollers and the microcomputer systems using EPROMs (electronically programmable read-only memories) and high-density RAMs (random-access memories). The pertinent software or firmware may take the nonvolatile form of mass-memory as tape cassettes, floppy-disks, and magnetic-bubble memory, or in some instances CCD (charge-coupled device) memory. With this kind of memory, the complexity of the control system is not an insurmountable problem for the design of the relevant digital filters, which are associated with the computer-controlled digital control system—once the requisite experience is gained for developing such software at the assembler or compiler level. Actually, this is the procedure presently adopted in the case of the satellite and other data communications systems. The approach via digital filters incidentally eliminates the requirement of evolving complex analog transfer functions, because the digital filters function on the principle of maximizing the signal-to-noise ratio (SNR) of the data-formatted control information or minimizing the integral squared error. The computer-processed data are finally decoded and phase-discriminated against reference pulse-data to provide the control information. The power plant, if any, is usually controlled via an ac carrier-servo, operating off the chopper-processed dc control information. The problems of theoretically handling the several nonlinear elements in practical control systems are mainly bypassed by the substitute applicability of the high-resolution analog-to-digital converters operating at $\pm\frac{1}{2}$ least-significant-bit at 10 and 12 bits.

Where simpler control systems are encountered, as the second approach, the LSI linear integrated circuits using the integrated injection logic (I^2L) as phase-lock loop (PLL) chips will take over the task to successfully accomplish the control function at minimal power dissipation and space requirements. In the case of the PLL, the requisite precision of stability is achieved by the provision of a compensation network external to the PLL-chip. The manufacturers themselves may provide the necessary data and aids for the design of the compensation networks. Telemetry and time-division multiplex data-communications via telephone networks, fiberoptic cables, and synchronous satellites will contribute as indispensable links for the provision of complex worldwide open-loop digital control systems. Even at this writing, remotely controlled digital control systems in defense and aerospace applications such as radar, remotely piloted vehicles (RPVs), aircraft attitude control, and satellite TT&C (telemetry, tracking, and control) are already commonplace; however, they are all high-budget items. But in the near future, with the advent of the low-budget microcomputer systems, such facilities will gradually become highly popular in low-cost commercial applications. Computers, control systems, and TDMA (time-division multiple-access) communications networks will make the C^3 (command, control, and communications) systems that will literally unite the world technologically.

In order to acquaint the reader with the basic foundations of the data-processing systems in this connection, the current chapter presents a brief insight into the concepts and techniques available for interfacing audio, video, and wide-band pulse information with modern high-capacity mini- and-micro-computer systems that can be linked to central large-scale computers via communications networks. Both LSI microcontrollers (as multiplexers) and microprocessors (as CPUs) will play a dominant role in future complex digital control systems. It is difficult to identify control and communications systems as separate entities in various applications. Direct Digital Control (DDC) using software of a high-level language in real-time high-speed multiprocessor systems is the order of the day.

5.1.1 Bits, Bauds, Bandwidth, Information, and Redundant Bits. The *bit*, as abridged from the binary digit, is a "1" or "0" (or a square pulse at 50% duty-cycle). It is the basic binary coding unit used in the computer machine (*object*) language. This is irrespective of the various other codes—such as octal, hexadecimal, decimal, unit-distance (Gray), redundant error-checking, and correction codes—used for software in the computer interface circuits for programming in low-level or high-level *source* languages appearing in assembler and compiler logic-packages. Also, as a tag or a *flag* or a specific timing, the individual digit may take the form of a rectangular pulse.

The term *bit/sec* is used as a measure of the number of information bits per second. It is derived from Shannon's theorem, which states that if a continuous function of time is sampled at a constant rate slightly in excess of twice the highest frequency component in the signal function, then the function can be reconstructed exactly. The only penalty is that the delay and amplitude changes effective in the reconstructed signal must be appropriately compensated. The minimum sampling-rate "twice-the-bandwidth in hertz per second" is called *Nyquist rate.*

If there are 100 bit/sec, the pulse repetition interval PRI is 1/100 = 0.01 sec, and the frequency is 1/0.01 = 100 Hz. And, as a square pulse (at 50% duty-cycle), the pulse-duration is 0.01/2 = 0.005 sec or 5 ms.

A *baud* is the unit used for signaling speed in data communications; it is the reciprocal of the shortest signal element: 1/0.005 = 200 bauds in the preceding example.

If T is the period of a return-to-zero train of pulses, the baud-rate, as shown, is twice the bit-rate. Incidentally, the baud is related to the bandwidth as bit/sec of transmission-speed, signal-power, and noise; that is, the baud-rate determines the type of the transmission channel required.

The teletype-unit code of 7 bit/character uses 25-msec *start*, 30-msec *stop*, and five 20-msec information pulse/channel for a total of 155 msec for one character. Therefore, the corresponding bit-rate = 1/0.155 X 5 = 32.3 information bit/sec.

If the space between the stop of one character and the start of the next is 40 msec, the baud-rate is identical but the *rate of information flow* is reduced:

$$\frac{1}{(0.155 + 0.40)} \times 5 = 25.6 \text{ information bit/sec}$$

The information bits include the *parity-bits* generally used for *error-detection.* Channel-capacity in bit/sec: $C = B \log_2 (1 + SNR)$

$$\text{(i.e.) } SNR_{dB} = 10 \log (2^{C/B} - 1) \ldots \text{Shannon's relationship}$$

where

B is the bandwidth in hertz

$$SNR = \frac{\text{Average signal power}}{\text{Average noise power}}$$

Owing to limitations in transmission channels, bandwidth-limited processing becomes necessary in a binary system, as shown in the following diagram:

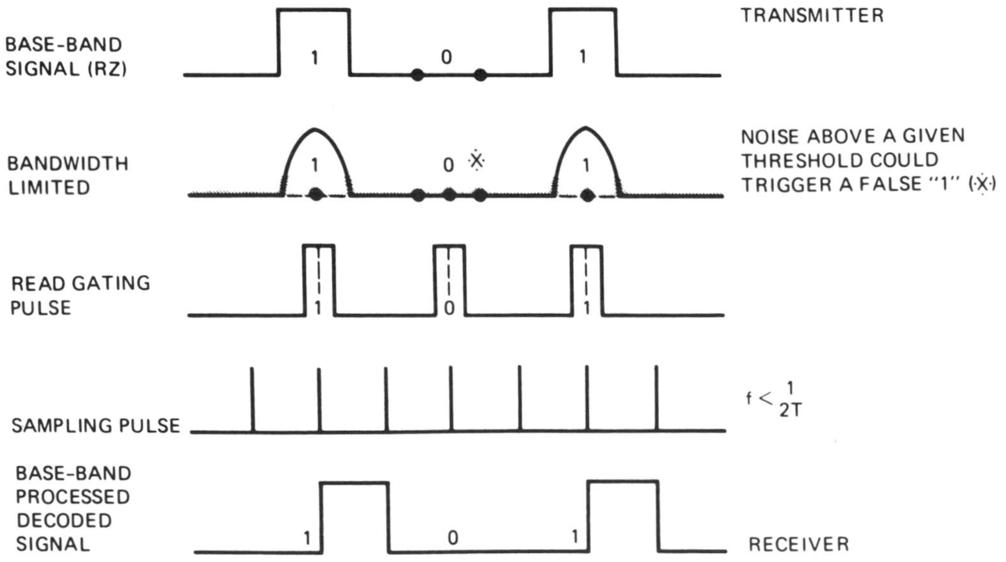

The noise-power increases as the square-root of bandwidth. This fact is taken into consideration when message enhancement is achieved by redundant-bit transmission in, for example, *error-correcting codes* with both information and redundant bits.

A parity-bit check with a *distance-2 code* will enable only the detection of one erroneous bit in a code word. "Distance-d" is the number of elements in which the code symbols (words) differ.

A distance of at least 3 ($d \geqslant 3$) is essential for the correction of one error-bit in a code symbol. For correction of two error-bits, d is 5. The popular Hamming code is a distance - 3 code capable of detecting and correcting one error-bit in a word.

Complement parity is another technique for reliability in data transmission. A command word of 8 or 16 bits is transmitted in practice in conjunction with its complement (of inverted 1's and 0's) on twisted pairs, via line-drivers and line-receivers for random or impulse noise-immunity.

5.2 PULSE-FORMATTING (QUANTIZERS AND ANALOG-TO-DIGITAL CONVERTERS)

The synchronizing clock-pulse at rates of up to 20 MHz for high-speed digital computers makes the basis for pulse-timing in synchronous digital processing systems. (Asynchronous digital computer systems operate without this basic synchronization by using alternative methods of logic.) The clock-frequency is divided by down-counters to provide other timings at lower rates. Figure 5-1a illustrates the various pulse formats normally encountered in practice. The sampling-time may be either the leading- or trailing-edge of the clock-pulse at several rates. Unlike the continuously varying analog, video, audio, and carrier signals, the digital data signal change in only discrete levels with rise- and fall-times, as governed by bandwidth in Hz/sec. The binary machine language used in digital computers naturally operates at two discrete levels, 1's and 0's (High and Low or True and False). A fast rise-time of course corresponds to higher bandwidth. Digital systems are comprised of either the directly coded *base-band* data signals on a network transmission line or a frequency- or time-division-multiplexed pulse-modulated signals; teletype (Baudot) and high-speed data communications systems are typical examples, respectively. Analog information is converted by a sample-and-hold circuit to a digital format by sampling the continuous signals at discrete timings, and converting each instantaneously sampled-level to a corresponding binary number to formulate a pulse-code modulation (PCM) system. The common PAM (pulse-amplitude modulation), PCM, PDM (pulse-width modulation), and PPM (pulse-period modulation) waveforms for the base-band and analog signals are illustrated in Fig. 5-1b.

5.2.1 Quantizers. The latest digital filter techniques in digital control systems naturally require the use of finite number of bits to express the filter state-variables as coefficient values. The continuous analog signal is therefore digitized in amplitude and timing by the sampling process. The process introduces samples of amplitude at several levels and involves quantization-noise and quantizing-errors.

An analog *ramp* (or sawtooth) signal is converted to a PCM format by momentarily holding each instantaneously sampled voltage level in a sample-and-hold circuit. The quantized step, quantum E_q is actually a pulse-amplitude modulated signal. A 7-bit code PCM signal represents (2^7=)128 possible amplitude levels or M channel-

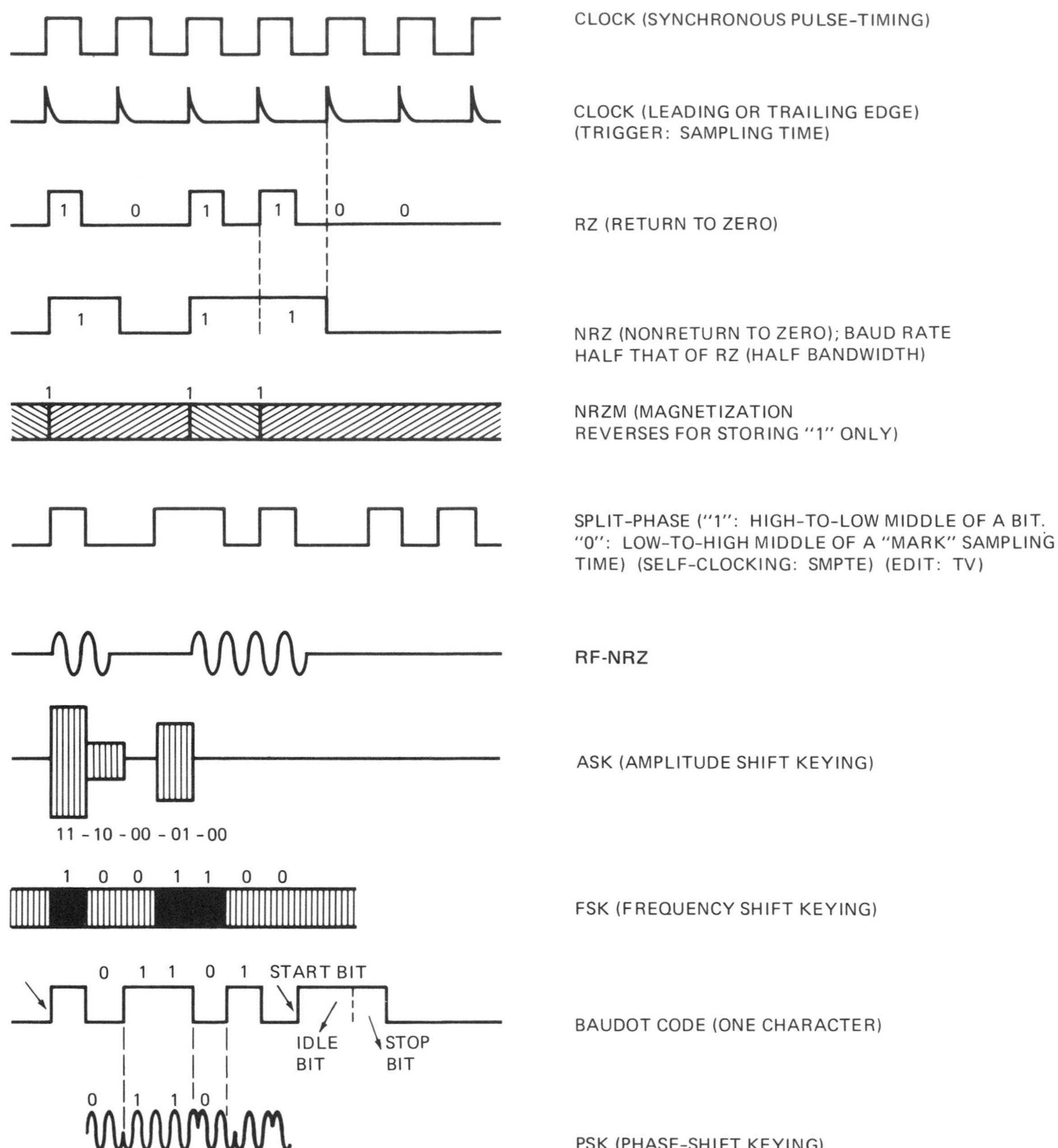

Fig. 5-1a. RZ, NRZ, NRZM, split-phase, RF-NRZ.

symbols or words. The code is a set of characters that, taken in order, correspond to the set of levels.

In data communications networks, the source-symbols transmitted at a rate,

$$R = \frac{1}{T} \log_e M = k/T \text{ bit/sec}$$

with k source-bits or $\log_e M$ binary signals (or 2^k "M-ary" channel-symbols = 2^7 in the above example), are processed in a *block encoder* to provide the channel signals.

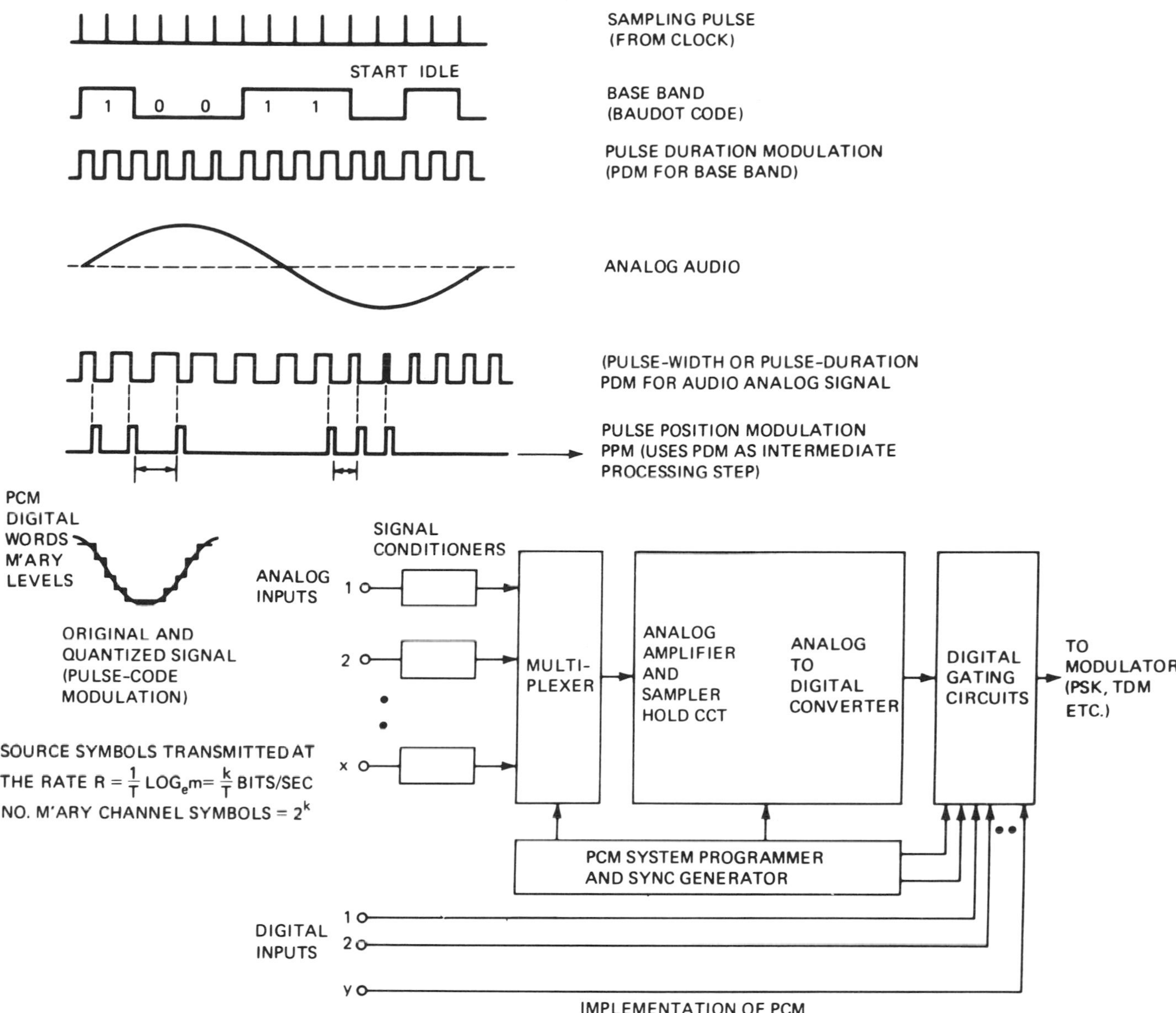

Fig. 5-1b. Base band and analog signals. PDM, PPM, and PCM.

The maximum quantizing error or noise for 128 levels in PCM is given by

$$E_q/2 = \frac{1}{2}\text{ quantum} = \frac{1}{2 \times 128} = 0.4\%$$

The quantized signal is thus 99.6% accurate, as far as the original analog information is concerned. The basic sample-and-hold circuit and a typical 4-bit quantizer with its binary formatting circuit are illustrated in Fig. 5-2.

The subjects of quantizing and coding in PCM are extensively treated by K. W. Cattermole in his book on pulse code modulation.* Although it is more complex than other pulse modulation methods, it is the most appropriate technique for digi-

**Principles of Pulse Code Modulation*, Iliffe Books, London, 1969.

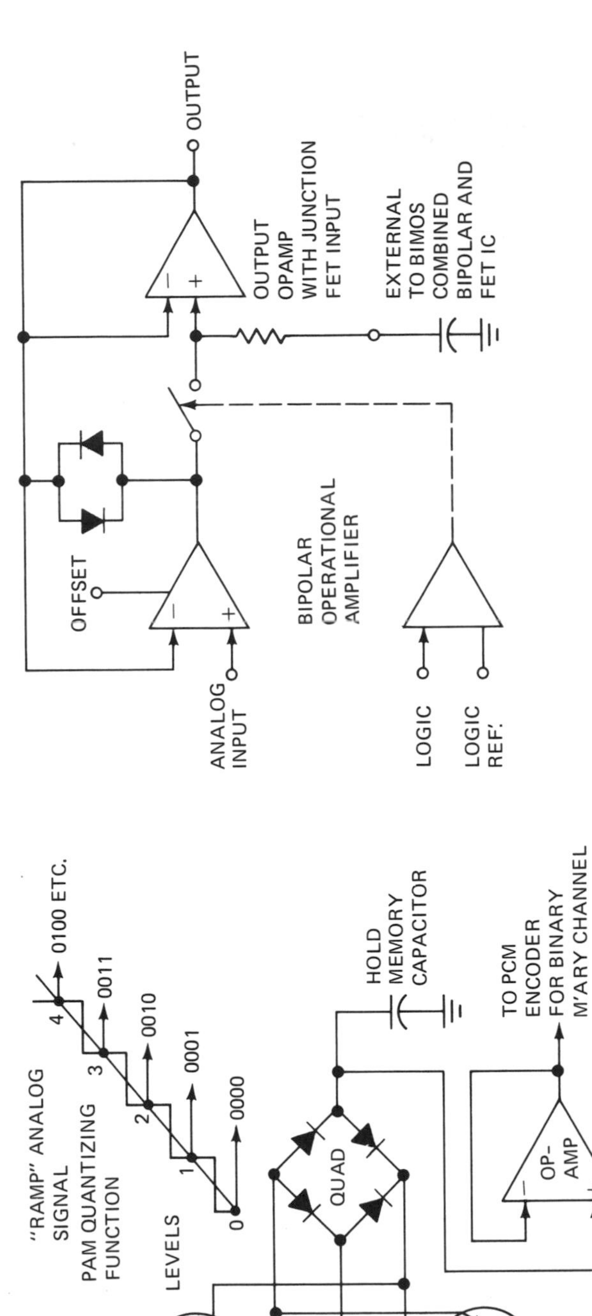

BIFET-MONOLITHIC SAMPLE AND HOLD CHIP (NATIONAL SEMICONDUCTOR CORP.)

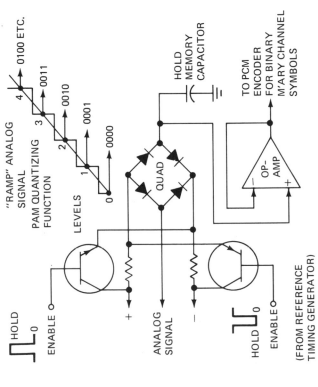

SAMPLE AND HOLD CIRCUIT

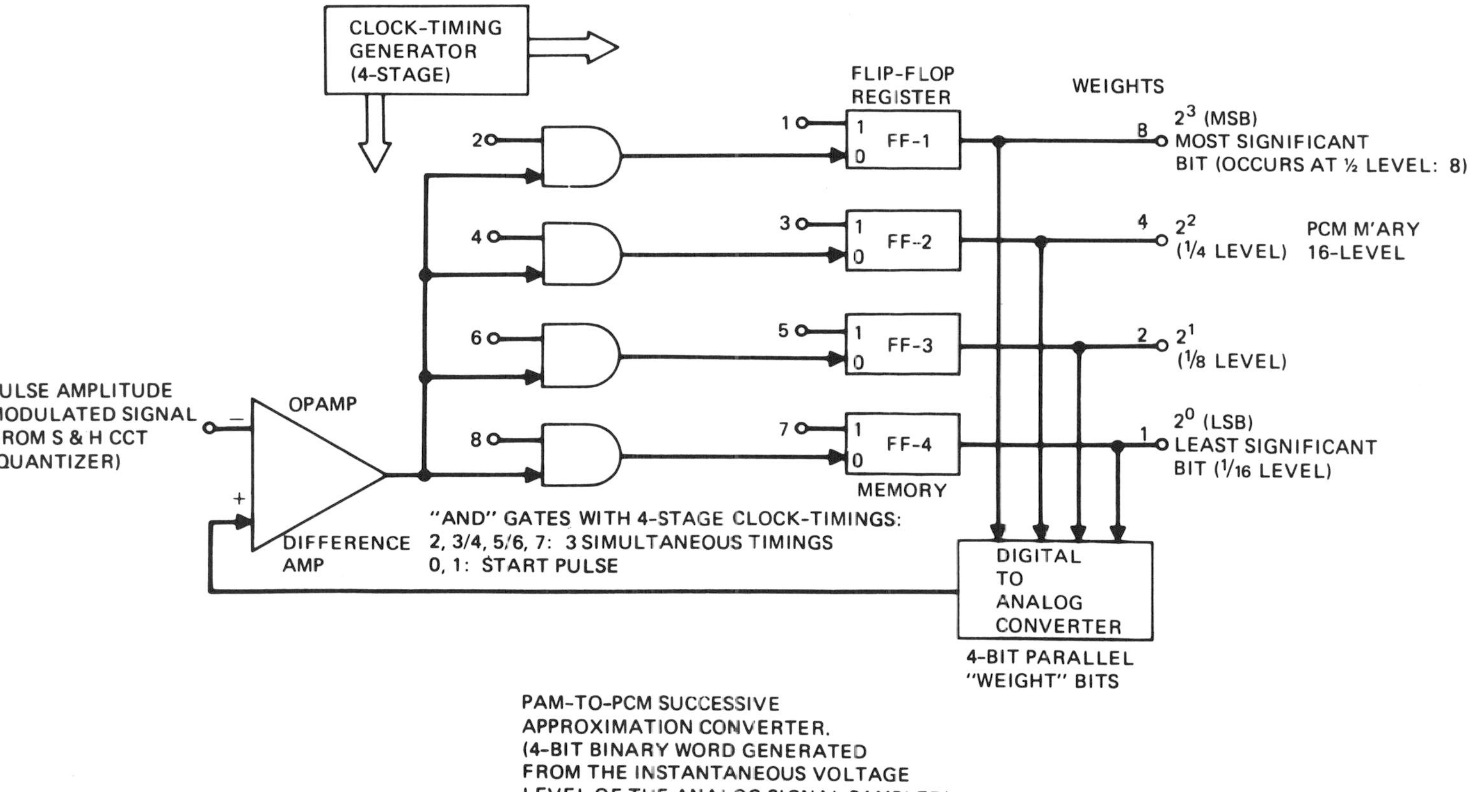

Fig. 5-2. Sample-and-hold circuit and 4-bit PCM encoder.

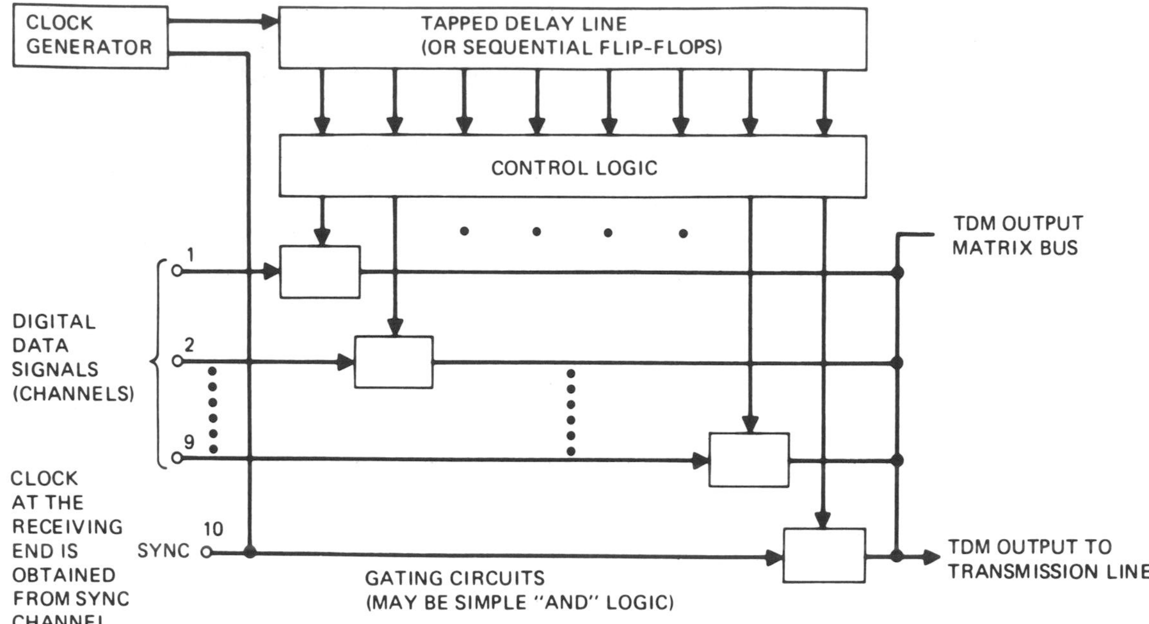
CLOCK GENERATOR
TAPPED DELAY LINE (OR SEQUENTIAL FLIP-FLOPS)
CONTROL LOGIC
TDM OUTPUT MATRIX BUS
DIGITAL DATA SIGNALS (CHANNELS)
1
2
9
CLOCK AT THE RECEIVING END IS OBTAINED FROM SYNC CHANNEL
10
SYNC
GATING CIRCUITS (MAY BE SIMPLE "AND" LOGIC)
TDM OUTPUT TO TRANSMISSION LINE
(SEE FIG. 5-35 FOR A TYPICAL DATA PACKET.)
A TDM MODEM WILL CONSIST OF A SWITCHER (COMMUTATOR) OF THE ABOVE CONFIGURATION AT BOTH THE CODING TRANSMITTING AND THE DECODING RECEIVING ENDS.
THE CLOCK RATE MUST BE AT LEAST EQUAL TO SAMPLING-RATE TIMES THE NUMBER OF CHANNELS SAMPLED (10 AS ABOVE).

Fig. 5-3. Concept of time-division multiplex.

tizing audio and video signals for modern computer-controlled time-division multiplex transmission. Each instantaneously sampled voltage level (*quantum*) of the ramp-function is held momentarily by the *sample-and-hold circuit*, as the conversion to PCM takes place with pulse-amplitude modulation as an intermediate step. The midpoint of each amplitude-step is the corresponding sampling time. Figure 5-3 illustrates the basic principle of the TDM (time-division multiplex) used for data transmission. The clock-rate must be at least equal to the sampling-rate times the number of channels being sampled, including the sync channel.

5.2.2 Analog-to-Digital Converter. The analog-to-digital (A/D) converter (ADC) is an encoder that accepts an analog signal A_i and reference voltage A_r, and provides an approximated digital output with a resolution of

$$\chi \equiv (A_{i/A_r})$$

The binary approximate form

$$\frac{A_i}{A_r} \approx (a_1 2^{-1} + a_2 2^{-2} + \cdots a_n 2^{-n})$$

The quantization error $\Delta A_i = A_i/r^n$ where r is the radix and n is the number of digits in χ. A sampling error is also present if the input-frequency is high relative to the conversion-rate of the A/D converter.

There are several types of A/D converters, including:

1. *parallel-feedback* with a comparator consisting of a summing circuit (for input and feedback analog signal from a D/A converter) and a threshold voltage to output an error of either polarity, which drives a bank of counters that register and store the updated count,
2. *successive approximation*, which makes n successive comparisons at each conversion between the input analog signal value and a time-dependent feedback voltage
3. *servo* with the digital output changing in such a direction as to reduce the error from a detector
4. *indirect converter*, which derives the digital format from an intermediate (partly or wholly) analog-step
5. *simple-ramp comparison* with a precise ramp generator and comparator.

The successive approximation A/D converter is the most popular type; at a clock frequency of 500 kHz and 12-bit words, the device using MSI requires 1 to 24 μsec for the conversion of small or large values of the input signal. A 12-bit successive approximation A/D converter is shown in Fig. 5-4. The timing generator output is properly sequenced to trigger bistable-latches and inhibit readout until conversion is complete.

5.2.3 Digital-to-Analog Converter. A simple system block-diagram of a high-speed current-output, 12-bit/3-decade, DAC-3 of Data Devices Corporation is shown in Fig. 5-5 along with the principle of a digital-to-analog (D/A) converter (DAC). The specifications of the DAC-3 follow:

High resolution: 12-bit or 3-decade BCD
Output current: up to 2 mA (full-scale)
Unipolar or bipolar output: (pin-programmable)
Choice of BCD, binary, off-set, or 2's complementary binary input format
Temperature coefficient: 15 ppm/°C
Linearity: ±0.0125%
Settling time: 500 nsec to 0.1%
Reference: built-in, temperature-compensated.

TTL, and DIP-pin compatible, Teledynes fastest DAC, type 4060, has a settling-time of 85 nsec. It is suitable for radar-pulse digitizing, video digitizing, simultaneous sample-hold systems, CRT displays, and waveform analysis.

5.2.4 Transducers. A transducer is a device that converts a change in some form of energy (such as heat, radiation, sound, pressure, motion, or angular velocity), or a natural phenomenon or event, into a change of a variable appearing as a measurable electrical parameter. And, transducers are in general affected by more than one of the preceding forms of energy.

An *active transducer* produces a variable output voltage or current that can be measured with or without external power, while a *passive transducer* produces a change in a passive parameter such as resistance/inductance/capacitance. All feedback analog/continuous or digital control systems derive their basic control information from some form of transducer, governing displacement, flow, force, humidity,

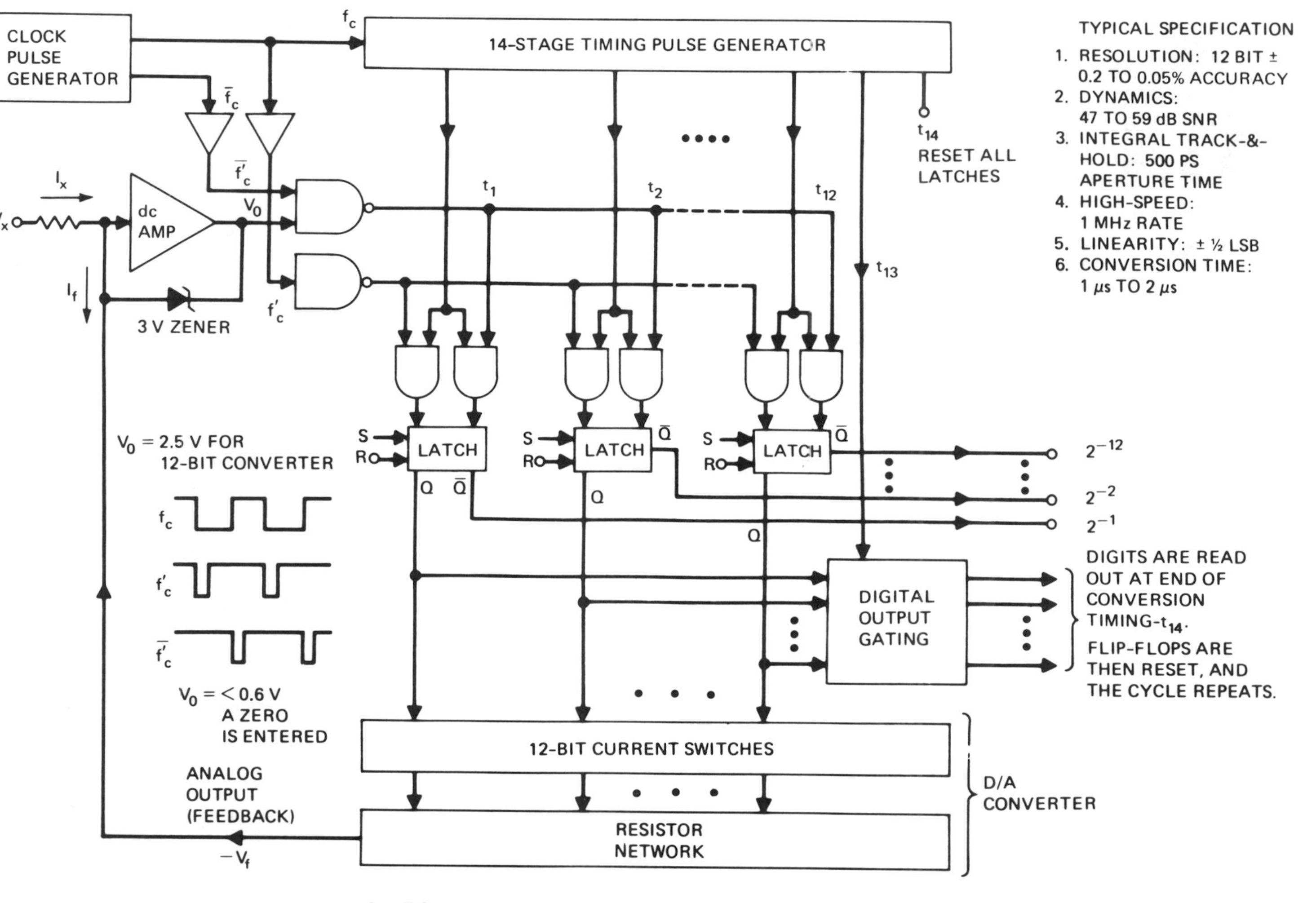

Fig. 5-4. Twelve-bit analog-to-digital converter.

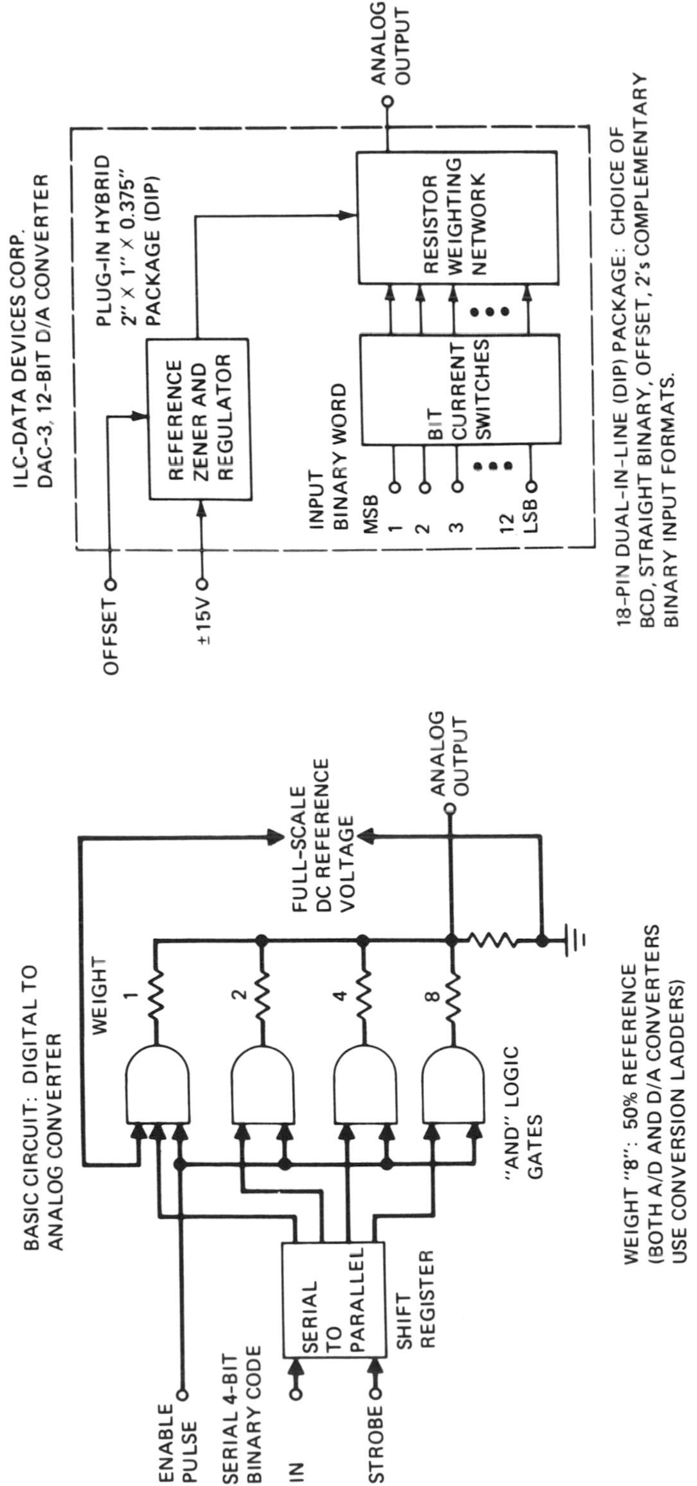

Fig. 5-5. High-resolution digital-to-analog converter.

level, light, mass, pressure, angular velocity, acceleration, strain, temperature, thickness, velocity, viscosity, etc. After sufficient amplification, this basic analog control information is converted to the digital format in digital systems.

Piezoelectric crystals, photoelectric (phototransistor), thermoelectric, magnetoelectric, electronic, and electrochemical and radioactive voltages/current generating devices are active transducers, while a variable R/L/C device, or a magnetostrictive element or a differential transformer make a passive type. All of these devices are mostly nonlinear beyond a linear range, and the operation of the control systems is generally approximated to the linear range or compensated appropriately.

5.2.5 SMPTE Control Time-Card. The Society of Motion Pictures and Television Engineers (SMPTE) has standardized videotape and audio time-control codes for the editing of the videotape and the synchronization of the audiotape. The code includes a biphase mark, with BCD time of day, BCD frame-count, and optional binary-word information. The code, consisting of 80 bit/television frame or 40 bit/field at 2400 (= 60 × 40) bit/sec, is recorded on the audio-cue track of the videotape recorder; the bandwidth is consistent with normal forward tape speeds. In editing, the code is used to search for the required edit timings during fast-forward and reverse shuttling of the tape, requiring a bandwidth approaching 100 kHz. The sync word (bits 4 through 79), with twelve 1 bits followed by 0 and 1 indicate *end of frame* and forward direction of tape. The reverse direction is indicated if twelve 1's are followed by two 0's.

Starting with bit-number 0, the 80 bits are sequenced thus: frame units (4), first binary group (4), frame tens including "drop-frame" (4), second binary group (4), seconds units (4), third binary group (4), seconds tens (3), unassigned address bit-number 27, fourth binary group (4), minutes units (4), fifth binary group (4), minutes tens (3), unassigned address bit-number 43, sixth binary group (4), hours units (4), seventh binary group (4), hours tens (2), unassigned address bit-number 58, fixed bit-0 (1), eighth binary group, and sync word from bit-number 64 to bit-number 0 of next frame. NTSC color video requires a drop-frame because the NTSC color frame-rate is 29.97002/sec in the place of the monochrome 30/sec. To allow for this difference, two frames are dropped every minute except every tenth minute.

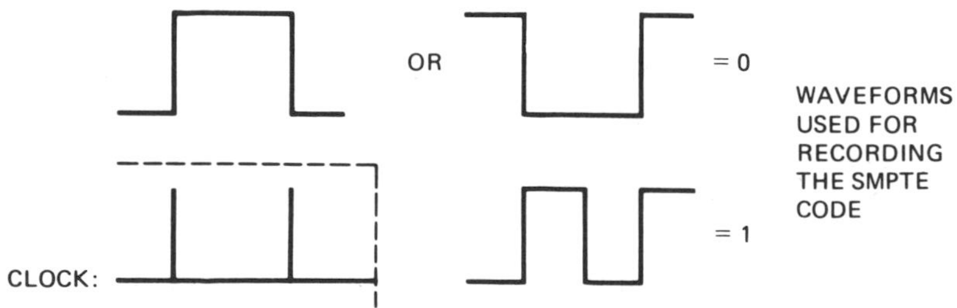

ANSI (American National Standards Institute), ASCII encoding chart is modified for television broadcast use in video character generators. Alphanumeric characters may be made to "crawl," usually from right to left across the bottom one-fourth of the picture to present statistics, or the display may be made to "roll" usually from bottom to top for information of "credits." Only capital letters are used for the ASCII decoding chart, and columns 6 and 7 are deleted. In column 1, row 3, CRSR abbreviates the word *cursor*—which when turned ON indicates the position of the next character to be typed on the keyboard; it appears as a white square on the monitor. The cursor control actually replaces BS (back space), HT (horizontal tabulation),

Table 5-1 ASCII Encoding Chart (7-bit code)

				Bits								
				$b_7 \rightarrow$	0	0	0	0	1	1	1	1
				$b_6 \rightarrow$	0	0	1	1	0	0	1	1
				$b_5 \rightarrow$	0	1	0	1	0	1	0	1
b_4 ↓	b_3 ↓	b_2 ↓	b_1 ↓	COL / ROW	0	1	2	3	4	5	6	7
0	0	0	0	0	NUL	DLE	SP	0	@	P		p
0	0	0	1	1	SOH	DC1	!	1	A	Q	a	q
0	0	1	0	2	STX	DC2	"	2	B	R	b	r
0	0	1	1	3	ETX	DC3	#	3	C	S	c	s
0	1	0	0	4	EOT	DC4	$	4	D	T	d	t
0	1	0	1	5	ENQ	NAK	%	5	E	U	e	u
0	1	1	0	6	ACK	SYN	&	6	F	V	f	v
0	1	1	1	7	BEL	ETB		7	G	W	g	w
1	0	0	0	8	BS	CAN	(	8	H	X	h	x
1	0	0	1	9	HT	EM	)	9	I	Y	i	y
1	0	1	0	10	LF	SUB	*	:	J	Z	j	z
1	0	1	1	11	VT	ESC	+	;	K	[	k	{
1	1	0	0	12	FF	FS	,	<	L	\	l	¦
1	1	0	1	13	CR	GS	-	=	M	]	m	}
1	1	1	0	14	SO	RS	.	>	N	^	n	~
1	1	1	1	15	SI	US	/	?	O	_	o	DEL

VT (vertical tabulation), FF (form feed), and GS (group separator) in the standard ASCII code. *Home* position is the upper left of the *page.* The function of the crawl and roll are selected on a separate group of control buttons on the keyboard unit. FLASH, OPEN, CLOSE, ETX (end of transmission), SND MSG, SND LINE, ERASE, RM (stop code enters memory), HOME, LOAD REQ, and CRSR are the major functions included in this chart. The ASCII Encoding Chart and the ASCII Decoding Chart for television are presented in Tables 5-1 and 5-2.

5.3 DIGITIZING AUDIO SIGNALS

Digitized audio (and video) signals can be conveniently "stored" in LSI (large-scale integrated) PROMs (programmable read-only-memories) and RAMs (random-access memories), so that the information can be read-out or manipulated on any convenient time-base desired. With LSI microprocessors and microcontrollers becoming extremely popular at economy prices, it is natural such information will gradually become a part of the worldwide PCM (pulse-code modulation) data control and communications systems.

A digitized voice channel of the telephone network with a frequency response up to 3000 Hz is shown in Fig. 5-6. Sampling at 8-kHz rate is converted to a 7-bit binary word from "0" on (2^7=)128 levels at 10-V peak audio level and 0.078-V

Table 5-2 ASCII Decoding Chart for Special Television Use

			Bits	$b_7 \rightarrow$	0	0	0	0	1	1	1	1
				$b_6 \rightarrow$	0	0	1	1	0	0	1	1
				$b_5 \rightarrow$	0	1	0	1	0	1	0	1
b_4 ↓	b_3 ↓	b_2 ↓	b_1 ↓	ROW \ COL	0	1	2	3	4	5	6	7
0	0	0	0	0	NUL		SP	0	`	P		
0	0	0	1	1	SOH	FLASH	!	1	A	Q		
0	0	1	0	2		OPEN	"	2	B	R		
0	0	1	1	3	ETX	CRSR	#	3	C	S		
0	1	0	0	4	EOT	CLOSE	$	4	D	T		
0	1	0	1	5			%	5	E	U		
0	1	1	0	6			&	6	F	V		
0	1	1	1	7	BEL		'	7	G	W		
1	0	0	0	8	CRSR LEFT		(	8	H	X		
1	0	0	1	9	CRSR RIGHT	EM	)	9	I	Y		
1	0	1	0	10	LF	ERASE	*	:	J	Z		
1	0	1	1	11	CRSR DOWN		+	;	K	[		
1	1	0	0	12	CRSR UP	CRSR HOME	,	<	L	~		
1	1	0	1	13	CR	CRSR NEW LINE	-	=	M	]		
1	1	1	0	14		SND	.	>	N	^		
1	1	1	1	15		SEND MSG	/	?	O	_		

quantum. The quantizing error is 0.39% of 10-V peak. *Signal-to-*(quantizing) *distortion ratio* (sdr) at maximum input level or full load is given by

$$\text{sdr} = 6n + 1.8 \text{ dB} = 43.8 \text{ dB (if } n = 7)$$

An sdr of 43.8 dB is approximately equivalent to 1.2% of *total harmonic distortion* (*thd*)—a fine tolerance from the distortion point of view. Where high-quality broadcast music is involved, 12- to 14-bit binary word is in order with an sdr of 85.8 dB (n = 14). When the signal is 40 dB below full level, the thd at 1% is still fine.

For broadcast music at 15-kHz bandwidth and a sampling rate of 33 kHz and 14-bit binary words, the transmission rate is 462 kb/sec (33000 × 14); the bandwidth achieved at an SNR of 40 dB is given by:

$$\text{BW} = \frac{\text{channel capacity}}{\log_2 \text{SNR}} = \frac{462 \text{ kb}}{\log_2 10000} = 35 \text{ kHz}$$

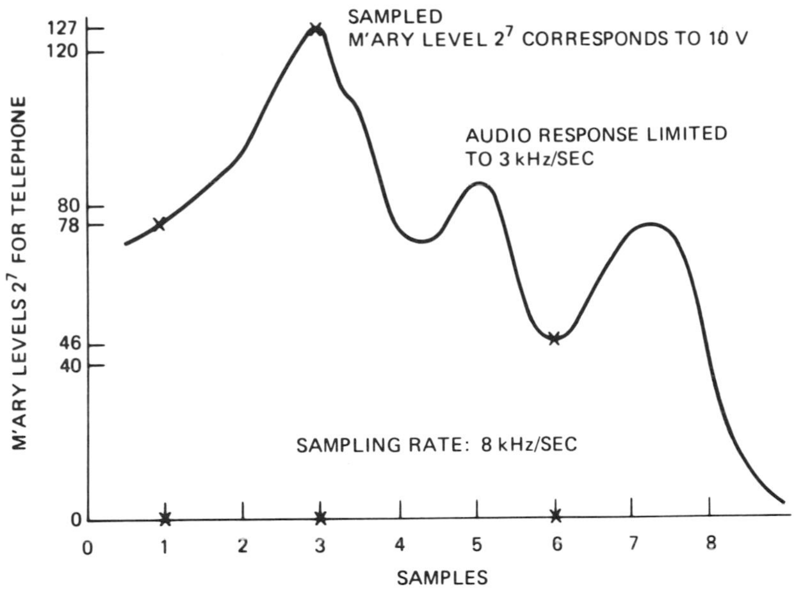

PCM: AUDIO M'ARY LEVELS IN BINARY WORDS REPRESENTING BINARY WORD HIGH'S POSITIVE-GOING AND LOW'S NEGATIVE-GOING:

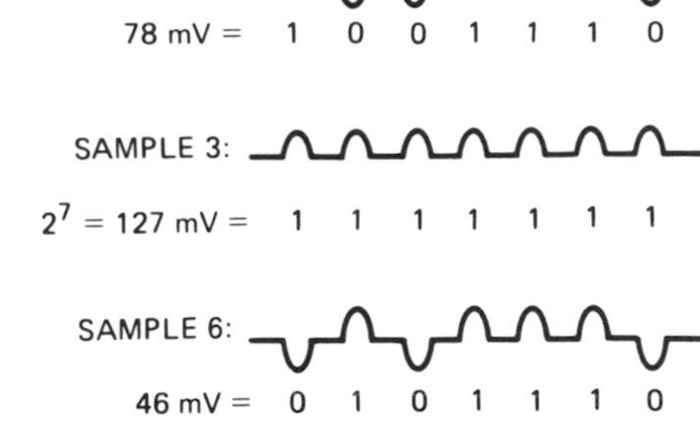

2^6 2^5 2^4 2^3 2^2 2^1 2^0

LEVEL	VOLTAGE
127	10 V
78	6.14 V
46	3.62 V
*1	0.078 V
0	0 V

*EACH QUANTUM LEAST SIGNIFICANT BIT (LSB) = 78 mV ∴ MAXIMUM QUANTIZING ERROR = 39 mV. (½ Q)

SIGNAL/DISTORTION = 6 X BITS/WORD + 1.8 dB; TOTAL HARMONIC DISTORTION = 1.2% (THD)
TELEPHONE: AN SDR OF (6 X 7) + 1.8 = 43.8 dB CORRESPONDS APPROXIMATELY TO 1.2% THD
TELEVISION: AN SDR OF (6 X 14) + 1.8 = 85.8 dB CORRESPONDS APPROXIMATELY TO 0.01% THD
(M'ARY LEVELS = 2^{14} FOR HIGH-FREQUENCY RESPONSE TO 15 kHz AND A CORRESPONDING HIGH SAMPLING RATE OF 33 kHz/SEC)

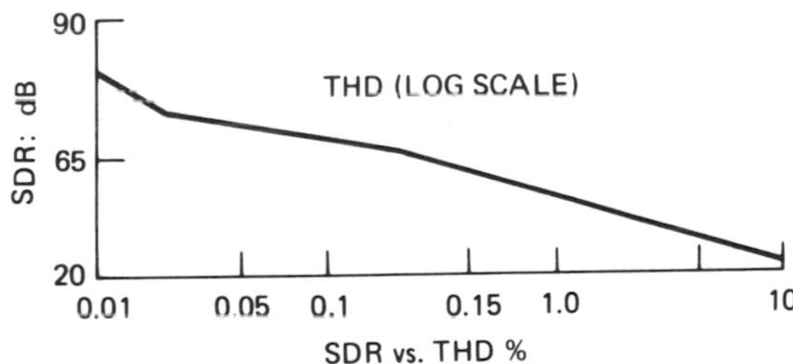

Fig. 5-6. Digitized voice channel in a telephone network. (*Courtesy of Sam's Publications*)

In practice, 10- to 12-bit M-ary levels are used for a channel bandwidth of 20 kHz. Bandwidth is further reduced by using such concepts as *companding* to truncate the three MSBs during soft passages and the three LSBs during loud passages, and automatically reinstating them during decoding. Program-modulated noise due to quantizing, etc., is eliminated by the usual technique of preemphasis and deemphasis prior to quantization. Another technique, *digital audio delay*, which is adjustable, is a common *special effects* feature in sound systems to minimize the effects of distracting echoes and loss of intelligibility.

5.4 DIGITAL AUDIO FOR TELEVISION

The accompanying diagram indicates how the binary audio data are carried by a 5.5-MHz digital subcarrier when audio accompanies the digitized video in a television transmission system. The color subcarrier 3.579545 MHz is divided by 104 to provide a 34.42 kHz sampling rate at a 13-bit word level for four audio input signals.

$$33.42 \text{ kHz} \times 13 \times 4 = 1.79 \text{ Mbit/sec}$$

The 5.5-MHz digital subcarrier furnishes four-phase phase-shift-keying modulation for the four audio channels. The encoder receives this signal along with a filtered video signal (of 4.5 MHz bandwidth), so that the base-band output is a 5.5 MHz subcarrier. The audio and video signals are separated at the decoder, and the four audio channels are separated by filtering.

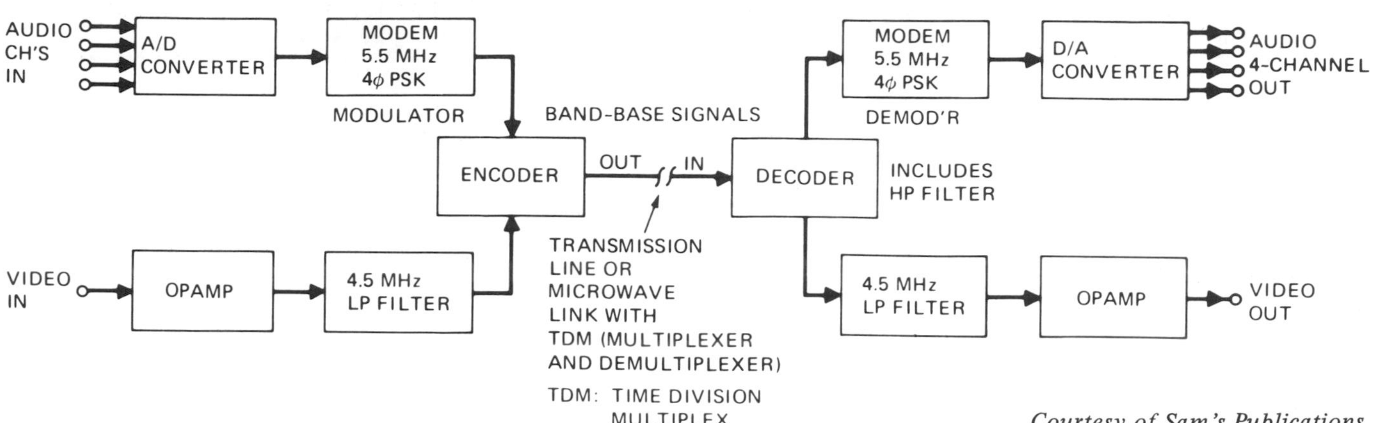

Courtesy of Sam's Publications

There is a possibility that the audio information may be inserted in future as digital data within the horizontal blanking intervals of the composite television signals with the sync content.

5.5 LSI VIDEO A/D CONVERTERS (TRW LSI PRODUCTS)

Digital television is presently an accomplished fact in the highly reliable, medium-cost LSI format. No doubt the processing will eventually become a popular low-cost technique when digital television becomes a mass medium around the world via synchronous domestic satellites. The following analog-to-digital converter is a timely new LSI component for the low-cost implementation of digital television:

1. TRW TDC 1007J is an 8-bit monolithic bipolar (TTL) fully parallel A/D converter, using 255 sampling comparators, combinatorial Exclusive-OR logic, and an output buffer for digitizing 7-MHz analog black-and-white or NTSC/PAL color-television signals or radar signals at rates from dc to 30 megasamples/sec (MSPS). The combinational logic is involved in the conversion of unit-distance code to binary. The sample-and-hold circuit is built in; recovery from a step-input occurs within 20 nsec with the sampling comparators of 40-MHz bandwidth. The aperture jitter is hardly 30 psec at a differential phase of 0.5° and a differential gain of 1.5%.

Peak-signal to rms noise at 2.438-MHz input is 54 dB, and noise power ratio (at dc to 7 MHz white noise-bandwidth) is 38 dB. The 64-pin LSI chip operates through a temperature range of −55° to +130°C, and requires a drive of 1-V peak-to-peak composite video signal (or −0.5-to-5.5-V radar pulse) and provides an 8-bit binary or two's complement output, using ±15- and +5-V power supplies. The system organization of the converter is shown in the accompanying block diagram. The cost of the plug-in PC-board is $550, complete with the chip, selectable input impedance and amplitude, and adjustable offset for unipolar or bipolar input.

2. For lower-cost closed-circuit television and other applications, a 6-bit resolution TRW TDC 1014J is available in a 24-pin ceramic dual in-line package at a lower cost of $186 and power dissipation of $\frac{3}{4}$W. Compared to the 8-bit converter, the

6-bit chip uses 63 strobed sampling comparators; the other details are identical to those in the 8-bit chip; linearity in this case is specified as $\pm\frac{1}{4}$ least-significant-bit (LSB). The device requires a single command to digitize an analog waveform of amplitude between 0 and -1 V. The peak-signal to rms noise with 2.438-MHz passband input is 42 dB, and the noise power ratio for dc to 7-MHz white-noise bandwidth is 27 dB.

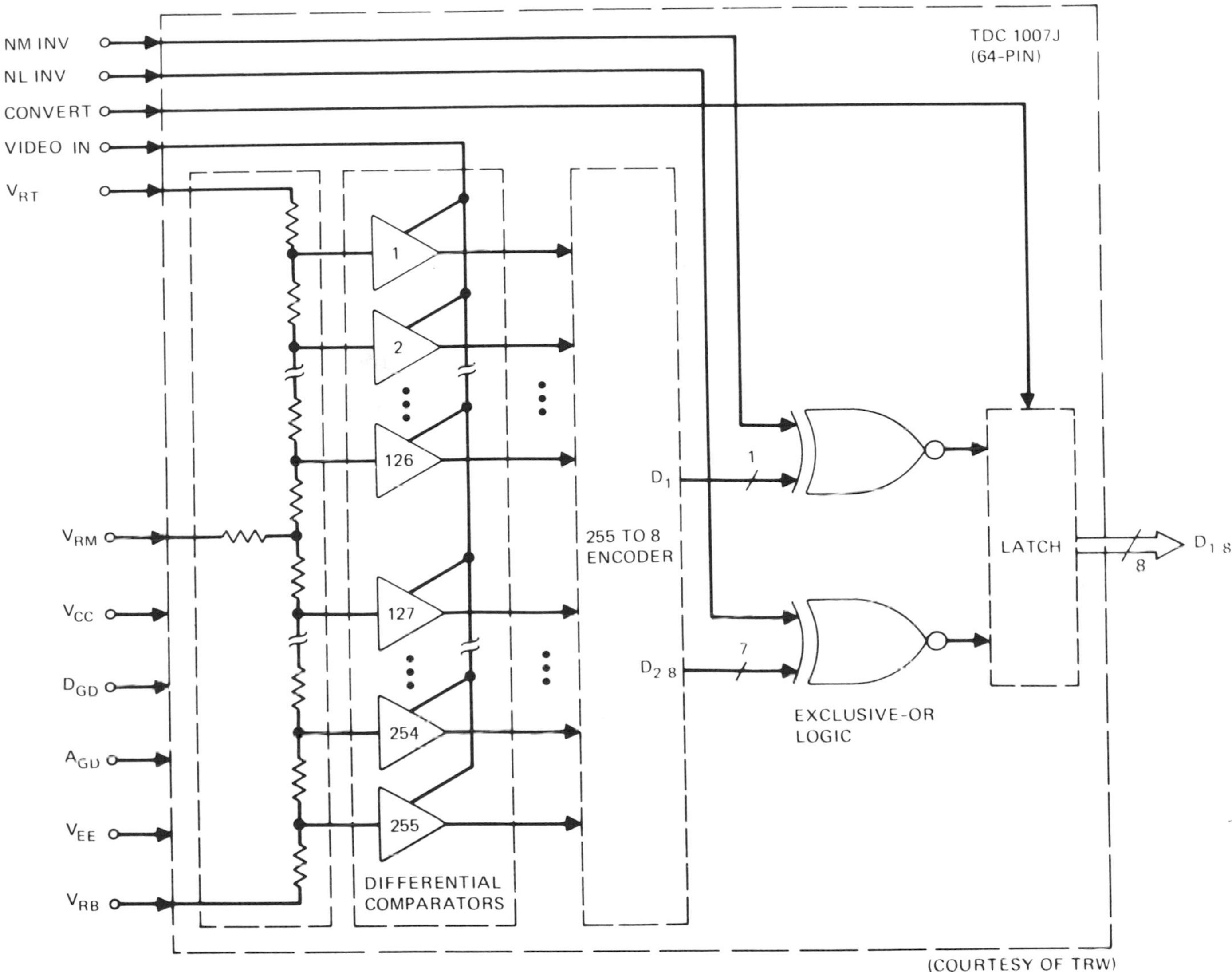

(COURTESY OF TRW)

5.6 DIGITIZED VIDEO SIGNALS

With the recent introduction of worldwide color television broadcasts via satellites, digitized video is an accomplished fact. Other digitized video applications include (1) automatic raster-and-color synchronizers and timers, (2) time-base correctors for video recorders, (3) image enhancers, (4) automatic international standards conversion, and (5) video bandwidth compressors and expanders. At this time, temporary storage via LSI RAMs at a clocked rate is an inherent processing technique in all of these applications. Sample-and-hold circuits to convert the low-pass filtered analog signal to a PAM signal, and the subsequent ADC to generate the format corresponding to the instantaneously sampled PCM signal, are the other two stages in this signal processing technique.

In the case of video, the whole synchronous picture information occurs at even multiples of the horizontal line-scanning frequency, and color information is "interleaved" at odd multiples of one-half the line frequency by starting with a color subcarrier at 3.579545 MHz. Unlike the audio, a smaller 8-bit binary word can be used at three times the color subcarrier in the case of video ($2^8 = 256$ levels) for very high-quality color picture of a bandwidth of 4.5 MHz. For closed-circuit television and VTRs, 6- and 7-bit (64 to 128 levels) binary words will do for peak-to-peak signal-to-RMS noise ratios from 43 to 49 dB.

Transmission rate at the above high-quality sampling rate:

$$10.7 \text{ MHz} \times 8 = 86. \text{ Mb/sec}$$

Quantizing aperture effect is generally frequency compensated by preemphasis and deemphasis before quantizing and decoding, respectively. COMSAT has experimentally demonstrated satisfactory international satellite telecasts with its DITEC-PCM and comb-filters at a transmission-rate of 33.6 Mb/sec in the place of 86 Mb/sec. The *teeth* of a luminance comb-filter have maximum response at the harmonics of the line-frequency (15.75 kHz) and nulls at the odd harmonics at half the line-frequency for color. And the *teeth* of the color comb-filter occur at the odd harmonics of half the line frequency for color, and nulls at the even harmonics of the line frequency for luminance. When four times the color subcarrier frequency is used for sampling the I and Q components of color, the samples on all the time-sequential picture lines are well-aligned for the comb-filters. However, by employing a Phase-Alternation Line Encoding (PALE) technique, the digital comb-filters perform equally well at three times the subcarrier (10.7 MHz) sampling-rate, using one-third fewer digits for the binary words. A simplified block-diagram of the PALE technique is illustrated in Fig. 5-7a. The phase of the sampling frequency is shifted by 180° on alternate horizontal scan lines during the breezeway at the leading-edge of the color-sync burst. When a signal is stored in a shift-register, the stored data are read out as a reconstructed PALE-clock, which is phase-synchronized with the samples on each scan-line. Switching transients and ringing at the output of the digital-to-analog converter are minimized by the *resampling* process using a filter, as shown in Fig. 5-7b. The high-frequency performance is redeemed by aperture compensation.

Digital-comb-filtering and a matrix network recover the in-phase I and quadrature Q color information as shown in Fig. 5-7c. They are superimposed on the dc luminance component M. The sampling is shown at the 10.7-MHz rate. The I and Q signals are digitally extracted with the following matrix:

$$I = A - C; Q = B - D; \text{and } M = \tfrac{1}{2}(A + C) \text{ or } \tfrac{1}{2}(B + D)$$

5.7 DIGITAL TIME-BASE CORRECTOR FOR VIDEOTAPE RECORDERS

The combined digital and analog open-loop monochrome automatic time-base corrector (MATC) and the color automatic time-base corrector (CATC) and the associated adaptive closed-loop PLL systems of the quadruplex color videotape recorder (Q-CVTR), using voltage-controlled variable delay-lines, can presently be replaced by digital time-base correctors. The Ampex TBC-800 digital time-base corrector is an example. The digital accessory incidentally includes a tape dropout compensator and an optional digital velocity compensator for color to eliminate the effects of

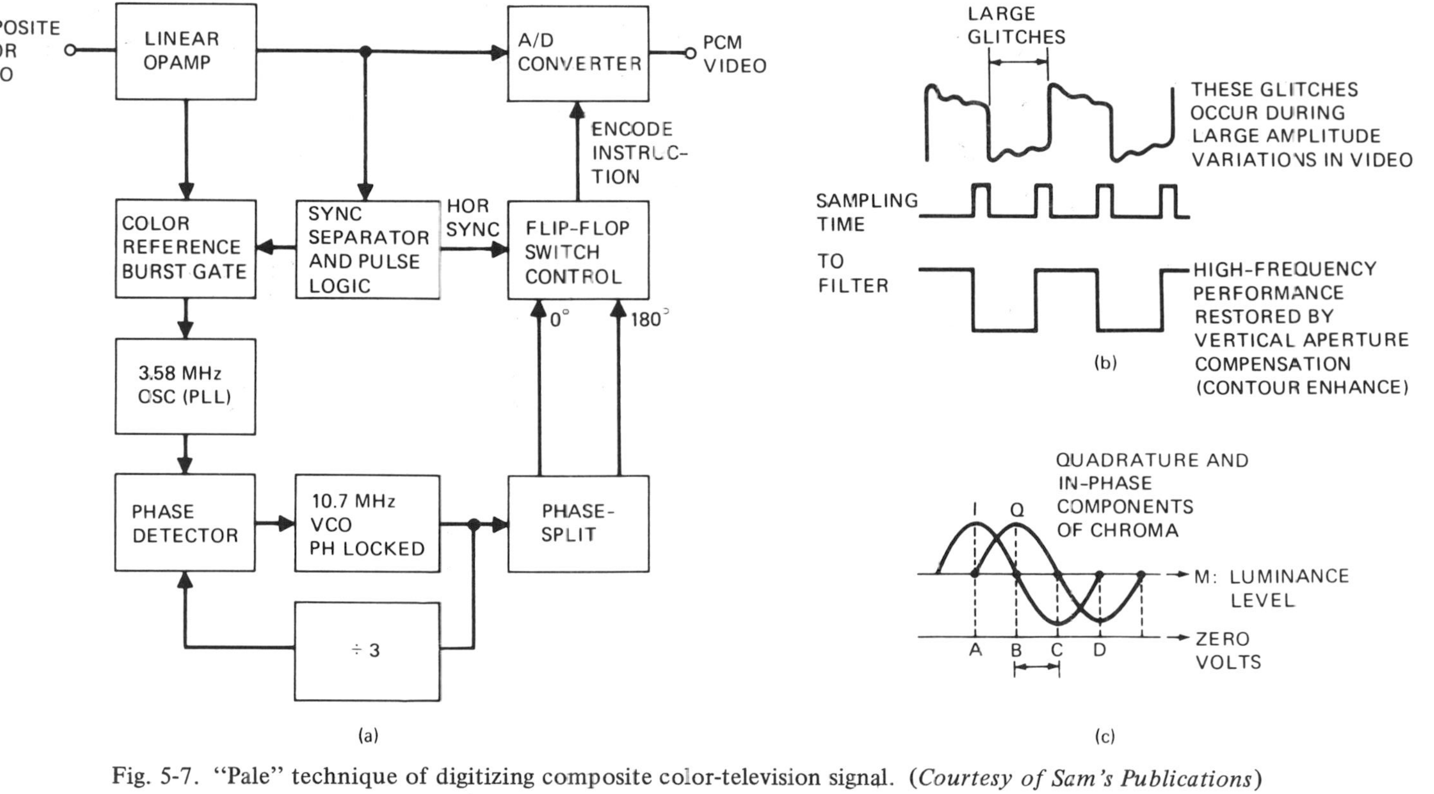

Fig. 5-7. "Pale" technique of digitizing composite color-television signal. (*Courtesy of Sam's Publications*)

color-banding off old videotapes. This major digital accessory also includes a facility for digitally phase-locking (by using quantizers) the digital studio sync generator to a remote composite color video signal to enable special pictorial effects between the local color camera/tape and remote nonsynchronous color video signals.

The digital time-base corrector has another unique advantage. The accessory enables the monochrome and color time-base corrector facility for not only the broadcast-quality quadruplex color videotape recorders but also the economy mobile and portable helical-scan tape recorders (Sony, Ampex, etc.). The helical-scan tape recorders frequently require a correction "window" to ± 1 horizontal line (± 63.5 μsec) as compared to less than 1 μsec for the Q-CVTR. Figure 5-8 illustrates the concept of the digital time-base corrector. Built-in test signals are incorporated to enable automatic testing of the digital memory and the A/D and D/A converters. For speed and accuracy, A/D conversion is done in two feedback sequences, first for the first four MSBs and then for the four LSBs to form the 8-bit word representing one analog sample. Each horizontal line is made up of 683 words; and these 8-bit words are processed at exactly three-times the off-tape subcarrier rate (approximately 10.7 MHz). When these 8-bit words are formatted into 24-bit words, they make the off-tape 3.597545 MHz color subcarrier rate.

The digital data are thus loaded into the digital LSI, RAM memory at a rate directly related to the off-tape sync and back-porch color reference burst—which naturally contain all the time-base errors of the tape recorders. The digital data are then read out of the RAM memory at the rate corresponding to the reference sync and color reference burst of the local studio digital sync generator in the TBC-800 accessory. This procedure enables the proper correction and time-synchronization of the helical-scan tape-recorder signals or remote television signals to the local television signals to enable synchronized color telecasts.

The latest television studio accessory equipment such as TBC-800 gradually facilitate the automatic mini or microcomputer-controlled telecasting capability with a minimum of operator errors on a preprogrammed basis. Most of the complexity in previous combined analog and digital sampled-data control signal-processing circuits is effectively substituted by introducing A/D converters and subminiaturized digital processing systems that use read-in and read-out of data in ROM/RAM storage memory; after read-out, the data are logically computed, and buffered out by means of the microcontrollers and the associated software and firmware.

5.8 INTRODUCTION TO USE OF ECONOMY MICROCONTROLLERS AND MICROCOMPUTERS IN COMPLEX DIGITAL CONTROL SYSTEMS

5.8.1 Data Acquisition. Data acquisition, which primarily comprises of input/output monitoring, alarm self-checking, and data logging on magnetic-tape is one of the first applications of real-time computers in complex control systems that process industrial plants. The CPU (central processing unit) instantaneously scans a set of analog input variables such as temperature, flow, position, pressure, weight, and angular velocity. Digital input subsystems are involved when the input variables are multivalued. The CPU works in conjunction with an operator's console and a disk file along with peripheral output devices such as keyboard line printers and tape. Presently, individual microcontrollers take over most of these tasks.

5.8.2 Direct Digital Control. Next, *direct digital control* (DDC), by means of the mini or microcomputer as the feedback element in the plant, will replace the large

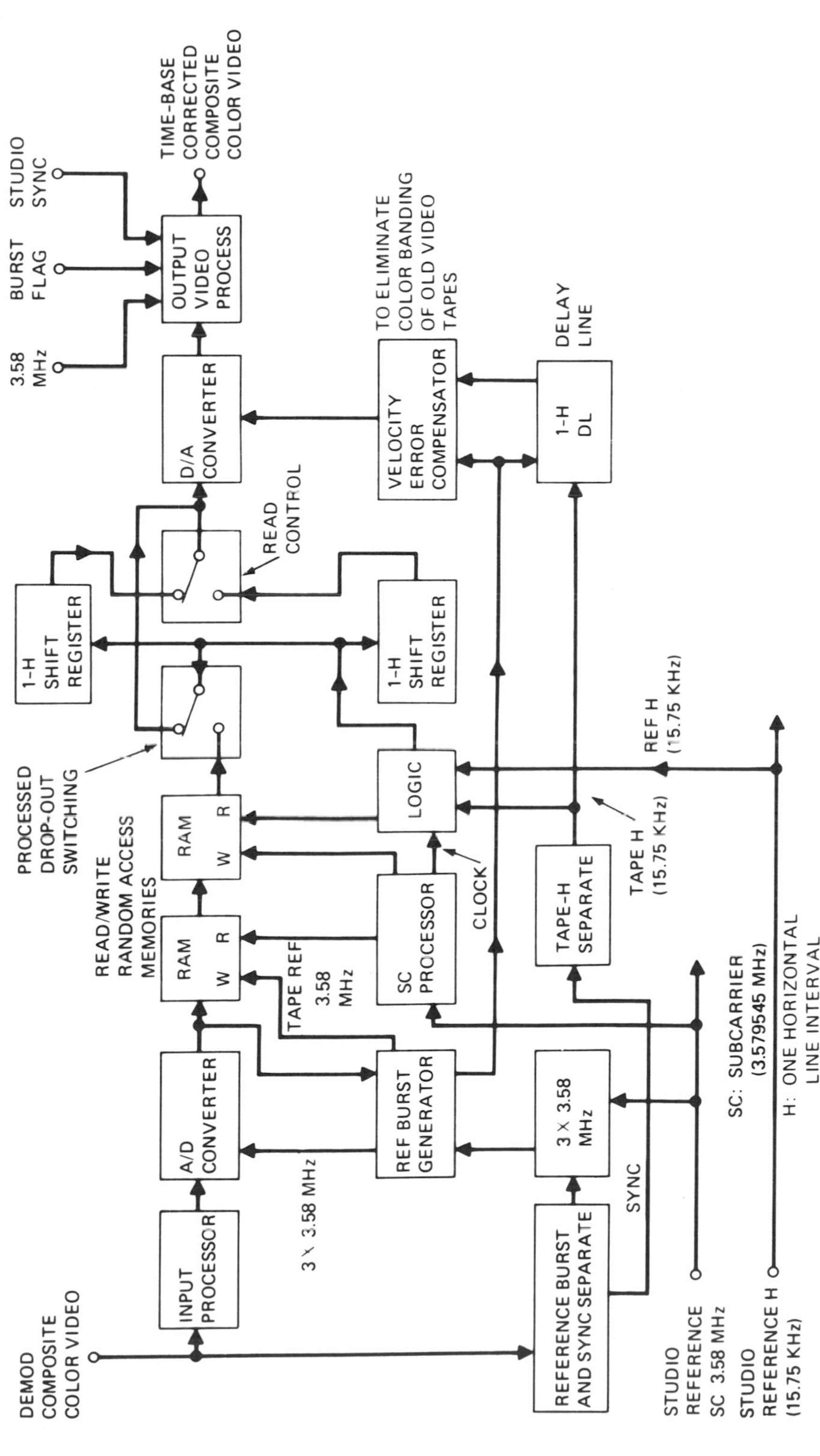

Fig. 5.8. System concept: digital time-base correction. (Reprinted from the March 1976 *SMPTE Journal* with the permission of the Society of Motion Picture and Television Engineers, Inc. Copyright © 1976 by the Society of Motion Picture and Television Engineers, Inc., 862 Scarsdale Ave., Scarsdale, NY 10583.)

number of conventional analog control devices. These analog controllers involve a multiplicity of feedback loops to regulate temperature, pressure, flow, position, and angular velocity. For accuracy, these controllers in practice work on a complex three-mode algorithm, an output signal proportional to (1) the error, (2) the time-integral of the error, and (3) the time-derivative of the error. A single high-cost large-scale computer that could time-share among many such feedback loops, using extensive software was the requirement, and the amount of programming involved was prohibitively expensive, while the down-time of the former hardware in large-scale digital computers added to the inherent problems in a complex industrial control system. The latest revolutionary LSI microprocessors have completely changed the situation toward a reliable, practical low-cost DDC system at the present moment in any industrial system-complex. Software will appear in the form of electronically alterable programmable LSI read-only-memory, and high-speed LSI RAM storage-memory will work off the all-electronic mass-memory systems of the category of magnetic-bubble memory and beam-addressable devices, that require negligible space, but operate at high-speed.

In the computing hierarchy, a powerful parallel processing configuration of 10 to 20 microcomputer cards, employing the high-speed bit-slice LSI/I^2L (integrated injection logic) bipolar microprocessors operating at a mere 1.5 V (such as the Texas Instruments SBP-0400A) or the Intel 3000 series using a 16-bit ALU (arithmetic logic unit) computation, may be interfaced around a central host-minicomputer real-time operating system (having a cycle-time of about 145 nsec) to drastically reduce the data-processing and computation time needed to operate highly complex industrial control plants. A bit-slice microprocessor card will typically use a fast bipolar 256-word data-memory, a microprogram control unit, and a 512 $\times$ 280 bit micro-program instruction-memory. A 16-bit I^2L microprocessor can further simplify this real-time computation. Each microprocessor module performs the same program, but on different data, in a pipeline fashion. The central host minicomputer with an elaborate CPU may be programmed in one of the high-level languages like PASCAL/APL (advanced programming language) to evaluate the results from the microcomputers in the surrounding network and adjust the situation to a new set of variables or commands according to the predetermined software. Also, from the viewpoint of cost estimate, the hardware and firmware required with the mini-micro general-purpose DDC setup would be hardly 5-to-10% of the former large-scale general-purpose digital computer systems, once the initial investment is made to develop the comparatively expensive software in a high-level language.

5.8.3 Hypercube Architecture of Microcomputers. A large network-array of microcomputers (each consisting of two Zilog-8000 microprocessors, one functioning as the task-processor and the other as the controller or multiplexer for input/output interface) is capable of exceeding two or three times the computing power and speed of the large-scale main-frame computers of IBM (International Business Machines) and CDC (Computer Data Corporation) at a fraction of power, area, and cost. IMS Associates Inc. (San Leandro, California) has developed a prototype system using sixteen 8080 microcomputers to simulate the computing power of one of the present high-cost medium-scale digital computers, as used in industrial plants. A 256-microcomputer network is on the drawing board. The *Hypercube* architecture is shown in Fig. 5-9. The individual microprocessor units (MPU) operate independently on separate parts of the complex control system on different algorithms. IMS's

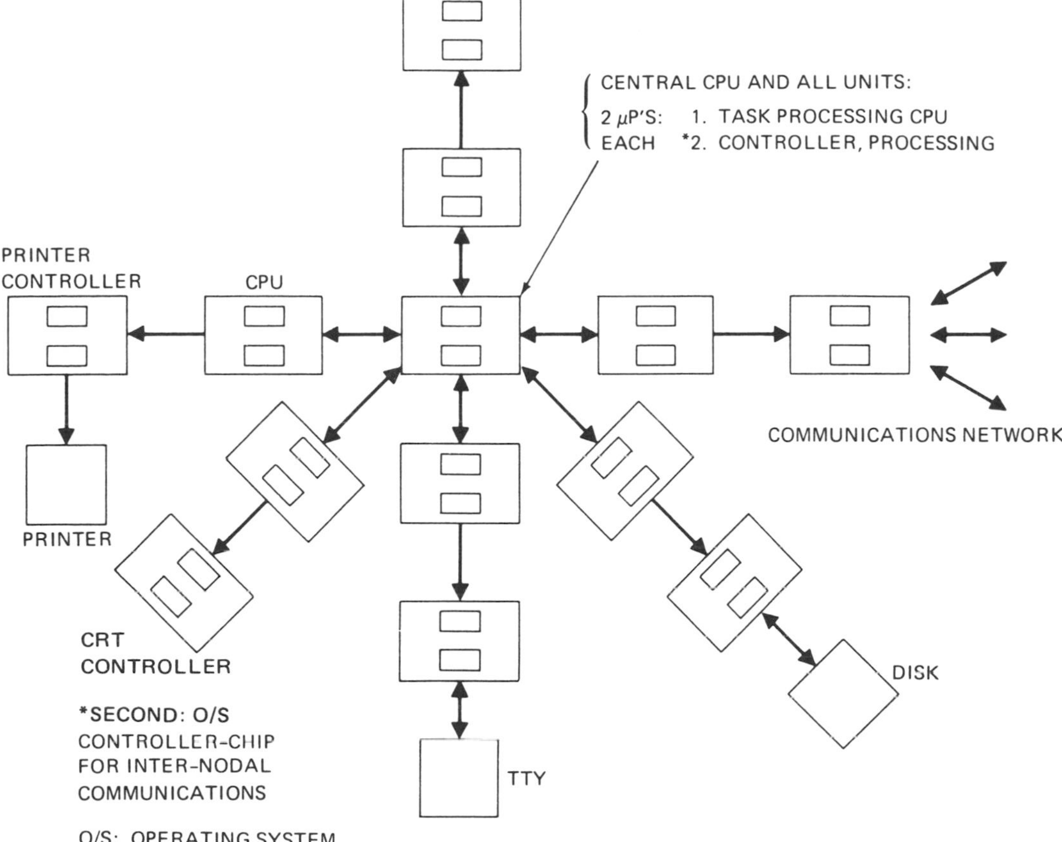

Fig. 5-9. *Hypercube* architecture of microcomputers. (*Courtesy of Electronics*)

Hypercube microcomputer system employs several nodes, each using a distributed program in a low-level assembly language. Each node is capable of handling 1 million instructions/sec, using a direct memory access (DMA) capacity of 2 M-bytes and 16-K-bytes of user program, which is extendable to 64-K bytes. Each subsystem located at each node is linked to eight other microcomputers in the adjacent nodes. One MPU of each node handles the user's program, while the second MPU oversees the communications tasks and Operating System software by using the DMA. Some nodes may operate on input/output only, or interface with a TTY or a CRT or a line-printer or a disk or a magnetic-bubble memory. This multiprogramming concept is more complex than handling successive batch programs on the same central processing unit (CPU). Some satellite telemetry and complex simulation systems, large-scale data-base systems, and data communications networks are being planned at this time under private and public research-and-development contracts.

5.9 MICROPROCESSORS AND MICROCOMPUTER SYSTEMS PRESENTLY AVAILABLE FOR DIGITAL CONTROL APPLICATIONS

The types of single-chip LSI microprocessors, such as CPU, and single-chip stand-alone LSI microcomputers (that include a clock generator with external crystal, ROM and RAM, and I/O) number about 40 at present. For flexibility of system design in various applications, the microprocessor as a CPU is, naturally, the preferred technique. That is, many of the logic and circuit designers will progressively become system designers using LSI microprocessors, LSI memories, LSI field-programmable logic arrays FPLAs, and single-chip dedicated microcomputers. Of course, the manufacturer's applications engineers will necessarily play an important role in this function during the learning stage. Several chip- and system-design houses having logic-design capability have already commenced putting together custom LSI chips for I/O interface, etc. in NMOS, or in CMOS or I^2L at competitive prices; this implementation is practical within a reasonable period of 3 to 6 months.

The single-chip microcomputers, 8-bit and 16-bit, from about a half dozen manufacturers are mostly based on NMOS technology; however, Fairchild Semiconductors is presently exploiting a 16-bit bipolar isoplanar I^2L microcomputer chip (9440) as a prelude to its successful feasibility and demand in the 1980s. Incidentally, some of the more popular microprocessor types do have second-source suppliers.

5.9.1 Intel-8080. Intel N-Channel Microprocessor: 8080 with Intellec-8 Program Development Facility. As the most popular "workhorse" among the microprocessors, the 40-pin silicon-gate N-channel microprocessor is TTL-compatible. Processing speeds are up, 10 to 100 times, compared to those of former Intel P-channel microprocessors. It is operationally a parallel CPU with an instruction cycle-time of 2 μsec at the nominal clock-rate of 2 MHz, and 78 microinstructions, and clocked with two-phase, nonoverlapping clock. It has 14 control lines, and 8-line bidirectional data-bus; a 16-line bus is used for addressing the memory along with a 24-I/O line-section. All system controls are decoded on the chip. The CPU accesses up to 64-K bytes of memory, and operates up to 256 input and 256 output 8-bit channels; it has a provision for 8 interrupt levels.

High-speed I/O structure, memory, and control lines permit its use as a controller and a data-processing subsystem. Stack architecture enables the programmer to effectively process both subroutines and interrupts. Instructions are capable of handling strings of data along with decimal and double-byte arithmetic. Both decimal and binary data are handled with equal speed. Source programs can be written in either PL/M high-level system-oriented language or in a macroassembler language; programs written for the previous P-channel Intel-8008 may be compiled or assembled for the use of type-8080, which needs 20% fewer instructions.

The Intel-8080 microprocessor employs the programmable peripheral interface (PPI) 8255 for easy interface to printers, keyboards, displays, and motor drives. It has 13 options for memory circuits, such as 16-K bit ROMs, 8-K bit EPROMs, and 4-K bit RAMs (at high density and low cost) plus CMOS-RAMs for minimum power requirements. Type-M8080A is ruggedized for operation at Military Standards −55 to 125°C, type-8080A-1 is designed for a 1.3-μsec instruction cycle-time in higher-speed real-time applications. Intel has a microcomputer hardware/software development system, Intellec-MDS, which employs an ICE-80 in-circuit emulator to simultaneously debug both software and hardware from the initial prototyping stage through production; it is supported by six comprehensive software packages, which include a macroassembler ICE-80, and a diskette operating system. This facility enables the direct availability of a versatile Intel microcomputer system with the computing power of the IBM-5100 or DEC-LS-11 microcomputer system.

Intel's MCS-80 system design-kit is comprised of a type 8080-A CPU, a crystal clock-generator, a system controller, a programmable communications interface (PCI), and a programmable peripheral interface (PPI). The system includes two 1-of-8

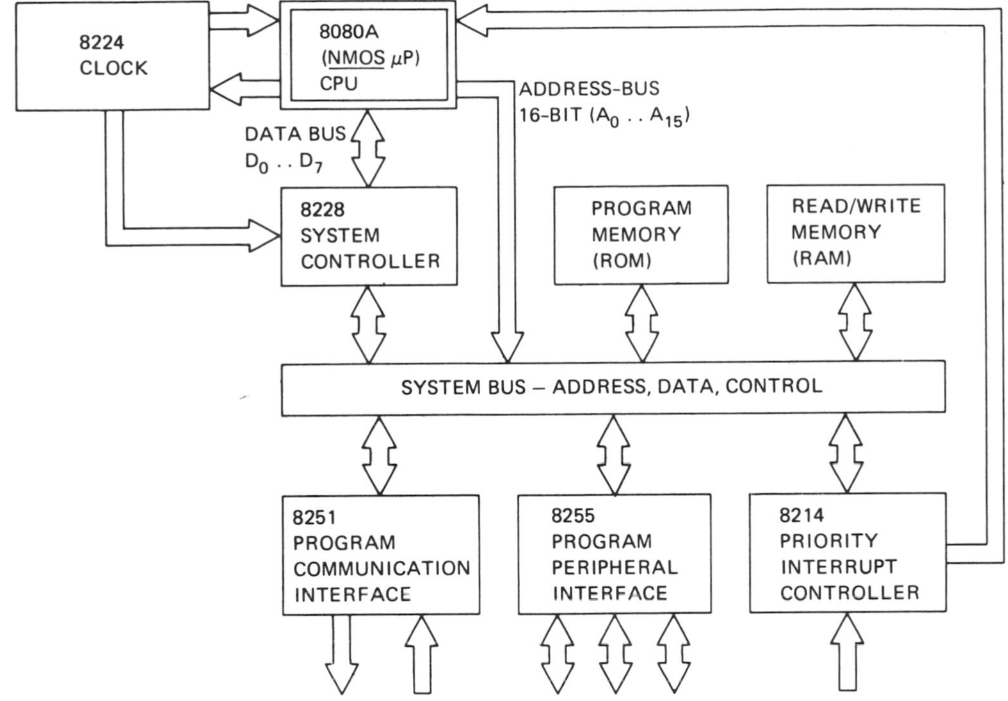

(COURTESY OF INTEL CORPORATION).

Fig. 5-10. Intel MCS-80 microcomputer system.

binary decoders, 256 bytes of static RAM as main memory, 2-K bytes of EPROM, and a PC board, at a total cost of $350. Figure 5-10 illustrates Intel's MCS-80 system design kit.

Gordon Moore and Robert Noyce of Intel pioneered the microprocessor concept in 1968 with their first PMOS, Intel-4004 microprocessor. Since then, Intel's technological track record in the advance of NMOS state-of-the-art in both the microprocessor and memory fields has been outstanding.

5.9.2 Intel SBC 80/10 Microcomputer-on-a-Card. The single-card microcomputer SBC 80/10 is an example of a kind of subsystem (or OEM–original equipment manufacturer– supercomponent), employing type-8080A microprocessor and one 8192-bit EPROM; it includes programmable LSI I/O interface, components to provide customized software for parallel I/O ports, and communications interface. This facility eliminates the need for inefficient specialized hardwired designs for custom applications, and results in the availability of a microcomputer-on-a-card, 6.75" X 12" at a price of $295, in the place of a superfluous minicomputer at a price of thousands of dollars, for most common general-purpose applications. The type-8080A CPU contains over 5000 transistors on a silicon-substrate chip, 0.164" X 0.19". The organization of this system version is shown in Fig. 5-11 along with the pertinent interface components for most applications. The central processor, using the interrupt-control (LSI-8228 status latches) and the bus-control logic, the crystal-controlled clock (LSI-8224), the high-current drivers for memory and I/O bus expansion, and other CPU-related control functions, is implemented with the NMOS 8080A CPU and two Schottky-bipolar LSI devices. The cycle-time is 1.95 μsec.

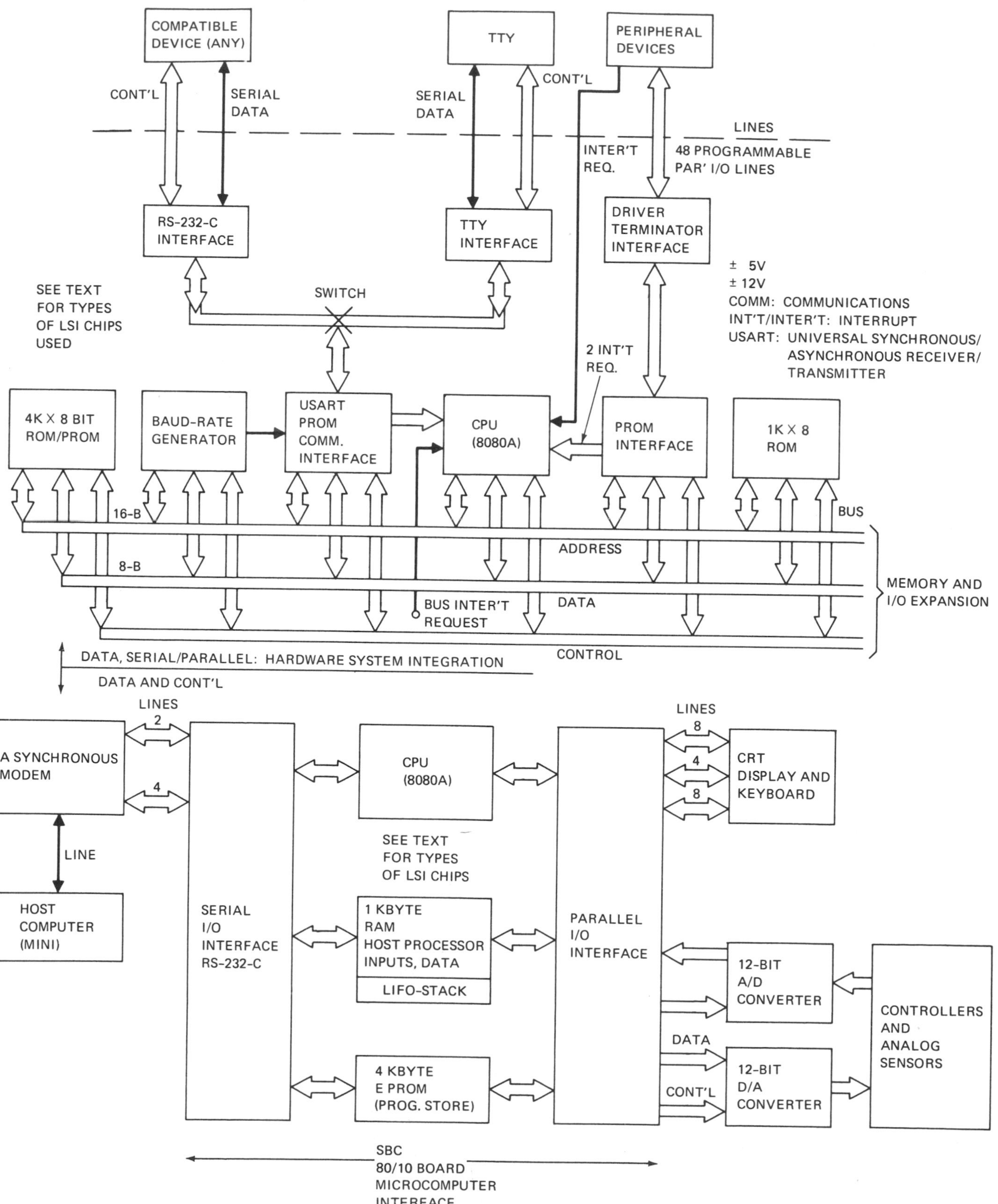

Fig. 5-11. Organization of Intel SBC 80/10 microcomputer on a card. (*Courtesy of Electronics*)

The read/write main memory on the card consists of 1-K bytes of static RAM (two LSI-8101). The card has provision for four sockets to plug in 4-K bytes of EPROMs (LSI-8708) or ROMs (LSI-8303) to facilitate control memory in 1-K byte increments.

The programmable peripheral interface (PPI-LSI, 8255) provides 48 software-configurable parallel I/O lines; sockets are provided for interchangeable quad-line drivers and terminators to enable the user's choice of sink currents, polarities, and so forth. The programmable synchronous and asynchronous communications interface (PCI-LSI, 8251) includes a variable baud-rate generator and a jumper-selectable RS-232C interface with TTY drivers and receivers.

Since active programs are normally stored in a nonvolatile memory to eliminate the need for reloading RAMs every time the system is turned on, formerly external CORE memories were used for this purpose. This requirement is now eliminated by packaging such dedicated active memory on the card itself by using plug-in 1-K byte EPROMs or ROMs. The programmable interface devices mentioned previously accommodate various operating modes and protocols for control of data transfer to a variety of external devices such as switches, motor drivers, bistable sensors, analog-to-digital and digital-to-analog converters, displays such as CRT/LED, keyboards, line printers, communications modems for the TTYs in centralized or distributed communications networks, cassettes, and other micro- and mini-computers.

5.9.3 Intel 8748/8048 Microcomputer-on-a-Chip. The latest LSI technology introduces a dedicated microcomputer on a single-chip as an all-in-all digital controller by itself. As anticipated, the Intel track record continues unabated as the leading contender in the state-of-the-art microcomputer technology. Not unlike the present low-cost stand-alone 4-bit calculator chips, the Intel type-8748 and type-8048 microcomputer twins have a built-in 8-K bit (1,024 × 8-bit) EPROM, to facilitate a periodic updating of the program in one case and to facilitate a mask-programmable ROM in the other case, respectively. The requisite software-development function is reserved for the electronically programmable prototype 8748, and the finalized application program is established for the regular large scale mask-programmable production chip 8048. This single-chip microcomputer thus accomplishes a stand-alone computer function in the built-in CPU, programmed ROM memory, 64 bytes of 9-bit scratch-pad data RAM memory, pulse-clocks and timers, and I/O interface for the dedicated application. In the case of the programmable version 8748, the program is altered or debugged by the technique of ultraviolet light-erasure. (The Prolog UV-Erase light system costs $150.)

The 8-bit CPU will provide the function of the ALU and the accumulator for all the binary and decimal arithmetic functions. The I/O function allows three 8-bit I/O ports and three test/interrupt ports—all directly controlled by the program instructions in the masked ROM area. The system details of the microcomputer-on-a-chip are shown in Fig. 5-12.

One could add flexibility of expansion to this stand-alone microcomputer by directly interfacing an *Expander* chip type-8243, to handle 16 additional I/O lines. At the same time, the program and data bus-oriented architecture of the chip facilitates further expansion of the data-processing capability by means of an external ROM/RAM complement using an external latch (8212)—as in the case of any micro- or mini-computer system.

Present multichip microcomputer designs have an inherent delay in the transfer of the data between the memory and the CPU, and the single-chip microcomputer

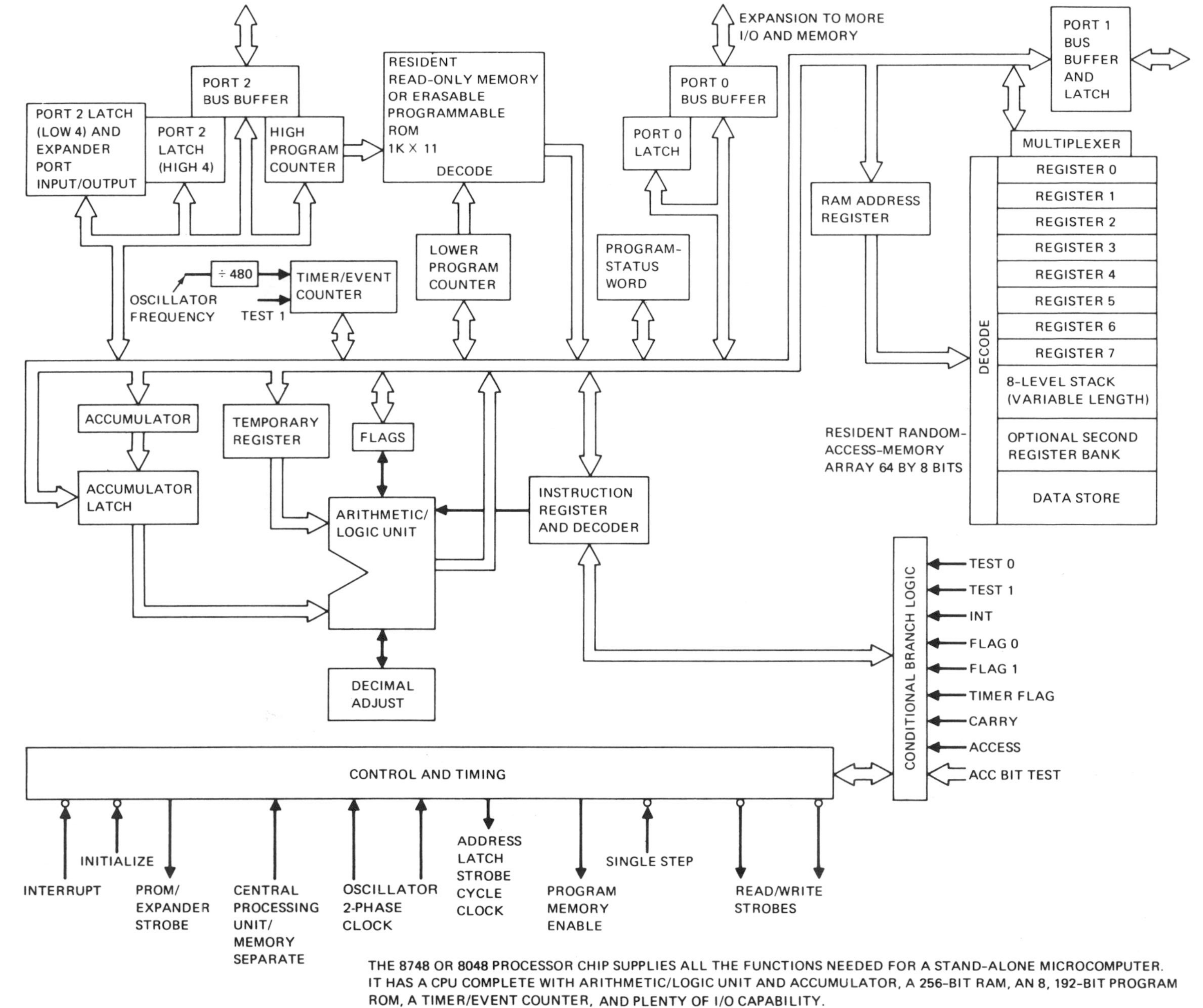

Fig. 5-12. Intel 8048 microcomputer on a chip. *(Courtesy of Intel)*

scores an advantage in eliminating that delay. Direct access to the 64 bytes of the 8-bit dynamic RAM allows the execution of indirect internal instructions, fetching the address of memory location and its contents, and the storage of the resulting information, all in one cycle-time of 2.5 μsec. The RAM consumes a mere 75 mW for decoding the sense circuits involved in this operation.

Advantages of the single-chip microcomputer include:

1. In multiple-chip design, some time-delay is involved in transferring data between the memory and the CPU; with single-chip design, the specified instruction cycle-time is all that matters.
2. The inclusion of respective data and program RAMs and ROMs simplifies the user's interface needs.
3. Since the eight-level stack scratch-pad RAM operation is included in the CPU function, no *refresh* is required in this respect, although the internal clock is used for refreshing the low-power access to the dynamic RAM main memory in a fraction of an instruction cycle. Indirect internal instructions, requiring multiple addresses for (1) fetching the memory location to be operated on, (2) fetching the contents of the addressed location, and (3) storing the results of the operation, can be executed in one instruction cycle-time.
4. The type-8748 with the EPROM can be operated on a special double-cycle instruction-cycle mode (called the *third-state mode*) for programming and convenient verification of the EPROM.
5. The instruction cycle of the 8748/8048 microcomputer takes place in five states: (1) instruction input, (2) decoding and program counter incrementing, (3) start of the program execution, and (4, 5) parallel pipeline operation with the next cycle's program address.

5.9.4 Microprocessor (Signetics Type 2650, North-American Philips). Five-volt microcomputer system. Interface on the chip; powerful instruction set; fixed instruction set of 75 (40% arithmetic); TTL-compatible; 576-bit ROM; 250-bits of register and 900 logic-gates on processor; 8-bit bidirectional tristate data-bus and separate address-bus; 32,768-byte addressing range; internal 8-bit parallel structure; seven 8-bit GP registers; 8-level on-chip subroutine, return-address stack; program-status-word for flexibility and processing power; separate adder for fast address calculation; low power.

The Signetics 2650 is a multisource microprocessor chip, priced at $21.50. It is available with external MOS/bipolar memories, programmable peripheral interface, communications interface, A-D converters, synchronous data-link converters, 16-K byte NMOS and bipolar RAMs, 4-K byte and 8-K byte NMOS EROMs, and 8-K byte bipolar PROMs. Development software in PL/μs high-level language compiler reduces programming effort and time. Type-2650/AS 1000/1100 assembler and type-2650/SM 1000/1100 simulator are also available in both 32- and 16-bit on GE and NCSS time-sharing BASIC. ANSI-standard FORTRAN IV is applicable. For development, prototyping facilities are provided by means of a twin floppy-disk mass-memory, a resident assembler, and a text editor.

The 2650A microprocessor, using a 77-instruction set, is much smaller in size as a 40-pin DIP; the cycle-time is therefore reduced from 2.4 to 1.5 μsec. The NMOS depletion-mode chip, costing less than $20, consumes 625 mW at 5 V. Two prototyping cards and kits, and a 4-K byte card are available along with 16/32-bit assemblers and compilers as software support. For further details, see Fig. 5-13.

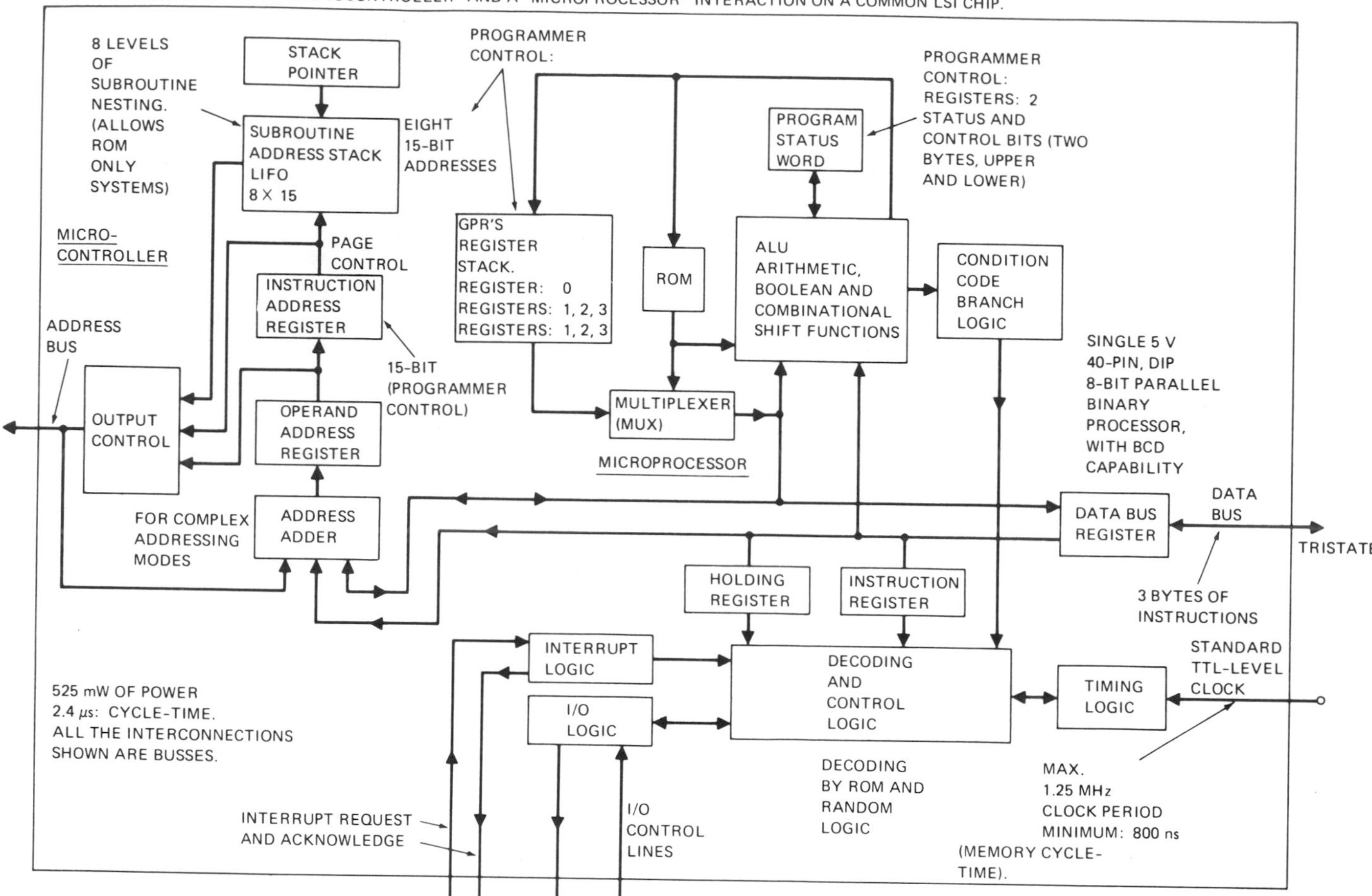

- DIRECT INSTRUCTIONS REQUIRE 2, 3 OR 4 PROCESSOR CYCLES. 4.8 TO 9.6 μs.
- OP. REQ. (OPERATION REQUEST) IS THE MASTER CONTROL SIGNAL TO COORDINATE EXTERNAL OPERATIONS.
- VECTORED INTERRUPT (8 LEVELS) (INTERRUPT SERVICE ROUTINE CAN BEGIN AT ANY ADDRESSABLE MEMORY LOCATION).
- INSTRUCTION SET. (75).
  THREE-LETTER MNEMONIC AND 6-BIT OPCODE).
  LOAD AND STORE (2)
  ARITHMETIC (3)
  LOGICAL (3: AND, OR, EXCL. OR)
  ROTATE COMPARE (3)
  BRANCH (7)
  SUBROUTINE BRANCH/RETURN (6)
  MISCELLANEOUS I/O (9)
  PROGRAM STATUS (5)

- REGISTER-TO-REGISTER INSTRUCTION: 1 BYTE. REGISTER TO STORAGE: 2 OR 3 BYTES LONG (2 BYTES: IMMEDIATE OR RELATIVE ADDRESSING TYPES).
- AUTOMATIC INCREMENTING OR DECREMENTING AN INDEX REGISTER.
- ALL BRANCH INSTRUCTIONS CAN BE CONDITIONAL.
- 1-, 2-, OR 3 - BYTE INSTRUCTIONS FOR THE VARIOUS ADDRESSING MODES.
- A MICROCOMPUTER SYSTEM (TTY) REQUIRES SEVEN MINIMAL NO. OF CHIPS:
  2,650 (A_0 . . A_9 ADDRESS BUS; D_0 . . D_7 DATA BUS, SENSE, FLAG, R/W, OPREQ, CLOCK, A_{10})
  2606 (256 X 4 RAM): QUANTITY 3; RANGE: 32-K BYTE.
  74123 CLOCK GENERATOR.
  7439 MISCELLANEOUS LOGIC (NAND).
  4049 (INVERTER).

NOTE: SIGNETICS HAS A SCHOTTKY BIPOLAR 2-CHIP MICROPROCESSOR (N-3002 CENTRAL PROCESSING UNIT AND N-3001 MICROCONTROL UNIT, USING STANDARD EXTERNAL TTL/ECL MSI SUPPORT CIRCUITS, FOR A MICROINSTRUCTION CYCLE-TIME OF 100 ns. USING 6 PROMS, INSTRUCTIONS MAY USE 48-BIT WORDS. 512-MICROINSTRUCTION CAPABILITY, WITH 9-BIT MICROPROGRAM ADDRESS REGISTER, ALLOWS 18-BIT PROCESSING.

Fig. 5-13. Signetics type 2650 microprocessor (Si-gate NMOS, ion-implanted). (*Courtesy of Signetics*)

Type 2650 is a comparatively faster depletion-mode NMOS unit; it has the capability of 8/16/24-bit instruction words. Internal address capacity is 8-K bytes with 75 instructions. However, it has an addressing capacity of 32-K words. The basic card uses type-2650 microprocessor, two 2606B ICs, one 2608 chip, one N74123, and one N7438A. The instruction formats used for the 2650A microprocessor are presented in detail in Table 5-3.

Table 5-3. Instruction formats—Signetics, type-2650 microprocessor *(Courtesy of Signetics)*

Addressing Modes	R = Register No. V = Value or condition X = Index register No. I = Indirect bit	* Index control 00 : Nonindexed 01 : Indexed with auto-increment 10 : Indexed with auto-decrement 11 : Indexed only		
(Z) Register Addressing (word: 1 byte)	Bits 7 6 5 4 3 2 1 0 Opcode (7–2) | R/V (1–0)	←8-Bit instruction		
(I) Immediate Addressing (word: 2 bytes)	Bits 15 14 13 12 11 10 9 8 Opcode (15–10) | R (9–8)	Bits 7 6 5 4 3 2 1 0 Data mask or binary value	←16-Bit instruction	
(R) Relative Addressing (word: 2 bytes)	Bits 15 14 13 12 11 10 9 8 Opcode (15–10) | R/V (9–8)	Bits 7 6 5 4 3 2 1 0 I (7) | Relative displacement $-64 < \text{Displacement} < +63$	←16-Bit instruction	
(B) Absolute Addressing: Nonbranch Instructions (word: 3 bytes)	Bits 23 22 21 20 19 18 17 16 Opcode (23–18) | R/X (17–16)	Bits 15 14 13 12 11 10 9 8 I (15) | *Index control (14–13) | Higher-order address (12–8)	Bits 7 6 5 4 3 2 1 0 Lower-order address	←24-Bit instruction
(A) Absolute Addressing: Branch Instructions (word: 3 bytes)	Bits 23 22 21 20 19 18 17 16 Opcode (23–18) | R/V (17–16)	Bits 15 14 13 12 11 10 9 8 I (15) | Page (14–13) | Higher-order page address (12–8)	Bits 7 6 5 4 3 2 1 0 Lower-order address	←24-Bit instruction
Indirect Addressing (word: 2 bytes)	Bits 15 14 13 12 11 10 9 8 (15) | Page (14–13) | Higher-order address (12–8)	Bits 7 6 5 4 3 2 1 0 Lower-order address	←16-Bit instruction	
(E) Miscellaneous Instructions (word: 1 byte)	Bits 7 6 5 4 3 2 1 0 Opcode	←8-Bit instruction		

5.9.5 SCP-234 Microcomputer System (RCA Avionics CMOS). Availability of small payload, low-power, flight computers enable overall spacecraft performance or capabilities such as enhancing

1. Command and control
2. Precision attitude control
3. EW system applications
4. Communications subsystems

The SCP-234 Microcomputer was specifically developed by Astro-Electronics Division of RCA for Aerospace applications. The unit, eight cards in all, weighs 8 lb at 300-in.^{3} space requirement and consumes 5 W of power. The computer uses CMOS-LSI, both for the CPU and the memory (RAMs). In one application, its read/write memory, with a write-protect feature, has a capacity of 16-K words of 16 bits each, plus 256 words of PMOS ROM control. The RAM is assigned areas by the jumper wiring, in blocks of 1024 words, to separate the data areas from the area of instructions and constants.

Arithmetic is binary 2's complement, fixed-point. (2's complement: subtract by inverting all bits and adding 1.) Since 15-bit precision is inadequate in precision control, software uses double-precision with two words without a speed penalty. Cued priority-interrupt system uses 16 levels, of which 14 are external. Single-word, 16-bit instruction execution is controlled by a ROM-microprogram. A total of 52 instructions are implemented: 12 arithmetic, 8 logical, 16 branch/skip, 13 load/store/transfer, 2 I/O, 1 control. ADD, in either the single- or double-precision format, takes 4.68 μsec.

I/O interface of the CPU is 16-bit parallel bidirectional-bus. The ROM contains the bootstrap loader; software can be written in a high-level language (SPL). A simpler assembler is also available for coding functions such as the command-control.

A block diagram of the attitude-control system in aerospace applications, given in

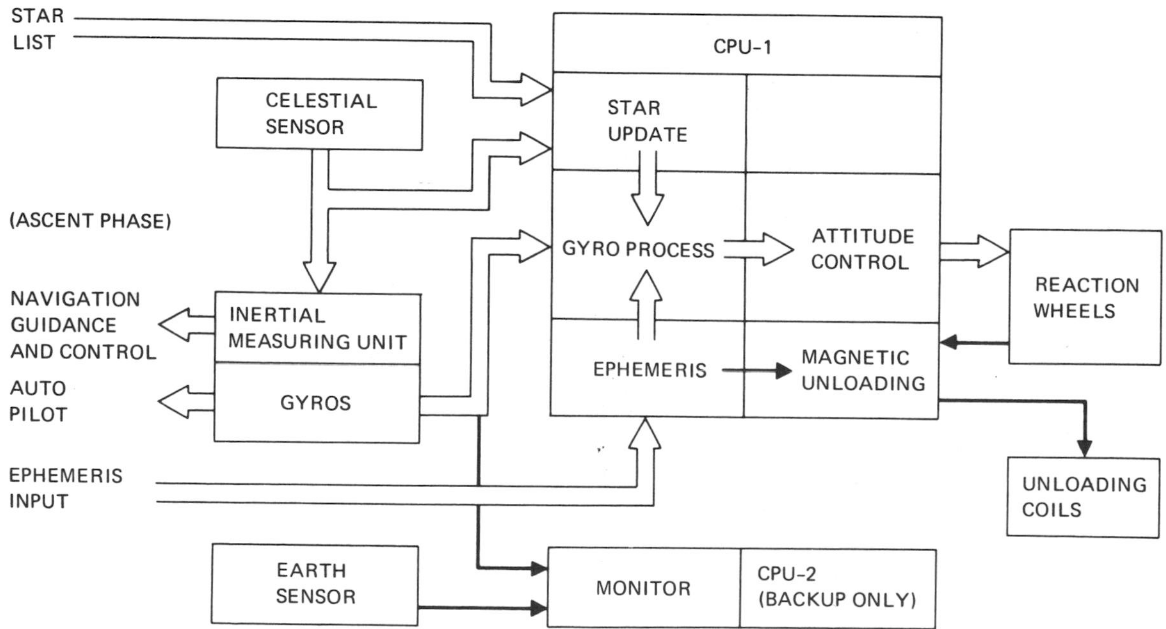

Fig. 5-14a. Application of the RCA Avionics, SCP-234 microcomputer system for attitude-control. (*Courtesy of IEEE Aerospace and Electronics Systems Society Newsletter*)

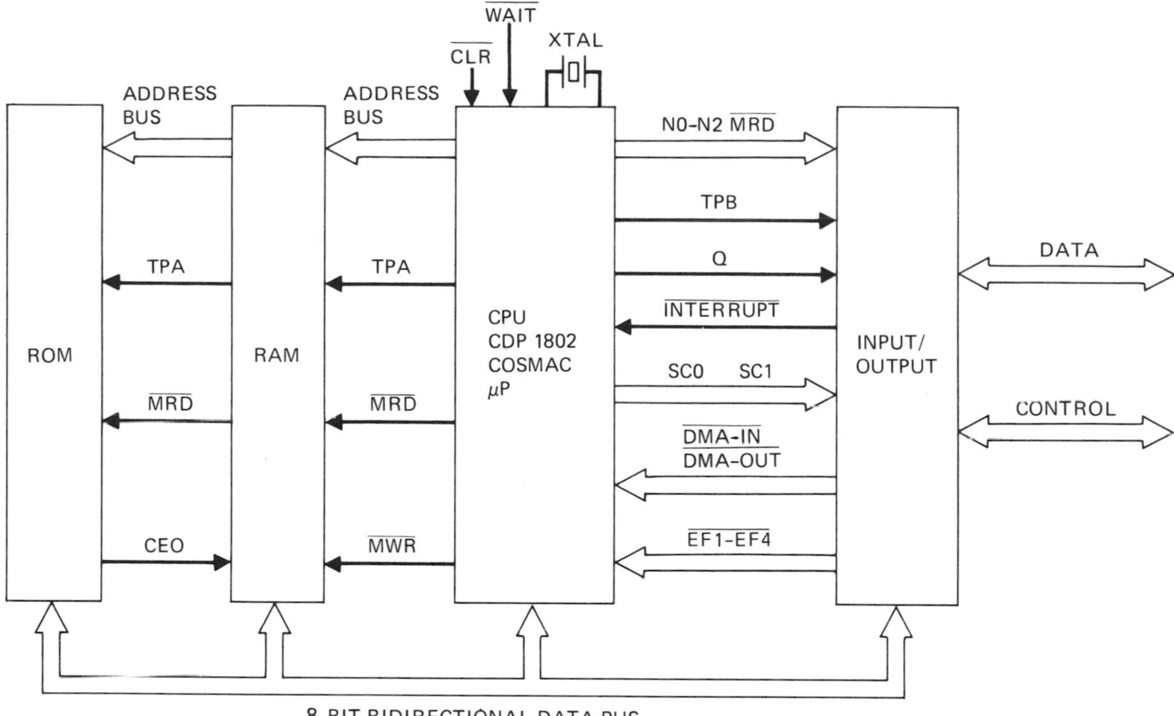

Fig. 5.14b. CDP-1802 Microprocessor. (*Courtesy of Electronics, McGraw-Hill*)

Fig. 5-14a, indicates the computer's functions. The CMOS microprocessor used in this application is illustrated in Fig. 5-14b.

5.10 RAM MEMORY

By using standby battery cells, low-power 1024-bit static CMOS-RAM chips are being evaluated and used at this time to replace the power-consuming higher-cost CORE memory.

MOS-LSI RAMs incorporate the following on-chip circuitry:

1. Address-decoder to select the desired cell(s) within the chip
2. Chip-select CS input signal, which activates the LSIs addressing and/or read/write circuitry
3. Write-amplifier to insert data into the address-selected memory cells
4. Provision to read or sense amplifiers to nondestructively read data out of the selected cells
5. *Open-collector* or *three-state* output buffers (see Fig. 5-15).

Static and dynamic RAMs need standby/"refresh" battery power only during computer off-time intervals; precharge signals and logic are also necessary to start operation. This facility will enable nonvolatility like the CORE memory.

Recent efforts in the memory field have accomplished the following tasks:

1. The bit-density of the leading NMOS dynamic RAMs is extended to 16-K and 64-K bits at the traditional 200-to-500-nsec access time.
2. New 1024-bit NMOS RAMs with 50-nsec access time are introduced as a substitute to 50-nsec higher-power ECL RAMs.

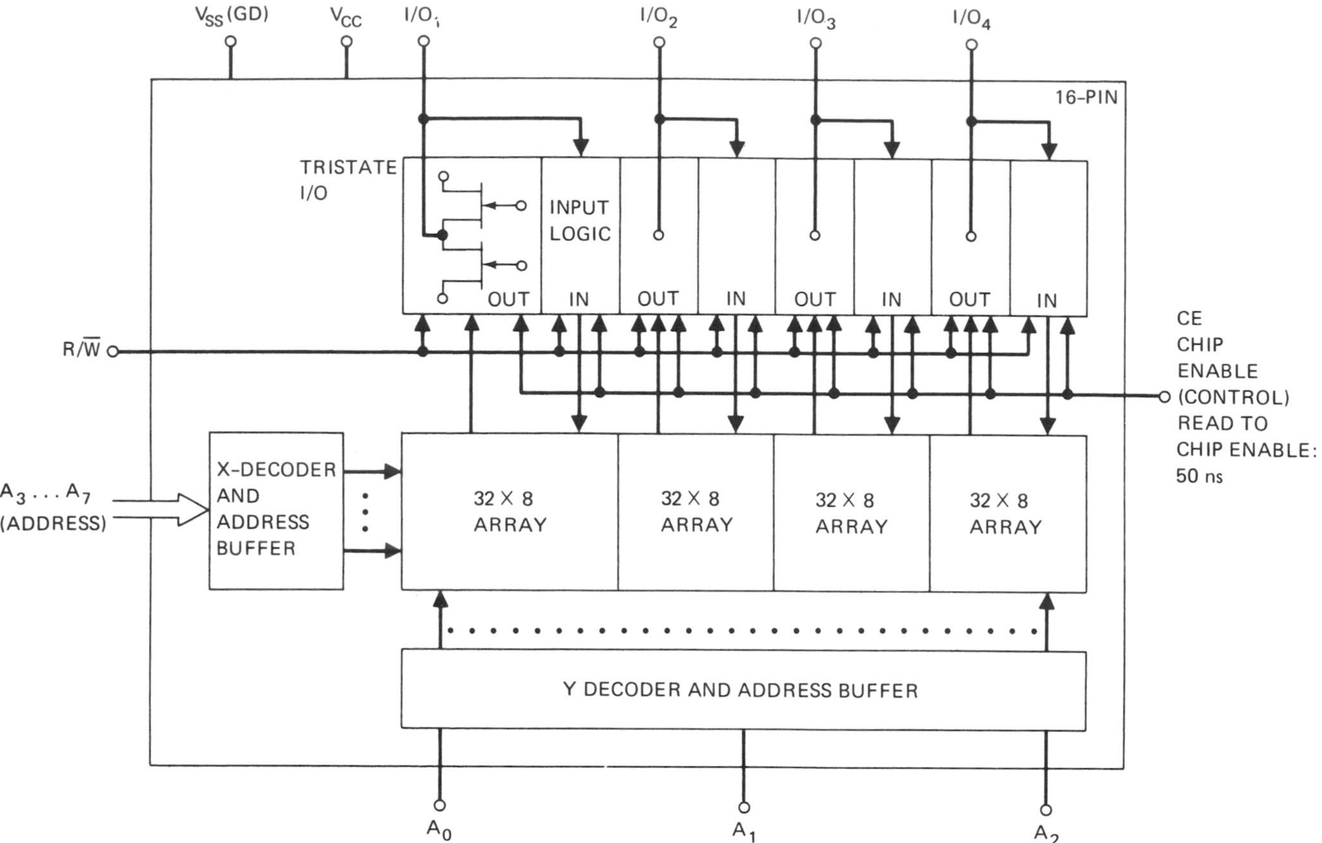

1. TYPE 2606 (SIGNETICS): NMOS, SI-GATE MAIN MEMORY.
2. ACCESS TIME < 750 ns. NO CLOCKS REQUIRED (STATIC): READ-CYCLE TIME: 500 ns.
3. INTERFACE SIGNALS TTL-COMPATIBLE. ADDRESS TO WRITE, TIME: 150 ns; WRITE CYCLE-TIME: 500 μs.
4. TRISTATE OUTPUTS ALLOW EASY EXPANSION OF MEMORY.
   (OPEN-COLLECTOR TECHNIQUE WITH EXTERNAL PULL-UP RESISTOR OF DIAGRAM 1(2).

Fig. 5-15. Static NMOS RAM, tristate output buffers (256 X 4 bit). (*Courtesy of Signetics*)

3. The available low-power CMOS, 4096-bit static RAMs are a substitute to the previous CORE memory, since they require just a little power from a standby battery.
4. I^2L, low-power high-density RAMs (both static and dynamic) are introduced as an alternative to NMOS, CMOS, and previous bipolar dynamic RAMs.

The 4096-bit dynamic NMOS RAMs are priced at 0.2 cents/bit, as compared to 0.5 cents/bit of CORE price. The static 1-K bit CMOS RAM is the most attractive memory because it consumes on standby 100 μW against 150 mW for a 1-K bit NMOS static RAM. A 1-K bit CMOS/SOS (silicon-on-sapphire) static RAM, with a 100-nsec access-time and just 1 mW of standby power, is available from RCA. Even a dynamic CMOS-RAM retains memory for weeks by activating the *refresh* circuits with a backup of four penlight batteries.

16-K bit RAM for main memory. The low-power, high-density, silicon-gate NMOS technology is developing further to reach the bipolar performance standards. Single and double polysilicon gate, V-notch, double diffusion, and charge coupling are the various techniques. (Polysilicon is the "metal" in the place of aluminum gate.)

TI's TMS-4070 is the first 0.5 W, 16,384-bit dynamic RAM that employs the sim-

Table 5-4. Types of RAM memory and possible density on chip.
(Courtesy of McGraw-Hill. From Carr, W. N., and Micze, J. P., 1972, **MOS/LSI Design and Applications.)**

Types of RAM Memory	Single or Multiphase Clocking Pulses	Need for "Refresh"	Area per Cell, mil^2
CMOS (6 transistor)	1	none	9
Static RAM (6 transistor) NMOS	1	none	13
Dynamic RAM (4 transistor) NMOS	2	parallel	11
Dynamic RAM (3 transistor) NMOS	2, 3, 4	serial	9
Dynamic RAM (1 transistor + capacitance) NMOS	2, 3	serial	3.5
CCD (LARAM)	1	none	1
I^2L (bipolar)	1	none	less than 3

pler NMOS silicon-gate single-level polysilicon approach. A 16-pin, single-chip, 16-K bit memory array requiring 1 mil^2/cell is no larger than the previously available 4-K bit RAM. The 16-K bit RAM will become the most favorite memory component in microcomputer systems. The double polysilicon approach with two separate polysilicon levels for transistors and capacitances, is the preferred production approach by other manufacturers, since it is a direct extension of the 4-K bit RAM process, and promises still higher memory capacity in a single chip. The high-density, double-poly technique approaches TTL performance in speed at a significant gain of three in power-delay product.

CMOS-on-sapphire, SOS RAM. The presently popular N-MOS RAMs are expected to run against an alternative preference for static CMOS-on-sapphire, where high speed and micropower are the criteria. RCA's NWS-5001 (1-K × 1 bit) at 150-nsec access time at 5 V, and its NWS-5040 (256 × 4 bit) at 120 nsec are two examples for applications in point-of-sales terminals, and automotive and telecommunications applications. A CMOS-on-sapphire, 512 × 8 bit ROM is also available from RCA for high-density memory applications. Table 5-4 indicates the various types of RAM memory and possible density of each type on a chip, as deduced from the area of a memory cell.

The RAM is both read-and-write. It could be partly low-power static and partly low-power dynamic; however, it is *volatile*, unlike the power-consuming nonvolatile core main-memory that presently has an access time varying from 300 nsec to 10 μsec. Besides the present NMOS/CMOS and bipolar I^2L, high-speed static semiconductor memory using multicell, power-consuming, bipolar TTL and ECL-10K are available; however, they are applicable to large- and medium-scale computers only. The latest ion-implanted bipolar semiconductor memories have access times of about 10 to 100 nsec, with gate-switching times of 1 to 2 nsec. A memory-instruction cycle-time is usually two to four times the access time. (Gate: logic gate; in the case of MOS, it is the "gate" electrode.)

The access time of the present high-density PMOS/NMOS and CMOS-on-sapphire is comparable to that of the high-speed CORE memory (300 to 1000 nsec), and hence their overwhelming popularity in the place of the CORE in practically all the microcomputer systems. As pointed out earlier, the high-density single-cell NMOS

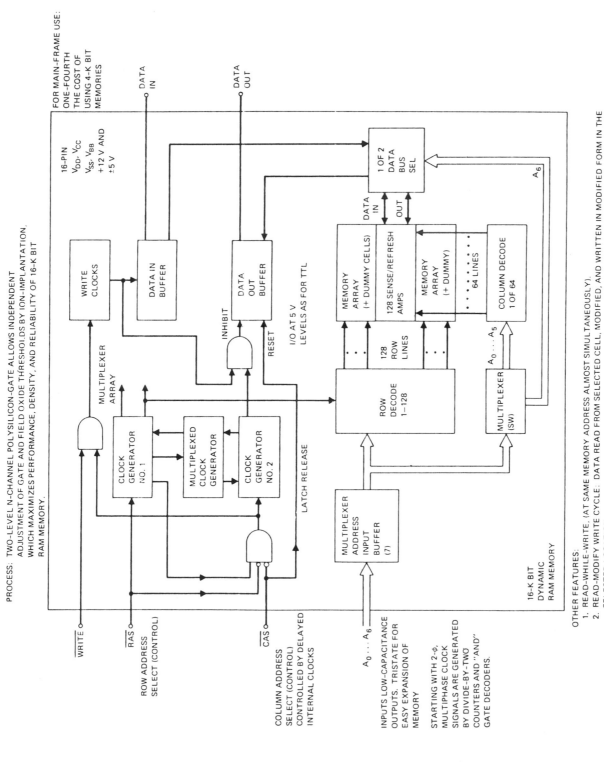

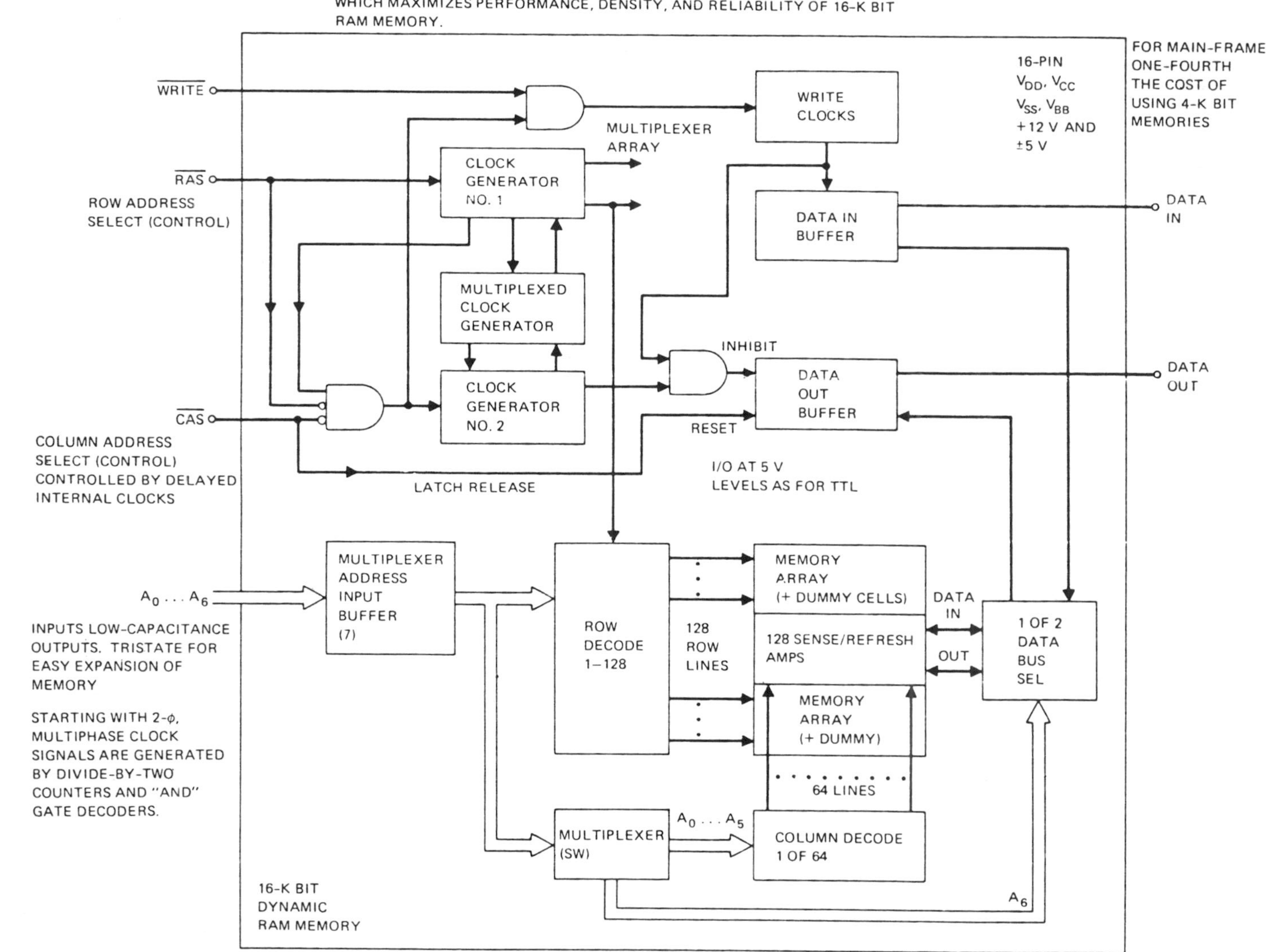

OTHER FEATURES:

1. READ-WHILE-WRITE, (AT SAME MEMORY ADDRESS ALMOST SIMULTANEOUSLY).
2. READ-MODIFY WRITE CYCLE: DATA READ FROM SELECTED CELL, MODIFIED, AND WRITTEN IN MODIFIED FORM IN THE SELECTED LOCATION.
3. PAGE-MODE: PAGE BOUNDARY EXTENDED BEYOND 128-COLUMNS IN A SINGLE RAM.
4. INTERNALLY ORGANIZED AS 2 X 8,192-BIT SUBARRAYS SEPARATED BY 128 SENSE AMPLIFIERS. TO THE USER, IT IS A 128 X 128 MEMORY (SINGLE TRANSISTOR CELL/BIT).
5. INTERNAL CLOCK GENERATORS ARE ACTIVATED BY CAS AND RAS CONTROL SIGNALS. (TESTING THE CHIP IS EASIER.)
6. SENSE AMPLIFIERS DO THE REFRESHING OF THE DYNAMIC MEMORY. (REFRESH EVERY 2 mS.)

Fig. 5-16. Latest high-density RAM main-memory, MOSTEK-MK4116, 16-K bit (volume production). Eight 16-pin chips to provide 16-K byte main-memory. (*Courtesy of Electronics, McGraw-Hill*)

memories (4K, 8K, 16K, 64K) are dynamic and volatile, and their storage function must be constantly replenished during the operation of the microcomputer by single or multiphase clock *refresh* circuitry. The auxiliary bulk or external mass-storage disk, drum, and floppy-disk memories can be also classified under *on-line* random-access memory, but their access time is very slow indeed, mostly in tens or hundreds of milliseconds. They are best suited to program libraries for software in high-level languages in the case of microcomputer systems. A high-density RAM is illustrated in Fig. 5-16.

5.11 INPUT/OUTPUT INTERFACE

5.11.1 Input/Output Interface: Typical Digital Data. Digital computer users conventionally establish requirements for the physical, functional, and electrical characteristics of the desired I/O Interface for transfer of digital data from the peripherals to the digital computer. It is the task of the Interface designer to provide the necessary hardware to implement this function in terms of the following bidirectional signals:

1. Parallel data transfer on twisted pairs up to 4-K 16-bit or 18-bit words on one cable. Binary voltage levels of 0-V(logical-1) and minus 3-V(logical-0) is one standard as an example.
2. Serial data transfer of words up to 64-K bit/sec on one twisted pair. (Bipolar ±3.25-V levels.)
3. External Interrupt Enable line
   External Interrupt Request line
   Input Data Acknowledge line
   Input Data lines
   External Function Request line
   External Function Acknowledge line
   Output Data Request line
   Output Data Acknowledge line
   Output Data line
   Input Enable, Output Enable, Control Frame, etc.
4. Drivers and receivers used for sending and receiving the signals on the twisted pairs are in practice differential amplifiers for the signal and its return. Conventionally, both the signal and its complement are sent on two adjacent twisted pairs.
5. Besides the preceding signals, in the case of synchronous digital processing, a digital clock and a strobe pulse-timing for the word accompany the above signals on separate twisted pairs. If provision is not made for a parity-bit in the regular words along with a sign-bit, as in the case of the 18-bit word computers, a separate twisted pair may be specified for the parity check-bit.

Typical I/O Interface *Handshake* signals of a computer and a peripheral device are shown in Fig. 5-17, along with typical pulse waveform timings. Figure 5-18 indicates bipolar high-speed real-time Microcontroller timings as a comparison (Signetics Microprogram control unit type N3001).

6. Some of the preferred standards of data transfer to a computer are specified in the following table.

Interface (I/O) parallel transfer/cable

Data Transfer Words/sec; TYPE	Bit 1	0	Logic
I: SLOW 41,667	dc	- 15 V	positive
I: FAST 250-K	0	- 3 V	positive
II: FAST 250-K	0	+3.5 V	negative

SERIAL: 10-M bits/sec/cable. Bipolar ±3.5 V.

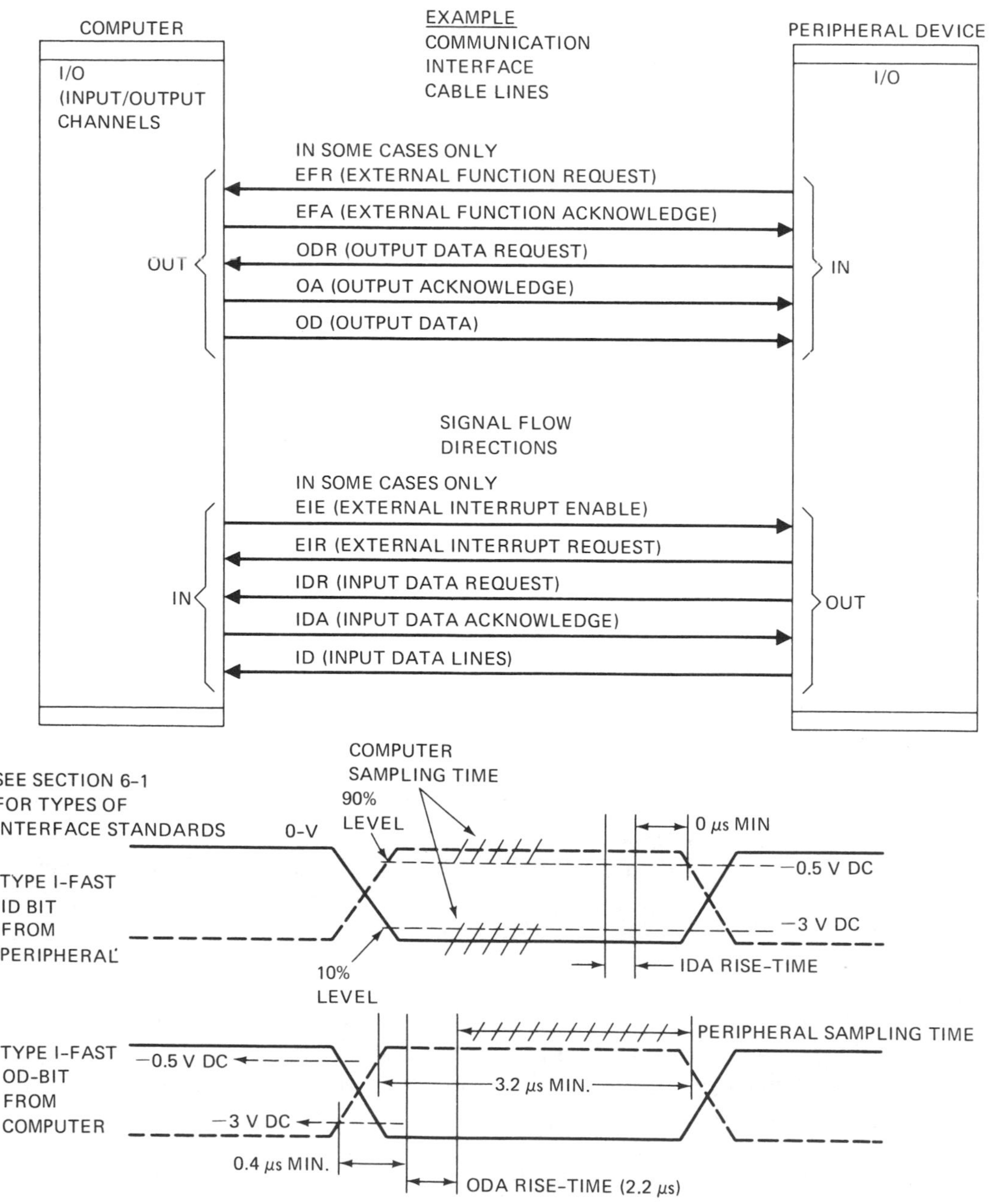

Fig. 5-17. Typical interface "handshake" signals of a computer and a peripheral device. Pulse waveform timings.

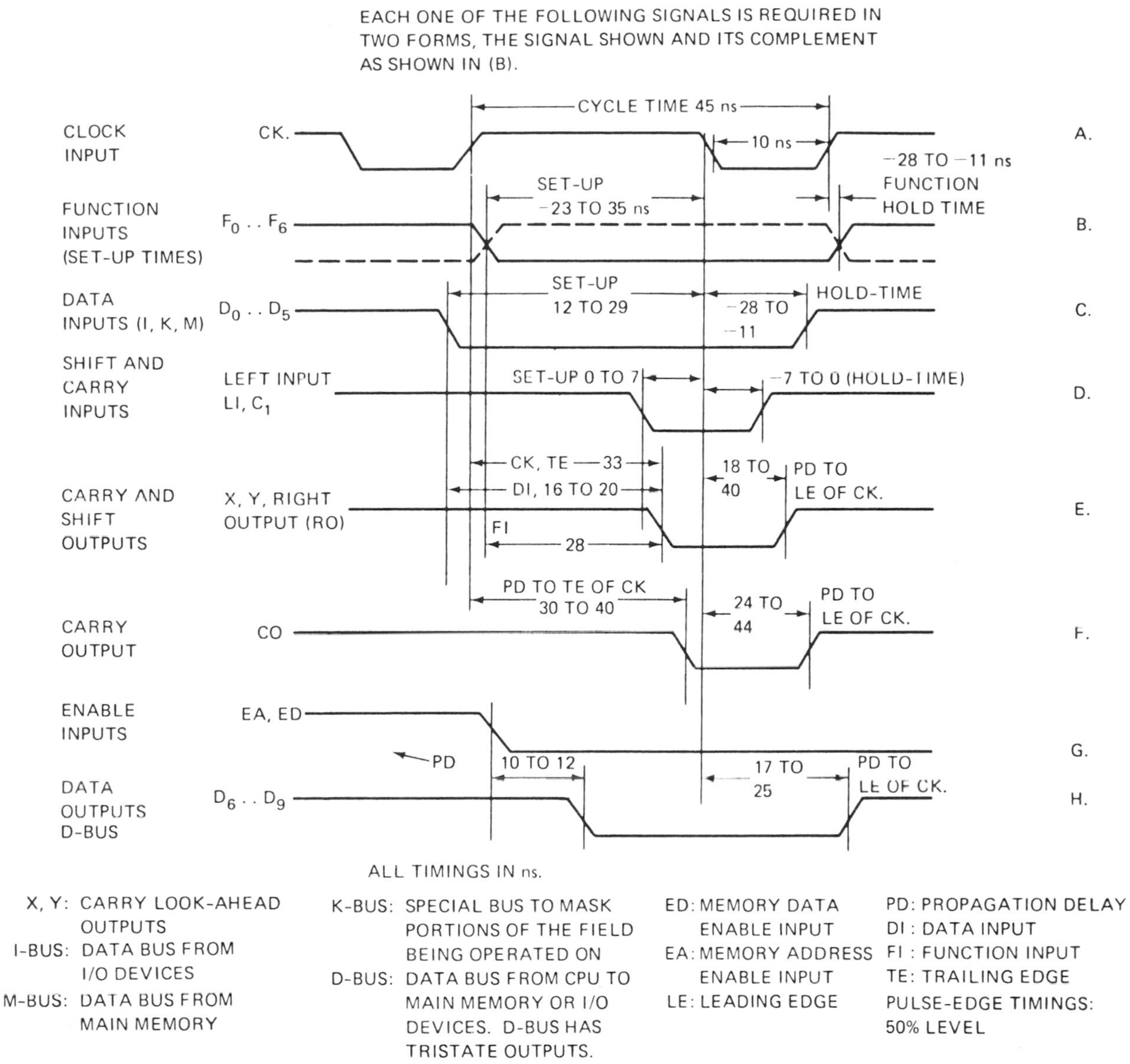

EACH ONE OF THE FOLLOWING SIGNALS IS REQUIRED IN TWO FORMS, THE SIGNAL SHOWN AND ITS COMPLEMENT AS SHOWN IN (B).

CYCLE TIME 45 ns

A. CLOCK INPUT — CK. — 10 ns

B. FUNCTION INPUTS (SET-UP TIMES) — $F_0 \ldots F_6$ — SET-UP −23 TO 35 ns — −28 TO −11 ns FUNCTION HOLD TIME

C. DATA INPUTS (I, K, M) — $D_0 \ldots D_5$ — SET-UP 12 TO 29 — HOLD-TIME −28 TO −11

D. SHIFT AND CARRY INPUTS — LEFT INPUT LI, C_1 — SET-UP 0 TO 7 — −7 TO 0 (HOLD-TIME)

E. CARRY AND SHIFT OUTPUTS — X, Y, RIGHT OUTPUT (RO) — CK, TE —33 — DI, 16 TO 20 — FI — 28 — 18 TO 40 PD TO LE OF CK.

F. CARRY OUTPUT — CO — PD TO TE OF CK 30 TO 40 — 24 TO 44 PD TO LE OF CK.

G. ENABLE INPUTS — EA, ED — PD — 10 TO 12

H. DATA OUTPUTS D-BUS — $D_6 \ldots D_9$ — 17 TO 25 PD TO LE OF CK.

ALL TIMINGS IN ns.

X, Y: CARRY LOOK-AHEAD OUTPUTS
I-BUS: DATA BUS FROM I/O DEVICES
M-BUS: DATA BUS FROM MAIN MEMORY
K-BUS: SPECIAL BUS TO MASK PORTIONS OF THE FIELD BEING OPERATED ON
D-BUS: DATA BUS FROM CPU TO MAIN MEMORY OR I/O DEVICES. D-BUS HAS TRISTATE OUTPUTS.
ED: MEMORY DATA ENABLE INPUT
EA: MEMORY ADDRESS ENABLE INPUT
LE: LEADING EDGE
PD: PROPAGATION DELAY
DI: DATA INPUT
FI: FUNCTION INPUT
TE: TRAILING EDGE
PULSE-EDGE TIMINGS: 50% LEVEL

Fig. 5-18. Bipolar *high-speed* real-time microcontroller (MPU) timings (compatible to latest high-speed minicomputer CPU timings). (*Courtesy of Signetics*)

5.11.2 Built-In Input/Output Section of a Microprocessor.

1. The I/O busses of the Microprocessor apply 4/8/16 bits in parallel, per port, to the bus leading to the peripherals via the Interface MSI chips (or LSI) as shown in Fig. 5-19.

2. There will be a number of I/O ports for each microprocessor.

3. Where the number of I/O ports are excessive, a multiplexer and a serial format are used between the I/O bus and the interface units. The I/O will generally use a serial to parallel data converter to communicate with the device in circuit at any instant.

4. Although all the peripheral Interfaces are connected in parallel, only one of them will be able to decode the device-address issued in the I/O instruction. The other Interfaces will not be able to decode that address, and hence they remain inactive.

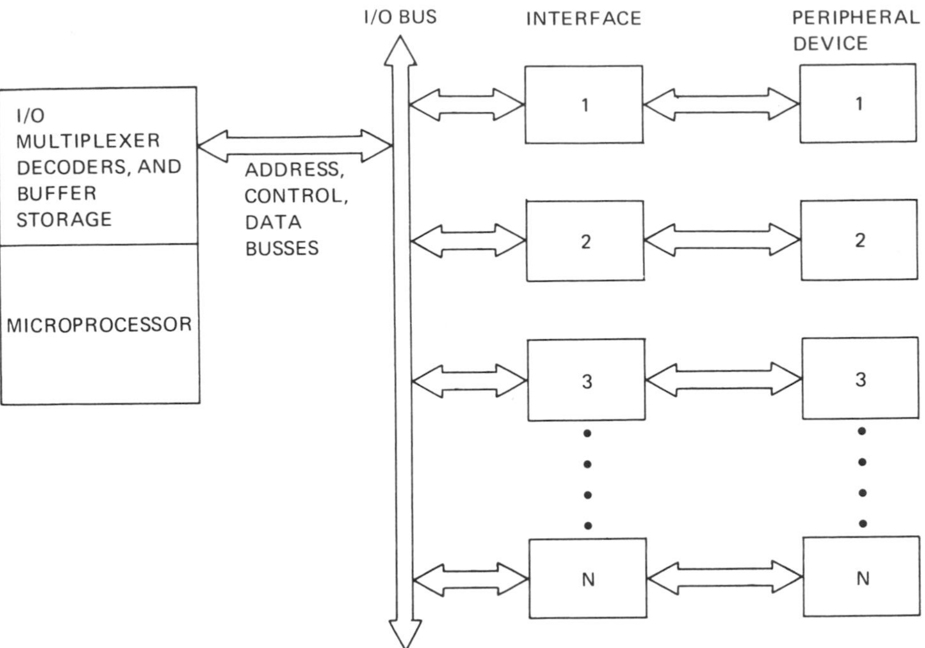

Fig. 5-19. I/O-bus and peripheral devices.

5. The general form of the I/O instruction word-format follows:

		OPERAND FIELD	
GROUP CODE	OPERATION CODE (OPCODE)	DEVICE ADDRESS CODE	FUNCTION CODE

The opcode specifies the instruction from one of the general groups of I/O instructions noted below:

INPUT: A word of data is read from the selected peripheral and placed in a working register.

OUTPUT: A word of data is transferred from the relevant working register to the selected peripheral.

CONTROL: A signal is issued by the microcomputer I/O to command a specific control function of the peripheral.

SENSE: The status of the peripheral Interfaces is tested. Conditional branches, which occur in the program, depend on this testing.

5.11.3 Multiplexer or Multiplexer Channel. The *channel controller* of a digital computer connects several peripherals, either directly or via a common or individual *device controller*. Each one of these may in turn take the form of a microprocessor, if they are connected to a central large-scale, medium-scale, or mini-computer. The multiplexer enables the computer to interface with not only the fast devices or peripherals but also slow devices such as TTYs and telephone line-conditioners or data-access units.

The multiplexer thus takes the form of a computer-within-a-computer. Actual data exchange between the main computer and the peripheral device or network terminal takes place at such a slow rate that it is possible to simultaneously service tens or even hundreds of such devices with a single channel-controller. A special microprocessor could be employed as a built-in CPU for the multiplexer and its control, while the multiplexer itself functions as an integral Interface for the main computer.

A peripheral may provide information in bytes or characters that differ in size (number of bits) from those of the words stored in the main memory of the computer. The multiplexer channel must do the necessary reformatting of the data or the instruction words from the peripheral device or the terminal.

Direct communication between the main memory and the multichannel controller can be also achieved by cycle-stealing without the awareness of the computer *control* system. The channel sneaks into the main memory and steals a memory cycle for the I/O; most often, the computer may not be held up at all when this transfer takes place.

Polled is synonymous with *multiplexed.* A digital multiplexer or encoder is in principle equivalent to a single-pole multiposition switch. It is a *combinational logic network* with 4, 8, 16, or more, input or data lines, one synchronized output line (and a complement of output-line), and i-control or data-select lines, where 2^i is the number of input lines. Combinational logic is realized by appropriate application of the input variables or binary constants to the data and data-select inputs. Conversely, a Decoder/Demultiplexer is a combinational network that handles serial input-data, i control inputs, and 2^i outputs, where each decoded output is unique. A combination logic network is the part of digital logic that does not include memory elements such as flip-flops.

A multiplexer or an encoder may in principle have any number of inputs and outputs; at any instant, several combinations may be true depending on the actual switching logic used. The corresponding *polling* configuration will then take the form of, for instance, a 4-pole, 4-throw switch. A decimal to binary-coded-decimal (BCD) code converter may be called a decimal to BCD/8-4-1 weight encoder or multiplexer since it can multiplex a 10-line input to a 4-bit binary equivalent on 10 $\times$ 4 outputs, if required. A typical open-collector 3-input, 4-bit digital multiplexer is illustrated in Fig. 5-20; it is analogous to a 4-pole, three-position switch.

5.11.4 Decoder/Demultiplexers. Corresponding to the multiplexing polling function just described, the multiple-line decoders do the reverse code-conversion process such as obtaining the 10-digit decimal equivalent on 10 output lines for a binary input bit-set of say four weighted lines (8-4-2-1), either pure binary or BCD, etc.

There are, however, several types of decoders. For example, a 2-input variable code produces four possible output variables (00, 01, 10, 11); and a 3-input octal code produces eight possible outputs. The device implementing the necessary logic in these cases to produce an output that indicates the state of the input variables is also a decoder. In this classification, a *majority decoder* using three input variables produces one true output only when two or three inputs are true; the corresponding "minority decoder" produces one true output when only one input is true; if an odd or even number of true inputs produce a true output, the decoders are odd or even, respectively. Similarly, special-purpose counter readouts, adders, subtractors, parity comparators, etc. can be classified as general decoding functions.

A binary coded 2-line to dual 4-output line low-power, high-speed ECL decoder/demultiplexer is shown in Fig. 5-21, as an example, along with its truth table of positive logic. This parallel decoder uses a special technique of internal emitter dotting

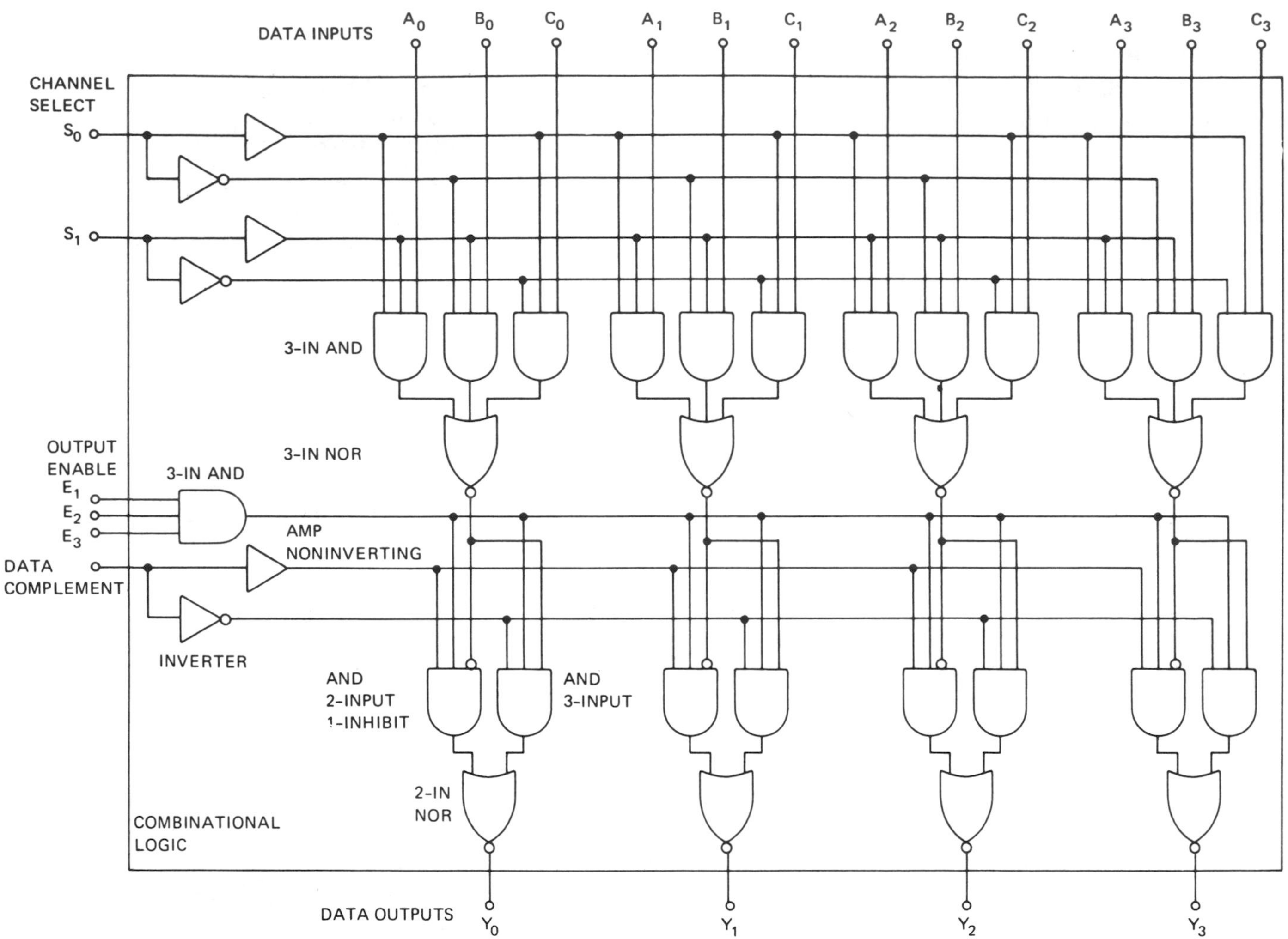

- FOR STORAGE, FOUR INVERTERS AND FOUR R-S FLIP-FLOPS ARE INCLUDED IN THE CHIP WITH A CLOCK INPUT.
- TWO-INPUT CHANNEL-SELECTION CODE DETERMINES WHICH INPUT SHOULD REMAIN ACTIVE.
- DATA COMPLEMENT INPUT CONTROLS CONDITIONAL COMPLEMENT CIRCUIT AT THE OUTPUT TO EFFECT EITHER INVERTING OR NONINVERTING DATA FLOW.
- OPEN-COLLECTOR OUTPUTS ALLOW EXPANSION OF INPUT TERMS.

- EXPANSION: OF ABOVE MSI CHIP TO FOUR-POLE 24-POSITION SETUP
  1. CONNECT ABOVE OUTPUTS TO THE OUTPUTS OF ANOTHER SIMILAR MULTIPLEXER.
  2. PROVISION IS MADE FOR USE OF A 3-BIT CODE (OUTPUT ENABLE E_1, E_2, $E_2 \equiv 2^0, 2^1, 2^2$ AND COMPLEMENTS) TO DETERMINE WHICH MULTIPLEXER IS SELECTED. THUS, EIGHT MULTIPLEXERS ARE USED TO EFFECT A FOUR-POLE 24-POSITION SWITCH.
  3. OPEN-COLLECTORS: USE COMMON COLLECTORS WITH EXTERNAL PULL-UP RESISTORS (ONE/4 OUTPUTS) AND USE OUTPUT ENABLE CODE.

Fig. 5-20. A typical open-collector, three-input, 4-bit digital multiplexer. (TTL MSI-chip; analogous to four-pole, three-position switch.) Type 8263, Signetics. (*Courtesy of Signetics*)

- OUTPUTS NORMALLY HIGH, WITH SELECTED OUTPUTS GOING LOW.
- EMITTER DOTTING DESIGN TECHNIQUE OF THIS CHIP ELIMINATES UNEQUAL PROPAGATION DELAY TIMES.

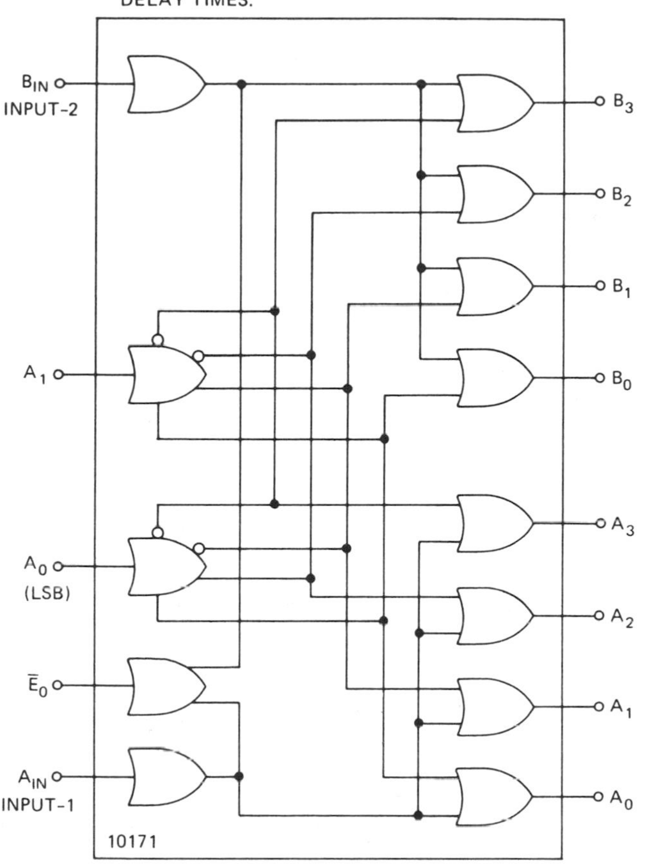

ECL 10,000 SERIES OPEN EMITTER OUTPUTS

- APPLICATIONS:
  DUAL 1 : 4 DECODER
  CROSSBAR SWITCH.
  HIGH FAN-OUT 1 OF 4 DECODER.
  MEMORY CHIP SELECT DECODER.
- 4 ns: ADDRESS TO OUTPUT.
- HIGH FAN-OUT DRIVES 8 OF 50 OHM LINES.
- HIGH IMPEDANCE INPUTS; INTERNAL 50-K Ω PULL DOWNS.
- ENABLE INPUT WHEN HIGH FORCES ALL OUTPUTS HIGH.

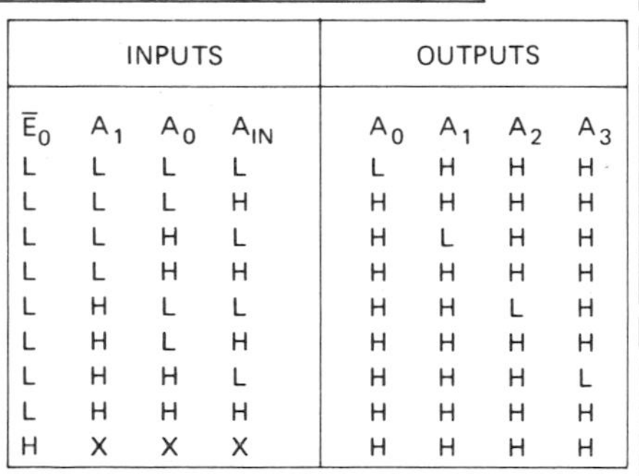

INPUTS				OUTPUTS			
$\bar{E}_0$	A_1	A_0	A_{IN}	A_0	A_1	A_2	A_3
L	L	L	L	L	H	H	H
L	L	L	H	H	H	H	H
L	L	H	L	H	L	H	H
L	L	H	H	H	H	H	H
L	H	L	L	H	H	L	H
L	H	L	H	H	H	H	H
L	H	H	L	H	H	H	L
L	H	H	H	H	H	H	H
H	X	X	X	H	H	H	H

X: DON'T CARE

TRUTH TABLE

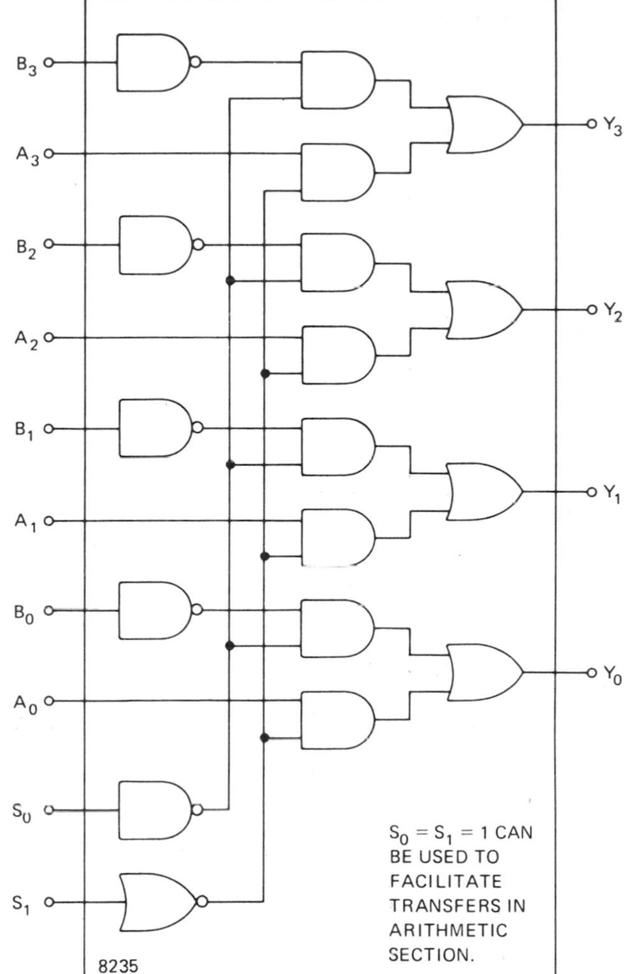

- USED AS A FACILITY FOR INPUT TO ADDERS, REGISTERS, ETC.
- OPEN-COLLECTOR OUTPUTS, WHICH CAN BE DIRECTLY WIRED TO OTHER FREE 4-BIT WORDS (100 WORDS CAN BE MULTIPLEXED).

TWO-INPUT 4-BIT DIGITAL MULTIPLEXER USED FOR GENERAL-PURPOSE DATA-SELECTION.
(SHOWN TO ILLUSTRATE THE REVERSE FUNCTION OF THE DECODER AT LEFT.)

Fig. 5-21. A typical two-input, dual four-output decoder/demultiplexer. Types 10171 and 8235. Signetics. (*Courtesy of Signetics*)

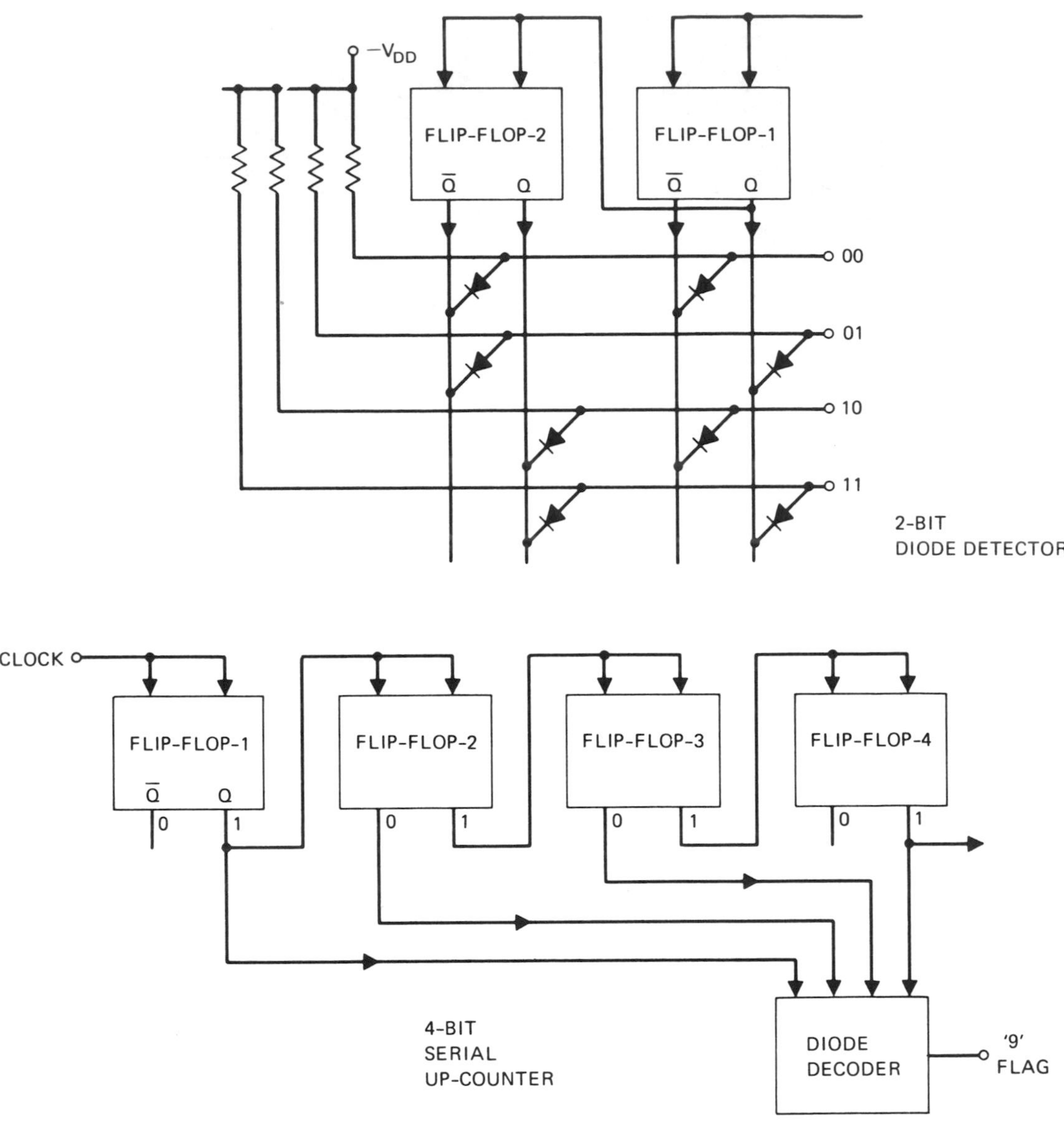

Fig. 5-22. A simple 2-bit diode decoder.

technique to eliminate unequal delay times, and can be used for demultiplexer, telephone cross-bar switching, fan-out for distribution of signals, and memory chip-select decoder applications. For direct comparison, an MSI/TTL 2-input 4-bit multiplexer, used in data applications, is shown adjacent to the decoder.

There are other possible variations of decoders. A simple diode matrix decoder can be represented by two flip-flops with two vertical output lines and matrixed-diodes for detection (or decoding) on horizontal lines. This matrixing feature can be extended to a large number of variables.

A simple diode decoder, for instance, for decimal 9, from a 4-bit serial up-counter can provide a *flag* for a subsequent ALU function such as *compare.* Figure 5-22 illustrates the configurations of a simple 2-bit diode decoder along with its two flip-flops.

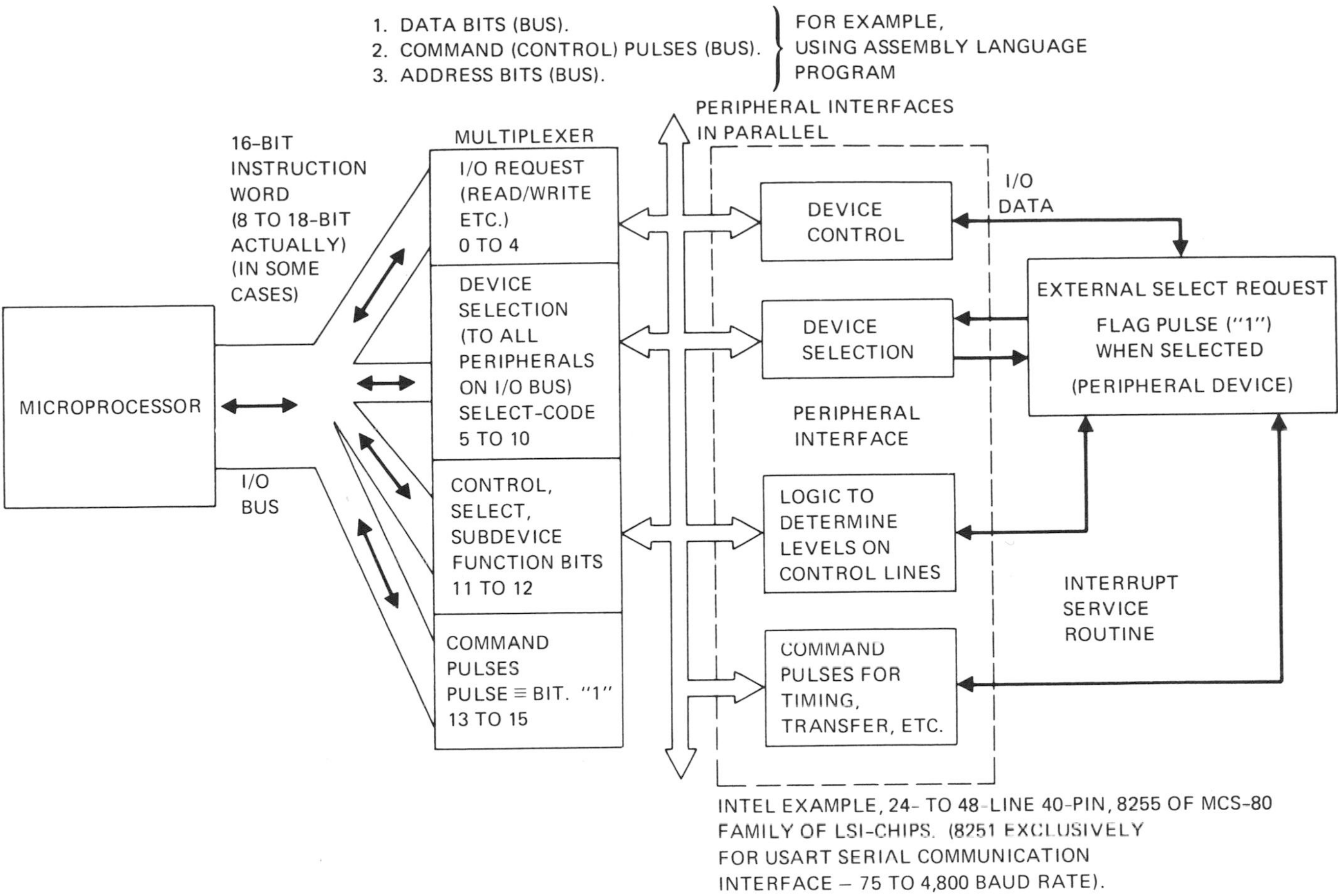

Fig. 5-23. I/O-bus interface data for several peripherals.

5.11.5 User's Line I/O Bus Interface. When a few peripheral devices are used in a microcomputer system, a bidirectional bus of 8 to 18 data-bit lines is adequate by including multiplexers for device selection via data lines. However, a common interface technique used for connecting a large number of peripheral devices, such as displays, A-D/D-A converters, printers, TTYs, etc., to a mini- or micro-computer of a wider scope is the so-called party-line I/O bus system. The peripherals send and receive data to the accumulator or memory by means of a parallel I/O data bus connected via appropriate logic to an I/O register in the microprocessor. Other device-select lines carry control-logic signals for device selection and synchronization of the data-transfers to the memory cycle-time. Figure 5-23 illustrates the simple interface system, and how a 16-bit I/O instruction word is allotted for device selection and control logic. Bits 0 to 4 request an I/O instruction word from the processor; bits 5 to 10 are device-address bits in a device-select code; bits 11 and 12 indicate special functions, if any; and bits 13 to 15 produce sequentially timed commands. A 16-bit I/O instruction word can then select any one of 2^{11} = 2,048 devices or device functions in four memory cycles.

5.11.6 Programmable Peripheral Interface of Intel. Intel LSI chip type 8251 is a typical example of the future course of direction in respect to the Interface of a microprocessor and a peripheral device-controller or a transmission link. (An

Interface design that includes the necessary decoding and buffering logic usually requires about 20 to 30 TTL or CMOS IC's.)

The Intel type 8251 LSI configuration is shown in Fig. 5-24, along with Intel MCS-80/85 system features. The type of transmission (whether serial synchronous or asynchronous), the length of the instruction word, and so forth have to be defined before starting the program. Each time, at start, the program loads the four control registers shown. A divide-by-four counter, along with a reset control-bit, selects (multiplexes) these registers in sequence to eliminate the need of four exclusive address words. For a dedicated application, the chip is preprogrammed, but the facility of control-ROM programmability makes the LSI chip highly flexible for any other application.

5.12 MICROCOMPUTER PERIPHERALS

- Peripheral device such as powerful source-program editing CRT-display (in place of a TTY)
- Line-printer hard-copy device for hard-copying the listings of the source-programs connected with debugging and documentation
- Mass-storage device, such as a cassette and/or floppy-disk, for rapid access to the assemblers of microprogramming and the compilers of high-level languages
- High-speed paper-tape system to handle paper taped programs.

All four are effective for the successful development of compatible software of minicomputer systems. Using available minicomputer hardware-interface, they are all applicable to the microprogrammable microcomputer systems. If efficient software development were attempted afresh for the microcomputer systems, such a development system would cost 75% of the whole microcomputer operating system. However, the latest high-speed miniperipherals would reduce such software development cost by at least 50%. As a long-range perspective, exclusive efficient software development is in order for the future microcomputer systems by using a few LSI programmable-logic array (PLA) chips in the place of the present bulky hard-wired logic designs. ROM-able macroassembly and high-level languages are foreseen.

For example, by using microprocessor-controlled CRT intelligent terminals, Raytheon has evolved a PTS-1200 distributed data processing system that performs the following six functions:

1. Source data-entry and preprocessing
2. File and record maintenance
3. Unattended two-way point-to-point or multipoint communications
4. Stand-alone batch-processing
5. Local report-printing
6. On-line IBM-3270 emulation

Some of the more common peripheral devices presently available for the use of the microcomputer systems are described in the following paragraphs:

1. Punched-card is widely used unique form of computer input with the following advantages:

   a. Each card is discrete as a unit-record.
   b. It provides the convenience of manipulation for addition, deletion, and rearrangement in records.
   c. It provides the facility of direct manual reference to individual records.

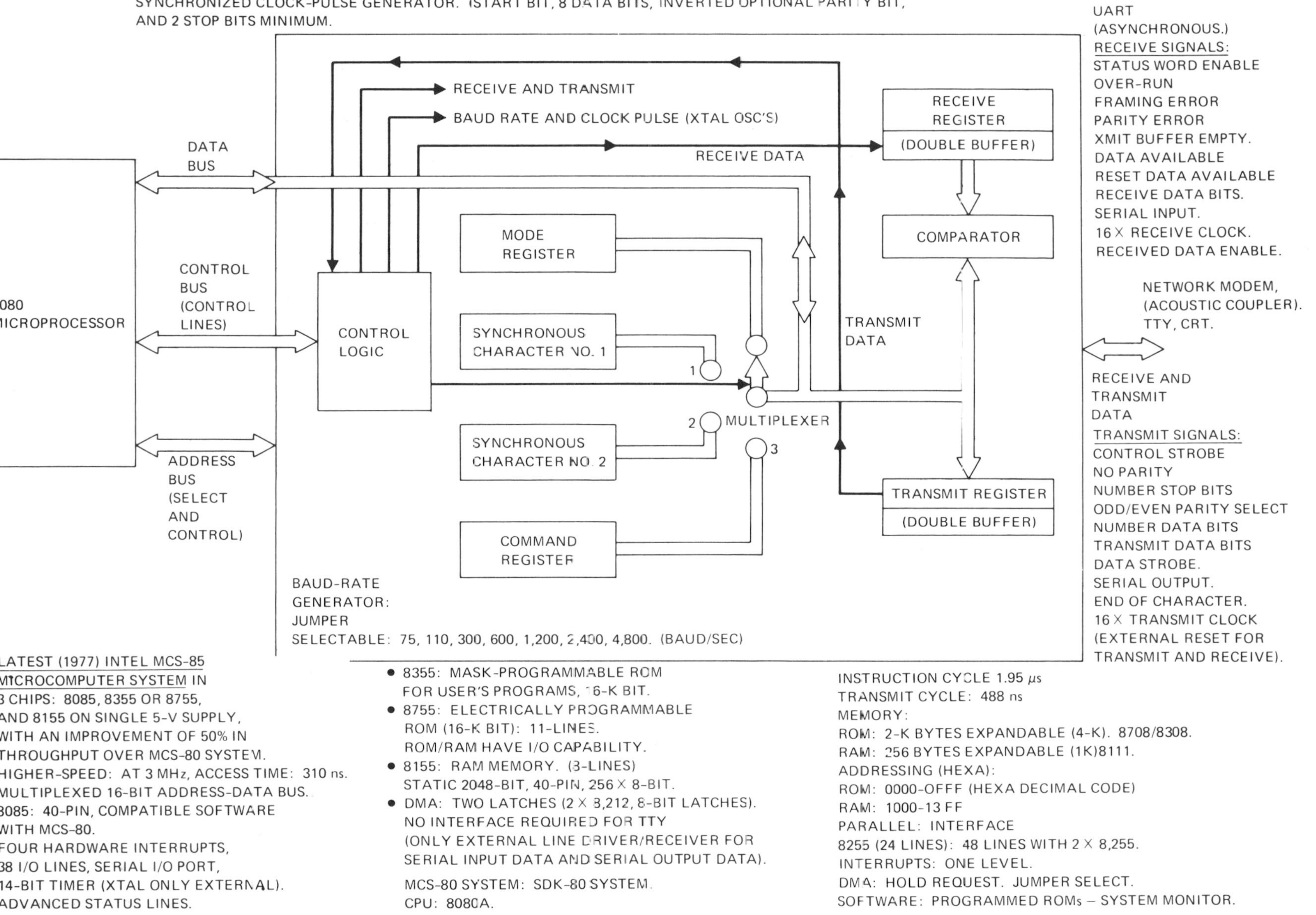

Fig. 5-24. LSI Interface, Intel type 8251 (for USART serial communications interface, EIA, RS-232-C). LSI chips used in Intel systems types MCS-80 and MCS-85.

The keypunch machine is a typewriterlike machine for punching the character code of a word as holes in a 12-bit × 80-column card. The IBM punched-card dimensions are specified as 3.25″ × 7.375″ × 0.0067″.

2. TI model-84 Card-Reader is a column-oriented device that reads punched-hole data in 80-column cards at approximately 400 cards/min. The data are transferred through fiber-optics to an array of photosensors. The Card Reader is complete with input hopper, feed mechanism, read station, stacker mechanism, output stacker, and timing mechanism with supporting drive-belt and motor, control-and-error electronics, and power supply.

3. TI model-306 line-printer is a medium-speed, impact printer that uses a standard 10-point type 5 × 7 dot-matrix for character generation and prints 80 characters/line/sec. Standard sprocketed paper is sprocket-fed, and paper widths from 4″ to 9.5″ can be used. One master and up to four copies are available. Standard print-format is 10 characters/in. and six lines vertical/in. The other specifications are:

Transmission rate: 100 to 9600 baud/sec
Code: ASCII, 64 characters printed
Character buffer: 80 characters (one line)
Paper feed: Adjustable to 9.5 in.
Dimensions: 12.75″ × 18.75″ × 23.5″; weight: 66 lb

4. Punched paper-tape has remained the most popular program-loading device for minicomputers. In view of the electronically alterable semiconductor programmable ROMs, program-loading by paper tape may not prove to be as popular for the microcomputer. Sophisticated microcomputer systems would prefer tape-cartridges or floppy-disk mass-memories; stand-alone special-purpose microcomputers would use PROMs and ROMs. However, where TTYs are involved as peripherals, the paper tape as a program-loader will continue in usage, especially in the extensive communications field. Teletypewriter types ASR-33 and ASR-35 have built-in 10 characters/sec paper-tape-punches and readers to feed the program tapes in the binary machine-language for occasional program preparation. Digital Equipment Corporation markets a 50-character/sec punch and a 300 characters/sec reader, using a fanfold paper tape that does not require rewinding (price: $300 to $3,000, depending on the tape handling speed).

In the case of the TTYs ASR 33/35, an 8-character paper tape in ASCII code appears with binary 1-2-4 weighted 3-bit groups, one group of holes on either side of the sprocket holes, and a third 2-bit group at the bottom for characters 7/8. The 212 line-feed and 215 carriage-return codes function as controls in the abovementioned TTYs.

5. Tape and CRT-keyboard interface (IBM-5100)

a. The large-capacity magnetic tape-storage employs plug-in capstan-driven data cartridges with file-protect features. Each cartridge holds up to 204,000 characters on a 300-foot, 0.25″ magnetic tape. Searching or rewinding are both performed at 40 ips.
b. The visual display-screen presents up to 1024 alphanumeric characters in 16 lines on a 5″ diagonal CRT.
c. A full-function keyboard and a 10-function calculator pad are provided.
d. An optional type-5103 desktop printer (12.25″ × 13.25″ × 23″. 60 lb, at a price of $3675) can be easily integrated into the portable computer system.

e. An optional auxiliary tape unit is also available as an added software facility. Each of these peripheral devices has a built-in controller and the necessary interface-electronics on a plug-in MSI printed-circuit board.

A back-plane card guide-assembly provides the facility for plug-in of the controllers and additional memory adjacent to the CPU. This system organization appears to be universal in all the existing and upcoming microcomputer systems.

Tape cartridges, reading and writing speed (IBM). The magnetic tape-cartridge unit driven at 40 ips has an effective reading-rate of up to 2850 alphanumeric characters/sec; however, in the record mode, the machine can write up to 950 characters/sec only, because, during the write-mode, a self-checking read-head is involved with its own time allotment for output information.

6. Video-RAM (V-RAM). As used in the IBM-5100 portable computer system, the video random access memory is a modular device that provides an interface between the microcomputer system and an ordinary TV-monitor of any picture size. The V-RAM functions like any other static-RAM; it is organized in 128 words of 8 bits each, and its output is presented as a conventional analog composite video signal that can be directly connected to any standard television-monitor. A single V-RAM has the provision to drive 25 television monitors. The device has a large repertoire of characters (ASCII capital and lower-case alphanumerics, symbols, and Greek alphabet) and operates at 5 V, 1 W. (One VLSI single-chip may provide this facility in a television receiver.)

7. Floppy Disk. The floppy-disk random-access memory used, for example, with the minicomputer of the Digital Equipment Corporation comes in either single-disk or dual-disk drive at a comparatively low cost. The disk-storage system is interfaced to DEC, LSI-11 microcomputer system as well, by means of one dual-height interface card that connects either drive to the system backplane. The floppy disk serves

a. as a mass-storage medium of 256-K or 512-K byte-capacity
b. for a high-level language programming such as the multiuser BASIC and FORTRAN-IV.

For this Operating System, 360 rpm, 3,200 bit/in., 77 tracks/disk, and a latency of 83 ms make the specification.

8. Tape Cartridge. As a substitute to the floppy-disk mass-storage, the IBM-5100 portable microcomputer has a system-integrated, miniature 3M built-in tape unit that reads up to 2850 characters/sec, and simultaneously writes and checks at 950 characters/sec. The tape unit uses a capstan-driven, 0.25-in. magnetic-tape, IBM data-cartridge that stores up to 204-K bytes. It searches and rewinds tape at 40 ips. As data and programs are developed, they can be stored off-line on these data cartridges for subsequent usage or for use in other software applications. Data stored on the cartridge are protected from accidental erasure by means of a file-protect feature. Also, data can be secured by removing the cartridge from the microcomputer unit. An optionally available auxiliary tape unit can be connected to the microcomputer system to speed up and simplify the preparation of file-updates of the cartridge plugged in the built-in tape unit. IBM, for example, supplies for its portable computer a library of ready-to-use cartridges for engineering, statistical, and business applications; the interface system is built in.

A typical specification of a magnetic-tape drive system for the mini- and microcomputers: minireel, single-capstan, 12.5–75 ips, 200–800 bpi (bits); 7-track non-return-to-zero interface; a phase-encoder tape-controller is required. Heads:

single-gap read/write and dual-gap write/read. (Nine-track tape-drive is also used on 0.5-in.-wide tape.) 7 tracks: Numeric 1, 2, 4, 8. Alphabet-Zone A, B. Even parity-bit C. Start-stop: 8 ms at 45 ips. Small reels with limited storage are called either *tape-cassette* or *tape-cartridge systems.* Limited storage on the reel necessitates multiple cassette drives. Incidentally, flexible disks can easily replace cassettes to provide faster access to data or instruction words. Figure 5-25 illustrates the inter-record gap and data record on a 7- or 9-track magnetic tape.

RZ and NRZ Magnetic Tape Recording. The coding method used for pulse-signal recording is called *return-to-zero* (*RZ*); and DC-level recording is called *non-return-to-zero* (*NRZ*). RZ recording requires one pulse to turn on from zero-to-saturation and another pulse to turn off from saturation-to-zero. As the tape can be positively or negatively saturated about zero, pulse-type RZ recording may have three signal levels. A pulse tape-cell is the significant requirement. The direction of magnetization of the tape-cell determines an area bit-content as a "1" or "0." NRZ recording is a dc-level change to signify plus-saturation level and minus-saturation level. There is no zero-level as such; hence, it is termed *NRZ.* If alternate "1s" and "0s" are to be written, level changes take place between cells, and the waveform resembles RZ recording. The corresponding IBM technique is called "NRZ-MARK" or "NRZ Modulo-2"; however, it is more sophisticated.

5.13 SOFTWARE PROCEDURE IN MICROCOMPUTER SYSTEMS

The programming techniques applicable to minicomputer systems do apply, although the actual number of instructions add up to need external ROM and RAM memories, in view of the limitation of instruction lengths to one-to-three 8-bit bytes. Programming experience with such high-level languages as BASIC, minicomputer FORTRAN, or APL cannot be directly utilized to apply to the low-level language required by most microprocessors. Software talent in program planning and in testing or debugging of minicomputers should be augmented with some familiarity of the details of the microcomputer hardware system and a specific programming technique for a specific microcomputer system.

Latest developments in hardware/firmware. With a drastic cut in engineering turn-around time, the hardware/firmware field is experiencing a kind of explosion in both developments and applications that reach most minicomputer levels of complexity. The latest microprocessors, mass-produced at low cost and custom-designed along economically feasible techniques, make the previous task of logic design much easier in microcomputer applications. The system and logic designers presently work with firmware of compilers, assemblers, editors, and debugging software related to special and custom hardware from microprocessor manufacturers; engineering changes are mostly made by altering a pattern here and there on a programmable electronically-erasable ROM or by plugging in an alternative ROM or PLA. Complexity of testing new microcomputer systems depends on the degree of assurance required and hence cost, whereas mass-produced dedicated systems enjoy an advantage in this respect.

Firmware allowing hardware/software trade-off. The hardware cost of computer systems is presently declining at a rapid rate, and LSI memories can economically provide computer functions that are normally performed by software. LSI chips (meeting Defense standards) with custom-designed gates (as field programmable logic arrays, or FPLAs) operate at speeds about twice as fast as those of LSI-ROMs, presently used for microprograms. Computational speed is basically the main reason for furnishing such hardware for software, especially for such complex computations

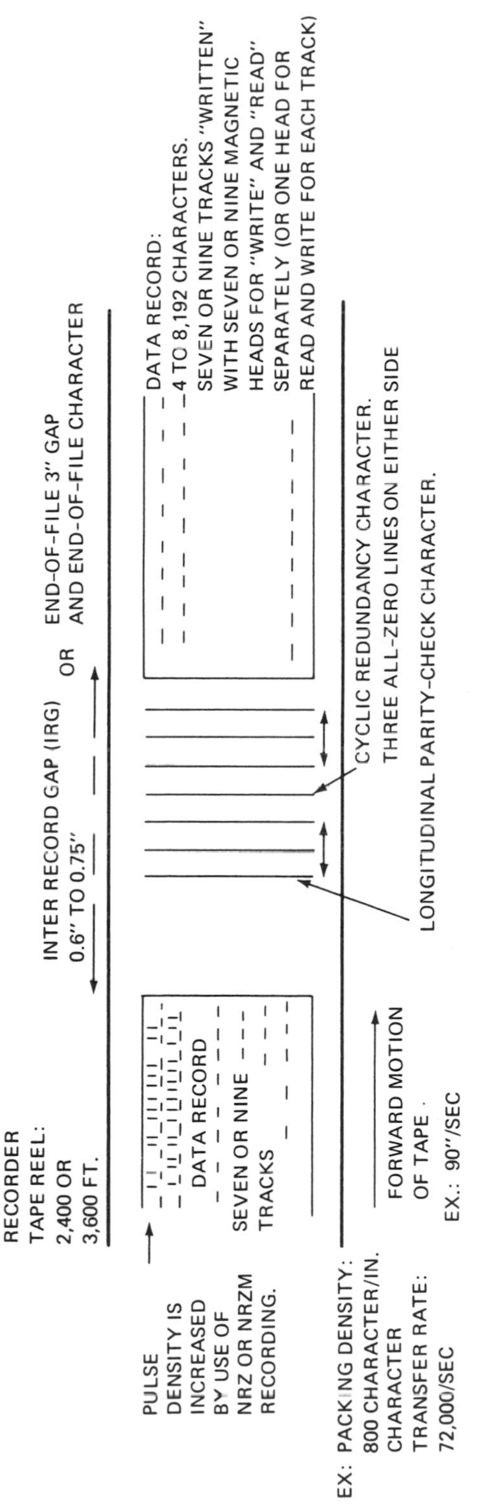

Fig. 5-25. Magnetic tape-Data record on 7/9-track tape.

as sorting, Fast Fourier Transforms in digital signal process, and floating-point arithmetic. For example, an algorithm for recognizing hand-printed characters, programmed in IBM PL-I high-level language, required 7 sec to run on an IBM-360/65; a hardwired processor or FPLA would take 7 msec to run the same algorithm. Memory management is best performed by hardware in the new micro- and mini-computing systems. A list is given below to indicate software features that can be presently implemented by hardware:

1. Memory allocation, memory reclamation, virtual memory management, paging, segmentation (absent segment interrupt), memory and data protection, stack operations, address generation, indexing, indirect addressing, storage protection.
2. Multiple-precision arithmetic, decimal multiply-and-divide, floating-point arithmetic, sorting, algorithms for data manipulation.
3. Program linking and binding, program relocation, data relocation, data structure, formal checking, character-string manipulation, data-type conversion.
4. Symbolic addressing, variable-field lengths (16, 32, 64 bits), variable data structures, alphanumeric field manipulation, context switching, emulation, queues, links, compilation, task-dispatching, next software-instruction fetch.
5. Interrupts, interrupt checking, trap catchers, peripheral data transfer, time-sharing supervision, text-editing, control command instructions, format checking.
6. Parity checks, error-control coding (detecting and correcting), automatic retry, automatic diagnosis.

5.13.1 Time-Share Computer Services. Time-share computer services offer proprietary programs, supplied by the microprocessor vendors, which allow the assembly of *source* high-level programs into *object* code suitable for execution in a microprocessor. The source program written in the assembly language of the microprocessor comprises the symbolic operation-code and the symbolic label that corresponds to the actual binary address at which the instruction Opcode will reside during program execution. This assembly language also keeps track of the start addresses of subroutines utilized by the main program. The object-code generated by the assembler loads the processor's instruction-PROM memory in the desired sequence.

In actual practice, the logical sequence of events in the program are established by a flowchart, and each step in the flowchart is converted to one or more instructions in the Assembly language. The instructions are keyed into a data-file in a time-shared computer, and the program or source-file is processed by the assembler program in the time-shared system. The error-free output-record renders the object-file in a suitable format for storage; the object-file of bit-patterns is "burned" by a vendor into usable ROM memory chips.

Central time-sharing computer. Real-time interaction between humans and machines in high-level language provides a continuous dialogue for quick results, instead of the one-program-at-a-time *batch-processing.* The time-sharing computer, therefore, serves as a powerful library for an entire community. The system organization of a time-sharing computer is illustrated in Fig. 5-26.

5.13.2 Memory Organization, Virtual Memory, and Bootstrap Loader. The microcomputer's memory is composed of several thousand words, each consisting of a number of binary digits or bits (e.g., 16 for a 16-bit word), and *the word at any ad-*

Software

Source Code	Translation	Object Code in Bits
High-level BASIC, FORTRAN, or APL	Compiler	Machine instructions and data addresses
High-level BASIC, FORTRAN, or APL	Interpreter (hardware or firmware)	Machine instructions and data addresses
Assembly Language (phrased in Mnemonic–alphanumeric commands)	Assembler	Machine instructions and data addresses

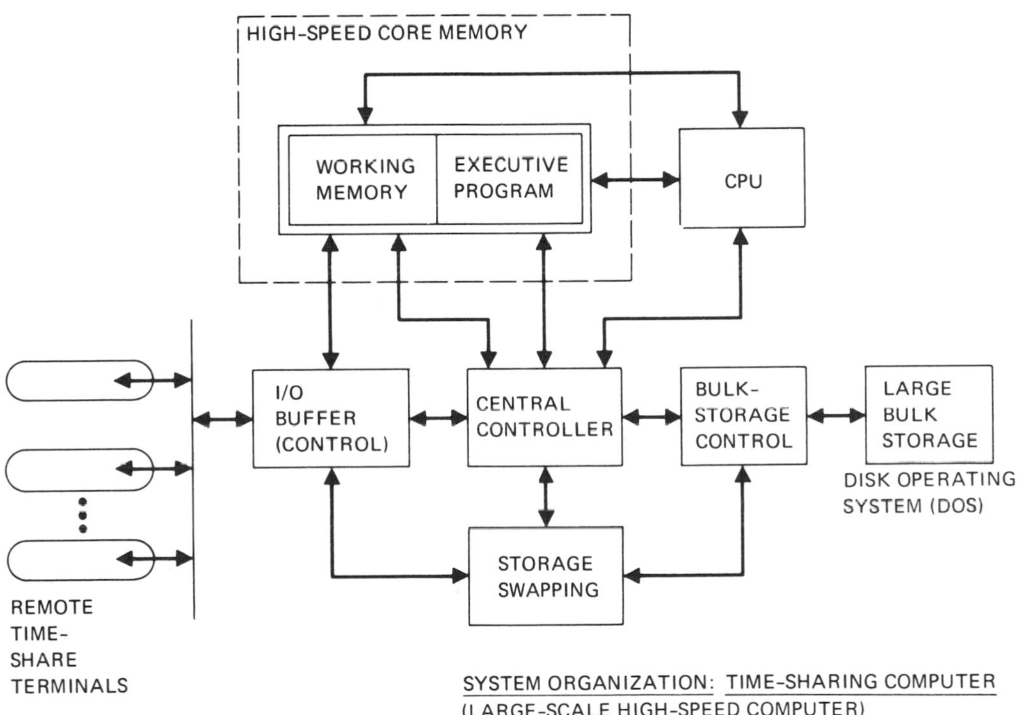

Fig. 5-26. Organization of a time-share system.

dress may represent either an operating instruction (for instance, of the stored program at address 55) *or a block of binary-coded-decimal data* such as the number 2730 in 3-bit octal (base 8) or 4-bit hexadecimal (base 16) code. The most significant bit 15 (MSB) of a word is often used to indicate sign (0 = +; 1 = –). The least significant bit of the word is numbered bit 0, as shown in the following diagram.

The computer handles one word at a time in a clock cycle-time; a word size may vary from 4 to 8, 12, 16, 18, 24, 32, or 64 bits. The microprocessor normally uses the memory under the jurisdiction of the control logic. But an I/O control or multiplexer can "sneak" into memory and steal a memory cycle without the awareness of the control, by way of a special *direct memory access* facility and "steal" a memory-cycle by not necessarily holding up the regular control circuitry by a memory-cycle.

Usually the program is stored in consecutive memory locations and is executed sequentially unless a "jump" or "branching" instruction comes along. It is actually

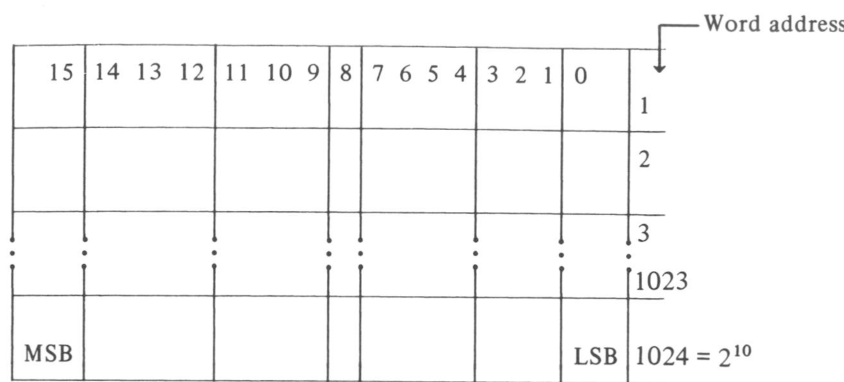

the sophisticated software, the logic of the program, that determines whether the contents of a memory word is data or an instruction. Firmware, via the medium of the latest LSI hardware in the form of the programmable logic array (PLA) and ROMs, is the fastest system of software presently available.

Virtual memory. Software programs, such as loaders, assemblers, and compilers, are generally provided by the microcomputer manufacturer. An Executive program, for instance, collects a sequence of software programs, such as write/edit/assemble or compile/and execute, into an operating system. The Executive program is also responsible for:

1. Job scheduling, such as sequencing the loading and execution of a user's program, for example, by way of teletype or the start punch-cards of a job-deck.
2. Monitoring programs and I/O controls for such utilities as debug. Larger Operating Systems using this Executive software have bulk-memory peripherals such as disk or drum or external RAM/ROM memory-stacks. The Executive software will transfer programs from its internal memory to the said external bulk-memory when out of use, loading each one back again when required. This kind of handling memory is termed *virtual memory*, since this configuration of the transferable software memory appears almost unlimited to the programmer.

Bootstrap loader. The bootstrap loader is a small program that enables at start in an empty memory the loading of a few instructions or data (for instance, for commanding a peripheral device to input the program instruction codes). It may typically consist of 5 to 10 instructions, which may be either read from an internal ROM or keyed in via:

1. The control panel switches.
2. The keyboard, if the microcomputer console has a built-in keyboard facility.

5.13.3 Assembler. The assembler programs are powerful but simple at the lowest level of program-development software. (The IBM BAL–Basic Assembly Language–is an example.) They allow the programmer to design programs up to several thousand words and to specify the instructions, addresses, and data by name in decimal, octal, or hexadecimal formats by using **mnemonics** (abbreviated symbolic words) for the instruction words, address names, and data numerical constants. The mnemonics are usually three to five characters long. Special user programs are usually developed by computer manufacturers for the specific hardware and peripherals involved.

The source-language assembler allows the inputting of a user program via the keyboard or TTY in the form of coded symbolic instructions. In addition, the assembler program will:

1. Translate these codes into the object binary machine-language
2. Calculate all the necessary memory reference addresses from the start address given
3. Convert all formats of data to the binary machine language.

Label is the equivalent of a memory address–where referenced only. Instruction mnemonic specifies type of operation. **Operand** is the operand part of the instruction word (a constant or a memory address).

The remarks under a separate column in the program-format are added for only understanding the program during *debugging.*

In the event an instruction word needs a modification, it can be typed in; the assembler recalculates the memory addresses. If the assembler is commanded to assemble the input program, it will list the user program on a line printer, if one is available.

Macroassembler. A Macroassembler allows the programmer to develop and specify blocks of instructions. As an example, if the programmer specifies a four-instruction sequence for interfacing an I/O device, the Macroassembler allows the programmer to define the sequence by a simple four-character mnemonic. When this mnemonic is encountered by the Macroassembler during the processing of a source program, it will automatically go through the four-instruction sequence. These user-defined multiple-words are called *macroinstructions.* So, this is a mix of the assembler and the high-level language programs.

5.13.4 Compilers for High-Level Languages Such as FORTRAN and BASIC. The Macroassembler is a start in the direction of a high-level language. The compiler translates the high-level language and provides the following simplifying software facilities to the programmer:

1. A mathematical operation can be specified in algebraic form, and the compiler converts it into a series of instructions in the form of binary words; it is the machine-language for actually operating the computer.
2. Real numbers can be conveniently handled in shifting floating-point notation. Since most computers are binary fixed-point machines, the compiler provides a floating-point feature.
3. Data can be specified in matrix (table) or subscript format to handle large amounts of data.
4. Macroassemblers can be indexed for a loop sequence, which requires several instructions at the start and end of each loop. Example: Program to evaluate $y = \frac{AX + B}{5.6}$, requires only four commands in FORTRAN.

   a. READ A, B, X
   b. Y = (A*X + B)/5.6
   c. PRINT Y
   d. END

In the machine code or Assembly language, a large number of program instructions are involved for this computation.

The compiler assigns the CPU-registers, calculates addresses, and assigns memory areas to programs and subroutines without the programmer ever knowing the working registers and actual instruction set. The demerit is that a much faster method of computing a function is not feasible by using a lesser number of internal machine instructions (as could be optimized by using Assembly language and mnemonics). At the same time, obviously the compiler needs high overhead of memory. Presently, the compilers are rendered "intelligent" by automatically rearranging in sequence of operations for the optimum throughput.

5.13.5 Interpreter. The interpreter differs from the compiler in that it translates the high-level instructions one by one, and executes each immediately to slow up the execution of the program. However, the program can be more conveniently modified on a one-by-one basis. This software technique is especially convenient in a programmable calculator.

5.13.6 Debugging a Program. Debugging is an essential utility routine during the generation of software for a specific application. A simple TTY debugger-program will include the ability to load memory from a paper-tape, and dump a section of the memory to a TTY printer in a readable format. It will have the ability to examine and modify a memory address or the contents of a memory address, and execute a program starting at a specific address. A *breakpoint* facility allows the programmer to stop the program at a specific step, by inserting a breakpoint at an address in the program. The debugging program then goes to that address, removes and stores the instruction, and then inserts a command that causes the computer to jump back to the debugging program; then, the stored original instruction is replaced to execute a program up to the point and then stop. This allows the programmer to examine the memory to see if any failures have occurred in the user's program. The debugging program also duplicates many front-panel hardware functions to check the relevant software aspects.

Generally, if a program is less than 200 words, support software may not be required. The program can then be generated directly in binary, octal, or hexadecimal format by entering it through the front panel and debugging it manually via the front-panel controls. For programs of 200 to 4000 words, a paper-tape assembler and a debug-package are sufficient. For multiprocessing of several programs, more sophisticated software packages are required. It is only during the debugging procedure that a system programmer, using the BASIC or FORTRAN, may have to deal with the machine-language instructions.

The use of available program-development software supplied by the manufacturer results in fewer program errors, because such automatic tools provide many design-checking features. Software diagnostics are called the debugging programs, and these diagnostic programs determine whether a system failure is due to hardware or software. The CPU hardware diagnostics are generally supplied by the manufacturer. When one designs an I/O interface, one must write corresponding diagnostics to debug the hardware design.

5.13.7 Text Editor. Text editor has the function of inserting or deleting information from words. If the memory-word outputted is 002350000, the editing procedure may involve reformatting this word by:

1. *Extracting* the meaningful part of the word
2. *Writing* or *masking* new characters over part of the word according to a *mask.*

After editing, the word may take the format "$23.50." The editing procedure may thus involve shifting bits or characters in a register with respect to the word, so that a decimal point, if any, can be inserted at the appropriate place. In general, delete, insert, copy, or mask-and-shift make the requisite manipulations.

5.13.8 Linkage Editor. In programs, *Load* instruction may have several functions, such as

1. Loading and relocating the machine-language program in the main memory, which may be CORE or RAM
2. Communicating with the various segments of the main program
3. Bringing in names, masks (or overlays), and addresses, and link to the subroutines from a subroutine library of look-in tables like ROMs.
4. Monitor the contents of the memory to see that its capacity bounds are not exceeded.

More flexible software systems distribute these functions on different passes by using a Linkage Editor, which may take a basic assembly-language program from a tape or disk and convert it into the corresponding *absolute* machine-language on a one-to-one instruction basis, and append or *link* subroutines from a relocatable library like an array of external static RAMs; at this stage, if the program exceeds the limits of the CORE/RAM main-memory, it is rejected.

5.13.9 Batch Processing. It is a programming technique where all information to be processed is coded and collected into separate *decks* or batches of cards *off-line* before actually processing through the computer in a sequence. This is the free-standing configuration, when a bulk-storage medium such as an *on-line random-access disk operating system* (DOS) is not the case. Microcomputer Systems, with *RAM* memories and floppy-disk mass-memories, make *real-time* random-access systems. When slow-access tape cartridges are used, software loading into the popular static RAM main-memory is off-line.

5.13.10 Stack Interrupt Processing. The stack is a reserved area of the main-memory where the CPU automatically sets aside the contents of the program-counter and the working registers when a program interrupt occurs via the I/O control system. The stack normally forces the CPU to return to resumption of processing from the interrupts in the same order that the sequence of the interrupts from the external devices occurred in the first instance.

5.13.11 Software Trends. As noticed in the case of the more sophisticated IBM-5600 and DEC-LSI-11 microcomputer systems, the software previously developed for minicomputers will be adapted for the direct use of microcomputer systems in the form of external miniature tape or floppy-disk facilities and internal RAM and ROM memory. Loading of the internal or extended external RAM memory from the tape or disk mass-memory software may be considered an off-line operation.

The distinction between the large-scale general-purpose digital computers and minicomputer systems with LSI hardware is presently determined by mostly software differences. A modern minicomputer operating system is as powerful as some of the original medium-scale general-purpose digital computer Operating Systems of IBM-360. The minicomputers are highly competitive in throughput, speed, access time, and so on, with latest medium-scale machines in the IBM-370 series too. (Large-scale IBM, CDC, and Burroughs Computers would be mainly reserved for the central multiprogrammed function of pipe-lined time-share and distributed network centers. There are several other large-scale main-frame manufacturers.)

The previous situation will be repeated for the eventual distinction between the latest microcomputers and the present medium-speed real-time minicomputer operating systems. For example, the word sizes in minicomputers have been increasing to 32 and 64 bits for handling larger word lengths during each memory cycle, to allow less programming effort and faster run-times for the same computational accuracy of the double-precision floating-point arithmetic. To keep in step, the latest 16-bit, bit-slicing microcomputers can now handle 4, 8, 16, and 24-bit word sizes in a memory cycle in medium-speed real-time applications, and they show signs of replacing some of the original minicomputer systems by exhibiting a more powerful architecture and performance. During the next few years, when the high-density, high-speed, low-power bipolar, Schottky I^2L-LSI processing reaches maturity, microcomputer Operating Systems are expected to reach the standards of the latest minicomputer systems in both performance and scope, and gain an advantage with the claimed higher MTBF of 12,000 hr (for Intel 8080) as compared to the average 500-hr figure for most minicomputers.

5.13.12 High-Level Language—BASIC. BASIC programming language is presently used on the teletype time-sharing computer systems (e.g., GE-265, GE-645). It is ideal for solution of scientific or business problems of moderate size and complexity. The BASIC consists of 31 commands using:

1. Variable name
2. Expression
3. Message, line number
4. Counter
5. Beginning value
6. Ending value
7. Step size
8. Numeric value
9. Alpha value
10. Dimension value
11. Operand
12. Relation
13. Function name
14. Function argument
15. Function definition

One uses a teletype machine with a built-in digital microcontroller and dials the computer's telephone number to hear a high-pitched tone for connection. The computer is told the problem in the proper form of "line statements" and asked to print the answer via the keyboard-printer. The computer prints back the answer in line statements. IBM-5100 microcomputer system uses BASIC in conjunction with its own keyboard and CRT output display, as do several other microcomputer systems at this moment.

5.13.13 High-Level Language—APL. As an alternative to BASIC, or as an advanced computational facility, the IBM-5100 *portable* computer system provides predeveloped APL software via its ROS/RAM internal memory system and external tape cartridges. The APL has a close similarity to the notation of vector algebra, but it is more simple and is further extended in applicability. APL contains a rich set of conveniently usable well-defined *primitives*—built-in functions that make it applicable over a wide area of arithmetic, trigonometric, and hyperbolic functions (and their inverses); matrix products and inverses, look-up lists or tables; set of relations; and logic.

APL (advanced programming language, IBM). The APL, an interactive high-level language used for problem-solving, enables a concise expression of complex mathematical statements.

1. It contains mathematical operators that allow the systems analyst to perform array processing functions, trigonometric and hyperbolic functions, and common arithmetic, logical, and relational operations.
2. It treats scalars, vectors, and matrix arrays with equal facility and lets one store data (or the results of computation) using a variable name, so that the current value is automatically substituted.
3. It permits previously defined programs to be used as a function and to perform operations on arrays of up to 63 dimensions.

The APL software is available on tape cartridges for various applications in statistics and mathematics—approximations to functions, advanced mathematical functions including calculus and linear equations, and matrix analysis. The main-memory storage in the form of fast-access 2-K bit NMOS RAMs is available in increments of 16-K bytes up to 64-K bytes (16,384 to 65,536 memory cells of words). Fourteen program command-keys for the BASIC and APL software permit direct inputting of the system-commands via the CRT keyboard in the case of the IBM-5100.

The set of primitive scalar functions in monadic and dyadic forms includes plus, negative, signum $[xB \longrightarrow (B > O) - \overline{B < O}]$, reciprocal, ceiling Γ, floor L, exponential, natural logarithm ⊕, magnitude (|), factorial (!), roll (?), pi-times (o), not (~), dyadic functions (all trigonometric functions), and ∧, or ∨, nand ⍲, Nor ⍱, less <, not greater ⩽, equal =, not less ⩾, greater >, and not equal ≠. The set of primitive mixed functions includes size (ρ A), reshape (V ρ A), ravel (, A ⟷ ∟, or first A integers), catenate (V, V ⟷ VV), index (V [A]), index generator (∟ S), index of (V ∟ A), take (V ↑ A), drop (V ↓ A), grade-up (⍋ A). grade-down (⍒A), compress (V/A), expand (V\A), reverse (ϕA), rotate (A ϕ A), transpose (V ⍉ A, ⍉A), membership (A ∈ A), decode (V ⊥ V), encode (V ⊤ S), and deal S?A, random deal of S elements from first A integers). Restrictions on argument ranks are indicated by S for scalar, V for vector, M for matrix and A for Any.

5.13.14 PL/1 (Programming Language/One), ALGOL, PASCAL, and FORTRAN.

PL/1, a user's high-level source program originated by IBM for large-scale computer users, aims at minimizing clerical work involved in writing an assembler program. The rigidly *coded* program with appropriate *syntax* of 60 (or 48) characters, divided into three groups, is keypunched and entered into the computer as data input for a standard *library program* preloaded in the computer as *PL/1 compiler*. (In IBM system 360 operating system 0/5-F level, and in tape/disk operating system T/DOS, PL/1 compilers are used.) The cards use columns 2–72 only; 73–80 are reserved for identifying the program. The computer outputs a listing of the program input and an equivalent program in machine language for subsequent debug or computation.

ALGOL 60. ALGOL 60 is a high-level language that was originated at Princeton University on a Control Data Corp. 1604 computer for translation to the Assembly-language of this machine; it is used more commonly in Europe. The language aims at separating the functions of *defining the language* and *translating to another*. The system operates in three phases. The first enables the translator's "diagramming program" to output syntax definitions pertinent to the particular ALGOL program, e.g., *defining identifiers*. The second phase enables DIAGRAM to output Assembly language program. The third phase is the actual assembly of the code. This compiler of CDC-1604 outputs about 300 assembly-language instructions per second.

PASCAL. This European high-level language is presently attracting the attention in academic circles because of its special capability to describe *data* and *procedure* in pidgin English in the case of complex programs, although the language is considered rather weak in I/O capabilities at the present stage. The program will define all its variables with a VAR statement. The start and completion of the procedure are denoted by BEGIN and END. Predefined procedures RESET, READ, WRITTEN signify I/O operations. The operator := assigns the statements corresponding to the common left-pointing arrow. (Incidentally, PASCAL is named for the famous seventeenth-century French mathematician and physicist.)

PASCAL is presently recognized as a more powerful high-level language than the former PL/M, PL/1 series of languages for processors ranging from the 8080 to the superfast Cray-1. It is specially encouraged as a structured language by the Institute for Information Systems of the University of California, San Diego (UCSD). Unlike the GO TO statements for branching and subroutines required in BASIC and FORTRAN, modern PASCAL allows the underlying algorithm to be solved on the spot without getting involved with the whole program. As 64-k bit and other large-capacity VLSI ROMs become available, software compilers will be put onto boards for convenient modularization in the near future. Bell Labs has recently come up with a C-programming language, which is an improved version of PASCAL.

For microprocessors, Kenneth Bowles of UCSD has compiled PASCAL in an intermediate machine-level program code called *P-code.* This code in turn gets easily interpreted in the individual target microcomputers; that is, a small *interpreter* is all that is required for adapting PASCAL to any low-cost microcomputer system. The PASCAL host *compiler* can then have extensions for strings, disk files, interactive graphics, system programming, editor, file manager, debugger, and various utilities. The code generated by the PASCAL compiler can be easily put into ROMs since the code is position-independent without any modification, and it is timely as a successful solution to compatible microcomputer software. American Microsystems, Inc. has already adopted the UCSD PASCAL as part of its MDC-100 microprocessor development software, since it is more powerful than BASIC, FORTRAN, PL/1, and the Assembly languages for defining *data structures.* With PASCAL's control constructs, programming time can be minimized by at least four times, and in 1977 a quarter of a million programs were successfully prepared. Texas Instruments has recently adapted PASCAL to handle fixed-point and decimal arithmetic capabilities and array processing presently used in PL-series languages. Currently, the trend is to standardize PASCAL for universal application in the near future. The code generated in PASCAL requires only twice the memory and execution time compared to the Assembly language; with low-cost high-capacity ROM memory on PC-boards, it is an insignificant factor, as compared to its advantages.

Following is an example of statement in PASCAL:

```
BEGIN
  READ(DATAFILE,ARRAY I);
  I:=I+1
END;
```

FORTRAN (Formula Translation) is developed for numerical methods, statistics, engineering applications, matrix and Boolean algebra, Monte Carlo techniques, and so forth. FORTRAN programming has several versions such as Data General, DEC 4-K, DEC 8-K, H.P. 4-K, H.P. 8-K, and General Automation FORTRAN for synchronous transmission of binary coded decimal (BCD) data in telecommunications. The form of FORTRAN that is used for minicomputers will be directly adapted for microcomputer system software by each manufacturer.

DEC 8-K Word FORTRAN. The language is similar to the standard ASA FORTRAN II, providing single-precision arithmetic only (and no mixed-mode arithmetic) and two array-dimensions. A two-pass compilation procedure is used to produce a relocatable binary-coded tape from the FORTRAN source statements. Maximum program size is from 200 to 300 lines of coding (excluding comments), depending on the size of the storage arrays and the number of the library-routines used. The software used for the compilation, loading, and running of the FORTRAN programs include an 8-K word compiler, plus assembler, linking loader, and a two-part FORTRAN library. This software runs on a DEC PDP-8/81 minicomputer with 8-K words of storage, using the floppy-disk facilities for the compilation, loading, and handling of the programs. DEC has adapted this software for use with its PDP-11/03 microcomputer system. Binary-coded tape copies (DEC minireels), available at $30 per set for the minicomputers, will be adapted for the new DEC microcomputer systems.

5.13.15 Routine Software Needs. The system manufacturer delivers an initial hardware with a free-standing Operating System, so named because it does not use slower bulk-storage devices such as tape and cards during real-time operation. The routine software consists of an Executive program, which controls:

1. Loaders, including a link-editor (linkage editor)
2. Text editor
3. Assembler
4. Debug monitor
5. I/O subsystem control software, for driving the peripheral devices via the appropriate interfaces
6. Utilities library of mathematical subroutines
7. High-level language translator such as a FORTRAN/APL/BASIC-compiler.

Assembly language programs use the I/O subsystem for all peripheral device communication and the utilities library for all arithmetic operations.

Once the Executive program is loaded into the memory via the memory data and address registers, a peripheral such as the TTY sends commands to the Executive Editor, and the Debug program initiates the execution of all subsequent system operations. See section 5.12.6, Debugging a Program.

Additional software in the mass-storage system is transferred to the compiler for source-to-object language translation, for occasional secondary applications that run when the microprocessor is otherwise idle during the data-processing intervals.

5.13.16 Executive Program. The Executive program brings together the various items of the software mentioned previously to make an operating system (O/S), which may include (1) a Disk-Operating System and (2) a Real-Time Operating System. With a disk assembler/compiler O/S, at the command of, for example, a TTY, the Executive will load all the other program modules of the O/S onto the bulk-storage unit and recall them when necessary, leaving the rest of the RAM main-memory free for user program. The user programs can be automatically transferred to the disk if they exceed the capacity of the main internal microcomputer memory. Also, too many infrequent user programs and data clutter up the disk storage.

5.13.17 Real-Time Operating System. As an addition to the above Disk O/S, the Real-time O/S:

1. Allows the execution of the programs to be scheduled by time of day.
2. Allows the programs to "queue up" for execution, waiting for the CPU to finish the more important jobs such as monitoring experiments and so on.
3. Every program is assigned a priority in an O/S; for example, emergency control programs (0), monitor experimental run (1), reduction of the data in between the runs and calculation of the settings for the next run (2), and compilation of some of the FORTRAN source programs and running these programs (3).

The CPU keeps track of time by way of a built-in crystal-controlled real-time clock (RTC), which generates interrupts at specified times. The RTC-interrupts, one by one, return control to the Executive, and the Executive checks if a higher-priority program needs immediate attention. Thus, a Real-time O/S provides maximum utilization of a computer system. With a memory cycle-time of 2 μsec, a microcomputer can execute half-a-million instructions/sec; however, it waits most of the time for the I/O devices transmit or receive data at rates of 1000 characters/sec. A TTY, for example, outputs 10 channels/sec, and an IBM Communications Interface Adapter operates at 800-bit (baud)/sec line-rate using an EIA RS-232C standard modem Interface for data communications.

As a typical example, in an Operating System, a computer, running intermittently, may take:

1. About 1 min at the start to monitor the instruments, during the early phase of an experiment at 1% of the busy-time.
2. Collect data during the next minute at 40% of the busy-time.
3. Process collected data during the next 2 min at 25% of the busy-time.
4. Set the starting conditions for the next run during the fifth minute at 10% of the busy-time.

5.14 PROGRAM DEVELOPMENT SYSTEMS

5.14.1 Interface Adaptor. A terminal digital controller (e.g., a microprocessor with its I/O Interface) using a TTY will be interconnected to a central large-scale computer system via a data-communication line or a telephone line by means of an Interface Adapter, which is otherwise called a front-end processor, data set, or modem; it is a terminal for conditioning the outgoing data and for accepting and recording the incoming data. The modem accomplishes its mod/demod functions by converting an input dc signal into an ac format and vice versa, mostly using PSK modulation (phase shift keying). The modem performs, in addition, automatic equalization for the line-losses in preferably a two-way full-duplex operation and provides the automatic level-limiting facility. The PSK Encoder/Decoder may have a data-rate capability from 75 to 9600 baud/sec, and several channels may be multiplexed on high-speed transmission lines at 40-K bit/sec. The low-speed data sets are generally asynchronous and compatible with one another. However, the high-speed modems above 32-K bit/sec are intended for synchronous operation, phase-locked to the multiple of a crystal-controlled reference frequency.

The high-speed synchronous units will phase-lock transmitting and receiving modems in a matter of hundreds or tens of milliseconds. The cost of a modem in general is about $1/baud. The baud represents a basic rate of transmission in pulse/sec, and the baud-rate is theoretically limited to twice the bandwidth in Hz. In low-speed systems, baud and bit rates are equivalent. Low baud-rate asynchronous modems can operate with about 20% distortion, while the high-speed synchronous modem is restricted to 7% distortion to enable minimal error-rate at the desired SNR. **Tymshare,** which effectively enables many users to use a central large-scale computer via terminal controllers such as TTY or CRT keyboard, requires the use of the Interface Adapter at each terminal. The computer Operating System actually services each user in sequence at its high data-speed, but the users operate as if they are served individually in one of the commonly used high-level languages such as FORTRAN or BASIC in serial 8-bit ASCII or IBM/EBCDIC code of characters.

5.14.2 Peripheral Interface of a Microcomputer System. Peripheral selection starts with the decoding of the I/O instruction in the digital computer.

Function selection (Example: writing and reading data, rewinding tape, back-spacing in a tape cassette or floppy disk.) Sensing the status of the conditions and service routines in the peripheral devices and their Interfaces is indicated by the error-checking schemes used in the peripheral with mostly an even or odd parity-bit (and occasionally additional correcting bits). The microcomputer addresses a peripheral device to interrogate via the control lines and takes the appropriate action on the sensed condition. A two-way data-transfer path between the high-speed com-

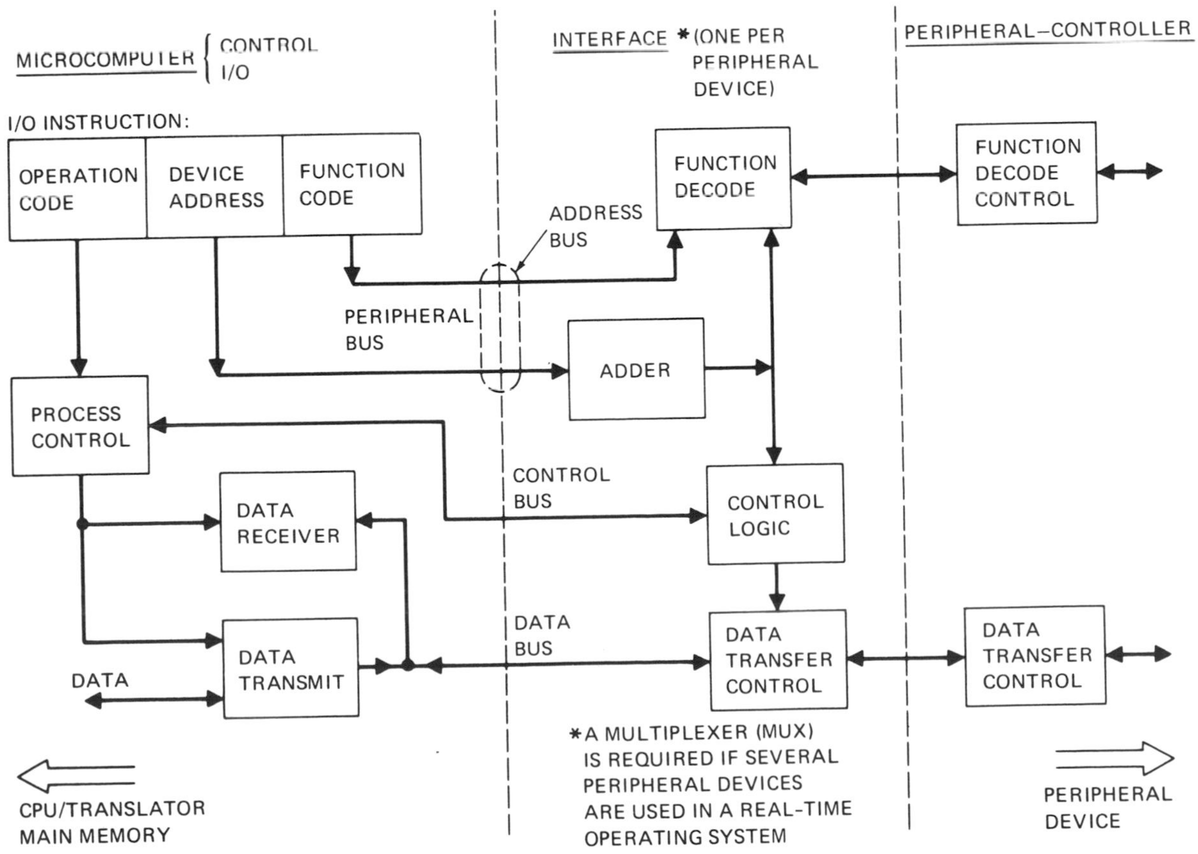

puter and the peripheral must be established without any loss of data due to the inevitable slow data-handling rate of the peripheral.

5.14.3 Addressing Modes. The microprocessor has the inherent ability to access a large amount of memory due to its high-speed cycle-time of the order of 1 or 2 μsec. One 16-bit program counter can address 64-K memory locations ($2^{16} = 65,536$). However, a regular 16-bit instruction word may actually specify the instruction in 8-bits and the memory-address in 8-bits to practically enable the addressing of only 256 locations ($2^8 = 256$). A few bits in the instruction are required for the opcode. To overcome this restriction, several *addressing modes* are used. Thus, the programming versatility increases with the number of modes available.

5.14.4 8080 Real-Time In-Circuit Emulator (muPro Inc.). muPro-80E:

Real-time program execution is achieved at a cycle-time of 350 to 650 nsec.

User programs may reside anywhere in the 64-K byte address space.

256 I/O device codes are available to the user's system.

Breakpoint is allowed on program/data/stack in PROM or RAM.

Comprehensive, fully transparent Control/Display Console is built-in.

Convenient memory and I/O Device allocation are feasible between the user's system and the Emulator, via four hexadecimal switches (multiplexers) on the Emulator.

Either the user's or the internal clock can be automatically selected.

Program-trace provides 64-instruction history, preceding a Halt or breakpoint.

Terminal is required for only software development.

The unit fits in a suitcase with its dimensions of 4.6″ H × 6.6″ W × 15″ D and weight of 18 lb.

5.14.5 muPro-80 Microcomputer Development System. The complete system consists of the CPU, using Intel-8080, memory as required, and a transparent control/display console that provides total debug capabilities. Automatic bootstrap is available from paper-tape or floppy-disk. The 64-K word memory-address system avoids the need for any monitor program. A total of 256 I/O device codes are available. Breakpoint can occur when the program is in ROM or RAM. There is a provision of eight card-slots on a mother-board for the CPU, hardware of multiply/divide option, memory refresh, 24-K word memory, floppy-disk interface, real-time clock, CRT/line printer/reader/punch interfaces. The chassis is extendable for full 64-K word memory, and additional I/O (serial, parallel, analog, and DMA). Total software support is available for a multiuser, multitask Disk Operating System (80-D), Resident BSAL-80 Block-structured assembly language, relocating/linking loader, text-editor, and BASIC high-level language. Dimensions: 4.6″ H × 6.6″ W × 15″ D; weight: 15 lb; price: $3950. A heavy-duty power supply system with regulated +5V at 15A and ±17V at 2A is included.

5.14.6 muPro-80D Multiuser/Multitask Disk Operating System. The dual-drive system includes additional memory as required. The system will allow more than one program to run at the same time and will therefore support multiple terminals or users. System development time is minimized by eliminating the need for paper tape. The controller supports two dual-drives (four diskettes). The disk operating system is high-speed and single-density with 250-K bytes/diskette and 100 ms maximum search-time from track 0 to track 76.

The RAM requirements are as follows:

muPro-80: 24-K bytes
Operating System: 8-K bytes
BSAL-80 Assembler: 8-K bytes (8-K Symbol Tables)
Text editor: 4-K bytes (12-K text buffer)
Linking-loader: 2-K bytes (14-K symbol table)

The muPro-80D, at a price of \$3765 provides comprehensive file management. Dimensions are identical to those of muPro-80. That is, the microcomputer system and the Disk Operating System together require 5.25-in. rack-mount side-by-side, with a total weight of 50 lb. Total price \$7715 (1977). Both the Development and Disk Operating Systems could be directly integrated without any need of redesign in CPU, memory, and I/O cards.

Comments. It will be noted, incidentally, that the latest low-cost microcomputer systems are gradually assuming the powerful real-time tasks and increased throughput of the original medium-scale and subsequent minicomputer systems. And, the solid-state RAM memory is slowly replacing the former power-consuming nonvolatile CORE memory. The starting time required for loading the RAM memory from the permanent or semipermanent software on disk and tape is, after all, not that significant.

The sectored software on Diskettes of this system is compatible with IBM-3740, with the powerful DMA data-transfer facility.

The source BSAL-80 software used in this system has a high-level language syntax with ALGOL-like commands (e.g., IF, THEN BEGIN, END ELSE BEGIN, END CONSTRUCTS); the object machine-language output is "relocatable" like that in larger computers, with automatic variable memory allocation for program and data. BSAL-80 includes latest software features such as intuitive statement formats, extensive Error Diagnostics (36 descriptive error messages), extensive Macroinstructions, Symbol crossreference table for the assembler, and the Linking-loader (which combines the relocatable modules into the absolute program with a Load map).

The Text Editor uses commands that are meaningful, such as ADD, DELETE, COPY, GATHER, FIND, MODIFY, LIST, REPLACE, KEEP, TEXT, and SET. This is all, of course, the result of the latest capability of turning software logic into low-cost subminiature LSI hardware. It conveniently enables sophisticated features of the nature of line and string modification, block-copying and moving, multiple commands per instruction word, single-character command option, local-edit during command or text-entry, and automatic instruction numbering by ADD, COPY, GATHER, and TEXT.

5.14.7 PROM Programmer (muPro-80P-2708). The PROM programmer enables on-card programming and thus avoids unnecessary chip-handling. Erasure and data verification each take less than 1 sec. Error conditions and program data from RAM, ROM, and emulator memories are displayed directly on the system control-display. It is transparent to CPU-timing, and no memory or I/O device codes are required. The accessory for programming up to eight type-2708 PROMs in about 5 min. costs \$775; and 8-K byte ROM card is included. (An independent programmer for the insertion and programming of eight chips in about 40 min. is priced at \$1115.)

5.14.8 VRAM-TV Monitor Display. The latest low-cost CRT digital computer display system is the VRAM-TV monitor display. One merely adds a card of hardware to a regular low-cost (under $100) television video monitor, operating at 15.75-kHz line-scan rate and 60-Hz interfaced-field (30-frame) vertical-scan rate. The VRAM-TV display is the compatible economy CRT-Terminal for the present economy microcomputer systems. Within a year or two, a few LSI chips will be available to enable home monochrome or color-television sets to interface with data-communications systems.

The system block-diagram of the VRAM (video random-access memory) technique that converts the computer digital format to the regular composite black-and-white or color television signal is shown in Fig. 5-27. Incidentally, the television signal is a composite of synchronizing pulses at the horizontal 15.75 kHz/sec rate and the vertical 60 Hz/sec rate and analog black-and-white or color video (picture) information. The color signal is compatible to the black-and-white information in that merely additional chroma (in green, red, and blue) and color-synchronizing information, amplitude-and-phase modulated on a color subcarrier of 3.58 MHz, is added to the black-and-white monochrome information. The recent IBM portable-computer type-5100 provides the VRAM-TV monitor facility.

Video random-access memory. VRAM operates like a regular RAM; data and address bits are similarly processed, and the location of an element of the picture on the television screen corresponds to the cell location of the VRAM address. As explained in Section 5.16, the digital part of the VRAM system processes the microprocessor instruction words and formats alphanumeric data; a ROM character-generator in conjunction with a shift register feeds the data to a video generator to obtain the analog (linear) video information for the television monitor. (I^2L) LSI chips would be ideal for processing in VRAM terminals. Special high-resolution television monitors with extended bandwidth could handle graphics that demand the configuration of line-segments along with alphanumerics in a 256 × 256 dot-matrix arrangement. The ROM character generator in the VRAM-TV will actually receive ASCII codes and output digital pulses that form letters in, for instance, a 7 × 9 format. The alphanumeric digital format may include all the present special features such as blinking, variable intensity, brightness reversal, and scrolling. The television sets may be conveniently switchable to the European line, field, and color frequency standards, if they are easily portable.

5.15 VENDOR'S INFORMATION ON MICROPROCESSORS (SOFTWARE AND HARDWARE)

The following example of the LSI chip MC-6800, the microprocessor unit/MPU of Motorola Semiconductor Products, illustrates the comprehensive information typically available from the various manufacturers.

The features of MC-6800 are:

- 8-bit parallel processing
- Bidirectional data-bus
- 16-bit address-bus—65-K bytes of addressing
- 72 instructions—variable length
- Seven addressing modes—direct, relative, immediate, indexed, extended, implied, and accumulator
- Variable-length stack
- Vectored restart
- Maskable interrupt vector

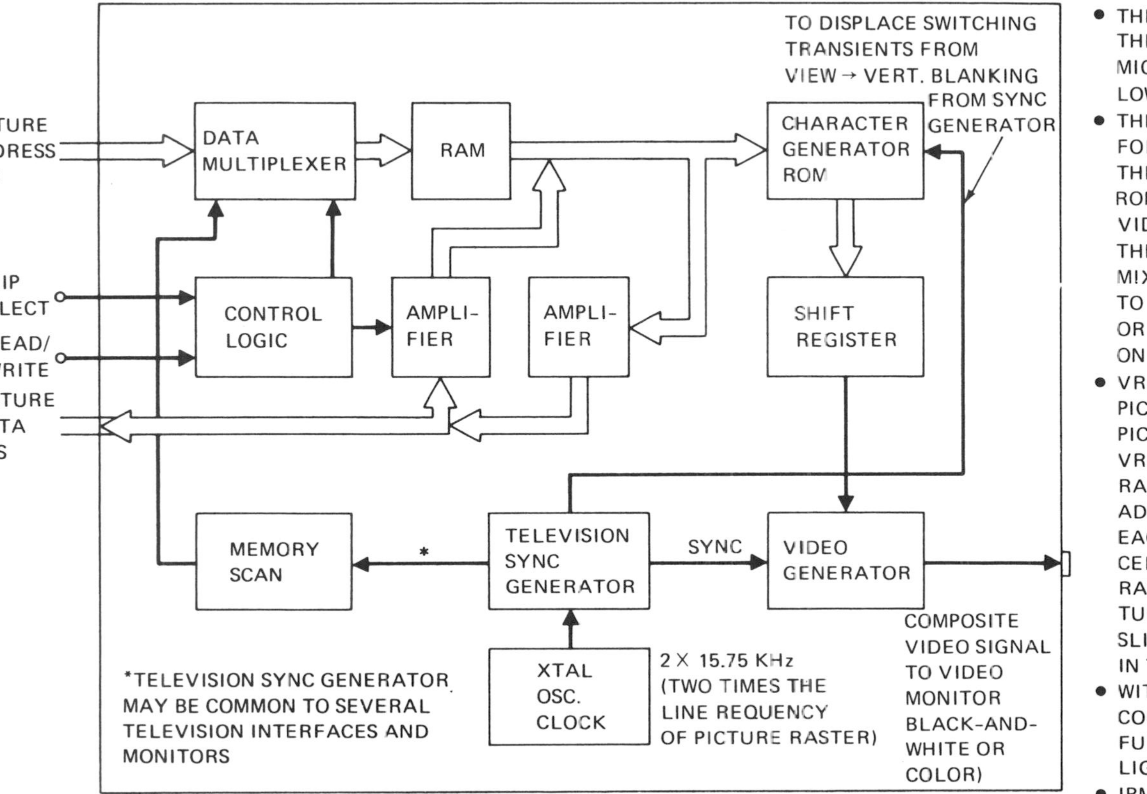

- THE ABOVE INTERFACE SYSTEM FOR A REGULAR TELEVISION RECEIVER (OR TELEVISION MONITOR) CAN BE REDUCED TO TWO LSI CHIPS (A MICROPROCESSOR AND AN INTERFACE CHIP).

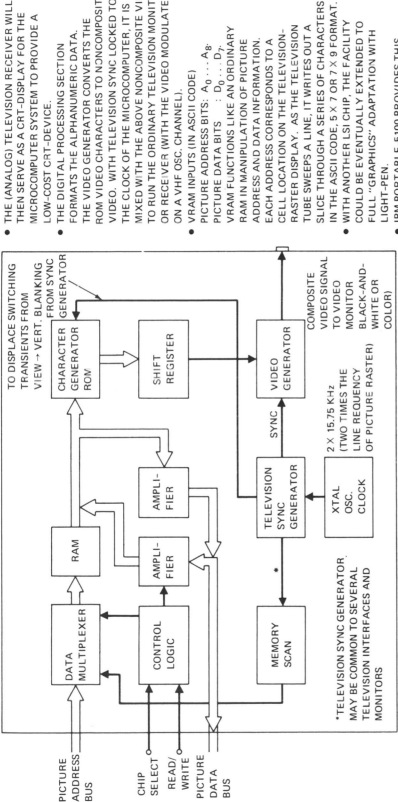

- THE (ANALOG) TELEVISION RECEIVER WILL THEN SERVE AS A CRT-DISPLAY FOR THE MICROCOMPUTER SYSTEM TO PROVIDE A LOW-COST CRT-DEVICE.
- THE DIGITAL PROCESSING SECTION FORMATS THE ALPHANUMERIC DATA. THE VIDEO GENERATOR CONVERTS THE ROM VIDEO CHARACTERS TO NONCOMPOSITE VIDEO. WITH TELEVISION SYNC LOCKED TO THE CLOCK OF THE MICROCOMPUTER, IT IS MIXED WITH THE ABOVE NONCOMPOSITE VIDEO TO RUN THE ORDINARY TELEVISION MONITOR OR RECEIVER (WITH THE VIDEO MODULATED ON A VHF OSC. CHANNEL).
- VRAM INPUTS (IN ASCII CODE)
  PICTURE ADDRESS BITS: $A_0 \ldots A_8$.
  PICTURE DATA BITS : $D_0 \ldots D_7$.
  VRAM FUNCTIONS LIKE AN ORDINARY RAM IN MANIPULATION OF PICTURE ADDRESS AND DATA INFORMATION. EACH ADDRESS CORRESPONDS TO A CELL LOCATION ON THE TELEVISION-RASTER DISPLAY. AS THE TELEVISION TUBE SWEEPS A LINE, IT WRITES OUT A SLICE THROUGH A SERIES OF CHARACTERS IN THE ASCII CODE, 5 X 7 OR 7 X 9 FORMAT.
- WITH ANOTHER LSI CHIP, THE FACILITY COULD BE EVENTUALLY EXTENDED TO FULL "GRAPHICS" ADAPTATION WITH LIGHT-PEN.
- IBM PORTABLE-5100 PROVIDES THIS VRAM-TELEVISION MONITOR FACILITY AS AN EXTENSION TO ITS SMALL-SCREEN CRT-DISPLAY PERIPHERAL.

Fig. 5-27. "VRAM-converter" for black-and-white or color television monitor or receiver. (*Courtesy of EDN, March 5, 1977*)

Table 5-5. Microprocessor instruction set—Alphabetic sequence (*Courtesy of Motorola Semiconductor Products, Inc.*)

ABA	Add Accumulators	CLR	Clear	PSH	Push Data
ADC	Add with Carry	CLV	Clear Overflow	PUL	Pull Data
ADD	Add	CMP	Compare		
AND	Logical AND	COM	Complement	ROL	Rotate Left
ASL	Arithmetic Shift Left	CPX	Compare Index Register	ROR	Rotate Right
ASR	Arithmetic Shift Right			RTI	Return from Interrupt
		DAA	Decimal Adjust	RTS	Return from Subroutine
BCC	Branch if Carry Clear	DEC	Decrement		
BCS	Branch if Carry Set	DES	Decrement Stack Pointer	SBA	Subtrace Accumulators
BEQ	Branch if Equal to Zero	DEX	Decrement Index Register	SBC	Subtrace with Carry
BGE	Branch if Greater than or Equal to Zero			SEC	Set Carry
		EOR	Exclusive OR	SEI	Set Interrupt Mask
BGT	Branch if Greater than Zero			SEV	Set Overflow
BHI	Branch if Higher	INC	Increment	STA	Store Accumulator
BIT	Bit Test	INS	Increment Stack Pointer	STS	Store Stack Register
BLE	Branch if Less or Equal	INX	Increment Index Register	STX	Store Index Register
BLS	Branch if Lower or Same			SUB	Subtract
BLT	Branch if Less than Zero	JMP	Jump	SWI	Software Interrupt
BMI	Branch if Minus	JSR	Jump to Subroutine		
BNE	Branch if Not Equal to Zero			TAB	Transfer Accumulators
BPL	Branch if Plus	LDA	Load Accumulator	TAP	Transfer Accumulators to Condition Code Register
BRA	Branch Always	LDS	Load Stack Pointer		
BSR	Branch to Subroutine	LDX	Load Index Register	TBA	Transfer Accumulators
BVC	Branch if Overflow Clear	LSR	Logical Shift Right	TPA	Transfer Condition Code Register to Accumulator
BVS	Branch if Overflow Set				
		NEG	Negate	TST	Test
CBA	Compare Accumulators	NOP	No Operation	TSX	Transfer Stack Pointer to Index Register
CLC	Clear Carry			TXS	Transfer Index Register to Stack Pointer
CLI	Clear Interrupt Mask	ORA	Inclusive OR Accumulator		
				WAI	Wait for Interrupt

Separate nonmaskable interrupt—internal registers saved in stack
Six internal registers—two accumulators, index register, program counter, stack-pointer, and condition code register
Direct memory addressing (DMA) and multiple processor capability
Clock-rates as high as 1 MHz
Simple bus-interface without TTL
HALT and single-instruction execution capability

A set of 72 instructions used for programming the Motorola type MC6800 microprocessing unit (MPU) are given in Table 5-5. The flowchart and bus-interface for this particular 8-bit parallel-processing MPU are shown in Figs. 5-28 and 5-29.

5.16 CRT LINE TERMINALS

5.16.1 CRT Display (Graphics). The CRT-display has a persistence-time ranging from 0.05 to 95 msec, depending on the phosphor used. For extended phosphor-life, medium persistence and a 60-Hz/sec *refresh* (rewrite) technique are commonly used with phosphor P-31 for minimum flicker. For telephone transmission at a data-rate of 2400 baud/sec, a local memory is used to store the data, and drive or refresh the CRT at an increased rate.

The low-cost, fixed-format, limited-symbol, 11-in. diagonal CRT-display provides

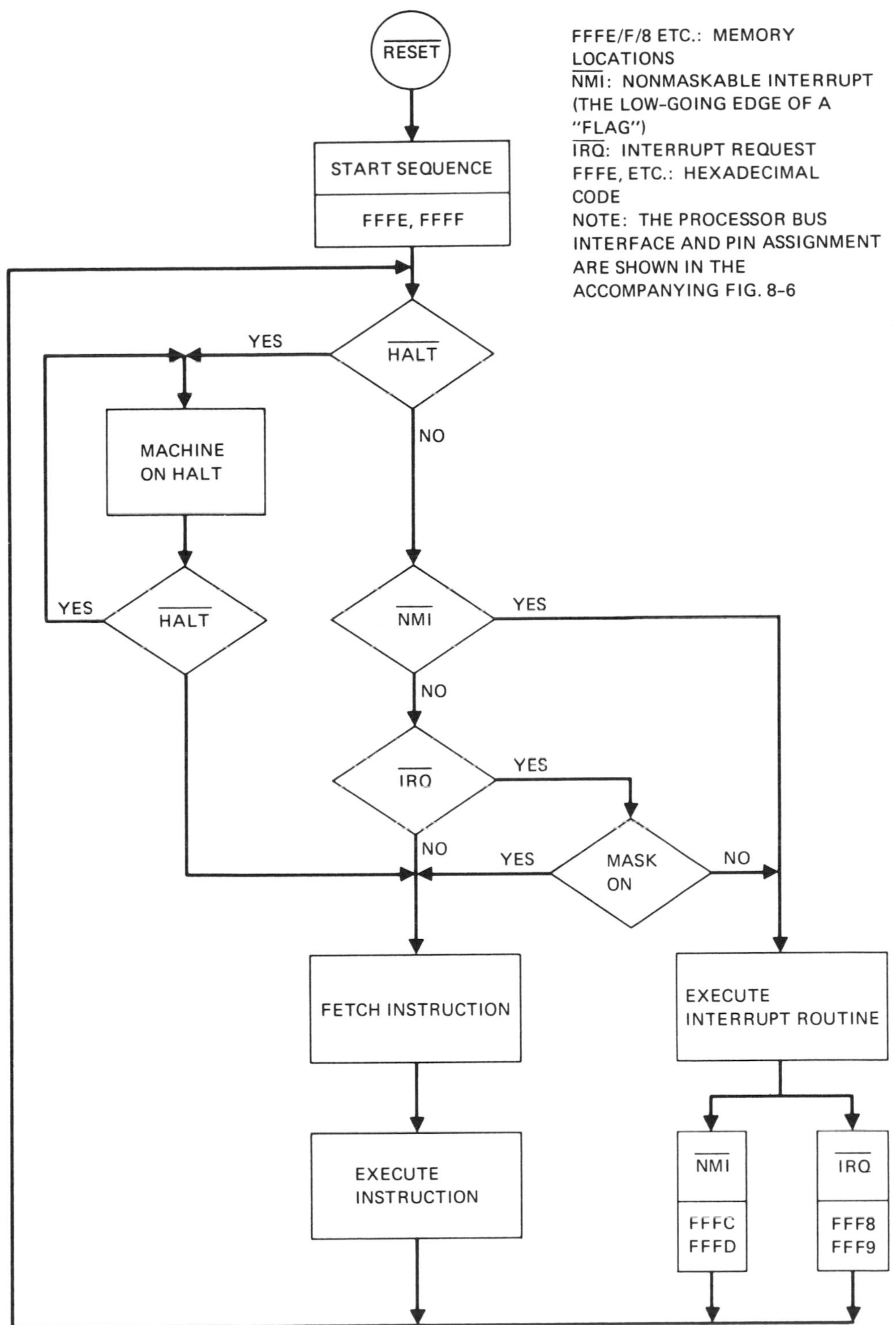

Fig. 5-28. MPU flowchart, MC-6800. (*Courtesy of Motorola Semiconductor Products, Inc.*)

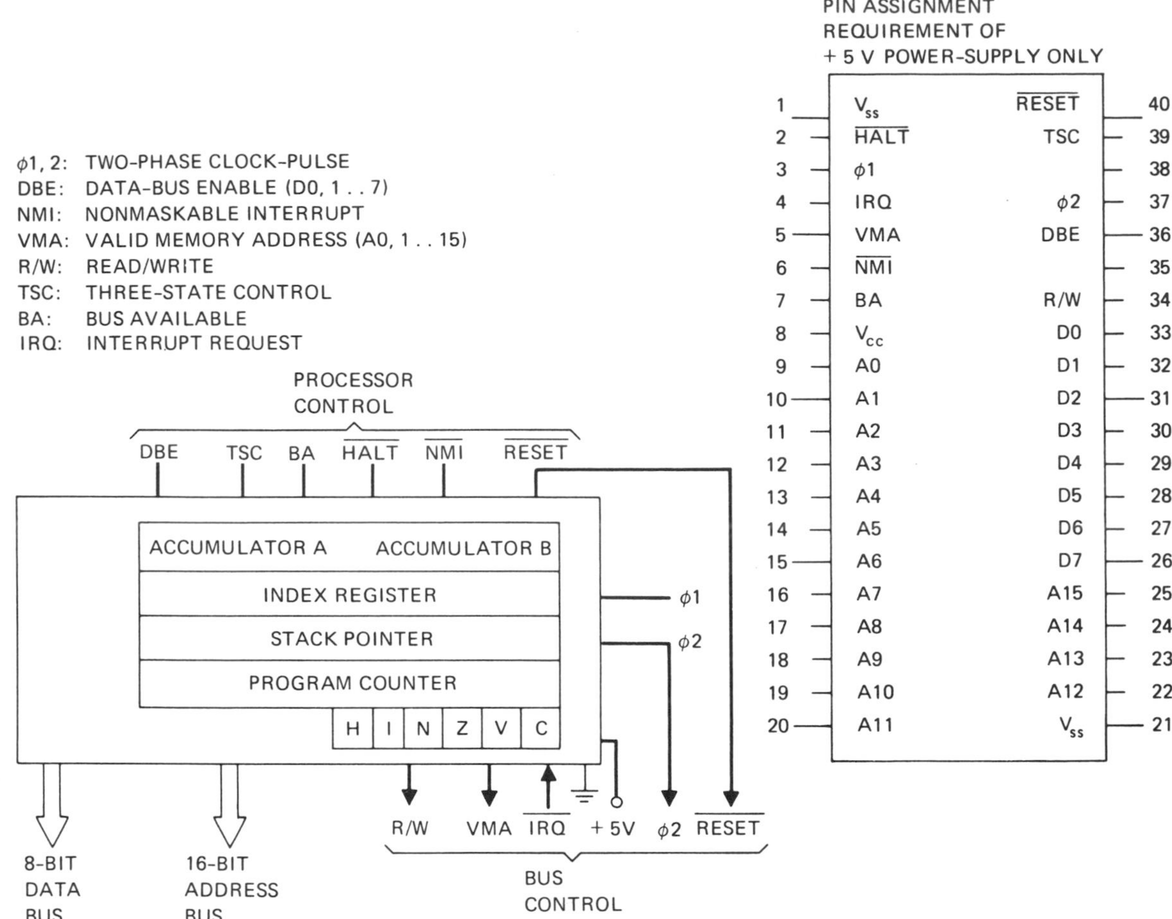

Fig. 5-29. Processor bus-interface, MC-6800. (*Courtesy of Motorola Semiconductor Products, Inc.*)

a fixed number of lines and characters per line (10 to 20 lines, 50 characters/line for a maximum of 1000 characters, in the absence of the interactive computer Graphics facility). The refresh-memory is a low-cost acoustic or magnetostrictive delay-line, limited to 8000 bits or 1000 characters for banking and recordkeeping applications. The characters are displayed and redisplayed each time the memory cycles, and a microprocessor-controller interprets input-commands from the associated keyboard and computes for positioning, spacing, line-feed, back-space, rub-out, erase, and cursor. A fixed-format "refreshed" CRT display system is illustrated in Fig. 5-30.

The higher-cost sophisticated alphanumeric and graphics CRT displays use a random-format 50 X 50 symbol-display with a local memory of approximately 64-K bits. If a 30-Hz refresh-rate is used, a refresh must be done in 33.3 msec. The 50 X 50 symbol would then require refreshing at about 13 μsec/symbol, while a 7 X 9 dot-matrix generates each symbol. Since this requires high-speed dot-generation at 200 nsec, the large-capacity refresh system is operated in a *stroke-generation mode* by X-Y analog deflection voltages. The characters are generated by line segments or strokes with additional software. This processing actually requires special hardware in the form of a video character generator or a vector generator.

The local memory allows the use of *light-pen* and *selective-erase* modes of operation. The light-pen is a light-sensing device to locate a graphic point on the CRT

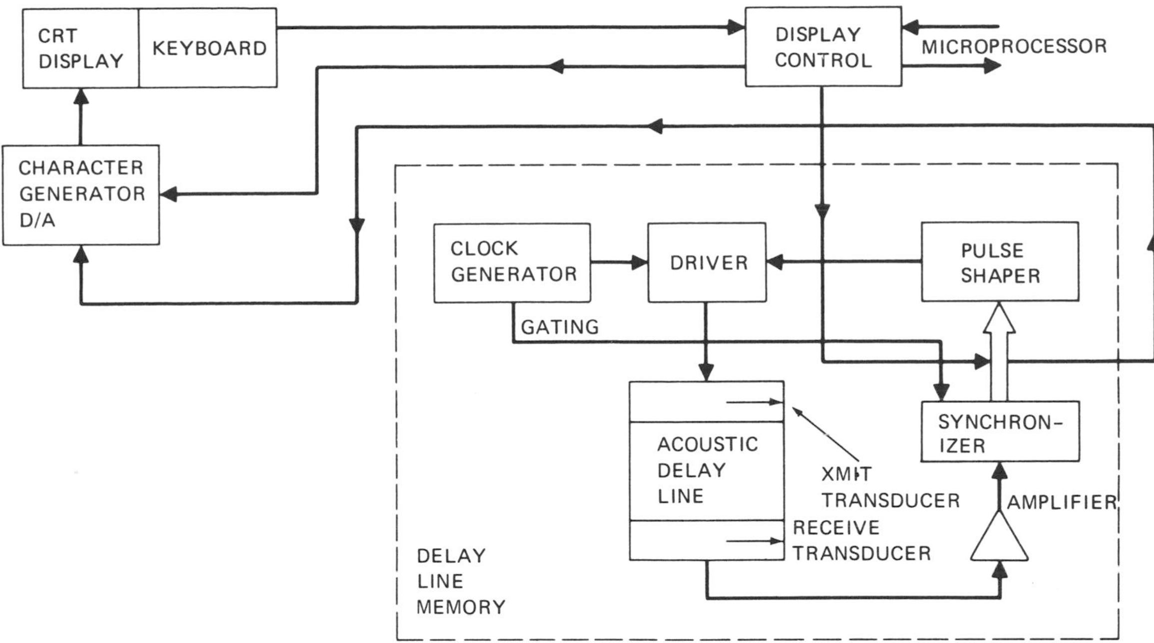

Fig. 5-30. Fixed-format refreshed CRT-display system. (*Courtesy of Tektronix Inc.*)

display and instruct the computer to perform a certain operation at that point. Some light-pen Graphics systems provide an error-sensing feedback mode; the CRT-beam follows the pen movements of simulated "writing." A light-pen system is shown in Fig. 5-31.

A selective-erase system can erase a character from the local memory without affecting the rest of the CRT display. A regular RAM-memory could be used for the local memory. Since the CRT terminal information is sent to the MPU controller (from the keyboard or the line), and not directly to the display, it is called *echo-plexing* or *half-duplexing*.

The direct-view nonflicker bistable *storage* CRT display is another low-cost, high-density computer display system. This system also requires the character/vector generator to convert the serial digital data into the required analog format. A dot-character generator and a vector-generator follow the system technique shown in Fig. 5-32.

5.16.2 Commercial Models of CRT Display or Video Display Terminal. A video display terminal (VDT) is commonly used for in-house time-sharing, data entry, inventory control, production control, and planning by the keyboard entry of data and commands. The Texas Instruments 913A keyboard VDT is a typical example, with a 15.5″ (H) × 12.8″ (D) × 19″ (W) display.

Besides the usual display controls, the cursor controls allow access up, down, left, right, and home (the tope upper left of the screen). The display terminal includes 14 keys on the keyboard for the entry of the commands. The display characters are formed by a 5 × 7 matrix. Twelve lines and 80 characters/line make the total screen-capacity for using 57 upper-case display characters and 32 control characters of the standard set of ASCII code. The screen can be filled in less than 20 msec. The cursor is positioned by the program written in a built-in ROM memory. The screen

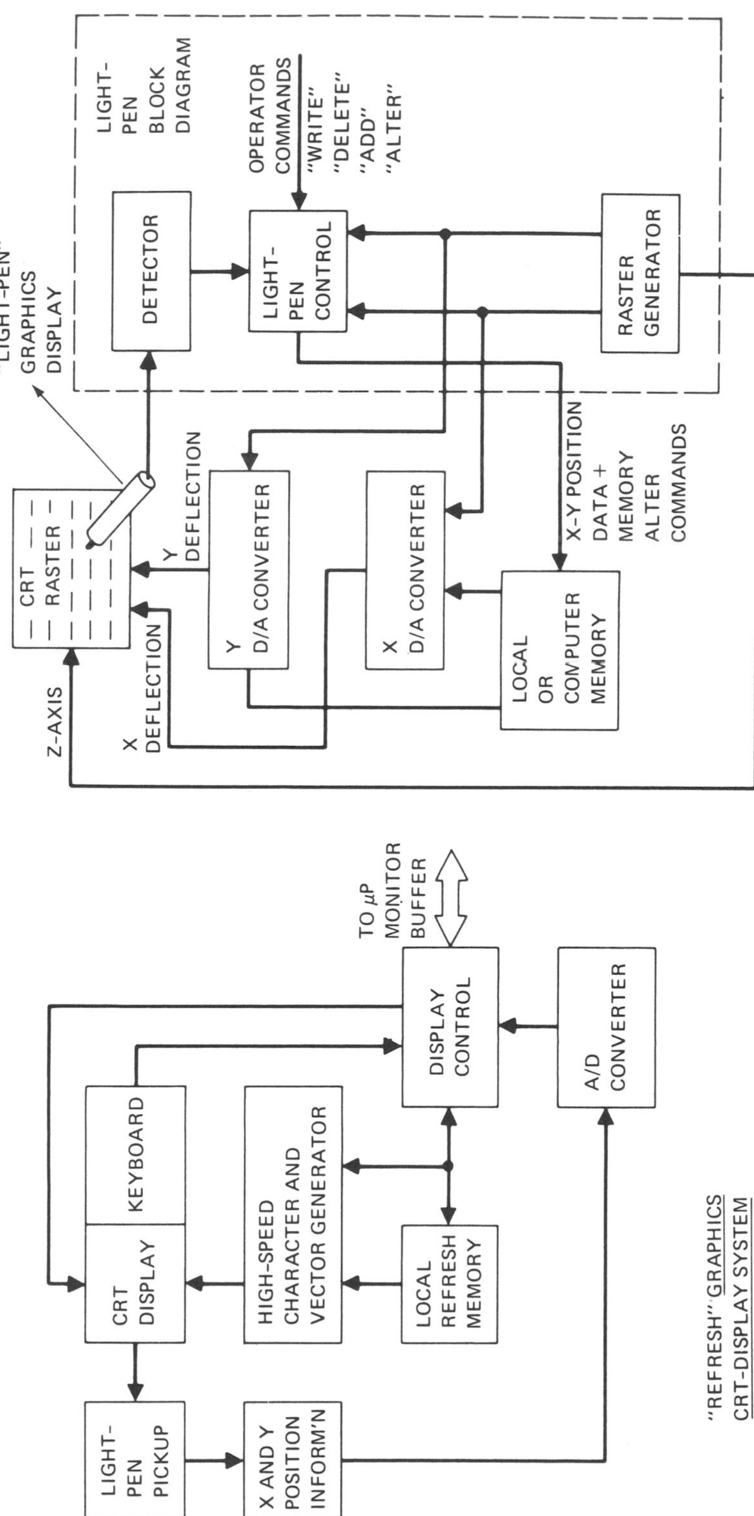

Fig. 5-31. "Refresh" graphics CRT-display system. (*Courtesy of Tektronix Inc.*)

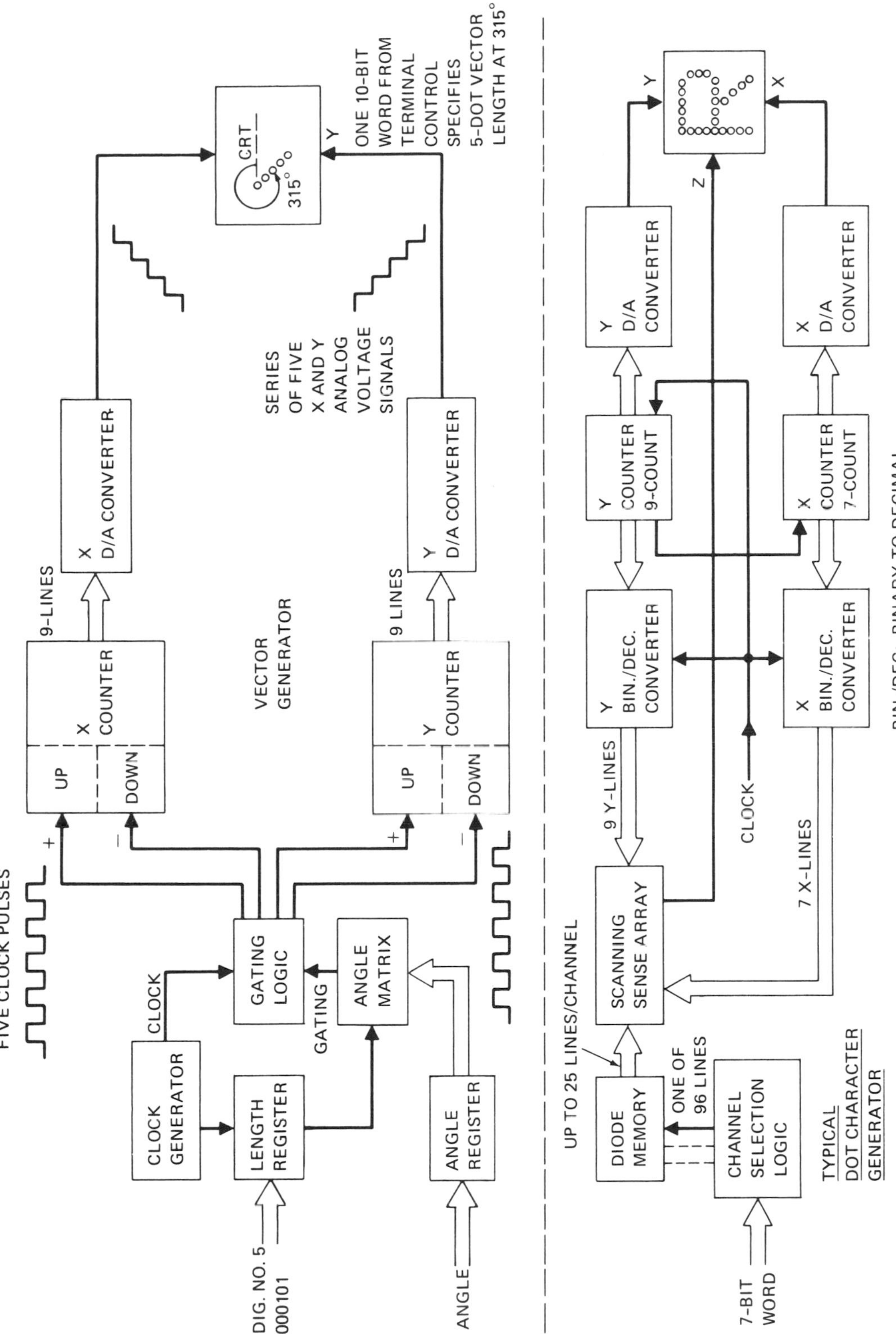

Fig. 5-32. Vector generator and dot-character generator. (*Courtesy of Tektronix Inc.*)

refreshes at 50 or 60 frames/sec under the control of the memory located in the controller.

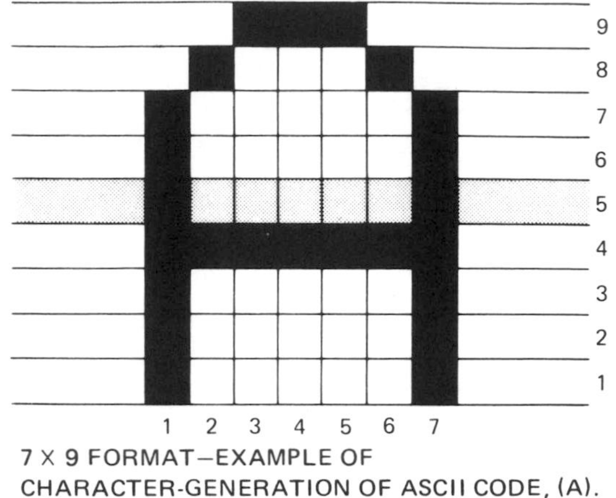

7 X 9 FORMAT—EXAMPLE OF CHARACTER-GENERATION OF ASCII CODE, (A).

5.16.3 Typical Microprocessor-Based On-Line CRT Display Terminal (Model JK-435, Matsushita). A microprocessor-based CRT terminal can meet the requirements of business, industrial, and scientific applications by the use of EPROMs or by merely plugging in preprogrammed ROMs. The various interface units and peripherals are shown in Fig. 5-33 for this on-line microcomputer system that employs a CRT keyboard display. The complete system is controlled by a single 16-bit microprocessor. The peripherals will of course be supplemented by individual microcontrollers.

I/O data is accepted from the keyboard via its interface or a remote computer via the requisite modem and the communications interface. The microprocessor uses:

1. The external PROM to process the incoming data into 11-bit slices for the display function, and the remaining 5-bit slices for the field-control function.
2. The RAM for temporary storage. The 11-bit data is transferred to the CRT-

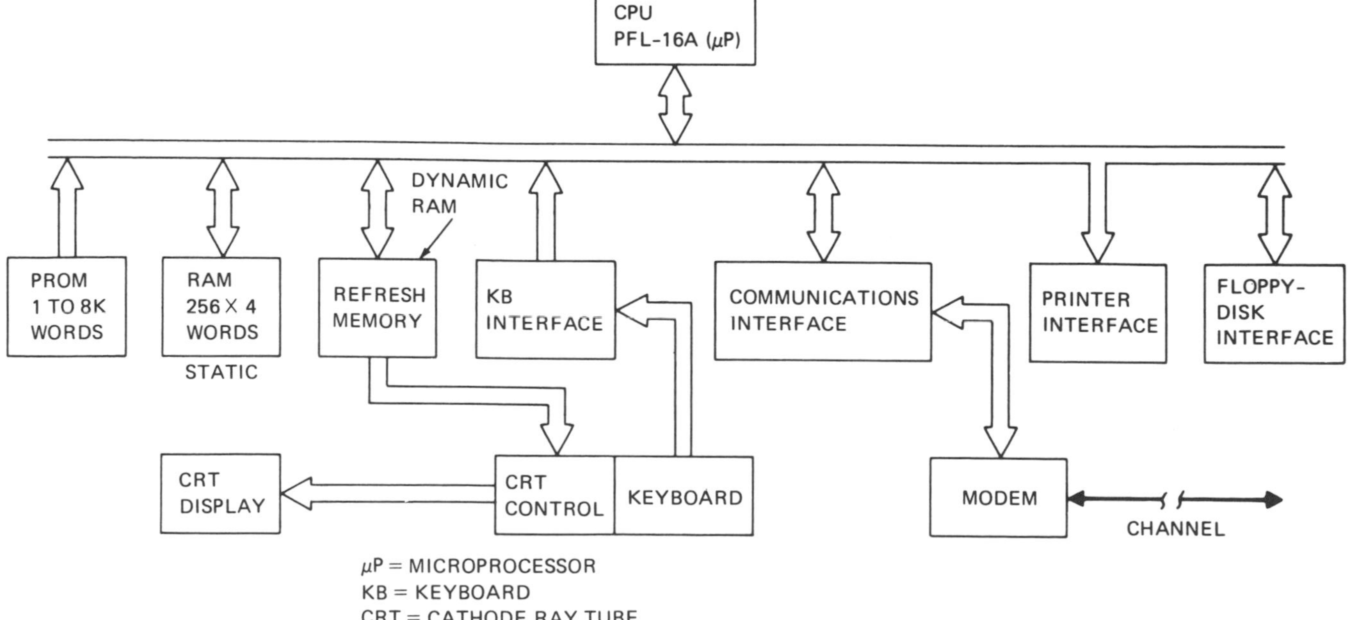

Fig. 5-33. Microprocessor-controlled CRT-Terminal. (*Courtesy of Electronics*)

refresh RAM-memory for the CRT readout. For printout, storage, or transmission, the microprocessor, under the command of the ROM program, picks up the data from the refresh memory to perform the necessary data processing.

The microprocessor PFL-16A has an instruction cycle-time of 3.3 μsec, and 33 basic instructions, five arithmetic registers, two index registers, three interrupt levels, and 256 I/O entries. As a facility for the display application, the microprocessor can operate in the direct, indirect, and relative addressing modes. With the index-modifier software used, the number of instruction steps in the ROM are minimized. Besides, the data is manipulated in bit, byte, or word to conserve ROM memory-capacity. Since one instruction corresponds to one word, programming and debugging procedures are simplified.

The ROM and RAM memory capacity can be extended to 8192 and 4096 words, respectively. With a P-39 phosphor and its 200-msec persistence, the retrace-rate of 44 Hz is flicker-free. The 14-in.-high CRT display can display 24/25 lines at 80 characters/line. The 11-bit display-slice consists of 8-bit character-data, 1 memory-protect bit, 1 cursor-display bit, and 1 graph-designation bit. The 128 different characters used in this display system include alphanumerics, punctuation marks, Japanese Katakana characters, and simple graphics.

The field control-data is implemented for the control of the next-field brightness, use of ruled lines, character-blinking, and erasure.

The keyboard has 16 command-function keys for character-insert, character-delete, or data transmission. The modem used in this system has provision for asynchronous data rates of 110, 200, 300, 600, and 1200 baud/sec, and synchronous data rates of 2400, 4800, 9600 baud/sec. An optional mini–line-printer, such as Texas Instruments Model 306, and Texas Instruments floppy-disk executive DX-10, can be appropriately interfaced to the PFL-16A. The interfaces in this system are hard-wired, but for dedicated applications, the interface units can be designed with preprogrammed LSI-PLA. An alternative ROM control-memory can be used for the optional printer or disk. The ROM *firmware* will control most of the CRT display functions, e.g., field-protect, field-erase, numeric field-control, tab, backtab, character-insert, character-delete, scrolling, and erase.

If the central computer employs peripherals of other makes, the CRT terminal will modify its firmware to meet the remote peripheral requirements.

5.17 FDM AND TDM IN DATA COMMUNICATIONS

The data user may choose to transmit the data either in a serial or a parallel mode. The choice usually depends on the original format of data to be transmitted or it may depend upon the optimization of the channel in use. For parallel transmission, the channel is band-split into narrow-band subchannels for Frequency Division Multiplex (FDM). Each bit of a character (e.g., ASCII 8-bit word) is transmitted over a separate narrow-band channel as shown in Fig. 5-34. The bit-rate of each narrow-band channel is reduced by a factor equal to the number of channel segments. By multiplexing, the incoming channels are contiguously put side-by-side in frequency. In serial transmission, each bit of an individual character is transmitted sequentially over a single channel.

Time Division Multiplex (TDM) is the latest preferred technique, since the concept is more appropriate to digital data communications, although FDM is easier to implement and less costly. A typical FDM multiplexer may put together twelve

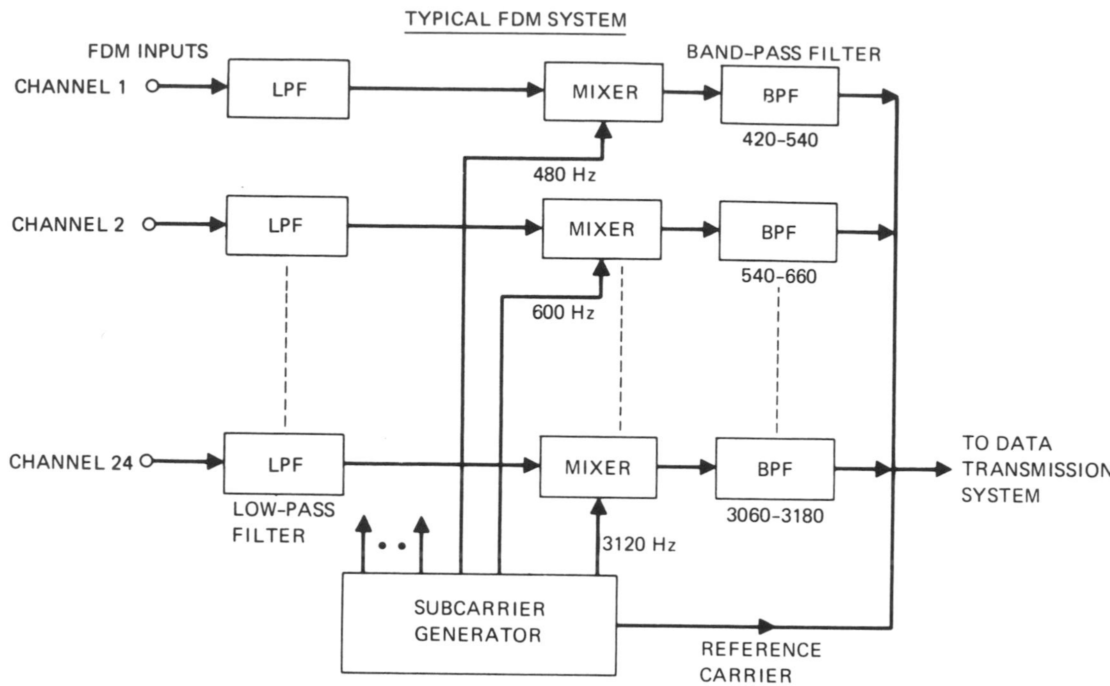

Fig. 5-34. Typical FDM system. (*Courtesy of IEEE Spectrum, February 1971*)

110-baud channels, or alternatively two 600-baud channels over one voice-grade 1200-baud line. A small part of the usable bandwidth must, however, be reserved as an allowance for channel *guarding.*

TDM is also termed *nonmessage switching*, and the relatively short time-intervals are repeatedly assigned to the same user; this is called *Time Division Multiple Access* (TDMA). The present FDMA satellite links are due for conversion to the TDMA at the next opporunity.

A typical character-interleaved TDM system is illustrated for the coded message (A), shown in Fig. 5-35. It is first stripped of its *start* and *stop* bits (B). The character of channel-1 is moved into the frame (C). Then, the frame synchronizing information is added for the next packets of data in respective channels (D). Although it is mandatory for TDM to reserve some message space for sync, framing, and channel-identifying bits, it represents the only practical way of formatting a *digital multiplexing technique* over a digital bit communications network, since the bits packed-together need not be put through D/A and A/D converters at either end. With TDM, low-cost microprocessor hardware will lower the costs too. The application of FDM/TDM data communications and color television by synchronous satellites is illustrated in Fig. 5-36.

5.18 FLOWCHART AND PROGRAMMING

The flowchart is a logical representation of the path of the processing or computation control to arrive at a set of instructons (a program) for obtaining the solution of a problem. A recursion or iteration procedure in the algorithm of the problem is indicated by a closed loop in the diagram. A simple linear path indicates that no control decisions are required. Each block in the flowchart presents the status of

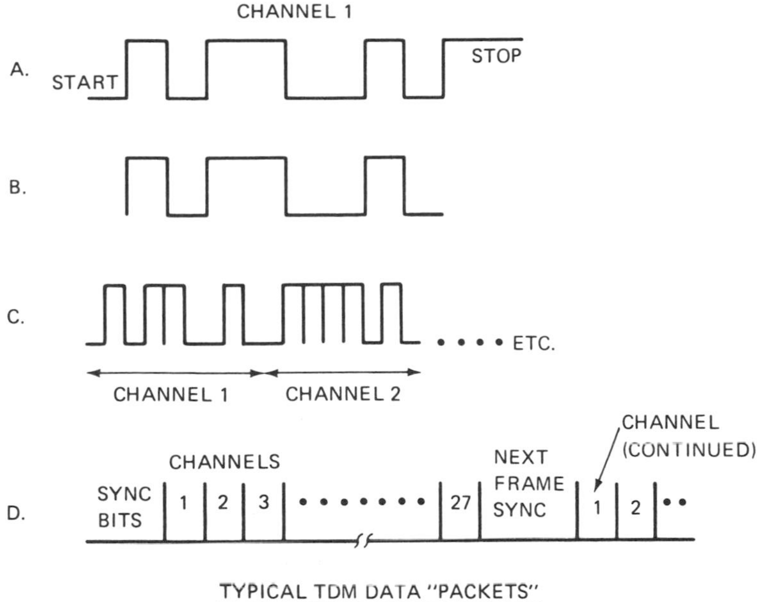

Fig. 5-35. Typical TDM data packets. (*Courtesy of IEEE Spectrum*)

the variables and constants in a logical sequence. An example of the said iterative procedure is shown in the following diagram when a control decision with an iterative loop is involved.

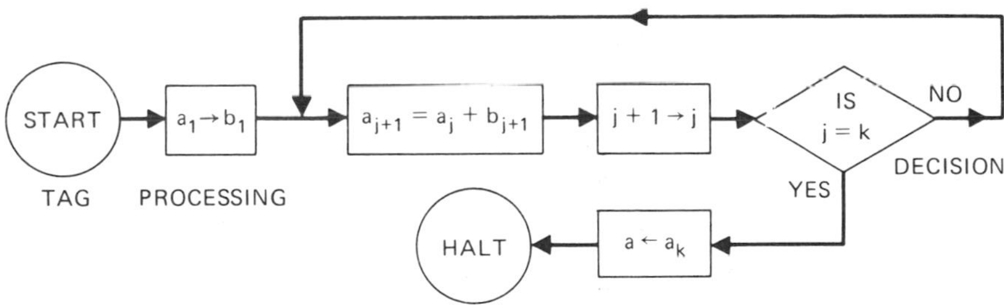

5.19 FAST FOURIER TRANSFORM

A series of oscillations, the frequencies of which are integral multiples of the fundamental frequency, can be represented by the Fourier series when the function X(t) is single-valued and finite and has a finite number of maxima and minima in the interval of one oscillation.

Frequency and Time Domains

1. Analog continuous Fourier transform pair, CFT

$$X(f) = \int_{-\infty}^{\infty} x(t)e^{-i2\pi ft}\,dt$$

$$x(t) = \int_{-\infty}^{\infty} X(f)e^{i2\pi ft}\,df$$

$$\left\{\begin{array}{l} -\infty < f < \infty \\ -\infty < t < \infty \\ i = \sqrt{-1} \end{array}\right.$$

$$X(j) = \frac{1}{N}\sum_{k=0}^{N-1} x(k)e^{-i2\pi jk}$$

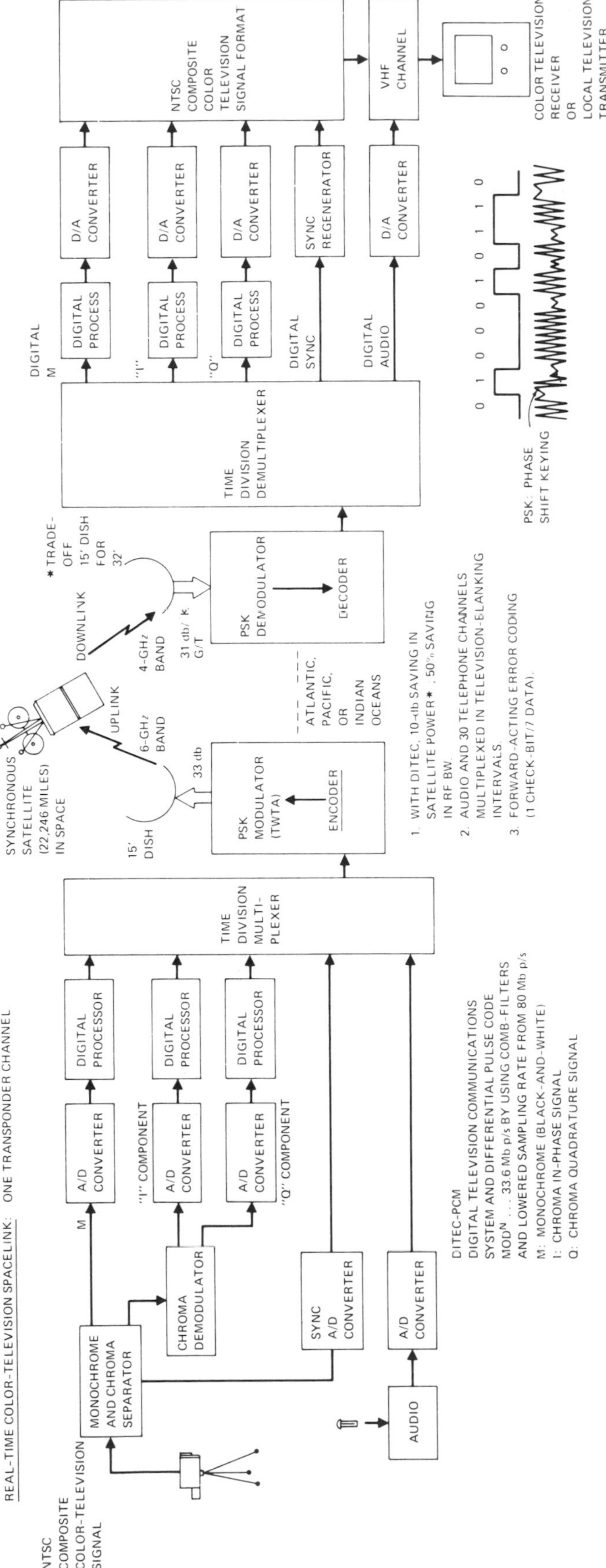

Fig. 5-36. Digitized real-time global color-television transmission via satellites.

2. Digital form for a complex series, DFT

$$x(k) = \sum_{j=0}^{N-1} X(j)e^{i2\pi jk} \qquad \begin{matrix} j = 0, 1 \ldots N-1 \\ k = 0, 1 \ldots N-1 \end{matrix}$$

Substitute $e^{2\pi i/N} \equiv W_N$ to obtain the following

3. Cooley-Tukey Fast Fourier Transform (FFT) algorithm:

$$X(j) = \frac{1}{N}\left[\sum_{k=0}^{N-1} x(k)^* W_N^{jk}\right]^*$$

*Complex conjugate operation.

$$X^*(j) = \hat{X}(j) = \sum_{k=0}^{N-1} A(k)W^{jk}$$

$$j = 0, 1 \ldots N-1; W = e^{2\pi i/N}; A = X^*/N.$$

When N = 8: j and k arc 0, 1, . . . 7

$$W^8 = [e^{2\pi i/8}]^8 = e^{2\pi i} = 1$$

For N = 8, a direct evaluation of the DFT algorithm requires 64 operations of complex multiply-and-add operations. But for FFT, the number of multiply-and-add operations are reduced to 24 only. Computation of N^2 operations in DFT are actually reduced to $\frac{N}{2}\log_2 N$ complex multiplications in FFT. When N = 1024, the simplification in computation is as much as 200 to 1. (G. D. Bergland, Bell Telephone Labs.)

The FFT technique has practically become conventional, not only in computing spectrograms for short-term power spectrums as function of time, but also for correlation of two time series and for digital filtering in convolution of two series of data samples. Using a mini- or micro-computer system and 8-K words of memory, a 1024-point transformation can be presently achieved in less than 1 sec without using the floating-point format. At this time, an accessory CCD processor appears ideal in a microcomputer system for high-speed FFT computation—since the basic operations needed to perform this computation are arithmetic and delay elements.

The Fast Fourier Transform is ideal for digital processing of waveforms in a literal sense, since the frequency components of a continuous waveform can be rapidly identified by first sampling it at finite intervals to enable the discrete DFT form of analysis in principle. The FFT algorithm is merely a rapid computational technique to translate the time-domain to the frequency-domain components. The closer the samples are taken, the more accurately the resulting digital series represents the original waveform. The sum of the finite series, considered as coefficients of successive harmonic frequencies, is an approximation in DFT. The coefficient of the first term is the average value of all the samples (i.e., their sum divided by N). Without FFT, each coefficient would require the summation of N real and N imaginary terms, each of which is a sample value and a trigonometric weight. For the real and complex conjugates, N^2 products would be required to compute the coefficients. The simplifying FFT algorithm requires the combining of the Fourier coefficients for two interleaved sets to obtain the coefficients of the composite set. A single set of coefficients for all the sets of the even- and odd-numbered samples.is the actual computational simplification. Hence, in the final analysis, the FFT digital signal processing, as stated previously, requires a total of N/2 real multiplications in regular DFT signal process. The Cooley-Tukey FFT approach is, in short, based on a set of nested multiplications using full-adders and delay elements—hence the popularity of the CCDs for simplifying the processing of waveforms by means of the FFT algorithm.

Special-purpose digital hardware, interfaced to a minicomputer or microcomputer, is currently available to compute 1024 real-point FFT in 139 msec. Complete real-time control software is provided along with the FFT processor. Some 1024 real-points correspond to 512 complex points, and actual FFT processor time is only 18 msec out of the overall computer time of 139 msec. With full 16-bit accuracy, Elsytec Company's FFT Array Processor 306-MFFT transforms from 16 to 16,384 real points. The processor is also capable of high-speed correlation and digital filtering, forward and inverse FFT, spectral magnitude and complex multiplication, and arithmetic processing of arrays or blocks of numbers. The control software needs only 1100 RAM cells. FFT calculations utilize block floating-point scaling. Complex array multiplications require 17 μsec per complete point.

5.20 SOLID-STATE MEMORIES

A digital control microcomputer system on one or two chips—that is where the large-scale integrated (LSI) solid-state memories are leading the microprocessors. The following types of internal main memory and external mass-storage memory will gradually replace the internal CORE memory and external disk and drum electromechanical memories, from the viewpoints of cost, size, power, reliability, maintenance, and performance as measured by cycle and access times. (New high-density nonvolatile CORE will continue in special applications.)

1. ROM (read-only memory), preprogrammed. 1 to 64-K bits per chip.
2. PROM (programmable ROM). 1 to 64-K bits per chip.
3. EAROM (electrically-alterable ROM). 1 to 64-K bits per chip.
4. RAM (bipolar I^2L, NMOS and CMOS-on-saphire random access memory). The higher capacity dynamic memory needs to be refreshed periodically by appropriate clock and *refresh* circuitry, since electrode charge-storage capacitances are involved. The static memory uses a larger number of transistors per storage cell and maintains the logic levels indefinitely, but unlike the CORE, it is nonvolatile in the absence of power. 1 to 16-K bits/chip (static); 1 to 64-K bit/chip (dynamic).
5. Nonvolatile MNOS (metal nitride-oxide memory). It is nonvolatile for all practical purposes, but the writing-speed requires milliseconds.
6. CCD Memory (charge-coupled device) or fast-access Fairchild LARAM (line-addressable RAM) up to 64-K bit capacity/chip.
7. FET memory of IBM, using electron-beam lithographic and ion-implantation techniques.
8. EBAM (electron-beam addressable memory) or BEAMOS (beam-addressable MOS mass memory of GEC).
9. CAM (content-addressable memory). This type of memory combines the comparator search logic at each bit position with the storage facility; it is ideal for search through data files and interrelate specific data against sets-of-data in files.
10. Magnetic-bubble mass memory. With its average 4-ms random access time

and 100-K bit/sec memory-transfer rate, it is even faster than the floppy disk. A single PC board can accommodate 92-K bytes of external nonvolatile RAM memory.

11. SOS-CMOS and I^2L chips have the advantage of very low power consumption and hence the ability to operate on battery cells for a considerable period.

5.21 DIGITAL FILTERS

The design of digital filters is comprehensively treated in Chapters 1 and 3 with examples. With the present trend of application of LSI microcomputers to complex digital control systems, digital filters naturally take up a position of prominence. So, in the context of the computer-controlled digital control systems, the subject of digital filters is given further exposition as a supplement to the topic of microcomputer control systems and microcomputer systems.

5.21.1 First-Order Digital Filter. Digital signals are sequences of numbers and hence discrete. Besides the basic arithmetic operations (addition, subtraction, and multiplication) that apply to digital networks, the delay element, as a shift register, is required to delay a pulse by one period T.

If the input to the network is a sequence, x(nT), and the output is y(nT), and if it is delayed by T to make it Y[(n - 1)T], the first-order digital filter is given by the first-order linear difference equation:

$$y(nT) = a_1 y(nT - T) + x(nT) \text{ and (enT) additive noise}$$

where for stability $|a_1| < 1$. If $|a_1| > 1$, the system is unstable. The impulse response of the digital filter is given by:

$$h(nT) = \sum_{n=0}^{\infty} a_1^n \delta(t - nT) = a_1^n$$

$$H(z) = \frac{a + az^{-1}}{1 - a_1 z^{-1}} \text{ or simply } \frac{1}{(1 - a_1 z^{-1})}$$

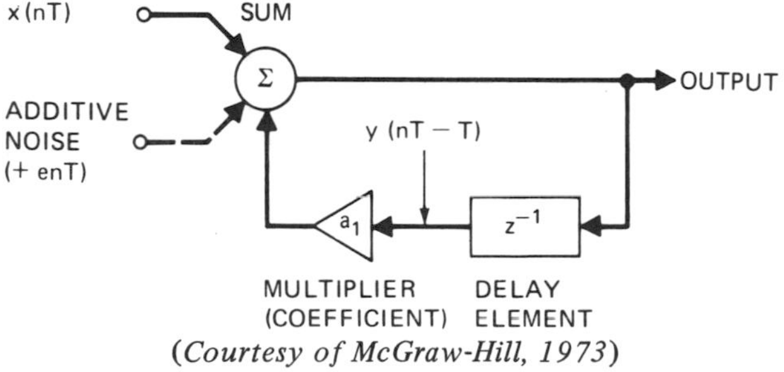

(*Courtesy of McGraw-Hill, 1973*)

For $n = 0$: $y(0) = a_1 y(-T) + x(0)$ $y(-T) = 0$ and $x(0) = 1$ are given initial
$n = 1$: $y(T) = a_1 y(0) + x(T)$ conditions. $y(0) = 1$
$n = 2$: $y(2T) = a_1 y(T) + x(2T)$, and so on for $n = 3, 4, \ldots$

If the input is a constant-amplitude pulse-train starting at $t = 0$, the initial condition $y(-T) = 0$.

$$x(nT) = 1, n \geqslant 0$$

$$y(0) = 1$$

$$y(T) = 1 + a_1$$

$$y(2T) = a_1(a_1 + 1) + 1 = 1 + K + K^2, \text{ etc.}$$

$$y(nT) = (a_1^{n+1} - 1)/(a_1 - 1). \qquad \text{System is unstable for } a_1 \geqslant 1.$$

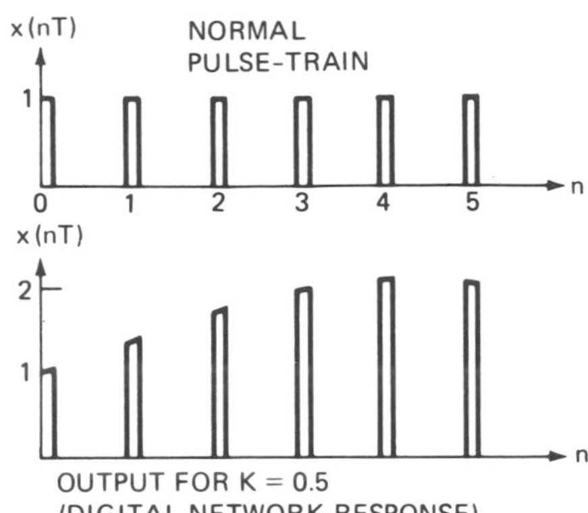

y(T) increases from 1 to $\dfrac{1}{1 - a_1}$. This is a steady-state as n increases. This response of the first-order digital filter yields a response similar to that of a first-order RC analog filter.

Using the z-domain analysis,

$$H(z) = \frac{1}{(1 - a_1 z^{-1})}$$

$$H^*(z) = \frac{z^{-1}}{(z^{-1} - a_1)}$$

and

$$H(z)H^*(z) = \frac{1}{(1 - a_1 z^{-1})} \cdot \frac{z^{-1}}{(z^{-1} - a_1)}$$

Now evaluate the output variance (σ) in the z-domain by calculating the residues of a contour integral defined from the discrete Parsevals theorem, with e(nT) as additive noise

$$\sigma_o^2 = \frac{E_o^2}{12} \cdot \frac{1}{2\pi j} \oint H(z)H^*(z) \frac{dz}{z}$$

with error $-\dfrac{E_o}{2}$ to $\dfrac{E_o}{2}$ for rounding, and $-E_o$ to 0 for truncation.

$$\sigma_o^2 = \frac{E_o^2}{12(2\pi j)} \oint \frac{z^{-1}}{(z^{-1} - a_1)(1 - a_1 z^{-1})} \cdot \frac{dz^{-1}}{z^{-1}}$$

This equation is evaluated by calculating the residue due to the singularities within the contour of integration (unit circle). The singularities (poles) of $H(z)H^*(z)$ occur at $z^{-1} = a_1$ and at $z^{-1} = 1/a_1$ on the z^{-1} plane. For $|a_1| < 1$, there is only one pole within the contour of integration, and this is at $z^{-1} = a_1$. The residue due to this pole $= \dfrac{1}{(1 - a_1^2)}$.

$$\sigma_o^2 = \frac{E_o^2}{12(2\pi j)} \cdot 2\pi j \ \Sigma \text{ (residue)} = \frac{E_o^2}{12(1 - a_1^2)}$$

If $a_1 = 0.9$, then: $\sigma_o^2 = 8.33E_o^2$ = average output noise power. In general, for first-order Kalman digital filters, the effect of a zero is to attenuate the noise.

5.21.2 Second-Order Digital Filter. A second-order digital filter is illustrated in two forms; (1) cascade series realization and (2) direct form realization. It can be shown to give an oscillatory response similar to that of the quadratic RLC analog filter. From the configuration shown, the output y(nT) is the sum of three signals applied to the summing network.

The difference equation for the circuit is

$$y(nT) = x(nT) + a_1 y[(n - 1)T] + a_2 y[(n - 2)T]$$

DIRECT FORM

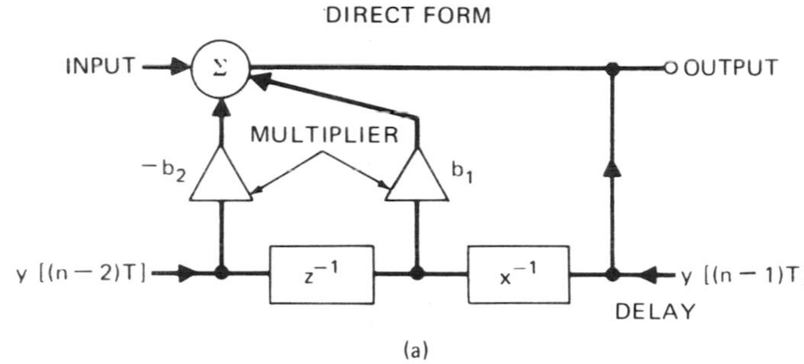

(a)

CASCADE FORM

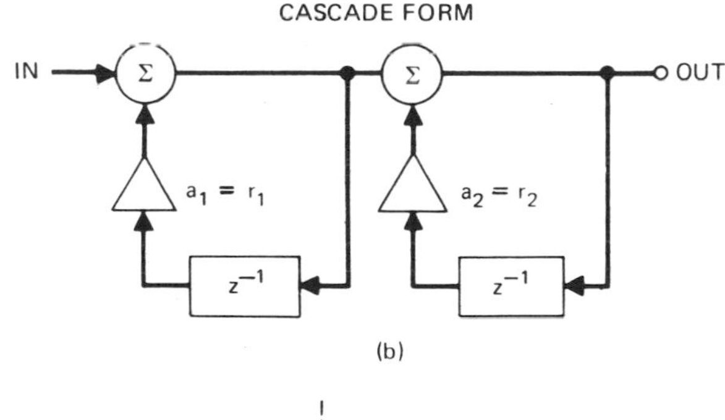

(b)

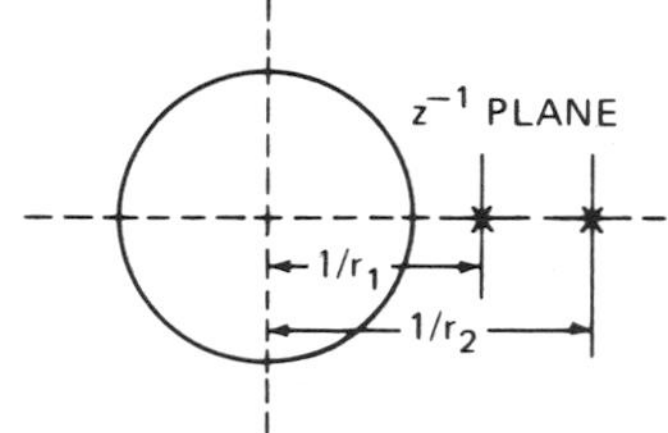

z^{-1} PLANE AND POLES OF SECOND-ORDER FILTER

(c)

(Courtesy of McGraw-Hill, *Digital and Analog Systems, Circuits, and Devices: An Introduction*, 1973)

The poles are located on the z^{-1} plane as shown. They are located at $z^{-1} = \dfrac{1}{r_1}$ and $z^{-1} = \dfrac{1}{r_2}$.

Direct-form case a. If $a_1 = 0.5$, $a_2 = -0.5$, $x(0) = 1$ and $x(nT) = 0$ for $n \neq 0$. That is, the input is a single pulse occurring at t = 0. Then, with the initial conditions y(-T) = y(-2T) = 0, we obtain the following sequence of values.

n	y(nT)	x(nT)	0.5y[n - 1)T]	- 0.5y[n - 2)T]
0	1	1	0	0
1	0.5	0	0.5	0
2	- 0.25	0	0.25	- 0.5
3	- 0.375	0	- 0.125	- 0.25
4	- 0.063	0	-0.188	+0.125
5	+0.156	0	- 0.032	+0.188
6	+0.110	0	0.078	+0.032
7	- 0.023	0	+0.055	- 0.078

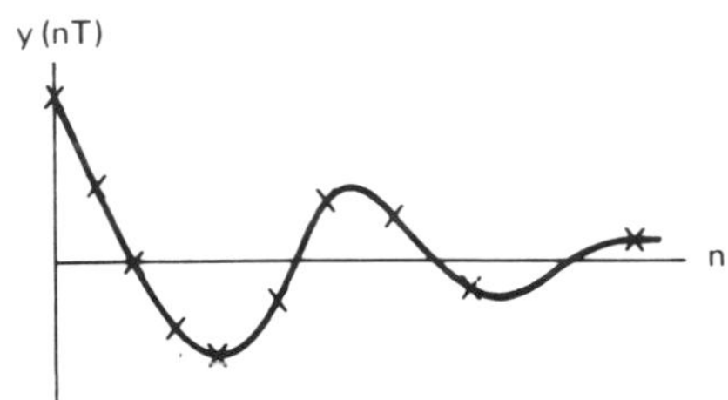

SECOND-ORDER FILTER y (nT) VERSUS n (NUMBER OF SAMPLES) (MORE VERSATILE THAN ANALOG RLC FILTER)

(d)

The plot y(nT) versus n shows the usual oscillatory response of the analog RLC network. The response, of course, ranges from nonoscillatory to unstable oscillations depending on the values of a_1 and a_2 chosen. That is, the actual coefficients chosen for the difference equation determine the optimum operation of the filter. By using the technique of *multiplexing with staggered pulses*, the second-order digital filter can be made more versatile than an RLC analog filter. The staggering time-interval between pulses will enable the processing of different pulses by a common digital filter. Also, higher accuracy is feasible with a digital filter, as compared to an analog one.

The response of a linear digital filter, according to the *principle of superposition*, comprises the *transient* and *steady-state* solutions when a sinusoidal signal is applied.

The transfer function

$$H(f) = \frac{Y_m \cdot \exp j\omega(n - N)T}{X_m \cdot \exp j\omega(n - N)T} = \frac{Y_m}{X_m}$$

The output of the filter

$$y(nT) = \mathrm{Re}[H(f)X_m \cdot \exp j\omega nT]$$

The second-order difference equation is given by

$$y(nT) = a_0x(nT) + a_1x(nT - T) + a_2x(nT - 2T) + b_1y(nT - T) + b_2y(nT - 2T)$$

Then,

$$H(z) = \frac{a_0 + a_1z^{-1} + a_2z^{-2}}{1 - b_1z^{-1} - b_2z^{-2}}$$

The pole positions are located at:

$$r_1 = \frac{b_1}{2} + \sqrt{\frac{b_1^2}{4} + b_2}$$

and

$$r_2 = \frac{b_1}{2} - \sqrt{\frac{b_1^2}{4} + b_2}$$

and zeros:

$$r_3 = \frac{-a_1}{2a_2} + \frac{1}{2a_2}\sqrt{a_1^2 - 4a_0a_2}$$

$$r_4 = \frac{-a_1}{2a_2} - \frac{1}{2a_2}\sqrt{a_1^2 - 4a_0a_2}$$

The amplitude and phase characteristics of the second-order digital filter would therefore depend upon the choice of the values of a_0, a_1, a_2, b_1, and b_2. The transfer function of the second-order digital filter with poles located at

$$z^{-1} = \frac{1}{r_1} \text{ and } z^{-1} = \frac{1}{r_2}$$

$$H_1(z) = \frac{1}{(1 - r_1z^{-1})(1 - r_2z^{-1})}$$ for cascade realization (b) $a_1 = r_1; a_2 = r_2$

or

$$H_1(z) = \frac{1}{1 - (r_1 + r_2)z^{-1} + r_1r_2z^{-2}}$$ for direct form (a) $b_1 = r_1 + r_2; b_2 = r_1r_2$

When the above equations are partially differentiated,

$$\partial r_1/\partial a_1 = 1; \partial r_1/a_1 = 0$$

$$\partial r_2/\partial a_1 = 0; \partial r_2/\partial a_2 = 1$$

$$\partial r_1/\partial b_1 = \frac{r_1}{r_1 - r_2}; \partial r_1/\partial b_2 = \frac{1}{r_2 - r_1}$$

$$\partial r_2/\partial b_1 = \frac{r_2}{r_2 - r_1}; \partial r_2/\partial b_2 = \frac{1}{r_1 - r_2}$$

Hence,

$$\left.\begin{aligned}\frac{\partial r_1}{\partial b_1} &> \frac{\partial r_1}{\partial a_1}\\ \frac{\partial r_2}{\partial b_2} &> \frac{\partial r_2}{\partial a_2}\end{aligned}\right\}\text{ if both } r_1 \text{ and } r_2 \text{ are of the same sign}$$

$$\left.\begin{aligned}\frac{\partial r_1}{\partial b_2} &> \frac{\partial r_1}{\partial a_2}\\ \frac{\partial r_2}{\partial b_1} &> \frac{\partial r_2}{\partial a_2}\end{aligned}\right\}\text{ for all realizable values of } r_1 \text{ and } r_2$$

Therefore, for most second-order digital filters, changes in b_1 and b_2 produce larger changes in r_1 and r_2 than changes in a_1 and a_2. That is, the direct-form configuration of the digital filter is more subject to the quantization effects of the coefficients.

Filters that have both poles and zeros are classified as *recursive* or *Infinite Impulse Response* (IIR) filters, and they are realized as usual in terms of adders, multipliers, and delay elements (shift registers). With all common factors cancelled in the following expression, the denominator coefficients are identically nonzero:

$$H(z) = \frac{a_0 + a_1 z^{-1} + a_2 z^{-1} + \cdots}{1 + b_1 z^{-1} + b_2 z^{-1} + \cdots} \tag{1}$$

Nonrecursive filters are represented by a polynomial in z^{-1} when all common factors are cancelled in Eq. 1.

$$H(z) = h_0 + h_1 z^{-1} + h_2 z^{-1} + \cdots \tag{2}$$

The denominator coefficients in Eq. 1 must be chosen in such a way that the transient response of the system does not increase without bound and become oscillatory; and of course it depends on the location of the poles of the system.

Digital filters can be synthesized from continuous filter data. A typical example is illustrated to give an idea. Cutoff frequency = 6 kHz; transition frequency = 8.8 kHz. Maximum passband attenuation = 1 dB; minimum passband attenuation = 30 dB at a sampling frequency of 32 kHz (beyond passband).

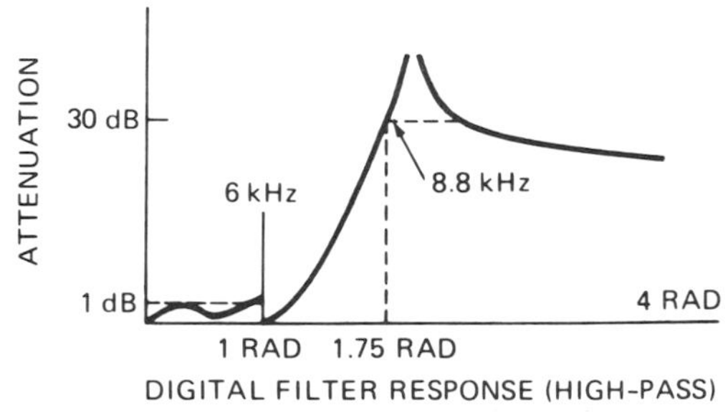

(e)

5.22 Use of PROM in Microprocessor or Microcontroller Applications (Prior to Burning in Program Instructions on ROM). A Signetics type 2650 microprocessor with 3-K bytes of ROM and 3-K bytes of RAM is illustrated in the digital microcomputer system shown in Fig. 5-37. Two input/output ports can be conveniently designed by means of a 32 X 8 bit programmable read-only memory as the decoding logic element and eliminate expensive hard-wired logic. The technique shown here is one of the most common implementation procedures one should get acquainted with in the application of microcomputer systems in digital control applications.

In the application shown in Fig. 5-37, the PROM is programmed to generate the chip-enable signals for the 6-K byte memory and provides the clock pulses to the I/O ports. The PROM actually decodes the control and address lines from the microprocessor to enable any one of the memory banks or I/O ports and functions as a kind of programmable logic array.

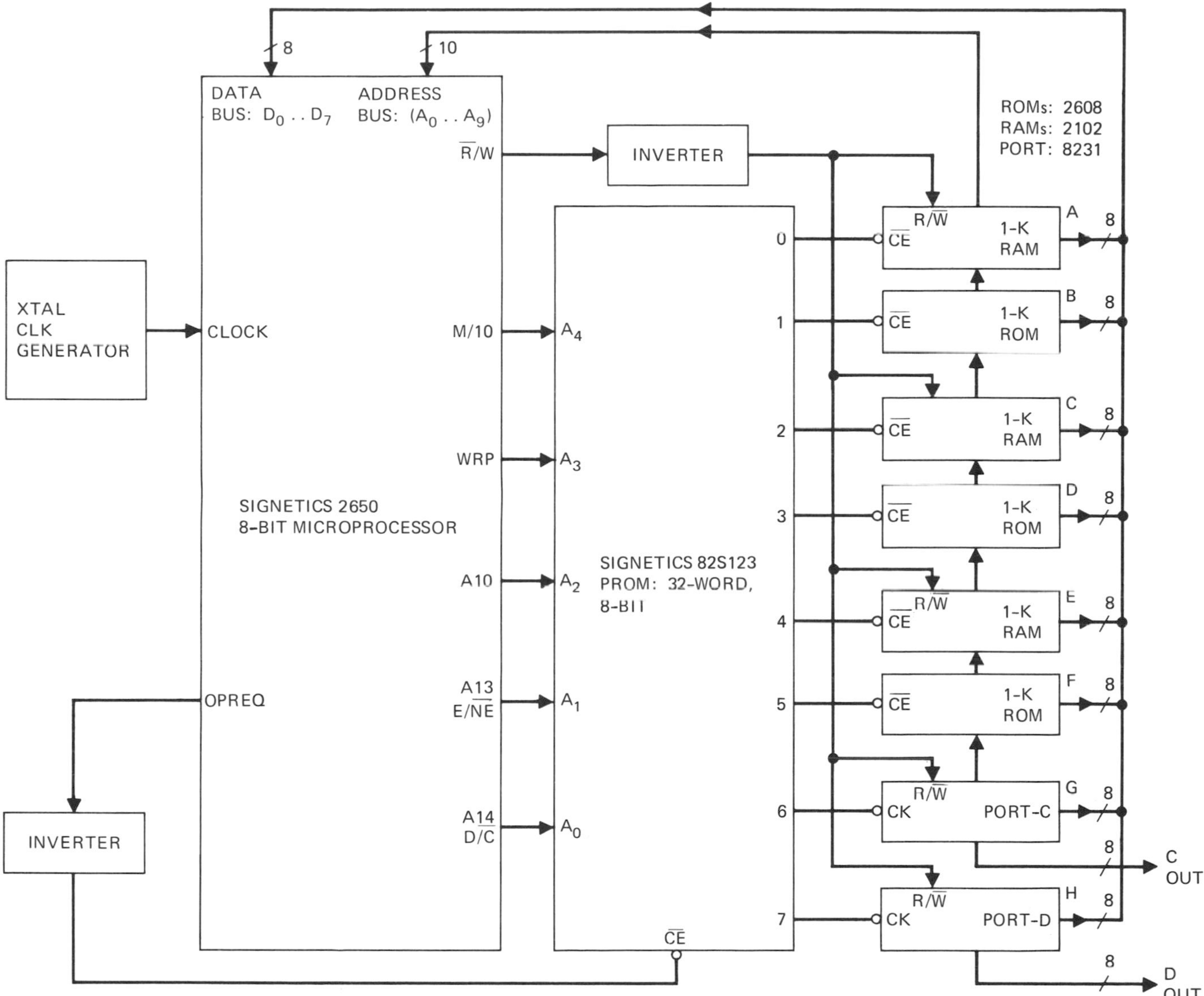

Fig. 5-37. PROM implementation in computer-controlled system. (Reprinted from *Electronics*, September 2, 1976: Copyright © McGraw-Hill, Inc., 1976.)

Signetics 2650 allows 18-bit processing by means of its 9-bit microprogram address-register and microinstruction capability. With its 77-instruction set and 8-level vectored interrupt, the operation request is the master control signal to coordinate external operations. The NMOS depletion-mode 40-pin DIP (dual-in-line package) has a cycle-time of 1.5 μsec, and consumes 625 MW at 5 V. It has the capability of 8/16/24-bit instruction words, 8-level on-chip subroutine, internal 8-bit parallel structure, and seven 8-bit general-purpose registers. A separate adder enables fast address calculation. The architecture reflects interfacing of a microcontroller and a microprocessor on a common LSI chip. Each RAM bank consists of eight 2108 1 K × 1-bit static RAMs, and each ROM bank consists of a single 2608 1-K byte ROM. Each of the I/O ports is an 8T31 8-bit bidirectional I/O interface structure. The PROM enables just one of the ROM/RAMs to read or write at a memory location command by the 10-address line memory-bus between the 2650 and the six memory banks. This bus can have 2^{10}(=1024) addresses, and the enable signals from the PROM under consideration can select these to any one of the six memory banks for 6-K bytes of memory locations for data via the 8-bit data-bus.

The two highest-order address lines that determine the 8-K byte page of memory and I/O operations are multiplexed on lines A14 (E/$\overline{\text{NE}}$) and A13 (D/$\overline{\text{C}}$) by the microprocessor. If E/$\overline{\text{NE}}$ (A1) is low, port D or C is enabled depending on whether D/$\overline{\text{C}}$ (A0) is high or low, when M/$\overline{\text{I,O}}$ line of the microprocessor selects the memory and I,O operation. That is, A4 must be high in any PROM input that enables one of the six memory banks, and A4 must be low to enable or clock either I/O port. The exclusive write-pulse (WRP) line at A3 must be high to enable any ROM for R/W. A2 must be high to enable a ROM and low to enable a RAM. The OPREQ, going to $\overline{\text{CE}}$ (chip-enable) of PROM must be high to enable any $\overline{\text{ROM/RAM/PORT}}$ D/$\overline{\text{C}}$. ROMs and RAMs are enabled at low (and hence $\overline{\text{ROM-A}}$ or $\overline{\text{ROM-B}}$) but I/O ports are enabled at high. For example, Port "C" is enabled at high. (R/W = read or write.)

Table 5-6 indicates the program that charts the operating conditions for ROMs, RAMs, or I/O ports. The coding of this program on the PROM is indicated in Table 5-7. As an example of clarification, words 0 through 15 all have A4 low for I/O operation, and words 16 through 31 have A4 high for memory operation. The first 10 address lines of the address bus A0 through A9 of the microprocessor and two page-address/I/O port lines are all thus decoded by this program. Undecoded A11/A12 are "don't care" lines, so that the same 1-K byte of information can appear at four places on one page, but the first three pages only are significant. (The ROM and RAM position on each page can be revised by merely recoding the PROM.) The coding in the blank rows is all redundancy and hence "don't care." If the PROM enables one of the I/O ports instead of a memory bank, the $\overline{\text{R}}$/W signal determines whether the port reads data on to the data-bus for write or reads the data off the bus.

PROMs can be used to drive 7-segment display to show the decimal value of a 4-bit input signal. A 32 × 8-bit programmed PROM can directly provide the drive

Table 5-6. ASCII enabling conditions for ROMs, RAMs, or I/O ports. (*Courtesy of McGraw-Hill*)

$(\overline{\text{RAM A}}) = (\text{OPREQ})\ (\text{M}/\overline{\text{IO}})\ (\text{WRP})\ (\overline{\text{A13}\cdot\text{E}/\overline{\text{NE}}})\ (\overline{\text{A14}\cdot\text{D}/\overline{\text{C}}})\ (\overline{\text{A10}})$
$(\overline{\text{ROM B}}) = (\text{OPREQ})\ (\text{M}/\overline{\text{IO}})\ (\overline{\text{A13}\cdot\text{E}/\overline{\text{NE}}})\ (\overline{\text{A14}\cdot\text{D}/\overline{\text{C}}})\ (\text{A10})$
$(\overline{\text{RAM C}}) = (\text{OPREQ})\ (\text{M}/\overline{\text{IO}})\ (\text{WRP})\ (\text{A13}\cdot\text{E}/\overline{\text{NE}})\ (\overline{\text{A14}\cdot\text{D}/\overline{\text{C}}})\ (\overline{\text{A10}})$
$(\overline{\text{ROM D}}) = (\text{OPREQ})\ (\text{M}/\overline{\text{IO}})\ (\text{A13}\cdot\text{E}/\overline{\text{NE}})\ (\overline{\text{A14}\cdot\text{D}/\overline{\text{C}}})\ (\text{A10})$
$(\overline{\text{RAM E}}) = (\text{OPREQ})\ (\text{M}/\overline{\text{IO}})\ (\text{WRP})\ (\overline{\text{A13}\cdot\text{E}/\overline{\text{NE}}})\ (\text{A14}\cdot\text{D/C})\ (\overline{\text{A10}})$
$(\overline{\text{ROM F}}) = (\text{OPREQ})\ (\text{M}/\overline{\text{IO}})\ (\overline{\text{A13}\cdot\text{E}/\overline{\text{NE}}})\ (\text{A14}\cdot\text{D}/\overline{\text{C}})\ (\text{A10})$
$(\text{PORT C}) = (\text{OPREQ})\ (\overline{\text{M}/\overline{\text{IO}}})\ (\overline{\text{A13}\cdot\text{E}/\overline{\text{NE}}})\ (\overline{\text{A14}\cdot\text{D}/\overline{\text{C}}})$
$(\text{PORT D}) = (\text{OPREQ})\ (\overline{\text{M}/\overline{\text{IO}}})\ (\overline{\text{A13}\cdot\text{E}/\overline{\text{NE}}})\ (\text{A14}\cdot\text{D}/\overline{\text{C}})$

Table 5-7. ASCII coding of 82S123 PROM. *(Courtesy of McGraw-Hill)*

	INPUTS					OUTPUTS								COMPONENT ENABLED	
						H	G	F	E	D	C	B	A		
WORD	A_4	A_3	A_2	A_1	A_0	7	6	5	4	3	2	1	0		H G
0	0	0	0	0	0	0	1	1	1	1	1	1	1	PORT C	0 1
1	0	0	0	0	1	1	0	1	1	1	1	1	1	PORT D	1 0
2	0	0	0	1	0	0	0	1	1	1	1	1	1	–	
3	0	0	0	1	1	0	0	1	1	1	1	1	1	–	
4	0	0	1	0	0	0	1	1	1	1	1	1	1	PORT C	0 1
5	0	0	1	0	1	1	0	1	1	1	1	1	1	PORT D	1 0
6	0	0	1	1	0	0	0	1	1	1	1	1	1	–	
7	0	0	1	1	1	0	0	1	1	1	1	1	1	–	
8	0	1	0	0	0	0	1	1	1	1	1	1	1	PORT C	0 1
9	0	1	0	0	1	1	0	1	1	1	1	1	1	PORT D	1 0
10	0	1	0	1	0	0	0	1	1	1	1	1	1	–	
11	0	1	0	1	1	0	0	1	1	1	1	1	1	–	
12	0	1	1	0	0	0	1	1	1	1	1	1	1	PORT C	0 1
13	0	1	1	0	1	1	0	1	1	1	1	1	1	PORT D	1 0
14	0	1	1	1	0	0	0	1	1	1	1	1	1	–	
15	0	1	1	1	1	0	0	1	1	1	1	1	1	–	
16	1	0	0	0	0	0	0	1	1	1	1	1	1	–	
17	1	0	0	0	1	0	0	1	1	1	1	1	1	–	
18	1	0	0	1	0	0	0	1	1	1	1	1	1	–	
19	1	0	0	1	1	0	0	1	1	1	1	1	1	–	
20	1	0	1	0	0	0	0	1	1	1	1	0	1	ROM BANK B (PAGE 0)	
21	1	0	1	0	1	0	0	0	1	1	1	1	1	ROM BANK F (PAGE 2)	
22	1	0	1	1	0	0	0	1	1	0	1	1	1	ROM BANK D (PAGE 1)	
23	1	0	1	1	1	0	0	1	1	1	1	1	1	–	
24	1	1	0	0	0	0	0	1	1	1	1	1	0	RAM BANK A (PAGE 0)	
25	1	1	0	0	1	0	0	1	0	1	1	1	1	RAM BANK E (PAGE 2)	
26	1	1	0	1	0	0	0	1	1	1	0	1	1	RAM BANK C (PAGE 1)	
27	1	1	0	1	1	0	0	1	1	1	1	1	1	–	
28	1	1	1	0	0	0	0	1	1	1	1	0	1	ROM BANK B (PAGE 0)	
29	1	1	1	0	1	0	0	0	1	1	1	1	1	ROM BANK F (PAGE 2)	
30	1	1	1	1	0	0	0	1	1	0	1	1	1	ROM BANK D (PAGE 1)	
31	1	1	1	1	1	0	0	1	1	1	1	1	1	–	

signals from a machine-generated 4-bit binary code to a $1\frac{1}{2}$ digit display of the numbers 0 to 15, and also include lamp test and inhibit commands. (Otherwise, a binary-to-BCD converter, and BCD-to-7 segment decoder/driver ICs are required.)

5.23 RELIABILITY AND MAINTAINABILITY OF DIGITAL CONTROL SYSTEMS

With each technological advance in the reduction of size, cost, and power consumption toward solid-state medium- and large-scale integration, the latest microcomputer systems have achieved progressively higher reliability in operation. As regards flexibility and computation capacity, the microcomputer is more powerful than the first- and second-generation vacuum-tube digital computers of the 1950s; the microcomputers perform at speeds enhanced by 4 to 5 orders of magnitude. They have gained a mean-time before failure (MTBF, the inverse of failure rate) of 12,000 hr after the commercial production-yield of the chips becomes stabilized.

For maintainability, Hewlett-Packard (HP) has come up with low cost special-purpose "546A Logic Pulser" and "547A Current Tracer" test-probe devices. By merely touching the Pulser to the circuit (pin) under test and pressing the *pulse* button, all circuits connected to the node (outputs as well as inputs) are briefly driven to their opposite state. No unsoldering of pins is required, and pulse injection is automatic; high nodes are pulsed low and low nodes high, each time the button is pressed. The Pulser is a programmed microprocessor, and generates a pulse or a pulse-train with high output-current capability to source or sink up to 0.65 A, in order to override IC outputs in either high or low state. Output pulse-width is limited so that the amount of energy delivered to the circuit under test is never excessive. The output is three-state so that the circuit under test is unaffected until the Pulser is activated. The Pulser/Tracer combination presents the digital circuit designer or troubleshooter a stimulus-response test capability; the Pulser acts as both a voltage and current source. (IC = Integrated circuit.)

The 546A Logic Pulser has six ROM-programmable output patterns (single pulser; pulse-trains of either 1, 10, or 100 Hz; or bursts of 10 or 100 pulses) to continually pulse a circuit, or provide a convenient means to insert an exact number of pulses into counters or shift registers. It provides bipolar TTL circuits $\leqslant$650 mA output current at $\geqslant$0.5-μsec pulse-width with a "high" of $\geqslant$3-V dc and "low" of $\leqslant$0.8-V dc; and CMOS $\leqslant$100 mA current at $\geqslant$5-μsec pulse-width with a $-$1-V dc "high," and $\leqslant$0.5-V dc "low"; power supply requirements for TTL/DTL: 4.5-to-5.5-V dc at 35 mA. CMOS: 3-to-18-V dc at 35 mA, protected to 25-V dc ($175).

The 547A Current Tracer locates low-impedance faults in digital circuits by pinpointing current sources or sinks. When several points in logic are stuck in one state by a short, the Tracer pinpoints the fault or hairline solder bridges on a node (even in multilayer boards) without any need of unsoldering IC pins. The Tracer senses the magnetic field generated by fast rise-time current pulses in the circuit, and displays transistions, single-pulses, and pulse-trains using a simple single-light indicator. Since it is not voltage-sensitive, the Tracer operates on all logic families (TTL, ECL, CMOS) on current pulses exceeding 1 mA, and repetition rates less than 10 MHz. The Tracer probe pinpoints the problem by tracing current flow (with the aid of light intensity) to the source or sink causing the node to be stuck. The sensitivity is 1 mA to 1 A. Single-step current transitions ($\leqslant$200 nsec at 1 mA), single pulses

$\geqslant$50 nsec, and pulse-trains to 10 MHz (and 20 MHz for current pulses $\geqslant$10 mA). Price: \$350. Power supply 4.5-to-18-V dc at $\leqslant$75 mA. (Overvoltage protection ±25-V dc, 1 min.)

5.24 DIRECT DIGITAL CONTROL WITH DUAL MICROCOMPUTER SETUP IN PLACE OF ANALOG CONTROLLER

Implementation of *direct digital control* facilitates computing, monitoring, storage, and analytic capabilities. Digital automation, using a central host minicomputer as a batch processor and several branch microcomputers, provides several advantages over the former methods of implementation that use Analog Controllers and auxiliary digital logic circuits.

Process industries generally use analog controllers for controlling variables such as flow, level, temperature, pressure, angular velocity, etc. Both pneumatic and electronic devices provide this type of control. Figure 5-38 illustrates a typical single-loop analog control system that computes a proportional/reset/and derivative-control operation. At the start, the coefficients of the three nodes are manually set for best response under normal operating conditions; if conditions change, the operator arbitrarily changes the set-points on the basis of experience. As the systems become more complex with inferential, feed-forward, and cascade processes, the analog controllers used for such continuous control loops rarely allow operator set-point adjustment, because the system is interacting with several variables preset on a predetermined basis. High production petrochemical processes started this way; however, combinations of special-purpose digital and analog equipment have been used to satisfy the demands of batch-operated discontinuous process systems, as a matter of compromise between several control constants at a limited efficiency.

The latest advances in direct digital control technology facilitate the use of a central minicomputer and several branch microcontroller-microcomputer combinations; they are in turn interfaced to high-precision A/D and D/A converters, transducers, and power plant to enable a modern complex digital control operating sys-

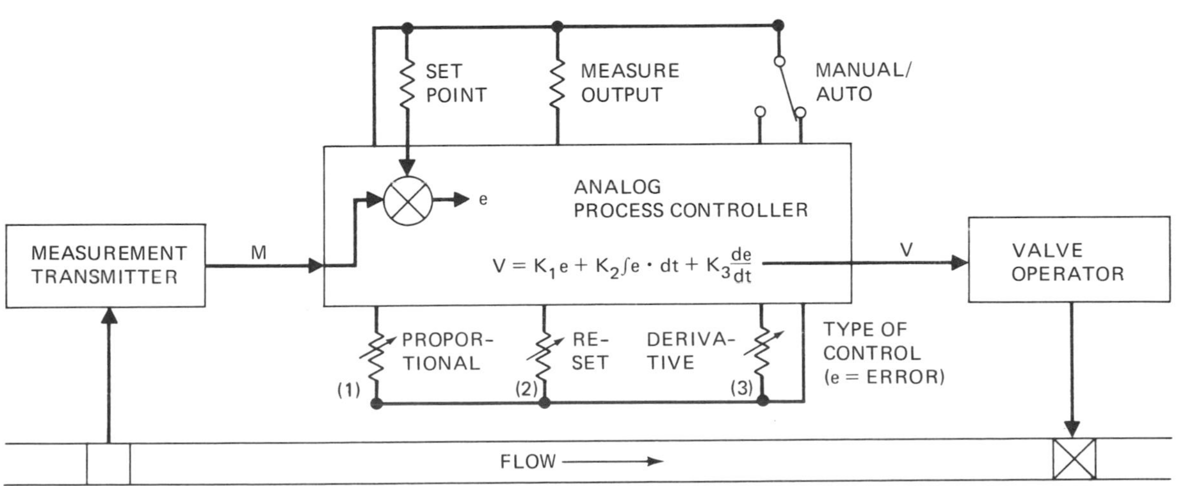

Fig. 5-38. Former single-loop analog feedback control system. (*Courtesy of AFIPS Conf. Proceedings, 1967*)

tem function at high efficiency and flexibility. This is a parallel to a modern data communications network centered around DOS large-scale computer. And several far-off production plants could be interfaced with one another via an identical data communications network, which could possibly include a domestic or international synchronous satellite.

Digital computers are ideal in the industrial process control field in view of their flexibility to store programs, calculate simple and complex process-algorithms, compute variables that are not directly measurable, and *monitor* the process locally or at remote headquarters according to a preplanned schedule. The digital computers have the multitasking and multiprogramming capability to perform complex computations at extremely high-precision. Changes could be incorporated into process dynamics, programs, and schedules in a flexible and economy fashion. The batch process systems that preceded the modern direct digital control systems required increased man-machine communication as merely an intermediate stage of conversion; otherwise, the finalized direct digital control systems require minimum of man-machine communication. Adaptive and optimized on-line digital control systems control set-points for economic or minimized production costs. Economic constraints relating to material inventory, choice, throughput, etc. could be conveniently developed stage by stage without any stoppages. In certain instances, although *mathematical models* could be developed more readily, the mechanism of transferring the model to a *process model* might be more difficult to achieve unless the exact digital process is arrived at on a heuristic design basis in *real-time*.

Some incidental functions of an on-line process computer are enumerated below:

1. Log operating data in engineering branches
2. Compute and display operator information
3. Process material-flow from inventory
4. Report on process statistic such as material used, fuel usage, throughput, programmers, personnel involved
5. Compute, display, and record unmeasurable variables such as Btu rate and mass flow
6. Monitor and alarm process limits
7. Record process events during unusual disturbances
8. Monitor and record changes in set-points, alarm limits, etc., made by operators or instruments
9. Report on demand operator-information such as trend recording, alarm status-report, loop set-point, and parameter data.

In the past, direct digital control presented some problems due to the inherent down-time percentages of the previous second- and third-generation computers, and analog controllers had to be substituted as alternatives under such circumstances. But with the latest availability of the lower-cost MSI and LSI mini- and microcomputers, direct digital control has eliminated this problem with its extremely high reliability—as far as the hardware is concerned. The former MTBF of approximately 2000 hr has promptly elevated to 12,000 hr, and the need for alternative analog controllers is completely eliminated. Instead, redundancy logic, appropriate digital filters, and automatically substituted low-cost standby microcomputers will furnish the requisite reliability in modern complex direct digital control systems.

Figure 5-39a illustrates a direct digital control (DDC) process-control system; Fig. 5-39b shows a parallel DDC system with a digital backup computer. The

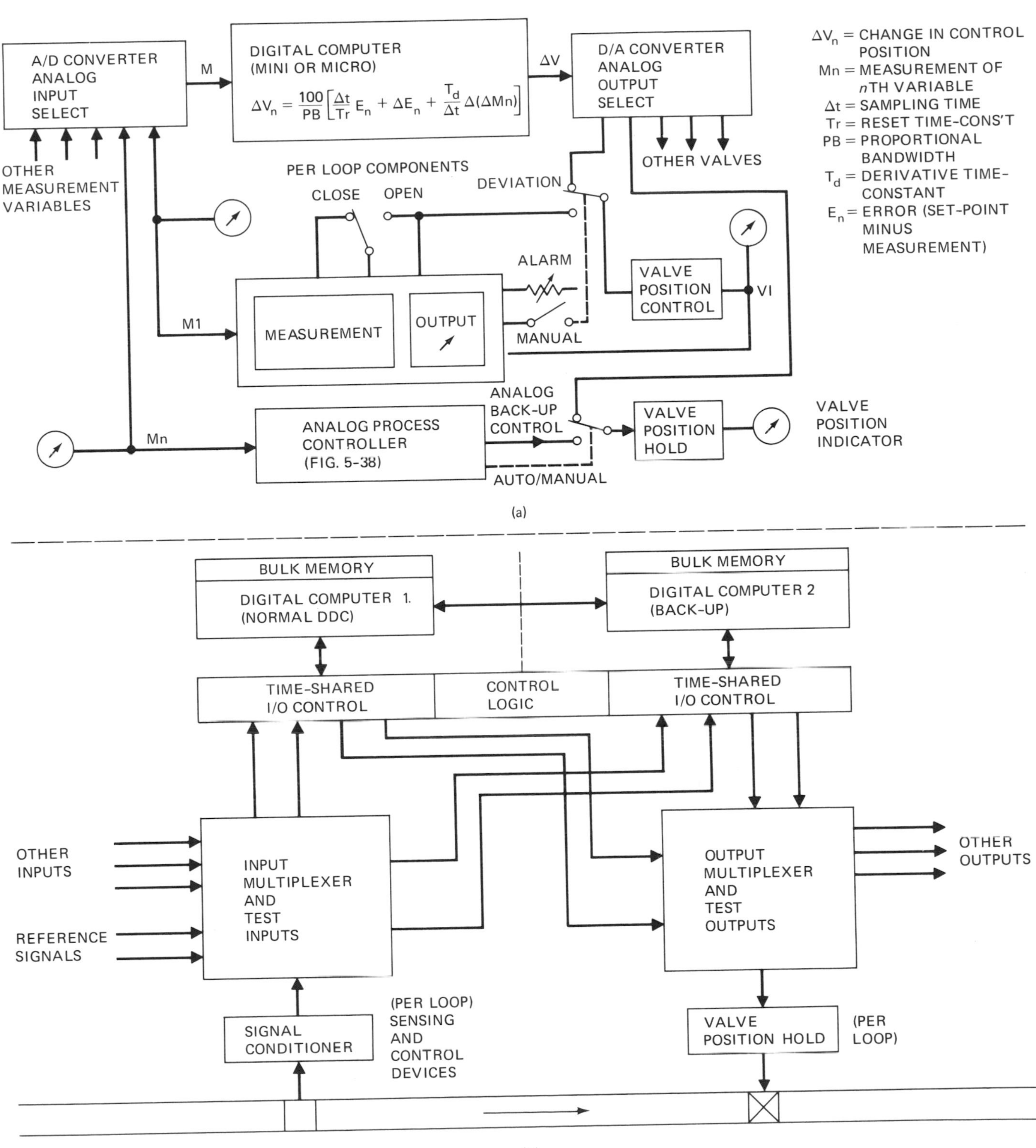

Fig. 5-39a. Direct digital control system.
Fig. 5-39b. Parallel direct digital control with backup digital computer.
(*Courtesy of IEEE Press, 1972*)

backup computer provides in addition time-shared analog and digital input/output equipment, which connects the computer to the various measurement and control variables; it backs up all interloop controls, as well as cascaded control facilities. The incidental functions enumerated in a previous paragraph are conveniently implemented by the backup computer. With its added diagnostic programs, the backup computer could improve the MTBF by at least 50%. The failed computer subsystem would be available for self-checking, while the backup computer subsystem takes over the direct digital process control.

Figure 5-40 illustrates one of the latest low-cost bipolar I^2L microcomputer systems that perform practically at the digital processing capability of the former medium-price minicomputers. Fairchild 9440 Microflame is, according to the manufacturer, the world's first 16-bit bipolar microprocessor CPU-on-a-chip that executes a minicomputer instruction set (that of the Data General NOVA 1200 minicomputer). Software, such as text editor, symbolic debugger, and high-level business BASIC, are also available.

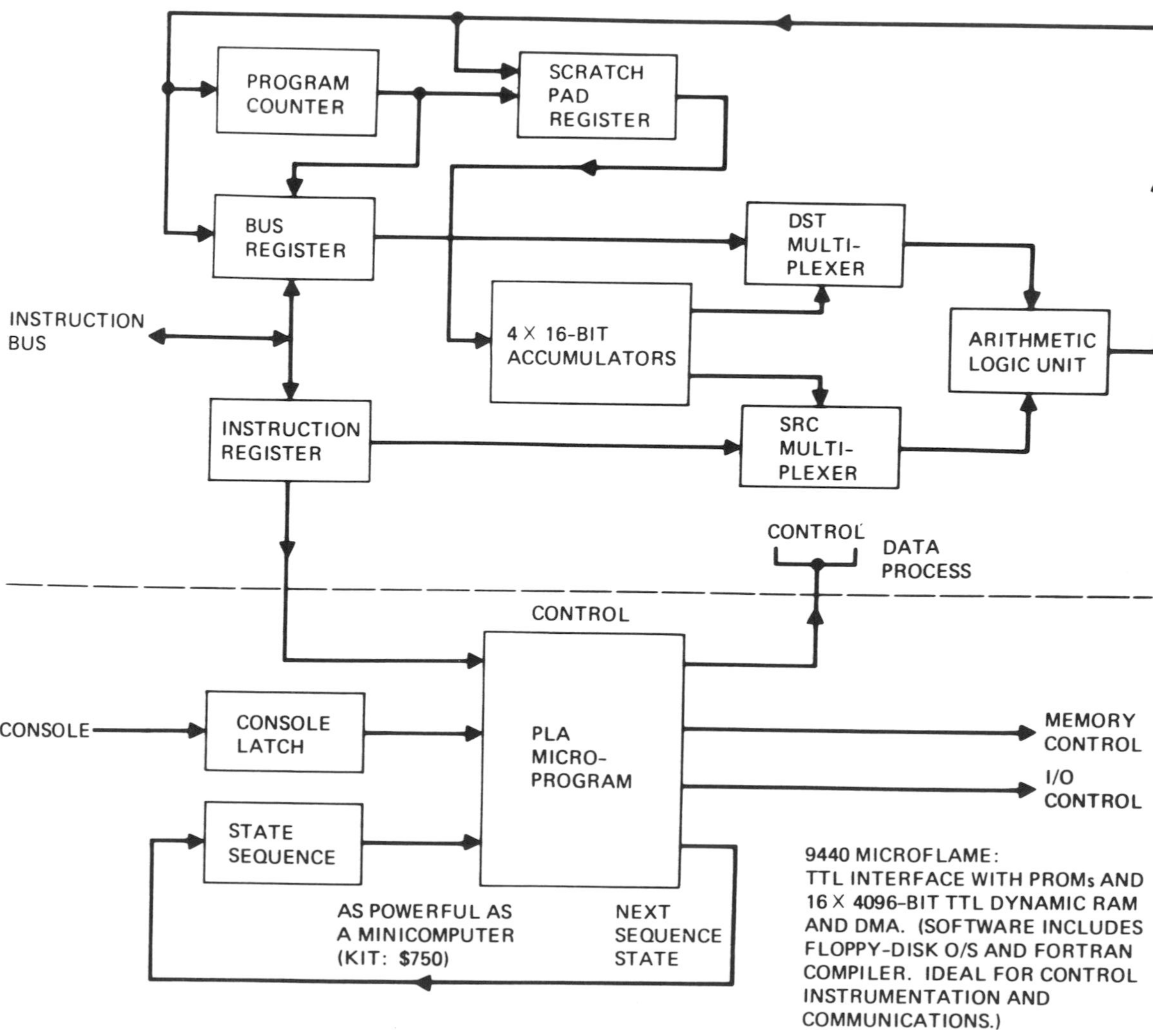

Fig. 5-40. Fairchild 9440 Microflame: 40-pin DIP. Latest 16-bit I^2L microprocessor. Executes Nova 1200 minicomputer's instruction set. High-speed bipolar microprocessor using internal PLA microprogram. (*Courtesy of Electronics, McGraw-Hill*)

Packaged in a 40-pin DIP, the microprocessor is ideal for applications such as control and instrumentation, telecommunications PBX and PABX switching systems, distributed intelligence, distributed multiprocessing, and front-end (Terminal) processing in data communications networks. Fairchild integrated real-time executive (FIRE) software is an initial software package for the CPU microprocessor. It is comprised of the requisite development aids such as diagnostics, bootstrap and binary loader, and an interactive entry and debugging program. Fairchild calls this I^2L chip isoplanar-I^2L (I^3L), and it combines bipolar high-speed and MOS packing density and interfaces directly for TTL-interface with other PROM and RAM memory chips.

The basic 9440 μFLAME microcomputer system comprises of 9440 CPU, sixteen 4-K bit TTL RAM dynamic memories (for 4-K byte memory) and a few SSI/MSI ICs for memory control, software manual, and instruction set ($750). Other software would include a floppy-disk operating system and a FORTRAN compiler, 16-K TTL dynamic RAM, a memory-control with refresh and DMA capabilities, an I/O control, and a hardware multiply/divide capability to make it as powerful as a modern basic minicomputer system.

To continue the topic of DDC, as in all control systems, each measurement in the control system is routed to the computer I/O buffer system individually (without interfering with other loops). The time-shared I/O system must have the capability to identify and diagnose any problem in any individual circuit on an automatic on-line diagnostic program. Output devices are normally expected to have a 1-hr battery backup facility in case of power failure. The diagnostic program must automatically isolate the failed control logic to inhibit the operation of the corresponding I/O control device.

The intercommunication link between the two parallel computers must continuously update the backup program-data and status on a periodic basis. The backup batch-process computer usually receives this information in seconds, and program changes made on-line must be transferred to the backup computer simultaneously. Of course, this is all the task of the real-time software devised for the system; it should permit updating and on-line diagnostics while time-sharing the real-time programs in disk mass-memory. When the backup computer is not on control, it could be used for program compiling, debugging, and problem *simulation* with test inputs and outputs. At the present time, with the advent of economy powerful microcomputer systems, direct digital control in industrial processing systems is an economic and attractive proposition, provided the generally more expensive software is devised in the form of, for example, bulk nonvolatile magnetic-bubble memory cards and solid-state 32/64-K bit LSI ROM/PROM and static RAM memory chips.

5.25 SOME BASIC CHARACTERISTICS OF DIGITAL FILTERS

5.25.1 Finite Duration Impulse-Response Digital Filters. Powerful optimization algorithms for computer programming are presently available to design digital filters. Although simplified *closed-form* solutions are not feasible in most cases, sophisticated *iterative programs* can be devised to yield optimum solutions. These iterative techniques are briefly described in this section, and the basic characteristics of the digital filters are defined.

1. **FINITE-DURATION IMPULSE RESPONSE (FIR).** The duration of the impulse-response h_n of the digital filter is considered finite.

$$h_n = 0, n > N_1 < \infty$$
$$h_n = 0, n < N_2 > -\infty$$

and

$$N_1 > N_2$$

INFINITE-DURATION IMPULSE-RESPONSE (IIR). The duration of the filter impulse-response h_n is infinite without any finite values for N_1 and N_2.

FIR filters with a linear-phase characteristic can be readily designed for a specific magnitude-frequency response to an arbitrary performance-index. (IIR filters are also capable of this feature.)

FIR filters are efficiently realized *recursively* by using a comb filter and a bank of tuned resonators; the filters are realized *nonrecursively* by using the FFT approach for high-speed *convolution*. Since the FIR nonrecursive filters contain only zeros in the finite z-plane, they are always stable.

Quantization, round-off, and accuracy problems in sharp-cutoff nonrecursive FIR filters are negligible, but such problems are inherent in recursive IIR filters.

2. **RECURSIVE DIGITAL FILTERS.** Both FIR and IIR filters belong to this category, when the *present filter output* is determined in terms of not only *past filter outputs* y_{n-1}, etc., but also *past and present filter inputs* x_n, x_{n-1}, etc.

$$y_n = F(y_{n-1}, y_{n-2}, \ldots, x_n, x_{n-1}, \ldots)$$

NONRECURSIVE DIGITAL FILTERS. Both FIR and IIR filters are feasible in this category too, when the *present filter output* is determined explicitly in terms of *past and present filter inputs only*.

$$y_n = F(x_n, x_{n-1}, \ldots)$$

However, nonrecursive realization of FIR filters, and recursive realization of IIR filters are more efficient in general and hence preferred.

3. **LSI IMPLEMENTATION OF DIGITAL FILTERS.*** Large-scale integration (LSI) at this point in time enables the implementation of multisection digital filters on a single-chip. Filter sections in hundreds, consisting of serial 2's complement adders, serial 2's complementers, serial multipliers, M-input multiplexers, shift-register delay elements, read-only memory coefficients, etc., can all be incorporated in LSI to furnish LSI digital filters for microprocessor-controlled digital control systems.

The filters can be easily modified to realize a wide range of filter forms, pulse transfer-functions, multiplexer facilities, and round-off noise levels by merely changing the contents of programmable read-only memories, timing signals, and length of shift-register delays. For example, LSI CCDs (charge-coupled devices) are ideal for simplified full-adder logic implementation. Analog-to-digital converters too are practical in single chip BIMOS implementation.

The transfer characteristics of the *canonical forms* of digital filters are represented in terms of their z-domain pulse transfer-functions. The corresponding direct, cascade, and parallel forms of digital filters are shown in Fig. 5-41.

a. *Direct form*

$$H(z) = \sum_{i=0}^{n} a_i z^{-1} \Big/ 1 + \sum_{i=1}^{n} b_i z^{-1} \cdots z^{-1} \text{ is the } \textit{unit-delay operator.}$$

*Reprinted, with permission, from *IEEE Transactions Audio Electroacoustics*, vol. AU-16, pp. 413–421, Sept. 1968.

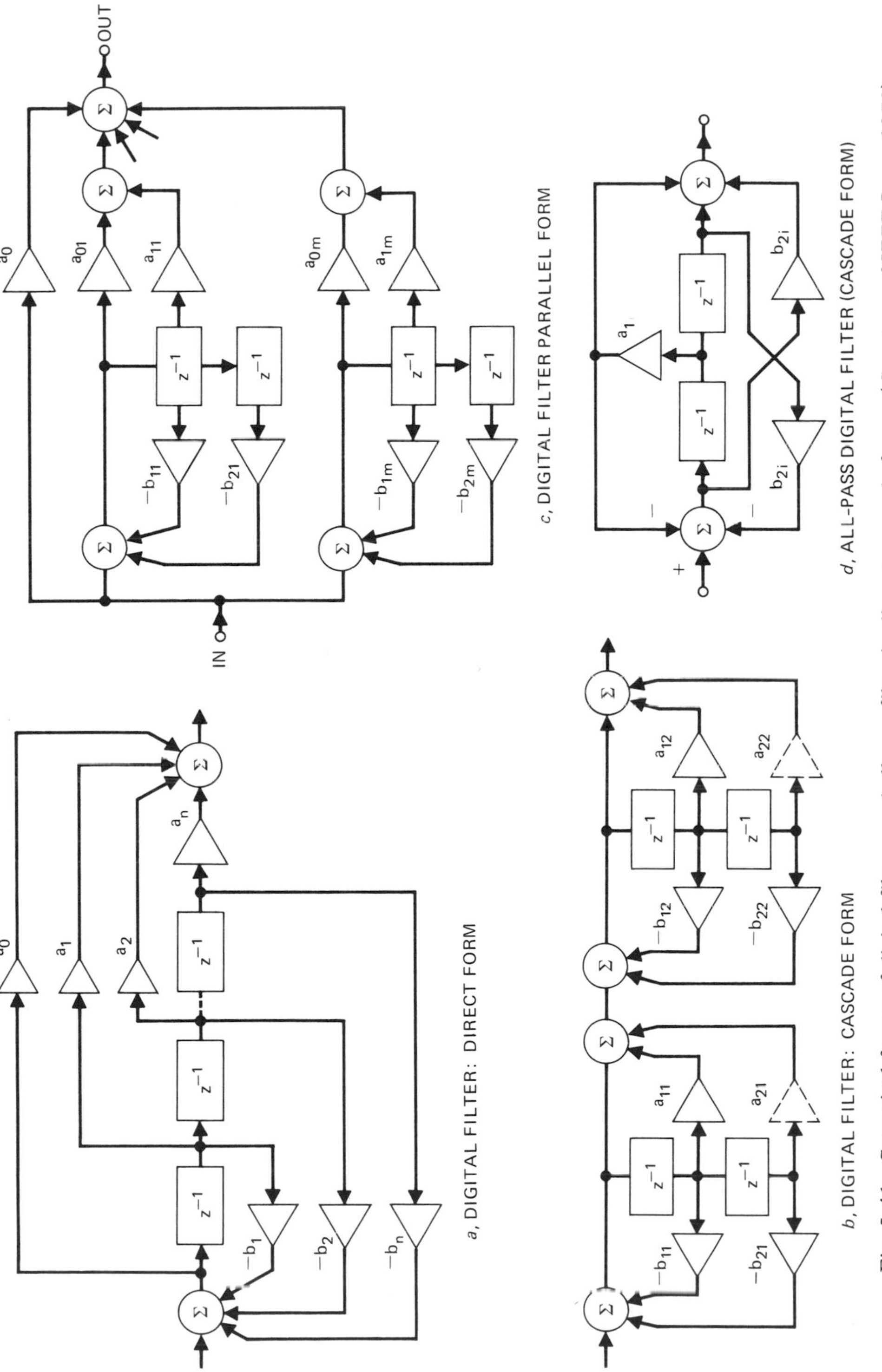

Fig. 5-41. Canonical forms of digital filters, and all-pass filter in discrete cascade form. *(Courtesy of IEEE Press, 1972)*

The canonical forms imply minimum number of adders, multipliers, and delays.

b. *Cascade form*

$$H(z) = a_0 \prod_{i=1}^{m} \frac{a_{2i}z^{-2} + a_{1i}z^{-1} + 1}{b_{2i}z^{-2} + b_{1i}z^{-1} + 1}$$

where m is the integer part of $\frac{(n+1)}{2}$. The numerator and denominator polynomials are factorized to produce H(z) of this second-order form with real coefficients, rather than a mixed set of first- and second-order factors for real and complex roots to simplify the implementation of the cascade form during multiplexing.

c. *The parallel form* is obtained by partial-fraction expansion of the direct form (a):

$$H(z) = c_0 + \sum_{i=1}^{m} \frac{c_{1i}z^{-1} + c_{0i}}{b_{2i}z^{-2} + b_{1i}z^{-1} + 1}$$

where $c_0 = \frac{a_n}{b_n}$. All three canonical forms are practically equivalent with respect to the extent of storage required (n-unit delays) and the number of arithmetic operations required, viz., (2n + 1) multiplications and 2n additions per sampling period.

The cascade form is, however, ideal because it requires significantly fewer multiplications for zeros on the unit circle, especially for band-pass and band-stop as well as low- and high-pass digital filters.

d. The *all-pass filter* (APF), which is an equalizer with unity-gain at all frequencies, is given by the following expression in the discrete form:

$$H_A(z) = \sum_{i=0}^{n} b_{n-1}z^{-i} \Big/ \sum_{i=0}^{n} b_i z^{-i}$$

With $a_i = b_{n-1}$ and $b_o = 1$, the direct form can be employed to implement the discrete APF of the following form as an approximation on the filter coefficients.

$$H_A(z) = \prod_{i=1}^{m} \frac{z^{-2} + b_{1i}z^{-1} + b_{2i}}{b_{2i}z^{-2} + b_{1i}z^{-1} + 1}$$

These second-order digital APF formats are shown in Fig. 5-41.

Recursive digital filters with arbitrary, prescribed magnitude characteristics can be presently designed, by employing the Fletcher-Powell optimization algorithm to minimize a square-error criterion in frequency-domain. This is a computer-aided design method.

The block-diagrams of the LSI hardware required for the preceding three canonical forms and the discrete APF are illustrated in Fig. 5-41 with appropriate description of the various configurations.

5.25.2 Equiripple or Minimax Responses of Digital Filters. A digital filter can be represented by a linear difference equation having constant coefficients and operating on equally spaced samples of a signal. The error signal is defined by the difference between the actual response of the digital filter and an ideal response—which could be either a time or frequency response. The response errors noted can be specified as either *equiripple or minimax.*

1. The *minimax* value of the error is the minimum of the maximum absolute value of the response error. The minimum is obtained by varying the filter's coefficients, and the maximum is obtained by sampling the error as often as sufficient.

2. An *equiripple* filter has more than one maximum of the absolute value of the error, and maxima are all equal to produce a ripple, and it may or may not have a minimax error. The converse is also true. Nonrecursive filters alone have both minimax and equiripple response errors.

5.25.3 Simplex Linear Programming. Nonrecursive filters having even-symmetric linear coefficients are determined by the *simplex method of linear programming.* These coefficients that minimize the maximum of the error in the complex response provide samples of error represented by

$$E_k = \sum_{q=-P/2}^{P/2} h_q \exp(-2\pi j x_k q) - F(2Wx_k) \qquad k = 1 \cdots N$$

h_q denotes the real coefficients of the digital filter, P is even, and $F(2Wx_k)$ denotes the complex response at a normalized frequency given by x_k, $|x_k| \leqslant \frac{1}{2}$, and 2W is the sampling rate. N must be larger than P; for example, N = 2P. The real and imaginary parts of the error E_k are specified to be less than some set of upper limits V_k.

Now, the Simplex method can provide a numerical solution of the following problem to determine the real coefficients h_q that would minimize Q (below unity) in

$$-QV_k \leqslant \text{Re}\,\{E_k\} \leqslant QV_k$$
$$-QV_k \leqslant \text{Im}\,\{E_k\} \leqslant QV_k \quad \cdots k = 1 \cdots N$$

By increasing the number of coefficients on a (P + 1) iterative basis, and repeating the linear programming process, we can minimize Q; *successive iterations* can obtain the smallest value of (P + 1) for which Q is less than unity. The *Simplex method of linear programming* may be thus applied to minimize the errors in time-response, which appear as *side-lobes* on both sides of the *main-lobe* of digital filters.

5.25.4 Nonlinear Programming (Time-Domain Synthesis and Constrained Optimization.* Both recursive and nonrecursive digital filters are applicable in the method devised by Athanassopoulos. It follows the linear programming approach. *Recursive* networks must have poles and unit-impulse response of infinite duration. *Nonrecursive* networks have only zeros and a finite duration unit-impulse response.

Recursive Case: If z^{-1} denotes the usual delay by one sample, the z-transform of the filter is chosen for (1) cascade and (2) parallel connection of elementary sections.

$$H(z) = g \prod_{i=1}^{s} \frac{1 + a_i z^{-1} + b_i z^{-2}}{1 + c_i z^{-1} + d_i z^{-2}} \cdots s = \text{No. of stages} \tag{1}$$

$$H(z) = g + \sum_{i=1}^{s} \frac{a_i + b_i z^{-1}}{1 + c_i z^{-1} + d_i z^{-2}} \cdots s = \text{No. of parallel sections} \tag{2}$$

*Reprinted, with permission, from *IEEE Trans. Audio Electroacoustics*, vol. AU-19, pp. 87–94, March 1971.

Error E_k is denoted by the difference between the actual response of the digital filter and the ideal values of the squares of the amplitude response:

$$E_k = |H \exp(2\pi j x_k)|^2 - |F(2Wx_k)|^2$$

2W is the sampling rate (greater than twice the highest frequency in the input waveform), and the error is a differentiable function of the coefficients a_i, b_i, c_i, and d_i, and g is for the square of the amplitude response.

If U_k and L_k are the real and positive upper and lower limits,

$$\left.\begin{aligned} G_k &\triangleq QU_k - E_k \\ H_k &\triangleq QL_k + E_k \end{aligned}\right\} \begin{aligned} &k = 1 \cdots N \\ &Q > O \end{aligned}$$

A compensation function is then selected to constrain errors of phase- or delay-response:

$$Q + \sum_{k=1}^{N} \frac{r}{G_k} + \sum_{k=1}^{N} \frac{r}{H_k}$$

Then a suitable software program, such as the conjugate gradient method of Davidon, Fletcher, and Powell, is chosen to obtain *unconstrained minimization* of the above compensation function with respect to Q, g, a_i, b_i, c_i, and d_i, $i = 1 \ldots s$. The minimization procedure is repeated for r/2 or r/5, until Q becomes nearly constant below unity. That is, r as a factor eventually becomes negligible compared to Q.

5.25.5 Special Computer Techniques, Digital Filters.

INTEGER PROGRAMMING TECHNIQUE. If the variables are integers, the corresponding program is available to provide *minimax error*, despite the quantization of the coefficients of a *nonrecursive filter*.*

MAPPING. If the coefficients of a recursive digital filter approximate a piecewise-constant frequency response, the coefficients can be determined by "mapping" a Laplace transform into a z-transform.†

WINDOW TECHNIQUES. Window techniques allow the measurement of Fourier transform or the power spectrum of the samples of a waveform by using the simplifying Fast Fourier Transform approach of discrete Fourier. The "windows" allow spectrum measurements with minimax leakage.

5.26 MICROPROCESSOR-CONTROLLED DIGITAL CONTROL SYSTEM TO IMPLEMENT DIGITAL FILTERS AND DIRECT DIGITAL CONTROL

Microprocessor manufacturers such as Intel have recently begun developing LSI software and making it commercially available in the form of ROMs, and the associated application control program support in the form of PROMs for the popular microprocessors available off-the-shelf. Thus, it is now feasible to plan and develop microprocessor-based direct digital control and ROM-programmed LSI digital filters for the subsystems in a complex digital control system. As a matter of fact, software packages for high-level languages such as PL/M and BASIC for certain network control and data communications and display applications are already available at moderately low prices.

*Dantiz, G. B. *Linear Programming and Extensions*, Princeton University Press, Princeton, New Jersey, 1963.

†Kaiser, J. F., *Systems Analysis by Digital Computer*, John Wiley & Sons, New York, 1966.

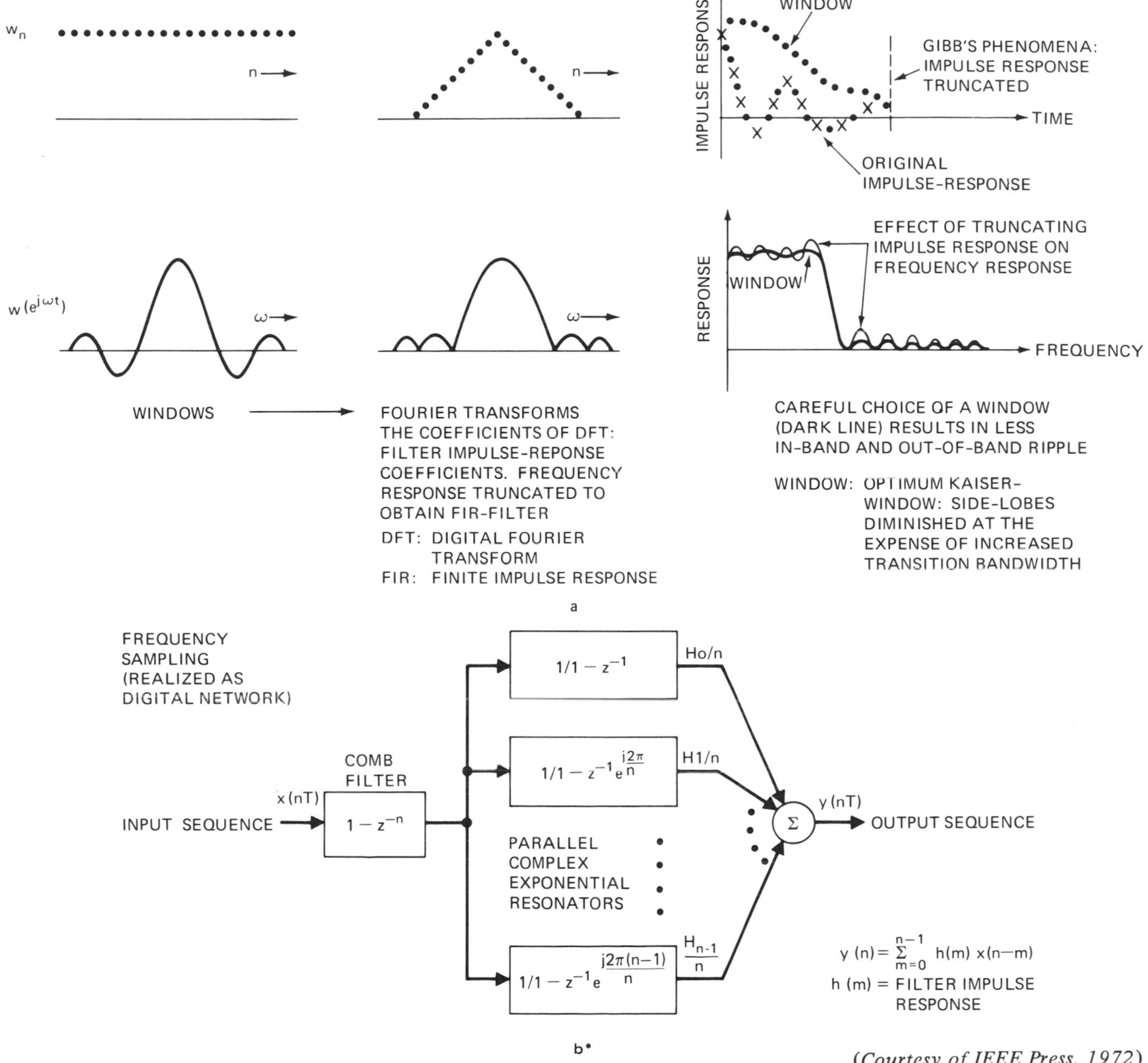

(Courtesy of IEEE Press, 1972)

As a facility in direct digital control and data communications for the I/O and support circuit areas, Signetics markets (1) LSI 2656 system memory interface SMI chip (2) 2651 programmable communications interface PCI chip, (3) 2655 programmable peripheral interface PPI chip, (4) 2657 direct memory access interface DMAI chip, and (5) 2652 USRT multiprotocol universal programmable synchronous receiver/transmitter chip for 8/16-bit microprocessors.

A general-purpose microcomputer system that could be programmed to demonstrate the concepts of direct digital control and digital filters is illustrated in Fig. 5-42a. It depicts the requisite functions such as (1) the input signal conditioner, the

*Reprinted, with permission, from *IEEE Trans. Audio Electroacoustics*, vol. AU-18, pp. 83–106, June 1970.

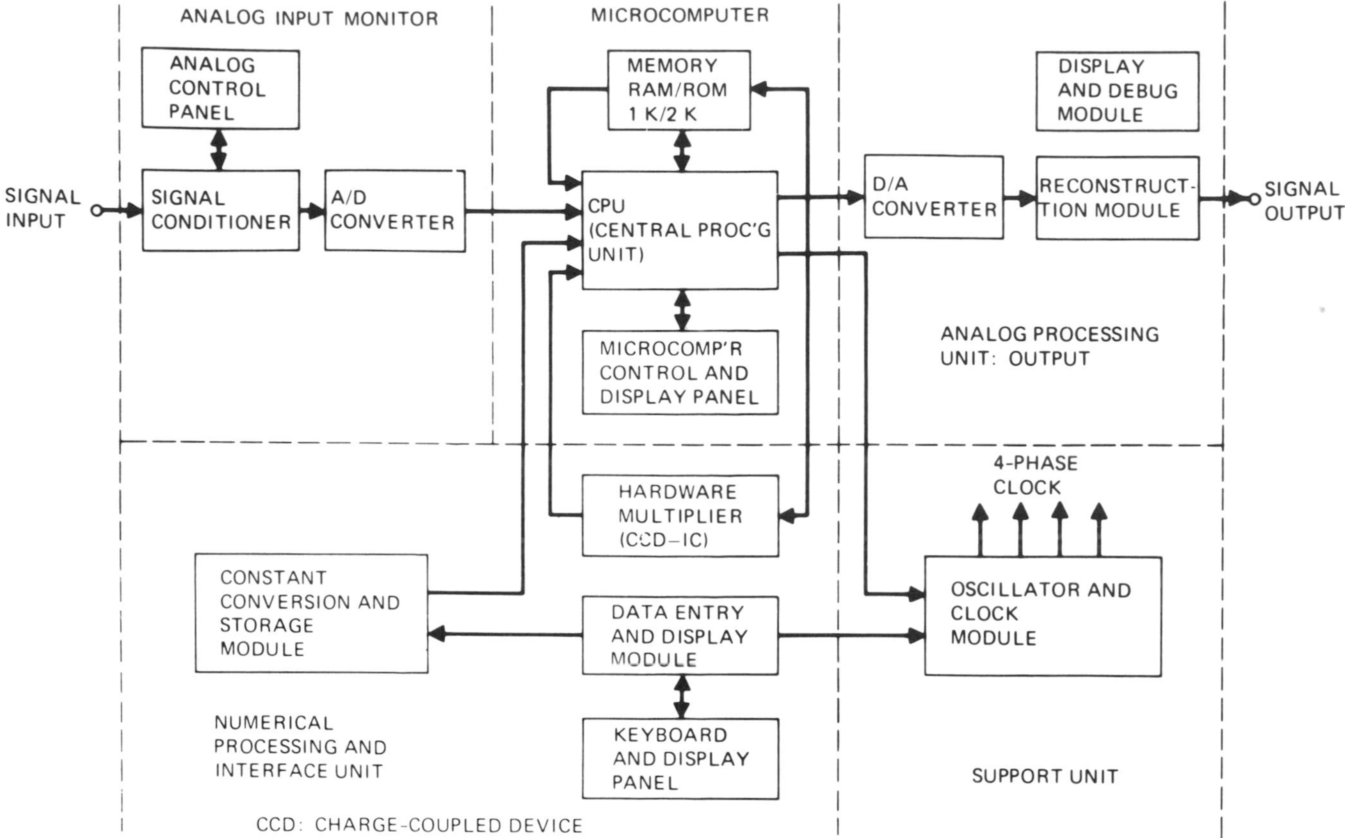

Fig. 5-42a. Microcomputer-controlled control system. System organization.

analog control panel, and the A/D conversion; (2) the central processing unit performing add/multiply/store filter-coefficients, generation of a constant delay-interval, and processing of signals and data within the system; and (3) the D/A conversion of the output to the appropriate analog format. The unit is programmable by means of the PROMs used, and the functions and parameters can be hence varied during the design stage. I/O buffers and registers are provided for loading and debugging the programs as also a turnkey capability for operating on a stored program.

Each control microcomputer subsystem may consist of a CPU, a RAM main memory, a control/debug/display panel, a numerical processing and interface unit, an I/O signal reconstruction unit, and an analog monitoring facility. Since the signals on the input data-bus are generally time-multiplexed, external latches, buffers, and multiplexers are required to control the flow of information. The data entry and display unit shown in Fig. 5-42b employs a decimal keyboard facility; it enables the loading of the digital filter-coefficients and the sampling frequencies into a BCD shift-register. The register contents are displayed on a BCD shift-register. The register contents are displayed on a seven-segment front-panel display. Other keys are provided for entering frequency scale-factors, RUN or HALT commands for the master reset of the microprocessor, and for converting the BCD coefficients into the binary machine code. The monitoring facility should have a warning signal to indicate if the sampling frequency selected is too high for completion of one cycle of the digital filter program.

Figure 5-42c illustrates an input signal-processor feeding the 8-bit digital data

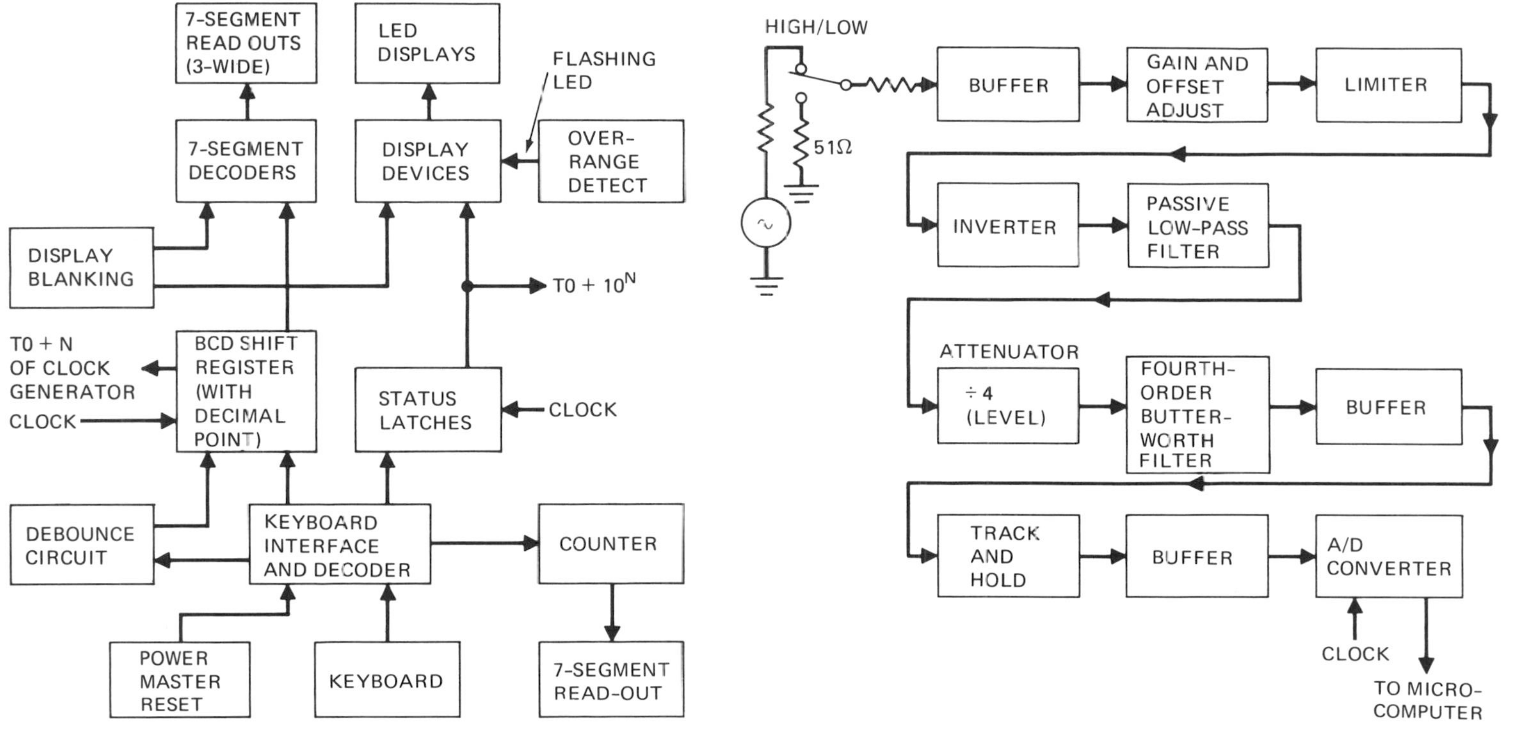

Fig. 5-42b. Data entry and display panel.

Fig. 5-42c. Input signal conditioning module. Output reconstruction module similar to digital-to-analog register, digital-to-analog converter, and de-glicher. (*Courtesy of IEEE Press, 1972*)

from the A/D converter. A fourth-order low-pass Butterworth filter may be used in both the input and output signal processing units to remove the sampling harmonics.

As a prototyping microcomputer subsystem, the unit described could be used for such functions as analog frequency conversion, automatic calibration and testing, automatic range measurement and nulling, convolution, correlation, covariance, data logging, digital filtering, direct open- and closed-loop digital control, pulse stretching, signal compression, signal scrambling, time compression and expansion, transient recording, waveform averaging, and waveform generation.

5.27 DIRECT DIGITAL CONTROL: MICROPROCESSORS AND APPLICABLE ALGORITHMS

Before the end of the century, low-cost mass production via automation in process control will be an accomplished fact throughout the world for the benefit of more and more people. The economies realizable by the use of microprocessors and multi-microprocessing systems have brought about this viable feasibility—notwithstanding the usual conservatism of the industry against rapid technological developments during the initial stages. Economy data communications via optical fiber links enable the facility of distributed control in plant automation not unlike the present central large-scale computer and the distributed network of remote terminals. As the computing power of the 16-bit microcomputer system goes on escalating, the tasks of the central large-scale computer and the terminals will be gradually taken over by economy multi-microprocessor systems for the various central and distributed control tasks. The central multi-microprocessor system will of course provide the supervisory pipeline control for interchange, monitoring, and communications while the distributed loops under individual microprocessor control perform the various subsidiary functions in the mass-production system. Where the distributed control tasks take a dominant role, parallel redundancy can be economically introduced for higher reliability. The superior computing power of the latest microprocessors incidentally add a high figure-of-merit compared to the presently used individual analog controllers in the distributed network. New innovations in digital control algorithms applicable to microprocessors with their LSI memory will enhance the capabilities of the distributed tasks without noticeable interaction to the operation of the other loops in the complex digital control system. It may be noted that the subsidiary analog controllers have been the order of the day since the former high-cost minicomputers are prohibitively expensive for the distributed tasks of process control; the high-cost demerit is not presently valid in the light of the suitability of economy microprocessor chips for the various stand-alone single-loop subtasks.

The to-and-fro data communications between the various process control tasks and the central supervisory command could take place either in parallel as a star-network, or sequentially along a serial-bus with requests for access to the bus of a coaxial nature or a fiber-optic time-division data-multiplex channel. The remote process control tasks may include, for example, 128 loops as in some of the complex Honeywell process control systems, or just a few under 10 as in Bristol and Measurex systems. *Single-loop* microcontrollers (using field-programmable PROMs) in place of the present proportional-integral-derivative (PID) analog controllers are immediately practical, and the PID-algorithms in these new direct digital controllers could, respectively, assume the following positional and velocity formats:

Sampled-data positional set-point,

$$u(k) = -k_p y(k) + k_q \sum_{i=0}^{k} [r(i) - y(i)] - k_d [y(k) - y(k-1)] \tag{1}$$

and the velocity incremental change of the manipulated variable,

$$\Delta u(k) = k_p [y(k-1) - y(k)] + k_q [r(k) - y(k)] + k_d [2y(k-1) - y(k-2) - y(k)] \tag{2}$$

y(k) is the sampled-data plant output, and k_p, k_q, and k_d are the proportional, integral, and derivative control gains, respectively. The set-point derivative-swing r(k) is effective in Eqs. 1 and 2 in the integral main-loop case only as shown in the sampled-data feedback control system of Fig. 5-43. The final output relates to the control value final output; in the case of velocity, the control value itself acts as the integrator. Variations such as PID error-squared criterion on gain, or integral and digital-filter compensation for nonlincarities and lead-lag action may be supplemented as well.

A few digital controller algorithms for direct digital control using microcontrollers have been devised.* These algorithms are discussed below to show the latest control application trends using modern state-space techniques for simple on-line discrete-time models based on step-response.

A typical unit-step response of process control is shown in Fig. 5-44 for the exponential sigmoid-shaped response characteristic. The pulse transfer-function is given by

$$G_p(z-1) = (g_1 z^{-1} + g_2 z^{-2} \cdots + g_n z^{-n})/(1 - pz^{-1}) \tag{3}$$

p is a decaying factor (0 to 1), and it satisfies the condition that the model steady-state gain matches the plant gain k_p:

$$p = 1 - \frac{g_n}{k_p - \sum_{i=1}^{n-1} g_i} \tag{4}$$

*Auslander, D. M., Takahashi, Y., and Tomizuka, M., Direct Digital Process Control: Practice and Algorithms for Microprocessor Application, *Proceedings IEEE*, February 1978.

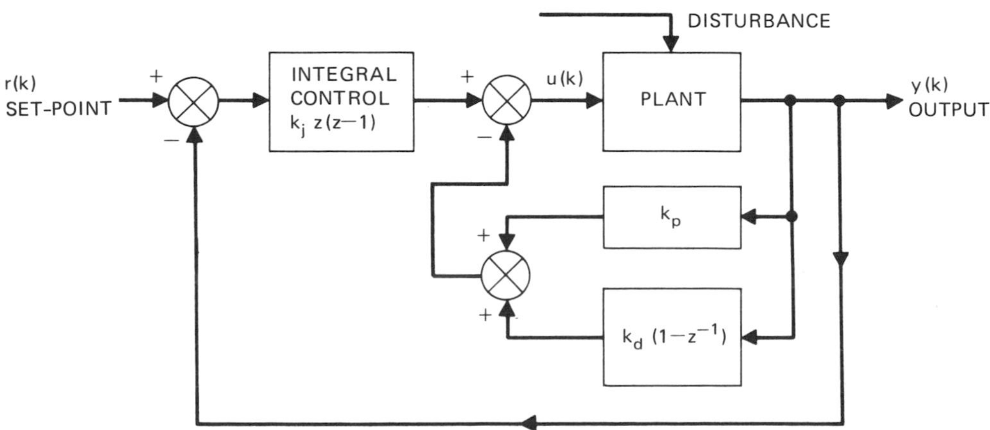

Fig. 5-43. PID digital controller. Reprinted, with permission, from the *Proceedings of the IEEE*, February 1978.

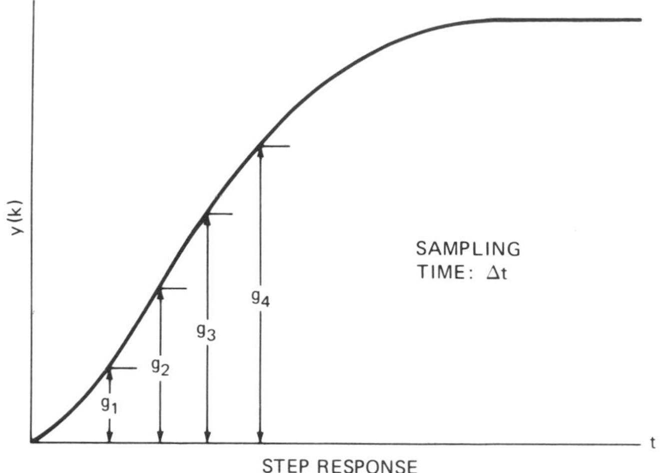

Fig. 5-44. Step response. (*Courtesy of IEEE Proceedings, February 1978*)

The approximation error is introduced from $k = (n + 1)$ and is made arbitrarily small by increasing the sampling period or by increasing the order of the mathematical model for a fixed sampling period.

The scalar input $h(k)$ and the scalar output $y(k)$ as an n-dimensional state-vector are related by

$$\text{State-vector, } x(k+1) = Px(k) + qu(k)$$
$$y(k) = cx(k) \tag{5}$$

where

$$P = n \times n \text{ matrix}, \begin{bmatrix} 0 & 1 & 0 & \cdots & 0 \\ 0 & 0 & 1 & \cdots & 0 \\ \vdots & & & & \vdots \\ 0 & 0 & 0 & \cdots & p \end{bmatrix}; q = n \times 1 \text{ matrix}, \begin{bmatrix} g_1 \\ g_2 \\ \vdots \\ g_n \end{bmatrix}$$

and

$$c = 1 \times n \text{ matrix}, [1 \quad 0 \quad 0 \quad \cdots \quad 0]$$

The discrete-process model (Eq. 5) is applicable to processes with all real eigenvalues and also those processes that have overshoot or oscillatory step-response. The method of *controller and observer design* algorithm, which follows, is an approximation of the preceding method of sigmoidal step-response.

a. *Finite-time settling observer algorithm (FTSO)*. Since it is extremely difficult to implement direct state-vector feedback based on complete state-vector measurement, an estimated $\hat{x}(k)$ of the state-vector is generated by an *observer* algorithmic model.

$$\overline{x}^0(k+1) = P\overline{x}(k) + qu(k) \tag{6}$$

This *estimated* state-vector is then *updated* by comparing the latest output $y(k + 1)$ with the estimated output $\overline{y}^0(k + 1) = c\overline{x}^0(k + 1)$.

$$\overline{x}(k+1) = \overline{x}^0(k+1) + f[y(k+1) - c\overline{x}^0(k+1)] \tag{7}$$

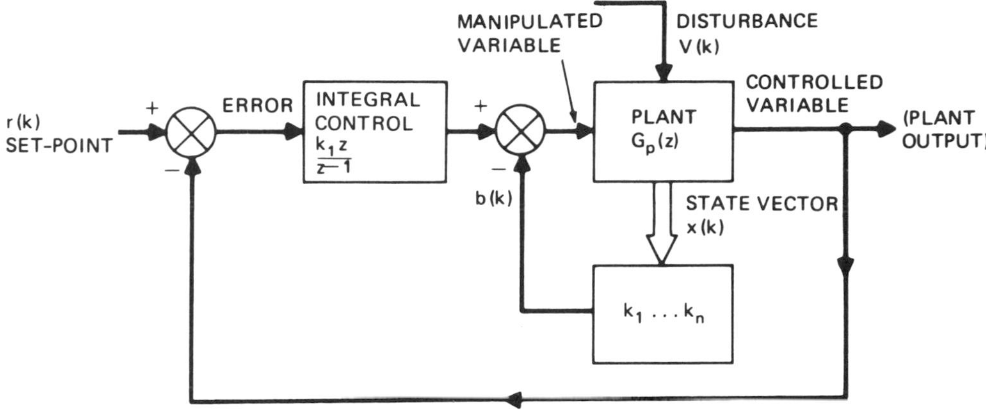

Fig. 5-45. Direct state-vector feedback control system. (*Courtesy of IEEE Proceedings, February 1978*)

If there is a measurable input disturbance, with an updating gain-vector given by $f' = [1\ p\ p^2 \ldots p^{n-1}]$, then an *exact estimation*, $\overline{x}(k) = x(k)$, is achieved for the *nth* order plant, viz., $y(k) = cx(k)$ within a minimum finite-time settling period $(n - 1)$ steps or less. This algorithm is called the *finite-time settling observer* (*FTSO*). If measurement noise is a problem, the gain-vector f may be chosen from optimal digital-filter theory. Since the FTSO has the feature of decoupling from the input control in the closed-loop format, digital control algorithms are derived assuming a direct state-vector feedback.

b. *Finite-time settling control algorithm* (*FTSC*). The state-variable feedback *controller* shown in Fig. 5-45 is represented by:

$$u(k) = k_q \sum_{i=0}^{k} [r(i) - y(i)] - \sum_{i=1}^{n} k_j x_j(k) \tag{8}$$

k_q and k_j are the control gains, and x_j are the *jth* state-variable defined in $y(k) = cx(k)$. The second state-variable feedback term in Eq. 8 is indicated by the minor-loop, and it replaces the proportional derivative terms in analog-controllers; the main-loop then stands for the integral control.

For an *nth* order system with integral control action, the order of the closed-loop characteristic equation is $(n + 1)$. The coefficients of the characteristic equation are represented by

$$k_1 = 0, k_2 = k_3 = \cdots k_{n-1} = k_q = \frac{1}{g_1 + g_2 + \cdots g_n}$$

The closed-loop poles or eigenvalues are equated with zero, to obtain $(n + 1)$ control gain terms for the FTSC.

For the pulse transfer function (Eq. 3), the control-gain k_n is given by

$$k_n = \frac{1}{g_n} [1 + p) - (g_1 + g_2 + \cdots + g_{n-1})k_q]$$

A digital-filter of the first-order $G_s(z)$ is used to "soften" the "initial jump" of the step-response.

$$G_s(z) = \frac{hz + (1 - h)}{z + \phi} \tag{10}$$

where h = 0.5 to 1, and ϕ is the desirable feedback gain for finite-time settling. With the above filter, the control-gain is modified to k'_n.

$$k'_n = \frac{pk_n + (1 - h)k_q}{h(p - 1) + 1} \tag{11}$$

The feedback gain, ϕ is given by

$$\phi = (1 + p) - h[k_q(g_1 + g_2 + \cdots + g_{n-1}) + k_n g_n] \tag{12}$$

The state-vector feedback via *observer* is shown in Fig. (5-46).

The accompanying flowchart is developed for the on-line computation by programming of a microcomputer in a third-order control plant, using the finite-time settling control algorithm (FTSC). The FTSO and FTSC gains are computed after determining g'_is and p.

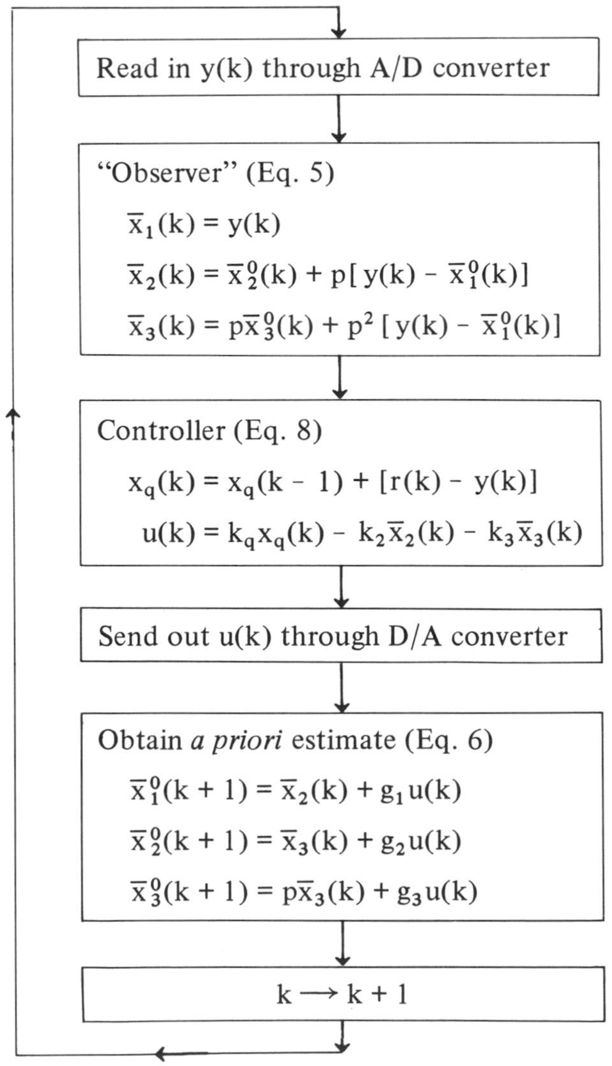

Read in y(k) through A/D converter
"Observer" (Eq. 5)
$\bar{x}_1(k) = y(k)$
$\bar{x}_2(k) = \bar{x}_2^0(k) + p[y(k) - \bar{x}_1^0(k)]$
$\bar{x}_3(k) = p\bar{x}_3^0(k) + p^2[y(k) - \bar{x}_1^0(k)]$
Controller (Eq. 8)
$x_q(k) = x_q(k - 1) + [r(k) - y(k)]$
$u(k) = k_q x_q(k) - k_2\bar{x}_2(k) - k_3\bar{x}_3(k)$
Send out u(k) through D/A converter
Obtain *a priori* estimate (Eq. 6)
$\bar{x}_1^0(k + 1) = \bar{x}_2(k) + g_1 u(k)$
$\bar{x}_2^0(k + 1) = \bar{x}_3(k) + g_2 u(k)$
$\bar{x}_3^0(k + 1) = p\bar{x}_3(k) + g_3 u(k)$
$k \longrightarrow k + 1$

c. *Optimal Controller for Linear Quadratic Problem.* The state-variable feedback

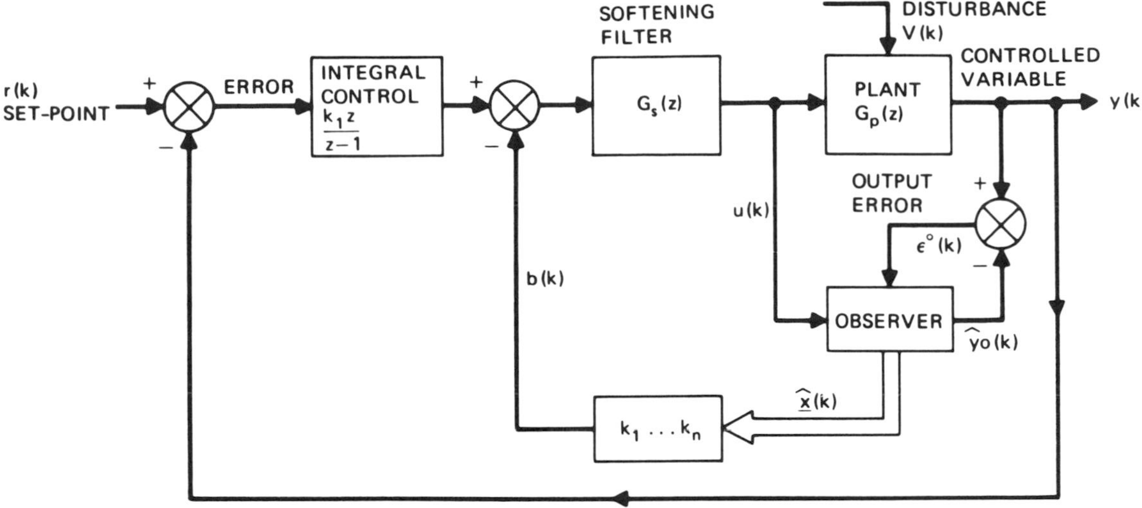

Fig. 5-46. State-vector feedback via *observer.* (*Courtesy of IEEE Proceedings, February 1978*)

gains can be optimized by optimal control theory, starting with the cost-functional J to be minimized:

$$J = \sum_{k=0}^{\infty} \{[r(k) - y(k)]^2 + w[\Delta u(k)]^2\} \tag{13}$$

The first term is the squared-error between the set-point and the plant output, and the second term is the squared increment in the manipulated variable. Assuming $r(k) = r_0$ = constant and ignoring the intial state vector for optimal control, the algorithm for optimal control is derived:

$$u_{opt}(k) = k_q \sum_{i=0}^{k} [r - y(i)] - \sum_{j=1}^{n} k_j x_j(k) \cdots \begin{cases} j = 1, 2, \ldots n \\ \text{and } k_q = k_1', k_j = k_{j+1}' \end{cases} \tag{14}$$

The matrix Ricatti equation converges to give the solution of the optimal control gains. The flowchart procedure for on-line computation is similar to the one for the previous case. With the weighting factor w = 0.01, good response is obtained for both step-reference and step-disturbance inputs.

5.28 SIMULATION MODELS

Simulation in principle aids the optimization of both hardware and software for choosing the right design of a complex computerized control system. The *multi-associative processor* (MAP) developed at the University of Colorado makes a fine example with its eight control units and 1024 processing elements. The complexity of the simulation model depends upon the desired system and its cost.

Alternative simulation models are based on the following design techniques:

1. prediction by way of the mathematical analysis of average independent variables, provided their *variance* is small
2. precise *queuing* models that allow the service and request procedures to be described by *probability distributions* for random values of independent variables

3. expensive *discrete* simulation models that include the preparation of *programs* for the various components
4. expensive experimental development of *several prototypes* that incorporate a minimum amount of detail.

Each of these techniques has its own merits and demerits, and only high-cost military and aerospace control systems can afford actual simulation models. For example, the current state-of-the-art of proving programs is not practical when large programs are involved; at the same time, it is comparatively simpler to test, alter, and retest a simulation model.

The MAP system illustrated in the accompanying block diagram allows the following system features to be tested:

1. the design of the main memory with respect to the various control units
2. the choice of the number of shared data busses with minimum conflict

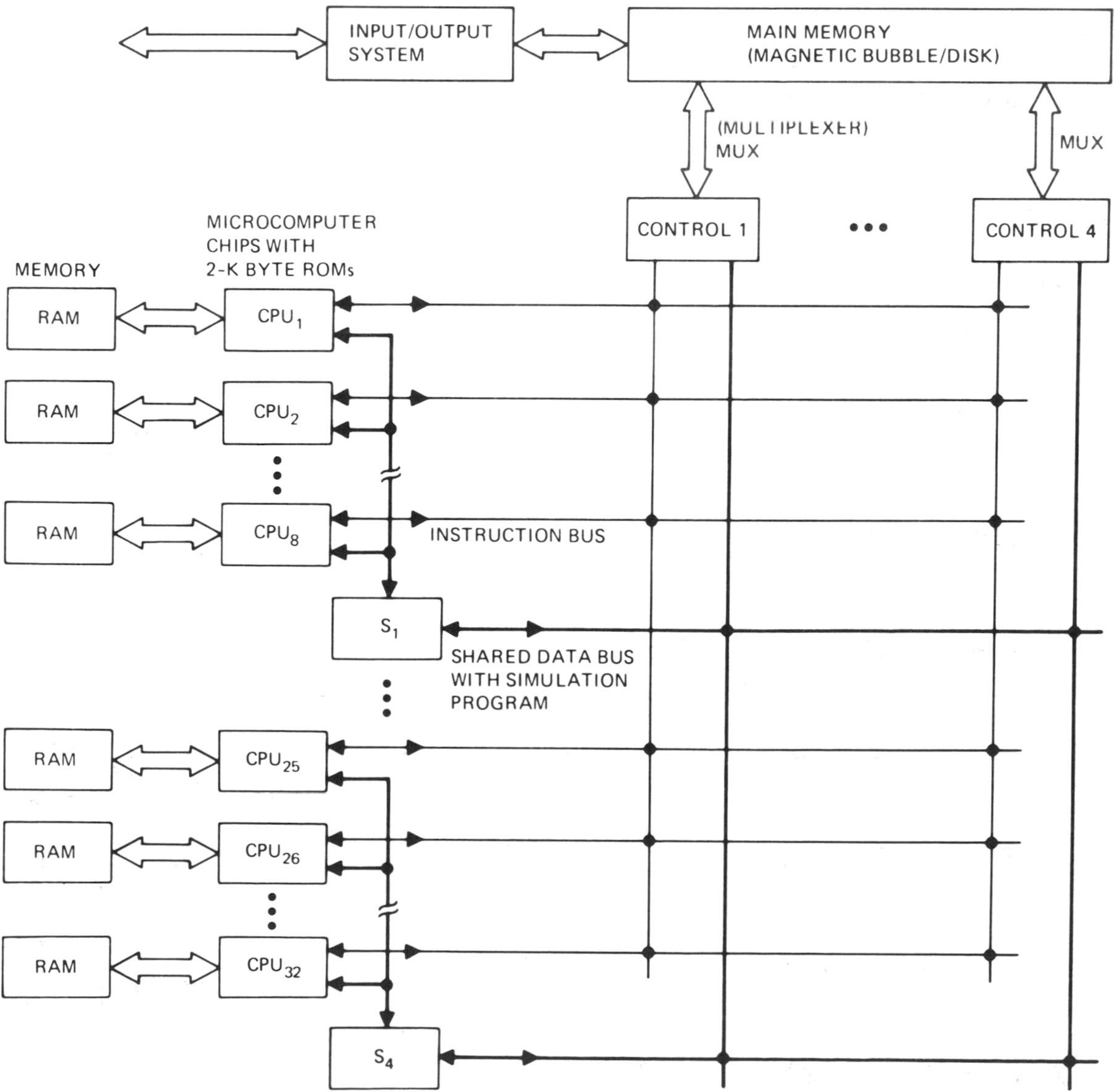

3. the selection of facilities needed in an operating system
4. testing the individual MAP programs

Each control unit, with its access to several CPUs, simulates an independent computation to allow several computations to proceed in parallel. Two-level simulation models are also devised: (1) to isolate single programs and components and (2) to minimize conflict for shared resources.

5.29 TRADE-OFFS BETWEEN HARDWARE IMPLEMENTATION BY LSI AND RESTRUCTURED SOFTWARE

The following example describes the latest design trend in systems and control engineering applications. Intel 8253 LSI programmable counter/timer chip is architecturally organized to control three active intervals of time with three independent 16-bit counters (binary or BCD) at dc to 2 MHz. It operates in several modes to generate different waveforms and timing requirements. As a typical systems-method of its application as a feedback controller for dc motor speed, R. D. Grappel of the Massachusetts Institute of Technology's Lincoln Laboratory improvised this chip to enable a microcomputer (1) to control the power supplied to the motor-load on a proportional basis and (2) to generate a motor-parameter such as revolutions per minute to be digitally fed back to the microcomputer.

It is obvious from the technique illustrated that the simple form of control algorithm, depicted by the direct implementation of hardware in Fig. 5-47, naturally results in a minimization of time and effort on computer software. Most of the tasks involved were formally assigned to the expensive software implementation for this specific digital control system. The flowchart for the CPU interrupt-handler and the main program are included in Fig. 5-47.

Timer 1 of 8253 provides overall system control and generates periodic interrupts to command the computer that a new power pulse is required. It is programmed to provide a 1-sec interval between interrupts by using a 10-kHz clock. Timer 2 is triggered to generate the power pulse. Timer 0 is gated to count feedback pulses from the motor load; it acts as a gated counter to feed back the state of the load to the computer; the computer interrupt-software modifies this count to a new count for timer 2.

A disk-encoder with 10 perforations and a lamp/photocell arrangement are used as a transducer to indicate motor speed. A single-shot integrated circuit 74121 shapes the photocell signals for timer 0. The motor speed-range from 600 to 6000 rpm translates to 100 to 1000 pulse/sec. The pulse generator of timer 2 drives a buffer (7407) and an opto-isolator. The motor is driven by a Darlington (2N6037) capable of handling loads up to 100 W.

At start, the first task of the microcomputer is to initialize the three timers. One interrupt/sec is received to read timer 0, and the corresponding count is compared to that in the program to increment/decrement the preset count of timer 2 within the set limits. Then, the microcomputer reloads timers 2, 1, and 0 for returning control to interrupt.

With the technique illustrated, one microcomputer (having cycle-times of the order of 1 to 2 μsec) can attend to several control tasks in parallel by using different control algorithms. The LSI hardware technology is thus instrumental at this time in

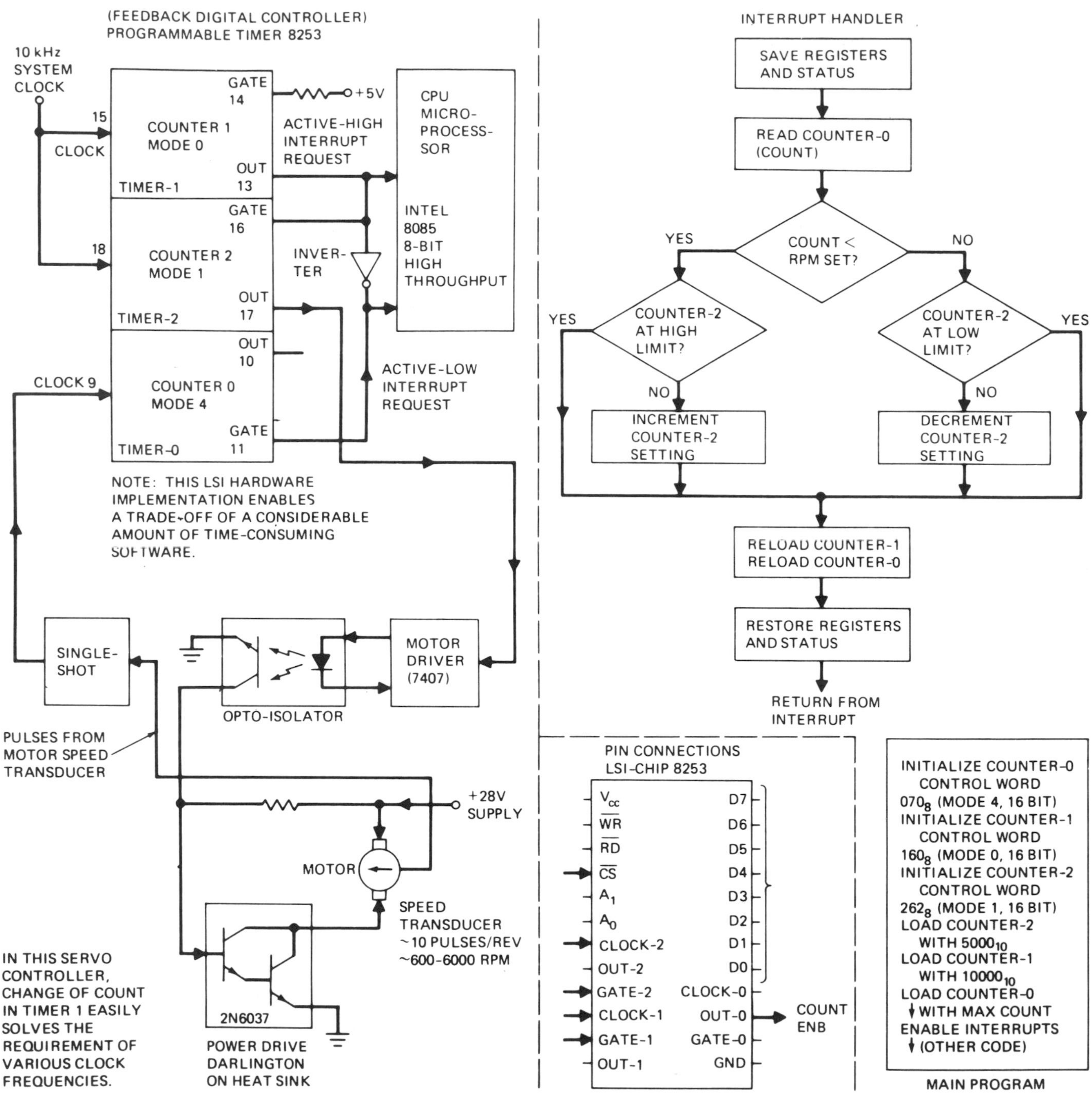

Fig. 5-47. Feedback controller. Intel-8253 as a programmable interval-timer in motor-speed application. (*Courtesy of EDN, Cahners Publishing Company*)

economizing fairly complex digital control systems by minimizing space and comparatively more expensive and time-consuming software demands.

5.30 MICROPROCESSORS TAKE OVER THE TASKS OF AVERAGE MINICOMPUTERS IN DIGITAL CONTROL

5.30.1 Zilog-8000, 32/16-bit scaled-NMOS Microprocessor (Zilog, Inc., Cupertino, California). The Zilog-8000 is the first VLSI (*very large-scale integrated*) microprocessor that is architecturally *register-oriented* instead of byte to rival the former average minicomputers in trends of application. Multiple registers reduce the instructions necessary for execution in order to simplify programming. At the same time, Z-8000 fits into the applications of the present 8/16-bit microprocessors. The VLSI feature means 17,500 transistors in 5833 gates at a power dissipation of 1.5 W, compared to 4800 transistors in 1600 gates of the LSI Intel 8080 (1974) at 1.2-W dissipation. (Intel's VLSI, 16-bit 8086 microprocessor is comprised of 20,000 transistors.) The Assembly language notation, the 81 distinct operation-code instructions, data types, and addressing modes all amount to a powerful set of 414 instructions, compared to a corresponding 65 of 8080. Therefore, the throughput of Z-8000 is as much as 10 times that of the present microprocessors. The VLSI Motorola scaled-NMOS 68000 microprocessor, expected during 1979, boasts 68,000 transistors; it can handle 32-bit arithmetic operations like any minicomputer and interface with 1 megabyte of random access memory.

With 24 registers of 16 bits each and a minimum need of memory reference, the Z-8000 may be considered as a *fourth-generation microprocessor* in the lineage 8080, Z-80, Z-80A/Intel 8085, and Z-8000.

A large majority of the instructions can use any of five main addressing modes with 8-bit byte and 16- or 32-bit data words.

With a direct-memory-addressing capability of 8 megabytes, the Z-8000 organizes memory into a set of 128 segments of up to 64-K bytes each, in order to simplify programming. As in large-scale and minicomputers, the Z-8000 accomplishes *dynamic relocation* and memory protection by using an auxiliary *memory-management chip* in a *data-base setup*. The NMOS chip is processed in VLSI scaled-down depletion-load silicon-gate technology, requiring a 5-V supply and a single-phase 4-MHz clock for timing.

To meet the requirement of compatibility with both the present series of microprocessors and new minicomputer applications, Z-8000 is available in (1) a 40-pin version that uses 16 lines to address a single segment of 64-K bytes and (2) a 48-pin version that uses 23 lines to address memory-segmented random-access memory of 8 megabytes. Real-time operating system and application programming features are isolated for maximum throughput under system- and normal-operating modes as in modern larger computer systems. At least three clock-cycles are mandatory for one memory-cycle. That is, Z-8000 requires memory devices with a cycle-time of 750 nsec and an access-time of 430 nsec. The architecture of Z-8000 is illustrated in Fig. 5-48. The organization of the registers and the chip-interface signal requirements are shown in Fig. 5-49. The throughput of this microprocessor is boosted by means of a look-ahead instruction decoder and accelerator on its internal bus. The systematic instruction-set allows an instruction to begin actuation as it enters the instruction register.

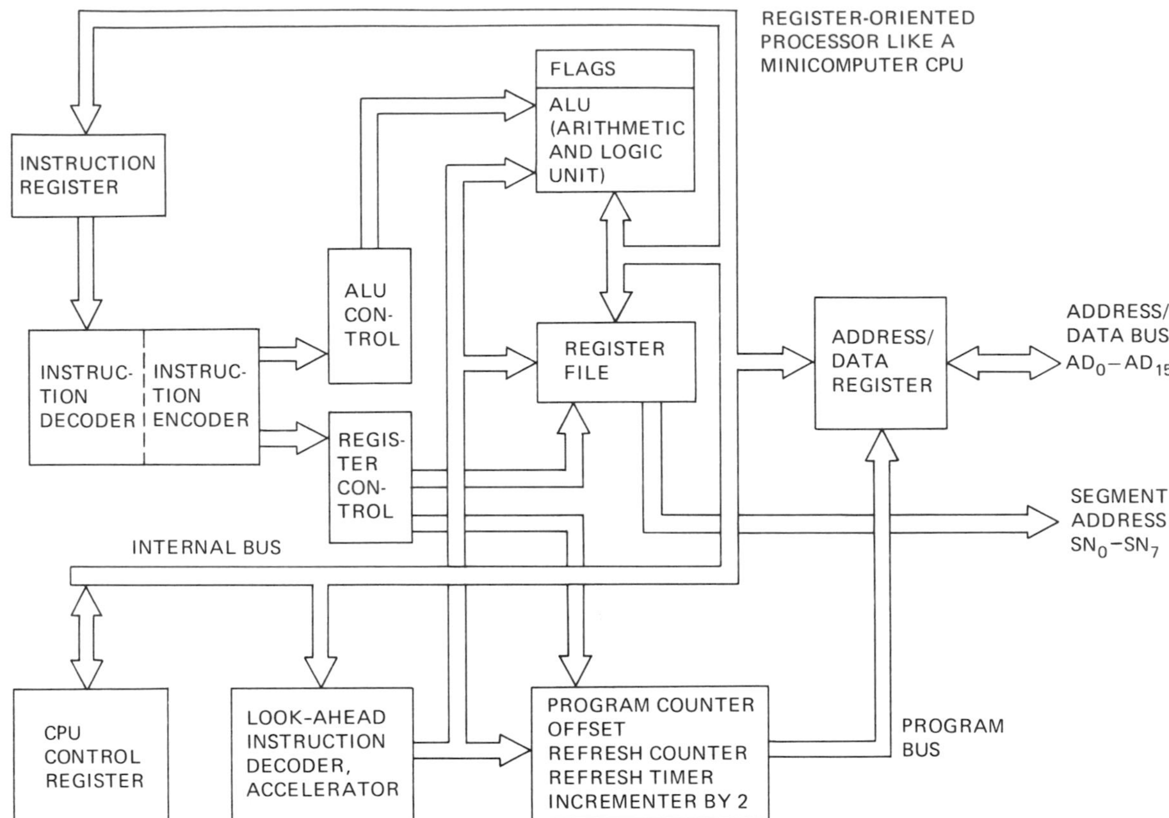

Fig. 5-48. Z-8000 Architecture. (Reprinted from *Electronics*, December 21, 1978: Copyright © McGraw-Hill, Inc., 1978.)

5.30.2 Intel VLSI 16-bit 8086 Microcomputer (Gordon Moore). The processing capability of the Intel 16-bit 8086 microcomputer, with its clock-rate of 5 MHz or alternatively 8 MHz, enables execution speeds about 10 times faster than those of 8080A-based computation, using roughly 10% to 25% of shorter programming effort. Its architecture aims at bridging the gap between 8- and 16-bit microcomputers, because the 8-bit 8080 software package and program development scheme could be easily adapted.

The advantages include signed 8- and 16-bit arithmetic, multiply and divide, interruptible bit-string operations, and improved bit manipulation. Minicomputer features such as reentrant code, position-independent code, dynamically relocatable programs, direct-address capability up to 1 megabyte of RAM or disk/bubble-memory, and flexibility for multiple-processor operation are available.

The VLSI high-performance NMOS, using 29,000-transistor, scaled-down, depletion-load, silicon-gate fabrication (H-MOS), enables propagation delays of 2 nsec per gate. Memory with cycle-times of 500 to 800 nsec is suitable, and access-time of data is from 300 to 460 nsec. Powerful 8- or 16-bit register structure, unlimited levels of interrupts, efficient input/output interface, and a new segment register-file contribute to a high computational throughput.

For control and timing, a bus-interface unit (BIU) maintains an optimized 6-byte fetch-ahead instruction queue for overlapping the execution and fetching of instructions. The instruction set supports stack management in block-structured high-level language of PL/M-86, which is an extension of the former PL/M language developed for 8080/8085 microprocessors.

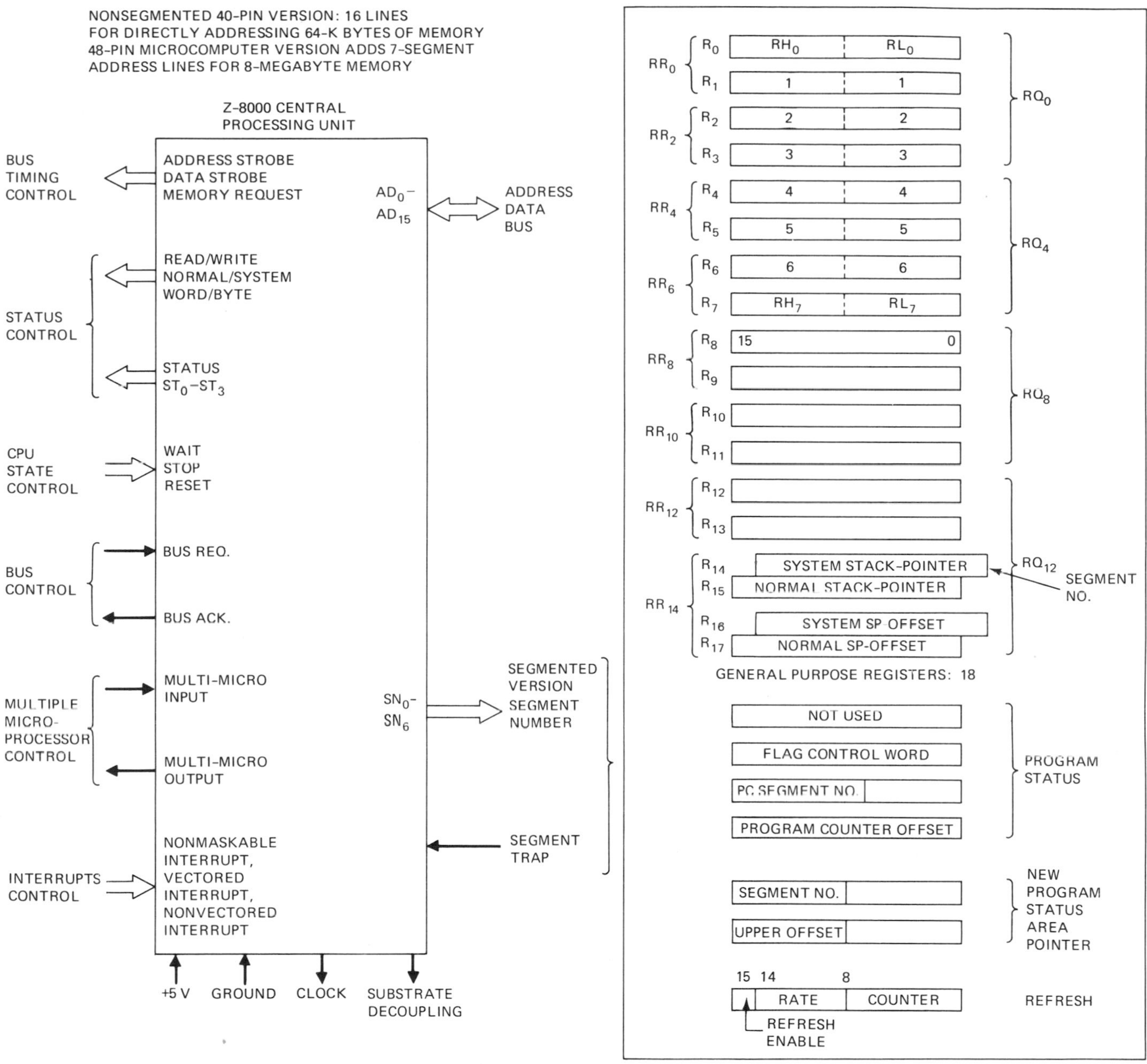

Fig. 5-49. Z-8000 Microprocessor: Nonsegmented version (40-pin); minicomputer-like 48-pin version, and register organization. (Reprinted from *Electronics*, December 21, 1978: Copyright © McGraw-Hill, Inc., 1978.)

The register-structure, the BIU, and execution unit (EU) of 8086 are illustrated in Fig. 5-50a and b. In the minimum-mode system (Fig. 5-50c), the access-times required of memory and I/O devices for the high-speed 8-MHz CPU operation are 265 nsec from receipt of address and 130 ns from receipt of read- or write-enable, respectively.

5.30.3 Intel-2920, Real-Time VLSI Microcomputer for Linear Signal Processing in Control Systems. A programmable general-purpose microcomputer, Intel-2920,

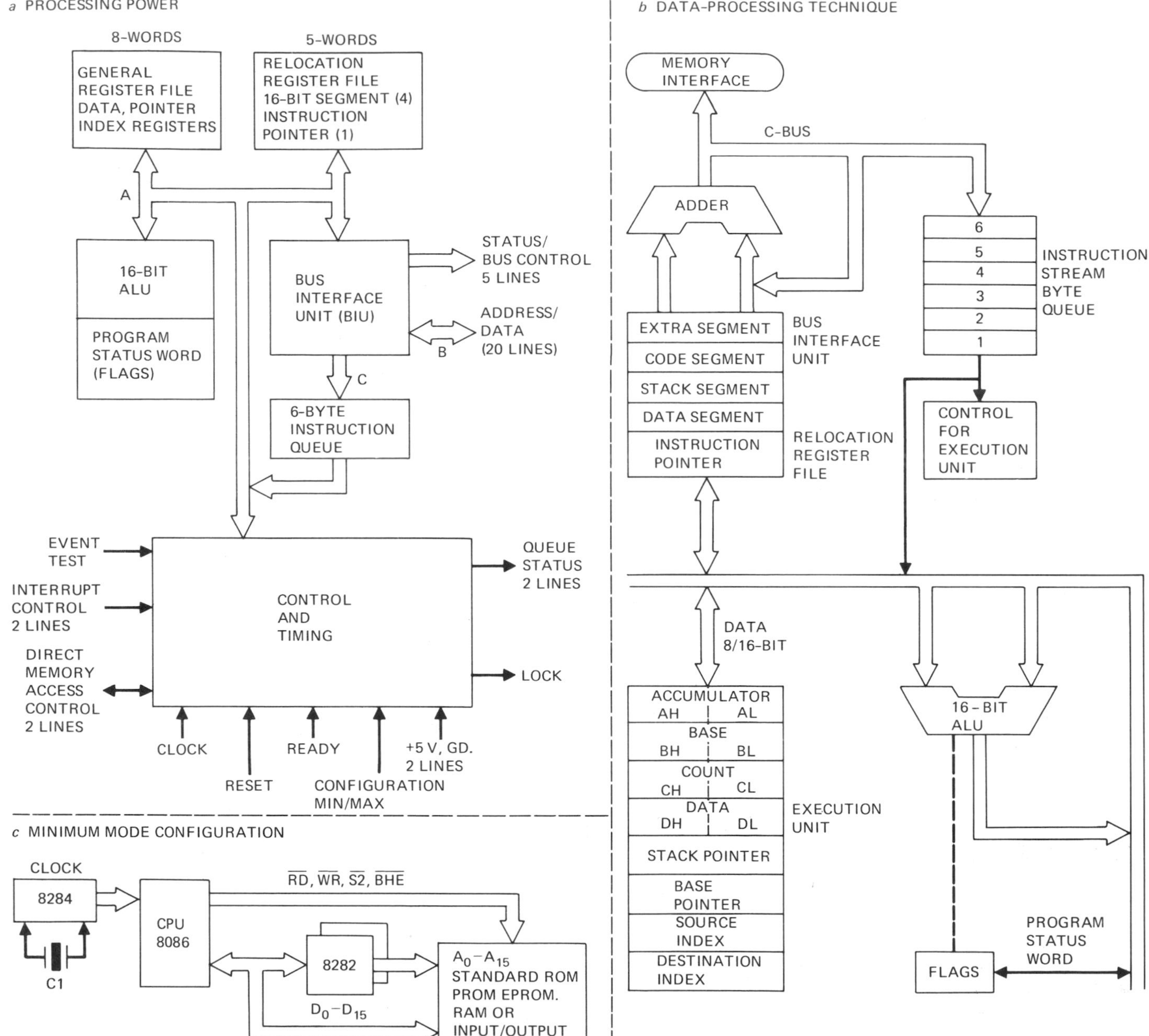

Fig. 5-50. *a*, INTEL 8086 CPU processing power. *b*, Processing data. *c*, Minimum mode microcomputer configuration needs 11 components to form a complete system. Large systems feasible with more components. (Reprinted from *Electronics*, February 16, 1978: Copyright © McGraw-Hill, Inc., 1978.)

with built-in EPROM program control, 40-word RAM working memory, control and arithmetic logic units, and A/D and D/A converters, is a timely arrival for popularizing the direct application of VLSI microcomputers in automatic control and data communications.

It can be programmed with 192 instructions in 24-bit words. The RAM working memory uses 25-bit words. The digital section enables parallel instructions to boost throughput and interface with the analog section by way of a data-register, which actually takes the role of a memory location.

With the availability of this VLSI chip, complex programmable devices such as modems, equalizers, tone generators and receivers, process controllers and motor or servo-motor drives can be readily designed without a large number of passive components and linear-amp integrated circuits.

The system organization of this NMOS VLSI microcomputer is shown in Fig. 5-51. The technique of architecture in this chip maintains total program control, and execution time remains constant. The program execution time is determined by the fixed sampling-rate that is mandatory with this digital-control chip. The high-speed, number-crunching, pipelined architecture of the CPU using a much simpler algorithm (in place of the usual shift-and-add multiplier) facilitates, in particular, the needed high-capacity computation between real-time sampling in processor control applications. External LSI digital filters and additional 2920 microcomputer chips can be provided for more complex digital control systems. The simulation of the microcomputer chip as a digital filter is interpreted in Fig. 5-51.

From the hardware point of view, the design of fairly complex digital control systems, requiring dynamic memory relocation, memory protection, and multitasking programmability, is economically feasible at the present stage. A new breed of combined hardware and software professionals are needed in order to minimize the current high cost of software. Companion VLSI chips are available for "memory management;" they enable VLSI 16/32-bit microcomputers of the category of Intel 8086, Zilog 8000, Motorola 68000, etc., to make full use of 8-megabit external addressing capability for such complex digital control systems. High-capacity 64-kilobit and 131.072-kilobit (28-pin) ROMs, having an initial access-time under 450 nsec are presently available to speed up most processors. External real-time high-capacity working memory is also readily available with 5-V dynamic 64-kbit RAMs, which use a single pin to control self-refreshing for both active and idle states. All this economy hardware naturally facilitates the scope of long-term development of software for plant digital automation in the near future.

Power tools, toys, food-processing appliances, floppy-disk drives, radio-controlled automobiles, etc., use millions of new universal motors. Single-chip digital microcomputer chips costing $2.00 can replace bulky analog feedback control circuitry hitherto used for speed-control. The microcomputer, besides adjusting speed, checks motor current against a list of maximum current values, stored in a high-capacity ROM lookup table right on the chip, to limit instantaneous long-duration surge and peak currents. This makes it feasible to use a smaller economy motor for most applications.

5.31 DISTRIBUTED DATA PROCESSING

5.31.1 Distributed Processing Networks. As a concept, distributed processing networks involve application and/or *data-base processing activity at two or more network nodes. Distributed data processing (DDP)* is generally the term connected with the system integration of *computer networking* and *data-base management* associated with *time-share*, using high-level programming languages BASIC, ALGOL, PL/1, APL, FORTRAN, and the latest block-structured languages such as PL/M or PL/Z (Zilog) or PASCAL/C (all ROM-able in the present microcomputer systems). The concept involves an economically flexible and reliable computer system that distributes the power of a *central* large-scale network computer to "Node" computer

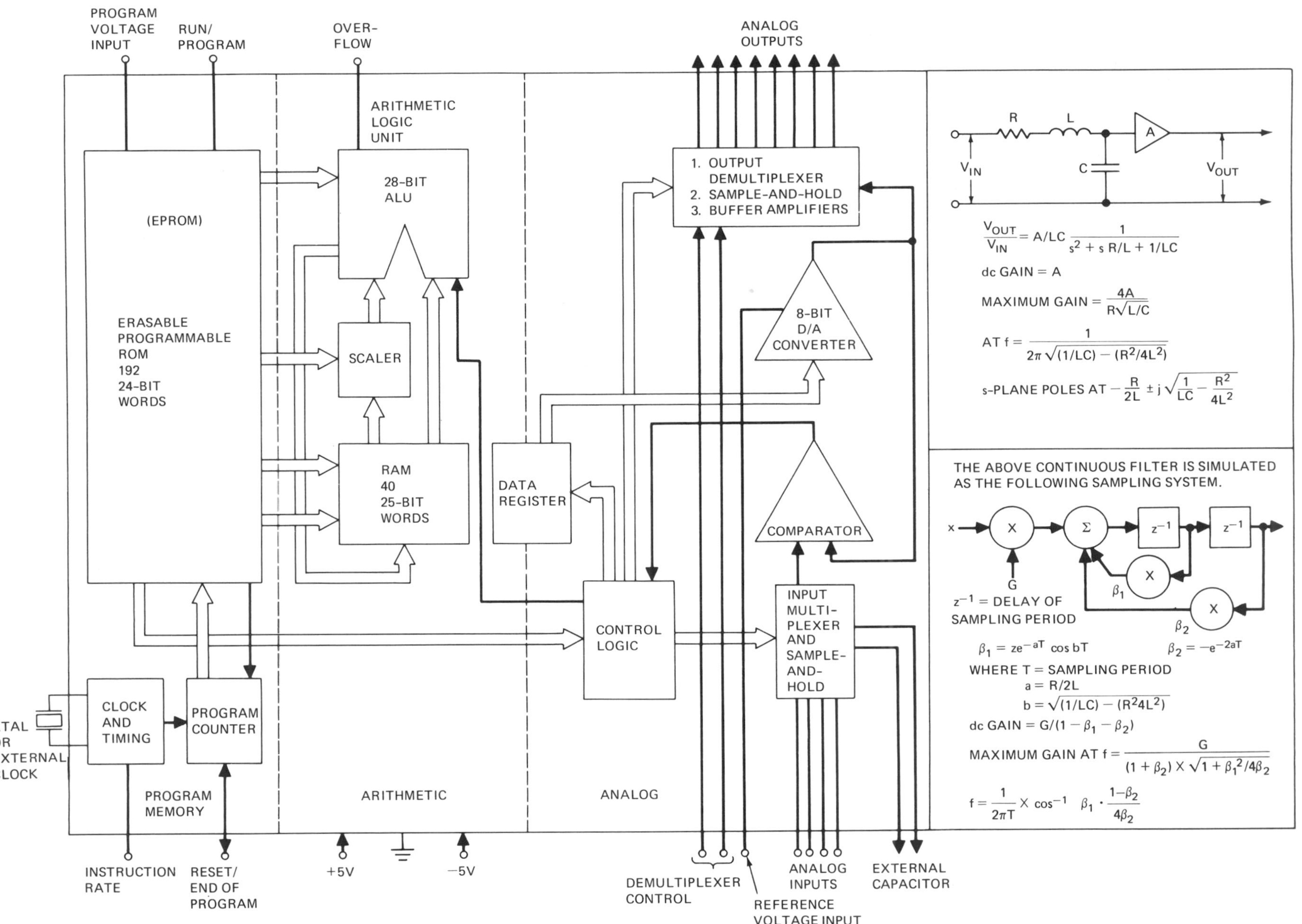

Fig. 5-51. Intel-2920. NMOS microcomputer for real-time signal processing in control systems. (Reprinted from *Electronics*, March 1, 1979: Copyright © McGraw-Hill, Inc., 1979.)

terminals where some raw information is generated and processed in *real-time*; the customer's host mini- or micro-computers at the terminals share resources such as software and data stored in libraries of disk and other peripherals. Such systems naturally reach a high degree of sophistication as a (difficult-to-implement) standardization takes place in ultimately deciding on a common high-level language, and they minimize undesired additional hardware at the terminals for necessary translation. The Nodes in turn may serve subscriber terminals such as CRTs and TTYs with built-in microcontrollers.

For example, IBM's latest distributed processing unit 8100 offers more stand-alone programming and processing power in a network without requiring a system/370 mainframe as the host computer. This facilitates on-the-spot computing in order to minimize data communication costs in the system network architecture and upgrade reliability via decentralization, since one unit's failure does not shut down the entire system. Today's 8-in. double-sided double-density floppy disk in a PDP-11 low-cost microcomputer system can store over a megabyte of data (1 megabyte of main memory was unheard of in the high-cost medium-scale computers of the early 1960s). Present real-time economy operating systems have become versatile enough to process concurrent tasks.

Network topologies can vary in DDP. However, network control may be hierarchical under a central or host computer; up to five levels of satellite processors are not uncommon. The goal is a network operating system that can allocate network resources.

Latest network software has an *automatic routing capability* that selects the shortest operable links between two points at data rates as fast as 56 kilobit/sec, using modems that allow synchronous speeds up to 9600 baud/sec over dial-up voice-grade lines. The increasingly powerful mini- and micro-computers at the remote terminals do not have to transmit raw data to the host computer for processing over expensive communication lines. Presently, a remote mini- or microcomputer in a systems network architecture (SNA) has the capability to develop, compile, and execute programs. That is, the hierarchical aspect from a central system/370 computer is eliminated by a set of remote 8100s (concentrators) that use central disk application programs in real-time in a peer network. IBM's synchronous data link control uses time-division-multiplex packet-switching techniques at data communications rates up to 56 kilobit/sec. In addition to IBM, other decentralized distributed data-processing networks are:

Burroughs	Honeywell
Computer Automation	Modular Computer Systems
Control Data	NCR
Datapoint	Sperry Univac
Digital Equipment	Sycor
General Automation	Tandem Computers
Hewlett-Packard	

Most nations have their own DDP networks. As far as standardization of high-level languages is concerned in the case of the microcomputer systems, it is still not too late to decide upon a common block-structured language such as PASCAL or Bell Telephone's C (modified PASCAL) for full implementation in an exclusive network in the near future. Zilog's distributed data base processor uses a Z-80 microprocessor to sort out the protocols and talk with other computers in a "virtual network."

As a typical example of DDP, the DECnet architecture of Digital Equipment Corporation shown in the following diagram employs three protocols: (1) data access for input/output, (2) network services, and (3) digital data communications message for supporting communications.

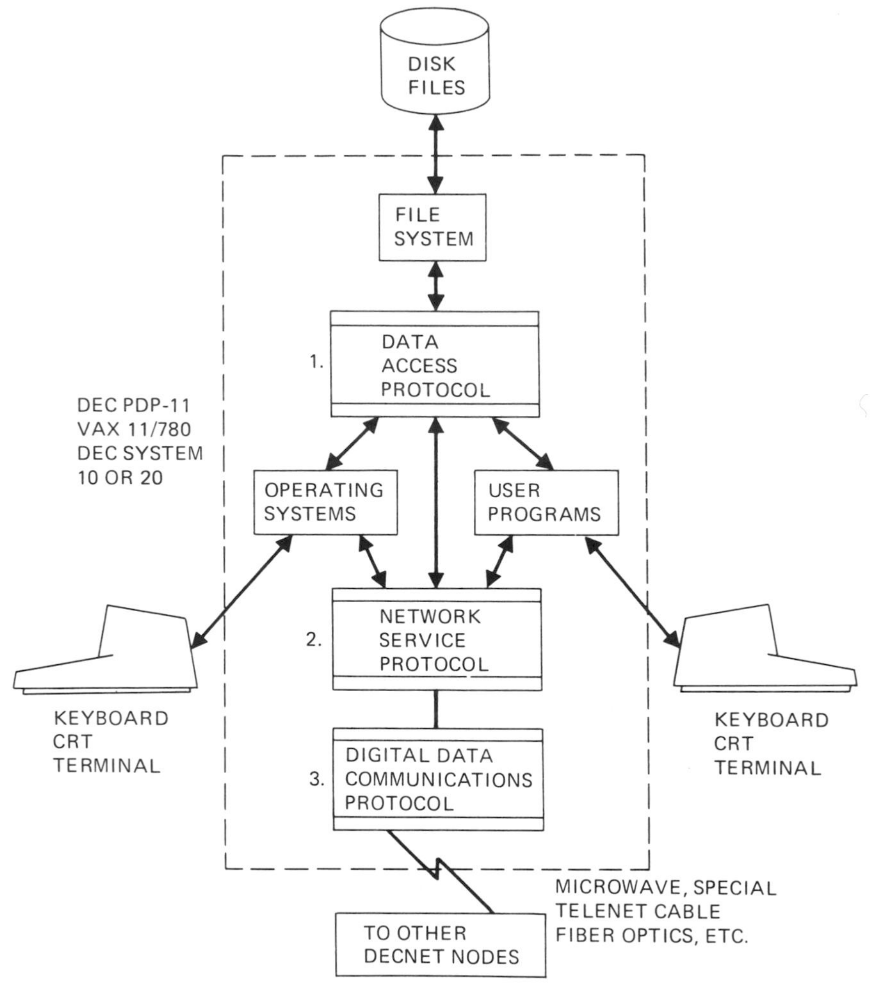

DEC PDP-11, VAX 11/780, DEC SYSTEM 10 OR 20

5.31.2 Distributed Digital Processing Techniques. The following techniques are encountered in distributed digital processing:

1. *Bit rates of physical channels*
   a. *Asynchronous channels* up to 1200 baud/sec

Telex: single-speed of 50 baud/sec. Digital information is transmitted in a 5-bit Baudot code between compatible subscriber-teletypewriter terminals in a Telex network, which provides international message-transfer service via public telegraph circuits.

TWX: Arbitrary codes or transmission rates under 150 baud/sec. The TWX (teletypewriter exchange) network of Western Union provides dial-up service between compatible teletypewriters. It uses analog transmission over the public telephone network.

Baud: Baud is the unit of signaling speed. It is the number of discrete conditions or signaling events per second on a transmission line. Baud is the same as bit for binary digital transmission. It may be a group of two bits in 4-phase modulation as dibit (00, 01, 10, 11).

b. *Synchronous channels* at higher bit rates

2.4, 4.8, 9.6, 19.2, 56, 112, 224, 1344 kilobit/sec

The switching time for the lower-speed channels of the Datran network is a fraction of that on the telephone network; it is therefore called a fast-connect network with full-duplex (FDX) two-way transmission. The channels are digital end-to-end, and the users should not need modems. However, modems must be used on analog telephone lines to convert the data transmitted in PSK analog form.

Other bit rates are:

64 Kb/sec: Digitized telephone channel (CCITT)
1.544 Mb/sec: T1 carrier wire-pair cable
6.312 Mb/sec: T2 carrier wire-pair cable
50-Mb/sec: Satellite transponder
274-Mb/sec: T4 carrier channels in various transmission systems
800-Mb/sec: Satellite throughput, cable television, optical-fiber transmission
16-Gb/sec: WT4 waveguide transmission system

2. *Vocoder:* 4.8 Kb/sec PCM (pulse code modulation) is processed to convey intelligence with enough information to synthesize a voice for human perception only.

3. *Picturephone:* 1-MHz bandwidth video signal is carried by a bit stream of 6.3 Mb/sec, using a 2-MHz sampling rate. For standard PCM, this gives 3 bits/sample, and hence eight discrete amplitude levels can be reconstructed to present a grainy picture. (As a comparison, presently 4.2-MHz bandwidth color television signals are transmitted using 8 to 10 bits per sample.) The coarse picture-phone signals and voice are improved by using differential PCM and delta modulation.

4. *Message-switching:* Non-real-time people-to-people message traffic is stored at the switch nodes and then transmitted onward to its destination on a *store-and-forward* centralized (star/tree) switching basis in a fraction of an hour. A long message sent in a single transmission is filed for possible retrieval in future.

5. *Packet switching:* The international standards organization for telephony and telegraphy (CCITT) defines a *packet* as a group of binary digits (including data and call control signals), which is switched as a composite whole. The data, call-control signals, and possibly error-control information are arranged in a specific format. *Packet switching* is defined as the transmission of data by means of addressed packets, whereby a transmission channel is occupied for the duration of transmission of the packet only. The channel is then available for use by packets being transferred between some other data-terminal equipment. (The data may be formatted into a packet or divided and then formatted into a number of packets for transmission and multiplexing purposes.)

Packet-switching is high-speed, and it is primarily intended for *real-time distributed data processing* in computer networks using several terminals. Long messages are segmented into small slices: 1008 bits per packet on the United States Telenet system; the Datapac system in Canada processes 255 bytes of 8 bits each. The segmented packets in a message can be *queued* in the main memory of the switching nodes and transmitted from node to node. At the destination, the whole message is reassembled. The computers that the network serves are called *host computers*.

A methodology in which the selected rates vary with the conditions of traffic in the network is called *adaptive routing.* The ARPA (Advanced Research Project Agency of the Department of Defense) and Telenet network use adaptive routing, with each node sending a service message every half-second. If, due to some error, the packet is not delivered at the correct destination, it is returned to the origin after a certain count in its field. Traffic congestion is prevented by controlling the *input* node, using control messages such as "ready-for-next-message (RFNM)."

6. *Frequency Division Multiple Access (FDMA):* FDM is *frequency division multiplex* in which the available transmission bandwidth is split by linear-phase filters into narrower bands as separate channels. FDMA implies that separate narrow-frequency bands are allocated to different users (as in broadcasting), and a transmitting or receiving terminal must be tuned to the frequency that is assigned to it.

Multiple-access techniques (used in satellite data communications) can operate as (1) analog fixed-channel assignments to the various users or (2) with channels assigned according to demand, as *demand-assignment.* When the user makes a request, a channel is allocated to the user after a short time-lag depending on the traffic in the channel circuits. It is *demand-assigned multiple access* (DAMA) as used in cable television (CATV). In cases where optical-fiber CATV cables are used, the high-speed digitized bit-rate will allow video (picture) communication between homes and terminal stations.

7. *Time Division Multiple Access (TDMA):* TDM (time-division multiplex) is the procedure in which high-capacity PCM channels are shared between user terminals by time division (time-slots)—as in digital radio and microwave links, CATV cables, waveguides (and fiber-optics), and satellites. If there are many access points for the time-slots by several terminals (in the place of only point-to-point send-receive multiplex terminals), the technique becomes *time-division multiple-access* (TDMA). Unlike synchronous TDM, TDM can have different aggregate bit-rates for incoming and outgoing lines. In general, *multiplexing* refers to static channel-assignment schemes in which given time-slots (or frequency bands) on a shared channel are assigned on a fixed, predetermined (*a priori*) basis with mostly balanced input/output capacities. Dedicated time-slots or subchannels for each port in the sharing group are categorized as synchronous time-division multiplex (STDM). *Asynchronous or statistical TDM* dynamically allocates time-slots on a statistical basis and increases transmission line efficiency by providing time-slots for ports actively transmitting data. The terminals naturally use digital computers for this purpose. STDMs share a synchronous communication line by cyclically scanning incoming data from I/O ports, gating bits, or characters and interleaving them into *frames* off a single high-speed multiplexed data pulse-train.

A *concentration* technique describes a methodology in which a number of ports dynamically share a limited number of output subchannels on a *demand* basis. As the aggregate I/O bit-rates need not be matched in a concentrator, the digital mini- or micro-computer used in a concentrator at the terminal node will generally use software algorithms based on statistics and queuing. Therefore, a statistical TDM may be referred to as a *multiplexer-concentrator* or simply a *dynamic multiplexer.* Incidentally, a line-switching concentration is sometimes referred to as *space-division multiplex.* The following block-diagram points out the features in a modern satellite TDMA system.

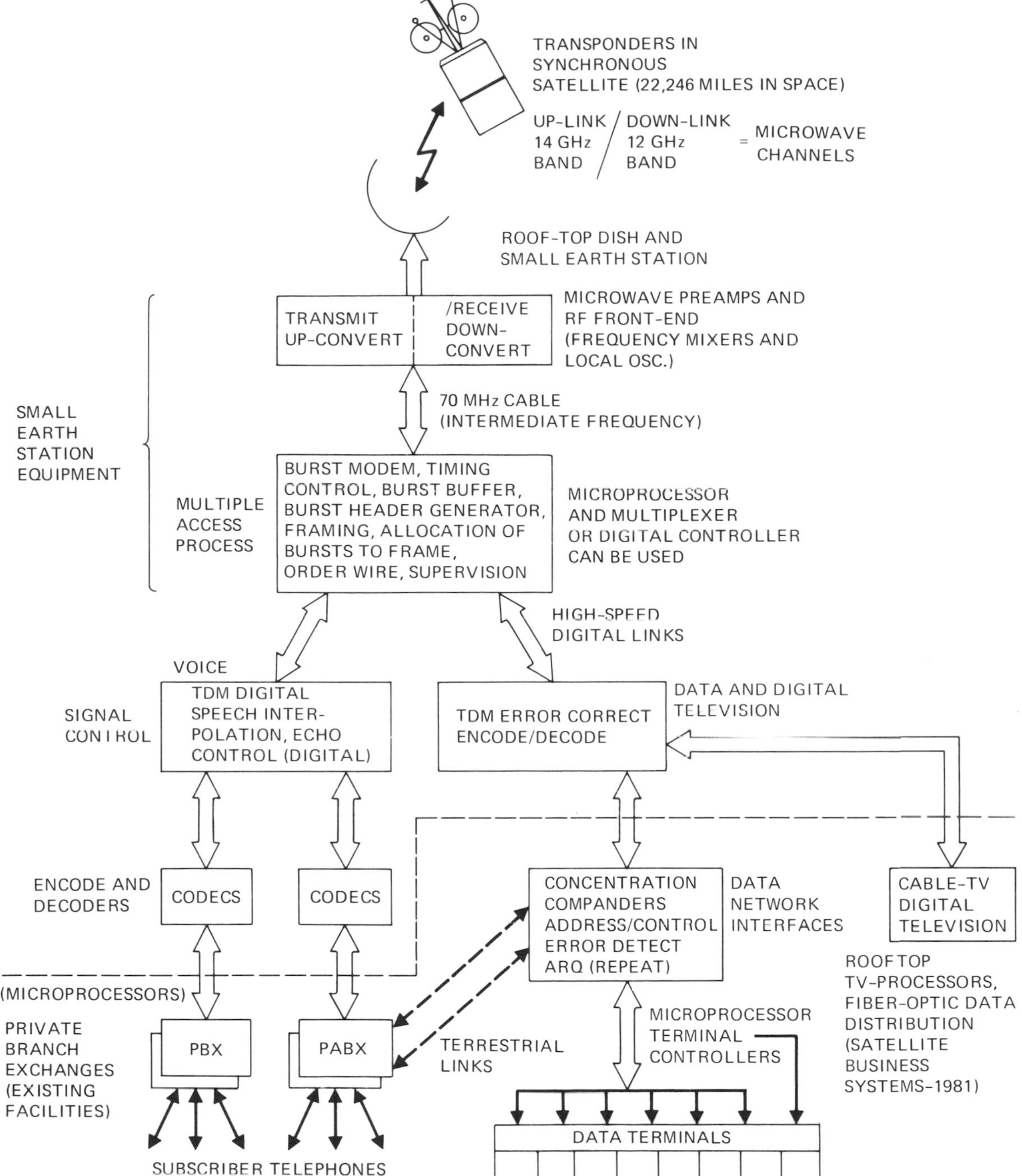

5.32 STATUS OF DIRECT DIGITAL CONTROL

Data General's data acquisition and control subsystem (DG/DAC) is a new *complete* I/O modular subsystem that performs both A/D and D/A, and digital I/O multiplexing in any mix, up to 256 lines. *Modularity* allows ready expansion to several units to meet complex requirements, catering up to 1000 lines. As shown in the following block-diagram, two microcomputers, using two VSLI microprocessors of the class of Zilog, register-oriented, 32/16-bit Z-8000 CPU chips, and associated software, may be incorporated in a system to develop a complex process-control function. For system interface, two LSI microcontrollers would perform the above tasks of DG/DAC as I/O multiplexers, A/D and D/A converters. Should one computer fail, the other would conveniently take over critical operations with no interruption or data loss. While one computer is handling critical process operations, the secondary computer would take over lower priority tasks such as program development.

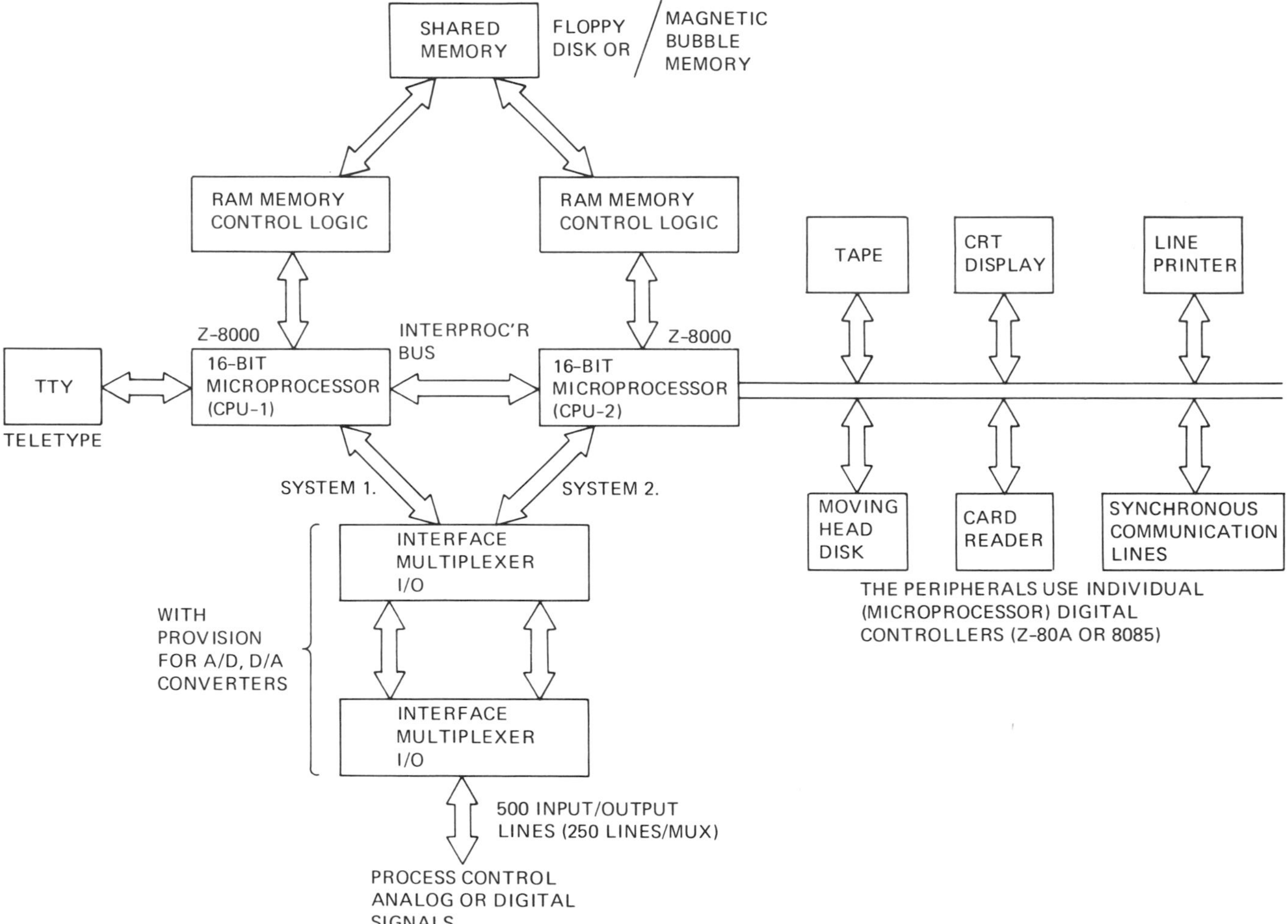

With large-scale mass production, the hardware is comparatively low-cost. However, the initial capital investment in software/firmware is both costly and time-consuming; only affluent nations could afford it. *But*, once the software for a specific process is established in an internationally standardized high-level language code, subsequent mass- and working-memory for software of dedicated process applications will naturally assume the form of low-cost mass-produced hardware as compact magnetic bubbles/CCDs, ROMs/RAMs/PLAs, and similar, more advanced, high-density, long-term and short-term storage memory. This is going to be the way of automation and mass-production in the twenty-first century.

World-wide automation and satellite data communications will be able to distribute basic living comforts (presently restricted to a very small percentage of the world population) to the remote corners of the earth during the twenty-first century (irrespective of the ideological differences):

1. if the affluent nations and the labor unions observe patience, restraint, and a

give-and-take philosophy in solving temporary international problems such as unemployment in the face of universal automation and successfully avoid atomic warfare and other forms of self-destruction, and

2. if nations work hard to prevent a population explosion and maintain a demographical zero population growth, and

3. if nations and individuals observe dedication against self-indulgent life-styles, alcohol and drug addiction, and explore ways to use the ever-increasing amount of leisure time more constructively.

With unlimited and safe fusion energy on the horizon, the remaining 20 years of the current century can prepare the way for this challenging goal in the evolution of *Homo sapiens*, and it is hence a magnificent opportunity for the present generation.

5.33. STATUS OF MICROCOMPUTER SOFTWARE WITH HIGH-LEVEL LANGUAGES (HLL-SOFTWARE)

5.33.1 Intel PL/M Software Engineering.* PL/M is a high-level language exclusively developed to simplify the programming of Intel 8008/8080/8085/8086 microcomputer systems using *either 8-bit or 16-bit words*. (A corresponding PL/Z software is available for the higher throughput chips Z-80/Z-80A/Z-8000 of Zilog.)

Along the lines of the Fortran-IV/V and PL/1 high-level languages, by using *structured design techniques* for writing *correct* programs, the PL/M *compiler* facilitates the automatic control of the internal registers, main memory, and stacks, and the external memory and peripherals in disk and tape operating systems via the input/output interface. With the use of high-capacity 64-K bit RAMs, PROMs, and PLAs, reliable PL/M software development, checkout, portability, and documentation for program maintenance and modification are presently feasible in microcomputer systems as well—and at comparatively low cost.

The characteristics and modular-software development methodology are explained with simple explanatory definitions in Section 5-33.3. The PL/M compilers look after the details of machine and assembly language programming, while the programmer concentrates on the effective software and logic design. A PL/M compiler may be directly outputted into the Intel series of 8- and 16-bit *simulator programs* for interactive, symbolic debugging, or may be punched to paper tape in hexadecimal (hex) format for loading into an Intellec microcomputer development system. The PL/M compilers are written in ANSI standard Fortran IV, and are designed to *run on timeshare* on any large-scale computer system using minimum 32-bit integer format (word size).

5.33.2 PL/M Features.* Salient features of the PL/M include:

1. A sequence of *declarations* and *executable statements* make the program.

2. The declarations allow the control of *memory allocation* and define simple *textual macros* and *procedures*. As a *block-structured language*, the procedures may involve secondary declarations and their subsidiary functions, which in turn might include further procedures.

3. The procedural facility reflects *modular programming* as a sectional *division into subprocedures* such as keyboard input, binary-to-decimal conversion, and output printing. The subprocedures facilitate easy formulation and debugging, and make a convenient *library* for direct incorporation into *other programs*.

4. The *data* are made up of (1) BYTE, variable or constant, as an 8-bit identity

and (2) ADDRESS, variable or constant, as a 16-bit (2-byte) word. The programmer can declare variable *names* or *vectors* (*arrays*) to represent BYTE or ADDRESS.

5. The executable statements specify computation such as arithmetic, boolean algebra, and operators BYTE and ADDRESS for comparison. Combined, these *relational operators* and *operands* make the *algebraic* EXPRESSIONS such as A/(B - 1)*C—which are actually the PL/M ASSIGNMENT *statements*. They compute the result and store it in a *memory location* defined by a *variable name* such as Q in the statement,

$$Q = A/(B - 1)*C$$

The computation to the right of equal sign is done first and saved in a memory location labeled by Q.

6. Conditional tests, branching, loop control, and procedure invocation with parameter passing make similar statements. Powerful *control block-structures* specify the *flow of program execution*. Basic *I/O statements* to read and write BYTES to and from the microprocessors may be defined by other procedures for more complex I/O operations.

7. A *compile-time macro* facility in the compiler provides a method of *automatic text-substitution* of a *character string* for a *symbolic name*. (A *string* is an arbitrary sequence of characters or alphanumerics.)

8. Programs are written *free-form*, with column-independent freely inserted *spaces*. The *character set* is a subset of both ASCII and EBCDIC codes. The following represent the valid PL/M alphanumeric and special characters:

ABCD . . . WXYZ 012 . . 789 $ =./( ) +-'*, <> : ;

Any other characters are merely invalid, and they are substituted by respective spaces. Special characters and combinations of special characters have special meanings in a PL/M program.

9. An *identifier* (up to 31 characters in length) names variables, procedures, macros, and statement labels. The identifier must be part alphabetic first and then alphabetic or numeric. Dollar signs are for readability only and are ignored by compiler. There are, however, 34 RESERVED WORDS, as part of PL/M language, but *not* used by programmer as identifiers. (An example of an identifier: X GAMMA LONGIDENTIFIERWITHNUMBER3.) Blanks may or may not be used in PL/M statements of identifiers and reserved and special characters. For readability, explanation, and documentation, a comment in the form shown is permissible; the delimiters enable the compiler to ignore the comment.

/*THIS IS A COMMENT*/

5.33.3 Structured Programming and Top-Down Design in PL/M.† Former programming techniques of general-purpose digital computers were without any standardized discipline, and hence most of those operating systems had too many errors. The situation has changed for the better with the latest techniques, just in time for the high-level language (HLL) programming of the new generation of a multitude of low-cost microcomputer systems. The PL-series of HLLs, and PASCAL in its modified forms are gradually taking a predominant role in fairly efficient and economic software engineering applications in these new systems, which apparently have an explosive character in demand and high-quality software. The latest programming techniques under consideration relate to two powerful software techniques: *struc-*

†All mnemonics copyright Intel Corporation 1975.

tured programming and *top-down design*. They enable error-free programming with a minimum of experience.

Structured programming is designed in terms of four *control structures—sequencing, selection, repetition*, and *subroutine* or *function invocation*—without the need of any elaborate flowcharts and the confusing, excessive GO TO statements common in most languages previously. Next, the *top-down design* relates to a hierarchical procedure or collection of statements in the form of *modules*, shown in the following multilevel hierarchical *tree*. Ready *documentation* is provided by mixing comments right inside the actual programs written in the high-level languages under consideration.

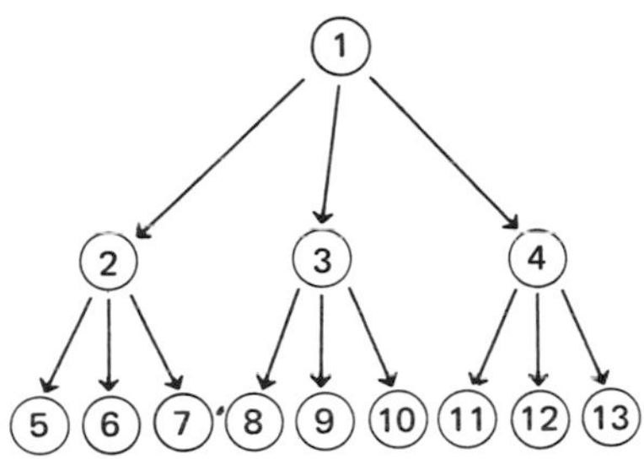

Entire program calls on *modules* 2, 3, and 4 to perform its *function* as *functional blocks* on level 2.

Independent modules in level 3 implement software with operations available in the HLL used (5 . . . 13) . . . (last statement section).

5.33.3.1 Definitions

1. *Data structure* can be organized into structures, arrays, stacks, strings, trees, plexes, graphs, queues, deques, etc.

2. *Structure* is a PL/ high-level-language term that refers to a *block of memory*. It can be organized into several *levels*. For example, if a 4-BYTE location is named PIECE, it can be DECLARED with four contiguous addresses into (level 1) PIECE TYPE; (level 2) PIECE COLOR; (level 3) PIECE POSITION ROW; (level 3) PIECE POSITION COLUMN. The three levels are actually termed *offsets* 0, 1, 2 in this example. A structure is designated by a POINTER, which in turn is the name of a WORD; COUNTER is the corresponding name of a BYTE. An address calculation is carried out at the *translation time* into the machine code at the *run-time* of the computer.

3. *Linear array* is similar to the structure; however, with the structure, the offset is known at the translation time, whereas the *array*, as a group of bytes or words, must be calculated at the run-time with extra access-time. An array will consist of several PIECES (or blocks of BYTES).

4. A two-dimensional array stored by *rows* or *columns* is declared a TABLE.

5. A *queue* is a linear list whose ends are labeled *front* and *rear;* it is implemented by a circular list (or *circular array*).

6. A *deque* (pronounced "deck") is a double-ended queue. A *bottomless stack* implements the deque with an *overflow* into tape or disk storage.

7. A *string* is a sequence of *characters*; it is stored in the computer as a linear array of character codes. One large array is preferred for all the strings in a program. The computer itself allocates space from this array for newly formed strings. A string is denoted by its *length* and the *address* of its leftmost byte, which together make a *descriptor*—it is declared as LIKE STRING. The length of the strings, called SIZE, is limited to the range, *null-string* 0 to a maximum length of 255. This limit

facilitates the sensing of when a *garbage collection* (or recovery of unused space) is required.

8. A *plex* is a *set of memory-blocks* interrelated with *pointers*. These memory-blocks are called *nodes*, and, since a memory-block is also a *structure*, a node is defined by a structure declaration. In general, one such *structure declaration* will do if all the nodes of a plex have the same structure.

9. A *chain* is a *linked list* implemented as a plex. If the last element of the list is linked back to the first, it is a *ring*. A chain may be searched by a *sequential search*.

10. A *computer tree* branches downward with the "root" of the tree at the top, and "leaves" at the bottom; the tree represents (1) hierarchical, (2) nested, (3) branching, and (4) converging structures. A collection of trees is called a *forest*. A tree in which each node has exactly two branches is called a *binary tree*. If every node that is not a "leaf" has two branches, left and right, it is called a *complete binary tree*. A *preorder traversal* represents node to branches left to right, and the inverse traversal from branches to the node is *postorder*. Compact bit tables require a single bit for each entry. Complicated programming, involving *shifting and masking operations*, is required to extract the desired bits from the bytes or words in which they are stored. Algorithms are *recursive* when the data structure being manipulated is itself recursively defined.

11. Maps, networks, mazes, signal flowcharts, schematic diagrams, state transition diagrams, syntax diagrams, and so forth are all known as *graphs;* they consist of *points* or *nodes* connected by lines or *arcs*. If the arcs have arrowheads, they are *directed graphs* with initial and terminal nodes.

12. A *record* is a block of data referring to a single entity. The record is usually a structure, and the components of the structure define the components or *fields*. Each entity will have certain properties or *attributes*. The information stored refers to an entity. When a program has *constructed an internal entity* as an aid to its manipulation, compilers make frequent use of such *internal constructs*.

13. A *binary search* requires an average of $\log_2 N - 1$ probes, and a maximum of $\log_2 N + 1$ probes, where N is the length of the list and $\log_2 N = \log N/\log 2$.

14. *Hashing* is a special technique used to store the records of a list in *random order*. The hashing function converts a key into an address or more commonly a *subscript*. The technique is ineffective after the list is about 80% filled. To avoid collisions as much as possible, the hashing function should distribute the stored records throughout the table and should *avoid clustering* them all in one or two areas. A hash table consists of a *prime area* and an *overflow area*.

15. *Bubble sort* is the simplest of the *exchange-sorting routines* to program for lists stored in *internal main memory* of a computer. It is the least efficient because the time required is proportional to N^2, if it is the number of records to be sorted. The *shell sort* is more efficient since the time required is proportional to $N^{1.2}$. The more complex *Quicksort* is the most efficient sorting technique with the corresponding figure $N*\log_2 N$.

The *external sorting* techniques for tape and disk memory are comparatively slow. *Merging* is the technique used for sequential access to files. Given two or more *ordered sequences* of records, we can merge them into a single sequence that is also ordered.*

*For further information, see Neil Graham, *Microprocessor Programming for Computer Hobbyists* (Blue Ridge Summit, Pennsylvania: Tab Books, 1977).

The following tables present (1) a sample PL/M program, (2) PL/M vocabulary, (3) special characters, (4) reserved words, (5) predeclared identifiers, (6) list of instructions for the original Intel 8-bit 8080 microprocessor, and (7) a typical set of macro-assembly programs with mnemonics.

Typical HLL software associated with microcomputer systems.

PL/M PROGRAMMING
PL/M Reserved Words

RESERVED WORD	USE
IF THEN ELSE	conditional tests and alternative execution
DO PROCEDURE INTERRUPT END	statement grouping and procedure definition
DECLARE BYTE ADDRESS LABEL INITIAL DATA LITERALLY BASED	data declarations
GO TO BY GOTO CASE WHILE	unconditional branching and loop control
CALL	procedure call
RETURN	procedure return
HALT	machine stop
ENABLE	interrupt enable
DISABLE	interrupt disable
OR AND XOR NOT	boolean operators
MOD	remainder after division
PLUS	add with carry
MINUS	subtract with borrow
EOF	end of input file (compiler control)

```
PL/M PROGRAMMING
PL/M Special Characters


SYMBOL  NAME               USE

  $     dollar sign        compiler toggles,
                           number and identifier spacer
  =     equal sign         relational test operator,
                           assignment operator
  :=    assign             imbedded assignment operator
  .     dot                address operator
  /     slash              division operator
  /*                       left comment delimiter
  */                       right comment delimiter
  (     left paren         left delimiter of lists,
                           subscripts, and expressions
  )     right paren        right delimiter of lists,
                           subscripts, and expressions
  +     plus               addition operator
  -     minus              subtraction operator
  '     apostrophe         string delimiter
  *     asterisk           multiplication operator
  <     less than          relational test operator
  >     greater than       relational test operator
  <=    less or equal      relational test operator
  >=    greater or equal   relational test operator
  <>    not equal          relational test operator
  :     colon              label delimiter
  ;     semicolon          statement delimiter
  ,     comma              list element delimiter
```

```
PL/M PROGRAMMING
PL/M Pre-declared Identifiers


CARRY
DEC
DOUBLE
HIGH
INPUT
LAST
LENGTH
LOW
MEMORY
OUTPUT
PARITY
ROL
ROR
SCL
SCR
SHL
SHR
SIGN
STACKPTR
TIME
ZERO
```

PL/M PROGRAMMING
ASCII codes

The ASCII (American Standard Code for Information Interchange) was adopted by the American National Standards Institute, Inc. (ANSI) in 1968. The standard itself, as distinct from the summary here presented, is available from ANSI, 1430 Broadway, New York, NY 10018, as USAS X3.4-1968. A previous version of this standard was adopted by the National Bureau of Standards as a Federal Information Processing Standard (FIPS 1). ASCII is a seven-bit code, which we are representing here by a pair of hexadecimal digits.

00	NUL	20	SP	40	@	60	
01	SOH	21	!	41	A	61	a
02	STX	22	"	42	B	62	b
03	ETX	23	#	43	C	63	c
04	EOT	24	$	44	D	64	d
05	ENQ	25	%	45	E	65	e
06	ACK	26	&	46	F	66	f
07	BEL	27	'	47	G	67	g
08	BS	28	(	48	H	68	h
09	HT	29	)	49	I	69	i
0A	LF	2A	*	4A	J	6A	j
0B	VT	2B	+	4B	K	6B	k
0C	FF	2C	,	4C	L	6C	l
0D	CR	2D	-	4D	M	6D	m
0E	SO	2E	.	4E	N	6E	n
0F	SI	2F	/	4F	O	6F	o
10	DLE	30	0	50	P	70	p
11	DC1	31	1	51	Q	71	q
12	DC2	32	2	52	R	72	r
13	DC3	33	3	53	S	73	s
14	DC4	34	4	54	T	74	t
15	NAK	35	5	55	U	75	u
16	SYN	36	6	56	V	76	v
17	ETB	37	7	57	W	77	w
18	CAN	38	8	58	X	78	x
19	EM	39	9	59	Y	79	y
1A	SUB	3A	:	5A	Z	7A	z
1B	ESC	3B	;	5B	[	7B	[(braces)
1C	FS	3C	<	5C	\	7C	! (bar)
1D	GS	3D	=	5D	]	7D	] (braces)
1E	RS	3E	>	5E	^	7E	(tilde)
1F	US	3F	_	5F	_	7F	DEL

PL/M PROGRAMMING
Grammar of the PL/M Language

VOCABULARY

	terminal symbols	nonterminals
1	!	<program>
2	;	<statement list>
3	HALT	<statement>
4	ENABLE	<basic statement>
5	DISABLE	<if statement>
6	IF	<assignment>
7	THEN	<group>
8	ELSE	<procedure definition>
9	DO	<return statement>
10	CASE	<call statement>
11	INTERRUPT	<go to statement>
12	<number>	<declaration statement>
13	PROCEDURE	<label definition>
14	<identifier>	<if clause>
15	)	<true part>
16	(	<expression>
17	,	<group head>
18	END	<ending>
19	:	<step definition>
20	RETURN	<while clause>
21	CALL	<case selector>
22	GO	<variable>
23	TO	<replace>
24	GOTO	<iteration control>
25	DECLARE	<to>
26	LITERALLY	<by>
27	<string>	<while>
28	DATA	<procedure head>
29	BYTE	<procedure name>
30	ADDRESS	<type>
31	LABEL	<parameter list>
32	BASED	<parameter head>
33	INITIAL	<go to>
34	=	<declaration element>
35	:=	<type declaration>
36	OR	<data list>
37	XOR	<data head>
38	AND	<constant>
39	NOT	<identifier specification>
40	<	<bound head>
41	>	<initial list>
42	+	<variable name>
43	-	<identifier list>
44	PLUS	<based variable>
45	MINUS	<initial head>
46	*	<left part>
47	/	<logical expression>
48	MOD	<logical factor>
49	.	<logical secondary>
50	BY	<logical primary>
51	WHILE	<arithmetic expression>
52		<relation>
53		<comp>
54		<term>
55		<primary>
56		<constant head>
57		<subscript head>

PL/M PROGRAMMING
Sort Program

A SORTING PROGRAM

Now we construct an example program using expressions, do-groups, and subscripted variables. Suppose a vector A contains a set of numbers in an arbitrary order, and we wish to sort them into ascending order.

```
/* INITIAL ORDERING OF 'A' IS ARBITRARY */

DECLARE A(10) ADDRESS INITIAL
  (33, 10, 2000, 400, 410, 3, 3, 33, 500, 1999);

                /* BUBBLE SORT */

/* SWITCHED = (BOOLEAN) HAVE WE DONE ANY
              SWITCHING YET THIS SCAN? */
DECLARE (I, SWITCHED) BYTE, TEMP ADDRESS;

SWITCHED = 1;       /* SWITCHED=TRUE MEANS NOT DONE YET */
DO WHILE SWITCHED;

  SWITCHED = 0;           /* BEGIN NEXT SCAN OF A */
  DO I = 0 TO 8;
    IF A(I) > A(I+1) THEN
      DO;                 /* FOUND A PAIR OUT OF ORDER */
      SWITCHED = 1;       /* SET SWITCHED = TRUE */
      TEMP = A(I);        /* SWITCH THEM INTO ORDER */
      A(I) = A(I+1);
      A(I+1) = TEMP;
      END;
  END;
  /* HAVE NOW COMPLETED A SCAN */

END /*WHILE*/;
/* HAVE NOW COMPLETED A SCAN WITH NO SWITCHING */

EOF
```

This program scans the vector A, comparing each adjacent pair of elements. When it finds a pair out of order, it swaps them. It does this repeatedly, until it completes an entire scan of A without having swapped any pair. Then it is done.

The variable SWITCHED keeps track of whether we have done a swap yet, this time through the array. So we zero it each time we start a new scan, and set it each time we do a swap.

PL/M PROGRAMMING
Interrupt Processing

```
DECLARE KEYMAX LITERALLY '72';
DECLARE KEYBUFF (KEYMAX) BYTE, KEYPTR BYTE;
DECLARE OVERFLOW LABEL;

KEYBOARD$PROCESS: PROCEDURE INTERRUPT 3;
   DECLARE CHAR BYTE;
   KEYPTR = KEYPTR+1;
   IF KEYPTR > KEYMAX THEN GO TO OVERFLOW;
   IF (CHAR := INPUT(5)) = '$' THEN RETURN;
   KEYBUFF(KEYPTR) = CHAR;
END KEYBOARD$PROCESS;

KEYPTR = .(KEYBUFF);
ENABLE;
/* MAIN PROGRAM */
...

OVERFLOW:
/* KEYBOARD BUFFER OVERFLOW */
...

EOF
```

In this example, KEYBOARDPROCESS operates on the global variables KEYPTR and KEYBUFF each time RST 3 is executed. If KEYPTR exceeds KEYMAX then control is transferred to the outer block label OVERFLOW and the saved machine state is discarded -- control never returns to the interrupted process. If KEYPTR does not exceed KEYMAX then the value of input port 5 is read and stored into CHAR. If the value of CHAR is ASCII dollar sign, then the interrupt procedure returns immediately to the interrupted process. Otherwise the value of CHAR is placed in the vector KEYBUFF and control returns to the interrupted process.

The 8080 interrupt mechanism is disabled by the occurence of an interrupt, and may be explicitly enabled with an ENABLE statement inside the interrupt procedure. Interrupts are enabled by a return from an interrupt procedure. Caution should be exercised when enabling interrupts inside an interrupt procedure: two activations of the same interrupt procedure must never be in process simultaneously, since there is only one data area for both activations. This exclusion can be accomplished by specifically disabling the interrupt source, or by establishing a priority of interrupts with external circuitry. The safest method is to leave interrupts disabled during all interrupt processing.

Interrupt procedures may contain nested non-interrupt procedures. On completion of a call, these nested procedures return to their point of call inside the interrupt procedure in which they are defined; it is only the RETURN's at the outermost interrupt procedure level which cause the machine state of the interrupted process to be restored.

Instruction set, Intel 8080 microprocessor.

Summary of Processor Instructions
By Alphabetical Order

Mnemonic	Description	Instruction Code[1] D_7	D_6	D_5	D_4	D_3	D_2	D_1	D_0	Clock[2] Cycles
ACI	Add immediate to A with carry	1	1	0	0	1	1	1	0	7
ADC M	Add memory to A with carry	1	0	0	0	1	1	1	0	7
ADC r	Add register to A with carry	1	0	0	0	1	S	S	S	4
ADD M	Add memory to A	1	0	0	0	0	1	0	1	7
ADD r	Add register to A	1	0	0	0	0	S	S	S	4
ADI	Add immediate to A	1	1	0	0	0	1	1	0	7
ANA M	And memory with A	1	0	1	0	0	1	1	0	7
ANA r	And register with A	1	0	1	0	0	S	S	S	4
ANI	And immediate with A	1	1	1	0	0	1	1	0	7
CALL	Call unconditional	1	1	0	0	1	1	0	1	17
CC	Call on carry	1	1	0	1	1	1	0	0	11/17
CM	Call on minus	1	1	1	1	1	1	0	0	11/17
CMA	Compliment A	0	0	1	0	1	1	1	1	4
CMC	Compliment carry	0	0	1	1	1	1	1	1	4
CMP M	Compare memory with A	1	0	1	1	1	1	1	0	7
CMP r	Compare register with A	1	0	1	1	1	S	S	S	4
CNC	Call on no carry	1	1	0	1	0	1	0	0	11/17
CNZ	Call on no zero	1	1	0	0	0	1	0	0	11/17
CP	Call on positive	1	1	1	1	0	1	0	0	11/17
CPE	Call on parity even	1	1	1	0	1	1	0	0	11/17
CPI	Compare immediate with A	1	1	1	1	1	1	1	0	7
CPO	Call on parity odd	1	1	1	0	0	1	0	0	11/17
CZ	Call on zero	1	1	0	0	1	1	0	0	11/17
DAA	Decimal adjust A	0	0	1	0	0	1	1	1	4
DAD B	Add B & C to H & L	0	0	0	0	1	0	0	1	10
DAD D	Add D & E to H & L	0	0	0	1	1	0	0	1	10
DAD H	Add H & L to H & L	0	0	1	0	1	0	0	1	10
DAD SP	Add stack pointer to H & L	0	0	1	1	1	0	0	1	10
DCR M	Decrement memory	0	0	1	1	0	1	0	1	10
DCR r	Decrement register	0	0	D	D	D	1	0	1	5
DCX B	Decrement B & C	0	0	0	0	1	0	1	1	5
DCX D	Decrement D & E	0	0	0	1	1	0	1	1	5
DCX H	Decrement H & L	0	0	1	0	1	0	1	1	5
DCX SP	Decrement stack pointer	0	0	1	1	1	0	1	1	5
DI	Disable Interrupt	1	1	1	1	0	0	1	1	4
EI	Enable Interrupts	1	1	1	1	1	0	1	1	4
HLT	Halt	0	1	1	1	0	1	1	0	7
IN	Input	1	1	0	1	1	0	1	1	10
INR M	Increment memory	0	0	1	1	0	1	0	0	10
INR r	Increment register	0	0	D	D	D	1	0	0	5
INX B	Increment B & C registers	0	0	0	0	0	0	1	1	5
INX D	Increment D & E registers	0	0	0	1	0	0	1	1	5
INX H	Increment H & L registers	0	0	1	0	0	0	1	1	5
INX SP	Increment stack pointer	0	0	1	1	0	0	1	1	5
JC	Jump on carry	1	1	0	1	1	0	1	0	10
JM	Jump on minus	1	1	1	1	1	0	1	0	10
JMP	Jump unconditional	1	1	0	0	0	0	1	1	10
JNC	Jump on no carry	1	1	0	1	0	0	1	0	10
JNZ	Jump on no zero	1	1	0	0	0	0	1	0	10
JP	Jump on positive	1	1	1	1	0	0	1	0	10
JPE	Jump on parity even	1	1	1	0	1	0	1	0	10
JPO	Jump on parity odd	1	1	1	0	0	0	1	0	10
JZ	Jump on zero	1	1	0	0	1	0	1	0	10
LDA	Load A direct	0	0	1	1	1	0	1	0	13
LDAX B	Load A indirect	0	0	0	0	1	0	1	0	7
LDAX D	Load A indirect	0	0	0	1	1	0	1	0	7
LHLD	Load H & L direct	0	0	1	0	1	0	1	0	16
LXI B	Load immediate register Pair B & C	0	0	0	0	0	0	0	1	10
LXI D	Load immediate register Pair D & E	0	0	0	1	0	0	0	1	10
LXI H	Load immediate register Pair H & L	0	0	1	0	0	0	0	1	10
LXI SP	Load immediate stack pointer	0	0	1	1	0	0	0	1	10
MVI M	Move immediate memory	0	0	1	1	0	1	1	0	10
MVI r	Move immediate register	0	0	D	D	D	1	1	0	7
MOV M, r	Move register to memory	0	1	1	1	0	S	S	S	7
MOV r, M	Move memory to register	0	1	D	D	D	1	1	0	7
MOV $_{r1,r2}$	Move register to register	0	1	D	D	D	S	S	S	5
NOP	No operation	0	0	0	0	0	0	0	0	4
ORA M	Or memory with A	1	0	1	1	0	1	1	0	7
ORA r	Or register with A	1	0	1	1	0	S	S	S	4
ORI	Or immediate with A	1	1	1	1	0	1	1	0	7
OUT	Output	1	1	0	1	0	0	1	1	10
PCHL	H & L to program counter	1	1	1	0	1	0	0	1	5
POP B	Pop register pair B & C off stack	1	1	0	0	0	0	0	1	10
POP D	Pop register pair D & E off stack	1	1	0	1	0	0	0	1	10
POP H	Pop register pair H & L off stack	1	1	1	0	0	0	0	1	10
POP PSW	Pop A and Flags off stack	1	1	1	1	0	0	0	1	10
PUSH B	Push register Pair B & C on stack	1	1	0	0	0	1	0	1	11
PUSH D	Push register Pair D & E on stack	1	1	0	1	0	1	0	1	11
PUSH H	Push register Pair H & L on stack	1	1	1	0	0	1	0	1	11
PUSH PSW	Push A and Flags on stack	1	1	1	1	0	1	0	1	11
RAL	Rotate A left through carry	0	0	0	1	0	1	1	1	4
RAR	Rotate A right through carry	0	0	0	1	1	1	1	1	4
RC	Return on carry	1	1	0	1	1	0	0	0	5/11
RET	Return	1	1	0	0	1	0	0	1	10
RLC	Rotate A left	0	0	0	0	0	1	1	1	4
RM	Return on minus	1	1	1	1	1	0	0	0	5/11
RNC	Return on no carry	1	1	0	1	0	0	0	0	5/11
RNZ	Return on no zero	1	1	0	0	0	0	0	0	5/11
RP	Return on positive	1	1	1	1	0	0	0	0	5/11
RPE	Return on parity even	1	1	1	0	1	0	0	0	5/11
RPO	Return on parity odd	1	1	1	0	0	0	0	0	5/11
RRC	Rotate A right	0	0	0	0	1	1	1	1	4
RST	Restart	1	1	A	A	A	1	1	1	11
RZ	Return on zero	1	1	0	0	1	0	0	0	5/11
SBB M	Subtract memory from A with borrow	1	0	0	1	1	1	1	0	7
SBB r	Subtract register from A with borrow	1	0	0	1	1	S	S	S	4
SBI	Subtract immediate from A with borrow	1	1	0	1	1	1	1	0	7
SHLD	Store H & L direct	0	0	1	0	0	0	1	0	16
SPHL	H & L to stack pointer	1	1	1	1	1	0	0	1	5
STA	Store A direct	0	0	1	1	0	0	1	0	13
STAX B	Store A indirect	0	0	0	0	0	0	1	0	7
STAX D	Store A indirect	0	0	0	1	0	0	1	0	7
STC	Set carry	0	0	1	1	0	1	1	1	4
SUB M	Subtract memory from A	1	0	0	1	0	1	1	0	7
SUB r	Subtract register from A	1	0	0	1	0	S	S	S	4
SUI	Subtract immediate from A	1	1	0	1	0	1	1	0	7
XCHG	Exchange D & E, H & L Registers	1	1	1	0	1	0	1	1	4
XRA M	Exclusive Or memory with A	1	0	1	0	1	1	1	0	7
XRA r	Exclusive Or register with A	1	0	1	0	1	S	S	S	4
XRI	Exclusive Or immediate with A	1	1	1	0	1	1	1	0	7
XTHL	Exchange top of stack, H & L	1	1	1	0	0	0	1	1	18

NOTES: 1. DDD or SSS – 000 B – 001 C – 010 D – 011E – 100H – 101L – 110 Memory – 111 A.
2. Two possible cycle times, (5/11) indicate instruction cycles dependent on condition flags.

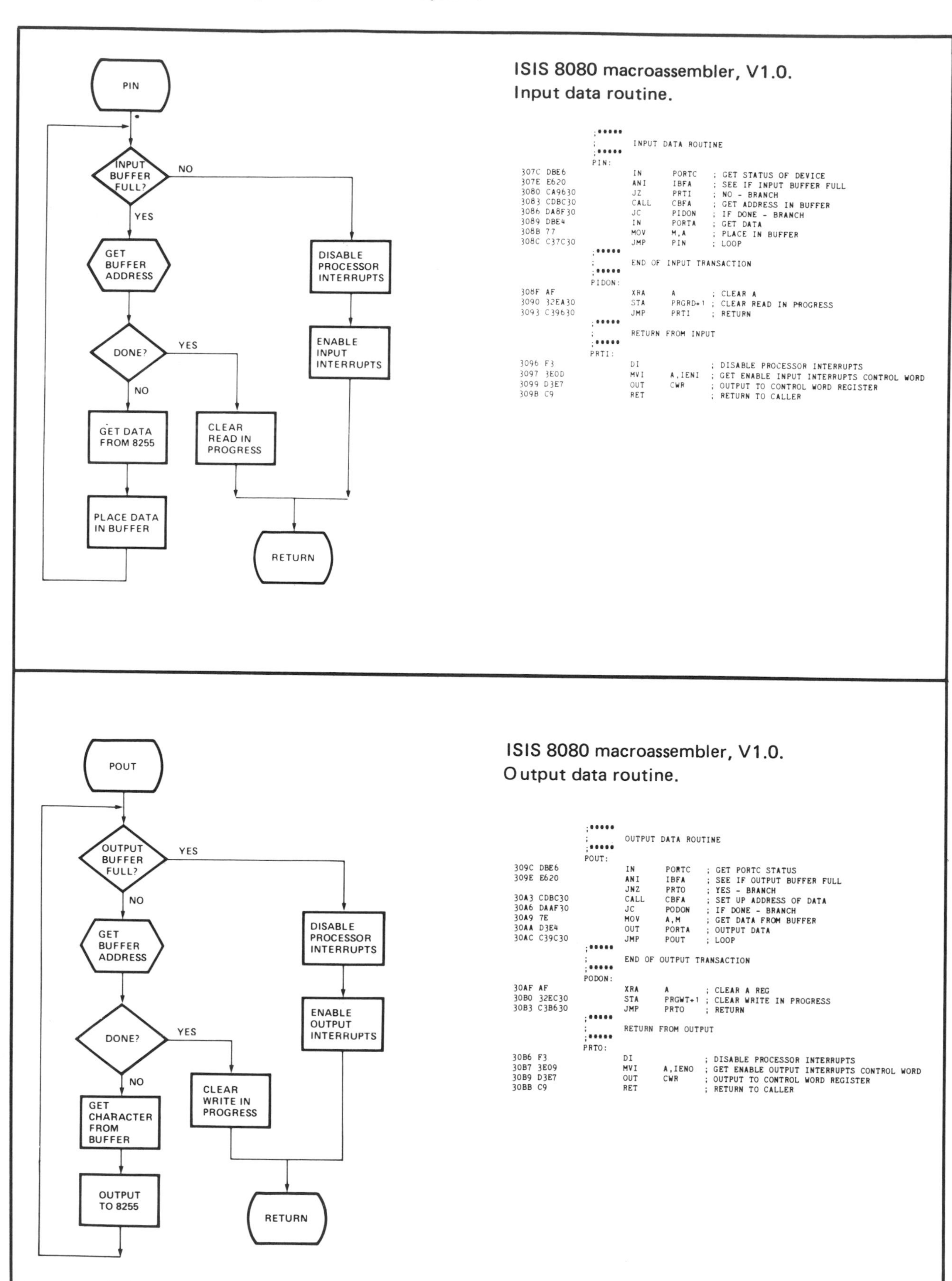

ISIS 8080 macroassembler, V1.0.
Input data routine.

```
               ;*****
               ;        INPUT DATA ROUTINE
               ;*****
               PIN:
307C DBE6               IN      PORTC   ; GET STATUS OF DEVICE
307E E620               ANI     IBFA    ; SEE IF INPUT BUFFER FULL
3080 CA9630             JZ      PRTI    ; NO - BRANCH
3083 CDBC30             CALL    CBFA    ; GET ADDRESS IN BUFFER
3086 DA8F30             JC      PIDON   ; IF DONE - BRANCH
3089 DBE4               IN      PORTA   ; GET DATA
308B 77                 MOV     M,A     ; PLACE IN BUFFER
308C C37C30             JMP     PIN     ; LOOP
               ;*****
               ;        END OF INPUT TRANSACTION
               ;*****
               PIDON:
308F AF                 XRA     A       ; CLEAR A
3090 32EA30             STA     PRGRD+1 ; CLEAR READ IN PROGRESS
3093 C39630             JMP     PRTI    ; RETURN
               ;*****
               ;        RETURN FROM INPUT
               ;*****
               PRTI:
3096 F3                 DI              ; DISABLE PROCESSOR INTERRUPTS
3097 3E0D               MVI     A,IENI  ; GET ENABLE INPUT INTERRUPTS CONTROL WORD
3099 D3E7               OUT     CWR     ; OUTPUT TO CONTROL WORD REGISTER
309B C9                 RET             ; RETURN TO CALLER
```

ISIS 8080 macroassembler, V1.0.
Output data routine.

```
               ;*****
               ;        OUTPUT DATA ROUTINE
               ;*****
               POUT:
309C DBE6               IN      PORTC   ; GET PORTC STATUS
309E E620               ANI     IBFA    ; SEE IF OUTPUT BUFFER FULL
                        JNZ     PRTO    ; YES - BRANCH
30A3 CDBC30             CALL    CBFA    ; SET UP ADDRESS OF DATA
30A6 DAAF30             JC      PODON   ; IF DONE - BRANCH
30A9 7E                 MOV     A,M     ; GET DATA FROM BUFFER
30AA D3E4               OUT     PORTA   ; OUTPUT DATA
30AC C39C30             JMP     POUT    ; LOOP
               ;*****
               ;        END OF OUTPUT TRANSACTION
               ;*****
               PODON:
30AF AF                 XRA     A       ; CLEAR A REG
30B0 32EC30             STA     PRGWT+1 ; CLEAR WRITE IN PROGRESS
30B3 C3B630             JMP     PRTO    ; RETURN
               ;*****
               ;        RETURN FROM OUTPUT
               ;*****
               PRTO:
30B6 F3                 DI              ; DISABLE PROCESSOR INTERRUPTS
30B7 3E09               MVI     A,IENO  ; GET ENABLE OUTPUT INTERRUPTS CONTROL WORD
30B9 D3E7               OUT     CWR     ; OUTPUT TO CONTROL WORD REGISTER
30BB C9                 RET             ; RETURN TO CALLER
```

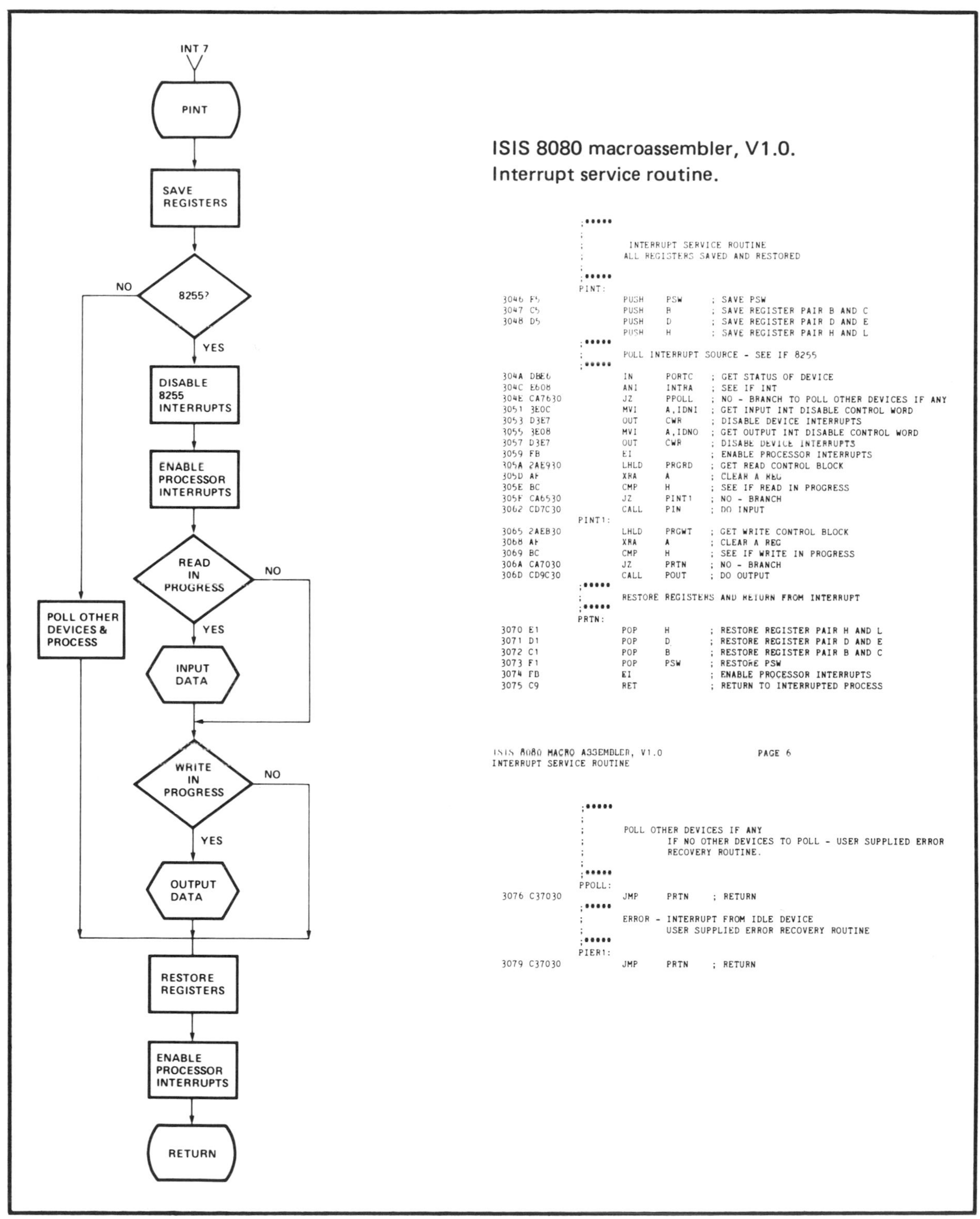

ISIS 8080 macroassembler, V1.0.
Interrupt service routine.

```
                 ;*****
                 ;
                 ;          INTERRUPT SERVICE ROUTINE
                 ;         ALL REGISTERS SAVED AND RESTORED
                 ;
                 ;*****
                 PINT:
3046 F5                    PUSH    PSW      ; SAVE PSW
3047 C5                    PUSH    B        ; SAVE REGISTER PAIR B AND C
3048 D5                    PUSH    D        ; SAVE REGISTER PAIR D AND E
                           PUSH    H        ; SAVE REGISTER PAIR H AND L
                 ;*****
                 ;         POLL INTERRUPT SOURCE - SEE IF 8255
                 ;*****
304A DBE6                  IN      PORTC    ; GET STATUS OF DEVICE
304C E608                  ANI     INTRA    ; SEE IF INT
304E CA7630                JZ      PPOLL    ; NO - BRANCH TO POLL OTHER DEVICES IF ANY
3051 3E0C                  MVI     A,IDNI   ; GET INPUT INT DISABLE CONTROL WORD
3053 D3E7                  OUT     CWR      ; DISABLE DEVICE INTERRUPTS
3055 3E08                  MVI     A,IDNO   ; GET OUTPUT INT DISABLE CONTROL WORD
3057 D3E7                  OUT     CWR      ; DISABLE DEVICE INTERRUPTS
3059 FB                    EI               ; ENABLE PROCESSOR INTERRUPTS
305A 2AE930                LHLD    PRGRD    ; GET READ CONTROL BLOCK
305D AF                    XRA     A        ; CLEAR A REG
305E BC                    CMP     H        ; SEE IF READ IN PROGRESS
305F CA6530                JZ      PINT1    ; NO - BRANCH
3062 CD7C30                CALL    PIN      ; DO INPUT
                 PINT1:
3065 2AEB30                LHLD    PRGWT    ; GET WRITE CONTROL BLOCK
3068 AF                    XRA     A        ; CLEAR A REG
3069 BC                    CMP     H        ; SEE IF WRITE IN PROGRESS
306A CA7030                JZ      PRTN     ; NO - BRANCH
306D CD9C30                CALL    POUT     ; DO OUTPUT
                 ;*****
                 ;         RESTORE REGISTERS AND RETURN FROM INTERRUPT
                 ;*****
                 PRTN:
3070 E1                    POP     H        ; RESTORE REGISTER PAIR H AND L
3071 D1                    POP     D        ; RESTORE REGISTER PAIR D AND E
3072 C1                    POP     B        ; RESTORE REGISTER PAIR B AND C
3073 F1                    POP     PSW      ; RESTORE PSW
3074 FD                    EI               ; ENABLE PROCESSOR INTERRUPTS
3075 C9                    RET              ; RETURN TO INTERRUPTED PROCESS


ISIS 8080 MACRO ASSEMBLER, V1.0                  PAGE 6
INTERRUPT SERVICE ROUTINE


                 ;*****
                 ;
                 ;         POLL OTHER DEVICES IF ANY
                 ;                 IF NO OTHER DEVICES TO POLL - USER SUPPLIED ERROR
                 ;                 RECOVERY ROUTINE.
                 ;
                 ;*****
                 PPOLL:
3076 C37030                JMP     PRTN     ; RETURN
                 ;*****
                 ;         ERROR - INTERRUPT FROM IDLE DEVICE
                 ;                 USER SUPPLIED ERROR RECOVERY ROUTINE
                 ;*****
                 PIER1:
3079 C37030                JMP     PRTN     ; RETURN
```

5.33.4. Microcomputer Programming with High-Level Languages. High-level languages such as PL/1, BASIC, FORTRAN, APL, PL/M, PL/Z, ALGOL, and PASCAL require less coding time, and new complex systems can be built expeditiously in neat, available system modules. However, they all require a comparatively large amount of memory. With low-cost 64-K bit LSI RAM/ROM memory chips and external magnetic bubble/CCD memories around, the possible waste of hardware memory space is not a problem anymore. The number of assembly instructions generated by a single HLL-statement may be as high as 12, and that is the potential range of software costs saved in the case of complex programming tasks. Logical system procedures and thinking are facilitated when one does not have to keep track of what every accumulator and register hold at any instant in any particular computer. English language statements such as DECLARE, IF, THEN, ELSE, DO, BEGIN, REPEAT, CALL, PLUS, BYTE, etc., ease the programmer's task of translating thinking processes the computer can understand via the *compiler* and any subsidiary *interpreter*.

The costly *flow-charting phase* of Assembly language programming in low-cost firmware is mostly eliminated. The costly *debugging* of assembly language instructions in mnemonics is also avoided. The block-structured PASCAL, with its data-handling versatility, is presently becoming popular with *ROM-able* mass-storage. For example, in most microcomputer applications, (HLL) PASCAL or BASIC coding would use one ROM for holding the interpreter (intermediate P-code) and another ROM for the mnemonic instructions in hexadecimal code; in most applications, the comparatively slower access-times are not a problem at all. The standardization of a powerful, universal high-level language such as one of the present PASCAL modified versions would be ideal from the viewpoint of inexpensive software for the multiplicity of microcomputer applications presently under consideration. The high-technology of microelectronics has cut the cost of LSI microcomputer system hardware enormously; given a chance, it will likewise do so during the 1980s for the present time-consuming and expensive software/firmware that make the real-time operating systems "go."

As far as the latest status of high-level languages for microcomputers is concerned, PASCAL (and its modified versions, especially the "C" language of Bell labs), as a block-structured programming language in the style of Algol, enjoys a leading position as a more powerful HLL than BASIC, FORTRAN, and PL/I. PASCAL programs are first *compiled* into the *intermediate P-code*, which in turn gets interpreted on the various types of *target microprocessors*. The native assembly languages of each target unit will be coded on a single ROM interpreter. Then the PASCAL compiler system-package developed at the University of California at San Diego (UCSD) will take over the task for the simple PASCAL program written for any application. Kenneth Bowles of UCSD claims that his PASCAL programs run on the Intel 8080, Zilog Z-80, Motorola 6800, Western Digital MCP-1600, Texas Instruments 9900, Rockwell 6502, and American Microsystems custom-unit microprocessors. The compiler has extensions for strings, disk files, interactive graphics and system programming, text-editor, file-manager, debugger, and utilities. The PASCAL programs are comprised of three levels of "modules": the "root" names the program and specifies the variables. The "functional block" follows as the body of the program. The block is divided into six "leaves." The first four declare the labels, constants, data types, and variables. The fifth names and precedes an actual "procedure" (or function). The sixth "statement" section will consist of the executable code for the named "procedure," *Labels* identify statements for reference. *Constants* equate numbers with names for use throughout the program (example, $\pi = 3.14$). *Data types* can be numerous. *Structured types* can be defined to include arrays, records, sets, and files (see *Definitions*, p. 409). Each variable named will be followed by its type; procedures can be written within procedures; and statements for each must be preceded with "BEGIN" and terminated with "END." *Operators* are defined for multiply, divide, add, subtract, logical, and relational. *Numerous control statements* are allowed. PASCAL is similar to other *block-structured programming languages*, but it is more powerful and elegant; it is potentially international for standardization, since the originator of this language, Nicklaus Wirth of Switzerland, named it PASCAL to honor the seventeenth-century French mathematician. For further information on PASCAL, see Section 5-13.14. We can see on the horizon the eventual demise of the Babel of computers—that is, if the international body politic is willing. Where secrecy is mandatory, *cryptography* is easily accomplished with the aid of protective ROM-software that could be translated to the standard HLL.

The *major advantages of a standardized HLL* include:

1. Assembly language coding produces a significant number of errors, since it is difficult to keep track of what is in each index register. The problem of debugging is minimized by the use of a high-level language.

2. The modular software engineering in a block-structured HLL improves program reliability and reduces costs.

3. The programmer feels that the compiler design itself executes the flowcharts.

4. After some experience with it, the HLL firmware enables the checkup of a product design much sooner than does other firmware.

5. Code written in a HLL is *portable*, since it is not machine-dependent like the assembly language. This portability is cost-effective since utility routines such as stack and queue processing, floating-point arithmetic, and number-conversion packaging are common in most applications.

6. Maintenance in firmware is simplified since programming in HLL is simpler to write. In practice, about 25% development time is saved.

7. The programmer becomes familiar with the efficient code-constructs when the HLL compiler furnishes, with its sophisticated macroinstruction capability of benchmark routines and subroutines, a listing of the assembly-language object code along with the source HLL-code. This kind of learning on the part of the HLL programmer (by way of CRT-presentation in particular) substantiates the efficiency realized to an extent of at least 30% in a short time.

8. The HLLs presently require expensive disk-based operating systems for their use. The latest and the next generation of extra high-capacity bubble-memory and ROMs will hold complex interpreter and compiler HLL software packages at the economy price levels of mass-produced VLSI chips—and that at error rates, orders of magnitude better than disk.

6 Conclusion

The combined continuous and sampled-data digital control systems in the quadruplex color videotape recorder have furnished a comprehensive signal processing medium to illustrate the impracticality of following a theoretical approach for expeditiously designing a modern complex commercial digital control system. As modern digital automation steps in to successfully implement these systems in practically every field of the industry and research, complex systems of this category with a multiplicity of phase-lock loops are frequently encountered. We have seen that the three multirate, multiloop, nonlinear, interacting sampled-data feedback control systems used in the modern high-quality television broadcast color videotape recorder—viz., (1) the high-precision headwheel sampled-data feedback control system, (2) the multispeed capstan sampled-data feedback control system, and (3) the vacuum-guide sampled-data feedback position-control system—make an interesting, sophisticated control-engineering example. The development and design of this successful, highly complex industrial application actually furnishes a meaningful replica to a considerable number of techniques encountered in general control theory, which, on its academic theoretical base, has been expanding in several directions. Chapter 1 summarized briefly the various theoretical approaches presently available.

As observed in Chapter 4, since the interacting control systems are intricately involved with the magnetics and the signal-processing electronics of the videotape recorder, and hence profoundly complex from the viewpoint of generating satisfactory transfer functions for direct application of available design techniques, they were originally developed as a commercial proposition, loop-by-loop, and facility-by-facility on a heuristic measure-and-optimization basis according to the basic principles of linear and nonlinear control systems. It was anticipated that this particular control system-complex in an up-to-date color-telecasting facility in the television industry would stimulate further theoretical interest on the subject, because it is an application as significant as the high-budget Defense radar in sampled-data control systems. For instance, the automatic compression of the state-space limit-cycle to ± 2.5 nsec in the multidimensional high-precision headwheel nonlinear feedback control system, on a loop-by-loop adaptive control basis at several rates of sampling, is an outstanding contribution as an application to modern control theory. The above-specified figure-of-merit or performance-index of this adaptive, interacting nonlinear control system is a prerequisite for the faithful reproduction of color from videotape during television broadcasts.

The *heuristic measure-and-optimize* technique, as used for the step-by-step development of these control systems in the laboratory, is considered justifiable for the following reasons. The development and design of the associated digital controllers is not akin to the conventional *trial-and-error* technique employed in closed-loop continuous control systems for the analysis and synthesis of suitable compensation networks of appropriate bandwidth. The terms *measurement* and *optimization* are more in line with the developments in modern control and optimal theory, where the measured variables through the various stages of a system, and the optimization of the system parameters and coefficients for the desired performance-index share a prominent role in state-space correlation techniques to satisfy some integral squared-error criterion, etc. The approach is partially oriented in the direction of modern (Kalman) digital filter theory—on a corresponding basis.

The heuristic measure-and-optimize techniques in the laboratory are explained by citing several PLL examples in Chapter 2. They illustrated a few actual development and design problems, and outlined how one proceeds in the laboratory to tackle such problems with the aid of specialized measuring equipment and the implied theoretical knowledge. These examples will therefore provide a channel of communication between the control engineer who is primarily expected to meet the program schedules in the laboratory, and the student of control theory. This practical successful approach is a versatile performance testing procedure (as a parallel to debugging of software in modern computer-control applications), and this specialized effort obviously depends not only on the state-of-the-art with the latest available components and instruments in the laboratory but also on the advances in the applicable theory—although it is difficult to closely interrelate the two.

Some insight into the heuristic measure-and-optimize techniques employed in the design of these complex combined analog and digital control systems is provided by comprehensive theoretical analyses of simple mathematical models in several outstanding cases. Some theoretical examples in this connection along different concepts merely provide an exposition of the corresponding theoretical aspects. So, as a whole, Chapter 3 provided a channel of communication between control theory in general and the circuit development engineers in the laboratory for a comprehensive understanding on their part. And, Chapter 4, which described the development and design of a modern complex digital control system, is an engineering-parallel to the various applications of theory exposed in Chapter 3. As an incidental exposition of the system environment, the emphasis placed on the subjects of international color television and magnetic tape recording is considered appropriate for a clearer understanding of the control systems involved and their operational functions.

Prior to the advent of the presently available, practical low-budget mini- and micro-controller or microcomputer systems, the scope of this book would have been necessarily limited to the theoretical intepretation of some of the more prominent features and peculiarities in the combined linear/nonlinear analog-cum-digital control system-complex. The effort would have been mainly the implementation of a systems engineering application in *applied research* as a parallel to the more common theoretical treatment on an academic base. The underlying idea of such an approach is that, for a class of complex interacting control systems of the character described, involving tens or hundreds of system variables:

1. it is neither practical nor absolutely necessary (before the development project is commended) to dwell into a rigorous and extraordinary theoretical effort by devising approximate mathematical models and develop a host of inadmissible or non-applicable theoretical solutions of *necessary and sufficient conditions* for the overall system stability
2. on the other hand, it is advantageous and time-saving, from the point of view of practical digital-controller systems development, that the circuit development engineer is fully conversant information-wise (if not application-wise) with the possible theoretical implications of the circuit techniques or novelties he or she attempts to accomplish the final evolution of the stable control system-complex. In complex cases, theory, as always, can at best provide the necessary guidelines and hence be complementary to the state-of-the-art developments in practice. With this basic philosophy, the author has entertained an optimistic orientation that adaptive automatic control of a class of highly complex nonlinear sampled-data control systems is entirely feasible on a stage-by-stage basis by applying heuristic measure-and-optimize techniques in the laboratory. This is despite the fact that the excessive number of control variables and the uncertainties, such as temperature and humidity effects, wear-and-tear of mechanical parts, and variations in pulse-widths and delay-elements, incidental in the former digital controllers of the individual phase-lock loops in the multiloop system far surpass the bounds and scope of the theoretical developments in control systems, and the design applications of analysis and synthesis to simple independent mathematical models.

At the present time, the activity in both theory and practice in the field of control is astounding, and the outlook is broad-based and gratifying because there is a prompt and concerted effort to coordinate the progress of theory and practice. From a historical point of view, in most cases the creative aspects of discovery and invention do exhibit a lead on the theory, because of the reality of instantaneous observation as direct cause and effect. However, the extent of this lead is a matter of opinion in some cases. On the whole, it is perhaps fair to say that technological and theoretical advances have, all along, contributed to one another. For instance, using the latest methods of invariance and the state-space optimal control, as briefly pointed out in Chapters 1 and 3, complex multidimensional interacting discrete control systems of the character of those in the quadruplex color videotape recorder can be studied as specialized research programs by reducing them to a set of unconnected one-dimensional control processes. This could be a step in the right direction. And surprisingly, this is exactly what has been happening since the latest mini or micro digital computers and their sophisticated highly-expensive and time-consuming software have taken over the individual tasks of the various loops in a pipeline fashion.

As a result of this new development, Chapter 5 was exclusively devoted to the latest status and application of the single-chip microprocessors and microcomputer systems, because they are all set to play a major role in complex digital control systems. The minicomputer has been in use during the last decade in fairly high-budget control system applications in industrial processing. But, the latest revolutionary microcomputer has radically altered the situation. Quite independent of the actual limitations of the budget allocations available, the large-scale integrated (LSI) microcontrollers and microcomputers, using mostly predesigned readily-available solid-state software or firmware, will gradually take over the tasks of control and digital automation in the near future. As a direct consequence, (Kalman) digital filters, both recursive and nonrecursive, will fit in exactly with the digital format as a complementary development in modern complex digital control systems. First the transducer, then the LSI analog-to-digital converter appear, and then the regular signal processing is all accomplished in the digital format by the microcomputer and its firmware to control the final power-plant, if any, on an automatic basis. In some cases, the actuating motor itself will act as the decoding device, if an ac carrier-servo is finally involved in the system; otherwise, LSI digital-to-analog converters will supplement the control function where necessary. The digital filter is an essential tool since all data are considered as corrupted by noise.

During the last decade, in the area of signal processing, it developed that one could manipulate waveforms digitally in ways that would be totally impractical with the alternative continuous representations. The advantage is obvious; when a digitized (sampled) waveform is stored in a digital memory (for example, a high-capacity solid-state LSI random-access RAM memory), one gains freedom from the constraints of time. The former delay-lines have achieved this function in continuous control systems, but they are expensive, cumbersome, and difficult-to-design for linear-phase operation. The digital shift-register takes over the task of the delay-element in digital signal-processing. Once the transient waveform is buffered into a computer RAM memory via an analog-to-digital converter or a quantizer of the requisite digital format, the waveform-data may be considered to have any associated time-base as desired for a certain function. For example, it can be restored to the continuous form on a new time-base, even a time-reversal. The flexibility in digital format is simply amazing. Digital filters are used in data communications, picture processing, biomedical applications, and vibration analysis.

From the spectrum-analysis point of view, digital filters, as the original nonrecursive type, had a limitation of prohibitive computing time. So, signal-processing involving spectral-analysis in the frequency domain required a bank of appropriate analog filters to make it practical. But recursive filtering techniques are the order of the day, since the Cooley-Tukey FFT (*fast Fourier transform* technique) was developed in 1965; it is presently a valuable algebraic tool to perform signal processing on waveforms in either the time or frequency domain, something impractical in continuous systems. Thus, the digital computer can be directly interfaced with new digital processing components and instruments such as digital filters, signal generators, spectrum analyzers, correlation measurements, and displays. That is, the digital computer in its new low-budget, low power-consuming, and high-speed micro LSI-form is presently a universal tool in practically every digital control application. On the theoretical side, several efficient algorithms are readily available for filtering and spectral analysis, and for representing waveforms by discrete time-series and modifying them in both time and frequency domains. Digital filtering actually became an important tool in *computer simulation* of speech-compression techniques and computation of maximum and minimum phase components in *pattern recognition.* Linear-phase digital filters are presently available as off-the-shelf MSI hardware components for filtering of Doppler-shifted radar returns, and for maximally flat Butterworth and equiripple Chebyshov filters for low-pass, high-pass, band-pass, and band-reject. (*Recursive realization* of a digital filter: the current filter output is obtained explicitly in terms of past filter outputs, *as well as* in terms of past and present filter inputs. *Nonrecursive realization* of a digital filter: the current filter output is obtained *explicitly* in terms of only past and present inputs; they have *finite duration impulse response* for linear-phase and stability [FIR].) However, FIR filters are suitable for nonrecursive (FFT) and recursive (step-recovery diode and YIG-tuned COMB) filters.

At one time in the past, prior to the advent of the minicomputer, there was some initial apprehension that theory and practice pertinent to complex sampled-data control systems were drifting apart; but the problem has gradually phased out due to the timely arrival of the LSI microcomputer systems and digital filters, and their applicability to complex digital control systems for practical checkup of complex algorithms via especially interactive-graphics with CRT-display.

Appendix A
Digital Frequency Synthesis of Local Oscillator in Super Heterodyne Radio/Radar/Television Receiver

A-1 DIGITAL FREQUENCY SYNTHESIZERS

Digital Frequency Synthesizers are mainly used for Test Signal Generators and frequency control of transmitters and receivers in aviation, military, broadcast, consumer-product (CB-radio), and international satellite communications applications. The present technology of the digital frequency synthesizers enables (1) fine accuracy in local-oscillators to set a large spectrum of *contiguous frequencies with small-increments* by using a single reference-crystal and (2) ease and simplicity with minimal errors unlike the high-cost application of individual crystal-control of transmitters and receivers.

A-2 DIRECT OR ANALOG FREQUENCY SYNTHESIS

Analog mixing techniques involve sum and difference heterodyne frequencies and original frequencies. Frequency selective circuits are required to select the sum or difference frequencies. *Mixing and selecting* as often as possible to get the required number of output frequencies from crystal-oscillator sources is the criterion. Continuous frequency-coverage is costly and bulky—thousands of (oven-controlled) crystals for small increments (it is impractical!). But using two mixers and three crystal-oscillators, by starting with 10 alternative crystals per oscillator, is perhaps practical from the viewpoint of simplification in switching, and reduction in size and weight several times. However, 30 crystals in a high-precision signal generator are too costly. But the technique of harmonic selection, using the basic oscillators, varicap harmonic-generators, and several frequency-selective circuits, actually involves only three basic stable-oscillators in practical standard test-instruments. The oscillator-frequency is that of the step-size being selected. For example, 10-kHz frequency increments will require 10-kHz basic oscillators. Since a bulky 10-kHz crystal-oscillator is not convenient, a 1- to 10-MHz master oscillator-frequency is divided into 100-kHz, 10-kHz, and lower basic frequencies. Thus, only one thermostat-controlled stable-oscillator, packaged in an oven with a heater-element, may be used as the source to derive the other requisite frequencies.

Disadvantages. Selection of frequencies to be mixed is a major problem, since we have to select by means of filters only the desired frequency with some conversion-loss out of several frequency-components, such as sum, difference, and odd- and even-harmonic spurs, appearing at the mixer output. If the filtering is inadequate for the requisite signal-to-noise ratio, we face high-level spurious frequency-components that produce problems in the overall system using the frequency-synthesizer. We could minimize this problem by careful selection of the mixer frequencies, appropriate shielding, and properly designed filters. At times, the filters may require tunability to select the desired frequency by switching, which further introduces mechanical switching problems. In critical applications, a *voltage-controlled oscillator* (VCO) may be required to continuously track the output of the frequency-synthesizer by means of a phase-lock loop (PLL). The well-designed PLL incidentally facilitates the effective isolation of the output from phase-noise and spurs in the synthesizer. Thus, the direct synthesizer does have many applications at this time, since most of the more difficult problems are minimized to allow for reasonably compact equipment size.

A-3 DIGITAL FREQUENCY SYNTHESIZERS

1. *Digital frequency synthesizers* have all the desirable properties of the direct synthesis but without most of its problems. Use of digital integrated-circuits, and LSI (large-scale-integrated) chips makes it simpler to design, produce, and test at lower cost and higher performance. The PLL, with its *voltage controlled oscillator*, phase-detector, low-pass filter, and a single heterodyne down-counter in the feedback path, is the basis of the digital frequency-synthesis. A varicap diode across a resonator converts the changing-current in the input common-base circuit of a specially designed transistor-multivibrator, and changes the frequency with the variation in dc voltage. The phase-detector compares the appropriately divided VCO frequency from a programmable-divider with the reference-frequency to develop an error-voltage that is proportional to the frequency/phase difference. In turn, it is low-pass filtered, and dc amplified for VCO control. The LP filter characteristics used adjust the transient properties of the PLL loop.

The PLL tracks an incoming signal for phase- and frequency-changes to provide a reverse-polarity error. After a minimal delay determined by the time-constant of the LP filter, a constant phase-difference is established in order to allow the phase-detector to provide the VCO with the required voltage that in turn holds the VCO at the desired frequency.

In general, the PLLs are employed (1) to detect signals at a detectable level, if they are submerged in noise and (2) as wide-band linear FM-detectors. In these linear-circuit applications, LSI-PLLs are available with external provision for the LP filter and linearity-control. The entire PLL is integrated on a single silicon substrate as a chip, and it is a production component at low cost. In time, the PLL-chips will attain the same popularity as the linear operational amplifier (opamp) ICs. In future, LSI-PLL chips may be available for still higher frequencies in the GHz bands by using the latest MESFET (metal semiconductor field-effect transistor) technology.

2. *Programmable-Divider* is the most important section in the digital-synthesizers.

Division of the input frequency F_{in} by N may be manually selected by thumb-wheel switches that activate digital integrated-circuits. A number of techniques are available for the up/down converters used for this purpose.

Basically, a variable-modulo counter moves through N states in a closed loop. If an output pulse is generated for one of these states, it will require N input pulses to produce an output pulse. This results in $F_{in} \div N$. It consists of a shift-register (SR) counter (or standard flip-flop binary-counter) that is so designed by way of feedback that it could jump certain states depending upon the state of the control lines.

Circuits to explain the principle

a. In Fig. A-1a, the input clocks a BCD up-counter. When the counter reaches the number programmed on the control lines, the comparator produces an output that resets the counter to zero. The reset-pulse is used as the divided output-pulse, producing ÷N.

b. In Fig. A-1b, the clock-pulse clocks the BCD down-counter and causes it to count down to zero. When the count reaches zero, the NOR gate detects that state and allows the preset-input (entering the BCD control-number) into the counter. The output of the divider is taken from the NOR-gate for ÷N.

In practice, in the case of the shift-register counters encountered, the input control requires a proper coding of the counts with respect to the actual frequency synthesized.

A-4 BASIC DIGITAL FREQUENCY SYNTHESIZER TECHNIQUES

1. The simplest technique requires a PLL as shown in Fig. A-2 and a programmable-divider. The feedback-loop counts to make $F_{ref} = F_{out}/N$, to make the VCO-out = N.(F_{ref}). By changing N, the F_{out} is changed. If the step or increment frequency is F_{ref}, then

$$(N + 1) = F_{out} + F_{ref}$$

Since the F_{step} is fairly low and stable oscillators at low frequencies are difficult to design for extremely fine accuracy (as required for the final-tolerance of the high

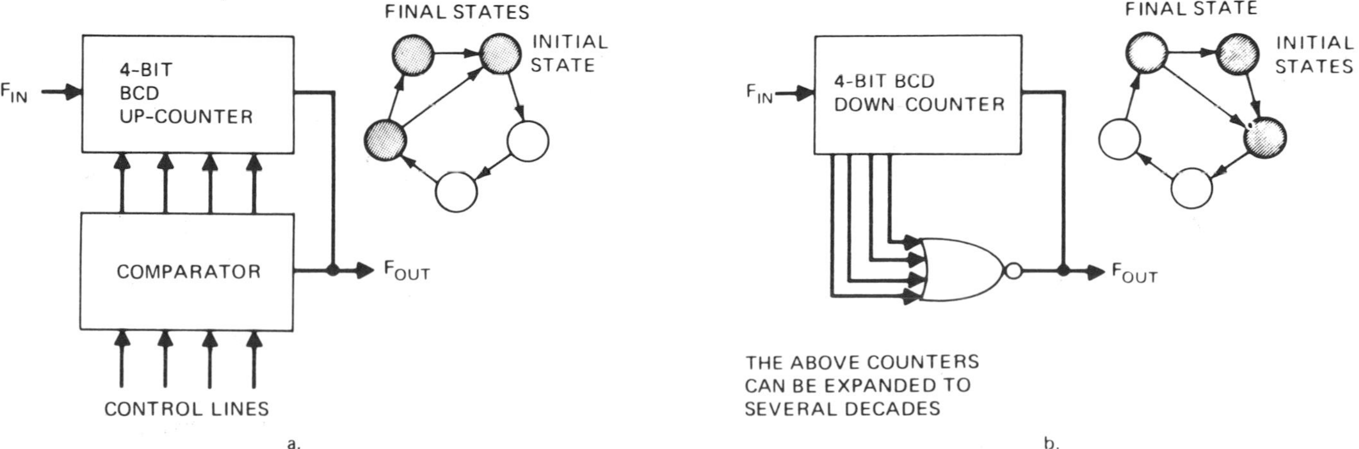

Fig. A-1. Initial and final states of a binary-coded-decimal up-and-down converter.

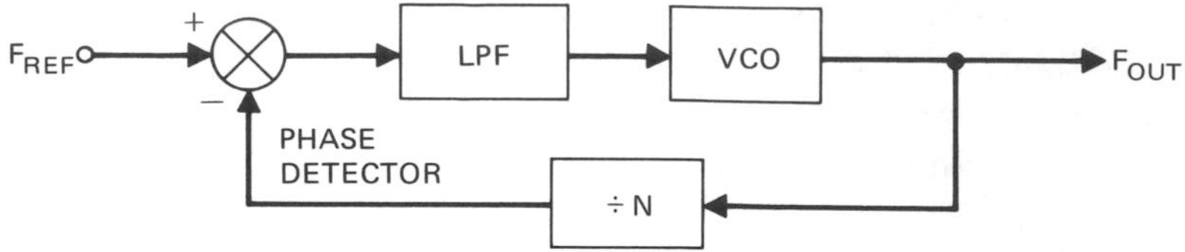

Fig. A-2. Basic digital feedback loop.

output-frequency), the F_{ref} is scaled-down from some higher frequency crystal-oscillator such as 1-MHz for good stability and frequency accuracy.

Example:

$$2.01 \text{ MHz} = N.F_{ref}$$

$$N = 201$$

$$F_{out} = 201 \times 10 \text{ kHz} = 2.01 \text{ MHz}$$

2. *Heterodyne down-conversion for high-frequency bands.* Until 1972, programmable-dividers were limited to 50 MHz due to the slower propagation-time of the former integrated-circuits. (This limitation has been moving up into the GHz bands as faster lines of logic are developed in medium-scale integration. Presently, it is extended to 10 GHz by using the latest MESFET MSI-technology.)

Example: Synthesis of 100–125 MHz for F_{step} = 10kHz. The technique used in (1) in the previous example is not suitable for frequencies above 50 MHz, because the programmable-divider N fails to count above 50 MHz. The *heterodyne down-conversion* shown in Fig. A-4 is the appropriate technique. The F_{out} is mixed or heterodyne down-converted to produce an intermediate frequency in the frequency-range of the IC-logic used.

$$F_{out} = 100.1 \text{ MHz; Mixer output} = (100.1\text{–}90) \text{ MHz} = 10.1 \text{ MHz}$$

$$N = 1010 \text{ for a divider output-frequency} = 10.1 \text{ MHz}/1010 = 10 \text{ kHz}$$

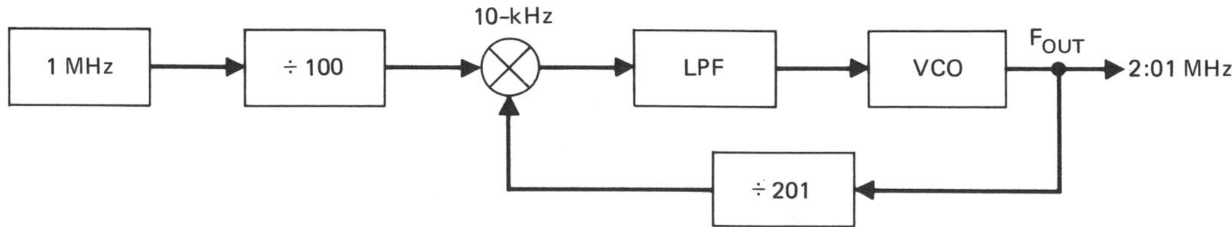

Fig. A-3. Typical digital phase-lock loop.

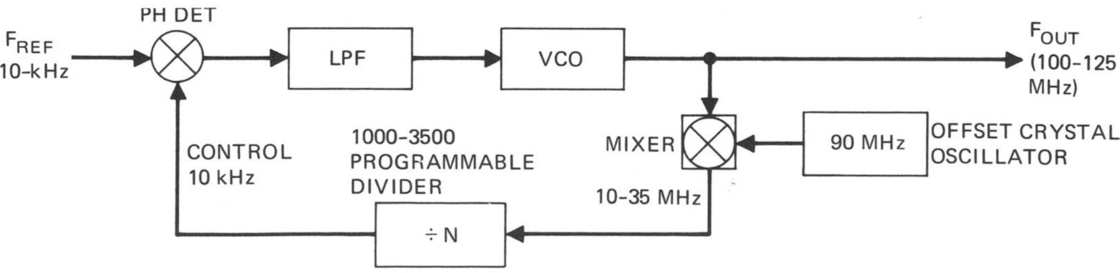

Fig. A-4. Phase-lock loop, Heterodyne down-conversion.

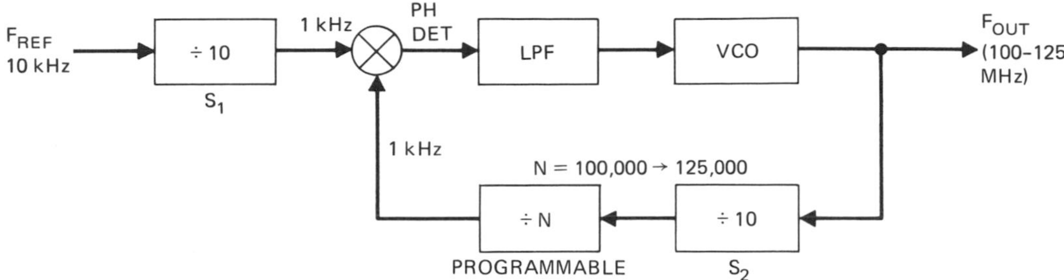

Fig. A-5. Phase-lock loop, scaling technique.

3. *Scaling Technique for high-frequency bands.* Scaling-down F_{ref} results in too large a time-constant for the low-pass filter to maintain the stability of the PLL. However, this procedure has the advantage of lowering the F_{step} to as low as 1 kHz to allow a F_{ref} of 1 kHz in a stable phase-lock loop. The effectively longer lockup time may be objectionable in some applications, but not necessarily so in some other applications. The scaler S_2 used in the feedback path is practical, because fixed frequency-dividers with simpler configuration are feasible up to as high as 600 MHz with the latest advances in semiconductor MSI technology using high-speed logic.

4. *Pulse-Swallowing Technique for Minimizing F_{step}* minimizes the high-speed logic by prescaling the programmable-divider's input, without any need of reducing F_{ref} to maintain the desired lower channel-spacing, such as 10 kHz in the high-frequency band under consideration.

The technique basically involves a special-purpose *modulo-2 prescaler* controlled by a "swallow-counter" M. The modulo-2 prescaler is not as easily implemented as a regular single-modulo prescaler, but it is more effective and simple as compared to a fully programmable high-speed counter. (The modulo-2 integrated-circuit chip, in conjunction with a special-purpose counter control-logic IC, makes the implementation fairly simple.)

At the start of the count-cycle, the modulo-2 prescaler is set to divide by 11. After M pulses are counted by the pulse-swallower M, the prescaler is shifted by means of an enable-pulse "High" to divide in ÷10 mode. It remains in this mode until ÷N down-counter is reset, and a "zero-detect" indication is given. Then the above dual-count cycle repeats.

The overall divide-ratio from the input of the prescaler to the input of the phase-detector is given by:

$$\text{Overall Divide-Ratio, } R = 11M + 10(N - M)$$

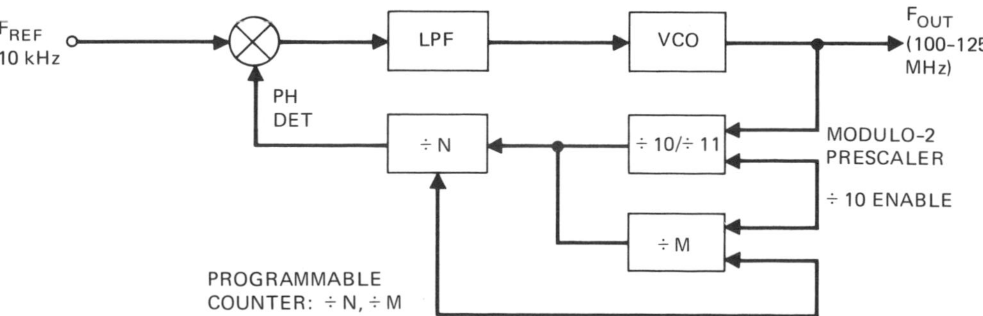

Fig. A-6. Phase-lock loop, pulse-swallowing technique.

Example: 100–125 MHz, 10-kHz spacing.
For 112.35 MHz:

$$R = 112.35\text{ MHz} \div 10\text{ kHz} = 11235$$

then, M = 5, and N = 1123 M $\longrightarrow$ LSD (least significant digit)

$$R = 11M + 10(N - M)$$

N $\longrightarrow$ MSD (most significant digit)

$$= (11 \times 5) + 10(1123 - 5) = 11235.$$

This simple count-procedure is not feasible for the overall Divide-Ratio if the simpler counter control-logic IC implementation is adapted. This specific implementation requires the facility of a special programming-code for the proper correspondence between the frequency and the count-digits.

The low-pass filter at the output of the phase-detector bypasses F_{ref} and the harmonic-content, so that it may not frequency-modulate the VCO and hence produce a spurious frequency-content (spurs) in the output. If F_{ref} is low, the slew-rate of the loop will be low, causing longer loop-lockup time when the frequency is changed. Incidentally, the filter characteristics will be similar to those used in the usual digital feedback control systems.

5. *Digital-to-Analog Converters:* If the VCO requires a relatively large voltage to cover the desired frequency-range, and if the phase-detector output is not adequate, a D/A converter may be used to "presteer" the VCO to the approximate F_{out}, by using frequency-programmable switches for the most significant digits as input to the D/A converter. The D/A converter need not be accurate because it is needed to adjust the VCO-frequency within the lockup range of the PLL.

6. *Receiver Offset.* In the case of the synthesizers used in transceiver applications, the synthesizer can be designed to produce the local-oscillator (LO) injection-frequency by changing the "divide-ratio" of the programmable-divider as required. For example, if the F_{ref} = 10 kHz and IF = 10.7 MHz, increasing the divide-ratio by (10.7 MHz/1010 kHz =) 1070 will result in a synthesizer output that is 10.7 MHz higher than the transmit-frequency. This specific procedure of changing the divide-ratio makes use of the principle explained in Section A-3 (2) for feedback to two initial states.

By suitable logic, the two final states differing by the desired offset in the divide-ratio can be decoded; using one would result in $F_{transmit}$, whereas the other would result in the LO injection-frequency. Switching between the two final states is a fairly simple proposition in the case of digital ICs. For dividers using the principle of feedback from two final states to the initial state, the circuit design manipulation requires the insertion of an adder/subtractor (a single LSI-chip) between the BCD control lines and the comparator. The adder/subtractor is merely disabled during transmit, and enabled during receive to allow the synthesizer to produce the LO frequency.

7. *Synthesizer for FM transceiver* application. The synthesizer shown in Fig. A-7 employs the "pulse-swallowing" technique, along with a (÷2) prescaler to produce frequency channels incremented at 10-kHz steps by using a 5-kHz F_{ref}. It has the LO offset provision and the Transmit/receive switching circuits.

The Frequency-Comparator circuit checks the F_{ref} using the output of the program-divider to produce an output when they are equal; only then, does the gate circuit open in the buffer to prevent the off-frequency operation of the FM transmitter.

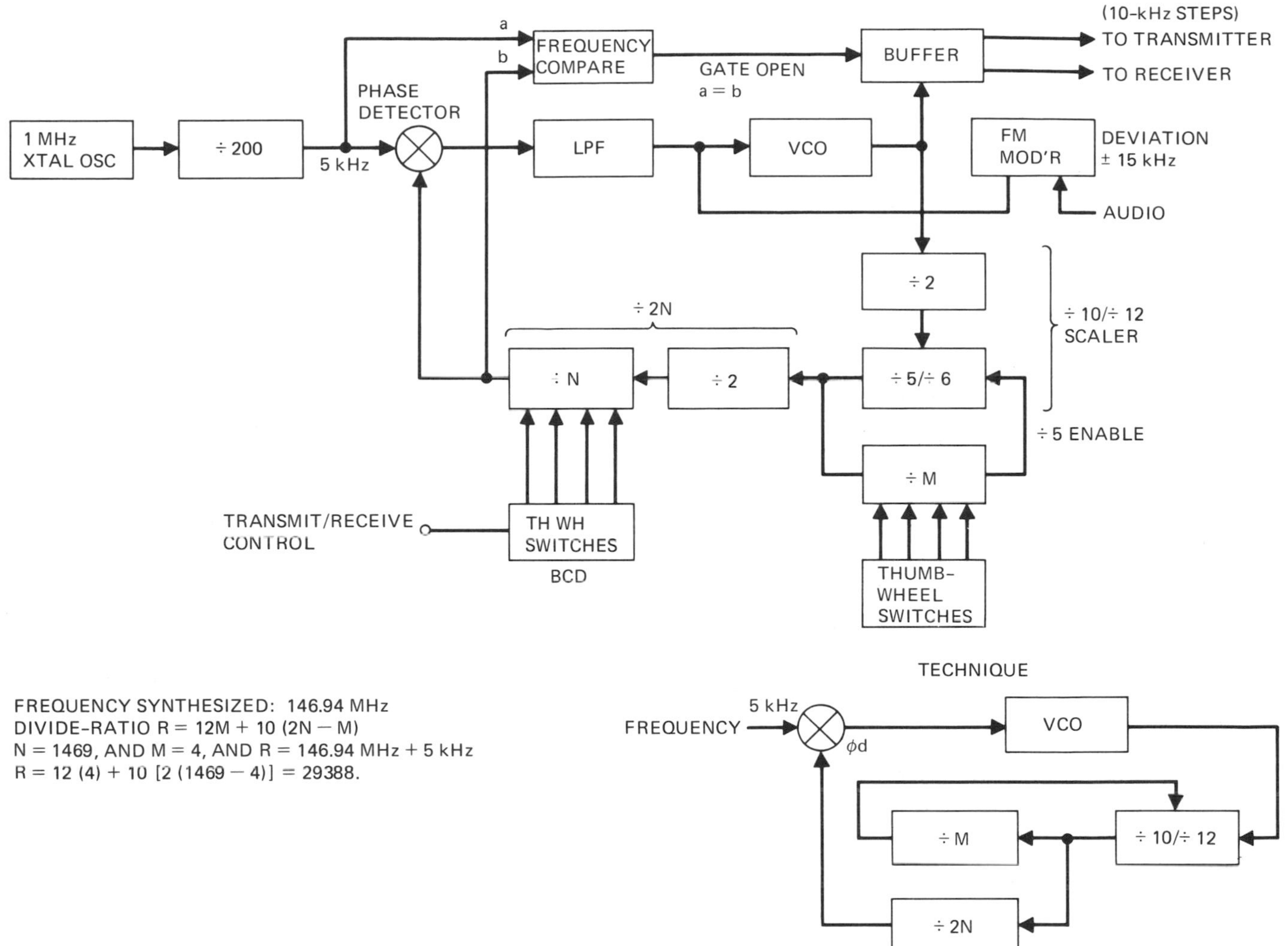

Fig. A-7. Typical synthesizer used in a transceiver application. (*Courtesy of Motorola*)

The programmable-divider configuration allows the direct encoding of the divider by means of the BCD-encoded thumbwheel switches. This prevents cumbersome switching manipulations and gives a direct read-out in frequency rather than just a channel number. The thumbwheel switches are standard BCD-encoded off-the-shelf components. The operating frequency is directly read out from the switches.

The FM-Modulator adds a small amount of the audio information (AF) to the VCO control-line during the Transmit-mode to obtain the FM output-frequency. Since FM-modulation of the VCO is used, it is essential that the low-pass (LP) filter cutoff frequency be below the audio-band (300 Hz to 3 kHz) to prevent the loop from attempting to cancel the modulation as drift or off-frequency operation.

Changing M by 1 results in a divide-ratio change (R) by 2, since F_{ref} is 5 kHz, and channel separation is 10 kHz. A change of one in N results in a change of 20 for R for the same reason.

The synthesizer output in the Receive-mode is 10.7 MHz below $F_{transmit}$. Then this output is used as the LO to produce the intermediate frequency IF of 10.7 MHz. Since F_{ref} is 5 kHz, R is changed by 10.7 MHz ÷ 5 kHz = 2140.

The desired frequency-change results in a change of $N = \frac{2140}{20} = 107$, and N is subtracted by 107 during the Receive-mode.

Example: If the operating Frequency is 146.94 MHz

$$N = 1469, M = 4, \text{Receive-Mode}, N = 1469 - 107 = 1362$$

$$R = 12(4) + 10[2(1362) - 4] = 27{,}248$$

$$\text{Receive local-oscillator frequency} = 136.24 \text{ MHz}$$

The VCO will have a voltage-range from 0.9 to 2.3 V available to produce the frequency-range in this application from 133.8 to 148 MHz, to cover the lowest LO to the highest $F_{transmit}$. Since the ratio between the highest and the lowest frequency output is only 1.1 to 1, it is fairly simple to design a VCO. Its F_{out}/F_{IN} characteristic must be as linear as possible for consistent modulation capability through the band. The frequency-stability of the VCO is reliable without any long-term drift, since the phase-lock loop takes care of this aspect.

8. *High-speed scalers.* The MSI emitter-coupled logic (ECL) provides the highest speed, and it is used in the scalers to divide at frequencies as high as 175 MHz in the case of the low-cost Motorola integrated-circuits. Matched lines at 50-ohm characteristic-impedance are required. They may be operated with the matching-network resistors on the input to the bias circuit for ac-coupled operation. The IC will have a high-input impedance by requiring approximately 0.8-V peak-to-peak transition as a trigger. Shift-register counters are avoided because the time-delay of the additional gates will be involved as a problem to the high speed required.

Plessey and Fairchild semiconductors have a series of fixed ÷2, ÷4, ÷5, and ÷10 count MSI ECL scalers operating from 40 to 600 MHz. They operate at - 5.2 V, and a positive ground-plane must be used for the circuit layout to prevent damage if the two output emitter-followers are inadvertently shorted to ground. The signal source is in practice ac-coupled with a 1000-pF mica capacitor. If the input is interrupted, a 15-K resistor must be used from the input to the negative voltage bus to prevent any possible oscillation under no-signal conditions. The input may be sinusoidal, but below 40 MHz the circuit operation depends on the slew-rate of the input rather than the amplitude. A square-wave input with a slew-rate of approximately 100 V/μsec will permit correct operation at the lower frequencies. The output voltage-swing can be increased by the addition of a dc-load to the output emitter-followers. Pull-down resistors of 1.5-K to the negative-bus provide an increase of 25% in the output voltage-swing. Leads and connections should of course be kept very short. These scalers can be conveniently outputted into high-speed TTL circuits.

9. Example of *Programmable-Divider.* A programmable-divider, suitable for example in the FM-transceiver synthesizer presented in Fig. A-7, is described as follows. A modulo-10 (decade) counter is the basic building-block of the programmable-divider. A Motorola MC-4016-P integrated-circuit chip is commonly used because of its special frequency-extended operation up to 25 MHz; the counter is built with TTL logic. A number is preset into the counter, the clock is enabled, and the counter cycles to zero, at which instant the preset number is effective. The operating speed is determined by (1) the toggle-rate (25 MHz) of the least significant bit (or the first flip-flop in the four flip-flop chain) and (2) the logic required to detect the zero and the preset conditions that override the effects of the clock-

input. The preset information enables the counter to start counting, and it is not the modulus of the counter.

Truth Table

Q_8 D	Q_4 C	Q_2 B	Q_1 A	Output Input
0	0	0	0	0
1	0	0	1	9
1	0	0	0	8
0	1	1	1	7
0	1	1	0	6
0	1	0	1	5
0	1	0	0	4
0	0	1	1	3
0	0	1	0	2
0	0	0	1	1
0	0	0	0	0

Clocking Transition: 0-to-1.

In some applications, a binary-counter is substituted in the place of the BCD decade-counter for dividing through 1599 instead of 999; this is allowed because the most significant digit counts down once, and only once, and does not recycle.

In the logic-diagram of the programmable-divider shown in Fig. A-8, the B output

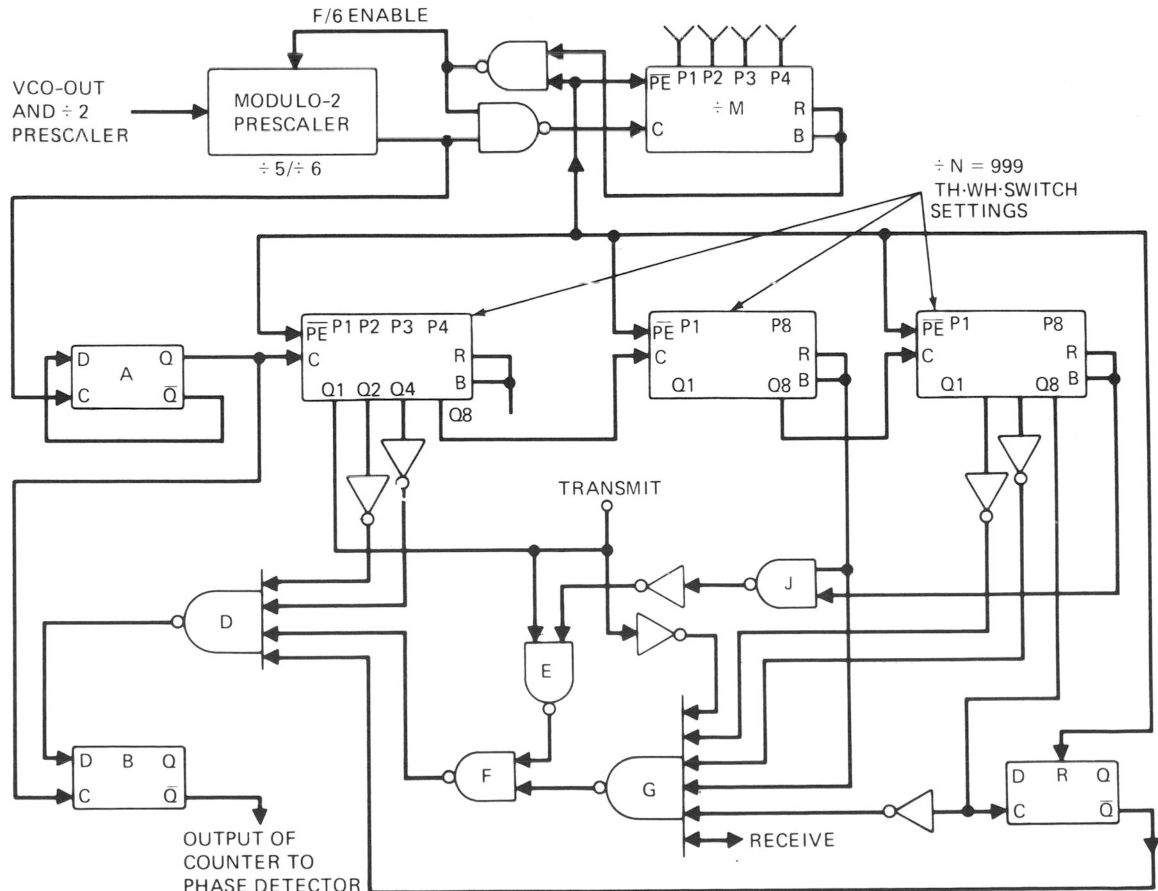

Fig. A-8. Programmable-divider for the frequency synthesis in a transceiver. (*Courtesy of Motorola*)

goes "high" when the counter reaches zero. By using the output on the ÷M count, the ÷N counter consists of three MC-4016-P chips and logic. The counters may be loaded by an external flip-flop (if the operation of the decade-counter used is not frequency-extended) to ensure that the circuit will operate reliably at this speed. The loading-circuit skips one clock-pulse while loading; hence, it is necessary to decode the 0001-state of the counter to set this flip-flop. Gates designated D, E, F, and J perform this operation during the Transmit mode. During the Receive mode, the desired number is 107 + 1 = 108; it is decoded by the gates D, F, and G.

The MSD of ÷N is permanently programmed as "1" to avoid a switch. The second MSD is also set up this way, but the synthesizer can be used to monitor the frequencies outside the band in use if desired, and the switch is therefore included. Suitable logic on the programming control-lines P_1, P_2, P_4, and P_8 of the counters would allow the use of two exclusive sets of thumbwheel switches for the Transmit and Receive modes.

The above logic circuitry can be modified to allow the scanning of a number of channels on Receive (for example, in surveillance or monitoring applications) and dwell on any channel for the duration of the activity of that channel. An LED frequency readout in digital-display can be included as part of this scanning facility when the scanner dwells on an active channel; a blinking LED indicator may signify that the scanning of the selected channels under scrutiny is in progress.

A-5 COMMON VHF/UHF RADIO-FREQUENCY BANDS OF INTEREST IN FREQUENCY SYNTHESIS IN SURVEILLANCE RECEIVER

These are enumerated in the following along with the corresponding local-oscillator frequencies. The approximate range of the step-frequency tuning is ±15% in any band.

RF Band (MHz)	Local-Oscillator (MHz)	Step-Frequency Increments (kHz)*
I. 20–35	90–105	10
II. 35–61	157–183	10
III. 61–107	257–321	50
IV. 107–187	481–561	50
V. 187–327	841–981	150
VI. 327–572	1471–1718	250
VII. 572–1000	2572–3000	250

*Step-frequency increments above 3 GHz are usually chosen as 0.5 or 0.625 MHz.

The digital frequency synthesis of the preceding rf bands is feasible with the present Motorola low-cost MSI components up to band II, namely, 157 to 183 MHz. The higher bands of interest can be conveniently derived therefrom by the simple technique of harmonic multipliers at the output of the basic phase-lock loop described below for band II. As an alternative, for the synthesis of the higher rf bands, the heterodyne harmonic-multiplier technique could be successfully used (Fig. A-4).

The block-diagram of the PLL includes the programmable-divider and the special-purpose high-speed MSI logic components presently available at average low-cost. At the rate the silicon monolithic microprocessor LSI components are being developed, there is a likelihood that the complete phase-lock loop for the desired rf band could be fabricated into two or three LSI chips and a few hybrid microwave compo-

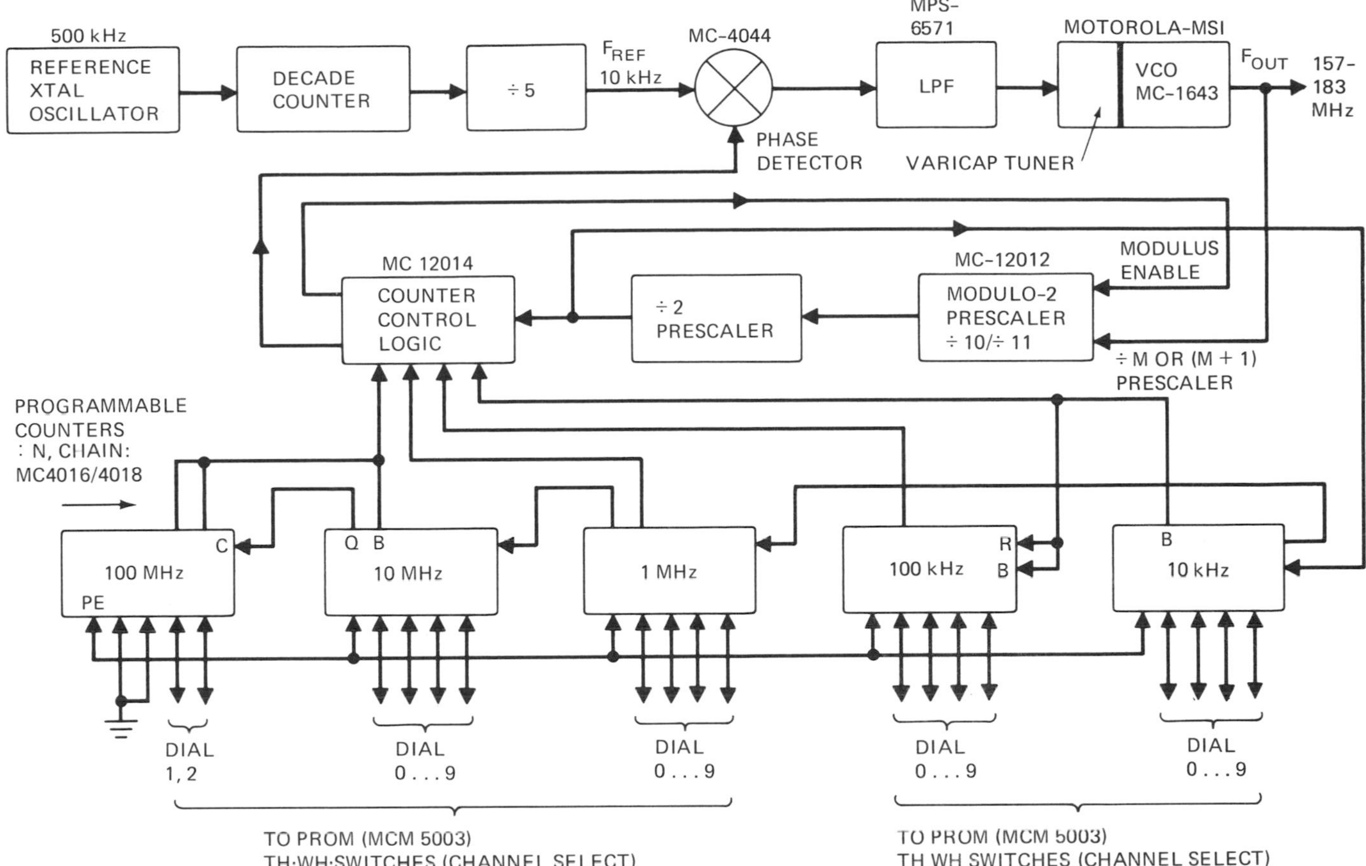

Fig. A-9. Local-oscillator frequency synthesizer for the radio-frequency band 157–183 MHz. (*Courtesy of Motorola*)

nents, to allow further reduction in cost and space. For example, a lower frequency-band with lower frequency-increments could be frequency synthesized starting with a standard LSI-PLL unit at the midband frequency-range by the simple signal processing technique of frequency-division.

A detailed schematic circuit diagram for the frequency-synthesis of the 157- to 183-MHz band at 10-kHz step frequency-increments is shown in Fig. A-9. The figure includes the programmable-memory ROMs (read-only-memories) for correspondence between the desired-frequency dialing and the actual counts of the modulo-control counter for the lower significant digits and the programmable-counter for the higher significant digits.

The phase-detector used in the PLL is illustrated in Fig. A-10; it is a digital frequency phase-detector with a digital-to-analog charge-pump. In the same integrated-circuit, a Darlington-pair of transistors are used as a VCO drive-amplifier. It is used in conjunction with a high-beta MPS 6571 transistor to make an active low-pass loop-filter.

The phase-frequency detector provides a pulse on one of the two terminals when F_{ref} or the phase is lower than that of the VCO-frequency divided by the programmable-count in the PLL. When the opposite is true, the other terminal provides this pulse. The charge-pump uses these signals to add or subtract a small charge from the capacitor in the active-filter to change the output error-voltage appearing at the input of the type MV-1404 varicap-diode that is connected to the

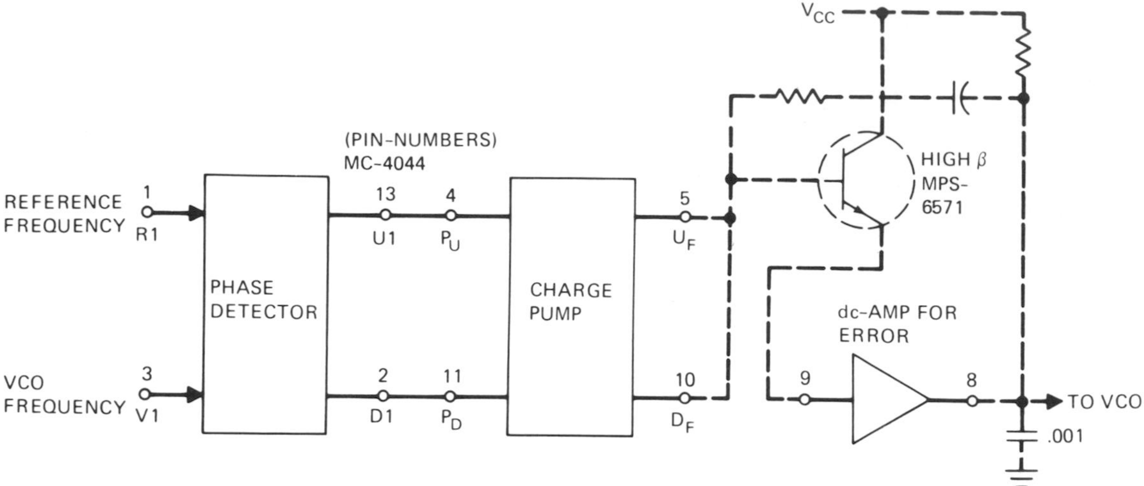

Fig. A-10. Medium-scale-integration phase-detector chip MC4044/Fairchild.

rf tuner of the LC-oscillator in the VCO. RC multivibrators have generally proven spectral purity; hence, the need for a higher-Q, LC-tuner for the VCO, type MC-1648. The latter consists of an emitter-coupled pair of transistors biased through the LC tank-circuit to cause degeneration. An automatic gain-control is added by means of an emitter-follower from the tank-circuit.

In the type MC-1648, VCO, shown in Fig. A-11, a cascode transistor-pair is used to couple from the above emitter-follower, since any additional load on the tank would decrease the Q of the oscillator and degrade spectral-purity. The cascode transistor setup provides the requisite transfer to a differential-pair of transistors to produce in turn a square-wave compatible with the existing high-speed ECL logic. The output is duly buffered from the frequency determining components so that the loading does not produce any frequency-shift.

The actual LC-tuner will of course require modification for higher-frequency VCO operation. As an example, a YIG-tuned oscillator for a higher-frequency output in GHz would require a Gunn voltage-controlled oscillator.

Thus, PLL digital frequency-synthesis is the ultimate answer, now that the monolithic large-scale-integration and the hybrid-IC technology have provided the

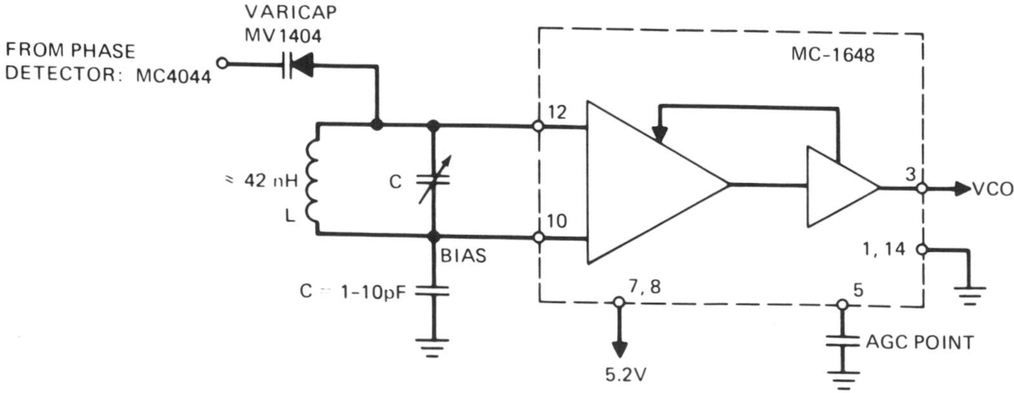

Fig. A-11. Voltage-controlled oscillator used in the VHF radio-frequency band.

solution for the reduction of cost, size, and power requirements in respect to hardware. Rigid full-spectrum utilization of the UHF and microwave bands in surveillance and consumer applications is another beneficiary. With new microwave components like digital MESFET-LSI, step-recovery diodes, and YIG (Yttrium-iron-garnet) resonator-filters, it is now feasible to employ digital frequency synthesis for minimal step frequency-increments such as 250 kHz right into the GHz frequency-bands. Mobile UHF and microwave two-way communication, for minimal frequency-separation in grid-zoning operations, is another beneficiary by packing more users into a narrow spectrum; mobile units would use limited power for transmission and automatically change frequency as they move from one area to the other in the zones established. Above all, digital frequency synthesis would allow remote-computer control of communications through LSI microprocessors.

A-6 SINGLE-CHIP PHASE-LOCK LOOP DEVICE (SIGNETICS)

As a parallel to digital integrated circuits and the latest monolithic LSI, the linear operational amplifier and the balanced differential-amplifier readily lend themselves to monolithic fabrication techniques. These amplifiers are also frequently fabricated by hybrid methods for close tolerances; however, where broad tolerances permit, mass-production monolithic techniques are applicable in most cases for video (pulse) bandwidths from dc to 100 MHz, with the corresponding propagation-delays down to 5 nsec at this time, because the use of capacitors and high-value resistors can be restricted to a minimum in fabrication.

Exceptional input-balance afforded by inherent match between differential pairs, and close match in temperature-coefficients of the components fabricated from the same material, assure stable electrical characteristics and noise immunity over a broad temperature range. Since the ratios are held constant output-to-input isolation is feasible. Also, simplified feedback configurations contribute to higher performance ratings. Economy low-power bipolar I^2L processing can be conveniently adapted by balanced differential-amplifier approach in linear monolithic fabrication techniques.

The linear monolithic approach has facilitated manufacturers like Signetics to produce phase-lock loop (PLL) chips on a mass-production basis. The medium-scale integrated (MSI) circuit chip is basically an electronic servo loop consisting of a phase detector, a low-pass filter, a dc amplifier, and a voltage-controlled oscillator (VCO). The VCO-phase makes it capable of locking or synchronizing to a reference frequency. If the VCO-phase changes, indicating the incoming frequency is changing, the phase-detector output voltage linearity increases or decreases just enough to keep the VCO-frequency the same as the incoming frequency, and preserve the locked condition. The average voltage applied to the controlled oscillator via a low-pass filter is a function of the frequency of the incoming signal. The dc amplified LP filter output voltage is the demodulated output when the incoming signal is frequency-modulated. The PLL facility enables independent center-frequency and bandwidth adjustment, high noise-immunity, high selectivity, high-frequency operation, and center-frequency tuning by means of a single external component.

NE560B, 562B, and 565 are general-purpose PLL chips operating to 30 MHz and locking within a range of 40% and a frequency-drift of ±600 Hz/°C, at a power dissipation of 0.2 W. The accompanying simplified block-diagram shows an external divider (2 to 16). Type 561B has an additional quadrature phase-detector (multiplier) for AM modulation, and it is linearized for FM and AM demodulator function. The PLL chips are presently used in applications such as FM demodulation, frequency synthesizing, frequency synchronization, signal conditioning, and AM demodulation.

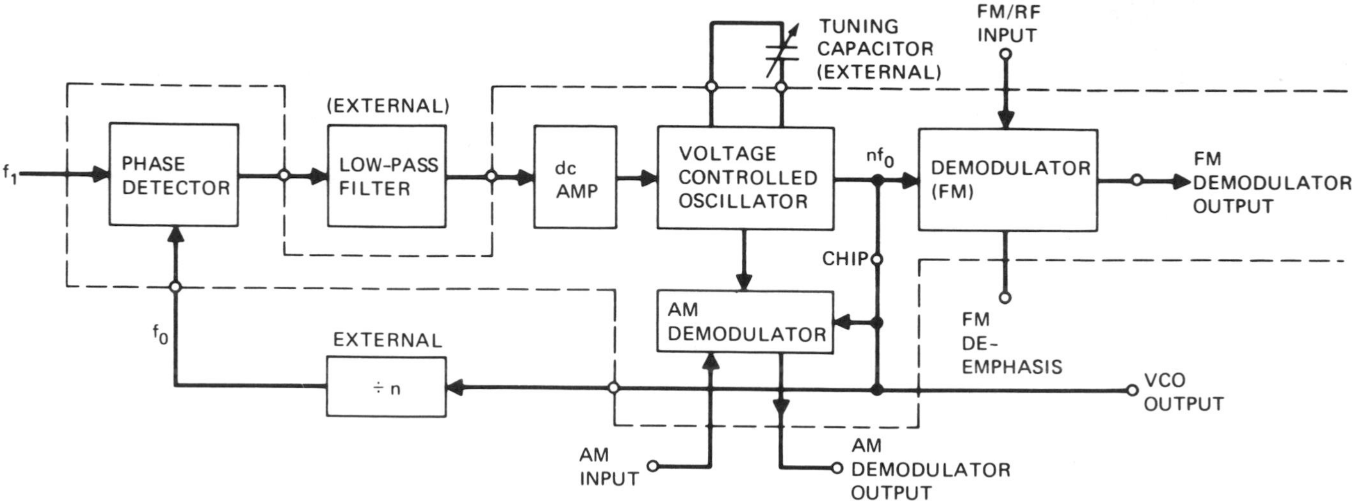

Typical PLL Chip. Permission to reprint granted by Signetics Corporation, a subsidiary of U.S. Philips Corp., 811 East Arques Avenue, Sunnyvale, CA 94086.

Appendix B
High-speed Phase-lock Loop for Improving Sensitivity of a Typical Computer-Controlled Radar Receiver

B-1 DESIGN OF PHASE-LOCK LOOP DEMODULATOR IN DIRECTION-FINDING INTERFEROMETER RECEIVING SYSTEM

The phase-lock loop (PLL) is an increasingly common application in many high-sensitivity communication systems. It is used in two different modes:

1. to *track* the carrier-signals that drift in frequency as a result of mistune or motion/Doppler-shift
2. to follow a modulated-carrier frequency- or phase-change so as to *improve the signal-to-noise ratio* (SNR) performance of a receiver, if the high-frequency *pass-band at the intermediate-frequency* (IF) (e.g., 34 MHz) *is greater than the modulation bandwidth*. The broader pass-band implies a *spread-spectrum* and hence improved SNR (see Appendix C-27).

In superheterodyne receivers, the scanned/tuned rf-signal alone is present, and a passive circuit or a crystal-filter establishes the correct IF by automatically bringing the frequency of a local-oscillator (LO) arbitrarily close to the incoming rf signal-frequency. In the direction-finding interferometer receiving system, the instantaneous frequency measuring system (IFM) and the voltage-controlled oscillator (VCO) combination make the LO set-on system fast and accurate, and the function of the PLL in the receiver is not one of tracking the incoming carrier-frequency. Instead, its function is to follow the incoming pulse-modulation as primarily a frequency-discriminator, as mentioned in (2), and to improve the SNR performance of the receiver by virtue of an added PLL-filter that attempts to simulate the received signal-modulation by way of a *high-speed phase-comparator*. In short, the PLL-demod in this application is merely a modified *matched filter*.

The receiver under consideration employs a 34-MHz wide-band IF frequency-discriminator. The pass-band of this device must be limited to pass only the main-lobe signal of the pulse-modulation, while remaining consistent with the minimum pulse-width requirements (approximately 100 to 200 nsec) of the radar signals encountered). To meet these requirements, a band upper-limit and hence a threshold-SNR of at least 12-dB at the activity-antenna of the interferometer antenna system are mandatory as far as the desired signal sensitivity-threshold (- 65 dBm) of the present systems are concerned.

If a PLL frequency-discriminator, employing an optimized narrow-band filter (satisfying a lockup time below 100 nsec) is incorporated as an added complement to the digital fine-tune facility (in the place of the conventional phase-comparator) and the following *low-gain quantizer* (for the formulation of the *fine-frequency word*), the threshold-SNR of 12-dB can be further extended for an improved *dynamic-range* of sensitivity. As a secondary contribution toward improved SNR, the new phase-comparator could employ *in-phase and quadrature synchronous phase-detectors*, using the coho IF-VCO reference-frequency from the PLL. The conventional IF in these radar receivers is 60 MHz.

B-2 PHASE-LOCK LOOP THEORY

The theory of the operation of the PLL as a demodulator was presented for a *linearized model* by R. Jaffee and E. Rechtin (1955). W. J. Gruen (1953, automatic frequency control) and D. Richman (1954, color television synchronizing techniques) used the linearized model for developing the *tracking characteristics under random noise conditions*, according to the applicable *superposition-principle* under these conditions. The linear PLL-model in principle restricts that the phase-error be constrained below 1 radian, lest dropout should occur; above 1 radian, the mode of operation could be treated under Gruen's AFC System analysis. The best justification for the validity of the assumption of a linear model, under conditions of noise in the vicinity of the phase-lock, is the actual experimental evidence in PLL tracking- and demod-applications for improved SNR. In fact, when the phase-lock is effective, and the SNR figure is high, it acts as a linear feedback system. In the PLL-demod application, the feedback loop is required to reproduce the phase-modulation of the pulse-envelope. Then a large extent of the general theory developed for tracking can be directly applied.

The exact model of a PLL is naturally *nonlinear*, and A. J. Viterbi, J. R. Rowbotham, and F. W. Sanneman have presented solutions for an approximated second-order PLL-network in limited cases to provide insight into the nature of the *loop-dynamics* such as the *lockup time and the threshold.* No general nonlinear analytical technique is practical to estimate the exact pull-in of signals under conditions of noise. In view of the justification of valid experimental evidence, the results of the linear-model analysis are applied in any feasibility study concerning the PLL-demod in the digital fine-tune subsystem of the IF local-oscillator.

However, Richman has obtained an analytical solution to the nonlinear problem of finding the time-interval required for the *pull-in of phase* under the following situation (NTSC Monograph No. 7, *IRE*, 1954). During the pull-in, the phase repeatedly goes through its entire range of values allowing the phase-detector to provide a positive and negative differential-voltage in an alternating sequence. This has the effect of slowing or speeding up the rate of change of the phase-difference between the incoming real-time IF and the VCO reference subcarrier on consecutive half-cycles of this differential-voltage. By virtue of the feedback-loop, the average frequency-difference is modified so that the pull-in may occur under the assumed conditions. Therefore, the results of the linear-model analysis are complemented by the technique used by Richman to obtain the *lockup time for the pull-in* in the case of this high-speed PLL configuration.

Richman's original analysis then proceeds to the excellent improvement the *dc quadricorrelator synchronous phase-detectors* allow in respect to the high-speed lockup time during the pull-in mode, while simultaneously (and automatically) permitting an independent choice of *stability* against the dynamic phase-jitter under the *hold-in mode during the dwell-period.* The proposed quadricorrelator technique is described in Section B-5 with a presentation of the system used and its details.

B-3 INCORPORATION OF PLL IN IF CENTER-TUNE CIRCUIT

The phase-shift control voltages at the output of the FM-demodulators, in the Center-Tune subsystem in the Activity-channel of the interferometer direction-finding receiving systems, enable (1) discrimination of the activity real/image conditions, (2) determination of the IF center-frequency as a center-tune flag in a flat-spectrum over a defined-band, and (3) pulse-edge time-differentiation of a linear "chirp" or FM-deviation or pulse time-modulation. The latter function is further developed in some systems as a separate spur-detection technique (relative to an SNR-threshold) for inhibiting the possible reading of a spur as a regular frequency-word in a computerized receiving system.

The degree of accuracy obtainable from the present envelope phase-detectors in a phase-comparator naturally depends on the contamination due to noise (known as *phase-jitter*), and the phase quantization-errors in the subsequent signal-processing circuits. From the viewpoint of noise-inhibition, the quadricorrelator PLL introduced in Section B-5 (Fig. B-3) is more effective. The dc quadricorrelator is a powerful tool as a *tracking-filter*; an outstanding merit is that it minimizes phase-jitter while simultaneously improving SNR.

The threshold-levels of weak emitter signals, practical in the present activity-channels, occur at a carrier-to-noise (CNR) level of 12 to 18 dB, depending on the microwave-band in use at any instant. For pulse-amplitude modulation, the CNR is proportional to SNR. The PLL demodulator, as a tracking-filter, is evaluated to improve the feasible threshold-levels to an extent of 5 to 6 dB of SNR in the present systems, since its bandwidth is defined independent of the frequency-band it tracks. Besides, it is adaptive in the sense that noise-bandwidth decreases with reduced carrier-levels. The noise-free IF subcarrier chosen as compatible to the preferred PLL variation (namely, the dc quadricorrelator) enables the then feasible loop-bandwidth to satisfy the conditions for a fast acquisition-time (or pull-in) and faithful *hold-in.* Filter-tracking takes place within a narrow noise-bandwidth of the PLL at a damping-coefficient that is suitable for the transmission of a high-speed step-function at the input of the system.

The dual filtering-modes effective in the quadricorrelator on an automatic basis allow the primary in-phase synchronous frequency/phase-detector look for the zero-beat; when it is realized, it acts to reduce the loop noise-bandwidth. Simultaneously, the FM-discriminator of the digital center-tune flag provides the high-precision analog information for pretuning the VCO of the LO setup so that the pull-in range of the frequency-interval is restricted to a narrow-band to allow minimum acquisition or pull-in time.

The loss of the phase-lock in the presence of excessive noise (or the absence of a signal from an emitter) at the input of the quadrature synchronous phase-detector—unlike the working of the present envelope-detectors at low SNR—automatically provides a clear-cut flag. Noise outputs a zero, and signal-plus-noise furnishes a 1-V step-function. This facility can be successfully adapted for suppressing the display of a disturbing *false alarm-rate* due to the incidence of excessive electromagnetic interference or jamming, as the displayed phase-information hops in integral-multiples of 2π-rad and skips cycles.

B-4 HIGH-SPEED PHASE-LOCK LOOP ANALYSIS

The basic PLL is composed of a phase-detector, an active filter element F(s), and a feedback connection to the phase-detector via the voltage controlled oscillator VCO.

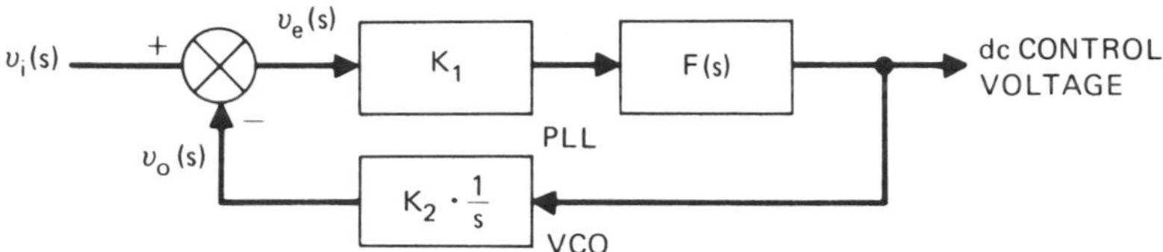

The general second-order differential equation used for the linear model will be derived and a specific solution for the requisite parameters of the PLL will be analyzed by using the relevant data, in this application of the PLL to the center-tune FM-Demodulator. Phase-error is given by

$$\nu_e(s) = \nu_i(s) - \nu_o(s) \tag{1}$$

Using Mason's rule for gain, the closed-loop transfer-function is obtained:

$$H(s) = \frac{\nu_o(s)}{\nu_i(s)} = \frac{K_1 K_2 F(s)}{[s + K_1 K_2 F(s)]} \tag{2}$$

The loop filter-element used in this design is an active-lag type, as illustrated below:

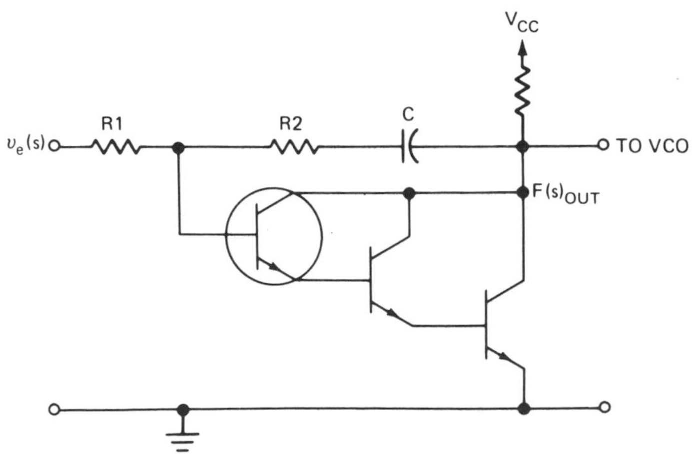

The transfer-function of the above filter is represented by

$$F(s) = \frac{\tau_2(s) + 1}{\tau_1(s)} \tag{3}$$

where

$$\tau_1 = R_1C \text{ and } \tau_2 = R_2C$$

The complete closed-loop transfer-function then takes the following form:

$$H(s) = \frac{K_1K_2\left(\frac{\tau_2 s + 1}{\tau_1 s}\right)}{s^2 + \left(\frac{K_1K_2\tau_2}{\tau_1}\right)s + \left(\frac{K_1K_2}{\tau_1}\right)} \tag{4}$$

The characteristic equation of this second-order system is represented by the differential equation:

$$s^2 + 2s\zeta\omega_n + \omega_n^2 = 0 \tag{5}$$

where ζ is the damping-factor, and ω_n is the natural loop-frequency. Substituting the RC values for τ_1 and τ_2,

$$H(s) = \left(\frac{K_1 \cdot K_2 \cdot R_2}{R_1}\right)\left[\frac{s + \frac{1}{R_2C}}{s^2 + \left(\frac{K_1 \cdot K_2 \cdot R_2}{R_1}\right)s + \left(\frac{K_1 \cdot K_2}{R_1C}\right)}\right] \tag{6}$$

Comparing the equation in the denominator with the characteristic equation of the loop,

$$\omega_n^2 = \frac{K_1K_2}{R_1C} \qquad \omega_n = \sqrt{\frac{K_1K_2}{R_1C}} \tag{7}$$

and

$$\zeta = \frac{R_2C}{2}\sqrt{\frac{K_1K_2}{R_1C}} \tag{8}$$

The 3-dB bandwidth of the closed-loop can be obtained by substituting $j\omega$ for s in Eq. 6.

Setting $H(j\omega)^2 = \frac{1}{2}$ and $10 \log \frac{2}{1} = 3$ dB,

$$\omega_{3\,dB} = \sqrt{\left[\frac{-B}{2} + \sqrt{\frac{B^2}{4} - C}\right]} \tag{9}$$

where

$$B = \left[-\left(\frac{K_1K_2R_2}{R_1}\right)^2 - \frac{2K_1K_2}{R_1C}\right] \tag{10}$$

and

$$C = -\left[\frac{K_1K_2}{R_1C}\right]^2$$

since

$$\zeta = \frac{R_2C}{2}\sqrt{\frac{K_1K_2}{R_1C}}, \text{ and } \omega_n = \sqrt{\frac{K_1K_2}{R_1C}} \tag{11}$$

$$\omega_{3dB} = \omega_n \sqrt{(2\zeta + 1) + \sqrt{(2\zeta^2 + 1)^2 + 1}} \tag{12}$$

Now, the loop gain-constants K_1 and K_2 are obtained as shown in the following diagram for the phase-detector and the VCO, respectively.

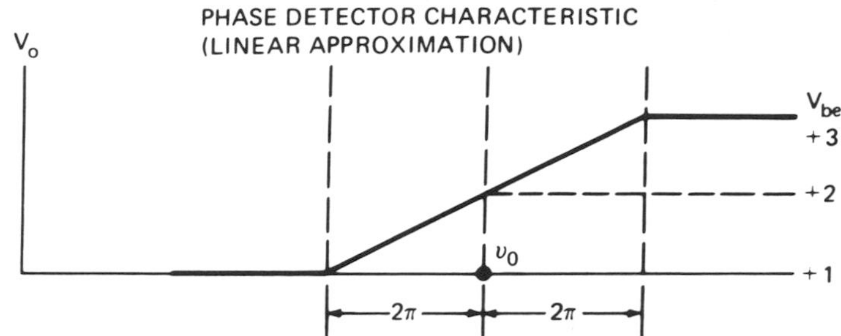

The phase-detector output-voltage is proportional to the phase-difference between the negative-going pulses applied to the phase-detector.

$$V_p = K_1 (\nu_i - \nu_o) \tag{13}$$

From the figure above,

$$V_p = \frac{V_o(\nu_i - \nu_o)}{2(2\pi)} \tag{14}$$

Substituting the expression for V_p from Eq. 13,

$$K_1 (\nu_i - \nu_o) = \frac{V_o(\nu_i - \nu_o)}{2(2\pi)}$$

$$K_1 = \frac{V_o}{2(2\pi)} \tag{15}$$

Consider as a typical example the following diagram of one of the latest phase-detector MSI chips such as the Motorola MC-4044, with a built-in charge-pump (as explained in Appendix A).

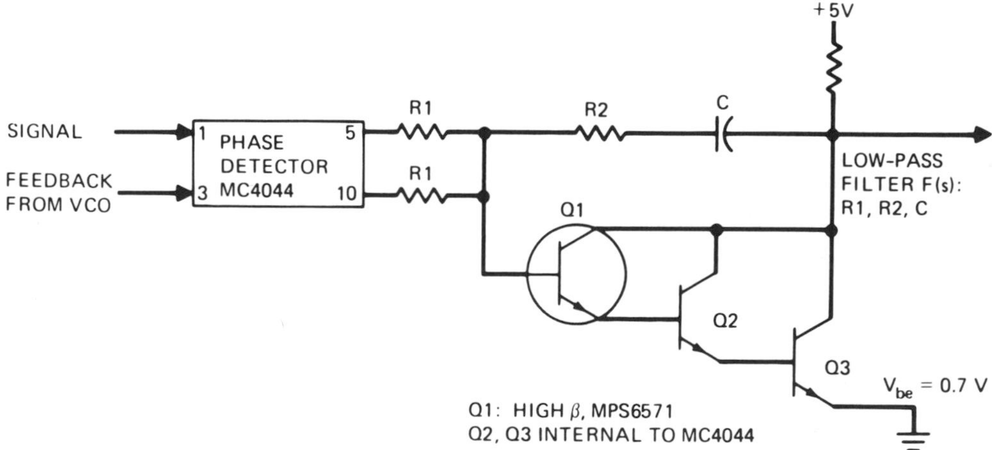

If V_o of Eq. 15 is the change in voltage at the summing junction of R_1's, through a phase-change of $\pm 2\pi$.

The phase-detector gain-constant,

$$K_1 = \frac{2V_{be}}{2(2\pi)} = \frac{0.7}{2\pi} = 0.11 \text{ V/rad} \tag{16}$$

since

$$v_o = 2V_{be} \text{ in the above circuit.}$$

The VCO gain-constant can be calculated from the data sheets or by measuring the change in the output-frequency of the VCO over a change in the input control-voltage.

From the manufacturer's specification of the VCO:

For an input excursion of 1.8 to 5 V, the frequency of the VCO changes by (77 - 42 MHz) = 35 MHz.

$$K_2 = \frac{(77 - 42)10^6 \cdot 2\pi}{(5 - 1.8)} = 6.86 \times 10^7 \text{ rad/sec/V} \tag{17}$$

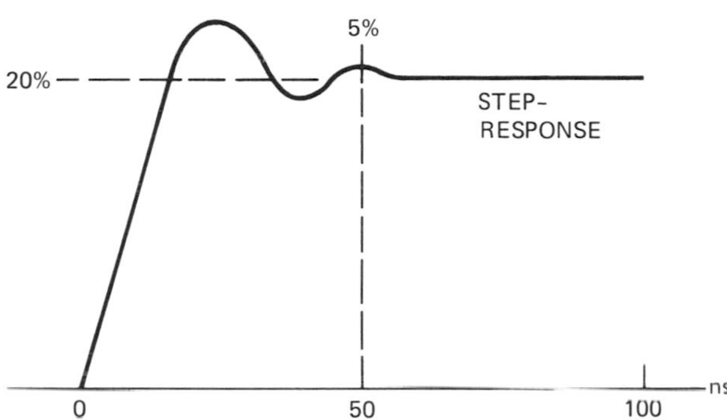

Time-interval t_s is defined as the time required for the VCO control-voltage to be within 5% of its final-value subsequent to a step-change in frequency. Assume t_s = 50 nsec, and an overshoot of 20%.

Using the linear second-order model, the PLL may be designed for a chosen damping-factor ζ of 0.707. (It has been experimentally established that a satisfactory balance between the transient-error and the output-noise occurs when the loop damping-constant $\zeta = 0.707$.)

Now consider a root-locus plot with an active lag-filter F(s).

For a damping factor $\zeta = 0.707$, the complex poles are located at ±45°.

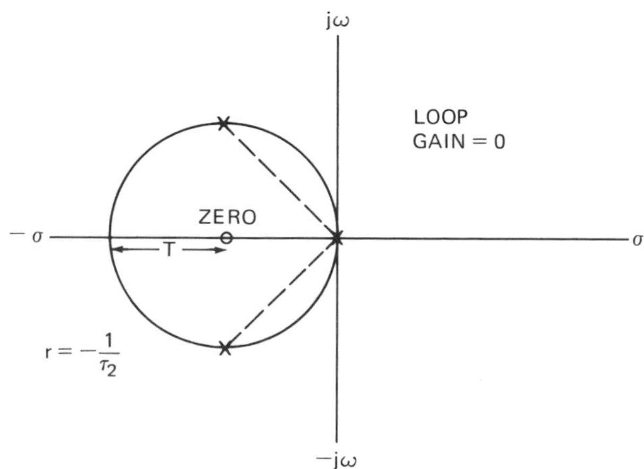

The closed-loop poles will be located $\pm 45°$ about the negative-real axis, and the root-locus circle will pass through the origin of the s-plane. One of the poles will be located at the origin when the loop gain is zero, and they traverse a circle with a radius of $1/\tau_2$, as the gain increases.

Two poles are located at $(-1 \pm j)$, and the zero is located at $-1/\tau_2$.

With the complex loop-poles above and below the system-zero for a damping-factor of 0.707, the overshoot in general will be approximately 20%.

Then the product of the natural-frequency ω_n and the settling-time t_s of 50 nsec for a 5% approximation of the final-value is represented by

$$\omega_n t_s = 5 \tag{18}$$

$$\text{Natural frequency, } \omega_n = 5/50 \times 10^{-9} = 100 \times 10^6$$

For the characteristic-equation, $s^2 + (2\zeta\omega_n)^2 s + \omega_n^2 = 0$, the denominator of the complex function, according to Eq. 6 is given by

$$s^2 + \left(\frac{K_1 K_2 R_2}{R_1}\right) s + \left(\frac{K_1 K_2}{R_1 C}\right) = 0 \tag{19}$$

The complex roots of the characteristic equation above can be formed into real and imaginary terms:

$$(s + b)^2 + \omega^2$$

i.e.,

$$s^2 + 2bs + (b^2 + \omega^2) \tag{20}$$

Comparing Eqs. 19 and 20 and the second-order differential equation (Eq. 5),

$$\omega_n^2 = b^2 + \omega^2 = \frac{K_1 K_2}{R_1 C} \tag{21}$$

$$R_1 C = \frac{K_1 K_2}{\omega_n^2}$$

With $\zeta = 0.707$ and $b = \omega$ in Eq. 21,

$$\omega_n^2 = 2b^2, \tag{22}$$

$$b = \omega_n/\sqrt{2}$$

Also, from Eqs. 19 and 20,

$$2b = \frac{K_1 K_2 R_2}{R_1} \tag{23}$$

$$R_2 = \frac{2bR_1}{K_1 K_2}$$

The natural-frequency $\omega_n = 100 \times 10^6$, and $K_1 = 0.111$ V/rad, and $K_2 = 6.86 \times 10^7$ rad/sec/V. From Eq. 21,

$$R_1 C = \frac{0.111 \times 6.86 \times 10^7}{(100 \times 10^6)^2}$$

Choose C for a physically realizable R_1. A capacitor of 100 pF is chosen for C.

$$R_1 = \frac{0.111 \times 6.86 \times 10^7 \times 10^{10}}{10^{16}} = 7 \text{ ohm} \tag{24}$$

$$b = \omega_n/\sqrt{2} = 100 \times 10^6/\sqrt{2} = 71.5 \times 10^6$$

$$R_2 = \frac{2bR_1}{K_1K_2} = \frac{2 \times 71.5 \times 10^6 \times 7}{0.111 \times 6.88 \times 10^7} = 133 \text{ ohm} \tag{25}$$

$$K = K_1 \cdot K_2 = (6.86 \times 0.111) = 7.6 \times 10^6 \text{ Hz/sec} \tag{26}$$

According to Richman's conclusion, the frequency derived is the filter cutoff frequency for the quadricorrelator (see Eq. 2 of Richman's optimum design technique). Therefore, for a settling-time of 50 nsec, the *low-pass active lag-factor* is chosen with the values:

R_1 = 7 ohm, R_2 = 133 ohm, and C = 100 pF. The preceding analysis is done with a signal input alone. In the linear model of the PLL considered, the *principle of superposition* holds good, and Viterbi has shown that a secondary white-noise (Gaussian) input of spectral-density N_o will have the same amount of noise-power at its output as the variance of the loop phase-noise. For the second-order PLL to function properly as a discriminator, the loop noise-bandwidth B_n is defined for an ideal low-pass filter of band-width B_n by the integral,

$$B_n = \int_o^{j\infty} \frac{H(s)^2}{2\pi j} \, ds \tag{27}$$

provided the phase-error is maintained small compared with 1 rad. For the VCO phase-jitter, the simplest parameter to compute is the variance of the VCO noise-jitter σ_ν^2:

$$\sigma_\nu^2 = \Phi \int_{-j\infty}^{j\infty} \frac{H(s)^2}{2\pi j} \cdot ds = 2\Phi B_n \tag{28}$$

Where Φ represents the (statistical) noise spectral-density normalized by the signal power A^2. Then, $\Phi = \dfrac{N_o}{A^2}$.

Since in this particular application of the PLL as a demodulator, both the modulation and the interference are random, the only measure of phase-error that can be calculated is the mean-square-error expressed as the sum:

$$\nu_e^2 = \frac{1}{2\pi j} \int_{-j\infty}^{j\infty} |\nu_i(s)|^2 \cdot |1 - H(s)|^2 + \Phi|H(s)|^2 \, ds \tag{29}$$

The loop is optimized by the criterion of minimization of ν_e^2. F. W. Lehan and R. J. Parks have used the Wiener optimization technique for minimization to finally arrive at the following relation for the loop noise-bandwidth B_n.

$$\omega_n = \tfrac{4}{3} \cdot \sqrt{2} B_n \tag{30}$$

For the requisite loop natural-frequency ω_n of 100×10^6, the PLL noise-bandwidth,

$$B_n = \frac{100 \times 10^6 \times 3}{4\sqrt{2}} = 53.2 \times 10^6$$

R. W. Sanders has analyzed the communication efficiency for the same linear model in the case of the PLL as a demodulator along the above procedure. When the 3-dB passband of the system is made equal to B_d, the data-bandwidth (which in the case of the Real-IF center-tune flag) is specified in this system as 14 MHz. Sanders has shown the loop noise-bandwidth B_n to be:

$$B_n = B_d \frac{3\pi}{2\sqrt{2}} = 14 \times 10^6 \times 9.45/2\sqrt{2} = 46.8 \times 10^6 \qquad (31)$$

B_n is a key parameter in PLL design. In the results shown, 46.8 MHz is used for the loop noise-bandwidth of the primary PLL System.

Richman's optimum design technique. In regard to television color synchronizing-burst (which is considered as a close-parallel to the radar pulse-envelope of Real-IF center-tune flag), Richman has treated the problems of the *pull-in* range and the *lockup* time (of color) in the presence of noise on the basis of a nonlinear system under a restricted set of conditions. Richman's analysis is used in the following treatment to come up with the final results. For instantaneous pull-in, and a lockup timing-accuracy of ±10 nsec, the analysis employs an antihunt RC-filter like that used in Richman's analysis.

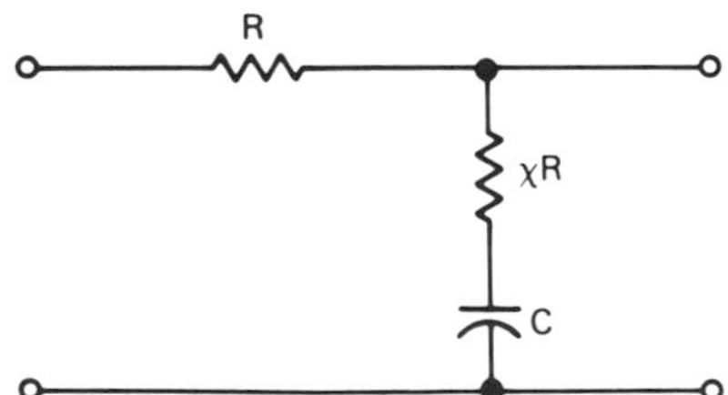

Let f_i be the initial frequency-difference before the pull-in. The *pull-in* is defined as the maximum f_i for which the final synchronized-phase $\theta(t)$ is less than $\theta_s(t)_{max}$.

$$f_i < |K| \cdot |\theta_s(t)_{max.}| \qquad (1)$$

$$K = K_1 \cdot K_2 = \left[\frac{(volt)}{rad} \cdot \frac{rad}{(sec)(volt)} = \frac{1}{sec} = \frac{1}{T}\right] = \text{Frequency (Hz/sec)} \qquad (2)$$

The phase in the steady-state is given by:

$$\text{Sin}\ \varphi = f_i/K \qquad (3)$$

where

$$\varphi = \theta_s(t)_{max} - \theta_s(t)$$

The maximum value of K Sin φ is K, and thus a steady-state can exist only if

$$f_i \leqslant K \qquad (4)$$

The maximum pull-in of the system is therefore K cycles, as explained by the dimension given in Eq. 2. And the value of K is obtained in the present analysis as 7.6 MHz.

For small φ and large K,

$$f_i/K = \text{Sin}\ \varphi \approx \varphi$$

and synchronization occurs if $|f_i| < K$. Richman has shown that the maximum pull-in range for automatic frequency-lock can be expressed by

$$|f_i| \leqslant K\sqrt{2m}, \text{ where } m = \frac{\chi}{\chi + 1} \qquad (5)$$

Then, the phase angle,

$$\sin \varphi \approx \varphi = \frac{f_i}{K} = \frac{K\sqrt{2m}}{K} = \sqrt{2m} \text{ rad}$$

Now, the *hold-in* range for the automatic frequency-correction (AFC) is defined by noting that, when once locked, the open-loop oscillator-frequency may drift by K cycles without any loss of synchronization when the loop is closed.

The ratio of the pull-in to the hold-in range, r, is then given by:

$$r = \frac{K\sqrt{2m}}{K} = \sqrt{2m} \tag{6}$$

Richman has also shown that the pull-in is practically instantaneous if

$$f_i < mK \tag{7}$$

The impulse noise-immunity I for the pull-in and the hold-in is determined by the loop noise-bandwidth B_n. It is so related to B_n that the peak phase-deviation in the output is inversely proportional to B_n:

$$I = \frac{1}{B_n}$$

This is a measure of the noise-immunity. Richman's analysis finally concludes with the following results. For a large K that allows the maximum possible pull-in and hold-in ranges at high phase-accuracy,

$$B_n = 1.25\sqrt{2\pi K\alpha} \tag{8}$$

where $\alpha = \dfrac{1}{(\chi + 1)T_L}$, and T_L = lock-up time, RC. If the pull-in time must be instantaneous, as in our present requirement (approximately 50 nsec),

$$\varphi_{max} = m \tag{9}$$

In the present center-tune PLL application, for a timing-accuracy of ±10 nsec when receiving a radar-pulse of 200-nsec duration at maximum phase-deviation φ_{max},

$$\varphi_{max} = \frac{2\pi \times 10 \times 10^9}{200 \times 10^9} = 0.314 = m$$

Since

$$m = \frac{\chi}{\chi + 1} = 0.314, \ \chi = 0.45$$

$B_n = 1.25\sqrt{2\pi K\alpha}$, where the pull-in range K = 7.6 MHz

$$\alpha = \frac{1}{(\chi + 1)T_L} = \frac{B_n^2}{1.25^2 \times 2\pi \times K} = \frac{46.8 \times 46.8 \times 10^{12}}{1.56 \times 6.28 \times 7.6 \times 10^{12}}$$

Lock-up time, T_L, for the conventional PLL

$$= \frac{1.56 \times 6.26 \times 7.6 \times 10^6 \times 10^9}{1.45 \times 46.8 \times 46.8 \times 10^{12}} = 24 \text{ nsec}$$

The interval of 24 nsec may be considered as the *instantaneous high-speed phase-lock.* In the quadricorrelator situation, after the first in-phase lock-in, the performance is automatically transferred to the quadrature synchronous phase-detector employing a *narrow-band filter.* This would naturally boost the basic 3-dB SNR improvement afforded by the coherent synchronous phase-detectors by an additional 2 to 3 dB—due to reduced bandwidth.

The pull-in range of the dc quadricorrelator is given by Richman's expression $\frac{2B_n}{\pi}$.

If B_d, the data-bandwidth provided by the quadrature-filter is assumed to be 7.6 MHz, that is, the cutoff frequency of F(s) obtained earlier from the PLL System analysis:

$$\text{Noise-bandwidth, } B_n = B_d \frac{3\pi}{2\sqrt{2}} = \frac{7.6 \times 10^6 \times 9.45}{2\sqrt{2}} = 23.4 \times 10^6$$

The improved pull-in range by using the dc quadricorrelator:

$$\frac{2B_n}{\pi} = \frac{46.8 \times 10^6}{3.14} = 15 \text{ MHz} \tag{10}$$

And the lockup time T_Q in this case is given by the expression for the signal + noise $\approx \frac{5}{B_n}$

$$T_Q = \frac{5 \times 10^9}{23.4 \times 10^6} = \underline{\underline{214 \text{ nsec}}}$$

For an R_Q of 133 ohm,

$$C_Q = \underline{\underline{0.0015}}\ \mu\text{F}$$

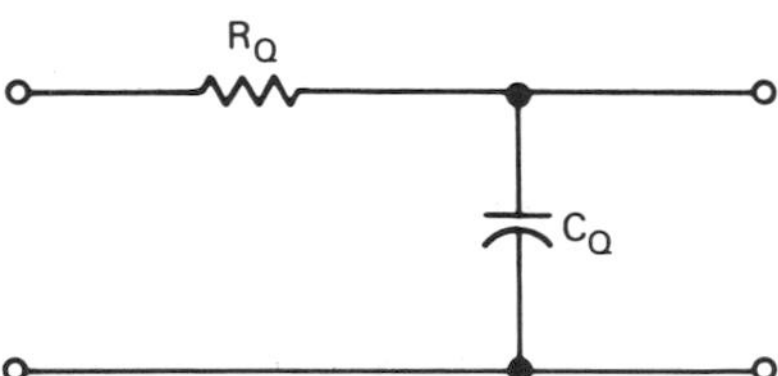

When the quadrature synchronous phase-detector is integrated with the in-phase synchronous phase-detector, T_L, the in-phase pull-in lockup time in the presence of noise is modified from 24 nsec to the following value:

$$T_L \approx \frac{4}{B_n} = \frac{4 \times 10^9}{46.8 \times 10^6} = 85 \text{ nsec}$$

That is, with the complete dc-quadricorrelator setup, the primary in-phase lock-up time is actually 85 nsec to satisfy the 50-to-100 nsec requirement.

Conclusion

- The use of coherent synchronous in-phase and quadrature phase detectors assures a 3-dB improvement of SNR due to the effective *averaging-out* of the signal against noise (besides that provided by a narrow-band tracking filter). Presently, television microwave links work at 8-dB SNR, as compared to the 12 dB commonly used in radar systems.
- The use of a unique dc quadricorrelator (as used in high-sensitivity color television

systems) enables the automatic switch-in of a narrow-band filter after a high-speed PLL lockup. The technique provides extended pull-in and hold-in frequency ranges, and high phase-accuracy to minimize phase-jitter.

- The quadrature synchronous phase-detector exclusively furnishes a technique to suppress or minimize the rate of false-alarm display due to EM Interference and jamming. This is an incidental dividend of this technique.
- Above all, the PLL-System proposed in this feasibility study as a replacement to the previous envelope-detectors, will enable a jitter-free LO frequency to the receiving system; it will incidentally furnish immunity against the effects of spurs.
- The dc quadricorrelator PLL is considered to be a powerful tool in communications and direction-finding systems.

B-5 DC QUADRICORRELATOR PHASE-LOCK LOOP TECHNIQUE

B-5.1. In the radar IF center-tune system application, the DC quadricorrelator PLL (DCQ-PLL) technique is proposed to provide automatic IF 60-MHz synchronization as an aid to the main VCO in the local-oscillator frequency generating system, for the Activity Image Phasing-Mixers (IPMs) of the Interferometer Direction-Finding system, as the radar emitter-frequencies are step-scanned by the digital computer through the various microwave-bands. That is, the device operation is simultaneously concerned with an improvement in the instantaneous frequency-measuring (IFM) capability of the system at SNR levels below the presently-practical average - 65 dBm sensitivity threshold-level. The facilty would therefore enable the acquisition of emitter signals below the common SNR levels.

The Frequency-Discriminator function involved is carried out by an analog/digital IF signal-processing system to generate the center-tune flag for the IF-window scanned. A coarse video phase-quantization is implemented for the formulation of the digital fine frequency-words in the IFM system. The DCQ-PLL subsystem is primarily applicable as an alternative to the coarse frequency/phase-discriminator used in the above implementation of the video phase-quantization. See Fig. B-1.

B-5.2. In the present direction finding systems, the *FM/phase-demodulator* consists of the system setup shown in Figs. B-2 and B-3.

At any incoming IF signal-frequency, the phase-relationships between the coarsely delayed (50.53 nsec) and undelayed outputs can be arbitrarily established for the relative phase-difference as a function of frequency.

$$\text{Phase-delay } \varphi = 2\pi d/\lambda = Kf \ldots \text{ where d is the delay}$$

As the incoming Real-IF center-tune signature varies, the phase-relationship will cycle periodically in multiples of 2π rad. The cyclic-frequency is then represented by the reciprocal of the time-delay. The delay-line and the direct reference-channel, respectively, drive the pair of in-phase and quadrature vector-detectors. The latter is comprised of a correlator or frequency-discriminator with *crystal video-detectors.* The correlator assembly is generally called a *phase-comparator* in the present system, and it converts the relative-phase of the delayed output with respect to the undelayed reference-output into *two orthogonal components.* As the IF is varied over a range equal to the reciprocal of the time-delay d, the vector-detectors describe four curves which, when differentially combined, generate the *sine- and negative cosine-function outputs.* The following *low-gain quantizer* synthesizes the coarse phase-data in these

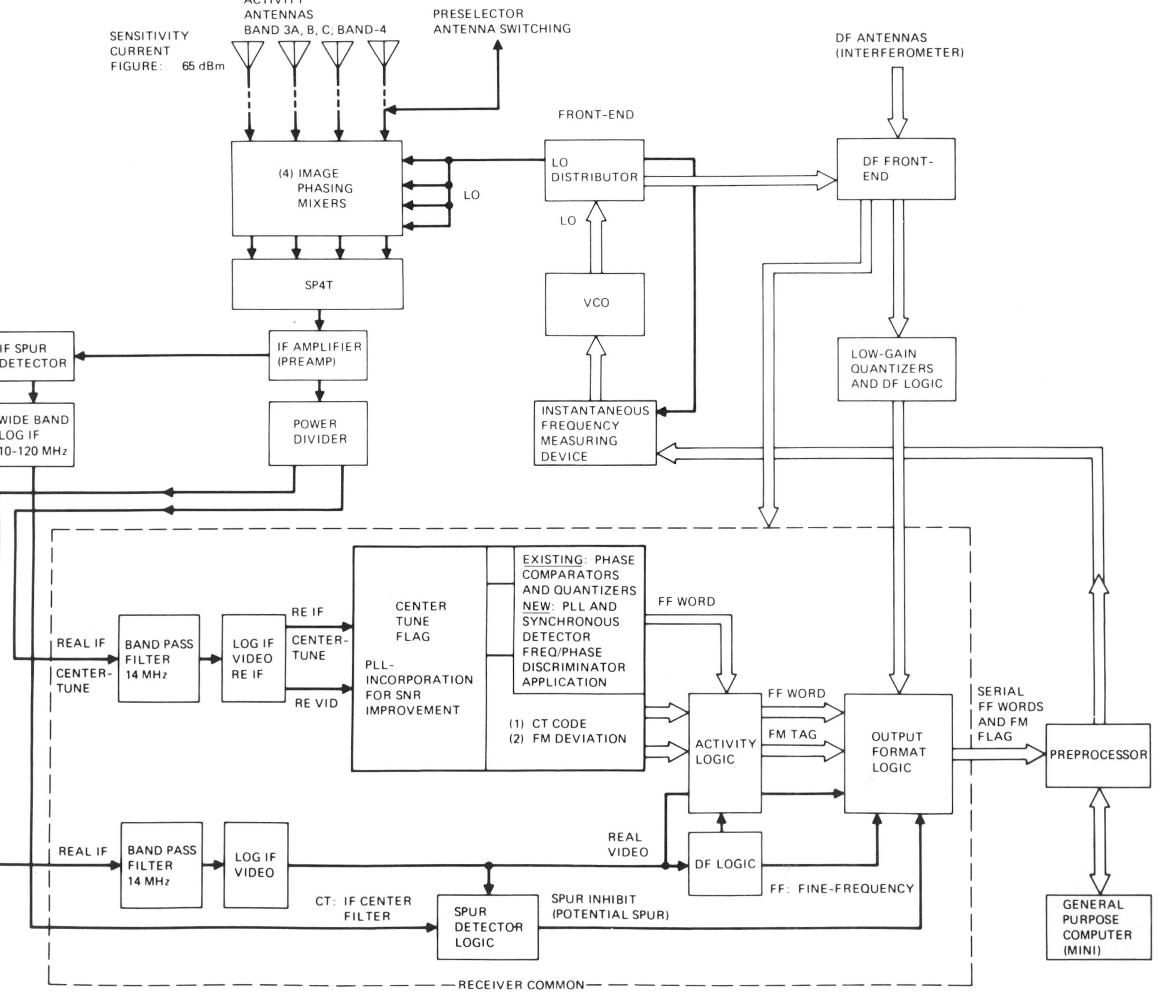

Fig. B-1. IF center-tune subsystem–interferometer superheterodyne computerized radar receiving subsystem.

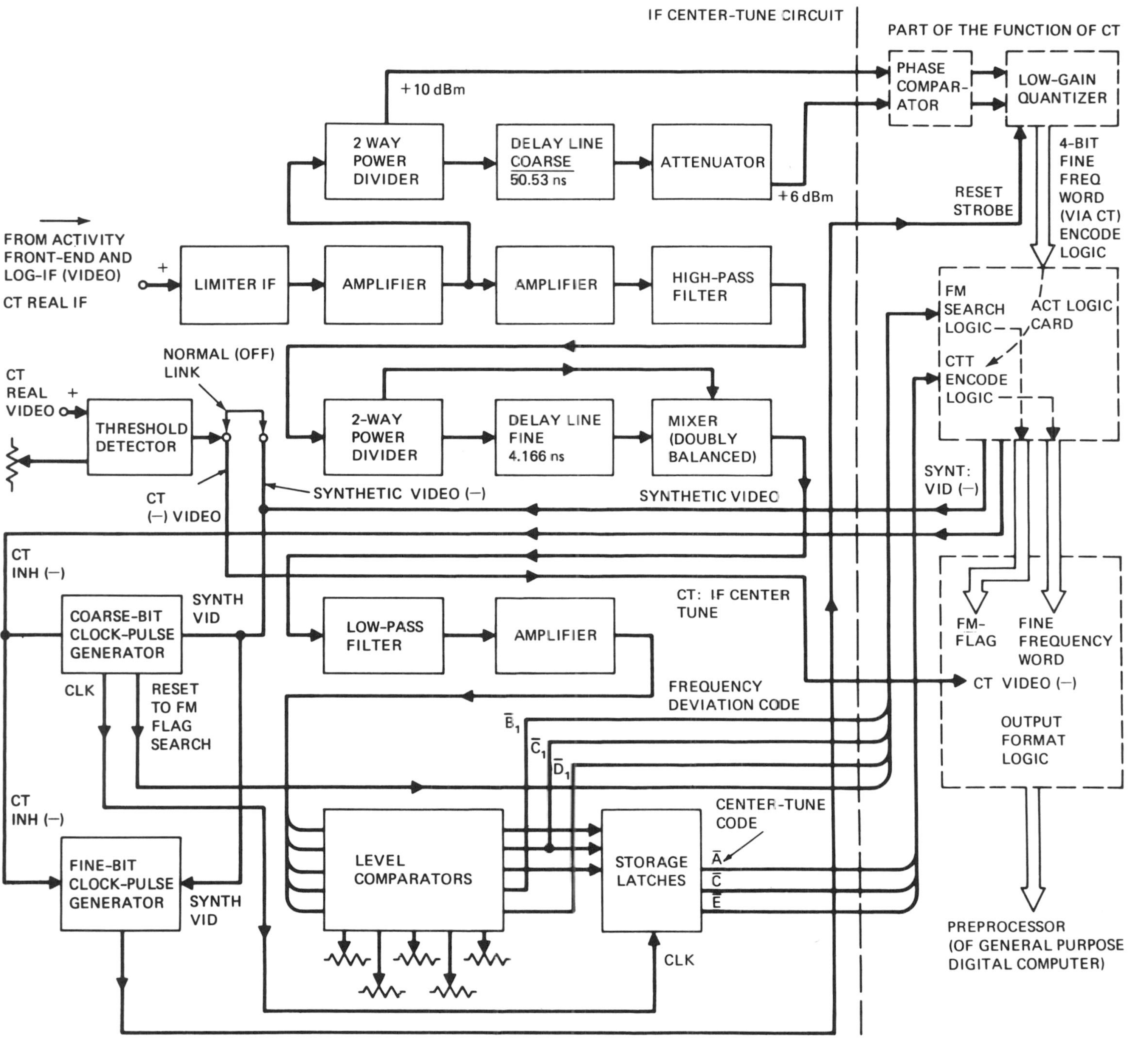

Fig. B-2. IF center-tune circuit for the incorporation of a phase-lock loop using a frequency/phase discriminator.

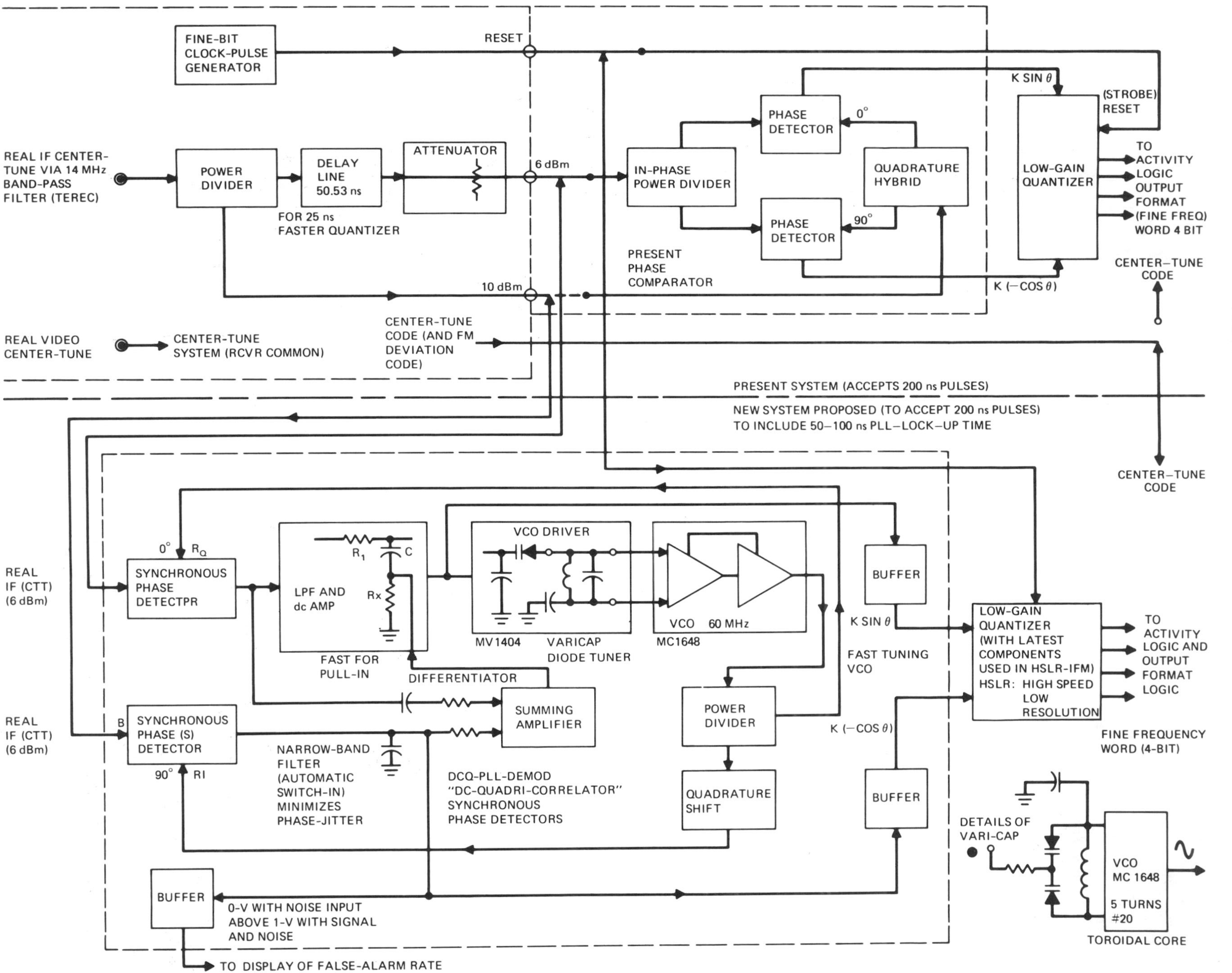

Fig. B-3. Phase-lock loop frequency-discriminator; DCQ-PLL synchronous demodulators for signal-to-noise ratio (and hence sensitivity) improvements.

curves to generate the corresponding digital binary-code for the subsequent formation of the fine frequency-word in the activity-logic connected with the emitter interception. This is done in conjunction with the fine center-tune code, which is generated simultaneously inside the center-tune circuit by means of a fine 4.166-nsec delay-line, a balanced-mixer/phase-detector and level-comparators. The center-tune circuit of the receiver incidentally generates an FM-deviation code for the formulation of the FM-indicator flag in the signal-processing circuits connected with the activity-logic.

B-5.3. In order to improve the SNR, the *DCQ-PLL demodulator using synchronous phase-detectors* is proposed as an alternative to the phase-comparator system. The lower-half of Fig. B-3 illustrates the basic integration of the DCQ-PLL demodulator into the conventional system. In view of the advances in MSI components, it is feasible to minimize the propagation-delay involved in the following low-gain quantizer for compensating against any trade-off between the lockup time of the PLL and the SNR improvement. It may be interesting to note in this context that the present radar receiving systems require an rf pulse-width of 100 to 150 nsec. The rise-time may be as low as 30 nsec. So, the applicable PLL-demodulator should primarily meet the requirement of a lockup time between 50 to 100 nsec. This is a valid proposition for a DCQ-PLL demodulator operating at 60 MHz.

The term *dc quadricorrelator* is derived from *quadr*ature *i*nformation-*correlator* and has been generally applied to synchronization systems that make use of the correlation existing between a pair of measurements at phase-quadrature in the presence of noise. When the phase-lock is effective, the phase-jitter is minimal, and the average cosine-output of the quadrature phase-detector approaches the value of unity as a transition from zero. The smoothed dc quadrature-output is called the *correlation output*, and it is suitable for applications such as automatic gain control (AGC) and generation of a flag to indicate the phase lockup condition.

B-6 SUMMARY

B-6.1. The dc quadricorrelator is a two-mode synchronization system that performs automatic frequency-/phase-control as a PLL, either as a tracking system or as a frequency-/phase-correlator (PLL-demodulator). The two modes of operation refer to independent pull-in and hold-in capabilities on an automatic switching basis, as the emitter signal-frequencies are scanned and intercepted in a computer-controlled IFM interferometer receiving system. The dc quadricorrelator technique simultaneously allows the incorporation of synchronous phase-detectors in the IF center-tune circuit, as an adjunct to a PLL-Demodulator. This naturally facilitates the added inherent advantage of *coherent integration of synchronous phase-detectors* at an improved SNR figure.

B-6.2. If V and P are the voltage and power of a threshold signal, and V_n and P_n are the rms-voltage and power of noise, respectively, and N_s is the total number of cycles in an AM pulse-envelope of an emitter-signal, it can be shown for fixed- or moving-targets that, for a train of pulses at the output of a linear second-detector, the *average noise-fluctuation* is reduced in accordance with

$$V^2 \approx V_n^2/\sqrt{N_s} \text{ or } P \approx P_n/\sqrt{N_s}$$

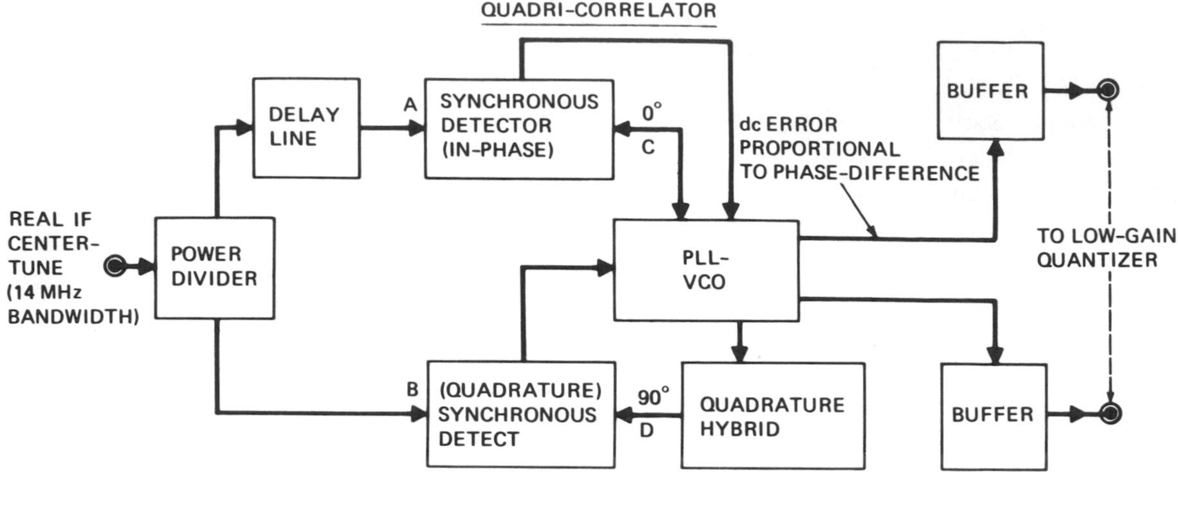

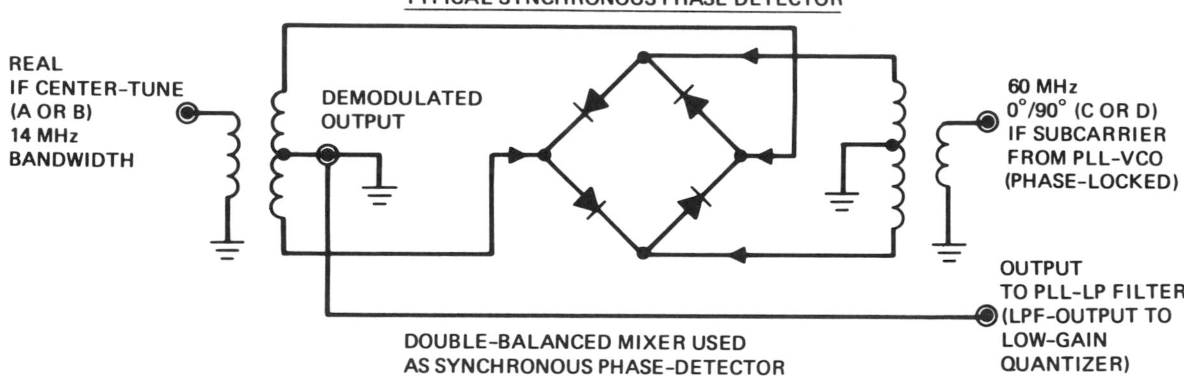

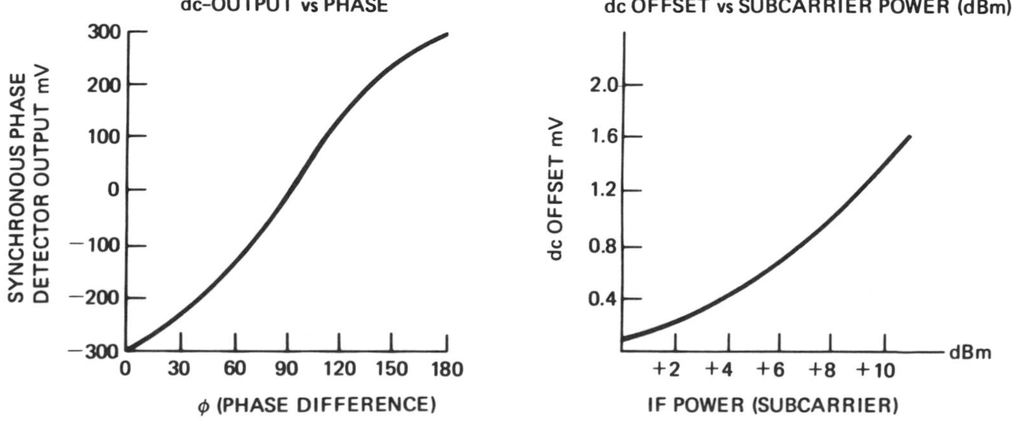

TWO DOUBLE-BALANCED MIXERS USED AS A PAIR OF SYNCHRONOUS PHASE DETECTORS (ONE IN-PHASE AND THE OTHER QUADRATURE) IN A dc-QUADRI-CORRELATOR OF A PLL-DEMODULATOR AS PART OF THE CENTER TUNE SUBSYSTEM OF THE RECEIVER FOR THE FORMULATION OF THE FINE FREQUENCY-WORD.

WHEN TWO SIGNALS OF IDENTICAL FREQUENCY BUT DIFFERENCE PHASE ARE APPLIED TO THE INPUTS OF A DOUBLE-BALANCED MIXER, THE OUTPUT (AT THE USUAL-IF PORT) WILL PROVIDE dc VOLTAGE PROPORTIONAL TO THE PHASE-DIFFERENCE, APPLICABLE (AFTER DUE PROCESS) TO THE VARICAP-DIODE TUNER OF THE 60 MHz VOLTAGE-CONTROLLED OSCILLATOR (VCO)

IN THIS PHASE DETECTOR APPLICATION, THE BALANCED CONFIGURATION REDUCES INPUT SIGNAL INTERACTION AND ELIMINATES THE NEED FOR TIME-CONSUMING NULL ADJUSTMENTS. (FOR OPTIMUM RESULTS, THE TWO INPUT SIGNALS, REAL-IF (CENTER-TUNE) AND THE PHASE- LOCKED IF SUBCARRIER SHOULD BE APPROXIMATELY OF THE SAME AMPLITUDE. (INPUT LEVELS OF 0 TO 7 dB ARE OPTIMUM FOR THE BALANCED PHASE CHARACTERISTIC OF THE OUTPUT.)

B-4. Synchronous phase detectors. Phase-lock loop demodulator (DC-quadricorrelator).

However, A. G. Emslie (1944) has shown that (1) by using coherent integration and letting the signal beat with a phase-locked subcarrier so as to present a reference phase-relation to the successive signal pulses, and (2) if the amplitude of the coherent continuous-wave is large compared with that of the rms Gaussian noise-distribution, the preceding condition will take the following form:

$$P \approx P_n / N_s$$

whether or not the detector is linear. The preceding coho-dependence has been checked experimentally, and the *injection of the coho does increment the threshold signal-power by approximately 3 dB*. (In fact, it is this very condition that makes the operation of a superheterodyne-converter approximate to a linear process in its function.) So, the incorporation of the synchronous phase-detectors with the phase-locked 60-MHz subcarrier will by itself provide a boost of 3 dB to allow a figure of 9-dB SNR in the place of the present 12-dB limit at the activity-antenna of the receiving system.

Appendix C
Basic Concepts of Feedback Control Systems

C-1 SERVOMECHANISM

The servomechanism is a power amplifying device in which the signal actuating the output is equal to the difference between the input and the output. Alternatively, a servomechanism is a feedback control system in which the *controlled variable* is basically a mechanical position.

C-2 FEEDBACK CONTROLLER

The feedback controller is a control system that tends to maintain a prescribed relationship between an input command variable and a feedback controlled variable by comparing the transfer functions of these variables and using the error relationship so obtained *as a means of control for establishing a* feedback control system with a *stable* controlled output.

C-3 OPEN-LOOP CONTROL SYSTEM

In the open-loop control system, the input command signal is applied manually by setting a dial position, and the output (amplified or otherwise) is normally predetermined by *calibration*.

Advantage. Simple and stable with high gain.

Disadvantage. The controlled output depends on validity of calibration, load changes, and component changes with temperature, humidity, and lubrication, etc.

C-4 CLOSED-LOOP CONTROL SYSTEM

In a closed-loop control system, the actual output is measured, and a *proportional* signal is fed back to the input for comparison against the input to obtain the desired output by means of the *actuating error signal rather than the calibrated signal.* They are specified under different classifications as (1) on-off controllers set by a thermostat, (2) step-controllers with one or more timing mechanisms to preset time, (3) continuous analog feedback control systems, and (4) sampled-data or digital control systems using digital filters.

Advantages. Variations in load or major components have little affect on accuracy. No periodic calibration is necessary.

Disadvantages. Conditionally stable with increase or decrease in gain, or unstable with increased gain.

C-5 STABILITY

Stability is the characteristic of a system or element whose response to *a stimulus* (such as an impulse) subsides when the stimulus is removed. In other words, the transients that result due to the application of the impulse gradually disappear within a finite interval.

If the system is stable, any bounded input or excitation will result in a bounded output response.

The stability of the system depends upon the system only, and not the driving-point function.

If the system is *unstable*, any type of excitation, even noise, will cause the system to *oscillate* at random.

Mathematically, the system is stable if it has *no singularity in the right-half complex plane*, except an isolated single pole with a finite residue at the origin.

A *nonlinear* system may be specified as stable if an amplitude-limited oscillation or error remains within a prescribed magnitude.

C-6 CHARACTERISTIC EQUATION AND STABILITY

$1 + KGH = 0$ is the *characteristic equation* if K is the gain-constant, G is the forward loop transfer function, and H is the feedback transfer function. This is the basis of investigation for stability in elementary feedback control systems by common *frequency response* techniques such as Bode, Nyquist, Routh-Hurwitz, and the *time-response* technique *root-locus.* The exact roots of the high-order polynomial characteristic equation determine the degree of stability of the system represented by the equation.

The *loop ratio* is the transform ratio or frequency response of the primary feedback (B) to the actuating signal (E) under linear conditions:

$$\frac{B}{E}(s) = KG(s)H(s)$$

The *closed-loop transfer function* is the transform ratio of the controlled variable C to the reference input R

$$\frac{C(s)}{R(s)} = \frac{KG(s)}{1 + KG(s)H(s)}$$

The *actuating signal ratio*

$$\frac{E(s)}{R(s)} = \frac{1}{1 + KG(s)H(s)}$$

C-7 CLASSIFICATION OF FEEDBACK CONTROL SYSTEMS

If H = 1 in the characteristic equation, 1 + KGH = 0

Example:

$$KG = \frac{K_1 \left[\left(s + \frac{1}{T_1}\right)\left(s + \frac{1}{T_3}\right)\right]}{s^n \left[\left(s + \frac{1}{T_2}\right)\left(s + \frac{1}{T_4}\right)\right]} \tag{1}$$

where

$$K = \frac{K_1 T_2 T_4}{T_1 T_3}$$

Type "0" system, if n = 0 in Eq. 1. A constant value of the controlled variable requires a constant actuating error signal, and results in a limited position error. (Examples: speed regulator, voltage, current, and temperature regulators.) The *static (position) error coefficient* K_p is given by:

$$K_p = \lim_{s \to 0} \frac{C(s)}{E(s)}, \text{ for constant output}$$

Type "1" system, if n = 1 in Eq. 1. A constant rate of change of the controlled variable requires a constant actuating error signal, to furnish a *lagging* error at constant velocity. No static position error results in this case, as in type "0" system.

$$\text{Static Velocity Error Coefficient} = \frac{\text{Velocity of output}}{\text{Applied error}}$$

$$K_v = \lim_{s \to 0} \frac{sC(s)}{E(s)} \text{ for constant velocity output}$$

Examples: position control systems, controllers (dc, hydraulic motors).

Type "2" System, if n = 2 in Eq. 1. A constant acceleration of the controlled variable requires a constant actuating error signal with constant error in acceleration only. (No position or velocity error.)

$$\text{Acceleration Error Coefficient } K_a = \frac{\text{Acceleration of output}}{\text{Applied error}}$$

$$K_a = \lim_{s \to 0} \frac{s^2 C(s)}{E(s)} \text{ for constant acceleration output}$$

Examples: Torque motors, position control with a pilot motor driving a control element, that in return controls the speed of the main drive motor.

Dynamic Error Coefficients

$$E(s) = \frac{R(s)}{1 + G(s)}$$

Using Maclaurin series (or division),

$$E(s) = \frac{1}{(1 + K_p)} R(s) + \left(\frac{1}{K_1}\right) sR(s) + \left(\frac{1}{K_2}\right) s^2 R(s) + \cdots \tag{2}$$

If $(1 + K_p) = K_0$, then K_0, K_1, K_2 are called *dynamic error coefficients.* They are the reciprocals of the coefficients of various derivatives. They indicate proportionality between dynamic error components and input derivatives. They are actually steady-state error coefficients, since the transient terms vanish during the series expansion. The dynamic error coefficients involve the time constants of the system; therefore, both high-gain and low system-time-constants are equally important for the dynamic performance of a control system.

C-8 STEADY-STATE ERROR AND TRANSIENT ERROR

The *steady-state error* is the error that remains after the transient has expired. A nonhomogeneous linear differential equation provides the particular solution, which is the steady-state component of errors. This error, which is an input-output variable, is one of the measures of servo performance; the driving functions must be determined to obtain the steady-state error.

The *transient error* is the difference between the system error at any time and the steady-state system error for a specified positive stimulus. (The transient- or time-response is the output as a function of time, following a prescribed input under specified conditions of operation.)

Driving Functions. Three types of driving functions are commonly used in basic feedback control systems.

A *step* function is used in position control systems:

1. Type "0": $R(s) = A/s$ for step function
2. Type "1": A *ramp* function is used in position control, or a *step* function is used in velocity control.

$$r(t) = v.t. \text{ Hence } R(s) = v/s^2$$

3. Type "2": A *step* function is used in acceleration control

$$r(t) = \tfrac{1}{2}\, at^2. \text{ Hence } R(s) = a/s^3$$

C-9 LINEAR CONTROL SYSTEM

Most linear control systems may be approximated to a second-order linear differential equation with constant coefficients.

$$\frac{d^2y}{dt^2} + 2\zeta\omega_0 \frac{dy}{dt} + \omega_0^2 y = f(t) \tag{1}$$

The corresponding Laplace transform equation:

$$F(s) = s^2 + 2\zeta\omega_0 s + \omega_0^2 \tag{2}$$

The two parameters ζ and ω_0 are sufficient to describe a second-order differential equation. When ζ, the *damping factor,* is small, the system gets more oscillatory. When ω_0 (the *natural frequency*) is large, the system reaches steady-state in a longer time-interval. Thus, the speed of response of the control system depends on both ω_0 and ζ. Only by changing the components can the (undamped) natural frequency ω_0 be changed.

In a closed-loop frequency response function containing one pair of complex roots, the corresponding Laplace transfrom equation is the characteristic equation.

Thus,

$$\frac{C(s)}{R(s)} = \frac{G(s)}{1 + G(s)} = \frac{K\omega_0^2}{s^2 + 2\zeta\omega_0 s + \omega_0^2} = \frac{K\omega_0^2}{(s + \sigma_0 + j\omega_d)(s + \sigma_0 - j\omega_d)}$$

where

ω_0 = natural frequency
ζ = damping ratio
$\sigma_0 = \zeta\omega_0$ = damping exponent
$\omega_d = \omega_0\sqrt{1 - \zeta^2}$ = natural damped (oscillatory) frequency.

For the characteristic (polynomial) equation $1 + GH(s) = 0$, with $H = 1$. Using partial fractions, with known R(s)

$$C(s) = \frac{A_1}{s} + \frac{A_2}{s + a_2} + \frac{A_3}{s + a_3} + \cdots$$

Transient response,

$$C(t) = \Sigma A e^{-\alpha t} + \Sigma M e^{-\beta t} \cos(\omega t + \theta)$$

$$= \text{Exponential term} + \text{damped sinusoid}$$
$$\text{(with } s = -\alpha) \qquad \text{(with } s = -\beta \pm j\omega_d)$$

The conjugate pairs become *damped sinusoids.* For higher degree differential equations, it is difficult to find the roots by using the root-locus techniques. Hence, the frequency response techniques of analysis are more popular.

C-10 LINEAR AND NONLINEAR CONTROL SYSTEMS

Linear systems. Analysis and design of systems are concerned with linear differential equations and their solutions. The characteristic (polynomial) equation will have roots, and they determine the transient response of the system. The *principle of superposition* holds, and the Fourier integral gives the formal relation between the *time-domain* and the *frequency-domain*, justifying the application of the common frequency-response techniques of analysis and design.

Nonlinear systems. There is no concept of roots for a nonlinear differential equation in nonlinear systems, and hence the transient response of the system cannot be defined. The principle of superposition does not hold, and there is no formal relationship between the time and frequency domains. If frequency-response techniques are used, the results are empirical as some linear-equivalent. However, *numerical methods* (using digital computation) can be used for approximating a single solution of transient-response to a specific input disturbance only. The problem of organizing a general theoretical technique still remains, although the best control signal-processing results can be achieved with nonlinear systems (by using heuristic measure-and-optimize techniques).

Types of nonlinearity. Amplifier saturation; motor sensitivity-limit due to coulomb friction; dead-band in overmodulation; hysteresis in motor characteristics; motor velocity and acceleration limits due to magnetic saturation; inherent motor nonlinearities due to multiphase characteristic; relay-switching; delay-time; granularity of follow-up potentiometer (helipot); backlash in gearings; and affects of time-varying (slow and rapid) system characteristics relative to actual response time. For a nonlinear control system, simultaneous signal levels and the operating range of all inputs must be specified in order to define the actual system performance.

Other nonlinear phenomena. Jump-resonance in lightly damped control systems when sinusoidally excited; limit cycle or bounded oscillation as explained by the phase-plane technique; subharmonic generation in systems with backlash and magnetic hysteresis; and intermodulation effects on gain with affect of noise on systems with saturation.

The *describing function* approach, using a complex quantity that gives both the amplitude and the phase relationships between the input and the output, is commonly used as a modification of the frequency response technique, such as Nyquist. The describing function as a function of amplitudes, quasi-linearizes the frequency response equations. This approach is feasible only if the output is periodic at a specific fundamental frequency, and the nonlinear element is *not time-varying.* (Other techniques such as phase-plane, piecewise linearization, Lyapounov's Second Method, and modern statistical control approaches are explained in Chapter 1.)

C-11 FREQUENCY RESPONSE OF A FEEDBACK CONTROL SYSTEM

The frequency response of a feedback control system (or the transfer function of an element) is determined by (1) the steady-state ratio of magnitude and (2) the difference-in-phase of the sinusoidal output with respect to the input. The range of frequency and the conditions of operation and measurement must be specified. In the time-domain, the transient response is determined by inputting a unit-step test waveform. The frequency-domain specification includes (1) the 3-dB down bandwidth, (2) dc performance at zero frequency, (3) freedom from resonant peaks, and (4) freedom from noise-component frequencies in a specific band.

At normalized gain, the maximum peak in transient response is limited to 2, and the peak frequency-response is limited to 1.3 to 1.5.

C-12 CONVOLUTION INTEGRAL AND TIME RESPONSE

The time response of a control system to an arbitrary driving function is best calculated by means of the convolution integral:

$$C(t) = \int_{-\infty}^{t} f(\tau)y(t-\tau)\,d\tau$$

where $f(\tau)$ is the input driving function, and $y(t)$ is the *weighting function* or characteristic time response to a unit impulse.

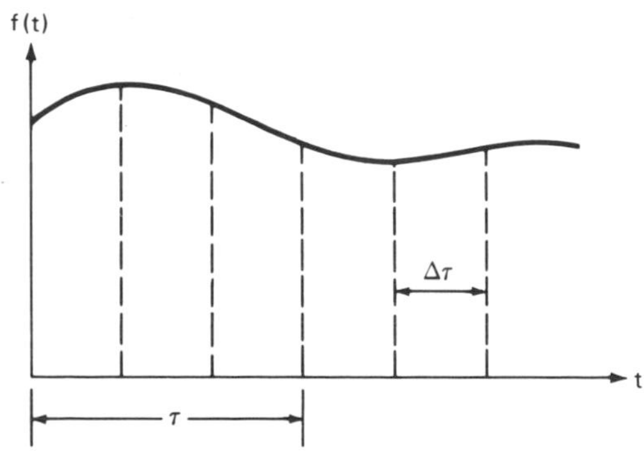

The weighting function y(t) is the exclusive inverse $\mathcal{L}$-transform of the transfer function f(t) and depends on the system only. If the weighting function is stable, all the time constants of the exponentials will be negative.

$$y(s) = \int_o^\infty y(t)e^{-st}\,dt$$

$$\underset{s \to 0}{y(o)} = \int_o^\infty y(t)\,dt = 1$$

i.e., if y(o) = 1, total weight = 1

$$c(t) = y(t) \cdot f(t)$$

To evaluate this equation, the arbitrary input f(t) is approximated by means of a series of impulses. If the impulse response y(t) is known, the sum of these responses to the impulses approximating the input signal constitutes the total time response.

$$\text{Response at time } t_1 = C(t_1) = \sum_{\tau = \tau_1, \tau_2, \ldots t_1} f(\tau)\,\Delta\tau\, y(t_1 - \tau)$$

C-13 NODAL AND MESH ANALYSES

Circuit analysis is generally approached by the method of nodal or mesh analysis. In nodal analysis, the algebraic sum of the currents at a junction (or node) is equal to zero (Kirchoff's first law). This concept enables the determination of an unknown voltage, given driving voltages and impedances.

Mesh analysis uses the voltage summations about a closed loop as equivalent to zero (Kirchoff's second law). An unknown current can be determined in terms of known voltages and impedances.

C-14 MINIMAL PROTOTYPE RESPONSE FUNCTION IN DIGITAL CONTROL

The minimal prototype response function is defined such that

$$1 - k(z) = (1 - z^{-1})^m$$

where k(z) = C(z)/R(z), and the resultant order of k(z) in z^{-1} is minimum phase.

Thus, if the system is to follow unit-step, ramp, and acceleration inputs without steady-state error, then k(z)'s for minimal prototype response are given by

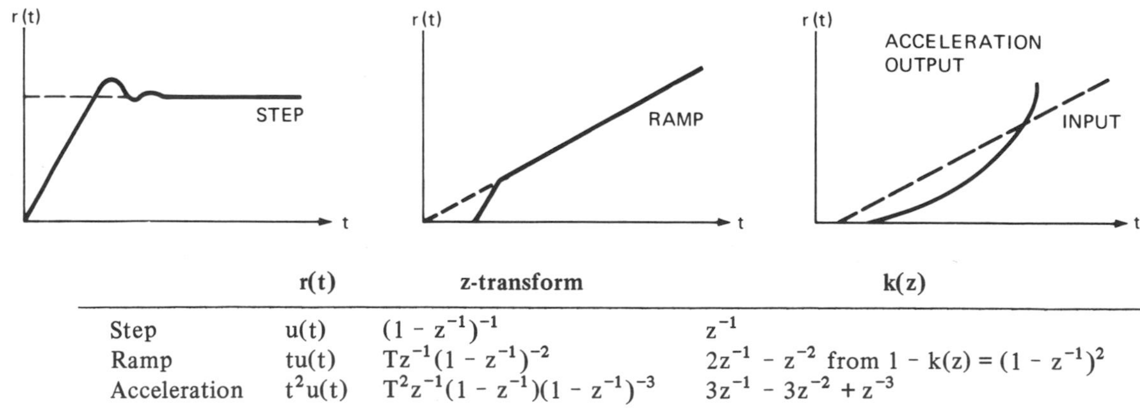

	r(t)	z-transform	k(z)
Step	u(t)	$(1 - z^{-1})^{-1}$	z^{-1}
Ramp	tu(t)	$Tz^{-1}(1 - z^{-1})^{-2}$	$2z^{-1} - z^{-2}$ from $1 - k(z) = (1 - z^{-1})^2$
Acceleration	$t^2u(t)$	$T^2z^{-1}(1 - z^{-1})(1 - z^{-1})^{-3}$	$3z^{-1} - 3z^{-2} + z^{-3}$

If a minimal prototype function is used, the system will respond without any error for a lower-order function in the steady-state (at the sampling instants).

Ripple-free system. In obtaining minimum finite settling-time at sampling instants, the servere shocks, which result in the plant, produce substantial ripple in continuous output, even though the output is correct at the sampling instants. In order to obtain ripple-free response, the feed-forward transfer function must generate a continuous output function that is closely similar to the input function. To meet this objective, the overall pulse transfer function must contain as its zeros all the zeros of the input pulse-transfer-function G(z), and not just the zeros of G(z) that lie outside the unit-circle in z-plane.

C-15 APPLICATION OF LAPLACE TRANSFORM TO CONTINUOUS FEEDBACK CONTROL SYSTEM

$$\frac{C(s)}{R(s)} = \frac{G(s)}{1 + G(s)H(s)}$$

where

$$C(s) = \frac{\text{Numerator}(s)}{\text{Denominator}(s)} = \frac{a_n s^n + a_{n-1}s^{n-1} + \cdots a_1 s + a_0}{b_m s^m + b_{m-1}s^{m-1} + \cdots b_1 s + b_0}$$

$$m \geqslant n$$

The solution $\mathcal{L}^{-1}[C(s)] = c(t)$ is obtained for simpler mathematical models by reference to Laplace transform tables. In general, it is necessary to express C(s) as sum of partial fractions with constant coefficients; their inverse transforms are then obtained from tables and added to yield the output time-response solution c(t).

An input function r(t) is represented by

$$A\left[\lim_{a \to o} \frac{u(t) - u(t - a)}{a}\right]$$

	Input, as F(s)	Response function, C(s)	Time Response as c(t)
Unit impulse: $A = \int_{-\infty}^{\infty} f(t)\,dt$	A	$A/(s\tau + 1)$	$\frac{Ae^{-t/\tau}}{\tau}$
Step: A = u(t)	A/s	$A/s(s\tau + 1)$	$A(1 - e^{-t/\tau})$
Ramp: A = A(t)	A/s^2	$A/s^2(s\tau + 1)$	$At - A\tau(1 - e^{-t/\tau})$

C-16 SAMPLED-DATA SYSTEMS

1. Systems that operate on data obtained at discrete intervals of time are called *sampled-data systems.*
2. The information obtained at a particular instant is called a *sample*.
3. Normally, the intervals are equally spaced in time, and the amplitude of the sample is proportional to the amplitude of the signal.
4. If the continuous or analog elements are linear, the sampled-data system is linear, and the superposition theorem is valid.
5. Notwithstanding the regular time-discontinuities, we can apply the solution of

a linear difference equation with constant coefficients (in the place of the linear differential equation) for analysis and synthesis.

6. The sampler acts as a pulse modulator of the input and generates a pulse train.

7. The high-frequency content inserted by the sampling process is attenuated by a *linear digital filter.*

8. The information contained in the input signal may be recovered with reasonable fidelity if the *sampling frequency* is at least twice the highest frequency component in the input signal (Shannon's *sampling theorem*).

The value of $\Delta\tau$, the maximum time-interval at which the values of the function are read: $\Delta\tau = \dfrac{1}{2f_s}$

Approximate Spectrum:

$$\Phi_{ap.}(\omega) = \frac{1}{2\pi}\int_{-\infty}^{\infty} u(\tau)\phi(\tau)\cos\omega\tau \cdot d\tau$$

The true spectrum is given by the Fourier transform of $\phi(\tau)$:

$$\Phi_{true}(\omega) = \frac{\tau}{\pi}\int_{-\infty}^{\infty} \Phi(\omega - \alpha)\frac{\sin\alpha\tau}{\alpha\tau}\, d\alpha$$

where $\dfrac{\tau}{\pi}\dfrac{\sin\alpha\tau}{\alpha\tau}$ is the transform of $u(\tau)$

C-17 BASIC FEATURES OF SAMPLED-DATA CONTROL SYSTEM

1. A digital computer can be used as part of the controller for compensation and processing instructions. The input data are fed in a sampled form via *quantizers* or analog-to-digital converters.

2. Simpler, low-power control elements can be used.

3. Advantage of finer *performance-index* in process-control of plants with inherent nonlinearities such as dead-time, hysteresis, etc.

4. Advantage of pulse-data information in most modern control systems. Input information is usually available in discrete samples in complex digital control systems of the classification of guided-missiles, telemetry, color videotape recorders, radar systems, etc. Thus, where digital sensors are already involved, sampled-data digital control is the obvious solution for higher accuracy.

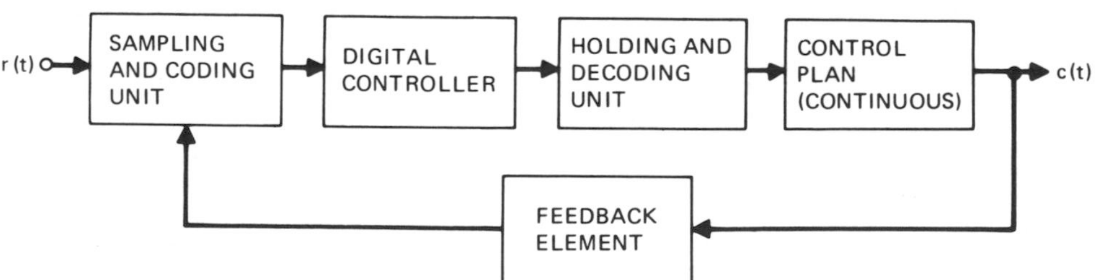

5. Sampled-data systems result in systems having dynamical performance; they cannot be matched by continuous systems from which they are derived. Real-time dynamic systems use the past samples, and data reconstruction by data-hold or digi-

tal filters is a *process of extrapolation* for one sampling interval by using the preceding sample.

6. *Sampling and quantization*. Sampling and *coding* operation is symbolized by a switch. The quantizing effect is practically ignored assuming that quantization is infinitely fine. The effect of sampling is described by linear difference equations, and they can be solved by *recursion formula* and numerical approximation. The quantizing effect, wherein the variables are generally quantized in amplitude (or phase), is much more difficult to account for, since it is described by a nonlinear equation. The unique property of a sampling process is that the output contains a small periodic component *ripple*, and it is possible to minimize or control the magnitude of this component.

7. The *z-transforms* in sampled-data control systems are analogous to Laplace-transform in continuous control systems. For systems with lumped constants and described by linear difference equations with constant coefficients, the z-transformation gives expressions that are rational polynomial ratios in the variable z. This variable is complex and is related to complex frequency s used in $\mathcal{L}$-transform by the relationship $z = e^{sT}$. The transfer function, mapping, inversion, etc. also hold good in z-transform theory.

8. The *digital controller* receives a sequence of numbers equally spaced in time; the controller processes the digital data in *real-time* into a command signal. It can be an active or passive compensating device. Its performance is described by a *pulse-transfer-function*, which is preceded and followed by synchronous samples. It enables the original continuous transfer function to achieve an overall stabilizing characteristic in the sampled-data system.

C-18 THE MINIMUM PHASE NETWORK

The minimum phase network is represented by a transfer function whose magnitude and phase are uniquely defined, and the specification of either one will signify the specification of the other. That is, for a specified magnitude characteristic, the phase minimum is possible for the transfer function at all the frequencies. The transform of a minimum-phase transfer function is one with no poles or zeros in the right half complex s-plane.

C-19 PHYSICAL MEANING OF MAKING s TRAVERSE THE $j\omega$-IMAGINARY AXIS IN THE COMPLEX s-PLANE

The statement implies the implementation of the *steady-state response* of the open-loop transfer function G(s)H(s). If the input is denoted by ($A \sin \omega t$) under steady-state condition, $B \sin (\omega t + \theta)$ corresponds to the output function. This specific condition allows the transients to die out to zero for a stable system with a gain-change of B/A and a phase-shift of θ, when the open-loop transfer function G(s)H(s) is represented as a vector with a corresponding magnitude and phase.

The concept is best illustrated by considering the Laplace transform of $\sin \omega t$:

$$\mathcal{L}[\sin \omega t] = \omega/(s^2 + \omega^2) = s/(s + j\omega)(s - j\omega)$$

The transform of this sinusoid is a pair of points on the $j\omega$ axis, with a distance of $\pm\omega$ from the origin. As the frequency of the sinusoid varies, ω varies, and s traverses the $j\omega$-axis.

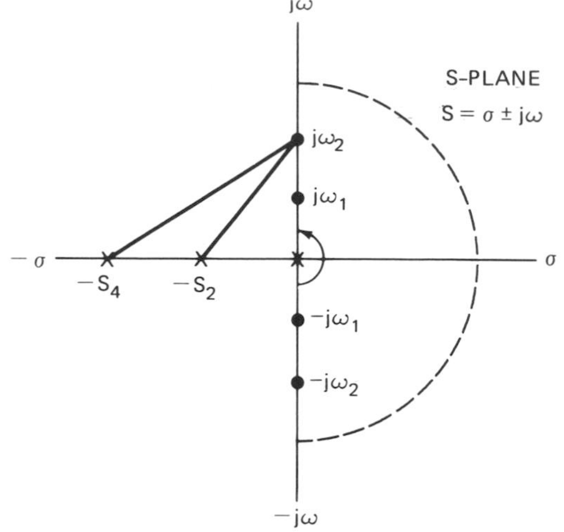

Origin of Nyquist plot. Physical systems have zero response to frequencies approaching infinite, since there are more poles and hence more jω elements in the denominator polynomial than in the numerator polynomial of the open-loop transfer function. Hence, the large semicircle (at ∞) in the s-plane in the right half-plane maps into a point at the origin in the KGH-plane. If there is a *singularity* at the origin on the jω axis in the complex s-plane, it is bypassed by a small circle about the origin in the counterclockwise sense. In the Nyquist plot, this minute circle corresponds to an infinite semicircle in the right-half plane; this semicircle is described in the clockwise direction for every bypassed s(pole) at the origin.

Incidentally, the *unit circle of the Nyquist criterion* corresponds to the *zero-dB line in the Bode plots.* The Nyquist diagram is a closed polar-plot of a loop transfer function, from which the *stability* of the representative system is determined. For a single-loop system, it is a *mapping* on the GH(s) plane of the s-plane contour that encloses the entire right-half of the s-plane. In the case of the Bode magnitude-phase plots, a *corner frequency* of each factor of a transfer function is the frequency at which the lines *asymptotic* to its log-magnitude characteristics intersect.

Relationship of pulse transfer locus of Nyquist to the frequency response function. Each point on the pulse transfer locus represents a point on the unit-circle of z-plane.

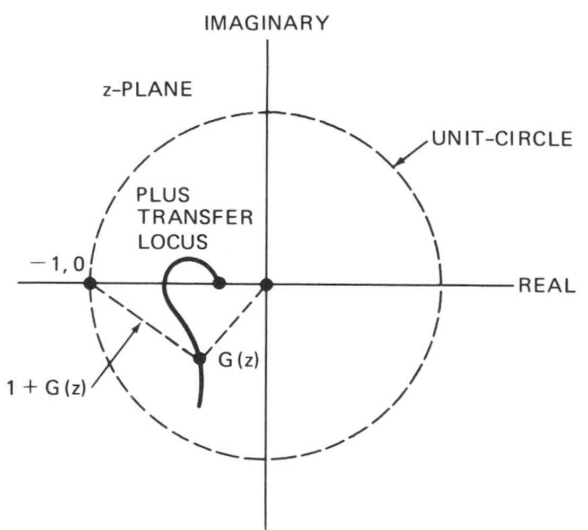

Viewed from frequency-domain, each point on the unit-circle represents a particular frequency.

For a unity feedback system, the pulse transfer function:

$$K(z) = G(z)/[1 + G(z)]$$

If the pulse transfer locus shown in the preceding diagram passes close to the critical point (- 1, 0), the function 1 + G(z) becomes very small, and the frequency response function hence becomes large. This peak in frequency reflects the fact that a pole lies close to the unit-circle, and the transient response will be relatively oscillatory (with ringing). In the time domain, we therefore get a satisfactory *transient response* with minimum overshoots by keeping the locus as far as possible away from (- 1, 0).

C-20 TYPES OF PERFORMANCE-INDEX OR CRITERIA OF GOODNESS IN CONTROL SYSTEMS

1. The open-loop gain gives a measure of the servo performance. Velocity error-constant, $K_v = 1000$ (good servo if stable); not satisfactory if too low. Position-error constant K_p also may be used.
2. In practice, most feedback control systems behave like second-order systems with a damping-ratio ζ and a natural-frequency component ω_n. They indicate damping and stability of the system. Speed of response depends on both the factors.
3. *Gain-margin* and *phase-margin* used in Bode/Nyquist plots (frequency-response methods) give a measure of stability and performance.
4. Minimum integral squared-error criterion is common for *deterministic* inputs such as step, sine, impulse (transient inputs).

a. ISE = $\int_0^\infty e^2(t)\,dt$, where $e^2(t)$ is mean-squared-error.

b. We use the following criterion for inputs of *stochastic* (random) nature. Mean-squared-error (MSE) of filtered outputs is also used, or time-weighted integral-squared-error: ITSE = $\int_0^\infty te^2\,dt$

c. Integrated absolute error: IAE = $\int_0^\infty |e|\,dt$

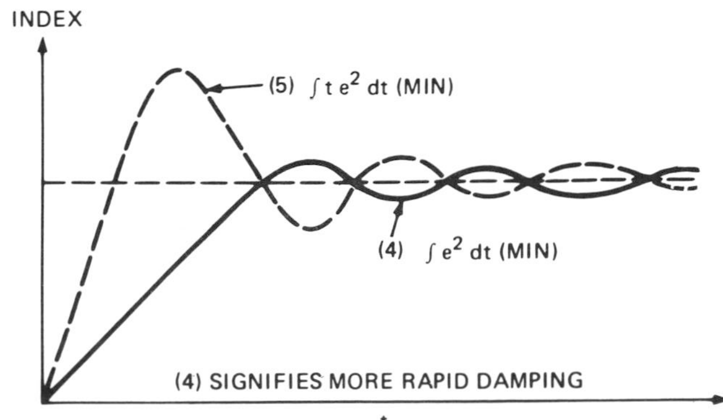

d. Time-weighted integral absolute error: ITAE: $\int_0^\infty t|e|\,dt$. This is the best performance criterion for optimal design of computer-controlled digital systems.

C-21 TYPES OF CONTROL PROBLEMS

A *regulator problem* is involved when some control aspect will have to be maintained constant in the presence of load disturbances. (Example: position, velocity or a combination thereof.)

A *servo problem* is involved when some control aspect of the system is a known or arbitrary function of time. (Example: Elevator, deceleration canceling, etc.; limited class of arbitrary functions such as those of an aircraft autopilot.)

A *navigation problem* is involved when a definite destination has to be reached despite wind, etc.

In a *ballistic missile and guidance problem*, a fixed trajectory path is required for the ballistic missile. No further guidance is involved after a present trajectory path; hence, it is more difficult than an aircraft or missile-guidance problem from the ground.

An *optimization problem* involves the achievement of the servo functions when the control aspect has to be optimized for the best performance-index.

An *adaptive control problem* involves the optimization problem in the presence of changes in environment.

C-22 POWER PLANT IN CONTROL SYSTEMS

1. The *amplidyne* (*metadyne or rotating amplifier*) is a constant voltage device. In short, it is a modified dc generator in which the armature flux is put to work by short-circuiting the armature and drawing off the current through an external load via a set of brushes and a compensating winding.

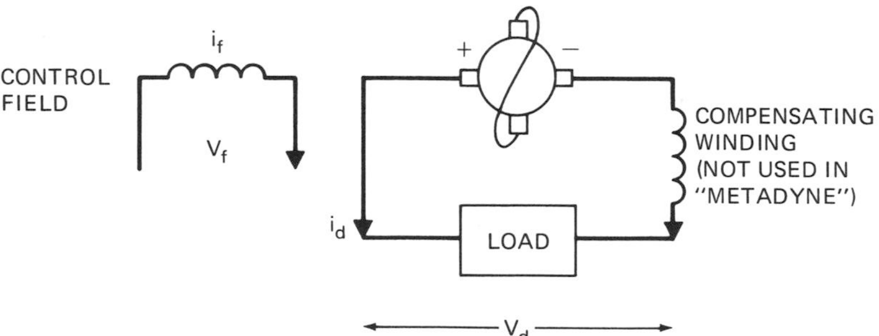

The compensating winding neutralizes any tendency of the magnetomotive force of the load current to produce a second armature reaction-reflux in opposition to the control excitation-flux.

$$\text{Amplidyne power gain} = V_d i_d / V_f i_f = \frac{10{,}000\ \text{W}}{1\ \text{W}} = 10{,}000$$

The excitation $V_f i_f$ is reduced from 100 to 1 W with the short-circuited armature in this power component.

2. *Ward-Leonard motor generator.* The dc motor is powered from a dc generator, whose output is controlled by a servo amplifier by changing the field current of the dc generator.

3. The *two-phase induction motor* is designed to provide maximum torque at stall, since in most applications the normal operation is a zero-speed region for position control. (See Guide-servo, p. 257.)

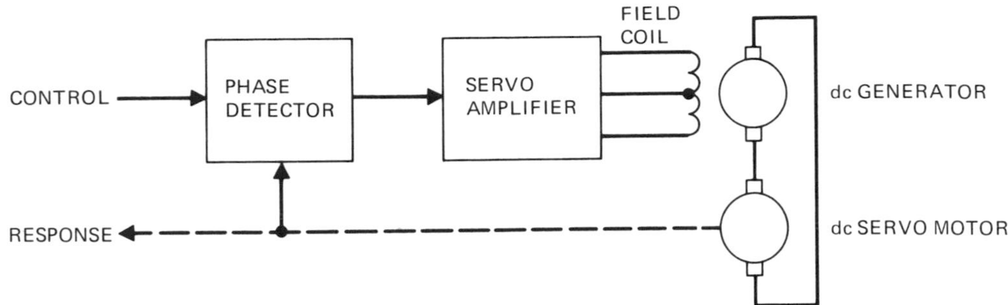

Slope and torque speed defines inherent viscous damping within the motor at a particular speed. Motor temperature affects motor impedance and so alters apparent motor gain. In high bandwidth applications, a second high-frequency breakpoint (for a secondary time-constant) is necessary to give an additional 90° phase-lag. Since the magnetic fields in the two coils (for reference and control excitation) are 90° out of phase, motor will rotate in a direction to correct error through the response-linkage to the input control-transformer. The rotating magnetic-field induces voltage in the rotor by transformer action to produce the required 90° out-of-phase magnetic field in the rotor with respect to the stator field. Maximum acceleration of a low inertia ac induction motor may reach 15,000 rad/sec^2.

Basic time-constants for 400-Hz motor: 10 to 30 msec.

For a 60-Hz motor, the time-constant is considerably lower. Lower motor time-constant is implemented by a rotor of small diameter to provide low rotor-inertia. A long rotor enables greater torque. For velocity servos subject to rapidly changing input rates, a high torque/inertia ratio is provided.

$$\frac{E_o(s)}{E_i(s)} = \frac{\text{Torque Coefficient}}{\text{Damping}} = K/s(sT + 1)$$

Gear drive. For minimum inertia, the motor-pinion is appropriately machined; external compensation will be unnecessary if internal damping and special gear-train are provided for the motor. (See Chapter 4 for three- and two-phase hysteresis asynchronous synchronous HAS motors of headwheel and capstan servos in the tape recorder.)

4. The *ac tachometer* is a special two-phase induction motor, designed so that

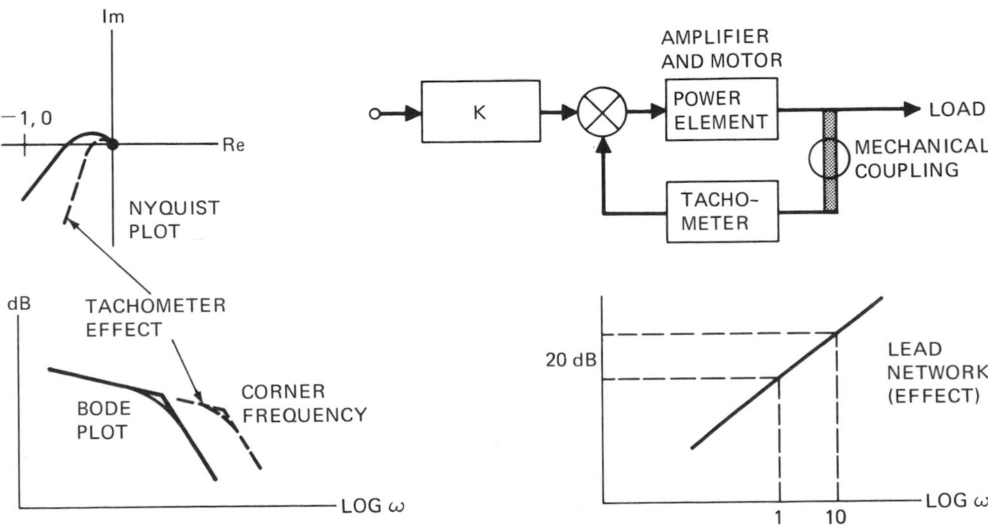

when one phase of stator is energized from constant amplitude ac voltage source, the other stator-phase produces a voltage of amplitude proportional to the speed with which the tachometer is driven by mechanical coupling to the power element of an angular position-control system.

As a feedback compensator, the tachometer shifts the corner-frequency (the pole near the origin) farther away to improve stability and transient response. The output is proportional to the time-rate of change of input.

Transfer function G(s) = Ks, where K is the sensitivity.

$$\text{Control error } e = K\,\frac{d\theta}{dt}$$

(Sometimes, a small dc generator with a permanent magnet is used as a tachometer.)

Example: A motor-tachometer (E-576, Electro-craft Corp.) is designed for high-performance dc instrument servos for precision velocity and position control systems with bandwidths up to 200 Hz. Both motor and generator armatures are assembled on the same shaft, but isolated magnetically and electrically. The motor-tachometer is ideal for low-speed digital tape capstan-servos and positioning drives in microfilm data retrieval system. Motor damping-factor is 4 oz-in./K rpm, with electrical time-constant of 2 msec. Tachometer linearity is 0.2% maximum deviation in either direction of rotation between 0 and 2000 rpm. Ripple is 5% peak-to-peak at 1000 rpm with 200-Hz RC-filter (5 K across tacho/RC: 12 K, 0.13 μF.) Dominant ripple-frequency 11 Hz/rev. Optimum load impedance: 5 KΩ.

5. There are presently two versions of the *moving-coil permanent-magnet dc motor: flat disk* (or *printed*) motor and shell (or cup) armature type. The moving-coil motor is characterized by the absence of rotating iron in the low-inductance, low time-constant armature. It provides high efficiency and good commutation; low inertia is not a requirement in its use. The armature time-constant is less than 0.1 msec. There is no reluctance torque-effect and hence no magnetic cogging.

The *printed* version uses an armature fabricated by photoetching. Stamped segments are arranged and joined to form a continuous conductor-pattern and a commutating surface. An eight-pole configuration provides a flux across an airgap of about 0.1 in. Current-flow is radial across the disk surface, while the rotary forces act tangentially. The outer end-turns provide a large inertia.

The *shell*-type motor with a cylindrical armature has the highest torque-to-

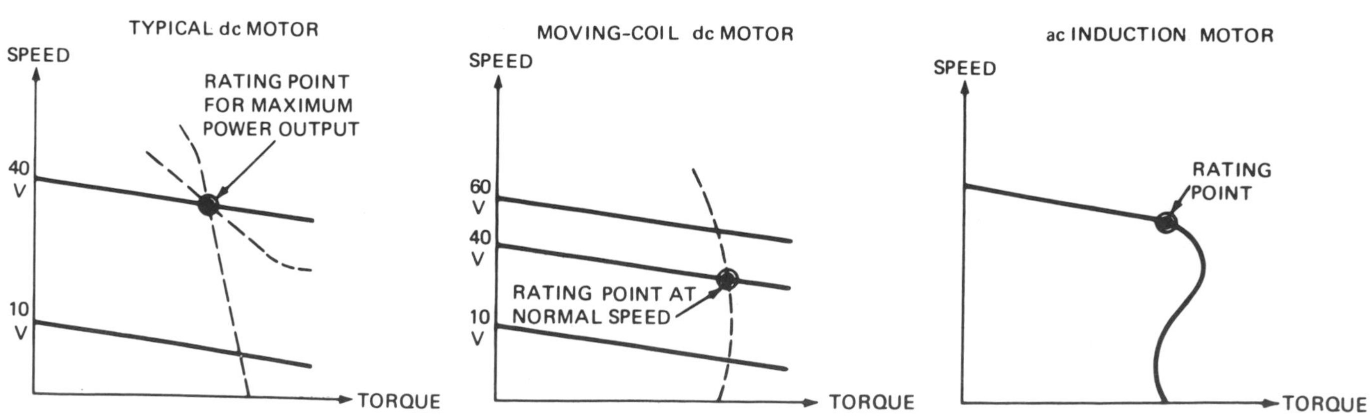

moment of inertia to provide a high acceleration capability of 1 million rad/sec^2. The Electro-craft M-1600 moving-coil servo motor is an example. The moving coil dc motor can handle a start-stop duty-cycle of several hundred cycles/sec. It is applicable to tape-transport capstan-servo operating at speeds of 250 in./sec, and 250 start-stop cycle/sec. Other applications include line printers, optical character readers, incremental motion drivers, phase-locked servos, machine tool drives, and video recorders.

M-1600 specification (uncooled moving-coil motor)
Maximum safe speed at rated load: 4.5 K rpm (7 K, no load)
Maximum continuous stall-torque: 60 oz-in.
Rated power output: 150 W
Rated power (continuous): 45 kW/sec
No-load acceleration at maximum pulse-torque: 1.7 $\times$ 10^6 rad/sec^2
Electrical time-constant: 0.1 msec
Damping factor: 1 oz-in./K rpm

6. The *brushless dc motor* (*BLM*) allows (1) the minimization of the effects of winding inductance and (2) control of torque and ripple as they relate to the turn-on and turn-off behavior of current in a segment of a conventional permanent-magnet dc motor. That is, the BLM dc motor should have the same linear torque-speed characteristics of the permanent-magnet (PM) dc motor, while its armature voltage is held constant. The nonlinear discontinuous nature of the step-motor's low-speed characteristic eliminates its use in velocity control applications. In the step-motor, the excitation voltage is held constant as the step frequency is varied. In the BLM case, torque is produced by the interaction of magnetic fields produced by a PM-mounted rotor shaft-and-hub assembly and by a dc current in the winding of an external stator structure. In a conventional PM dc motor, the stator and rotor situa-

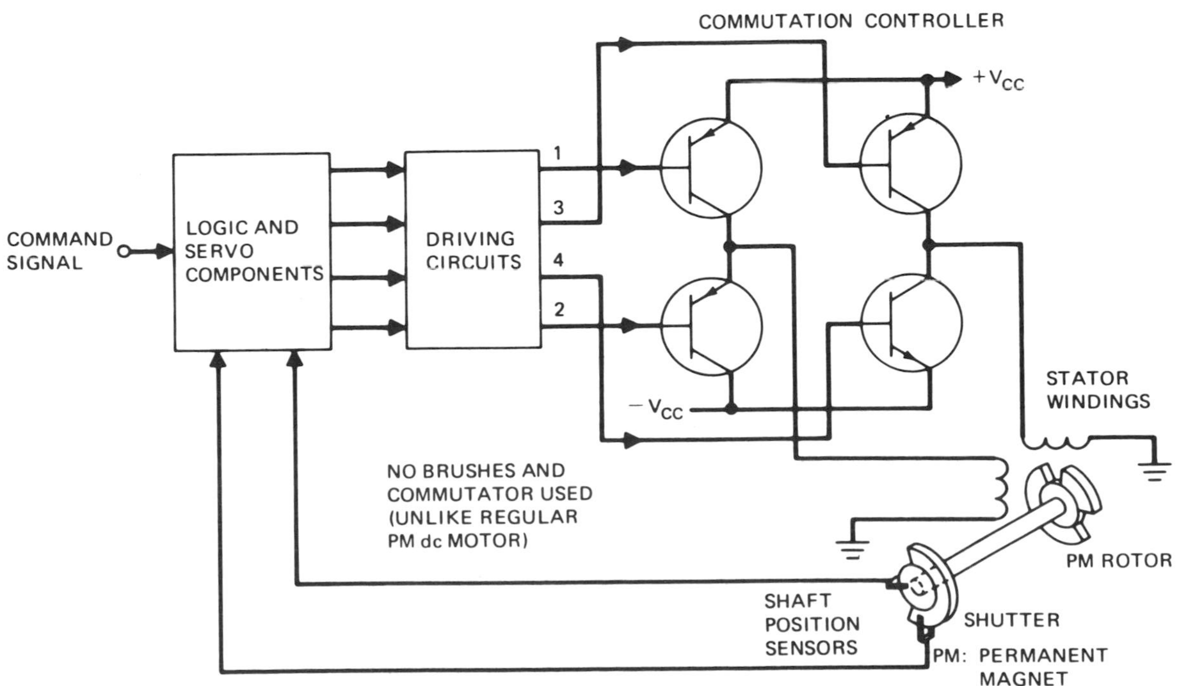

(From Electro-Craft Engineering Handbook, *D.C. Motors, Speed Controls, Servo Systems,* 4th Edition.)

tion is reversed. Electro-craft BLM-341, high-performance brushless dc servo motor specifications are:

a. Applicable to wide range of servo applications for computer peripherals and machine tool drives
b. Rated load torque: 80 oz-in.
c. Rated speed: 3600 rpm (no load: 4530 rpm)
d. Rated output power: 0.28 hp
e. Electrical time-constant: 1.85 msec
f. Armature inertia: 0.016 oz-in.-sec²
g. Acceleration at stall: 20,000 rad/sec²
h. Two of three phases energized sequentially by electronic commutation.

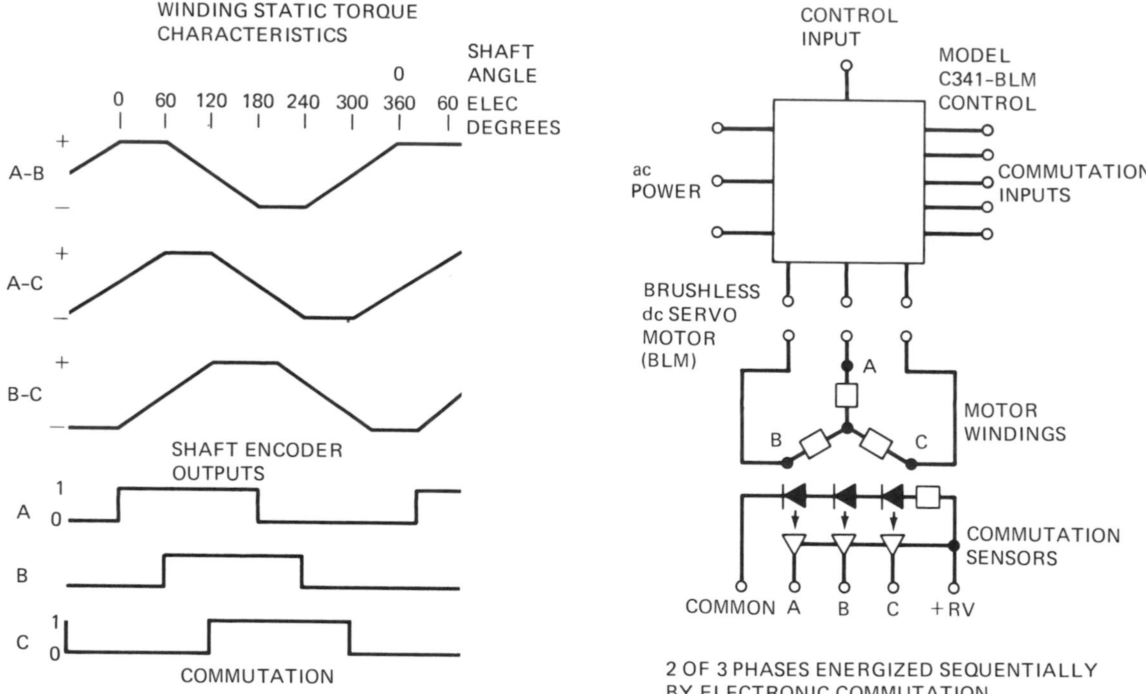

(From Electro-Craft Engineering Handbook, *D.C. Motors, Speed Controls, Servo Systems,* 4th Edition.)

7. *Synchro and synchro differential generator and differential motor* facilitate ditital/analog servo systems such as remote control of radar antennae, etc.

a. Torque transmission: synchro generator and motor.
b. *Synchro generator and control transformer* producing error signal: voltage indication or control voltage of servo system. The torque transmission is for rotating shaft of motor or synchro receiver; usually a damper flywheel is essential for the motor.
c. The *differential generator* (synchro) differs from ordinary synchro generator in that it transmits sum or difference of two signal inputs (one fed mechanically and the other electrically from another generator).
d. *Control transformer:* signal produces a voltage output varying in phase and magnitude, instead of a synchro receiver (motor) turning a rotor one way or the other.

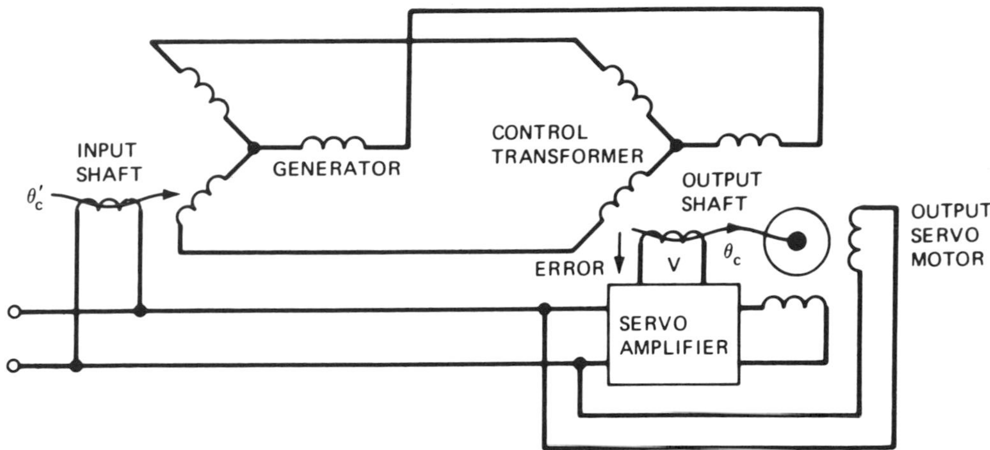

8. A *magnetic amplifier* is a device employing a saturable reactor, in combination with a dry-type rectifier to achieve power amplification. (High-permeability magnetic materials with gapless construction are used.) Magnetic amplifiers are used with feedback too.

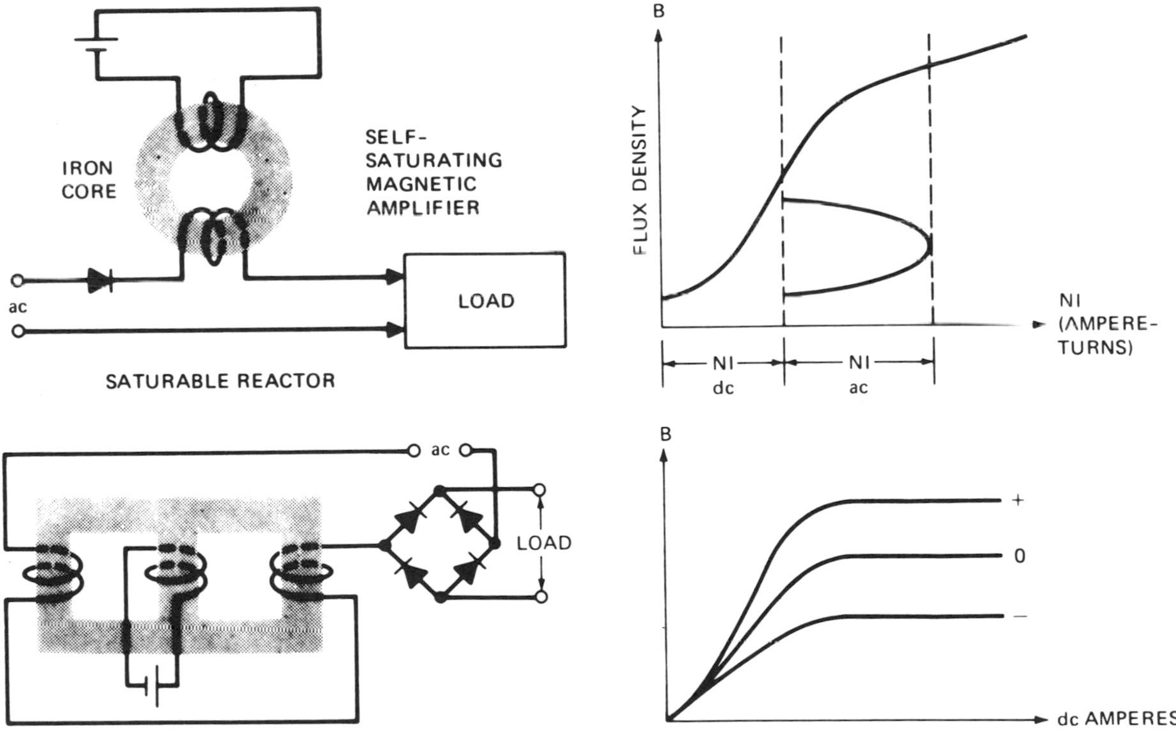

9. *Gyrotron*: A microwave oscillator (or traveling wave tube amplifier) with a built-in PLL, operating at 35 GHz as an electron-cyclotron, develops 100-kW output. The gyrotron is basically a vibrating tuning-fork maintained in continuous, constant-amplitude vibration by a suitable driving system consisting of a pick-off transducer, an amplifier, and a driving coil:

$$L = dH/dt = I\,d\omega/dt + \omega\,dI/dt$$

where I = moment of inertia. In principle it operates like a *rate gyroscope*, but with no wearing parts such as the spin-bearings. The periodic variation in inductance L results in a voltage proportional to the applied angular velocity ω. The gyrotron is used in "Tokamak" fusion research.

C-23 TYPES OF CONTROL SYSTEMS AND COMMON TRANSFER FUNCTIONS

1. *Proportional control system.* The output controlled function is proportional to the input error. The transfer function G(s) is of the form:

$$K/s(sT_1 + 1)$$

2. *Proportional plus derivative system.* The output controlled function is proportional to a linear combination of input error and first time-derivative of input. The transfer function G(s) is of the form:

$$\frac{K(sT_1 + 1)}{s(sT_2 + 1)}$$

3. *Proportional plus integral system.* The output controlled function is proportional to a linear combination of input error and first time-integral of input. The transfer function G(s) is of the form:

$$K(sT_1 + 1)/s^2(sT_2 + 1)$$

K is a constant representing system gain, viscous friction, and load inertia.

4. *Galvanometer:*

$$G(s) = K/s^2(sT + 1)$$

Transfer function G(s) represents the ratio of the position of the galvanometer element and signal current, and K is the torque coefficient.

5. *Gyroscope:*

$$G(s) = \frac{\text{Angular Velocity}}{\text{Signal Current}} = \frac{K}{s(sT + 1)}$$

6. *dc motor, speed control:*

$$G(s) = \frac{\text{Velocity of Motor, rad/sec}}{\text{Applied Voltage}} = \frac{K}{(sT + 1)}$$

7. *dc motor, position control:*

$$G(s) = \frac{\text{Output Position}}{\text{Applied Voltage}} = \frac{K}{s(sT + 1)}$$

Where K is the reciprocal of the voltage constant of the motor in V/rad/sec.

8. *dc generator and motor for position control:*

$$\frac{\text{Output Position}}{\text{Control Voltage of Generator}} = \frac{K}{s(sT_1 + 1)(sT_2 + 1)}$$

9. *Stabilizing network for rate signals: phase lead (differentiator):*

$$\frac{E_o(s)}{E_1(s)} = G(s) = sT/(sT + 1)$$

E_i E_o

10. *Integral signals:* phase lag.

$$G(s) = 1/(sT + 1)$$

11. *Rate and integral signals:* Lead-lag

$$G(s) = \frac{(T_1T_2)s^2 + (T_1 + T_2)s + 1}{(T_1T_2)s^2 + (T_1 + T_2 + T_{1,2})s + 1} \quad \begin{cases} T_1 = R_1C_1 \\ T_2 = R_2C_2 \\ T_{1,2} = R_1C_2 \end{cases}$$

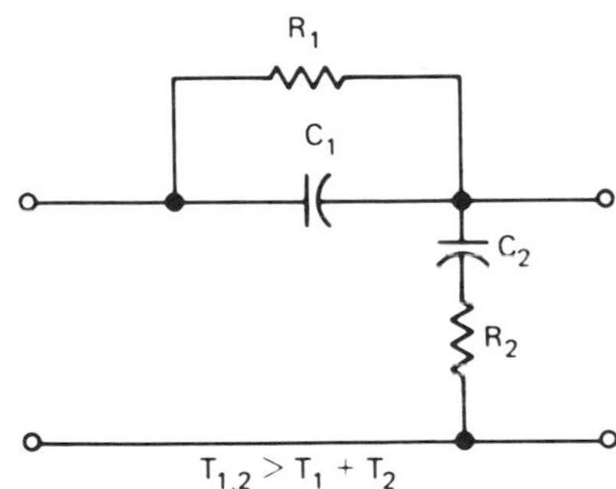

12. *ac, two-phase induction motor:*

$$G(s) = \frac{K}{s(sT + 1)}$$

C-24 BASIC CONCEPTS OF ELECTRICAL NETWORKS

1. *Superposition.* If a system is linear, the system (output) response to several inputs will be the sum of the response to each input individually.

2. *Thevenin's theorem.* The effect of any impedance element in an electrical circuit may be determined by replacing all the voltage sources by a single equivalent voltage source, and all other impedances by a single impedance in series with the impedance of interest.

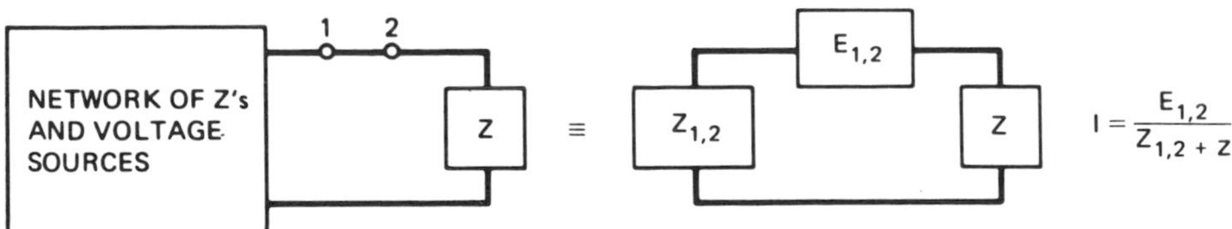

3. *Norton's theorem* is best defined by the three following equivalent circuits in terms of a constant-current generator.

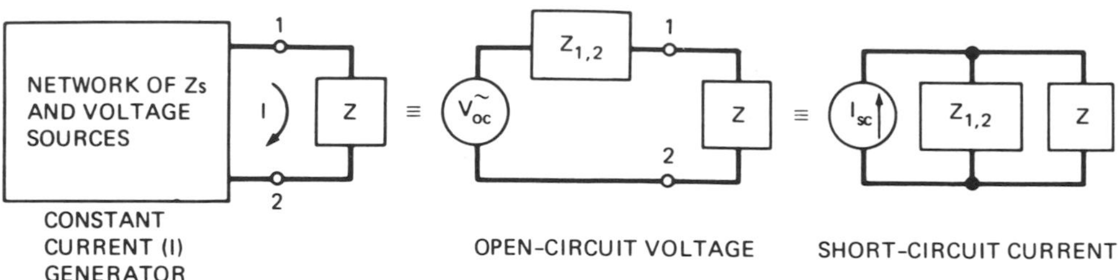

4. The *transfer function* of an element or a system is the ratio of the transform of the output to the transform of its input under the conditions of zero initial energy storage. The dynamic properties of the system are defined to mathematically represent the frequency response or the transient response to a specified input. The transfer functions are always represented in Laplace transform form.

The closed-loop feedback control system, consisting of the feed-forward and feedback transfer functions of the above character, is accurate and predictable in performance, with (1) less dependence upon system component characteristics, (2) less sensitivity to output load disturbances, (3) faster response to input commands, and (4) reliable stability under adequate gain and phase margins.

C-25 AC SERVO SYSTEMS

1. As compared to a dc servo system where the signals are proportional to the instantaneous amplitude, in an ac (carrier) servo system, the signals are modulated carrier, and information is modulation. The envelope consists of the modulation signal in an amplitude modulation system.

2. Due to the simplicity of using a chopper or synchro as a constant-frequency, constant-magnitude modulation, and a two-phase motor itself as a demodulator, ac carrier systems are generally suppressed carrier systems. (Frequency and phase modulation are also feasible for better null detection.)

3. *Advantages of ac carrier servo systems*

a. Use of sensitive, high-precision, low power-consuming, low-level sensors.
b. Inexpensive, easily produced servo amplifiers and power elements with less maintenance requirements.
c. An ac two-phase motor acts as an ideal demodulator with a prescribed static and dynamic relationship between the driving voltage-envelope and the output shaft position. (Sometimes, an ac tachometer can be improvised as a modulator.)

4. *Types of ac carrier servo*

a. When the elements in both the input and the output functions are modulated carriers, it is *type-1* ac servo.
b. When the modulators output a modulated ac carrier with an input at the signal frequency, it is *type-2* ac servo.
c. When a demodulator has a modulated carrier as an input, and a signal frequency as an output, it is *type-3* ac servo.

5. *Open- and closed-loop ac carrier servo systems*

Example: Open-loop ac carrier system

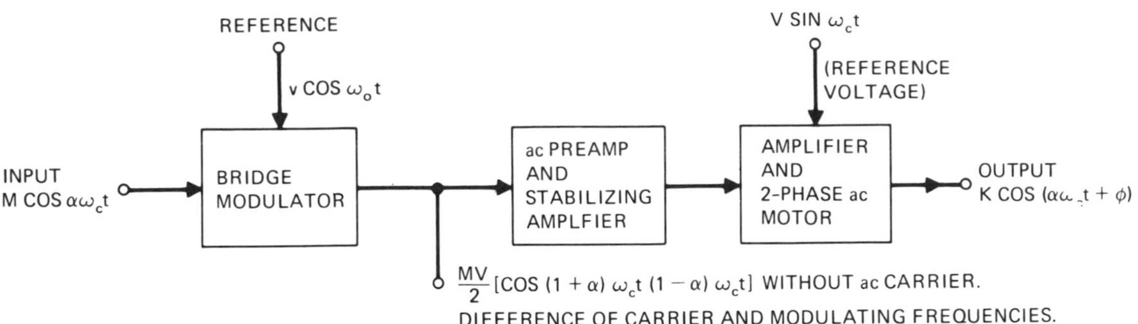

Open-loop servo parameters

Example: Typical suppressed carrier feedback control system using tachometer stabilization

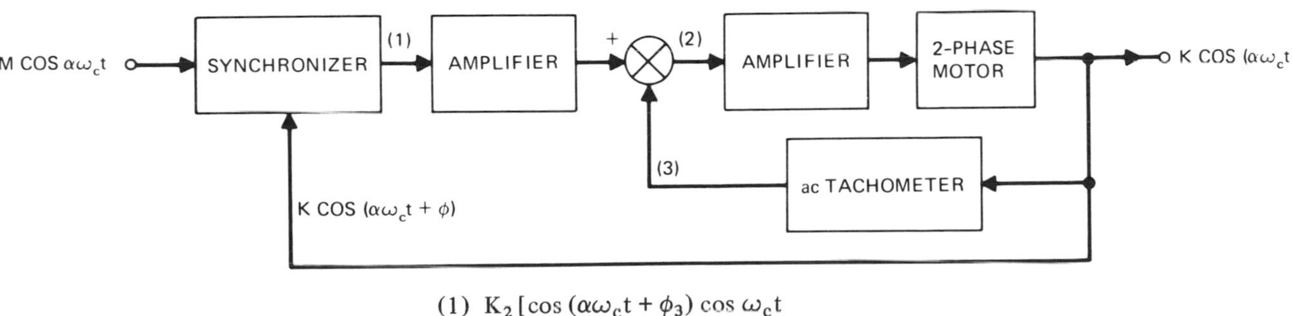

(1) $K_2\,[\cos(\alpha\omega_c t + \phi_3)\cos\omega_c t$
(2) $K_3 \cos(\alpha\omega_c t + \phi_2)\cos\omega_c t$
(3) $-K_1\alpha\omega_c t\sin(\alpha\omega_c t + \phi_1)\cos\omega_c t$

C-26 PHASE PLANE TECHNIQUE FOR NONLINEAR SYSTEMS

1. This is a graphical method used for *nonlinear* systems in the place of the analytical methods for linear systems.

2. The *phase-plane trajectories* give an overall picture of the types of *transient response* possible with various *initial conditions*, and a series of these trajectories show the effect of parameter variation, so that the plots indicate the trends and thus lead to the optimal conditions for design of a nonlinear system.

3. While the independent variable time is not explicit in the plot, it is possible to compute an accurate transient response from the phase-plane plot, and thus obtain a special solution from a general solution.

4. Practical handling of the problems for phase-space concepts require plotting in the phase-plane for the first- and second-order linear or nonlinear differential equations of motion in two dimensions only. This is of course the limitation of this technique—to *second-order systems* with a single degree of freedom. The approach is limited to autonomous systems where time does not appear as a parameter in any coefficients of the system. Impulse, step, and ramp inputs are the driving-point functions allowed in the procedure.

5. *Definitions.* The *phase-plane* has the coordinates of *velocity* (ordinate) *and position* (abcissa) for plotting the solutions of the differential equations on this sys-

tem of coordinates. The *locus* of a solution to the differential equation is called a phase trajectory or simply trajectory. The technique is sometimes called the method of *Isoclines* to draw a *phase-portrait* consisting of a series of solutions as phase trajectories.

C-27 SPREAD-SPECTRUM TECHNIQUES

A transmission system in which the transmitted signal is spread over a broader frequency band than the regular minimum bandwidth required is called a *spread-spectrum system.* A *coded signal-format* is the means for accomplishing this technique in modern communications, navigation, and test systems. Naturally, the advent of MSI and LSI microprocessors at this juncture is mainly responsible for the increasing applications of the spread-spectrum techniques in data communications to achieve (1) *selective addressing* capability, (2) *code division multiplexing for multiple-access* in fiber-optic and satellite communications, (3) low-density power-spectra for signal-secrecy and message-screening from eavesdroppers, (4) ranging with high-resolution, and (5) rejection of interference and jamming.

In view of the wide spectra generated by code modulation, the power transmitted in any narrow band is far less than that of the conventional frequency band involved for regular transmission of the base-band information. A signal is normally described in the time or frequency domain and transforms or mathematical operators are available for conversion from one domain to the other. The latest Fast Fourier Transform (FFT) algorithm is merely a rapid computational technique to translate the time-domain waveforms (displayed on an oscilloscope) to the frequency-domain components (displayed on a spectrum analyzer).

When the code signal-format is selectively chosen for low cross-correlation, minimum interference occurs between users in a network since receivers set to use different codes respond to transmitters sending the respective codes. Thus, several signals can be simultaneously transmitted at the same frequency without any incidence of ambiguity. A spread spectrum system, for example, distributes the few-kHz voice base-band signal over a band that may be several MHz-wide, by using a wide-band encoding process.

The simplest spread-spectrum technique in principle is FM with a deviation-ratio greater than 1; it is not only the information bandwidth, but the amount of modulation that is involved in the bandwidth of the FM signal. Besides, a signal-to-noise advantage is inherent in the mod-demod process. If δ is the deviation ratio, the signal-to-noise advantage or process gain = $\delta^2 \cdot (S/N)$ information.

In practice, spread-spectrum techniques incorporate the following types of code-modulation in PLL-controlled synchronous communications systems:

1. A digital *code sequence* is used with a much higher bit-rate than the information bandwidth in *direct sequence* version.
2. The carrier frequency is shifted in discrete increments in a pattern according to a code sequence—*frequency hoppers*; the transmitted frequency jumps within a preset range according to a code sequence. In *time-frequency hoppers*, the code sequence determines both frequency and time of transmission.
3. "Chirp" or pulsed-FM modulation in which the carrier is swept over a wide band during a radar pulse-interval.

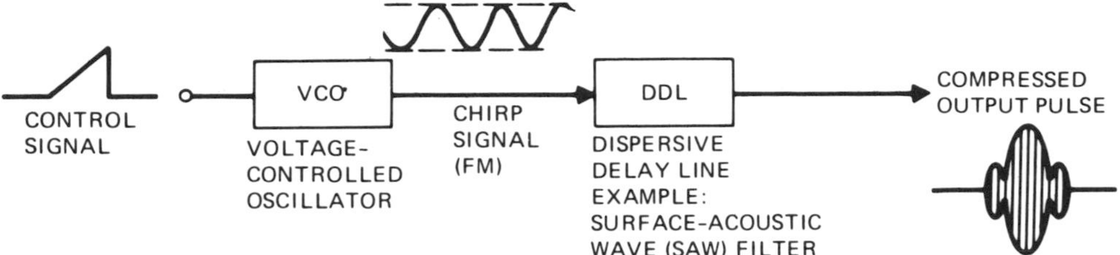

Channel capacity, as defined by Shannon, is the basis for the technique. If W is the bandwidth in hertz, and S/N is the signal-to-noise power ratio,

$$\text{Capacity in bit/sec, } C = W \log_2 (1 + S/N)$$

$$C/W = 1.44 \log_e (1 + S/N)$$

After the expansion of $\log_e$,

$$S/N = \frac{C}{1.44\ W}$$

Hence, bandwidth $W \approx NC/S$.

If, as an example, the noise power is 10 times the signal, and the audio bandwidth is 5 kHz or 5 kb/sec, the digital code-sequence must be transmitted with a bandwidth,

$$W = 5 \times 10^3 \times 10/1.44 = 33 \text{ kb/sec}$$

This bandwidth results in an equivalent rf bandwidth on modulation to allow the system to output error-free information in an excessively noisy environment.

The process gain G_p = rf bandwidth/information rate $= \frac{33}{5} = 6.6$. The widely used direct-sequence signal bandwidths are assumed equivalent to the bandwidth of the main lobe in the power spectrum of $(\sin x/x)^2$. This results in an rf bandwidth that is two times the system code clock-rate (generally designated R_c). A typical process-gain curve and code-sequence generator are illustrated. The clock-rate is the frequency or bit-rate reference used to set the rate of code generation in the spread-spectrum system.

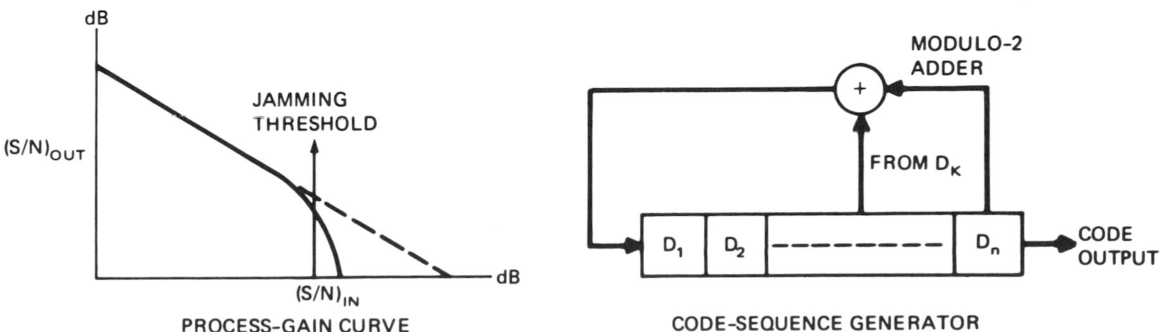

C-28 SURFACE ACOUSTIC WAVE FILTERS

The surface acoustic wave (SAW) filters are generally used as (1) dispersive delay-line filters for radar pulse compression, (2) phase-coded tapped delay-line matched filters for phase-shift-keying signals in spread-spectrum communications, (3) contiguous-channel IF filters in the VHF/UHF bands (preferably with a 5% pass-band) for instantaneous frequency measurement, and (4) tuned band-pass IF filters. The dispersive delay-line has the potential for enabling high-resolution frequency identification between time-coincident signals, separated by as little as 2 MHz in VHF/UHF bands.

The key element of the device is the *piezoelectric* substrate, the *surface* of which propagates an acoustic wave that is generated and translated by *interdigital* input-output transducers. When a radio-frequency signal is applied to the input transducer terminal of an ST-cut quartz,* the surface-wave property of this device permits access to the acoustic wave along the *entire path on* the surface, unlike the bulk acoustic delay-line devices. The electric field induced in the gap excites the wave via the piezoelectric properties of the substrate when the surface wavelength is commensurate with the radian frequency. The amplitude response of the acoustic device can be controlled by adjusting the *interdigital electrode overlap*—this is called *apodization.*

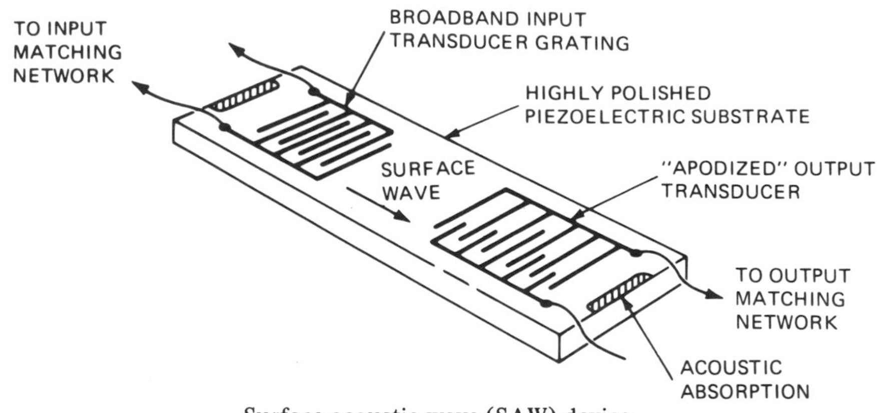

Surface acoustic wave (SAW) device.

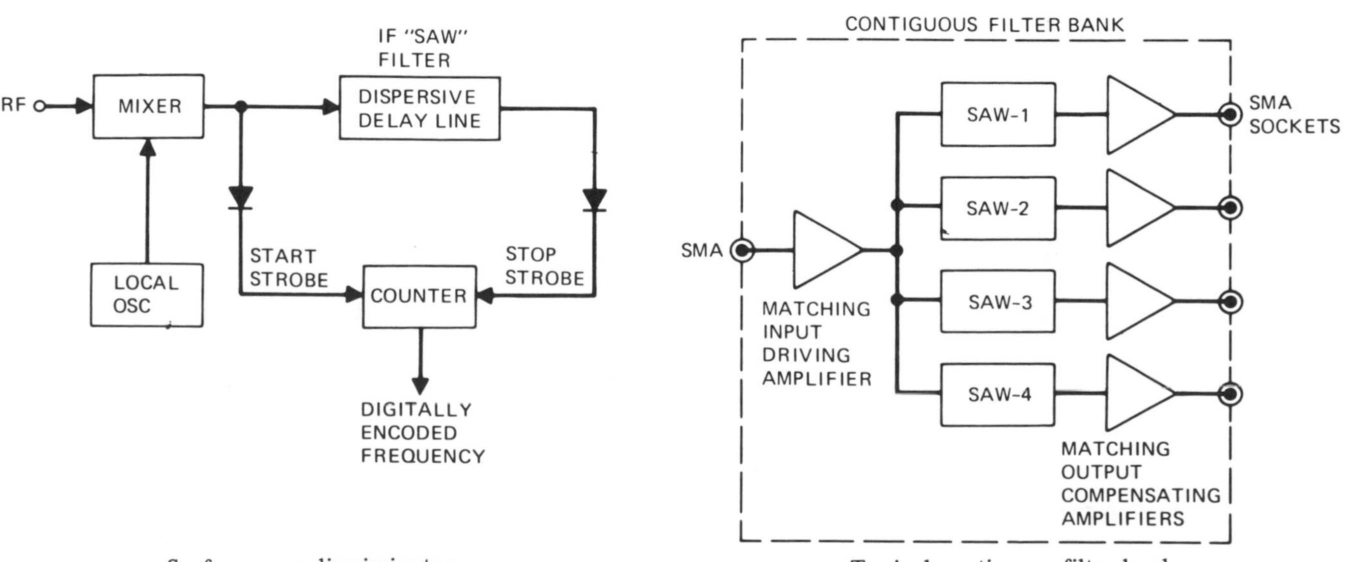

Surface wave discriminator

Typical contiguous filter bank

*The *quartz crystal* resonator, SiO_2, is cut with an orientation of the surface normal to the ST-axis direction to allow the propagation of wave in the X-axis direction for enabling zero temperature-coefficient at the frequency of resonance.

The transducers are represented by individual transfer functions separated by a broad-band time-delay of exponent $(-j\omega\, dV)$ where d is the distance between the centers of the transducers and V is the rate of surface-wave propagation. The frequency response is mainly controlled by the apodization overlap of the output interdigital transducer. For bandwidths less than 5%, the ST-cut quartz with a zero temperature coefficient at 25°C is the preferred substrate material, because it is temperature stable over the 100°C range without the need of a temperature-controlled oven, relatively insensitive to second-order perturbation effects, and effectively superior for triple-transit suppression of over 50 dB.

As an example of its application, the surface-wave discriminator shown in the block-diagram on p. 484 down-converts the received signal by means of a tuned local oscillator to an intermediate frequency of the surface-wave dispersive delay-line, and applies it simultaneously to the SAW delay-line and the threshold circuit. The threshold signal is detected and used to initiate the frequency-determining circuit such as a digital counter. The signal is also used to gate out any electromagnetic feed-through. The radio-frequency signal applied to the delay-line is delayed in time an amount that is proportional to the received frequency. This signal is then detected and used to stop the counter. The count obtained during this interval is therefore directly proportional to the received frequency.

In practice, a filter bank may consist of four SAW filters that use a common 50-ohm matched-input driving amplifier, and individual output amplifiers matched to the 50-ohm transducers of the SAW filters. The gain of each output amplifier is adjusted to equalize and compensate the insertion loss in each filter.

A typical specification of a SAW filter follows:

1. Frequency: 10 to 1000 MHz
2. Fractional bandwidth: 0.5% to 120%
3. Delays: 0.03 to 100 μs
4. Insertion loss: 8 to 50 dB (compensated)
5. Velocity (ST-quartz): 3.157×10^{-5} cm/sec
6. Pulse-width: $>$100 nsec
7. Side-lobe rejection: 45 dB
8. Shape factor: 1.27 (40-dB attenuation within 2 MHz of 3-dB cutoff frequency)
9. Frequency-stability: $<\pm 60$ kHz through 0° to 100°C
10. Dynamic range: $<$50 dB
11. Linear phase deviation: ±5°
12. Size of a four-channel filter bank: $2 \times 2 \times 0.5$ in. approximately

C-29 PATTERN RECOGNITION

The subject of pattern recognition is generally treated under *pattern analysis and machine intelligence.* In principle, *adaptive pattern recognition* involves several basic themes of *learning*, and a processing model features various methods of physical realization such as (1) a feature-vector defined in space, (2) the division of space into several regions, and (3) the space visualized as a structure. Where the knowledge of the structure is incomplete, a learning process in the form of heuristic *design* samples is effective.

The present state-of-the-art in pattern analysis and machine intelligence by way of digital computers may be subdivided into the following classifications: (1) statistical pattern recognition, (2) linguistic pattern recognition, (3) image coding and processing, (4) shape and texture analysis, (5) biomedical pattern analysis and information systems, (6) remote sensing, (7) speech recognition, (8) industrial and defense applications of pattern recognition and image processing, (9) character and text recognition, (10) semantic information processing, (11) theorem proving, (12) natural language analysis, (13) robotics, and (14) inherent complex digital process control. Microprocessors are ideal for applications in pattern recognition.

When a picture is scanned to obtain a two-dimensional array of points, usually with an excess of disturbing data, three classes of pattern-recognition algorithms—"template-matching," object-detection, and relationship-detection—are applied. The term *matching* may include the aspects of classification, identification, and recognition.

Although many theories are put forward for specific situations and partial problems, a general applicable approach to problems of pattern recognition as a whole does not exist. Each problem, depending on its nature and complexity, has to be tackled individually, using the numerous available algorithms in the fields of statistics and probability, decision theory, linguistics, information and communication theory, artificial intelligence, and so on. And finally, a judicious combination of heuristic measure-and-optimize experimentation in the laboratory is inevitably involved in arriving at a reliable and appropriate design solution. Where funding is available, the design of an initial simulation model is effective.

In conclusion, some of the actual problems in pattern recognition that are being tackled at this time are as follows:

1. *Medical data:* chromosome classification, blood-cell analysis, neuron identification, and cardiogram and encephalogram analysis.
2. *Reading symbols:* reading printed and written text, and drawings, for administrative purposes and for the blind.
3. *Detection of human environment:* recognition of faces, spoken language, fingerprints, clouds, objects in landscape and environment (for recognition by robots and tactile transducers), and remote sensing by infrared sensors.
4. *Scientific research:* analysis of bubble-chamber pictures in nuclear physics, interpretation of aerial and satellite photographs, and CRT waveforms.
5. *Industry:* inspection of the shape of objects to enable sorting, trimming circuits for the appropriate response-curve, detection of distortion in loudspeakers, and television tubes.
6. *Defense applications:* detection of planes by radar, and identification of surface and underwater moving objects.

Appendix D
Measure-and-Optimize Techniques

The methodology implies the heuristic laboratory techniques using special instrumentation for optimizing designated parameters to meet the requirements of an engineering specification at a certain state-of-the art. As the technology advances in sophistication along with further discoveries and inventions, the specifications and the key performance-index of a system, however complex, will improve to meet higher standards.

The effort is primarily based on an iterative biofeedback process, which the experienced engineer intuitively develops. A sophisticated interactive-graphics real-time operating system using an analog/digital computer as a simulation model is a close parallel. Invention is a possible logical or logistic end-result, and discovery is a probable by-product.

Some state-of-the-art development and design assignments successfully accomplished throughout the author's professional career are listed here to reveal the potential of this usual engineering development and design in the laboratory (while theory mostly provides only guidelines).

D-1

The control and electronics engineering contributions in projects associated with the Q-CVTR:

1. Development and design of a transistorized "switch-lock" capstan sampled-data feedback control system for the "pixlock" facility in the early broadcast-quality RCA vacuum-tube-version of the Q-CVTR. It was the very first solid-state control system in the Q-CVTR. A simulated tape capstan model was designed as a preparation for developing this control system (1959).

Advance development and the first demonstration ever of a switch-lock control system that operated without the need of a 30-Hz edit pulse-train over and above the regular 240-Hz control-track signal.

Advance development of a multirate 240/30-Hz capstan servo.

Reference pulse generator (30-Hz digital controller) for the switch-lock feedback-control system.

2. Preliminary advance development of a solid-state "line-lock" feedback control system for the headwheel servo. (1959)

3. Development and design of the first high-precision, solid-state, vacuum-guide, sampled-data position feedback-control system for the first solid-state Q-CVTR. Development of a special solid-state error-simulating device for the advance development of the guide servo (since the Q-CVTR, as a whole, was not available until the final stage). (1961)

4. Custom development and design of the first 6-channel time-division-multiplex radar antenna synchro-signal recording facility for the solid-state Q-CVTR. The hardware used for the RCA solid-state digital computer was improvised for this function. (1962)

5. Custom development and design of an automatic "pixlock" timing simulator for testing the monochrome timing corrector (MATC) control system in the Q-CVTR. (1962)

6. Advance development of an economy Q-CVTR without the regular 4-channel FM-switching on playback. (1961)

7. Automatic self-checking and indicating system for the three servo systems and other accessory devices in the Q-CVTR. (1961)

8. Selective improved video master-erase system for the electronic editing facility in the Q-CVTR. (1963)

9. A color stabilizing amplifier accessory to reprocess a distorted remote color video signal for video-recording purposes. (1963)

10. Remote-control facility for the Q-CVTR. (1962)

11. Advance development of an automatic tape tension control. (1963)

12. A digital controller for synchronizing group-delay measuring equipment to the Q-CVTR. (1964)

13. A special device for synchronizing a video-frequency sweep generator to the individual FM channels in the Q-CVTR. (1964)

14. System modification of two early "cannibalized" experimental Q-CVTRs for broadcase-quality performance. (1964)

15. Quadruplex burst stairstep video signal generator for comparing the performance of the four channels in the record/playback system of the color tape recorder. (1964)

16. High-precision ($\pm 0.1°$ differential phase, $\pm 0.1\%$ differential gain, $\pm 0.1\%$ k-factor), combined video and color processing and distribution unit in the recording system of the Q-CVTR, with available linear-phase low- and high-band filters, and video preemphasis networks for American and International television standards. (1965)

17. Selective line-trigger generator for the precise superimposition and measurement of k-factor by comparing the display of delay compensated 0.1-μs (half-amplitude width) sine-squared pulses on alternate television lines, at the input and output of the Q-CVTR or any video-processing section therein. That is, two corresponding points on any two alternate lines in the picture, one in the input signal and the other in the output signal, are displayed as sharp sine-square shaped pulses, and superimposed for direct comparison on a two-channel oscilloscope by virtue of its electronic switching facility on alternate fields. (1965)

18. The first intermodulation test instrument for measuring the intermodulation products in the FM section of the Q-CVTR. (1965)

D-2

The author was engaged in advance television development and design of studio color television equipment from 1949 to 1970. In addition to the preceding projects associated with the Q-CVTR, some noteworthy contributions in the field of color television are included in the following list:

1. Remote-controlled television-studio electronic video-switching system, with special effects, on an automatic frame-by-frame basis (hybrid relay and vacuum-tube version for black-and-white television, first ever, 1952; Pye Research, Cambridge, England).
2. The first "flyweel" sync for fringe-area television receivers in England, 1950. (There was no television broadcasting in Europe at that time.) A PLL system.
3. The first successful television stabilizing amplifier (for distorted network signals) in the European market (1953). It was later redeveloped for color signals (1954) to make it a color-processing amplifier, the first ever for regenerating a standard color signal from a distorted remote composite color video signal. (Special effects between a test transmission from the British Broadcasting Corporation in London and laboratory NTSC-color signals were demonstrated in Cambridge; a color subcarrier of 2.6 MHz was used with pleasing results in 405-line television.)
4. The first video special-effects equipment in England. (1951)
5. Development and design of both the early British and international high-definition NTSC color multiplexing equipment in England, using high-precision automatic chroma balance for the first time. (1953–1954)
6. High-precision, 18-channel camera-switching system (the first of its kind as regards precision, group-delay, etc., in color television transmission), 1956. This was a unique measure-and-optimize technique based on a theoretical background of high-frequency transmission lines. A paper on this subject was published by the author (A resonant coaxial-stub as an automatic equalizer, *IRE* (*IEEE*) *Transactions*, March 1960).
7. The first automatic vertical-interval color reference test signal equipment for the coast-to-coast NBC television network. (1957–1958)
8. The first high-reliable solid-state television synchronizing pulse distribution equipment in the laboratory (1959) that needed no maintenance for a decade or so.
9. The first scan-and-color conversion of American studio color television equipment to European standards for television studios in Rome–conversion of NTSC to CCIR standards, prior to PAL. (This equipment was demonstrated by RCA in Moscow in 1959.)
10. While program manager in advance development work for North American Philips: a single-line color-bar test signal during vertical blanking interval as a built-in accessory of color television cameras (Built-in single-line color-bar test signal during vertical-interval for color television cameras, *SMPTE*, 1968).
11. System analysis and design of low-cost, single-head, helical-scan color videotape recording system. (1968)
12. The *first* mono-drive, one-piece, broadcast-quality color television camera with internal sync-regeneration (for remote signals) using plumbicons and digital/

linear MSI. (A paper was presented along with a successful demonstration at NEB Exhibition, Washington, D.C., 1969.)

D-3

Communications systems analysis and design activity in the India federal government (1939–1949, as eventually Class I Technical Officer):

1. All-around activity in an international broadcasting system (1939–1942).
2. System development and design of three-channel carrier and automatic telephone exchanges, and high-frequency measurements of transmission lines. (1944)
3. Aeronautical (ground and aircraft) radio communications systems and navigational aids, including antennas–MF/HF/VHF–(1943–1949).

D-4

While a product manager in consumer electronics and a principal member of technical staff: computerized microwave radar receiver analysis/design and systems effectiveness (1973–1978):

1. VHF police-band FM scanning monitor receiver and front-end VHF/UHF preamplifiers.
2. Digital-coded paging system.
3. Center-tune-tag high-speed PLL system.
4. High-speed low-resolution instantaneous frequency measuring (IFM) device–white paper on system analysis/design.
5. A comprehensive article on K-band (18–40 GHz) IFM system design, with sensitivity analyses and cost estimates of three different versions. (1976, as a *first.*)
6. Computation, measurement, and paper on intermodulation products and spurs in microwave up-and-down converters.
7. A 2-channel mixed-base IFM receiver using contiguous SAW (surface-acoustic-wave) filters, etc.

Appendix E
Latest Developments in Digital Television

Table E-1. Specifications of synchronizing signal generator used for transmission of color television signals.

Details of Pulse Waveforms (see Fig. 4-4)	NTSC (United States)	CCIR (Europe) (Color: prior to CCITT PAL-Modification)
1. Horizontal frequency	15.73425 kHz	15.625 kHz
2. Picture frequency	29.97 Hz/sec	25 Hz/sec
3. Field frequency	59.94 Hz	50 Hz
4. Aspect ratio of CRT-display	4 X 3	4 X 3
5. Interlace	2:1	2:1
6. Field duration (V)	16.67 msec	20 msec
7. Horizontal duration (H)	63.6 μsec	64 μsec
8. Scanning during active periods	Left to right (H) Top to bottom (V)	Same
9. System capable of operating		
a. independent of supply frequency	Yes	Yes
b. locked to supply frequency	Yes, 60 Hz	Yes, 50 Hz
c. genlocked to remote sync	Yes	Yes
d. locked to external control frequency for color	31.46852 kHz (from frequency standard)	31.25 kHz (from frequency standard accessory)
e. locked to internal crystal frequency	94.5 kHz	93.75 kHz
10. Vertical blanking duration	13H to 21H (826 to 1334 μsec) Set at 21H.	18H to 22H ± 11.7 μs (1160 to 1417 μsec) Set at 21H.
11. First equalizing-pulse group interval	3H, 6 pulses	2.5H, 5 pulses

Table E-1. (continued)

Details of Pulse Waveforms (see Fig. 4-4)	NTSC (United States)	CCIR (Europe) (Color: prior to CCITT PAL-Modification)
12. Prevertical sync blanking interval	3.02H	2.52H
13. Vertical sync-pulse group interval	3H, 6 pulses	2.5H, 5 pulses
14. Second equalizing-pulse group interval	3H, 6 pulses	2.5H, 5 pulses
15. Vertical-drive leading edge delay on vertical blanking	None	None
16. First equalizing-pulse delay on vertical blanking	None	1.3 μs
17. Vertical-drive duration	Half vertical blanking duration. Set at 9H.	500–600 μs (3% of V)
18. Horizontal blanking width	10.2 to 11.5 μsec	11.5 to 12 μsec
19. Front-porch duration	Minimum of 1.27 μsec	1 to 1.5 μsec
20. Horizontal sync duration	4.45 to 5.1 μsec	4.5 to 5.2 μsec
21. Equalizing-pulse-width	2.54 μsec	2.3 to 2.6 μsec
22. Back-porch duration	4.45 to 5.7 μsec	5.5 to 6 μsec
23. Vertical sync serration pulse-width	3.82 to 5.1 μsec	4.5 to 5.8 μsec
24. Vertical sync pulse-width	26.7 to 28 μsec	24.2 to 25.9 μsec
25. All rise-and-fall times	0.2 to 0.4 μsec	0.2 to 0.4 μsec
26. Horizontal drive duration	Half horizontal blanking duration	7 to 8 μsec (11% of H)
27. Advance of horizontal drive on horizontal sync (leading edges)	Minimum 1.27 μsec	1.6 μsec
28. Pulse distribution a. sync b. blanking c. vertical drive and d. horizontal drive	Negative, 3.5 to 8 V peak-to-peak into 75 ± 5 ohm	Negative, 4 V ± 0.5 V peak-to-peak into 75 ± 5 ohm
29. Grating output: The non-composite signal is connected via the "test" position of the processing amplifier to the colorplexer for the addition of sync.	a. Dot pattern b. Test pattern c. Cross-hatch (V–H) d. V e. H	a. Dot pattern b. Test pattern c. Cross-hatch (V–H) d. V e. H
30. Remote sync, for genlock	−4 ± 0.5 V	−4 ± 0.5 V

Table E-2. Colorplexer (color encoder) with automatic chroma balance and aperture compensator): comparative transmission specifications.

Transmission Characteristics	NTSC (United States)	CCIR (Europe)
1. Channel bandwidth	6 MHz	7 MHz
2. Spacing between the vision and the sound carriers	4.5 MHz	5.5 MHz
3. Type of transmission	Vestigial side-band (-1.25 and 4.5 MHz about the vision carrier)	Vestigial side-band (-1.25 and 5.5 MHz about the vision carrier)
4. Color subcarrier	3.579545 MHz ± 5 Hz	4.429687 MHz ± 5 Hz
5. Vision modulation	Amplitude	Amplitude
6. Sound modulation (FM)	Frequency deviation ± ±25 kHz, 75 μsec, preemphasis	Frequency deviation ±50 kHz, 50 μsec, preemphasis
7. Minimum level of carrier as percent of peak-carrier	⩽15% at maximum luminance	10% minimum
8. Ratio of vision to sound effective isotropic radiated power (e.i.r.p.)	2/1 to 2/3	5/1
9. Approximate gamma of radiated signal.	0.45	0.5
10. Video bandwidth	4 MHz	5 MHz
11. Colorplexer: Monochrome, "M"	8 MHz	8 MHz
12. I-channel bandwidth (small-area color)	Less than 2 dB down, 1.3 MHz At least 20 dB down, 3.6 MHz	Less than 2 dB down, 1.6 MHz At least 20 dB down, 4.4 MHz
13. Q-channel bandwidth (large-area color)	Less than 2 dB down, 400 kHz Less than 6 dB down, 500 kHz At least 6 dB down, 600 kHz	Less than 2 dB down, 500 kHz Less than 6 dB down, 500 kHz More than 6 dB down, 750 kHz
14. Envelope delay (group-delay) distortion	-0.17 ± 0.05 μsec at 3.579545 MHz	-0.1 ± 0.02 μsec at 4.429687 MHz
15. Power unit	115 V, 60 Hz (The transformer is suitable for operation on 50-Hz supply.)	230-V, 50 Hz, 230/115-V, 50-Hz stepdown transformer is required.
16. Phase shifter of the color signal analyzer.[a] The RG-59U coaxial cables used for this purpose are cut on a strictly proportional basis for the new frequency and calibrated.	3.579545 MHz 6 in. per 1° shift	4.4296875 MHz 4.85 in. per 1° shift
	No. *in.*	*No.* *in.*
-5(2°)	RG-59U: 12	RG-59U: 9.7
-4(3°)	RG-59U: 18	RG-59U: 14.55
-3(4°)	RG-59U: 24	RG-59U: 19.4
-2(5°)	RG-59U: 30	RG-59U: 24.25
-1(6°)	RG-59U: 36	RG-59U: 29.1
0(7°)	RG-59U: 42	RG-59U: 33.95
1(8°)	RG-59U: 48	RG-59U: 38.8
2(9°)	RG-59U: 54	RG-59U: 43.65

Table E-2. (continued)

Transmission Characteristics	NTSC (United States)	CCIR (Europe)
$3(10^\circ)$	RG-59U: 60	RG-59U: 48.5
$4(11^\circ)$	RG-59U: 66	RG-59U: 53.35
$5(12^\circ)$	RG-59U: 72	RG-59U: 58.2
5°	RG-59U: 32	RG-59U: 26.25
10°	RG-59U: 61.5	RG-59U: 48.5
20°	RG-59U: 122.5	RG-59U: 99
30°	RG-59U: 182	RG-59U: 147
50°	RG-59U: 301.5	RG-59U: 243.5
90°	RG-59U: 540	RG-59U: 436.5

[a]The color signal analyzer is used for measuring the phase of the chroma components in a color-bar test signal in order to align the color encoder appropriately. (A vectorscope is used alternatively for continuous display of the color-dots in a vector-display of colors.)

Table E-3. EBCDIC and ASCII codes. (Delete MSB "1" for Standard ASCII.)

Character	(IBM) EBCDIC Representation	(Teletype and CRT) ASCII Representation
Blank	0100 0000	1010 0000
. Period, decimal point	0100 1011	1010 1110
< Less than	0100 1100	1011 1100
(Left parenthesis	0100 1101	1010 1000
+ Plus sign	0100 1110	1010 1011
| Logical OR	0100 1111	
& Ampersand	0101 0000	1010 0110
$ Dollar sign	0101 1011	1010 0100
* Asterisk, multiplication	0101 1100	1010 1010
) Right parenthesis	0101 1101	1010 1001
; Semicolon	0101 1110	1011 1011
¬ Logical NOT	0101 1111	
- Minus, hyphen	0110 0000	1010 1101
/ Slash, division	0110 0001	1010 1111
, Comma	0110 1011	1010 1101
% Percent	0110 1100	1010 0101
_ Underscore	0110 1101	
> Greater than	0110 1110	1011 1110
? Question mark	0110 1111	1011 1111
: Colon	0111 1010	1011 1010
# Number sign	0111 1011	1010 0011
@ At sign	0111 1100	1100 0000
' Prime, apostrophe	0111 1101	1010 0111
= Equal sign	0111 1110	1011 1101
" Quotation mark	0111 1111	1010 0010
A	1100 0001	1100 0001
B	1100 0010	1100 0010
C	1100 0011	1100 0011
D	1100 0100	1100 0100
E	1100 0101	1100 0101
F	1100 0110	1100 0110
G	1100 0111	1100 0111
H	1100 1000	1100 1000
I	1100 1001	1100 1001

Table E-3. (continued)

Character	(IBM) EBCDIC Representation	(Teletype and CRT) ASCII Representation
J	1101 0001	1100 1010
K	1101 0010	1100 1011
L	1101 0011	1100 1100
M	1101 0100	1100 1101
N	1101 0101	1100 1110
O	1101 0110	1100 1111
P	1101 0111	1101 0000
Q	1101 1000	1101 0001
R	1101 1001	1101 0010
S	1110 0010	1101 0011
T	1110 0011	1101 0100
U	1110 0100	1101 0101
V	1110 0101	1101 0110
W	1110 0110	1101 0111
X	1110 0111	1101 1000
Y	1110 1000	1101 1001
Z	1110 1001	1101 1010
0	1111 0000	1011 0000
1	1111 0001	1011 0001
2	1111 0010	1011 0010
3	1111 0011	1011 0011
4	1111 0100	1011 0100
5	1111 0101	1011 0101
6	1111 0110	1011 0110
7	1111 0111	1011 0111
8	1111 1000	1011 1000
9	1111 1001	1011 1001

E-4. Transmission of black-and-white and color television signals, latest developments

1. PICTURE-PHONE (American Telephone & Telegraph). Video bandwidth: 1 MHz; bit-rate/sample: 3; equivalent pulse code modulation (PCM) bit-rate used on telephone pairs with digital repeaters (Bell Telephone T-2 carrier): 6.312 Mb/sec, that is, 3 X 2 X the video bandwidth. (The bit-rate may be modified to the T-1 carrier, 1.544 Mb/sec, by using 4 frames/sec slow-scan television.)
2. DIGITAL TELEVISION BY SATELLITES in geosynchronous orbit (Satellite Business Systems). Rooftop 15-ft dish Small Earth Station is used. Microwave band: 11.7 to 12.2 GHz; color television signal of compatible black-and-white signal video bandwidth: 4.2 MHz; differential PCM data rate: 43 Mb/sec by frequency-division multiple-access (FDMA) or time-division multiple-access (TDMA). In the near future, TDMA at a rate of 274 Mb/sec is feasible at 18 GHz to carry several color television channels simultaneously. The SBS satellite is effective in 1981.
3. CABLE TELEVISION (CATV). Fiber-optic cable, LD-4 can presently carry six 43 Mb/sec digital color television channels at a multiplexed 274 Mb/sec digital rate from rooftop Small Earth Stations for international telecasts or coast-to-coast television reception in America. Helium-neon laser for long life or lower-cost LED laser can be used at this time to transmit the modulated digital color television signals via fiber-optic cables to homes within a radius of 1 mile. This kind of distribution (without any repeaters) makes it a Community Antenna System. 274 Mb/sec is the Bell Telephone T-4 carrier.

Bibliography

Acker, D. E. et al. Digital Time-base Correction for Video Signal Processing, *SMPTE Journal*, March 1976.

Allan, R. The Microcomputer Invades the Production Line, *IEEE Spectrum*, January 1979.

Aniebone, E. N. and Brathwaite, R. T. A Review of Analog-to-Digital Conversion, *Computer Design*, Vol. 8, pp. 49–54, December 1969.

Auslander D. M. et al. Direct Digital Process Control: Practice and Algorithms for Microprocessor Application. *IEEE Proceedings*, February 1978.

Ball, C. J. Communications and Minicomputer, *Computer*, Vol. 4, pp. 13–21, November 1969.

Barden, W., Jr., *How to Program Microcomputers*, Radio Shack, Tandy Corp., 62–2012, Howard W. Sons & Co., 1977.

Barna, A. and Porat, D. A. *Integrated Circuits in Digital Electronics*, John Wiley & Sons, New York, 1973.

Bellman, R. On Adaptive Control Processes, *Trans. IRE Automatic Control*, Vol. AC-4, pp. 1–9, November 1959.

Bellman, R. On the Application of the Theory of Dynamic Programming to the Study of Control Processes, *IRE Proceedings*, *Symposium on Nonlinear Circuit Analysis*, April 1956, pp. 199–213.

Belove, C., Schachter, H., and Schilling, D. L. *Digital and Analog Systems, Circuits, and Devices: An Introduction,* McGraw-Hill, New York, 1973.

Bennet, W. R. Spectra of Quantized Signals, *Bell System Technical Journal*, July 1948.

Bergland, Bell Telephone Labs. A Guided-Tour of the Fast Fourier Transform, *IEEE Spectrum*, Vol. 6, pp. 41–52, July 1969.

Bingley, F. J. Colorimetry in Color Television, *Proceedings IRE*, January 1954, Vol. 42, No. 1, pp. 48–58.

Bisset, S. LSI Tester Gets Microprocessors to Generate Their Own Test Patterns, *Electronics*, May 25, 1978, pp. 141–145.

Blacksher, R. PROM Decoder Replaces Chip-Enabling Logic, *Circuits for Electronics Engineers*, edited by Weber, S., McGraw-Hill Publications Co.

Blanchard, A. *Phase-Locked Loops*, Wiley-Interscience Publications, John Wiley & Sons, 1976.

Bogner, R. E. and Constantinides, A. G., eds., *Introduction to Digital Filtering*, John Wiley & Sons, New York, 1975.

Bohacek, P. K. and Tuteur, F. B. Stability of Servomechanisms with Friction and Stiction in the Output Element, *Trans. IRE Automatic Control*, Vol. AC-6, No. 2, May 1961, pp. 222–227.

Booth, T. L. *Digital Network and Computer Systems*, 2nd Edition, John Wiley & Sons, New York, 1978.

Brookner, E., ed. *Radar Technology*, Artech House, Dedham, Mass., 1977.

Brown, G. H. Mathematical Formulations of the NTSC Color Television Signal, *Proceedings IRE*, January 1954, Vol. 42, No. 1, pp. 66–71.

Burner, H. B. et al. A Programmable Data Concentrator for a Large Computing System, *IEEE Transactions Computers*, Vol. C-18, pp. 1030–1038, November 1969.

Busby, A. K. Principles of Digital Television Simplified, *SMPTE Journal*, July 1975.

Cattermole, K. W. *Principles of Pulse Code Modulation*, Iliffe Books, London, 1969.

Cawlan, N. Structure and Applications of FPLAs, Integrated Circuits Applications, *Electronics Engineering Times*, pp. 61–82.

Cerni, R. N. Transducers in Digital Process Control, *Instrumentation Control Systems*, Vol. 37, pp. 123–126, September 1964.

Chaffee, J. G. The Application of Negative Feedback to FM Systems, *Bell System Technical Journal*, Vol. 18, July 1939, pp. 403–437.

Chang, S. S. L. *Synthesis of Optimum Control Systems*, McGraw-Hill Book Co., New York, 1961.

Cheng, C. E. et al. Microprocessor Includes A/D Converter for Lowest Cost Analog Interfacing (Intel), *Electronics*, May 25, 1978, pp. 122–127.

Chenoweth, D. L. et al. A Microprocessor-based Digital Filter Prototyping System. *IEEE Microcomputer Conference Record*, April 1977.

Considine, D. M. and Ross, S. D. *Handbook of Applied Instrumentation*, McGraw-Hill, 1964.

Conway, R. *A Primer on PASCAL*, Winthrop Publishers, Cambridge, Mass., 1976.

Cosgriff, R. L. *Nonlinear Control Systems*, McGraw-Hill Book Co., New York, 1958.

Cunningham, W. J. An Introduction to Lyapounov's Second Method, *Trans. AIEEE, Application and Industry*, No. 58, January 1962, pp. 325–332.

Davidoff, F. Digital Video Recording for Television Broadcasting, *SMPTE Journal*, July 1975.

Davis, S. The Hysteresis Motor for Synchronous Power, *Product Engineering*, December 1952, pp. 167–171.

Desoer, C. A. Pontryagin's Maximum Principle and the Principle of Optimality, *Journal of the Franklin Institute*, Vol. 271, January 1961, pp. 361–367.

Develet, J. A. A Threshold Criterion for Phase-Lock Demodulation, *Proceedings IRE*, Vol. 51, No. 2, February 1963, pp. 349–356.

Devereux, V. Application of PCM to broadcast Quality Video Signals, *The Radio and Electronic Engineer*, September 1974.

Dixon, R. C. *Spread Spectrum Systems*, John Wiley & Sons, New York, 1976.

Doll, D. R. *Data Communications: Facilities, Networks, and System Design*, John Wiley & Sons, New York, 1978.

Electro-Craft Corporation, Minnesota. *DC Motors, Speed Controls, Servo Systems*, 3rd Edition, 1975.

Ennes, H. E. *Digitals in Broadcasting*, Howard W. Sons & Co., 1977.

Fink, D. G. *Television Engineering*, McGraw-Hill Book Co., New York, 1952.

Fredendall, G. L. Delay Equalization in Color Television, *Proceedings IRE*, January 1954, Vol. 42, No. 1, pp. 258–263.

Friedman, J. et al. *Fortran IV*, John Wiley & Sons, New York, 1975.

Godbole, V. R. *Circuits for Electronics Engineers*, Edited by Weber, S., McGraw-Hill Publications Co.

Gould, R. G. and Lum, L. F., eds. *Communication Satellite Systems: An Overview of the Technology*, IEEE Press, New York, 1976.

Grabbe, E. M., Ramo, S., and Wooldridge, D. E. *Handbook of Automation, Computation and Control*, Vols. 1–3, 1959.

Graham, D., and Lathrop, R. C. The Synthesis of Optimum Transient Response: Criteria and Standard Forms, *Trans. AIEE, Part II, Applications and Industry*, Vol. 72, 1953, pp. 273–288.

Graham, N. *Microprocessor Programming for Computer Hobbyists*, TAB Books, Blue Ridge Summit, Pa., 1977.

Gruenburg, E. L. *Handbook of Telemetry and Remote Control*, McGraw-Hill Book Co., 1967.

Helms, H. D. Digital Filters with Equi-ripple or Minimax Response, *IEEE Transactions Audio Electroacoustics*, Vol. AU-19, pp. 87–94, March 1971.

Hirsch, S. *BASIC: A Programmed Text*, John Wiley & Sons, New York, N.Y., 1975.

Hostutler, D. Frequency Synthesis, Term Paper, Akron University, 1971.

Howell, D. A. A Primer on Digital Television, *SMPTE Journal*, February, 1975.

Hu, S. C. Microprocessors in Complex Control and Measuring Systems. *IEEE Microcomputer Conference Record*, April 1977.

Intel Corporation, *Intel 8080 Microcomputer Systems Users Manual*, Santa Clara, Calif., 1975.

Ivey, K. A. *A. C. Carrier Control Systems*, Ch. 6, John Wiley & Sons, New York, 1964.

Jackson, L. B. et al. An Approach to the Implementation of Digital Filters, *IEEE Transactions Audio Electroacoustics*, Vol. AU-16, pp. 413–421, September 1968.

Jennes, R. R. *Analog Computation and Simulation; Laboratory Approach*, Allyn and Bacon Series, 1965.

Jensen, K. and Wirth, N. *PASCAL–User Manual and Report*, 2nd Edition, Springer-Verlag Inc., New York, 1975.

Jurgen, R. K. Electronics in Medicine, *IEEE Spectrum*, New York, January 1979.

Jury, E. I. A Simplified Stability Criterion for Linear Discrete Systems, *Proceedings IRE*, Vol. 50, 1493 (1962).

Jury, E. I. *Theory and Application of the z-Transform Method*, John Wiley & Sons, New York, 1964.

Kadota, T. Analysis of non-linear sampled-data systems with pulse-width modulation. Thesis, University of California, 1960.

Kalman, R. E. and Bertram, J. E. General Synthesis Procedure for Computer Control of Single and Multi-loop Systems, *Trans. AIEE*, Vol. 56, Part II, 1957.

Kapur, G. K. *IBM 360 Assembler Language Programming*, John Wiley & Sons, New York, 1971.

Katzan Jr., H. *The IBM 5100 Portable Computer: A Guide for Users and Programmers*, Van Nostrand Reinhold, New York, 1977.

Korn, G. A. Digital Computer Interface Systems, *Simulation*, Vol. 11, pp. 285–298, December 1968.

Korn, G. A. *Minicomputers for Engineers and Scientists*, McGraw-Hill, New York, 1973.

Kranc, G. M. Input-Output Analysis of Multi-rate Feedback Systems, *Trans. IRE*, Vol. PGAC-3, November 1957, pp. 149–159.

Lindorff, D. F. *Theory of Sampled Data Control Systems*, John Wiley & Sons, New York, 1964.

Lindsey, W. C. and Simon, M. K. (editors). *Phase-Locked Loops and Their Applications*, IEEE Press, New York, 1978.

Lombardo, J. M. The Place of Digital Backup in the Direct Digital Control System, *AFIPS Conference Proceedings*, Vol. 30, 1967, Spring Joint Computer Conference, pp. 771–778.

Luther, A. C. Automatic Timing Correction for Modern Color Television Tape Recorders, *RCA Engineer*, Vol. 9, No. 5, February 1964, pp. 52–55.

Maegele, M. Digital Transmission of Two Television Sound Channels in Horizontal Blanking, *SMPTE Journal*, February 1975.

Manassewitsch, V. *Frequency Synthesizers, Theory and Design*, John Wiley & Sons, New York, 1976.

Martin, D. P. and Berland, K. S. IC's Interface Keyboard to Microprocessor, *Circuits for Electronics Engineers*, Edited by Weber, S., McGraw-Hill Publications.

Martin, J. *Future Developments in Telecommunications*, 2nd Edition, Prentice-Hall, Englewood Cliffs, N.J., 1977.

Melvin, D. K. Microcomputer Applications in Telephony (PABX and Intel 8085 Microprocessor), *Proceedings of the IEEE*, February 1978.

Monroe, A. J. *Digital Processes for Sampled-Data Control Systems*, John Wiley & Sons, New York, 1962.

Newton, G. C., Gould, L. A., and Kaiser, J. F. *Analytical Design of Linear Feedback Controls*, John Wiley & Sons, New York, 1961.

Ogdin, C. A. *Microcomputer Design*, Prentice-Hall, Englewood Cliffs, N. J., 1978.

Ogdin, C. A. *Software Design for Microcomputers*, Prentice-Hall, Englewood Cliffs, N.J., 1978.

Phister, M. *Logical Design of Digital Computers*, John Wiley & Sons, New York, 1958.

Polak, E. Minimal Time Control of a Discrete System with a Non-linear Plant, *IEEE Trans. Automatic Control*, AC-8, No. 1, January, 1963, pp. 49–56.

Puri, N. N. and Weygandt, C. N. Second Method of Lyapounov and Routh's Canonical Form, *Journal of the Franklin Institute*, Vol. 276, No. 5, November 1963, pp. 365–384.

Rabiner, L. R. Techniques for Designing Finite Duration Impulse-Response Digital Filters, *IEEE Transactions Communications Technology*, Vol. COM-19, pp. 188–195, April 1971.

Rabiner, L. R. et al. An Approach to the Approximation Problem for Non-recursive Digital Filters, *IEEE Transactions Audioacoustics*, Vol. AU-18, pp. 83–106, June 1970.

Rabiner, L. R. and Rader, C. M. *Digital Signal Processing*, IEEE Press, 1972.

Rader, C. M. and Gold, B. Digital Filter Design Techniques in the Frequency Domain, *Proceedings IEEE*, Vol. 55, pp. 149–171, February 1967.

Ragazzini, J. R. and Franklin, G. F. *Sampled Data Control Systems*, McGraw-Hill Book Co., New York, 1958.

Rao, G. V. *Television Reference Test Signals for Monochrome and Color*, M. S. (E. E.) thesis, Moore School, 1959, Philadelphia, Pa.

Rao, G. V. *Microprocessors and Microcomputer Systems*, Van Nostrand Reinhold Co., 1978.

RCA, Electronic Recording Products, I. B. 31622 (Camden, N. J.).

Richman, D. The DC Quadri-correlator, a Two-Mode Synchronization System, *Proceedings IRE*, January 1954, Vol. 42, No. 1, pp. 288–298.

Rossi J. P. Color Decoding a PCM NTSC Color Television Signal, *SMPTE Journal*, June 1974.

Shima, M. Two Versions of 16-bit Chip Span Microprocessor, and Microcomputer Needs, *Electronics*, McGraw-Hill, New York, December 21, 1978.

Shinners, S. M. *Control System Design*, Ch. 7, John Wiley & Sons, New York, 1964.

Signetics. Digital, *Linear and MOS Data Book and Applications*, Menlo Park, Calif., 1977.

Sippl, C. J. *Microcomputer Handbook*, Van Nostrand Reinhold, New York, 1976.

Smith, O. J. M. *Feedback Control*, Ch. 18, McGraw-Hill Book Co., New York, 1958.

Snyder, R. H. A Magnetic Tape Recording System for Video Signals, *Trans. IRE-PGBTS*, February 1957.

Stout, T. M. A Step-by-Step Method for Transient Analysis of Feedback Systems with One Nonlinear Element, *Trans. AIEE*, Vol. 76, Part II, January 1957, pp. 378–390.

Stuehler, J. E. *Manufacturing Process Control at IBM*, AFIPS Conference, 1970, pp. 461–469.

Sugarman, R. Computers: Our "Microuniverse" Expands, *IEEE Spectrum*, January 1979.

Tou, J. T. *Modern Control Theory*, McGraw-Hill Book Co. (Chs. 4, 7), New York, 1964.

Tou, J. T. *Digital and Sampled Data Control Systems*, McGraw-Hill Book Co., New York, 1959.

Tou, J. T. and Vadhanaphuti, B. Optimum Control of Nonlinear Discrete-Data Systems, *Trans. AIEE*, Vol. 80, 1961, pp. 166–171.

Truxal, J. G. *Automatic Feedback Control System Synthesis*, Ch. 9, McGraw-Hill Book Co., New York, 1955.

Tsypkin, Y. Z. *Sampling Systems Theory*, Vols. I & II, A Pergamon Press Book, The Macmillan Book Co., New York, 1964. (Translated by Allan, A. and Cochrane, I.)

Uimari, D. C. A Practical Microprocessor Design Example, Integrated Circuits Applications, 1976, Part I, pp. 79–103, Electronics Engineering Times.

Urkowitz, H. Analysis and Synthesis of Delay Line Periodic Filters, *Trans. IRE*, Vol. CT-4, No. 2, June 1957, pp. 41–53.

Wakerly, J. F. Circuit Steps Program for 8080 Microprocessor Debugging, *Circuits for Electronics Engineers*, Edited by Weber, S., McGraw-Hill Publications Inc.

Wentworth, J. W. The Technology of Television Program Production and Recording, *Proceedings IRE*, May 1962, Vol. 50, No. 5, pp. 830–837.

Wetmore, R. E. DATE: a Digital Audio System for Television, *SMPTE Journal*, March 1974.

Yourdon, E. *Design of On-Line Computer Systems*, Prentice-Hall, Englewood, Cliffs, N.J., 1972.

Index